Springer Collected Works in Mathematics

More information about this series at http://www.springer.com/series/11104

JACK KIEFER
Fall, 1957

Jack Carl Kiefer

Collected Papers I

Statistical Inference and Probability (1951–1963)

Editors
Lawrence D. Brown
Ingram Olkin
Jerome Sacks
Henry P. Wynn

Reprint of the 1985 Edition

Springer

Author
Jack Carl Kiefer (1924–1981)
University of California
Berkeley, CA
USA

Editors
Lawrence D. Brown
University of Pennsylvania
Philadelphia, PA
USA

Ingram Olkin
Stanford University
Stanford, CA
USA

Jerome Sacks
Duke University
Durham, NC
USA

Henry P. Wynn
The London School of Economics
London
UK

Published with the cooperation of the Institute of Mathematical Statistics

ISSN 2194-9875
Springer Collected Works in Mathematics
ISBN 978-1-4939-3497-3 (Softcover)
978-0-387-96003-6 (Hardcover)

Library of Congress Control Number: 2012954381

Springer New York Heidelberg Dordrecht London

Printed on acid-free paper

Springer Science+Business Media LLC New York is part of Springer Science+Business Media (www.springer.com)

Jack Carl Kiefer
Collected Papers I

Statistical Inference and Probability
(1951–1963)

Published with the co-operation of the
Institute of Mathematical Statistics
and edited by

Lawrence D. Brown
Ingram Olkin
Jerome Sacks
Henry P. Wynn

Springer-Verlag
New York Berlin Heidelberg Tokyo

Lawrence D. Brown
Department of Mathematics
Cornell University
Ithaca, NY 14853
U.S.A.

Ingram Olkin
Department of Statistics
Stanford University
Stanford, CA 94305
U.S.A.

Jerome Sacks
Department of Mathematics
University of Illinois
Urbana, IL 61801
U.S.A.

Henry P. Wynn
Imperial College
London, England

AMS Subject Classification: 62-XX

Library of Congress Cataloging in Publication Data
Kiefer, Jack, 1924–1981
Jack Carl Kiefer collected papers.
"Published with the co-operation of the Institute of Mathematical Statistics."
Bibliography: p.
Contents: 1. Statistical inference and probability, 1951–1963—2. Statistical inference and probability, 1964–1984—3. Design of experiments.
1. Mathematical statistics—Collected works.
I. Brown, Lawrence D. II. Institute of Mathematical Statistics. III. Title.
QA276.A12K54 1984 519.5 84-10598

ISBN 978-0-387-96003-6

9 8 7 6 5 4 3 2 1

Preface

Jack Kiefer's sudden and unexpected death in August, 1981, stunned his family, friends, and colleagues. Memorial services in Cincinnati, Ohio, Berkeley, California, and Ithaca, New York, shortly after his death, brought forth tributes from so many who shared in his life. But it was only with the passing of time that those who were close to him or to his work were able to begin assessing Jack's impact as a person and intellect.

About one year after his death, an expression of what Jack meant to all of us took place at the 1982 annual meeting of the Institute of Mathematical Statistics and the American Statistical Association. Jack had been intimately involved in the affairs of the IMS as a Fellow since 1957, as a member of the Council, as President in 1970, as Wald lecturer in 1962, and as a frequent author in its journals. It was doubly fitting that the site of this meeting was Cincinnati, the place of his birth and residence of his mother, other family, and friends. Three lectures were presented there at a Memorial Session—by Jerry Sacks dealing with Jack's personal life, by Larry Brown dealing with Jack's contributions in statistics and probability, and by Henry Wynn dealing with Jack's contributions to the design of experiments. These three papers, together with Jack's bibliography, were published in the *Annals of Statistics* and are included as an introduction to these volumes. For those who knew Jack, they serve us as a reminder of the impact he had and how dearly he was held. For those who didn't know Jack, they provide some insight into his character and the development of his research efforts.

Jack's career was centered first at Cornell University and later at the University of California in Berkeley. Jack's presence as an influential faculty member for 25 years at Cornell was only superseded by his magnetic affect on the statistical community at large, drawing any number of visitors and associates from nearby and far. Although at Berkeley only a short time, by the time of his death Jack had become a focus of research activity among students, faculty, and visitors. Cornell and Berkeley each paid tribute to Jack by holding symposia in his honor, shared by Jack Wolfowitz at Cornell and Jerzy Neyman at Berkeley.

The symposia helped express a sense of loss of the person and a sense of honor and respect for his achievements. Above all, they showed how current and vital his work was even after his death. For it was Jack's scientific work that heavily affected the profession at large. His collected papers, numbering over 100 and comprising more than 1600 pages, exhibit an amazing range of interests. They cover sequential analysis, nonparametric analysis, decision theory, multivariate analysis, inventory theory, stochastic processes, and design of experiments. Brought together, these papers are a testament to a pre-eminent scientific career.

Though these papers stand by themselves, and so many of the papers contain introductions or comments reflecting Jack's acute sense of where matters stood at the time they were written, we thought it useful to provide some overview of the works, how they related to research at the time, and their impact on subsequent work. Therefore, we asked a number of people to provide commentary on some of these papers. The response was overwhelming in that virtually everyone responded with thoughtful and perceptive comments. Though our intention was to help readers and scholars to see a large picture, by being selective we run the risk of perhaps neglecting certain issues, threads, and pertinent facts. There are many who were not asked but could easily have given equally responsive comments. We know that they will understand that no slight was intended on their appreciation of Jack's works.

Many of Kiefer's papers cover a wide spectrum of areas for which there is no neat rubric to describe them. For lack of a well-defined title we label these Statistical Inference and Probability (Volumes I and II). A second group of papers can very naturally be characterized as Design of Experiments, and these comprise Volume III.

A few papers, especially the early ones, could be labeled either as Design or as Statistical Inference and Probability. Because of the length of the Design Volume, we have placed these in Volumes I and II.

The Bibliography is the basic source of referencing, and Kiefer's papers are referred to by [·]. Other references follow standard citation style.

Since many of the commentaries are on groups of papers, it was difficult to assign a particular commentary to a single paper. The commentary section for papers on Statistical Inference and Probability is located at the end of Volume II. The commentary section for Design is at the end of Volume III. Some duplication of commentary was unavoidable; these serve to offer different perspectives on a particular paper or issue. Some papers which have no direct comment are discussed in the introductory articles by Brown and Wynn.

To the commentators and to all those, including Jack's family, who helped in the task of assembling and preparing these volumes for publication, we wish to express our deepest thanks.

January, 1984

Lawrence Brown
Ingram Olkin
Jerome Sacks
Henry Wynn

Contents

*Papers with asterisks have commentaries (see pages ix–xiv).

Bibliography of Jack Kiefer

A. Papers (Published)

[1] Almost subminimax and biased minimax procedures, (with P. Frank). *Ann. Math. Statist.* **22** (1951), 465–468. [MR 13 (1952) 143, Zbl 43 (1952) 346]. (I)

[2] The inventory problem: I. Case of known distributions of demand, (with A. Dvoretzky and J. Wolfowitz). *Econometrica* **20** (1952), 187–222. [MR 13 (1952) 856, Zbl 46 (1953) 376]. (I)

[3] The inventory problem: II. Case of unknown distributions of demand, (with A. Dvoretzky and J. Wolfowitz). *Econometrica* **20** (1952), 450–466. [MR 14 (1953) 301, Zbl 48 (1953) 371]. (I)

[4] Stochastic estimation of the maximum of a regression function, (with J. Wolfowitz). *Ann. Math. Statist.* **23** (1952), 462–466. [MR 14 (1953) 299, Zbl 49 (1954) 366]. (I)

[5] Sequential minimax estimation for the rectangular distribution with unknown range. *Ann. Math. Statist.* **23** (1952), 586–593. [MR 14 (1953) 487, Zbl 48 (1953) 121]. (I)

[6] On minimum variance estimators. *Ann. Math. Statist.* **23** (1952), 627–629. [MR 15 (1954) 241, Zbl 48 (1953) 120]. (I)

[7] On Wald's complete class theorems. *Ann. Math. Statist.* **24** (1953), 70–75. [MR 14 (1953) 998, Zbl 50 (1954) 140]. (I)

[7a] Correction of a proof. *Ann. Math. Statist.* **24** (1953), 680. (I)

[8] Sequential minimax search for a maximum. *Proc. Amer. Math. Soc.* **4** (1953), 502–506. [MR 14 (1953) 1103, Zbl 50 (1954) 357]. (I)

[9] Sequential decision problems for processes with continuous time parameter. Testing hypotheses, (with A. Dvoretzky and J. Wolfowitz). *Ann. Math. Statist.* **24** (1953), 254–264. [MR 14 (1953) 997, 1279, Zbl 50 (1954) 148]. (I)

[9a] Corrections to "Sequential decision problems for processes with continuous time parameter testing hypotheses". *Ann. Math. Statist.* **30** (1959), 1265. (I)

[10] Sequential decision problems for processes with continuous time parameter. Problems of estimation, (with A. Dvoretzky and J. Wolfowitz). *Ann. Math. Statist.* **24** (1953), 403–415. [MR (1954) 242, Zbl 51 (1954) 366]. (I)

[11] On the optimal character of the (s, S) policy in inventory theory, (with A. Dvoretzky and J. Wolfowitz). *Econometrica* **21** (1953), 586–596. [MR 15 (1954) 333, Zbl 53 (1956) 279]. (I)

[12] On the theory of queues with many servers, (with J. Wolfowitz). *Trans. Amer. Math. Soc.* **78** (1955), 1–18. [MR 16 (1955) 601, Zbl 64 (1956) 133]. (I)

[13] On tests of normality and other tests of goodness of fit based on distance methods, (with M. Kac and J. Wolfowitz). *Ann. Math. Statist.* **26** (1955), 189–211. [MR 17 (1956) 55, Zbl 66 (1956–1957) 123]. (I)

[14] On the characteristics of the general queueing process, with applications to random walk, (with J. Wolfowitz). *Ann. Math. Statist.* **27** (1956), 147–161. [MR 17 (1956) 980, Zbl 70 (1957) 366]. (I)

[15] Asymptotic minimax character of the sample distribution function and of the classical multinomial estimator, (with A. Dvoretzky and J. Wolfowitz). *Ann. Math. Statist.* **27** (1956), 642–669. [MR 18 (1957) 772, Zbl 73 (1959–60) 146]. (I)

[16] Consistency of the maximum likelihood estimator in the presence of infinitely many incidental parameters, (with J. Wolfowitz). *Ann. Math. Statist.* **27** (1956), 887–906. [MR 19 (1958) 189, Zbl 73 (1959–60) 147]. (I)

[17] Sequential tests of hypotheses about the mean occurrence time of a continuous parameter Poisson process, (with J. Wolfowitz). *Naval Res. Logist. Quart.* **3** (1956), 205–219. [MR 18 (1957) 833]. (I)

[18] Some properties of generalized sequential probability ratio tests, (with L. Weiss). *Ann. Math. Statist.* **28** (1957), 57–74. [MR 19 (1958) 333, Zbl 79 (1959) 354]. (I)

[19] Invariance, minimax sequential estimation, and continuous time processes. *Ann. Math. Statist.* **28** (1957), 573–601. [MR 19 (1958) 1097, Zbl 80 (1959) 130]. (I)

[20] Optimum sequential search and approximation methods under minimum regularity assumptions. *J. Soc. Indust. Appl. Math.* **5** (1957), 105–136. [MR 19 (1958) 1097, Zbl 81 (1959) 385]. (I)

[21] On the deviations of the empiric distribution function of vector chance variables, (with J. Wolfowitz). *Trans. Amer. Math. Soc.* **87** (1958), 173–186. [MR 20 (1959) #5519, Zbl 88 (1961) 113]. (I)

[22] On the nonrandomized optimality and randomized nonoptimality of symmetrical designs. *Ann. Math. Statist.* **29** (1958), 675–699. [MR 20 (1959) #4910, Zbl 92 (1962) 361]. Corrections (and comments), unpublished. (III)

[23] Optimum designs in regression problems, (with J. Wolfowitz). *Ann. Math. Statist.* **30** (1959), 271–294. [MR 21 (1960) #3079, Zbl 90 (1961) 114]. (III)

[24] *K*-sample analogues of the Kolmogorov–Smirnov and Carmér-V. Mises tests. *Ann. Math. Statist.* **30** (1959), 420–447. [MR 21 (1960) #1668, Zbl 134 (1967) 367]. (I)

[25] Asymptotic minimax character of the sample distribution function for vector chance variables, (with J. Wolfowitz). *Ann. Math. Statist.* **30** (1959), 463–489. [MR 21 (1960) #6642, Zbl 93 (1962) 156]. (I)

[26] Optimum experimental designs. *J. Roy. Statist. Soc., Ser. B* **21** (1959), 272–319. [MR 22 (1961) #4101, Zbl 108 (1964) 153]. (III)

[27] A functional equation technique for obtaining Wiener process probabilities associated with theorems of Kolmogorov–Smirnov type. *Proc. Cambridge Philos. Soc.* **55** (1959), 328-332. [MR 22 (1961) #8557, Zbl 96 (1962) 334]. (I)

[28] Optimum experimental designs V, with applications to systematic and rotatable

designs. *Proc. 4th Berkeley Sympos. Math. Statist. and Prob.* **1** (1960), 381–405, Univ. California Press, Berkeley, Calif. [MR 24 (1962) #A3765, Zbl 134 (1967) 366]. (III)

[29] The equivalence of two extremum problems, (with J. Wolfowitz). *Canad. J. Math.* **12** (1960), 363–366. [MR 22 (1961) #8616, Zbl 93 (1962) 156]. (III)

[30] Distribution free tests of independence based on the sample distribution function, (with J. R. Blum and M. Rosenblatt). *Ann. Math. Statist.* **32** (1961), 485–498. [MR 23 (1962) #A2989, Zbl 139 (1968) 363]. (I)

[31] Optimum designs in regression problems, II. *Ann. Math. Statist.* **32** (1961), 298–325. [MR 23 (1962) #A735, Zbl 99 (1963) 135]. (III)

[32] On large deviations of the empiric d.f. of vector chance variables and a law of the iterated logarithm. *Pacific J. Math.* **11** (1961), 649–660. [MR 24 (1962) #A1732, Zbl 119 (1966) 349]. (I)

[33] Two more criteria equivalent to D-optimality of designs. *Ann. Math. Statist.* **33** (1962), 792–796. [MR 25 (1963) #701, Zbl 116 (1965) 113]. (III)

[34] An extremum result. *Canad. J. Math.* **14** (1962), 597–601. [MR 26 (1963) #1968, Zbl 134 (1967) 369]. (III)

[35] Channels with arbitrarily varying channel probability functions, (with J. Wolfowitz). *Information and Control* **5** (1962), 44–54. [MR 24 (1962) #B2506, Zbl 107 (1964) 345]. (I)

[36] Minimax character of Hotelling's T^2 test in the simplest case, (with N. Giri and C. Stein). *Ann. Math. Statist.* **34** (1963), 1524–1535. [MR 27 (1964) #6331, Zbl 202 (1971) 495]. (I)

[37] Asymptotically optimum sequential inference and design, (with J. Sacks). *Ann. Math. Statist.* **34** (1963), 705–750. [MR 27 (1964) #893, Zbl 255 (1973) 62063]. (I)

[38] Local and asymptotic minimax properties of multivariate tests, (with N. Giri). *Ann. Math. Statist.* **35** (1964), 21–35. [MR 28 (1964) #2605, Zbl 133 (1967) 418]. (II)

[39] Minimax character of the R^2-test in the simplest case, (with N. Giri). *Ann. Math. Statist.* **35** (1964), 1475–1490. [MR 29 (1965) #6579, Zbl 137 (1967) 368]. (II)

[40] Optimum extrapolation and interpolation designs, I, (with J. Wolfowitz). *Ann. Inst. Statist. Math.* **16** (1964), 79–108. [MR 31 (1966) #2806, Zbl 137 (1967) 131]. (III)

[41] Optimum extrapolation and interpolation designs, II, (with J. Wolfowitz). *Ann. Inst. Statist. Math.* **16** (1964), 295–303. [MR 31 (1966) #2806, Zbl 137 (1967) 131]. (III)

[42] Admissible Bayes character of T^2-, R^2-, and other fully invariant tests for classical multivariate normal problems, (with R. Schwartz). *Ann. Math. Statist.* **36** (1965), 747–770. [MR 30 (1965) #5430; 50 (1975) #11567, Zbl 137 (1967) 36; 249 (1973) 62058]. (II)

[42a] Correction to "Admissible Bayes character of T^2-, R^2-, and other fully invariant tests for classical multivariate normal problems". *Ann. Math. Statist.* **43** (1972), 1742. (II)

[43] On a problem connected with the Vandermonde determinant, (with J. Wolfowitz). *Proc. Amer. Math. Soc.* **16** (1965), 1092–1095. [MR 32 (1966) #115, Zbl 142 (1968) 269]. (III)

[44] On a theorem of Hoel and Levine on extrapolation designs, (with J. Wolfowitz). *Ann. Math. Statist.* **36** (1965), 1627–1655. [MR 32 (1966) #3230, Zbl 138 (1967) 140]. (III)

[45] Multivariate optimality results. *Multivariate Analysis. Proceedings of an International Symposium* (ed. by P. R. Krishnaiah), (1966), 255–274, Academic Press, New York. [MR 37 (1969) #2372, Zbl 218 (1972) 448]. (II)

[46] Optimum multivariate designs, (with R. H. Farrell and A. Walbran). *Proc. Fifth Berkeley Sympos. Math. Statist. and Probability* **1** (1967), 113–138. [MR 35 (1968) #5099, Zbl 193 (1970) 171]. (III)

[47]. On Bahadur's representation of sample quantiles. *Ann. Math. Statist.* **38** (1967), 1323–1342. [MR 36 (1968) #933, Zbl 158 (1960) 370]. (II)

[48] Statistical inference, (panel discussion with G. A. Barnard, L. M. LeCam, and L. J. Salvage). *The Future of Statistics* (Proceedings of a Conference on the Future of Statistics held at the University of Wisconsin, Madison, Wisconsin, June 1967) (ed. by D. G. Watts), (1968), 139–160, Academic Press, New York. (II)

[49] Statistical inference. *The Mathematical Sciences*, (1969), 60–71, The M.I.T. Press, Cambridge. [Reprinted in *Math. Spectrum.* **3** (1970–71), 1–11]. (II)

[50] On the deviations in the Skorokhod–Strassen approximation scheme. *Z. Wahrsch. Verw. Gebiete.* **13** (1969), 321–332. [MR 41 (1971) #1117, Erratum 41, p. 1965, Zbl 176 (1969) 482]. (II)

[51] Old and new methods for studying order statistics and sample quantiles. *Nonparametric Techniques in Statistical Inference* (ed. by M. L. Puri), (1970), 349–357, Cambridge University Press, London. [MR 44 (1972) #3442]. (II)

[52] Deviations between the sample quantile process and the sample df. *Nonparametric Techniques in Statistical Inference* (ed. by M. L. Puri), (1970), 299–319, Cambridge University Press, London. [MR 43 (1972) #2808]. (II)

[53] Optimum experimental designs. *Actes du Congrès International des Mathématiciens, Nice*, **3** (1970), 249–254. [MR 54 (1977) #8993, Zbl 237 (1972) 62050]. (III)

[54] Iterated logarithm analogues for sample quantiles when $p_n \downarrow 0$. *Proc. Sixth Berkeley Symp. on Mathematical Statistics and Probability* **1** (1970), 227–244, Univ. California Press, Berkeley. [MR 53 (1977) #6696, Zbl 264 (1974) 62815]. (II)

[55] The role of symmetry and approximation in exact design optimality. *Statistical Decision Theory and Related Topics.* (*Proc. Symp.*) (ed. by S. S. Gupta and J. Yackel), (1971), 109–118, Academic Press, New York. [MR 50 (1975) #3447, Zbl 274 (1974) 62050]. (III)

[56] Skorohod embedding of multivariate rv's, and the sample df. *Z. Wahrsch. Verw. Gebiete.* **24** (1972), 1–35. [MR 49 (1975) #6382, Zbl 267 (1974) 60034]. (II)

[57] Optimum designs for fitting biased multiresponse surfaces. *Multivariate Analysis — III. Proceedings of the Third International Symposium on Multivariate Analysis* (ed. by P. R. Krishnaiah), (1973), 287–297, Academic Press, New York. [MR 51 (1976) #2188, Zbl 291 (1975) 62093]. (III)

[58] General equivalence theory for optimum designs (approximate theory). *Ann. Statist.* **2** (1974), 849–879. [MR 50 (1975) #8856, Zbl (1975) 62092]. (III)

[59] Discussion on the paper: Planning experiments for discriminating between models, by A. C. Atkinson and D. R. Cox. *J. Roy. Statist. Soc. Ser. B* **36** (1974), 345–346. (III)

[60] Balanced block designs and generalized Youden designs, I. Construction (patchwork). *Ann. Statist.* **3** (1975), 109–118. [MR 51 (1976) #4578, Zbl 305 (1976) 62052]. (III)

[61] Construction and optimality of generalized Youden designs. *A Survey of Statistical Design and Linear Models* (ed. by J. N. Srivastava), (1975), 333–353, North-Holland Pub. Co, Amsterdam. [MR 52 (1976) #15877, Zbl 313 (1975) 62057]. (III)

[62] Review of the paper: Bayesian analysis of generic relations in Agaricales, by R. E. Machol and R. Singer. *Mycologia* **67** (1975), 203–205. (II)

[63] Optimal design: Variation in structure and performance under change of criterion. *Biometrika* **62** (1975), 277–288. [MR 52 (1976) #2064, Zbl 321 (1976) 62086]. (III)

[64] Optimal designs for large degree polynomial regression, (with W. J. Studden). *Ann. Statist.* **4** (1976), 1113–1123. [MR 54 (1977) #11676, Zbl 357 (1978) 62051]. (III)

[65] Asymptotically minimax estimation of concave and convex distribution functions, (with J. Wolfowitz). *Z. Wahrsch. Verw. Gebiete.* **34** (1976), 73–85. [MR 53 (1977) #1829, Zbl 354 (1978) 62035]. (II)

[66] Large sample comparison of tests and empirical Bayes Procedures, (with D. S. Moore). *On the History of Statistics and Probability: Proc. of a Symp. on the American Mathematical Heritage* (ed. by D. B. Owen), (1976), 347–365, M. Dekker, New York. (II)

[67] Admissibility of conditional confidence procedures. *Ann. Statist.* **4** (1976), 836–865. [MR 55 (1978) #11454, Zbl 353 (1975) 62008]. (II)

[68] Asymptotically minimax estimation of concave and convex distribution functions. II, (with J. Wolfowitz). *Statistical Decision Theory and Related Topics, II* (ed. by S. S. Gupta and D. S. Moore), (1977), 193–211, Academic Press, New York. [MR 56 (1978) #1572, Zbl 418 (1980) 62031]. (II)

[69] Conditional confidence statements and confidence estimators. *J. Amer. Statist. Assoc.* **72** (1977), 789–827. [MR 58 (1979) #24638, Zbl 375 (1978) 62023]. (II)

[70] The ideas of conditional confidence in the simplest setting, (with C. Brownie). *Comm. Statist.—Theor. Methods.* **A6**(8) (1977), 691–751. [MR 56 (1978) #3993, Zbl 392 (1979) 62002]. (II)

[71] Conditional confidence and estimated confidence in multidecision problems (with applications to selection and ranking). *Multivariate Analysis—IV. Proceedings of the Fourth International Symposium on Multivariate Analysis* (ed. by P. R. Krishnaiah), (1977), 143–158, North-Holland, Amsterdam. [MR 58 (1979) #18810, Zbl 381 (1979) 62009]. (II)

[72] Comparison of rotatable designs for regression on balls, I (Quadratic), (with Z. Galil). *J. Statist. Plann. Inference* **1** (1977), 27–40. [MR 58 (1979) #24769, Zbl 394 (1979) 62058]. (III)

[73] The foundations of statistics—are there any? *Synthese* **36** (1977), 161–176. [MR 58 (1979) #31488, Zbl 375 (1978) 60005]. (II)

[74] Comparison of design for quadratic regression on cubes, (with Z. Galil). *J. Statist. Plann. Inference* **1** (1977), 121–132. [MR 58 (1979) #24770, Zbl 381 (1979) 62062]. (III)

[75] Comparison of simplex designs for quadratic mixture models, (with Z. Galil). *Technometrics* **19** (1977), 445–453. [MR 57 (1978) #17972, Zbl 372 (1978) 62058]. (III)

[76] Comparison of Box–Draper and D-optimum designs for experiments with mixtures, (with Z. Galil). *Technometrics* **19** (1977), 441–444. [MR 58 (1979) #3233, Zbl 369–389 (1978) 62087]. (III)

[77] Asymptotic approach to familes of design problems. *Comm. Statist.—Theory Methods* **A7** (1978), 1347–1362. [MR 82g (1982) 62106, Zbl 389 (1979) 62058]. (III)

[78] A Diophantine problem in optimum design theory. *Utilitas Math.* **14** (1978), 81–98. [MR 80b (1980) 62091, Zbl 391 (1979) 62055]. (III)

[79] Comment on paper: Pseudorandom number assignment in statistically designed simulation and distribution sampling experiments, by L. W. Schruben and B. H. Margolin. *J. Amer. Statist. Assoc.* **73** (1978), 523–524. (III)

[80] Extrapolation designs and Φ_p-optimum designs for cubic regression on the q-ball, (with Z. Galil). *J. Statist. Plann. Inference* **3** (1979), 27–38. [MR 81a (1981) 62074, Zbl 412 (1980) 62055]. (III)

[81] Sequential statistical methods. *Studies in Probability Theory* (ed. by M. Rosenblatt), (1978), 1–23, Studies in Math., 18, Math. Assoc. Amer., Washington, D.C. [MR 80m (1980) 62077, Zbl 412 (1980) 62056]. (II)

[82] Comments on taxonomy, independence, and mathematical models (with reference to a methodology of Machol and Singer). *Mycologia* **71** (1979), 343–378. (II)

[83] Optimal design theory in relation to combinatorial design. *Ann. Discrete Math.* **6** (1980), 225–241. [MR 82a (1982) 62107, Zbl 463 (1982) 62066]. (III)

[84] Designs for extrapolation when bias is present. *Multivariate Analysis — V. Proc. Fifth International Symp.* (ed. by P. R. Krishnaiah), (1980), 79–93, North-Holland Pub. Co., Amsterdam. [MR 81i (1981) 62130, Zbl 458 (1982) 62063]. (III)

[85] *D*-optimum weighing designs, (with Z. Galil). *Ann. Statist.* **8** (1980), 1293–1306. [MR 82g (1982) 62104, Zbl 466 (1982) 62066]. (III)

[86] Time- and space-saving computer methods, related to Mitchell's DETMAX, for finding D-optimum designs, (with Z. Galil). *Technometrics* **22** (1980), 301–313. [MR 81j (1981) 62147, Zbl 459 (1982) 62060]. (III)

[87] Optimum weighing designs, (with Z. Galil). *Recent Developments in Statistical Inference and Data Analysis. Proceedings of the International Conference in Statistics in Tokyo* (ed. by K. Matusita), (1980), 183–189, North-Holland Pub. Co. Amsterdam. [MR 82a (1982) 62108, Zbl 462 (1982) 62059]. (III)

[88] Optimum balanced block and Latin square designs for correlated observations, (with H. P. Wynn). *Ann. Statist.* **9** (1981), 737–757. [MR 82h (1982) 62122]. (III)

[89] The interplay of optimality and combinatorics in experimental design. *Canad. J. Statist.* **9** (1981), 1–10. [MR 82m (1982) 62167]. (III)

[90] Relationships of optimality for individual factors of a design, (with J. Eccleston). *J. Statist. Plann. Inference* **5** (1981), 213–219. [MR 83c (1983) 62115, Zbl 481 (1982) 62059]. (III)

[91] Optimum rates for non-parametric density and regression estimates, under order restrictions. *Statistics and Probability, Essays in Honor of C. R. Rao* (ed. by G. Kallianpur, P. R. Krishnaiah, J. K. Ghosh), (1982), 419–428, North-Holland Pub. Co., Amsterdam. (II)

[92] On the characterization of D-optimum weighing designs for $n \equiv 3 \pmod 4$, (with Z. Galil). *Statistical Decision Theory and Related Topics III* (ed. By S. S. Gupta and J. O. Berger), **1** (1982), 1–35, Academic Press, New York. (III)

[93] Eight lectures on mathematical statistics. (Chinese) *Advances in Mathematics (Beijing)* (Shuxue Jin Zhan) **10** (1981), 94–130. (II)

[94] Conditional inference. *Encyclopedia of Statistical Sciences*, Volume 2 (ed. By S. Kotz, N. L. Johnson, and C. B. Read), (1982), 103–109, Wiley-Interscience. (II)

[95] Construction methods for *D*-optimum weighing designs when $n \equiv 3 \pmod 4$, (with Z. Galil). *Ann. Statist.* **10** (1982), 502–510. [MR 83i (1983) 62139]. (III)

[96] Autocorrelation-robust design of experiments, (with H. P. Wynn). *Scientific Inference, Data Analysis, and Robustness* (ed. by T. Leonard and C. -F. Wu), (1983), 279–299, Academic Press, New York. (III)

[97] Comparison of designs equivalent under one or two criteria, (with Z. Galil). *J.*

Statist. Plann. Inference **8** (1983), 103–116. (III)

[98] Optimum and minimax exact treatment designs for one-dimensional autoregressive error processes, (with H. P. Wynn). *Ann. Statist.* **12** (1984), 414–450. (III)

B. Book Reviews

[99] Review of *The Advanced Theory of Statistics*, Volume 2, "Inference and Relationship," M. G. Kendall and A. Stuart. *Ann. Math. Statist.* **35** (1964), 1371–1380. (II)

[100] Review of *The Savory Wild Mushroom*, M. McKenney. (Second edition, revised and enlarged by D. E. Stuntz). *Quarterly Review Biology* **47** (1972), 342–343. (II)

[101] Review of *A Field Guide to Western Mushrooms*, A. H. Smith. *Quarterly Review Biology* **52** (1977), 91. (II)

C. Books, Edited Volumes, Lecture Notes

[102] *Sequential identification and ranking procedures, with special reference to Koopman–Darmois populations*, (with R. E. Bechhofer and M. Sobel), Statistical Research Monographs, Vol. 3, 1968, The University of Chicago Press, Chicago, Illinois. [MR 39 (1970) #6445, Zbl 208 (1971) 446].

[103] Jacob Wolfowitz, selected papers (ed. by J. Kiefer, with the assistance of U. Augustin and L. Weiss). 1980, Springer-Verlag, New York. [MR 83d (1983) 01080, Zbl 447 (1981) 62001].

[104] Lectures on design theory, (1974). Mimeograph Series #397, Department of Statistics, Purdue University.

[105] Contributions to the theory of games and statistical decision functions. Doctoral dissertation, 1952, Columbia University.

[106] Notes on decision theory (1953), (notes recorded by J. Sacks).

D. Papers (Unpublished)

[107] Note on asymptotic efficiency of M.L. estimators in nonparametric problems, (with J. Wolfowitz), (circa 1960). (II)

[108] Mathematics 371 Final Examination (with D. Kiefer), (1972). (II)

[109] Lecture notes on statistical inference, (1973), to be published by Springer-Verlag, 1984.

[110] *D*-optimality of the GYD for $v \geq 6$, (1974). (III)

[111] Optimality criteria for designs, (1975). (III)

Jack Carl Kiefer
(1924–1981)

J. SACKS
Northwestern University

On August 10, 1981, we learned of Jack Kiefer's death. The initial shock gave way to a dismay that has not been dissipated by time, so that the loss and sorrow is felt as deeply now as then. Though what I say comes from my own experience and emotion it is no exaggeration when I express a communal feeling that we have lost a central figure in our profession and in our lives. The personal memories, that continue to flash, of times and events peripheral to our work nonetheless connect with it and remind us that science is not a cold-hearted private activity.

Each of us who knew Jack was aware of a wonderful spirit which pervaded not only his being but also the contributions he made to our field. The complete integrity and honesty we find in his written work was ever present in his personal life. The principled and nondogmatic view he had of life was the same view he had of science. Everything he did was expressed with such grace and style that our despair is due not only to having lost an intellectual force but, even more, of having lost a presence and vitality which revealed what a creative spirit is about.

His dedication to distinguishing right from wrong in science, be it in setting straight statistical methodology in the classification of mushrooms, or in combatting misunderstanding of the implications of mathematical facts about experimental designs, was the same dedication he brought to actively working to right the human wrongs of the Vietnam War and of Soviet repression.

The spirit which he brought to local politics in Ithaca and Tompkins County was the same spirit that covered his continual and effective activities in the Institute of Mathematical Statistics, his department, college and university (perhaps with less success since he failed to win election to the New York State Assembly as the Liberal Party candidate in 1968 though, intelligent statistician that he was, Jack had assigned zero prior probability to that event).

Jack didn't stop at arbitrary boundaries but was compelled to push to the frontiers of anything he touched in order to understand what we knew and, most importantly, what it was that we did not know. His work in statistics stands as a testament to the scope of this quality—there is no shallow generalization, only deeper and deeper exploration no matter how intricate and formidable the task.

Even his hobbies would carry him well beyond casual dalliance. Jack's introduction to the fascination of mushrooms by Jacques Deny in the early 1960's grew rapidly into a passion and eventually brought him from amateur status to professional standing as a mycologist. A Dorothy Sayers mystery could lead him, along with a friend and colleague, Bob Walker, to construct a mechanical change ringer that rang all the peals of the Nine Tailors. Obliged to organize the Cornell Mathematics Colloquium one year, he would introduce each speaker with a limerick or poem specially written to suit the occasion.

Jack was born in Cincinatti in 1924 and grew up in an energetic and intellectual environment. In 1942 he went to MIT to study electrical engineering and economics. This was interrupted by military service during World War II. It was through economics and Harold Freeman at MIT that Jack became interested in statistics although he nearly went into show business, having written, produced, and directed student musical comedy shows. His masters' thesis at MIT on sequential search for the maximum of a function grew into the famous Fibonacci Search Algorithm which ultimately appeared in 1953.

In 1948 he went to the Department of Mathematical Statistics at Columbia where Abraham Wald was preeminent in a department that included Ted Anderson, Howard Levene, Henry Scheffe, and Jack Wolfowitz. He wrote his doctoral thesis in decision theory under Wolfowitz and went to Cornell in 1951 with Wolfowitz. From his arrival he was a central figure in a whirl of statistical and mathematical activity involving other statisticians such as Bob Bechhofer, Walt Federer, Phil McCarthy, and Lionel Weiss, and probabilists such as Kai-Lai Chung, Gil Hunt, and Mark Kac. Numerous visitors in the 1950's including Julie Blum (whose recent death we also mourn), Aryeh Dvoretzky, Willy Feller (who had been on the Cornell faculty earlier), Esther Seiden, Milton Sobel, Frank Spitzer (who later joined the Cornell faculty) added to the richness of the Cornell environment.

In 1957 Jack married Dooley Sciple and they had two children, Sarah and Daniel, born in 1960 and 1962.

For many years Jack suffered greatly from an arthritic hip condition and in 1971 underwent a delicate operation to insert artificial joints in both hips. The success of the operation enabled him to resume many of the activities that had been so severely curtailed.

Jack received any number of honors during his career. He was a Fellow of the Institute of Mathematical Statistics and the American Statistical Association, President of the Institute of Mathematical Statistics (1967), the Wald lecturer in 1962, and a Guggenheim fellow. Jack was elected to the American Academy of Arts and Sciences and to the National Academy of Sciences in 1975, two years after being named the first Horace White Professor of Mathematics at Cornell.

Jack served on numerous editorial boards of statistics journals and on a large number of committees of various institutions and organizations. It was universal knowledge in the profession that you could count on Jack to give immediate and incisive responses to requests for opinions on a paper, on individuals, or on almost any professional matter.

After 28 years at Cornell Jack left in 1979 to join the Statistics Department at Berkeley; he was a Miller Research Professor there at the time of his heart attack and death in 1981 at the age of 57.

These facts tell us how distinguished a figure he was in statistics, but they omit the quality and elegance which suffused his work and lectures. The facts do not describe how, in an hour's lecture, he could synthesize and compress a subject and with unusual clarity and perception bring the audience to the boundaries of the area. The facts do not tell how he would come to Northwestern, Purdue, or UCLA for a visit to lecture on inference or design and impress the audience about the contributions of others as if theirs were at least as important as his own which rarely, if ever, was the case.

The facts do not tell how devoted a husband and father he was—of how he would bring his son Daniel to Chicago for a crucial four-game series between the Chicago Cubs and Philadelphia Phillies, or how he would drive 25 miles to post a letter to his daughter Sarah in France so it would reach her a day earlier.

Although we can read the legacy of accomplishment that Jack left and recognize the consistency and force that make it a model of a lifetime of scientific inquiry, it does not tell us about the support and inspiration he leaves to his students and associates. His students (Corrie Atwood, Larry Brown, Ching-Shui Cheng, Ker-Chau Li, Gary Lorden, David Moore, Praesert Na Nagora, Bill Notz, Doug Robson, Sally Sievers Nerode, Jerry Sacks, Dick Schwartz, Josefa Lopes Troya, Yehuda Vardi, Gloria Zerdy) all know how encouraging and considerate he could be. They knew that you could bring him a manuscript one day and, astonishingly, have it back a few days later with detailed annotations and penetrating comments written in the margins. He never diminished their work but only imbued them with the spirit of excellence that was stamped on everything he himself did.

He was encouraging not only to his students but also to his associates and colleagues nearby and far away. His advice was continually sought and given about work and career, organizing conferences, editorial matters, and about anything professional. He was a prodigious correspondent whose letters never failed to be filled with all sorts of interesting items hastily written up the sides and across the top of the page. He would spend uncountable hours on the telephone with friends and colleagues—one wondered where he found time to do anything for himself.

The spontaneity and diversity of mind and wit that made being with him such fun, whether at the Purdue Decision Theory Conference, or at the Bay Meadows race track, or at a Greek night club in Chicago, will not be forgotten. All his characteristics and all the facts of his life portray an unusual man whose imprint will not vanish.

But for those who were close to Jack there was something beyond the admiration and respect we felt for him, beyond the gratitude for his help and the urge for approval by a leading intellect, and beyond the pleasure of interchange of thoughts and good times. Perhaps it lies in the regard and affection he had of our own individuality; perhaps it was his compassion always expressed with such sincerity. Ultimately, I think it can only be defined as the love he had for each of us and the love we had for him. Jack Kiefer was more than just a great statistician; he was part of the fabric of our lives. And that is why we miss him so.

The Research of Jack Kiefer Outside of the Area of Experimental Design

LAWRENCE D. BROWN[1]

Mathematics Department
Cornell University

This article is a survey of Jack Kiefer's published works in areas other than design of experiments. The approximately 50 articles can be divided into categories by subject matter. These categories will be discussed individually in the following review. At times Jack worked in several areas simultaneously and ideas from one area sometimes influenced work in another. Some of these influences are noted in the following survey, as the categories are discussed in roughly the order Jack began work in them.

Collaborators: The majority of Jack's earlier papers were written jointly with his former thesis adviser, Jack Wolfowitz. Six of these were a three-way collaboration including also Aryeh Dvoretzky. Jack was brimful of curiosity and good ideas and a desire to assist and share. In all, this lead to joint work with 21 different collaborators throughout his career. The extent of this collaboration can be noted from the bibliography. No attempt will be made here to unravel the who-did-what-and-why of these collaborations.

Foundations and Decision Theory

Kiefer [7][2] distills the first half of Jack's 200 page Ph.D. thesis into a 6-page article. The main results here establish the existence of minimax and admissible procedures under regularity conditions weaker than those assumed in the fundamental book of Wald (1950). Jack's methods and results have since become antiquated by those of LeCam and others (see especially LeCam (1955)).

[1] Preparation of this manuscript was supported in part by N.S.F. Grant MCS 820031.

[2] References in square brackets refer to the collected bibliography of the works of Jack Kiefer. Other references are collected at the end of this article.

What still remains very much of note is that this work signals a commitment to the philosophy of Wald's decision theoretic approach. This philosophy transcends the important concepts of loss, risk, admissibility, minimaxity, etc. It holds that one must weigh carefully—in a precise mathematical formulation—the frequential consequences of any statistical course of action.

A quotation from Dvoretzky, Kiefer, and Wolfowitz [2] describes this philosophy in relation to the loss function:

> "It may be objected that our method requires one to specify the [loss] function and that this function may be unknown or difficult to give. We wish to emphasize that the need for a [loss] function, W, is inevitable in the sense that any method which does not explicitly use a function W simply uses one implicitly. Thus one who selects a method of solving the ... problem which ostensibly has the advantage of not requiring the specification of W is simply relinguishing control of W and may be implicitly using a W of which he would disapprove .... It is difficult to see what advantage can accrue ... from deliberately burying [ones] intellectual head in the sand."

The philosophy applies equally to the formulation of a probabilistic model, specification of the action space, and so forth.

The minimax principle is one feature of Wald's theory. Kiefer, as well as Wolfowitz, accepted it only as, "a *possible* principle ... [which] ... might be the course of a very conservative statistician." (Italics, mine. Quotation from Wolfowitz (1951, p. 461).) Indeed, Jack's first published article, Kiefer and Frank [1], consists of three relatively elementary examples of problems where one would clearly prefer not to use the minimax procedure.

As I will note, the minimax principle reappears many times in various contexts in Jack's work, sometimes in rather surprising ways.

The minimax principle is of course a central element of Kiefer [20]. Theorems establishing the existence (under general regularity conditions) of minimax admissible procedures are called Hunt–Stein theorems. This paper unifies and extends the contemporaneous Hunt–Stein theorems. It remains a classic in spite of the fact that its methods—though not its results—can be considered inferior to those developed later by Huber. (Huber's work on the Hunt–Stein theorem remains largely unpublished. A brief sketch of the method in relation to testing problems appears in [45]. Another such sketch and some important related results appear in Bondar and Milnes (1981).)

The fact that this paper extends the Hunt–Stein theorem to sequential problems should be noted for its own merit, and also as a measure of Jack's wide interests and his determination to extend theoretical results to their statistically natural boundaries.

Jack's insistence on establishing clearly the frequential consequences of any statistical action persisted. In the series of papers [72], [73], [74] (with C. Browme), and [75] he applied this insistence to the frustrating dilemma of conditional confidence. Jack's results leave an optimistic feeling that a frequential discussion of conditional confidence is possible, useful, and even perhaps natural. But, as expressed in Brown (1978), I think that a finished satisfactory frequential theory will have to diverge at some point from that proposed in these papers.

This immersion in Wald's theory had earlier led almost naturally to a climax in Jack's career. Given an insistence on establishing *a priori* the frequential consequences of statistical procedures, it was natural to look at problems of experimental design, as indeed Wald himself had done in Wald (1943).

The first paper to result was the remarkable paper [23], which presents the possibility of using randomized designs and discusses optimality properties in this context. After this began the climactic series of papers—several written jointly with Jack Wolfowitz. Kiefer and Wolfowitz skillfully brought together and refined various existing optimality criteria for judging experimental designs. In the course of this they discovered a formal parallel with the classical minimax theorems. The development of this parallel and the years of interesting consequences belong to the essay to follow about Jack's work on experimental design. Here, I want only to emphasize how the initial achievements in design related naturally to Jack's general statistical outlook and to his formal decision theoretic background.

Before turning to the next major area I will mention Jack's commentary papers, [49] and [77], for these express a side of Jack's statistical personality very familiar to those who knew him, but not so evident in his published papers. He was extremely concerned with the question of how well procedures justified by elegant theories actually work in real life. For example, how well do asymptotically optimal procedures perform for experiments involving realistic sample sizes?—or, what happens in design and analysis when fitting a linear or quadratic regression to a response curve which is not linear or quadratic?

Sequential Analysis

Wald's decision theory was developed in tandem with his theory of sequential statistical procedures. Analogously, the second half of Jack's doctoral thesis was an attempt to construct optimal sequential procedures for problems whose parameter spaces have three or more points. It turns out that there is no precise characterization possible here like that of the familiar Wald–Wolfowitz theorem for the sequential probability ratio test (SPRT) for testing a simple null hypothesis versus a simple alternative. Thus this second part of the thesis was never directly adapted for publication. But, what Jack discovered there appears in a very significant role in Kiefer and Weiss [19], which concerns properties of the generalized SPRT, and also in Kiefer and Sacks [39], which is discussed in more detail below.

Aside from work in inventory theory, which will be described later, Jack's first published paper in sequential analysis involved a problem of a very different sort from those in his thesis. Kiefer and Wolfowitz [4] proposed what has come to be known as the "Kiefer–Wolfowitz procedure." This is a consistent procedure for sequentially locating the maximum of an unknown unimodal regression function. The stimulus for this work was of course the earlier paper of Robbins and Munro (1951), describing a consistent sequential procedure for locating the zero of a regression function.

The problem for the Kiefer–Wolfowitz procedure is statistical—the values of the regression function are observed with statistical error. What happens if the regression function can be observed exactly? Jack's involvement with this question goes back to his master's thesis, written at M.I.T., but the final results did not appear until his paper [8], with generalizations in [21]. There is still a problem to be solved—namely, where to place the observations in order to have the most precise final statement concerning the location of the maximum. This is not a classical type of statistical problem. Nevertheless Jack was able to cast it into a minimax formulation (Again the minimax principle!) and to use some clever analysis (as usual) to produce a precise description, involving the Fibonacci numbers, of ε-minimax rules and of asymptotically optimum rules. Even to the present there is no comparably complete solution to the statistical problem attacked by Kiefer and Wolfowitz—indeed, it is not even known whether in their general formulation there can exist a procedure which achieves the optimal stochastic rate of approach to the unknown maximum. (If not, maybe it is necessary to restrict the formulation somewhat?)

The pair of papers [9] and [10] by Dvoretzky, Kiefer, and Wolfowitz considers statistical problems concerning continuous time stochastic processes—for example, a Wiener process with unknown drift or a Poisson process with unknown intensity. The first paper contains a proof of the optimum property of the SPRT for such problems. The second paper treats questions of estimation. Primarily, it characterizes those situations where a fixed sample size procedure is optimal among all sequential procedures.

A feature of these papers which is of note—apart from the important basic results themselves—is the understanding that discrete time processes can be profitably approximated by continuous time ones. Such approximations are now an essential part of modern methodology for attacking sequential problems. Dvoretzky, Kiefer, and Wolfowitz wrote, "There are many cases … in which an exact determination of the optimal procedure is possible in the continuous case [but not in the discrete case]. Thus even when treating the discrete case the continuous case … may be used to derive approximations." ([9, p. 255]) My impression is that this understanding was an important motivating force and building block for a second climax in Jack's work, which I'll describe later.

An earlier place where this realization proved useful in Kiefer's work was in the monograph [16] on ranking and selection by Bechhofer, Kiefer, and Sobel. This manuscript was more than 10 years in preparation. It was one of the few projects on which Jack felt himself to be distinctly the junior author, but he worked on it very industriously, especially to create and perfect sequential arguments involving approximation by continuous time stochastic processes.

This brings us to the monumental paper of Kiefer and Sacks [39]. This paper was not the first to consider asymptotic sequential properties—there were important precursors by Chernoff (1959) and by Schwarz (1962)—but it was the first comprehensive treatment. Within the models considered it clearly settled the issue of the asymptotic relationship between expected sample size and probability of error for asymptotically optimal procedures. In fact it constructed such procedures—that is, procedures which are asymptotically Bayes (as the cost per observation approaches zero) simultaneously for all priors whose support is the full sample space. It is characteristic of Jack's work that this paper began as an attempt to settle an even broader question, the construction of asymptotically optimal procedures in problems involving a sequential choice of design. After they began work on this broader question Kiefer and Sacks realized that the existing formulation and results for the basic (nondesign) problem were inadequate for their purposes. After solving this basic problem they then turned to the original design question, and in the second half of [39] they solved this problem as well.

Here is one main result from the first half of this paper. Consider the problem of sequentially testing a composite null hypothesis, Ω_0, versus a composite alternative, Ω_1, with the cost of each observation being c. The potential observations are independently and identically distributed (i.i.d.) with density f_ω, some $\omega \in \Omega_0 \cup \Omega_1$. Introduce the Kullback–Liebler distance

$$\lambda_i(\omega) = \inf\left\{ E_\omega\left(\log\left(f_\omega(X)/f_{\omega_j}(X)\right)\right):\ \omega \in \Omega_i, \omega_j \in \Omega_j,\ j \neq i\right\}, \qquad i = 0,1. \quad (1)$$

Assume

$$\lambda_i = \inf\{\lambda_i(\omega):\ \omega \in \Omega_i\} > 0, \qquad i = 0,1. \quad (2)$$

Then, under suitable mild regularity conditions there is a procedure satisfying

$$\sup\{P_\omega(\text{terminal error}):\ \omega \in \Omega_0 \cup \Omega_1\} = c|\log c| \quad (3)$$

and

$$E_\omega(\text{terminal sample size}) \leq (1 + o(1))|\log c|/\lambda_i(\omega) \quad \text{uniformly for } \omega \in \Omega_i,\ i = 1,2. \quad (4)$$

This procedure is a modification of the weight function procedure introduced in Wald (1947). If G is any prior distribution supported on all of $\Omega_0 \cup \Omega_1$, then the Bayes procedure for G achieves exactly the bounds on the right of (3) and (4). It follows that the modified weight function procedure is asymptotically Bayes.

The power and elegance of such a result often leads one to overlook its deficiencies. The primary deficiency is that the given asymptotically optimal procedure does not seem to be satisfactorily near optimality for practically realistic values of c. Kiefer and Sacks noted this deficiency. Others have since attempted to provide a theoretical basis for defining an improved asymptotically optimal procedure. Lorden (1972), (1976) and Zerdy (1980), both former doctoral students of Kiefer, have investigated related questions and have made some progress on this issue. Another possible deficiency lies in the assumption (2), which specifies that the null and alternative hypotheses be effectively separate. Analogous results without this assumption have not been proved, but some progress may be seen in Bickel and Yahav (1971) and works cited there, and from a somewhat different perspective in Brown, Cohen, and Samuel–Cahn (1983).

Writing Style

In terms of general style [39], considered above, is an excellent example of Jack's craftsmanship—so this seems a good place to pause and consider some general characteristics of Jack's writing.

Jack was always bursting with useful ideas. In writing a paper he tried to share them all and to pass along a wealth of information on the relationship of results and methods to other important related works, on possible generalizations, and on alternate formulations and methods of proof.

This is immediately obvious to anyone who glances at this work, or at almost any other of Jack's major works. Some other sides of Jack's mathematical personality are not so immediately apparent. Jack organized long and complicated mathematical arguments in a startlingly logical and concise way. He had a special skill for pulling apart complex proofs and putting them together in a step by step fashion, and for condensing routine arguments to their bare outlines while, on the other side, displaying and explaining fully all technical or conceptual innovations. All this requires careful reading to see, mainly because difficult mathematics, even when presented brilliantly, is still difficult and demands careful study.

Inventory Theory and Queueing Theory

Many types of problems are essentially sequential in character. Jack's background in both Wald's decision theory and sequential analysis prepared him well to look at some of these problems; and he did so with important consequences jointly with Dvoretzky and Wolfowitz. The following description of the work in inventory theory is borrowed from the introduction to the research of Jacob Wolfowitz in Kiefer et al. [93].

"Dvoretzky, Kiefer, and Wolfowitz [2] on the inventory problem suggest a development which applies to the broader class of settings now referred to as discounted dynamic programming, in which an optimal policy is sought for adjusting a chance process so that the sum of discounted rewards over many time periods is maximized. A particularly simple form of policy suggested in an earlier work of Arrow, Harris, and Marshak (1951) is shown in Dvoretzky, Kiefer, and Wolfowitz [11] to be optimum under certain conditions. The case in which the chance law that governs the process is unknown and thus, in effect, has to be estimated as time periods pass $\cdots$ is treated in Dvoretzky, Kiefer, and Wolfowitz [3]."

Kiefer and Wolfowitz [12] was the first general systematic treatment of the multiserver queue. In this paper, "general results are obtained on the convergence in probability of waiting times and other quantities of interest in queueing systems. The methods yield results on random walks in Kiefer and Wolfowitz [14], such as characterization of random walks S_n based on i.i.d. summands for which $E(\max S_n)^k < \infty$.

Non parametric tests, and related distribution theory.

A colleague once told me that of all Jack's accomplishments he was most proud of those related to the Kolmogorov–Smirnov problem. Whether or not Jack ever expressed such an opinion, he had every right to be proud of his work in this area.

This work began with the joint paper of Kac, Kiefer, and Wolfowitz [13]. This paper first develops the distribution theory for tests of normality based on the Kolmogorov–Smirnov statistic and the Von-Mises statistic. Then the minimax principle again appears. It is convincingly demonstrated—via an asymptotic minimax formulation—that these tests are vastly preferable to the standard χ^2-test. Roughly, if the χ^2-test of size $\alpha < 1/2$ requires N observations (N large) to obtain a certain power β, against all alternatives at a distance (depending on α, β, N) from the null hypothesis, then the proposed Kolmogorov–Smirnov type test requires only $O(N^{4/5})$ observations.

The minimax principle is also a cornerstone of Dvoretzky, Kiefer, and Wolfowitz [15] and Kiefer and Wolfowitz [26]. These papers establish the asymptotic minimax property of the sample cumulative distribution function (CDF) as an estimate of the true CDF, in terms of the Kolmogorov–Smirnov distance and a variety of other measures of loss. [15] deals with univariate problems. [26] treats the multivariate case and, in passing, develops a streamlined heuristic argument and proof which is useful also in the univariate case. In a sense hinted in these papers as well as Kiefer and Wolfowitz [17], and made much more transparent by contemporary research, the sample CDF plays the role for this nonparametric problem that the maximum likelihood estimator plays for nice parametric problems.

This asymptotic minimaxity result for estimation is of a vastly superior nature compared to the crude order of magnitude argument of [13] for the testing problem. At the present time there is still no completely satisfactory asymptotic minimaxity result for the testing problem.

Kiefer and Wolfowitz returned to this asymptotic question in [69] and [70] where they give analogous results for the situation when the cumulative distribution function involved is known to be concave or, convex, or to satisfy certain other similar conditions.

In the univariate case the Kolmogorov–Smirnov statistic is distribution free (so long as the underlying distribution is continuous). In the multivariate case it is not. Kiefer and Wolfowitz were not the first to note this fact, but they were the first to do something concrete about it. In [22] they established a uniform bound on the tail of the Kolmogorov–Smirnov statistic.

Here is their crude bound. The (one-sided) statistic is $T_n = \sup\{\sqrt{n}\,(F_n(x) - F(x)):\ x \in \mathbb{R}^k\}$, where F denotes the underlying CDF on $\mathbb{R}^k$ and F_n denotes the sample CDF. Then, for some $\alpha = \alpha(k) > 0$ and $c = c(k) < \infty$,

$$P(T_n > r) \leq c e^{-\alpha r^2} \qquad \forall n, r, F. \tag{5}$$

This bound is certainly very crude but it did at least suffice to prove the existence of the limiting distribution of T_n as a Gaussian process. Although Kiefer and Wolfowitz established the existence of this limiting distribution, no one as yet has found any explicit form for it. Indeed there is only one nontrivial case where reasonably accurate bounds are known. This is when $F = U_2$, the uniform distribution on the unit square. Here the limiting distribution is that of a pinned Brownian sheet, and fairly close lower and upper bounds on the limiting distribution appear in Goodman (1976) and Cabaña and Wschebor (1982), respectively. In a classic paper, [29], Jack greatly improved this bound. He proved that for all $\varepsilon > 0$ there is a

$c = c(k, \varepsilon)$ such that

$$P(T_n > r) \leq c e^{-2(1-\varepsilon)r^2} \qquad \forall n, r, F. \tag{6}$$

This bound, (6), has since been widely generalized to other Gaussian processes, (see, for example Marcus and Shepp (1972) or Revesz (1976a, b)) but not until now basically improved except in the case $F = U_2$ mentioned above. This bound is of interest on its own merits and also because—as Jack noted—it enables one to establish a law of the iterated logarithm for the multivariate Kolmogorov–Smirnov statistic.

As a footnote I'm pleased to mention, because Jack would have been pleased to hear, that Robert Adler and I were recently able to capitalize on a remark in [22] to refine (6) to the bound: There is a $c = c(k)$ such that

$$P(T_n > r) \leq c r^{2(k-1)} e^{-2r^2} \qquad \forall n, r, F. \tag{7}$$

(A paper containing this result and a similar lower bound is currently in preparation.)

In a series of articles, more probabilistic than statistical in nature, Jack then pursued various distributional questions involving the one-dimensional sample CDF or—in [25] and [27] (with Blum and Rosenblat) the distance between several independent one-dimensional CDFs.

Jack was clearly fascinated as well by the search for the best methodology. Paper [25] is apparently the first non parametric test for the equality of several sample CDFs. (The test uses an appropriate generalization of the Kolmogorov–Smirnov statistic.) In this paper the methodology is to reduce the limiting question to one about several independent Brownian bridges. Paper [27] works out the limiting distribution of a statistic for testing independence, and this time the methodology is a characteristic function argument. Then, in [33], Jack published a largely expository paper describing how similar results could be derived via an argument based on the differential generator of the process.

Next comes an important paper, [48], on laws of the iterated logarithm for distributions involving the sample quantiles. Here is one basic result from this paper. Let $Y_{p,n}$ denote the sample pth quantile based on a sample of size n from a uniform distribution on $(0,1)$. Let $R_n(p) = Y_{p,n} - p + (F_n(p) - p)$, where F_n denotes the sample CDF. Let $\sigma_p^2 = p(1-p)$. Then

$$\limsup_{n \to \infty} \left[R_n(p) / (32/27)^{1/4} \sigma_p^{1/2} n^{-3/4} (\log\log n)^{3/4} \right] = 1 \quad \text{w.p.l.} \tag{8}$$

Related questions were profitably pursued also in [53], [54], and [56].

In the course of studying sample quantiles Jack first encountered still another methodology for investigating the sample CDF—the Skorohod–Strassen embedding, or strong invariance principle. [52] contains important results about this embedding.

Jack was methodologically prepared for a second major climax in his career. In addition, his statistical background provided the problem: the ordinary Skorohod embeddings supply a satisfactory approximation to the sample CDF for a single large n, but they do not provide the right joint distribution for several different large n at once. Jack's familiarity with the asymptotics of sequential analysis emphasized the importance of this joint distribution.

Here is the background for Jack's formulation. Let $X_1, \ldots$ be i.i.d. real random variables with continuous CDF F. Take F to be uniform on $(0,1)$ without loss of generality. Let $T_n(x) = n^{1/2}(F_n(x) - x)$. Let $\{B(x)\}$ denote the Brownian bridge on $[0,1]$ and $\{W(x_1, x_2)\}$ the Brownian sheet on $[0,\infty)\times[0,\infty)$. Note that $B(x) = n^{-1/2}[W(x,n) - xW(1,n)]$ for each $n \in (0,\infty)$. Brillinger (1969) had earlier proved that for each fixed n

$$|B(x) - T_n(x)| = O_p\left(n^{-1/4}(\log n)^{1/2}(\log\log n)^{1/4}\right) \qquad \forall x \in [0,1]. \qquad (9)$$

Jack now introduced in [58] the Gaussian process

$$K(x,t) = W(x,t) - xW(1,t) \qquad (10)$$

on $[0,1]\times[0,\infty)$. This process has since been called, "The Kiefer Process." (Note that $B(x) = n^{-1/2}K(x,n)$. The process K can be understood as a Brownian sheet conditioned to be zero on $\{1\}\times[0,\infty)$.) He then used all of his accumulated methodological knowledge and ingenuity. In a paper exploding with significant detail he established the useful uniform bound

$$\left|n^{-1/2}K(x,n) - T_n(x)\right| = O_p\left(n^{-1/6}(\log n)^{2/3}\right) \qquad \forall x, n \in [0,1)\times[0,\infty). \quad (11)$$

Jack conjectured the validity of a much better bound. A few years later Komlos, Major, Tusnady (1975) verified that

$$\left|n^{-1/2}K(x,n) - T_n(x)\right| = O_p\left(n^{-1/2}(\log n)^{2}\right) \qquad \forall x, n \in [0,1)\times[0,\infty). \quad (12)$$

(This result incidentally also improves on (9).) An analogous result for multivariate random variables (whose study was initiated in [22], as has been described) appears in Revesz (1976a, b).

Multivariate Analysis

Jack's activity in this area covers only a brief fragment of his career; although his work here has roots into, and connections with, other of his interests. These connections are especially strong in the important Giri and Kiefer paper [40] in which the local optimality criteria discussed are intimately related to those Jack had exploited in design papers such as [23] and [32]. The connections are perhaps weakest in the startling result in Kiefer and Schwarz [42], which establishes the Bayes character (for rather unusual appearing priors) of various standard multivariate tests.

Giri, Kiefer, and Stein [37] and Giri and Kiefer [41] are very complex proofs of minimaxity over invariant shells of Hotelling's T^2 and of R^2 only in the very simplest nontrivial case. The question for other sample sizes and dimensions resisted Jack's attempts. A partial solution appears in Šalaevskiĭ (1968, 1969). It is necessary to approach these multivariate problems on an individual basis because they necessarily fall outside the scope of a general Hunt–Stein theorem, such as the one Jack proved in [20]. Incidentally, nonparametric problems are also outside the scope of the Hunt–Stein theorem. Thus, for example, as Jack pointed out, it is still an open question whether the best invariant (or, "equivariant," depending on terminology) estimator of the population CDF is a minimax estimator, even though it was proved already in Dvoretzky, Kiefer, and Wolfowitz [15] to be asymptotically minimax.

Additional Comments

I have tried to describe how the minimax principle gently guided much of Jack's work. It could be said that Jack's professional life was also characterized by a form of this principle—he accomplished a maximum amount in a minimum of time.

His published works are a remarkable accomplishment. But it is not primarily through these that I remember him; and I know that my feelings in this are shared by many others.

Jack was above all a loyal friend, a delightful comrade and colleague, a statistician of unerringly high standards and aspirations for himself and the profession, and most important a deeply and sincerely warm and nice person.

REFERENCES

1. Arrow, K., Harris, T., and Marschak, J. (1951). Optimal inventory policy. *Econometrica* **19**, 250–272.
2. Bickel, P. and Yahay, J. (1971). Wiener process approximation to Bayes sequential testing problems. *Proc. 6th Berk. Symp. Math. Stat. and Prob.*, 57–82.
3. Bondar, J. V. and Milnes, P. (1981). Amenability: A survey for statistical applications of Hunt–Stein and related conditions on groups. *Z. Wahrsch. verw. Geb.* **57**, 103–128.
4. Brillinger, D. (1969). An asymptotic representation of the sample d.f. *Bull. Amer. Math. Soc.* **75**, 545–547.
5. Brown, L. D. (1978). A contribution to Kiefer's theory of conditional confidence procedures. *Ann. Statist.* **6**, 59–71.
6. Brown, L. D., Cohen, A., and Samuel-Cahn, E. (1983). A sharp necessary condition for admissibility of sequential tests. *Ann. Statist.* **11**, 640–653.
7. Cabaña, E. M. and Wschebor, M. (1982). The two parameter Brownian bridge: Kolmogorov inequalities and upper and lower bounds for the distribution of the maximum. *Ann. Prob.* **10**, 289–302.
8. Chernoff, H. (1959). Sequential design of experiments. *Ann. Math. Statist.* **30**, 755–770.
9. Goodman, V. (1976). Distribution estimates for functionals of the two parameter Wiener process. *Ann. Prob.* **4**, 977–982.
10. Komlos, J., Major, P., and Tusnady, G. (1975). An approximation of partial sums of independent R. V.'s and the sample DF, I. *Z. Wahrsch. und Verw. Gebiete*, **32**, 111–131.
11. LeCam, L. (1955). An extension of Wald's theory of statistical decision functions. *Ann. Math. Statist.* **26**, 69–81.
12. Lorden, G. (1976). 2-SPRT's and the modified Kiefer–Weiss problem of minimizing an expected sample size. *Ann. Statist.* **4**, 281–292.
13. Lorden, G. (1972). Likelihood ratio tests for sequential *k*-decision problems. *Ann. Math. Statist.* **43**, 1412–1427.

14. Marcus, M. B. and Shepp, L. A. (1972). Sample behavior of Gaussian processes. *Proc. 6th Berk. Symp. Math. Statist. and Prob. II*, 423–439.
15. Revesz, P. (1976a). On strong approximations of the multidimensional empirical process. *Ann. Prob.* **4**, 729–743.
16. Revesz, P. (1976b). Three theorems of multivariate empirical processes. *Lecture Notes in Math.* **566**, 106–126.
17. Robbins, H. and Munro, S. (1951). A stochastic approximation method. *Ann. Math. Statist.* **22**, 400–407.
18. Šalaevskiĭ (1968).
19. Šalaevskiĭ (1969).
20. Schwarz, G. (1962). Asymptotic shapes of Bayes sequential testing regions. *Ann. Math. Statist.* **33**, 224–236.
21. Wald, A. (1943). On the efficient design of statistical investigations. *Ann. Math. Statist.* **14**, 134–140.
22. Wald, A. (1947). *Sequential Analysis*. Wiley and Sons, New York.
23. Wald, A. (1950). *Statistical Decision Functions*. Wiley and Sons, New York.
24. Wolfowitz, J. (1951). On ε-complete classes of decision functions. *Ann. Math. Statist.* **22**, 461–465.
25. Zerdy, G. (1980). Risk of asymptotically optimum sequential tests. *Ann. Statist.* **8**, 1110–1122.

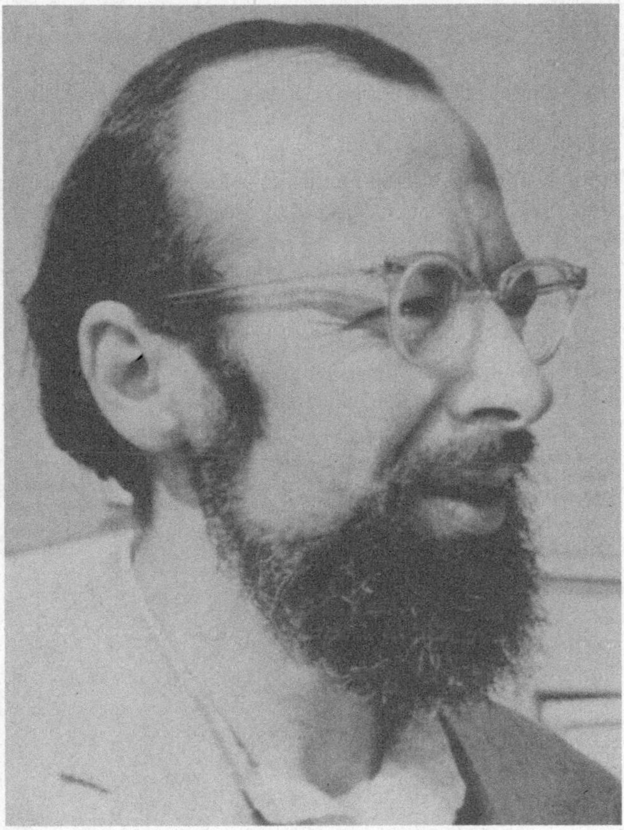

Jack Kiefer
Early Fall, 1970

Reprinted from THE ANNALS OF MATHEMATICAL STATISTICS
Vol. 22, No. 3, September, 1951

ALMOST SUBMINIMAX AND BIASED MINIMAX PROCEDURES[1]

BY P. FRANK AND J. KIEFER

Columbia University

Robbins [1] emphasized the notion of an "almost subminimax" procedure[2] and gave an example of such a procedure. The examples in this paper have been constructed with a view to simplicity and to the indication of the underlying mechanism which makes subminimax solutions exist in certain decision problems. At the same ime we point out another potentially undesirable property of a minimax procedure—biasedness.

All our examples fall within the following framework. A sample of one is taken from a population whose distribution is one of n given distributions: $F_1(x), F_2(x), \cdots, F_n(x)$. There are n decisions: $d_1, \cdots, d_n$. The weight function is $W(F_i, d_j) = 0$ if $i = j$ and $= 1$ otherwise. Instead of a finite number of F's, we may have a sequence of F's with a corresponding sequence of decisions. In all our examples each of the F's will be a uniform distribution over a finite interval of the x-axis, and our decision procedures will be randomized. These restrictions are made only for arithmetical simplicity.

With this setup, the risk when F_i is the true distribution is equal to the probability of not making decision d_i, which we will denote $r(F_i)$. We will not give an exact definition of an almost subminimax procedure, but just say that a procedure is almost subminimax if its maximum risk is "a little greater" than that of the minimax procedure (which risk is the same for all minimax procedures in our examples) and on the other hand its risk is "a lot less" than that of the minimax for "most of " the F's. Our examples will conform with this "definition" for almost any reasonable interpretation of the phrases in the quotes.

The first example will give an indication of the mechanism which makes a subminimax example possible. Let $F_1(x)$ be the uniform distribution on the interval $1 - a$ to 1, where $a > 0$ and small. Let $F_2(x)$ be the uniform distribution on the interval 0 to 1. An admissible minimax procedure to decide between d_1

---

[1] Research done under a contract with the Office of Naval Research.

[2] The examples of this paper fall into the framework of the definition in [1] of an "asymptotically subminimax solution" if each example is replaced by a sequence of examples whose a's approach zero. The present nomenclature was suggested as more suitable here.

and d_2 is to accept d_2 for $0 \leq x \leq 1 - a$ and for $1 - a < x \leq 1$ to accept d_1 with probability p_1 and accept d_2 with probability $1 - p_1$, where $p_1 = \frac{1}{1+a}$. With this procedure $r(F_1) = r(F_2) = \frac{a}{1+a}$.

Let us compare this procedure with the procedure which tells us to accept d_2 for $0 \leq x \leq 1 - a$ and to accept d_1 for $1 - a < x \leq 1$. For this procedure we have $r(F_1) = 0$, and $r(F_2) = a$. Thus we see that we can reduce the risk under F_1 to its absolute minimum 0, while increasing the risk under F_2 only slightly. (In fact, the ratio of the two risks under F_2, namely, a and $\frac{a}{1+a}$, approaches 1 as $a \to 0$.) This example, which may seem meaningless because $\frac{a}{1+a} \to 0$ as $a \to 0$, was given mainly to help in understanding the underlying mechanism in the almost subminimax example which follows. In the latter the maximum risk will be $> \frac{1}{2}$ for all a.

Let $F_3(x)$ be a uniform distribution from $-a$ to $1 - a$. In deciding between $F_2(x)$ and $F_3(x)$, an admissible minimax procedure is to accept d_2 for $1 - a < x \leq 1$, to accept d_3 for $-a \leq x < 0$, and to accept d_2 and d_3 with probability $\frac{1}{2}$ each for $0 \leq x \leq 1 - a$. The minimax risk is $\frac{1}{2}(1 - a)$. For a small, this is near $\frac{1}{2}$ and the two distributions are so intermeshed that there is little hope to disentangle them. When we now consider the problem of deciding between F_1, F_2, and F_3, we expect that the addition of F_1 can not do much to aggravate the difficulty already present in trying to decide between F_2 and F_3.

The following is an admissible minimax procedure for deciding between F_1, F_2, and F_3:

for $-a \leq x < 0$, we accept d_3;
for $0 \leq x < 1 - a$, we accept d_2 with probability p_2 and d_3 with probability $1 - p_2$;
for $1 - a \leq x \leq 1$, we accept d_1 with probability p_1 and d_2 with probability $1 - p_1$;

where $p_1 = \frac{1+a}{2+a}$ and $p_2 = \frac{1}{(2+a)(1-a)}$. For this procedure, $r(F_1) = r(F_2) = r(F_3) = \frac{1}{2+a}$.

Consider the alternative procedure which is exactly the same as the minimax procedure except that for $1 - a \leq x \leq 1$ we always accept d_1. For this procedure,

$$r(F_1) = 0; \quad r(F_2) = \frac{1}{2+a} + \frac{a}{2+a}; \quad r(F_3) = \frac{1}{2+a}.$$

Thus the alternative procedure reduces the risk from $\frac{1}{2+a}$ to 0 under F_1, increases it under F_2 by $\frac{a}{2+a}\left(\frac{a}{a+2} \to 0 \text{ as } a \to 0\right)$, and leaves it unaltered under F_3. The alternative looks more attractive.

This last example can be altered slightly so as to appear more striking. We can replace the distribution $F_1(x)$ by a sequence of distributions $F^1(x), F^2(x), \cdots,$ $F^n(x), \cdots$, where $F^n(x)$ is the uniform distribution on the interval

$$(1-a)+\frac{a}{n+1} < x \leq (1-a)+\frac{a}{n}.$$

Call this interval I_n, ($n = 1, 2, \cdots$). Corresponding to the distributions $F_2(x)$, $F_3(x)$, $F^1(x), \cdots, F^n(x), \cdots$, there are decisions d_2, d_3, $d^1, \cdots, d^n, \cdots$.

An admissible minimax procedure is described as follows:

for $-a \leq x < 0$, accept d_3;
for $0 \leq x \leq 1-a$, accept d_2 with probability p_2 and d_3 with probability $1-p_2$;
for $x \in I_n$, accept d^n with probability p_1 and d_2 with probability $1-p_1$;

where $p_1 = \frac{1+a}{2+a}$ and $p_2 = \frac{1}{(2+a)(1-a)}$. For this procedure,

$$r(F_2) = r(F_3) = \frac{1}{2+a} = r(F^j) \qquad \text{for } j = 1, 2, \cdots.$$

Consider the following alternative procedure:

for $-a \leq x < 0$, accept d_3;
for $0 \leq x \leq 1-a$, accept d_2 with probability p_2 and d_3 with probability $1-p_2$;
for $x \in I_n$, accept d^n;

where $p_2 = \frac{1}{(2+a)(1-a)}$. For this procedure,

$$r(F_2) = \frac{1}{2+a} + \frac{a}{2+a}; \qquad r(F_3) = \frac{1}{2+a}; \qquad r(F^j) = 0, \qquad j = 1, 2, \cdots.$$

If a is sufficiently small, the alternative procedure is certainly almost subminimax in the sense of our third paragraph: the maximum of the risk of the alternative procedure is only $\frac{a}{2+a}$ greater than that of the minimax procedure, and the alternative procedure has reduced the risk to zero for all except two of the distributions.

A decision procedure for deciding which of a class of distribution functions is the true distribution of X is said to be unbiased for F_i if $\text{Prob}(d_i \mid F_i) \geq$

$\text{Prob}(d_j \mid F_i)$ for all j. If a procedure is not unbiased for F_i, it will be said to be biased for F_i. In the next example every minimax procedure is biased for F_1.

Let $F_4(x)$ be the uniform distribution on the interval 0 to $1 - a$, with $a \leq \frac{1}{4}$. The problem is to decide between F_1, F_2, F_3, and F_4. An admissible minimax procedure is described as follows:

for $-a \leq x < 0$, accept d_3;
for $0 \leq x < 1 - a$, accept d_2 with probability p_2,
accept d_3 with probability p_3,
accept d_4 with probability $1 - p_2 - p_3$;
for $1 - a \leq x \leq 1$, accept d_1 with probability p_1,
accept d_2 with probability $1 - p_1$;

where $$p_1 = \frac{1 + a}{3}, \qquad p_2 = \frac{1 - a + a^2}{3(1 - a)}, \qquad p_3 = \frac{1 - 2a}{3(1 - a)}.$$

Thus, we have $\text{Prob}(d_2|F_1) = \dfrac{2 - a}{3} > \dfrac{1 + a}{3} = \text{Prob}(d_1|F_1)$.

This shows that the procedure is biased for F_1. By altering the procedure so that $p_2 = p_3 = \frac{1}{3}$ and $p_1 = 1$, we obtain a procedure which is unbiased for all F_i, and whose maximum risk is increased by only $\frac{2}{3}a$ over the minimax risk of $\dfrac{2 - a}{3}$.

The above example may be altered in the same way as the example of an almost subminimax solution so that there are infinitely many distributions for all but three of which the minimax solution is biased. In fact, it is possible to construct an example of a biased minimax solution for deciding among any number of distributions greater than two. It is impossible for a minimax procedure to be biased when there are only two distributions.

Along similar lines an example can be constructed for any $\epsilon > 0$ for a continuum of distributions, where any minimax procedure has constant risk of $\frac{2}{3} - \epsilon$ and is biased for all but three distributions, and where there exists an alternative almost subminimax procedure which is unbiased for all distributions and which reduces the risk to zero for all but three distributions where it is increased by less than 3ϵ.

REFERENCE

[1] H. Robbins, "Asymptotically subminimax solutions of compound statistical decision problems." *Proceedings of the Second Berkeley Symposium on Mathematical Statistics and Probability*, University of California Press, 1951.

THE INVENTORY PROBLEM: I. CASE OF KNOWN DISTRIBUTIONS OF DEMAND*

By A. Dvoretzky,[1] J. Kiefer, and J. Wolfowitz

The general solution of the inventory problem is developed. The solution applies also to a large class of problems in economics, biology, physics, etc., which can be put into the framework of a fairly general class of stochastic processes. In the present Part I of the paper the probability law of the stochastic process is assumed known; Part II will treat the case where this law is unknown.

1. Introduction

For simplicity of exposition and ease of understanding, this paper is written in the language of the inventory problem. However, the paper is actually devoted to the study of stochastic processes of a fairly general nature for which, at various instants of time, one may exert some degree of control over the distribution functions in question in order to achieve a desired result. Several applications of the methods and results of this paper to domains other than that of inventory control will be mentioned below.

The inventory problem is the general problem of what quantities of goods to stock in anticipation of future demand. Loss is caused by inability to supply demand (e.g., a store loses sales, soldiers in battle run out of ammunition) or by stocking goods for which there is no demand. Precise formulations of the problems are given below. It is already clear, however, that an optimum policy must strike a balance between overstocking and understocking.

Throughout Part I of this paper we shall assume that for the commodities in question the amounts demanded are chance variables with known distribution functions. In Part II we shall treat the problem under the assumption that the distribution functions are unknown.[2]

We shall always assume that the agency in charge of inventories (the ordering agency) can only order or not order goods, i.e., that returning or "dumping" of commodities is not permitted. However, if one desires to permit these actions, simple changes in the theory given below will suffice to take this into account; the chief change will be to allow $y < x$.

Throughout the paper we assume that the ordering agency must place a single order (which may be delivered in parts at several future times; see Section 5.3) for all of the commodities involved at every time point of the form of $n\tau$, where n is a nonnegative integer and τ a suitable time unit. (This, of course, includes the case where no order, i.e., a zero

* Research under contract with the Office of Naval Research.

[1] On leave of absence from the Hebrew University, Jerusalem, Israel.

[2] Part II of this article will appear in the July issue of Econometrica, 1952.

order, is placed.) By making τ sufficiently small this procedure can be made to correspond to any real physical situation, and in general approximates arbitrarily closely the "continuous" ordering procedure (where one can order at any time). The demand for the commodities during the time interval $[n\tau, (n + 1)\tau]$ will be assumed to manifest itself after the order has been placed. The general theory developed below permits that a demand for a commodity which is wholly or partially unsatisfied in one interval can be carried over to the next interval either wholly, partially, or not at all. Certain special cases treated below, however, involve restrictions on the carry-over of demand.

An optimum ordering policy depends, of course, upon the loss function W, which is a given function of all the quantities involved (including the demand, which is a chance variable), and measures the loss (or negative loss, i.e., gain) incurred in a certain set of circumstances characterized by the values of the arguments of W. In Sections 2, 3, 4, and 5 below we consider the situation where only one commodity is involved. In Section 6 we treat the case where more than one commodity is involved; the situation is formally little different from the one-commodity case, but the computational difficulties increase. In Section 2 the loss in the first interval only is considered. Section 3 treats the case where one considers the loss in N consecutive intervals. Section 4 discusses the case where the loss in infinitely many intervals is considered. In the case of stationarity (see also Section 5) a fundamental integral equation is derived the solution of which characterizes optimal policies. In Sections 2, 3, and 4 it is assumed that orders are filled instantly, without lag. In Section 5 we treat the situations of Sections 2, 3, and 4 in the presence of a lag in the delivery of goods ordered. Further generalizations are treated in Sections 5 and 6.

Throughout the paper we assume that the initial stock at the beginning of the first interval is given and not within the control of the ordering agency. (The initial stock at the beginning of an interval is the stock that the ordering agency possesses at the beginning of the interval. The starting stock is the initial stock plus the stock that arrives at the beginning of the interval. Thus the starting stock is the stock with which one starts the interval and before any demand has manifested itself.) If the initial stock at the beginning of the first interval is within the control of the ordering agency, one has only to choose it to be such that the total expected loss is a minimum.

As mentioned previously, the methods and results of this paper may be applied to many other problems in economics and other fields. For example, the problem of biological selection from a sequence of generations in order to obtain a population with desired characteristics falls naturally into the framework of Part II (and in special cases, Part I) of this paper. By taking τ to be sufficiently small, the setup of Part I may be used to treat the problems of designing control mechanisms with feedback to satisfy certain requirements. Similar problems in physics and other fields may obviously be treated by altering suitably the nomenclature used here.

Part I has many points of contact with the interesting papers of Pierre Massé,[3] and of K. J. Arrow, T. Harris, and J. Marschak,[4] (hereafter referred to as AHM). The reader is referred to both the above papers for further economic interpretations and to AHM for bibliography. Some of the problems treated here are formulated precisely in AHM, and most of the problems treated there are discussed below, in addition to other problems not treated in AHM. A principal difference in the treatment of the common problems is that AHM is limited to discussion of ordering policies of a certain special type (described in Example 1 below), while no such limitation is made by us. Several examples given below (e.g., Examples 2 and 5) show that an optimum ordering policy need not be of the type assumed in AHM.

2. CASE OF ONE TIME INTERVAL

2.1. As stated in the introduction, this section deals with a single time interval. Not only is this a highly important special case (cf. below), but also its correct understanding is necessary for the treatment of the general problem.

The case treated here is obviously a limiting case of those treated in Section 3. It occurs whenever there is a preponderant value attached to whatever happens in the immediate future. This seems to be the situation in certain problems of ammunition stocking.

2.2. We proceed to formulate the problem and the concepts involved.

Let $x (x \geqslant 0)$ be the *initial* stock at hand, i.e., the stock available before any ordering of new stock is done. Let $y (y \geqslant x)$ be the *starting* stock, i.e., the stock at the start of the time interval, after the *order* of $y - x$ additional stock has been filled. In this section we shall assume *instantaneous* fulfillment of orders. In Section 5 a time lag will be taken into account. If $y = x$ there is, of course, no ordering.

In this section we shall treat simultaneously the case where x is a continuous variable (e.g., stocking water) and the case where x assumes only discrete values (e.g., stocking tires). Whenever we are in the second (atomistic) case, the variables y and D (to be introduced below) are also restricted to integers. (The domains of x and y may also be restricted in any other consistent way throughout this paper.) We shall adopt the

[3] Pierre Massé, *Les réserves et la régulation de l'avenir dans la vie économique*, Paris: Hermann, 1946, Vol. 2, Avenir Aléatoire.

[4] K. J. Arrow, T. Harris, and J. Marschak, "Optimal Inventory Policy," ECONOMETRICA, Vol. 19, 1951, July, pp. 250–272.

terminology suggested by the continuous case; e.g., we shall speak of the curve $u = f(x)$, though in the atomistic case it degenerates to a sequence of points.

To every pair of numbers x, y with $0 \leqslant x \leqslant y$ there corresponds a real random variable $W(x, y)$ representing the *loss* that occurs when x is the initial and y is the starting stock. Following tradition, W represents loss; the gain is, of course, $-W$. The randomness may enter W through the *demand D* as well as in many other ways. For example, the cost of ordering and the value of the commodity to the consumer may also be chance variables, so that *speculation* on buying and selling prices may be involved in ordering. For simplicity, we shall assume in Sections 2–4 and 5.1–5.2 that the randomness enters W only through D; if one thinks of D as a vector, it is obvious that our results apply also when there are random elements other than the demand. The form of the setups of 5.3 and Section 6 will be seen explicitly to permit this more general treatment.

It may be objected that our method requires one to specify the function W and that this function may be unknown or difficult to give. We wish to emphasize that the need for a function W is inevitable in the sense that any method which does not explicitly use a function W simply uses one implicitly. Thus one who selects a method of solving the inventory problem which ostensibly has the advantage of not requiring the specification of W is simply relinquishing control of W, and may be implicitly using a W of which he would disapprove (if he knew it). It is difficult to see what advantages can accrue to the ordering agency from deliberately burying its intellectual head in the sand. Even if the function W is very difficult to obtain it seems preferable to make some attempt at an intelligent decision about it. A rough approximation or greatly simplified version of the underlying W may be preferable to completely ignoring this fundamental datum of the problem.

We assume, then, that the demand for the commodity during the time interval in which we are stocking is a nonnegative random variable D with a given distribution function,

$$F(d) = P\{D \leqslant d\}. \tag{2.1}$$

The function $F(d)$ may also depend on the quantities x and y. Given x, y, and that $D = d$, the loss is a specified, nonrandom function $W(x, y; d)$ of these quantities. For each pair x and y $(0 \leqslant x \leqslant y < \infty)$, $W(x, y; d)$ is assumed to be a Borel measurable function of d, so that $W(x, y; D)$ is a real-valued random variable.

Throughout this section we shall make the following assumption:

Assumption A: For all x and y with $0 \leqslant x \leqslant y < \infty$, *the expected value*

$$V(x, y) = \mathcal{E}\, W(x, y; D) \tag{2.2}$$

is finite, i.e.,[5]

$$\mathcal{E} \mid W(x, y; D) \mid < \infty. \tag{2.3}$$

Comparisons will be made on the basis of *expected* loss. Thus, if $y_1 > x$, $y_2 \geqslant x$, and $V(x, y_1) < V(x, y_2)$, it is preferable to order $y_1 - x$ rather than $y_2 - x$ when the initial stock is x.

An *ordering policy* is a function $Y(x)$ [with $Y(x) > x$] that associates with every x a value $Y = Y(x)$; the amount to be ordered is $Y(x) - x$.[6] A policy is *optimal* if, for every x,

$$V[x, Y(x)] = \min_{y \geq x} V(x, y). \tag{2.4}$$

Thus, if the minimum on the right side of (2.4) exists for every x, the problem of finding the amount to order is simply that of solving (2.4). In most practical applications, this minimum will exist for every x.[7] For simplicity, the remainder of this section will be presented for this case. However, the results of this section apply with obvious modifications when the minimum of (2.4) does not exist for every x. For example, an optimal policy will no longer exist for all x, but if $\inf_{y \geq x} V(x, y) > -\infty$ for every x, then for any $\epsilon > 0$ (no matter how small) one may obtain by our methods a policy $Y(x)$ which is *optimal to within* ϵ; i.e., for which

$$V[x, Y(x)] \leqslant \inf_{y \geq x} V(x, y) + \epsilon. \tag{2.5}$$

For those x for which $\inf_{y \geq x} V(x, y) = -\infty$, one may obtain similarly a policy $Y(x)$ for which the expected gain, $-V[x, Y(x)]$, is as large as desired.

[5] Actually it is sufficient to assume that the expected value of (2.2) exists for every x and y: i.e., that either $\mathcal{E}\{\mid W(x, y) \mid + W(x, y)\}$ or $\mathcal{E}\{\mid W(x, y) \mid - W(x,y)\}$ is finite for each x, y.

[6] One may also consider randomized ordering policies: a randomized ordering policy assigns to each x a distribution function $G(y; x)$ [with $G(x-; x) = 0$]. The amount $Y(x) - x$ to be ordered is then determined by an independent chance mechanism in such a way that $P\,[Y(x) \leqslant y] = G(y; x)$. For the sake of simplicity, Part I of this paper will be carried out using nonrandomized ordering policies as defined in the text above and in subsequent sections. Since the distribution of demand is assumed known, no randomized ordering policies are necessary. However, in Part II where the distributions are unknown, randomized ordering policies must be considered.

[7] A sufficient condition for the existence of this minimum is that, for every x, $V(x, y)$ is lower semi-continuous in y and that $\inf_{y \geq x} V(x, y) < \liminf_{y \to \infty} V(x,y)$.

2.3. Perhaps the simplest nontrivial example is a special case of one treated in AHM, pages 256 and 260. They assume that W has the functional form

$$W(x, y, d) = cy + c'(y, d) + c''(x, y), \tag{2.6}$$

with

$$c'(y, d) = \begin{cases} 0 \text{ if } d \leqslant y \\ A \text{ if } d > y \end{cases} \tag{2.7}$$

and

$$c''(x, y) = \begin{cases} 0 \text{ if } y = x \\ K \text{ if } y > x, \end{cases} \tag{2.8}$$

c, A, and K being positive constants. Here c may be interpreted as the carrying cost of a unit quantity of stock per unit time (assuming that this carrying cost is proportional to the size of the stock), $c'(y, d)$ represents the loss occurring whenever the demand exceeds the available supply (this loss A is assumed to be independent of the actual size of the excess of demand over supply), and $c''(x, y)$ represents the ordering cost [this cost occurs only when ordering is done (i.e., $y > x$), and K is assumed to be independent of the size of the order]. From (2.1) and (2.2) we have in this case

$$V(x, y) = cy + A[1 - F(y)] + \begin{cases} 0 \text{ if } y = x, \\ K \text{ if } y > x. \end{cases} \tag{2.9}$$

This function clearly satisfies the conditions of footnote 7.

A relatively simple generalization is to replace (2.8) by

$$c''(x, y) = K'(y - x) + \begin{cases} 0 \text{ if } y = x, \\ K \text{ if } y > x, \end{cases} \tag{2.10}$$

K' being a nonnegative constant. (This is the case $b_1 = 0$ treated on page 256 of AHM.) Our methods apply, however, to the most general case, where the functional form of W is not restricted to (2.6) (and, indeed, where other random elements in addition to the demand may affect the loss).

2.4. It is very useful to introduce the two auxiliary functions,

$$\lambda(x) = V(x, x) \qquad (x > 0), \tag{2.11}$$

and

$$\mu(x, y) = V(x, y) - \lambda(y) \qquad (y > x > 0). \tag{2.12}$$

$\lambda(x)$ is the expected loss when the initial stock is x and no ordering is done. It may be called the *unimproved expected loss*. $V(x, y)$ and $\lambda(y)$

both represent expected losses with starting stock y; however, when $y > x$, the second corresponds to the case when the initial stock was already y, while the first refers to the case when the quantity $y - x$ was ordered. Thus $\mu(x, y)$ represents the extra expense involved in ordering.

The solution of (2.4), i.e., the optimal ordering policy, can be restated in terms of μ and λ as follows: $Y(x) \geqslant x$ is a function of x for which

$$\lambda(y) + \mu(x, y) - \lambda(x) \tag{2.13}$$

achieves its minimum with respect to y. The minimum value of the expression in (2.13) is obviously nonpositive.

In the case of (2.6) we have

$$\mu(x, y) = c''(x, y) - c''(y, y), \tag{2.14}$$

so that no computation of expected values is involved. In the special case (2.10), (2.14) reduces to

$$\mu(x, y) = K'(y - x) + \begin{cases} 0 \text{ if } y = x, \\ K \text{ if } y > x. \end{cases} \tag{2.15}$$

This is typical of many problems where $\mu(x, y)$ is a function $\nu(x - y)$ of the difference $x - y$. In all such problems it is easy by plotting the function ν on transparent paper to have a graphical device for determining the optimal amount to be ordered.

In the case (2.8) treated in AHM, (2.15) takes the very simple form

$$\mu(x, y) = \begin{cases} 0 \text{ if } y = x, \\ K \text{ if } y > x. \end{cases} \tag{2.16}$$

In the general case the expression (2.13) suggests the following way (which is convenient graphically) of obtaining the optimal starting stock $y = Y(x)$ associated with any initial stock x. Plot the unimproved expected loss curve $v = \lambda(u)$ in the (u, v) plane. Plot for $u > x$ a second curve $v = \lambda(u) + \mu(x, u)$. (Under all realistic assumptions this second curve will lie above the first.) There are now two possibilities; either (i) no point of the second curve lies below the line $v = \lambda(x)$ or (ii) at least one point of that curve lies below this line. In case (i) an optimal policy would be to order nothing when the initial stock is x. If (ii) occurs then there exists a point $u = Y(x) > x$ at which the second curve assumes its absolute minimum. Then an optimal policy would be to order the amount $Y(x) - x$ when the initial stock is x.[8]

[8] If there are many points $Y(x)$ where the minimum is assumed, all such policies of ordering $Y(x) - x$ are equally good. A similar remark applies to case (*i*) when $v = \lambda(x)$ meets the second curve at one or more points.

2.5. *Illustrative examples.* We shall now consider several examples which will demonstrate the method of solution as well as the types of solutions which may be obtained (see especially Example 2).

EXAMPLE 1. In this example we assume that $V(x, y)$ is given by (2.9) with $A = 3$, $K = c = 1$. The distribution $F(z)$ of demand is given by

$$F(z) = \begin{cases} 0 & z < 0, \\ z & 0 \leqslant z \leqslant 1, \\ 1 & z > 1. \end{cases} \tag{2.17}$$

A simple calculation gives the following unique optimal ordering scheme: $Y(x) = 1$ if $x < \frac{1}{2}$, and $Y(x) = x$ otherwise. (One may actually take $Y(\frac{1}{2}) = \frac{1}{2}$ or $Y(\frac{1}{2}) = 1$).

The procedure of Example 1 illustrates ordering policies of the following type: There are given two numbers s, S, with $s \leqslant S$. Whenever $x < s$ one orders $S - x$ (i.e., $Y(x) = S$). Whenever $x > s$ one does not order (i.e., $Y(x) = x$). Our example above shows that such a procedure may be optimal. However, this need not always be so. To demonstrate this we give an example using the very simple scheme of (2.9) which falls within the case treated in AHM.

EXAMPLE 2. Let $c = 1$ in (2.9). The distribution $F(z)$ of demand is given by

$$F(z) = \begin{cases} 0, & z < 1, \\ \frac{1}{2}, & 1 \leqslant z < 2, \\ 1, & z \geqslant 2. \end{cases} \tag{2.18}$$

It is easy to verify that

$$V(x, x) = x + \begin{cases} A \text{ if } & x < 1, \\ A/2 \text{ if } 1 \leqslant & x < 2, \\ 0 \text{ if } & x \geqslant 2, \end{cases} \tag{2.19}$$

and that, for any $y > x$,

$$V(x, y) = y + K + \begin{cases} A \text{ if } & y < 1, \\ A/2 \text{ if } 1 \leqslant & y < 2, \\ 0 \text{ if } & y \geqslant 2. \end{cases} \tag{2.20}$$

Obviously, whenever $x > 2$, the optimal policy is $Y(x) = x$. If $1 \leqslant x < 2$, we have

$$V(x, y) - V(x, x) = \begin{cases} y - x + K - A/2 & \text{if } y > 2, \\ y - x + K & \text{if } x < y < 2; \end{cases} \tag{2.21}$$

while if $0 \leqslant x < 1$ we have

$$V(x, y) - V(x, x) = \begin{cases} y - x + K - A & \text{if } y > 2, \\ y - x + K - A/2 & \text{if } 1 \leqslant y < 2, \\ y - x + K & \text{if } x < y < 1. \end{cases} \tag{2.22}$$

It follows that the optimal policy is given by

$$(2.23)\qquad Y(x) = \begin{cases} 2 \text{ if } 2 > x > \max(1, 2 + K - A/2), \\ 2 \text{ if } 1 > x > 2 + K - A \text{ and } A > 2, \\ 1 \text{ if } 1 > x > 1 + K - A/2 \text{ and } A \leqslant 2, \\ x \text{ in all other cases.} \end{cases}$$

Any other optimal policy can deviate from (2.23) only in the obvious cases of ambiguity (e.g., $A = 2$ or $x = 2 + K - A/2 > 1$). In order to show more clearly the forms which the optimal $Y(x)$ may take for various values of K, we shall consider two specific values of A. If $A = 3$, we have: if $0 < K \leqslant \frac{1}{2}$,

$$(2.24\text{a})\qquad Y(x) = \begin{cases} 2 \text{ if } 0 \leqslant x < 2, \\ x \text{ if } x > 2; \end{cases}$$

if $\frac{1}{2} < K \leqslant 1$,

$$(2.24\text{b})\qquad Y(x) = \begin{cases} 2 \text{ if } 0 \leqslant x < 1, \\ x \text{ if } 1 \leqslant x < K + \frac{1}{2}, \\ 2 \text{ if } K + \frac{1}{2} \leqslant x < 2, \\ x \text{ if } x > 2; \end{cases}$$

if $1 < K < \frac{3}{2}$,

$$(2.24\text{c})\qquad Y(x) = \begin{cases} x \text{ if } 0 \leqslant x \leqslant K - 1, \\ 2 \text{ if } K - 1 < x < 1, \\ x \text{ if } 1 \leqslant x \leqslant K + \frac{1}{2}, \\ 2 \text{ if } K + \frac{1}{2} < x < 2, \\ x \text{ if } x > 2; \end{cases}$$

if $\frac{3}{2} \leqslant K < 2$,

$$(2.24\text{d})\qquad Y(x) = \begin{cases} x \text{ if } 0 \leqslant x \leqslant K - 1, \\ 2 \text{ if } K - 1 < x < 1, \\ x \text{ if } x > 1; \end{cases}$$

if $K > 2$,

$$(2.24\text{e})\qquad Y(x) = x;$$

where in each case of ambiguity we have assigned some specific value $Y(x)$. On the other hand, if $A = 1$, we have:

if $0 < K < \frac{1}{2}$,

$$(2.25\text{a})\qquad Y(x) = \begin{cases} x \text{ if } 0 \leqslant x \leqslant K + \frac{1}{2}, \\ 1 \text{ if } K + \frac{1}{2} < x < 1, \\ x \text{ if } 1 \leqslant x \leqslant K + \frac{3}{2}, \\ 2 \text{ if } K + \frac{3}{2} < x < 2, \\ x \text{ if } x > 2; \end{cases}$$

if $K > \frac{1}{2}$,

$$Y(x) = x. \tag{2.25b}$$

Equations (2.24) show that it may occur that the optimal $Y(x)$ can be characterized by a pair s, S for some values of K [$s = S = 2$ in (2.24a)] but not for others. Equation (2.25a) exemplifies the case where there exist triplets (S_1^i, S_2^i, S_3^i) $(i = 1, 2, \cdots, m)$ with $S_1^i < S_2^i \leqslant S_3^i < S_1^{i+1}$ and such that $Y(x) = S_3^i$ if $S_1^i < x < S_2^i$ and $Y(x) = x$ otherwise. However, (2.24b) and (2.24c) show that $Y(x)$ need not have this characterization, since $S_3^1 > S_1^2$ in these cases. An example possessing the properties demonstrated in Example 2 may similarly be constructed using a distribution function possessing a density function.

3. CASE OF FINITELY MANY TIME INTERVALS

3.1. In this section we shall treat the case when $N(>1)$ time intervals are to be considered. The same symbols as in Section 2 will be employed, with subscripts to denote the interval to which they refer. Thus x_i and y_i will be the initial stock and starting stock, respectively, at the beginning of the ith interval $(i = 1, \cdots, N)$. As in Section 2, we shall assume that the order $y_i - x_i$ is filled instantaneously.

Let

$$\begin{aligned} B_i &= (x_1, \cdots, x_i, y_1, \cdots, y_i, D_1, \cdots, D_{i-1}) \\ \beta_i &= (a_1, \cdots, a_i, b_1, \cdots, b_i, d_1, \cdots, d_{i-1}) \end{aligned}$$

be the generic designation of a sequence of $(3i - 1)$ nonnegative numbers such that $a_j \leqslant b_j$ for $1 \leqslant j \leqslant i$. The function $F_i(z \mid B_i = \beta_i)$ will denote the probability that the demand D_i in the ith interval will not exceed z, given that $B_i = \beta_i$. Thus we allow F_i to be a function of any or all of the quantities (in B_i) which are known at the beginning of the ith interval. Thus the various D's need not be independent. We assume that $F_i(z \mid B_i = \beta_i)$ is always known. Finally, we define

$$\begin{aligned} B'_i &= (x_1, \cdots, x_i, y_1, \cdots, y_{i-1}, D_1, \cdots, D_{i-1}), \\ \beta'_i &= (a_1, \cdots, a_i, b_1, \cdots, b_{i-1}, d_1, \cdots, d_{i-1}), \\ B''_i &= (x_1, \cdots, x_i, y_1, \cdots, y_i, D_1, \cdots, D_i), \\ \beta''_i &= (a_1, \cdots, a_i, b_1, \cdots, b_i, d_1, \cdots, d_i). \end{aligned}$$

It is clear that in this present setup the quantities $(x_2, x_3, \cdots, x_N)$ are functions of quantities which depend upon chance. Thus x_2 is a function of (at least) y_1 and D_1, x_3 is a function of (at least) y_2 and D_2, etc.

An ordering policy is a set of functions

$$\{Y_i(x \mid B''_{i-1} = \beta''_{i-1})\} \qquad (i = 1, \cdots, N)$$

which we shall designate generically as $Y(x)$, and such that in the ith interval when the past history is $B''_{i-1} = \beta''_{i-1}$, one orders the amount $Y_i(x_i \mid B''_{i-1} = \beta''_{i-1}) - x_i$ [$Y_i(x) \geqslant x$ for all i and x; $B''_0 = \beta''_0$ means that there is no restriction; always $x_i = \max(y_{i-1} - d_{i-1}, 0)$].

Given the initial and starting stocks x_i and y_i for the ith interval, and that $B''_{i-1} = \beta''_{i-1}$, the loss $W_i(x_i, y_i; d_i \mid B''_{i-1} = \beta''_{i-1})$ incurred if $D_i = d_i$ is a Borel measurable function of d_i, so that $W_i(x, y; D \mid B''_{i-1} = \beta''_{i-1})$ is for fixed x, y, and β''_{i-1} a real-valued random variable. Its expected value $V_i(x, y \mid B''_{i-1} = \beta''_{i-1})$ will hereafter be written $V_i(x, y)$ for the sake of readability, it being understood that V_i like W_i may also depend on all the quantities in B''_{i-1}. As in Section 2, we assume that Assumption A is satisfied for every β''_{i-1} and all i (see also footnote 5).

Let α_i be the present value (at some time before any ordering is done) of a unit loss in the ith interval. We shall assume, as seems reasonable, that $0 \leqslant \alpha_i \leqslant 1$. Actually, this assumption is unnecessary. Indeed, the introduction of the α's in this section is completely unnecessary, as the α_i could be absorbed in the V_i. However, the α_i are economically plausible (see AHM, pp. 261–262) and their introduction here will lead naturally to their use in Section 4, where they are essential.

In order to avoid circumlocutions we shall make the following assumption:

Assumption B: There is a value $\gamma > -\infty$ such that for all $i \leqslant N$ and all $x \leqslant y$ (and all β''_{i-1}), $V_i(x, y) \geqslant \gamma$.

This assumption is made only to avoid the circumlocutions required when the integrals which will appear below are equal to $-\infty$; it can easily be weakened. Henceforth in writing $V_i(x, y)$ we shall always assume that $y \geqslant x$.

We may now assume, without loss of generality, that the γ of Assumption B is $\geqslant 0$. For if γ were negative we need only add $-\gamma$ to every V_i and compensate the one placing the orders by giving him the sum $-\gamma \sum_{i=1}^{N} \alpha_i$ before the ordering begins.

We now define, for any ordering policy $Y(x)$ and for $i \geqslant 1$, the following:

$$A_{ii}(y_i \mid B'_i = \beta'_i) = \alpha_i V_i(x_i, y_i) \tag{3.1}$$

and inductively for $1 \leqslant j \leqslant i - 1$,

$$\begin{aligned} A_{ij}&(y_j \mid B'_j = \beta'_j) \\ &= \int_0^\infty A_{i,(j+1)}[Y_{(j+1)}(x_{j+1} \mid B''_j = \beta''_j)]\, d_{d_j} P\{D_j < d_j \mid B_j = \beta_j\} \\ &= \int_0^{y_j} A_{i,(j+1)}[Y_{(j+1)}(y_j - d_j \mid B''_j = \beta''_j]\, dF_j(d_j \mid B_j = \beta_j) \\ &\quad + \int_{y_j^+}^\infty A_{i,(j+1)}[Y_{j+1}(0 \mid B''_j = \beta''_j)]\, dF_j(d_j \mid B_j = \beta_j). \end{aligned} \tag{3.2}$$

(Here, as well as throughout the paper, finite limit points are included in the range of integration unless otherwise noted.) In (3.2) we have used the fact that

$$x_{j+1} = \max (y_j - d_j, 0). \tag{3.3}$$

The question of the measurability of the various functions that occur as integrands in (3.2) must now be discussed. Because of Assumption B, all the integrals which occur there will exist and be finite or $+\infty$, if each integrand is measurable. This is obviously the case if D_i, x_i, and y_i can take only values which are nonnegative integral multiples of some fixed positive constant. This is certainly true in any real situation. For mathematical simplicity, however, it is sometimes convenient to allow D_i, x_i, and y_i to be continuous variables. Throughout the rest of this paper we shall assume either that the above restriction (to integral multiples) is satisfied, or else that the F_i and $V_i(x, y)$ are such (see, e.g., Example 1) that F_i is a Borel-measurable function of all the quantities on which it depends and that in the constructions below the functions Y_i^* and Q_i^* (see Section 5) may be chosen to be Borel-measurable functions of all the quantities on which they depend. We shall also hereafter restrict ourselves to ordering policies for which the Y_i and Q_i are Borel-measurable functions of these quantities.

For any policy $Y(x)$ and any x_1, we define

$$A[Y(x) \mid x_1] = \sum_{i=1}^{N} A_{i1}(y_1 \mid B_1' = \beta_1') = \sum_{i=1}^{N} A_{i1}[Y_1(x_1)]. \tag{3.4}$$

We define a policy Y^* to be *optimum* if

$$A(Y^* \mid x_1) = \inf_Y A(Y \mid x_1) \tag{3.5}$$

for every x_1, and to be *optimum to within* ϵ for a given $\epsilon > 0$ if

$$A(Y^* \mid x_1) \leqslant \inf_Y A(Y \mid x_1) + \epsilon \tag{3.6}$$

for every x_1. As in Section 2, we have not made assumptions to insure the existence of a policy Y^* satisfying (3.5), and we shall consider the problem of obtaining an optimum ordering policy solved if, for any $\epsilon > 0$, we give a policy $Y^*(x)$ satisfying (3.6). We now proceed to the construction of such a policy.

Define

$$\bar{V}_j(x_j, y_j; Y_{j+1}, \cdots, Y_N) = \sum_{i=j}^{N} A_{ij}(y_j \mid B_j' = \beta_j'), \tag{3.7}$$

where we do not exhibit the dependence of $\bar{V}_j$ on the quantities in B''_{j-1}. We note that

$$\bar{V}_1(x_1, y_1; Y_2, \cdots, Y_N) = A[Y(x) | x_1]. \tag{3.8}$$

Let any $\epsilon > 0$ be given, and choose $y_N = Y^*_N(x_N \mid B''_{N-1} = \beta''_{N-1})$ so that

$$\bar{V}_N(x_N, y_N) \leqslant \inf_y \bar{V}_N(x_N, y) + \epsilon' \tag{3.9}$$

with $\epsilon' = \epsilon/N$. Suppose $Y^*_{(j+1)}, \cdots, Y^*_N$, with $j \geqslant 1$, have been defined. Choose $y_j = Y^*_j(x_j \mid B''_{j-1} = \beta''_{j-1})$ so that

$$\bar{V}_j(x_j, y_j; Y^*_{j+1}, \cdots, Y^*_N) \leqslant \inf_y \bar{V}_j(x_j, y; Y^*_{j+1}, \cdots, Y^*_N) + \epsilon'. \tag{3.10}$$

From (3.9) and (3.10) it follows that this construction gives the desired result (3.6).

In the case where $\alpha_i = \alpha^{i-1}$, $0 < \alpha < 1$, the $V_i(x, y)$ do not vary with i and do not depend upon the quantities in B''_{i-1}, the D_i's are independently and identically distributed, and N is large, the procedure of finding an optimum policy (within ϵ) may be greatly eased by use of the theory of Section 4.3 below. If

$$\frac{\alpha^N}{1 - \alpha} \sup_{x \geq 0} \lambda(x) < \epsilon, \tag{3.11}$$

then an optimum procedure for the case of 4.3 must be optimum to within ϵ for the present case of N intervals. A similar remark applies to the use of policies which are optimum to within ϵ for the case of 4.3, as well as to the case of stationarity without independence of 5.3.

3.2. It is illuminating to consider some simple examples wherein there are a finite number >1 of time intervals under consideration.

Example 3. We assume that there are two time intervals, that the two demands are independently distributed, that $\alpha_1 = 1$, and that $F_i(x)$ and $V_i(x, y)$ are for $i = 1, 2$ the same as $F(x)$ and $V(x, y)$ in Example 1 of Section 2. An optimum ordering policy for the second time interval is then given by that of Example 1, namely, $Y^*_2(x_2) = 1$ if $x_2 = \frac{1}{2}$, $Y^*_2(x_2) = x_2$ otherwise (again, one may take $Y^*_2(\frac{1}{2}) = \frac{1}{2}$ or 1). It is easy to verify that, with this definition of y_2,

$$\bar{V}^*_2(x_2) = \inf_{y_2} V_2(x_2, y_2) = \begin{cases} 2, & x_2 < \frac{1}{2} \\ 3 - 2x_2, & \frac{1}{2} \leqslant x_2 < 1 \\ x_2, & x_2 \geqslant 1. \end{cases} \tag{3.12}$$

A simple calculation shows that an optimum $Y^*_1(x_1)$ is given by $Y^*_1(x_1) = x_1$ whenever $x_1 > 1$; and that, if $x_1 < 1$, $\bar{V}_1(x_1, y_1; Y^*_2)$ attains its minimum for $y_1 > x_1$ at $y_1 = 1$. Hence, for $x < 1$, $Y_1^*(x) = 1$ if and only if

$$\begin{aligned} 0 &\leqslant \bar{V}_1(x_1, x_1; Y_2^*) - \bar{V}_1(x_1, 1; Y_2^*) \\ &= 1 + 2\alpha_2 - 2x_1(1 + \alpha_2) - \alpha_2 \int_{x_1}^1 \bar{V}_2^*(x)\,dx. \end{aligned} \tag{3.13}$$

If $x_1 \leqslant \frac{1}{2}$, (3.13) is always satisfied. If $\frac{1}{2} < x < 1$, (3.13) becomes

$$\alpha_2 x^2 + (2 - \alpha_2)x - 1 \leqslant 0. \tag{3.14}$$

Thus we have

$$\begin{aligned} Y_1^*(x) &= \begin{cases} 1 & \text{if } x_1 \leqslant \frac{1}{2} - \frac{1}{\alpha_2} + \sqrt{\frac{1}{\alpha_2^2} + \frac{1}{4}}, \\ x_1 & \text{otherwise;} \end{cases} \\ Y_2^*(x_2) &= \begin{cases} 1 & \text{if } x_2 \leqslant \frac{1}{2}, \\ x_2 & \text{otherwise.} \end{cases} \end{aligned} \tag{3.15}$$

It is interesting to observe that the interval where $Y_1^* = 1$ includes the interval where $Y_2^* = 1$. This is a hedge against an ordering cost in the second interval. As $\alpha_2 \to 0$ and the second interval diminishes in importance,

$$\frac{1}{2} - \frac{1}{\alpha_2} + \sqrt{\frac{1}{\alpha_2^2} + \frac{1}{4}} \to \frac{1}{2},$$

i.e., Y_1^* resembles Y_2^* more and more closely because the situation approaches that of Example 1 where there is no second interval. The above example illustrates also that the optimum procedure for an N-interval situation may be characterized by N pairs of numbers (s_i, S_i) $(i = 1, \cdots, N)$, the optimum policy being given by $Y_i(x_i) = S_i$ if $x_i \leqslant s_i$, $Y_i(x_i) = x_i$ otherwise $(i = 1, \cdots, N)$. The pairs (s_i, S_i) will usually not be the same for all i, and in fact the optimum policy (as in the case of Section 2) need not be characterized by such pairs, as we shall see in the next example. This example will also be used to illustrate certain other phenomena.

Example 4. Again, we suppose that there are two time intervals with independent demands. $F_i(x)$ and $V_i(x, y)$ are for $i = 1, 2$ the same as $F(x)$ and $V(x, y)$ in Example 2 of Section 2. We again put $c = 1$ and $\alpha_1 = 1$. For simplicity and ease of comparison in what follows, we shall put $A = 2K + 1$, with $K > 1$. For typographical ease we write

$\alpha_2 = \alpha (0 \leqslant \alpha \leqslant 1)$. It is easy to verify from (2.23) that for any $K > 1$, an optimum $Y_2^*(x_2)$ is then given by that of (2.24b) for $K = 1$, $A = 3$; namely,

$$
Y_2^*(x_2) = \begin{cases} 2 \text{ if } 0 \leqslant x_2 < 1, \\ x_2 \text{ if } 1 \leqslant x_2 \leqslant \frac{3}{2}, \\ 2 \text{ if } \frac{3}{2} < x_2 < 2, \\ x_2 \text{ if } x_2 > 2; \end{cases} \tag{3.16}
$$

$$
\bar{V}_2^*(x_2) = \begin{cases} 2 + K & \text{if } 0 \leqslant x_2 < 1, \\ x_2 + K + \frac{1}{2} & \text{if } 1 \leqslant x_2 \leqslant \frac{3}{2}, \\ 2 + K & \text{if } \frac{3}{2} \leqslant x_2 < 2, \\ x_2 & \text{if } x_2 > 2. \end{cases} \tag{3.17}
$$

Using the expressions for $V(x, x)$ and $V(x, y)$ given in (2.19) and (2.20), an exhaustive examination of the function $\bar{V}_1(x_1, y_1; Y_2^*)$ yields the following results for an optimum $Y_1^*(x_1)$:

If $1 \leqslant K \leqslant \dfrac{2}{\alpha}$,

$$
Y_1^*(x_1) = \begin{cases} 2 & \text{if } 0 \leqslant x_1 < 1 \\ x_1 & \text{if } 1 \leqslant x_1 \leqslant \dfrac{3}{2} - \dfrac{\alpha}{4} \\ 2 & \text{if } \dfrac{3}{2} - \dfrac{\alpha}{4} < x_1 < 2 \\ x_1 & \text{if } x_1 \geqslant 2. \end{cases} \tag{3.18a}
$$

If $\dfrac{2}{\alpha} < K < \dfrac{2}{\alpha} + \dfrac{3}{2}$,

$$
Y_1^*(x_1) = \begin{cases} 3 & \text{if } 0 \leqslant x_1 < 1 \\ x_1 & \text{if } 1 \leqslant x_1 < \max\left(1, \dfrac{5}{2} - \dfrac{K\alpha}{2} - \dfrac{\alpha}{4}\right) \\ 3 & \text{if } \max\left(1, \dfrac{5}{2} - \dfrac{K\alpha}{2} - \dfrac{\alpha}{4}\right) \leqslant x_1 < 2 \\ x_1 & \text{if } x_1 \geqslant 2. \end{cases} \tag{3.18b}
$$

If $K \geqslant \frac{2}{\alpha} + \frac{3}{2}$,

$$
(3.18c) \qquad Y_1^*(x) = \begin{cases} 4 & \text{if } 0 \leqslant x_1 < 1 \\ x_1 & \text{if } 1 \leqslant x_1 < \max\left(1, \frac{7}{2} + \frac{\alpha}{2} - K\alpha\right) \\ 4 & \text{if } \max\left(1, \frac{7}{2} + \frac{\alpha}{2} - K\alpha\right) \leqq x_1 < 2 \\ x_1 & \text{if } x_1 \geqslant 2. \end{cases}
$$

Comparing the formulas (3.18) with (3.16), we again find illustrations of the situation where the set of values of x_1 for which it pays to order is greater if there is a second time interval to follow than if there is only one interval. In fact, not only is this so, but in (3.18b) and (3.18c) we see that it also pays to order larger amounts, and that these amounts increase as K (and hence A) becomes larger relative to the carrying cost $c = 1$, as was to be expected. Again, for any fixed K, as $\alpha \to 0$, $Y_1^*(x_1)$ is given by (3.18a) and approaches the optimal policy for the case where there is only one interval. Finally, we note again that in an N-interval example it may occur that some or all of the optimum $Y_i(x_i)$ may not be characterized by pairs (s_i, S_i).

4. CASE OF INFINITELY MANY TIME INTERVALS

4.1. In Section 4 we shall treat the case where an infinite number of time intervals are under consideration. In 4.1 we study the general formulation of this case, the remainder of the section being devoted to a special case of great importance in applications, namely that of stationarity (see also 5.3) and independence.

We shall use the nomenclature of Section 3, the symbols D_i, x_i, y_i, $Y_i(x)$, $Y(x)$, $V_i(x, y)$, $F_i(z)$, and α_i having the same meaning as before (the unexhibited dependence of Y_i, V_i, and F_i being also as in Section 3), except that now the index i runs through all the positive integers. Throughout Section 4 we shall make Assumptions A and B with $\gamma = 0$[9] for all i (see also footnote 5). We shall also assume throughout Section 4 that the order $y_i - x_i$ is filled instantaneously. The functions A_{ij} are as in Section 3, and $\bar{V}_j$ is again defined by (3.7) except that the summation over i is now from j to ∞. An optimal policy will be one which minimizes $\bar{V}_1 = A(Y \mid x_1)$ for every x_1. As before we shall

[9] We have already explained that no loss of generality is involved in assuming $\gamma \geq 0$. (Even if Assumption B is weakened to $V_i(x, y) \geq \gamma_i$, it can be reduced to this case provided that $\Sigma\alpha_i \mid \gamma_i \mid$ converges.) Questions of measurability may be handled in a manner similar to that of Section 3.

consider our problem solved if for any given $\epsilon > 0$ we can find a policy Y^* which satisfies (3.6) for every x_1.

In the discussion of the paragraph which follows, we make the following two assumptions:

ASSUMPTION C.

$$\sum_{i=1}^{\infty} \alpha_i < \infty.$$

ASSUMPTION D.

$$\lambda_i(x) = V_i(x, x \mid B''_{i-1} = \beta''_{i-1}) < M < \infty$$

for all i, x, and all complexes β''_i.

(This assumption is not as restrictive as it may first appear: see the last paragraph of 4.3).

Let any $\epsilon > 0$ be given, and let $N = N(\epsilon)$ be such that $\sum_{i=N+1}^{\infty} \alpha_i < \epsilon/2M$. Consider the rule of action for all intervals after the Nth which puts $\tilde{Y}_i(x) = x$, $i > N$. Whatever the policy for the first N intervals, if the above rule is used for $i > N$, the corresponding $\tilde{V}_{N+1}$ is $< \epsilon/2$. For $i \leqslant N$, starting with the Nth interval and working backwards, we use the procedure of Section 3 to calculate a policy $(Y_1^{**}, \cdots, Y_N^{**})$ whose loss over the first N intervals is within $\epsilon/2$ of the infimum. Then $(Y_1^{**}, \cdots, Y_N^{**}, \tilde{Y}_{N+1}, \tilde{Y}_{N+2}, \cdots)$ is a policy which is optimal to within ϵ.

If N is large the above procedure will be computationally onerous. It can be considerably simplified in the case of stationarity and independence by use of Sections 4.2–4.5 (see also Section 5.3).

Assumptions C and D may be modified in many ways, the essential point being the existence for every $\epsilon > 0$ of a policy for all intervals after the Nth such that the corresponding $\tilde{V}_{N+1}$ is $< \epsilon$ for all possible past histories $\acute{\beta}_{N+1}$ resulting from the use of some policy which is optimum (to within ϵ) over the first N intervals. For example, the above method of "truncating" after N intervals and not ordering thereafter may be used if for each $\epsilon > 0$ one can show the existence of a finite value $B = B(\epsilon)$ such that (i) for no policy optimum to within ϵ does one order if $x_i \geqslant B$ nor order to an amount $> B$ if $x_i < B$ ($i = 1, 2, \cdots$, ad inf.) and such that (ii) there exists a finite-valued function $M(z)$ such that for every z

$$\sum_{i=1}^{\infty} \alpha_i \sup_{x \leq z} V_i(x, x) < M(z), \tag{4.1}$$

where for each i the supremum in (4.1) is taken over all β''_{i-1} as well as over all $x \leqslant z$. In this case, as well as in many other cases where Assumption D is not satisfied, it will be possible to use the method of "truncation" after N intervals as given above, if N is allowed to depend on x_1 as well as on ϵ. The reason that it is useful to give x_1 special treatment is that in many applications the number N corresponding to a given ϵ may be chosen so as to depend on x_1 but not necessarily on the subsequent history of the process, and in fact the smallest such N will depend on x_1 in such a way that $\lim_{x_1 \to \infty} N = \infty$.

4.2. As an important instance of the case considered in 4.1, we shall consider in the remainder of Section 4 a situation which could be described as one of "stationarity and independence," the first term signifying independence of time, and the second term indicating the independence of the chance variables of demand in different time intervals [for the case of stationarity without independence see equations (5.19) and (5.20) below]. More precisely, we shall assume that the loss function $W_i(x_i, y_i; d_i \mid B''_{i-1} = \beta''_{i-1})$ is a function only of x_i, y_i, and d_i, the function W_i being the same in every time interval; and that the chance variables of demand in different time intervals are independently and identically distributed, so that $V_i(x_i, y_i)$, the expected loss from the ith interval, is a function only of x_i and y_i, and it is appropriate to write $V(x_i, y_i)$ without a subscript on V. We shall also assume that $\alpha_i = \alpha^{i-1}$, where $0 < \alpha < 1$.

It is evident from our assumptions that nothing is to be gained by letting $Y_n(x)$ depend on any variables other than x. Moreover, it will become clear in what follows that, in seeking an optimum ordering policy, it suffices to find one for which the functions $Y_i(x)$ are the same function $Y_1(x)$; any ordering policy may be replaced by another of this type which is at least as good as the first in the sense to be described below.

We define $\ell(z)$ by

$$\ell(z) = \inf_Y A[Y(x) \mid z]. \tag{4.2}$$

Because of Assumption B with $\gamma = 0$, we have $\ell(z) > 0$. As in the case of Section 2, we shall carry on the remainder of our argument as if the infimum in (4.2) is actually attained, with obvious modifications applying when this is not the case. Note, however, that the existence of this minimum is not necessary for the validity of any of the discussion of 4.3. The discussion following (4.10) gives a bound on the extra loss which is incurred if the infimum of (4.2) is not attained for all x, and one orders at the ith stage an amount $y_i - x_i$ such that (4.8) is within ϵ of its infimum.

The distribution function[10] $F^*(v)$ of x_2, i.e.,

$$F^*(v) = P\{x_2 < v\}, \tag{4.3}$$

[10] The $<$ sign in (4.3) may be seen from (4.4) to be consistent with the $\leqslant$ sign in the definition of the distribution of demand given in (2.1).

is a function of y_1 only and is given as follows:

$$F^*(v) = \begin{cases} 1 & \text{when} \quad v > y_1 \\ 1 - F(y_1 - v) & \text{when } 0 < v \leqslant y_1 \\ 0 & \text{when} \quad v \leqslant 0. \end{cases} \tag{4.4}$$

From the independence and stationarity assumptions it follows that the present expected value (before any ordering is done, of the sum of the losses in all intervals except the first, incurred by entering the second time interval with a stock of amount x_2 and then proceeding in the optimum manner, is

$$\alpha \int_0^\infty \ell(v)dF^*(v) = \alpha \int_0^{y_1} \ell(y_1 - v)dF(v) + \alpha\ell(0)[1 - F(y_1)]. \tag{4.5}$$

We now readily conclude that $\ell(x)$ is a solution of the integral equation

$$\ell(x) = \inf_{y \geq x} \left\{ V(x, y) + \alpha\ell(0)[1 - F(y)] + \alpha \int_0^y \ell(y - v)dF(v) \right\}. \tag{4.6}$$

The function $\ell(x)$ is of fundamental importance for our problem. For the characterization of an optimal ordering policy, the function

$$h(y) = \lambda(y) + \alpha \int_0^y \ell(y - v)dF(v) + \alpha\ell(0)[1 - F(y)] \tag{4.7}$$

is of very great importance. An optimal ordering policy is defined as follows: At the beginning of the ith interval ($i = 1, 2, \cdots$, ad inf.), order the amount $y_i - x_i$, where y_i is the smallest quantity $\geq x_i$ for which [compare (2.13)]

$$h(y_i) + \mu(x_i, y_i) - h(x_i) \tag{4.8}$$

achieves its minimum.[11] Equation (4.6) can be written as

$$\ell(x) = \inf_{y \geq x} [h(y) + \mu(x, y)]. \tag{4.9}$$

Suppose $h(x)$ is not available but one employs an approximation $h_1(x)$, which is perhaps obtained by an iterative process such as will be described below in (4.21). If

$$| h_1(x) - h(x)| < \epsilon \tag{4.10}$$

for every x, then the loss incurred by using $h_1(x)$ instead of $h(x)$ will not exceed $\ell(x)$ by more than $\epsilon/(1 - \alpha)$. If one uses as y_i a quantity for

[11] We omit (compare the remark following (4.2)) consideration of the existence of a value for which (4.8) achieves its minimum, or of the existence of the smallest of such values. If $\lambda(x)$ is bounded in every finite interval the same is true of $\ell(x)$ and $h(x)$. If $\mu(x, y)$ and $h(x)$ are not finite, obvious modifications are necessary in the rule of (4.8).

which (4.8) is within ϵ of its infimum, this is equivalent to using an approximation $h_1(x)$ which satisfies (4.10).

We now define $If(x)$, the result of the operator I acting on the function $f(x)$ (until further notice, v, x, and y will be assumed to range over the nonnegative half-line) as

$$I\,f(x) = \inf_{y \geq x} \, [f(y) + \mu(x, y)]. \tag{4.11}$$

We define the operator G by

$$Gf(x) = \lambda(x) + \alpha \int_0^x f(x - v)dF(v) + \alpha f(0)[1 - F(x)]. \tag{4.12}$$

It is easy to verify that (4.6) may be written as

$$\ell(x) = IG\ell(x). \tag{4.13}$$

From (4.7) and (4.9), we see similarly that $h(x) = G\ell(x)$ satisfies the equation

$$h(x) = GIh(x). \tag{4.14}$$

In the next subsection we will give an iterative method of solving (4.13) and (4.14) and we will prove, under weak restrictions, that (4.13) and (4.14) have unique solutions which are bounded in every finite interval [and which must thus satisfy (4.2) and (4.7)]. Here we may remark that $\ell(x)$ satisfies

$$\ell(x) = I\ell(x) \tag{4.15}$$

whenever $\mu(x, y) + \mu(y, z) \geqslant \mu(x, z)$ for all x, y, and z such that $x \leqslant y \leqslant z$. This could be used to help gauge the closeness of approximation of functions used to approximate $\ell(x)$. However, the usefulness of (4.15) is limited by the fact that it obviously does not have a unique solution, so that it represents a necessary but not sufficient condition on the $\ell(x)$ of (4.2).

4.3. Since $\alpha < 1$, Assumption C is automatically satisfied in the present case. In this subsection we assume that Assumption D (as well as A and B) is also satisfied.

If Assumptions A, B with $\gamma = 0$ (see footnote 9), and D are satisfied, the functions $\ell(x)$ and $h(x)$ defined by (4.2) and (4.7) are bounded for $0 \leqslant x < \infty$ [since the total loss incurred by never ordering anything is no greater than the expression (4.23) below]. We shall now show that if the above assumptions are satisfied, (4.13) and (4.14) have unique bounded solutions, which hence must be the functions of (4.2) and (4.7).

Since the $\ell(x)$ of (4.2) satisfies (4.13), the latter has at least one bounded solution, say $\ell(x)$. Let $g(x)$ be any bounded measurable function over $0 \leqslant x < \infty$. Let

$$M = \sup_{x \geq 0} |\ell(x) - g(x)|. \tag{4.16}$$

Then $M < \infty$, and by (4.12)

$$\sup_{x \geq 0} |G\ell(x) - Gg(x)| \leqslant \alpha M. \tag{4.17}$$

It follows easily from (4.11) that

$$\sup_{x \geq 0} |IG\ell(x) - IGg(x)| \leqslant \alpha M. \tag{4.18}$$

Repeating the above argument n times, we obtain

$$\sup_{x \geq 0} |(IG)^n \ell(x) - (IG)^n g(x)| \leqslant \alpha^n M \tag{4.19}$$

for any positive integer n, where $(IG)^n f(x)$ represents alternate applications of the operators G and I (in that order) to the function f, n times for each. Since $\ell(x)$ is a solution of (4.13), $(IG)^n \ell(x) = \ell(x)$. If $g(x)$ were also a solution, then $(IG)^n g(x) = g(x)$. Since n can be arbitrarily large and $\alpha < 1$, (4.19) shows that $\ell(x) \equiv g(x)$, which was to be proved. A similar argument, employing $(GI)^n$, shows that, under the conditions we have postulated, there is a unique bounded solution of (4.14).

We notice also that, if $g(x)$ is any bounded measurable function over $0 \leqslant x < \infty$, (4.19) shows that

$$\lim_{n \to \infty} (IG)^n g(x) = \ell(x) \tag{4.20}$$

uniformly in x, and a similar argument shows that

$$\lim_{n \to \infty} (GI)^n g(x) = h(x) \tag{4.21}$$

uniformly in x. Equations (4.20) and (4.21) describe iterative processes which will lead, uniformly in x, to solutions of (4.13) and (4.14). Using (4.19) and an analogous argument leading to (4.21), it is simple to give a bound on the speed of approach of (4.20) and (4.21), namely

$$\sup_{x \geq 0} |J^n g(x) - k(x)| \leqslant \alpha^n \sup_{x \geq 0} \max [|g(x) - R|, |g(x)|] \tag{4.22}$$

where

$$R = \frac{1}{1-\alpha} \sup_{x \geq 0} V(x, x) \tag{4.23}$$

and $k(x) = \ell(x)$ or $h(x)$ according as $J = IG$ or GI. These bounds may be improved by improving the bound R on $\ell(x)$ and $h(x)$, and also by considering $|(IG)^n g(x) - k(x)|$ for a fixed x. For example, $\ell(y)$ is no greater than (4.23) with the supremum taken over those x for which $0 \leqslant x \leqslant y$.

We remarked in Section 2 that the domains of x and y may be restricted in a consistent manner throughout the paper. In that case the operators G and I will be modified in an obvious manner, and the theory above is entirely valid, except for these modifications. One restriction which will always occur in any application to a physical situation is that, for every i, $y_i \leqslant Z$, where Z is a finite positive number. In every real physical situation the total amount of stock must surely be bounded. This restriction need in no way affect $F(z)$. If such a Z is given or assumed, then $V(x, y)$ need be defined only for $0 \leqslant x \leqslant y \leqslant Z$, the functions $Y_n(x)$ which characterize an ordering policy are now defined for $0 \leqslant x \leqslant Z$ and satisfy the inequality $x_n \leqslant Y_n(x_n) \leqslant Z$ for $n = 1, 2, \cdots$, ad inf., and the infimum in (4.11) is taken over those y for which $x \leqslant y \leqslant Z$. If the existence of such a Z is not postulated, but if $\lambda(x)$ is bounded in every finite interval, and if to every Z_0 there corresponds a finite $Z_1 > Z_0$ such that, for every $Z_2 > Z_1$, the bounded solutions to (4.13) corresponding to $Z = Z_1$ and $Z = Z_2$ are identical for all $x \leqslant Z_0$, then the $\ell(x)$ which is uniquely defined for each x as the solution to (4.13) for Z sufficiently large, is obviously the unique solution to (4.13) (with no Z imposed) which is bounded in every finite interval. In the next subsection we shall see an example of this.

4.4. We shall now give an example of the solution to (4.13).

EXAMPLE 5. We assume that there is independence and stationarity, and that $F_i(x)$ and $V_i(x, y)$ are for $i = 1, 2, \cdots$, ad inf., the same as $F(x)$ and $V(x, y)$ in Example 2 of Section 2. We shall assume that $c = 1$, $A = 3$, and that $0 \leqslant K \leqslant 1 + \alpha/4$. We shall show that the solution to (4.13) for every $Z > 2$, and thus the $\ell(x)$ of (4.2), is given by

$$\ell^0(x) = \begin{cases} \beta & \text{if } x < 1 \\ x + \alpha\beta + \frac{3}{2} & \text{if } 1 \leqslant x < r + 1 \\ \beta & \text{if } r + 1 \leqslant x < 2 \\ x + \frac{1}{2}\alpha[\ell^0(x-1) + \ell^0(x-2)] & \text{if } x > 2 \end{cases} \tag{4.24}$$

where the last line of (4.24) defines $\ell^0(x)$ for $N \leqslant x < N + 1$, successively for $N = 2, 3, \cdots$, and where

$$r = \begin{cases} 0 & \text{for } 0 \leqslant K \leqslant \frac{1}{2} \\ \dfrac{K - \frac{1}{2}}{1 + \frac{\alpha}{2}} & \text{for } \frac{1}{2} \leqslant K \leqslant 1 + \frac{\alpha}{4}, \end{cases} \tag{4.25}$$

and

$$(4.26)\qquad \beta = \begin{cases} \dfrac{K+2}{1-\alpha} & \text{for} \quad 0 \leqslant K \leqslant \dfrac{1}{2} \\[2ex] \dfrac{K + 2 + \frac{5}{4}\alpha}{(1-\alpha)\left(1+\frac{\alpha}{2}\right)} = \dfrac{r+\frac{5}{2}}{1-\alpha} & \text{for} \quad \dfrac{1}{2} \leqslant K \leqslant 1 + \dfrac{\alpha}{4}. \end{cases}$$

The corresponding optimum policy, unique except at $x = r + 1$ or when $K = \frac{1}{2}$ or $K = 1 + \alpha/4$, will be shown to be

$$(4.27)\qquad Y(x) = \begin{cases} 2 \text{ if } 0 \leqslant x < 1 \\ x \text{ if } 1 \leqslant x < r + 1 \\ 2 \text{ if } 1 + r \leqslant x < 2 \\ x \text{ if } x \geqslant 2. \end{cases}$$

If $r = 0$, the first three lines of (4.24) and (4.27), respectively, simplify to $\ell^0(x) = \beta$ if $x < 2$ and $Y(x) = 2$ if $x < 2$. For simplicity, we shall assume in what follows that $Z \geqslant 2$ is an integer, although it will become clear that this restriction is unnecessary. By direct computation, one may verify that $\ell^0(2) \leqslant \ell^0(1)$ and that $\ell^0(2) < \ell^0(3) < \ell^0(4)$. For any integer $N > 4$, we then have inductively

$$(4.28)\qquad \begin{aligned} \ell^0(N) &= N + \tfrac{1}{2}\alpha[\ell^0(N-1) + \ell^0(N-2)] \\ &> N - 1 + \tfrac{1}{2}\alpha[\ell^0(N-3) + \ell^0(N-2)] = \ell^0(N-1). \end{aligned}$$

It may also be verified inductively that for integral $N \geqslant 0$, $\ell^0(x)$ is a nondecreasing (strictly increasing if $N \geqslant 2$) linear function of x in each of the intervals $N \leqslant x < N + r$ and $N + r \leqslant x < N + 1$, continuous at $x = N + r$ if $r > 0$. From this, (2.19), (2.20), (4.28), and the fact that

$$(4.29)\qquad G\ell^0(x) - \lambda(x) = \tfrac{1}{2}\alpha[\ell^0(x-1) + \ell^0(x-2)]$$

where we put $\ell^0(y) = \beta$ for $y < 0$, it follows that for $x \geqslant 2$, either

$$(4.30)\qquad IG\ell^0(x) = x + \tfrac{1}{2}\alpha[\ell^0(x-1) + \ell^0(x-2)] = \ell^0(x),$$

or else

$$(4.31)\qquad \begin{aligned} IG\ell^0(x) = \{x\} + K + \tfrac{1}{2}\alpha[\ell^0(\{x\} - 1) \\ + \ell^0(\{x\} - 2)] = \ell^0(\{x\}), \end{aligned}$$

where $\{x\}$ is the smallest integer $> x$. From our previous remarks, we see that (4.30) will hold for all $x > 2$ if and only if for every $x > 2$ which is not an integer we have

$$\ell^0(\{x\}^-) - \ell^0(\{x\}) \leqslant K. \tag{4.32}$$

By direct computation it may be shown that (4.32) holds for $\{x\} = 3$, 4. For any integer $N > 4$ we then have inductively

$$\begin{aligned} \ell^0(N^-) - \ell^0(N) &= \tfrac{1}{2}\alpha\{\ell^0[(N-1)-] - \ell^0(N-1) \\ &\quad + \ell^0[(N-2)-] - \ell^0(N-2)\} \leqslant \alpha K \leqslant K. \end{aligned} \tag{4.33}$$

We have thus proved (4.30) for all $x > 2$, the uniqueness of the last line of (4.27) following from the previously mentioned fact that for $2 \leqslant N \leqslant x < N + 1$, $\ell^0(x)$ is strictly increasing in x.

For $x < 1$, it follows from our previous remarks that either

$$IG\ell^0(x) = x + 3 + \alpha\beta \tag{4.34}$$

or else that

$$IG\ell^0(x) = K + \ell^0(2) = \beta. \tag{4.35}$$

In order that (4.35) hold for all $x < 1$ and that the first line of (4.27) be unique, it is necessary and sufficient that [with strict inequality for uniqueness of $Y(0)$]

$$0 \leqslant 3 + \alpha\beta - \beta. \tag{4.36}$$

From (4.26) and our restriction on K, it is easy to verify that (4.36) holds, with strict inequality except at $K = 1 + \alpha/4$.

Similarly, for $1 \leqslant x < 2$, either

$$IG\ell^0(x) = x + \tfrac{5}{2} + \alpha\beta \tag{4.37}$$

or else (4.35) holds. A necessary and sufficient condition on x that (4.35) hold (with $1 \leqslant x < 2$) is that (with ambiguity of $Y(x)$ only in the case of equality)

$$0 \leqslant x + \tfrac{5}{2} + \alpha\beta - \beta = \begin{cases} x - K - \tfrac{1}{2} \text{ for } 0 \leqslant K \leqslant \tfrac{1}{2} \\ x - r - 1 \quad \text{for } \tfrac{1}{2} \leqslant K \leqslant 1 + \alpha/4. \end{cases} \tag{4.38}$$

This completes the proof that (4.24) is the unique bounded solution of (4.13) for every $Z > 2$, and thus also satisfies (4.13) for $0 \leqslant x < \infty$. We see that (4.27) is the unique optimum policy, except at the points of ambiguity noted above.

$\leqslant \frac{1}{2}$, $r = 0$ and [see the remark following (4.27)] the optimum procedure is characterized by a pair of numbers s, S (with $s = S = 2$ in this case) such that the optimum policy at each stage is given by $Y(x) = S$ if $x < s$, $Y(x) = x$ otherwise. However, for $\frac{1}{2} < K \leqslant 1 + \alpha/4$, (4.27) is not characterized by such a pair. Comparing (4.27) with (2.24b) and (2.24c), we note in the latter two that for $\frac{1}{2} < K \leqslant 1 + \alpha/4$, the phenomenon mentioned in the examples of Section 3 is again present here: the set of values of x at which one should order is greater when there are more intervals to come, than when there is only one time interval. In the case of (2.24a), however, it turns out that the optimum procedure is the same in both cases. We note also that as $\alpha \to 0$, (4.27) approaches the corresponding forms of (2.24a) and (2.24b).

4.5. It is useful in the case of 4.2 to be able to give the loss function $A(Y \mid x)$ corresponding to any ordering policy Y for which $Y_i(x) = Y_1(x)$ for all i. For any such Y, writing $A(Y \mid x) = A(x)$ for convenience, an argument similar to that used to derive (4.6) shows that $A(x)$ must satisfy the integral equation

$$A(x) = TA(x), \tag{4.39}$$

where $Tf(x)$, the result of the operator T acting on the function $f(x)$, is defined by

$$\begin{aligned} Tf(x) = V[x, Y_1(x)] &+ \alpha f(0)\{1 - F[Y_1(x)]\} \\ &+ \alpha \int_0^{Y_1(x)} f[Y_1(x) - v]\, dF(v). \end{aligned} \tag{4.40}$$

If $V(x, y) < M' < \infty$ for all x and y, then $A(Y \mid x) < M'/(1 - \alpha)$ for every ordering policy Y. Noting the form of the operator T, it is easy to verify that if $\ell(x)$ is a bounded solution to (4.39) (such a solution must exist), then under the condition of (4.16), equations (4.18) and (4.19) hold if IG is replaced by T. Thus, (4.39) has a unique bounded solution which represents the loss function corresponding to Y. If $\ell(x)$ and $k(x)$ are replaced by $A(x)$ and if IG and J are replaced by T, we see similarly that (4.20) and (4.22) hold with $R = M'/(1 - \alpha)$. Thus, we have an iterative method for solving (4.39), as well as a bound on the speed of convergence of this method.

Remarks analogous to those in the last paragraph of Section 4.3 can also be made here.

5. LAG IN DELIVERY OF THE COMMODITY—STATIONARITY WITHOUT INDEPENDENCE

5.1. In 5.1 and 5.2 we shall consider the case where there is a fixed time-lag of T time intervals (T a positive integer) between the moment when an order is placed and the moment the ordered amount is received by the ordering establishment.[12] In 5.1 we shall treat the case of a finite number of time intervals (corresponding to Section 3), while in 5.2 we shall treat the case of an infinite number of time intervals (corresponding to Section 4). Throughout this section such questions as existence of minima will be omitted, since they may be handled in a manner similar to that of the previous sections (see also footnotes 8 and 11).

We suppose then that there are N time intervals ($N > 1$) to be considered and that $1 \leqslant T \leqslant N - 1$. Again x_i will represent the initial stock on hand at the beginning of the ith interval and before any stock has been received at that time. An amount q_i is ordered at the beginning of the ith interval for $i = 1, 2, \cdots, N - T$ (nothing is ordered thereafter, since it would not be received within the span of N intervals which concern us) and is received at the start of the $(T + i)$th interval, so that $y_{T+i} = q_i + x_{T+i}$ is the starting stock at the beginning of the $(T + i)$th interval, $i = 1, 2, \cdots, N - T$. Besides, there may be earlier orders $(q_{-T+1}, \cdots, q_0)$ arriving at the start of the 1st, $\cdots$, Tth intervals, respectively. Taking this into consideration, we see that the relation $y_i = q_{i-T} + x_i$ holds for $i = 1, \cdots, N$. We shall use the symbols C_i, C_i'', γ_i, γ_i'' to represent the past history in this case. These symbols will be defined in the same way as B_i, B_i'', β_i, β_i'', respectively, if the letter x is replaced by the letter q throughout these definitions and if, furthermore, the dependence on $q_{-T+1}, \cdots, q_0, x_1$ is exhibited for all $i > 0$. (It is clear that because of the time lag, the B_i, etc., will no longer represent the entire past history. It will be more convenient, especially in 5.2, to consider the past history in terms of y's, q's, and d's rather than in terms of x's, q's, and d's.) Again, equation (3.3) is always valid. An ordering policy is a set of $N - T$ nonnegative functions

$$Q(y) = \{Q_i(y \mid C''_{i-1} = \gamma''_{i-1})\}, \qquad (i = 1, \cdots, N - T)$$

such that in the ith interval when the past history is $C''_{i-1} = \gamma''_{i-1}$, one orders an amount $q_i = Q_i(y_i \mid C''_{i-1} = \gamma''_{i-1})$. The distribution function $F_i(z \mid C_i = \gamma_i)$ of D_i may depend on all the quantities in C_i.

Although the effect of ordering q_i is not entirely paid for until the $(T + i)$th interval, one may consider $W_i(y_i, q_i; d \mid C''_{i-1} = \gamma''_{i-1})$ to be the amount actually paid out in the ith interval, and this is a function of the quantities in C_i'' which for fixed γ''_{i-1}, y_i, q_i we assume to be a Borel-measurable function of d_i, so that $W_i(y_i, q_i; D \mid C''_{i-1} = \gamma''_{i-1})$

[12] The work of 5.1 may be carried out similarly in the case when the lag $T(i)$ between the placing of an order at the beginning of the ith interval and the time of its receipt is not independent of i.

is for fixed γ''_{i-1}, y_i, q_i, a real-valued random variable whose expected value $U_i(y_i, q_i) = U_i(y_i, q_i \mid C''_{i-1} = \gamma''_{i-1})$ may actually depend upon y_i, q_i, and all the quantities in C''_{i-1}. For the sake of readability, the dependence on γ''_{i-1} again will not be exhibited explicitly. (The reason we use the symbols U_i rather than V_i is that it will be convenient, especially in simple cases like that of (5.10) of 5.2, to treat this expected loss as a function of y_i and q_i, whereas in (2.9) and in Sections 4.2–4.5 it was a function of x_i and y_i.) The quantities α_i are as in Section 3. We assume throughout Section 5 that, replacing V_i by U_i, Assumption A and Assumption B with $\gamma = 0$ hold.

In analogy to the A_{ij}'s of Section 3, we define for any ordering policy $Q(y)$, the following:

$$R_{ii}(q_i \mid y_i ; C''_{i-1} = \gamma''_{i-1}) = \alpha_i U_i(y_i, q_i) \tag{5.1}$$

and inductively for $1 \leqslant j \leqslant i - 1$,

$$R_{ij}(q_j \mid y_j ; C''_{j-1} = \gamma''_{j-1})$$

$$= \int_0^\infty R_{i(j+1)} [Q_{j+1}(y_{j+1} \mid C''_j = \gamma''_j)] \, d_{d_j} P\{ D_j < d_j \mid C_j = \gamma_j\}$$

$$= \int_0^{y_j} R_{i(j+1)} [Q_{j+1}(y_j + q_{j-T+1} - d_j \mid C''_j = \gamma''_j)] \, dF_j(d_j \mid C_j = \gamma_j) + \int_{y_j^+}^\infty R_{i(j+1)} [Q_{j+1}(q_{j-T+1} \mid C''_j = \gamma''_j)] \, dF_j(d_j \mid C_j = \gamma_j), \tag{5.2}$$

where in (5.2) we have used the fact that

$$y_{j+1} = q_{j-T+1} + \max (y_j - d_j, 0). \tag{5.3}$$

In analogy to Section 3, we shall consider an ordering policy $Q(y)$ to be optimum if for every set $q = (q_{-T+1}, \cdots, q_0, x_1)$ it minimizes

$$J[Q(y) \mid q] = \sum_{i=1}^N R_{i1}(q_1 \mid y_1 ; C''_0 = \gamma''_0), \tag{5.4}$$

where we recall that $y_1 = q_{-T+1} + x_1$; and we shall consider the problem of obtaining a policy Q^* which is optimum to within ϵ, i.e., for which

$$J(Q^* \mid q) \leqslant \inf_Q J(Q \mid q) + \epsilon \tag{5.5}$$

for every q.

We shall consider our ordering problem solved if for any $\epsilon > 0$ we can give a policy $Q^*(y)$ which is optimum to within ϵ. We now proceed to the construction of such a policy. It will be noted that this construction is very similar to that of Section 3, except that one now begins by giving Q^*_{N-T} in place of Y^*_N. We define

$$\text{(5.6)} \quad \bar{U}_j(y_j, q_j; Q_{j+1}, \cdots, Q_{N-T}) = \sum_{i=j}^{N} R_{ij}(q_j \mid y_j; C''_{j-1} = \gamma''_{j-1}),$$

where we do not exhibit the dependence of $\bar{U}_j$ on the quantities in C''_{j-1}. We note that

$$\text{(5.7)} \quad \bar{U}_1(y_1, q_1; Q_2, \cdots, Q_{N-T}) = J[Q(y) \mid q].$$

Let any $\epsilon > 0$ be given, and choose $q_{N-T} = Q^*_{N-T}(y_{N-T} \mid C''_{N-T-1} = \gamma''_{N-T})$ so that

$$\text{(5.8)} \quad \bar{U}_{N-T}(y_{N-T}, q_{N-T}) \leqslant \inf_q \bar{U}_{N-T}(y_{N-T}, q) + \epsilon'$$

with $\epsilon' = \epsilon/(N-T)$. Suppose $Q^*_{j+1}, \cdots, Q^*_{N-T}$, with $j > 1$, have been defined. Choose $q_j = Q^*_j(y_j \mid C''_{j-1} = \gamma''_{j-1})$ so that

$$\text{(5.9)} \quad \bar{U}_j(y_j, q_j; Q^*_{j+1}, \cdots, Q^*_{N-T}) \leqslant \inf_q \bar{U}_j(y_j, q; Q^*_{j+1}, \cdots, Q^*_{N-T}) + \epsilon'.$$

From (5.8) and (5.9) it follows that this construction gives the desired result of (5.5).

We note that in many simple cases (see also 5.2), the calculations for the case $T = 1$ are of a much simpler form than for $T > 1$. The reason for this is that when $T = 1$, at the beginning of the ith interval ($i \leqslant N - 1$) the starting stock y_i includes all stock ordered in previous intervals; while for $T > 2$, at the beginning of the ith interval ($i \leqslant N - T$) one is faced with the knowledge that the quantities $q_{i-T+1}, q_{i-T+2}, \cdots, q_{i-1}$ will arrive at the beginning of the $(i + 1)$, $(i + 2), \cdots, (i + T - 1)$ intervals, respectively, and the ordering agency no longer has any control over these quantities. Even for simple loss functions, it will usually be the case that $Q_i(y \mid C''_{i-1} = \gamma''_{i-1})$ depends explicitly on $q_{i-T+1}, \cdots, q_{i-1}$, whereas (see 5.2) in simple cases when $T = 1$, Q_i may only depend on y_i.

We also note that the analogue in the present case to the last paragraph of 3.1 is valid here.

5.2. We shall now treat the case where there are an infinite number of time intervals. For brevity, we shall restrict ourselves to the case of independence and stationarity corresponding to 4.2–4.5, although it is clear from 5.1 that the general case corresponding to 4.1 may be treated similarly (see also 5.3). T, x_i, y_i, and q_i are as in 5.1, except that now $i = 1, 2, \cdots$, ad inf., and an order may be placed at the beginning of each interval. The distribution of demand in each interval is given by $F(z)$ and does not depend on the past history, and the demands are independent. The function $U(y_i\,; q_{i-T+1}, \cdots, q_i)$ is assumed to be a function only of its stated arguments, which may be written without a subscript i on U. We could also carry out the work of 5.2 with U a function of different arguments (this would be a special case of 5.3). However, the form given above is perhaps the simplest which will be useful for many applications. For example, the analogue to the case of (2.6), (2.7), and (2.10) of Section 2 is

$$U(y_i\,; q_{i-T+1}, \cdots, q_i) = cy_i + A[1 - F(y_i)] + K'_{q_i} + \begin{cases} K \text{ if } q_i > 0 \\ 0 \text{ if } q_i = 0. \end{cases} \tag{5.10}$$

(We remark that restricting U to the form $U(y_i\,; q_i)$ will not in general reduce the number of variables on which the function ℓ of (5.15) below depends.) It is also interesting to note that the loss function may be put into the form $U(y_i\,; q_{i-T+1}, \cdots, q_i)$ in some cases where it does not appear to have this form, by departing from the accounting convention mentioned in 5.1 where W_i was considered to be the amount paid out in the ith interval. For example, if there is a cost $c(x_{i+T}, q_i)$ of unloading and storing the received q_i when x_{i+T} is the initial stock, U will apparently not have the desired form if this cost is included in W_{i+T}. However, if $\alpha^T \mathcal{E}c(x_{i+T}, q_i)$ is included in W_i (the expectation being with respect to $D_i, \cdots, D_{i+T-1}$), we will achieve the desired form. With the above modifications in mind, Assumptions A and B will be assumed to hold throughout this section, where V is replaced by U, and again without loss of generality we assume $U \geqslant 0$.

We postulate that α has the same meaning as in 4.2. An ordering policy is a sequence of functions Q_i as in 5.1, except that now $i = 1, 2, \cdots$, ad. inf., and the assumptions of independence and stationarity make it unnecessary for Q_i to be a function of any variables other than $y_i, q_{i-T+1}, \cdots, q_{i-1}$ (if $T = 1$, there are no q_j's in this last set). In addition, all of the functions Q_i may for $i > 0$ be taken to be the same function Q_1. The R_{ij} are as in 5.1 and the definitions of (5.4) and (5.6) apply if the upper limit of summation is replaced by ∞. We note that, because of the form of dependence of U and Q, the quantity $\alpha^{1-j}\overline{U}_j$ does not depend on j, and for a given Q this quantity will be a function only of $y_j, q_{j-T+1}, \cdots, q_{j-1}$ and will be denoted by $\tilde{U}(y_j\,; q_{j-T+1}, \cdots, q_{j-1}\,; Q)$. We then define

$$\ell(y_j\,; q_{j-T+1}, \cdots, q_{j-1}) = \inf_Q \tilde{U}(y_j\,; q_{j-T+1}, \cdots, q_{j-1}\,; Q). \tag{5.11}$$

A policy Q^* is said to be optimum if

$$\bar{U}(y; q_1, \cdots, q_{T-1}; Q^*) = \ell(y; q_1, \cdots, q_{T-1}) \tag{5.12}$$

for every $y, q_1, \cdots, q_{T-1}$. Optimality to within ϵ is defined in the usual manner.

In deriving the fundamental integral equation satisfied by ℓ it is necessary to consider the situation from the viewpoint of the beginning of the Tth (or a later) interval (before any ordering is done) rather than from the first as in 4.2. The reason for this is that if $j \leqslant T - 1$, $\bar{U}_j$ depends on all or some of the quantities $q_0, \cdots, q_{-T+1}$, which are given in advance. This special form does not permit one to use the method of 4.2 to derive the integral equation for $\ell(y; q_1, \cdots, q_{T-1})$ where none of the arguments are so fixed. The fact that this point does not arise when $T = 1$ is a manifestation of the fundamental difference noted in 5.1 between the cases $T = 1$ and $T \geqslant 2$.

The derivation of the fundamental integral equation now proceeds along the lines given in 4.2 (except that we consider $\bar{U}_T$ rather than $\bar{V}_1$). The analogue of (4.3) and (4.4) is now

$$F^{**}(v) = P\{y_{T+1} < v\} = \begin{cases} 1 \text{ when } v > y_T + q_1 \\ 1 - F(y_T + q_1 - v) \text{ when } q_1 < v \leqslant y_T + q_1 \\ 0 \text{ when } v \leqslant q_1 . \end{cases} \tag{5.13}$$

The quantity corresponding to (4.5) is now

$$\begin{aligned} \alpha \int_0^\infty \ell(v; q_2, \cdots, q_T)\, dF^{**}(v) &= \alpha \int_0^{y_T} \ell(y_T + q_1 - v; q_2, \cdots, q_T)\, dF(v) \\ &\quad + \alpha \ell(q_1; q_2, \cdots, q_T)[1 - F(y_T)]. \end{aligned} \tag{5.14}$$

We conclude that, corresponding to (4.6), the fundamental integral equation which $\ell(y; q_1, \cdots, q_{T-1})$ must satisfy is given by

$$\begin{aligned} \ell(y; q_1, \cdots, q_{T-1}) = \inf_{q \geq 0} \Big\{ & U(y; q_1, \cdots, q_{T-1}, q) \\ & + \alpha \ell(q_1; q_2, \cdots, q_{T-1}, q)[1 - F(y)] \\ & + \alpha \int_0^y \ell(y + q_1 - v; q_2, \cdots, q_{T-1}, q)\, dF(v) \Big\}. \end{aligned} \tag{5.15}$$

In the simplest case of $T = 1$, this becomes

$$\ell(y) = \inf_{q \geq 0} \left\{ U(y; q) + \alpha \ell(q)[1 - F(y)] + \alpha \int_0^y \ell(y + q - v)\, dF(v) \right\}. \tag{5.16}$$

Although ℓ is a function of only one variable in this case, we note that the form of (5.16) is essentially different from that of (4.6). For example, the quantity over which the infimum is taken enters differently in each case.

An optimum policy may now be described as follows:[13] Let the expression in braces over which the infimum is taken in (5.15) be denoted by $h(y; q_1, \cdots, q_{T-1}, q)$. Given $y_1 = q_{-T+1} + x_1$, at the beginning of the first interval order the least amount q_1 for which $h(y_1; q_{-T+2}, \cdots, q_0, q_1)$ attains its minimum. Similarly, at the beginning of the ith interval, order the least amount q_i for which $h(y_i; q_{i-T+1}, \cdots, q_i)$ attains its minimum. (We recall that $q_0, \cdots, q_{-T+1}$ are given in advance and appear in $h(y_i; q_{i-T+1}, \cdots, q_i)$ for $i < T$.)

The expression in braces in the right member of (5.15) may be taken as the definition of the operator G on the function $\ell(y; q_1, \cdots, q_{T-1})$, taking a function of T variables into one (namely, h) of $T + 1$ variables. The infimum operation on $h(y; q_1, \cdots, q_{T-1}, q)$ defines the operator I. Thus, the functions ℓ and h satisfy the equations

$$\ell(y; q_1, \cdots, q_{T-1}) = IG\ell(y; q_1, \cdots, q_{T-1}) = Ih(y; q_1, \cdots, q_T) \tag{5.17}$$

and

$$h(y; q_1, \cdots, q_T) = GIh(y; q_1, \cdots, q_T) = G\ell(y; q_1, \cdots, q_{T-1}) \tag{5.18}$$

Note however that, unlike the case of (4.13) and (4.14), h and ℓ are no longer functions of the same number of variables.

We now pass to the considerations which were given in 4.3 for the case of no lag. In analogy to Assumption D, we make

Assumption E. There is a positive constant M such that

$$\sup_{y \geq 0} U(y; q_1, q_2, \cdots, q_{T-1}, 0) < M < \infty$$

for all $q_1, \cdots, q_{T-1}$.

Because of the way in which the quantity α enters into $G\ell(y; q_1, \cdots, q_{T-1})$, it is now easy to see that if $\ell(x)$ and $g(x)$ in (4.16) and what

[13] The analogue to footnote 11 applies here.

follows it are replaced by bounded functions $\ell(y; q_1, \cdots, q_{T-1})$ and $g(y; q_1, \cdots, q_{T-1})$, then the considerations of 4.3 regarding uniqueness apply. Thus, the unique bounded solution of (5.17) represents the function ℓ of (5.11); similarly, (5.18) has a unique bounded solution. Moreover, if V is replaced by U in (4.23) so that R becomes $(1 - \alpha)^{-1}$ multiplied by the M of Assumption E, the results of (4.22) on the speed of uniform approach of $J^n g$ to $k(= \ell$ or $h)$ for any bounded function g remain valid. In like manner, the result of the discussion following (4.10) may be derived in this case to give a bound on the additional expected loss incurred by using a function h_1 satisfying (4.10) (with h_1 also a function of $q_1, \cdots, q_T$ and the supremum being also over all q_i) or if an optimum policy does not exist and the amount q_i is such that

$$h(y_i\,; q_{i-T+1}, \cdots, q_i)$$

is within ϵ of its infimum. Finally, the analogues in the present case to 4.5 and to the final paragraph of 4.3 are also applicable.

5.3. Many simple generalizations of the previous results of this paper are possible. The quantity q_i may be delivered in parts at several future times (which may or may not be controlled by the ordering policy) and furthermore q_i as well as its parts may be chance variables (e.g., crops). The lags in delivery time for the quantities q_i (or their parts) may depend on i or may even be chance variables with a joint distribution function which depends on the values of the q_i and the past history. In the stationary case with any or all of the above modifications (without postulating independence), one obtains the integral equation

$$\ell(H) = \inf_Q \left[U(H, Q) + \alpha \int \ell(H')\, dF \right] \tag{5.19}$$

where H is the past history of the process, Q is the ordering policy, H' is the history resulting from H by the application of Q and a random element whose distribution F may depend on H and Q (but in a stationary way), and U and ℓ are the analogues of the corresponding quantities of 5.2.

For example, in the case of no lag if the distribution F of demand as well as the expected loss V in the ith interval depend only on d_{i-1}, x_i, and y_i, (5.19) becomes

$$\begin{aligned} \ell(x; d) = \inf_{y \ge x} \Big[& V(x, y; d) + \alpha \int_0^y \ell(y - z; z)\, dF(z \mid x, y; d) \\ & + \alpha \int_{y+}^{\infty} \ell(0; z)\, dF(z \mid x, y; d) \Big] \end{aligned} \tag{5.20}$$

where $\ell(x; d)$ is the infimum of the present expected loss in the present and all future intervals when x is the starting stock and d was the demand in the immediately preceding interval. It is to be noted that here we can carry over demand for the commodity from one interval to the next, either wholly or partially.

6. SEVERAL COMMODITIES, CONSUMERS, SOURCES OF SUPPLY—DISTRIBUTING POLICIES—FURTHER GENERALIZATIONS

6.1. We shall now treat the case where there are a finite number $m > 1$ of commodities being stocked.[14] Thus, the quantities x_i, y_i, and D_i are now replaced by vectors $(x_i^1, \cdots, x_i^m)$, $(y_i^1, \cdots, y_i^m)$, and $(D_i^1, \cdots, D_i^m)$; for example, y_i^j is the starting stock of the jth commodity in the ith time interval.

Due to the fact that it may be possible to substitute (with perhaps some decrease in satisfaction) a quantity of a commodity of which there is an excess for a quantity of one whose demand is greater than its starting stock, a slight reformulation of the problem is necessary. Letting x_i^j, y_i^j, and D_i^j be as defined above, we now let e_i^j $(0 \leqslant e_i^j \leqslant y_i^j)$ be the amount of the jth item actually supplied by the ordering agency in response to the demand vector d_i, and we define $e_i = (e_i^1, \cdots, e_i^m)$.[15] Thus, x_{j+1} now satisfies the equation

$$x_{j+1} = y_j - e_j \tag{6.1}$$

rather than (3.3), where the difference between two vectors is defined in the usual way. Thus, e_i^j may be greater than, less than, or equal to d_i^j. Let $R = (x_1, \cdots, x_i, y_1, \cdots, y_{i-1}, D_1, \cdots, D_{i-1}, e_1, \cdots, e_{i-1})$, and let ρ_i represent a particular set of values of the quantities in R_i. In the cases corresponding to those of Section 2 and Subsections 3.1 and 4.1, an ordering polity (Y, E) (now an *ordering and distributing policy*) is now a sequence of vector functions $Y_1, E_1, Y_2, E_2, \cdots$, where Y_i is a vector $(Y_i^1, \cdots, Y_i^m)$ of functions of the quantities in ρ_i and E_i is a vector $(E_i^1, \cdots, E_i^m)$ of functions of the quantities in ρ_i as well as of y_i and d_i; and where at the moment of ordering or distributing supplies in the ith interval, the functions Y_i^j and E_i^j of the past histories determine the values of y_i^j and e_i^j, respectively. The (scalar) loss function W_i now depends on y_i, d_i, e_i, and ρ_i. For a given policy E_i, the expected value V_i of W_i is a function of y_i and ρ_i.

The function W_i measures the losses discussed in previous sections, as well as the loss incurred by substituting various commodities for

[14] The case of a denumerable set of "commodities" may be treated similarly.

[15] More generally, e_i could be an $m \times m$ matrix representing the amount of each commodity distributed in substitution for every other commodity.

others which are demanded (the corresponding increase or decrease in satisfaction not necessarily being linear in the amounts substituted). The joint distribution $F_i(d_i \mid y_i, R_i = \rho_i)$ of the quantities $D_i^1, \cdots, D_i^m$ may now depend on y_i and on all the quantities in R_i.

In analogy to (3.1) and (3.2), one now defines for a given ordering policy (Y, E), the following:

$$A'_{ii}(y_i, d_i, e_i \mid R_i = \rho_i) = \alpha_i W_i; \tag{6.2}$$

and inductively for $1 \leqslant j \leqslant i$,

$$A_{ij}(y_j \mid R_j = \rho_j)$$
$$= \int_{\substack{0 \le d^i < \infty \\ i=1,\cdots,m}} A'_{ij}[y_j, d_j, E_j(d_j \mid y_j, \rho_j)]\, d_{d_j} P\{D_j < d_j \mid y_j, R_j = \rho_j\}; \tag{6.3}$$

and, for $2 \leqslant j \leqslant i$,

$$A'_{i(j-1)}(y_{j-1}, d_{j-1}, e_{j-1} \mid R_{j-1} = \rho_{j-1}) = A_{ij}[Y_j(\rho_j) \mid R_j = \rho_j], \tag{6.4}$$

where in (6.4) we have used (6.1) to express ρ_j as a function of y_{j-1} $d_{j-1}, e_{j-1}, \rho_{j-1}$. We define, in analogy to (3.7), $\bar{V}_j = \sum_{i=j}^N A_{ij}$ and, $\bar{V}'_j = \sum_{i=j}^N A'_{ij}$, and then work backwards as in Section 3 to define E_N^*, Y_N^*, E_{N-1}^*, $Y_{N-1}^*, \cdots, E_1^*$, Y_1^* which give a policy satisfying (3.6) [with $A(Y, E \mid x_1)$ defined by the last expression of (3.4)].

Analogues of the remainder of Section 3, as well as of the results of 2, 4.1, and 5.1, may be carried out by introducing modifications similar to those given above.

In the case of 4.2–4.5, W_i is a function of x_i, y_i, d_i, e_i which may be written without a subscript i on W. The distribution function F_i may also be written without a subscript i and does not depend on the past history, and although D_i^j and D_i^k need not be independent for $j \neq k$, D_i and D_k are independent for $i \neq k$. $Y_i(x_i)$ may be assumed to depend only on x_i (but not on i), and E_i may be taken to depend only on x_i, y_i, and d_i (but not on i). Let

$$\ell(x) = \inf_{Y,E} A(Y, E \mid x). \tag{6.5}$$

Define the following operators:

$$I_1 f(y) = \inf_{e \le y} [W(x, y, d, e) + \alpha f(y - e)],$$

$$I_2 f(x, y, d) = \int f(x, y, d)\, dF(d),$$

$$I_3 f(x, y) = \inf_{y \ge x} f(x, y).$$

Then it is not difficult to show that

$$\ell(x) = I_3 I_2 I_1 \ell(y). \tag{6.6}$$

Let $\ell_2(x, y)$ be the minimum total expected loss when x is the initial stock and y is the starting stock. Then

$$\ell_2(x, y) = I_2 I_1 \ell(y).$$

With the aid of $\ell_2(x, y)$ one obtains an optimal ordering policy in a manner exactly analogous to that described in the discussion accompanying (4.8).

To obtain an optimal distributing policy (see footnote 11) when x is the starting stock, y is the initial stock, and d is the actual demand in the interval under consideration, choose a value e for which $W(x, y, d, e) + \alpha\ell(y - e)$ attains its minimum.

A discussion analogous to that of (4.10) and of 4.5 may be carried out in a similar manner and the form in which α enters into (6.9) and (6.10) shows that the analogue of 4.3 is also valid in its entirety. Finally, 5.2 and 5.3 may also be applied in the present case.

6.2. Extensions to even more general cases may be treated by making modifications in Sections 2, 3, 4, and 5 in similar manner to 6.1. For example, x_i and y_i may be $m \times L$ matrices representing the initial and starting stocks of m different commodities at L different warehouses, D_i may be an $m \times S$ matrix representing the demand of each of S consumers for each of the m commodities, and e_i may be an $m \times S \times L$ array representing the distribution of each of the m items from each of the L warehouses to each of the S consumers, all in the ith interval (using the idea of footnote 15 this becomes an $m \times m \times S \times$ L array). One may also treat the case where there are several levels of ordering agencies, those on each level ordering from the next level in response to demand on themselves and then receiving and distributing orders with certain time lags, and similar generalizations.

With slight changes in interpretation our setup includes the important economic situation wherein some or all of the ordered commodities may be transformed in several stages of production (over one or several time intervals) into other commodities, the transformations perhaps including chance elements, and the prices of commodities (bought and sold) being also subject to chance variation as mentioned in 2.2.

The methods of this paper can be extended in an obvious way to include problems where there are additional constraints on the ordering policy. For example, the ordering agency may wish to minimize the loss corresponding to one commodity, subject to some restriction on the loss corresponding to another commodity.

One may also wish to consider the problem where, at each time inter-

val, there is a known probability that the process of ordering will terminate at the end of this interval. This probability need not be constant, and may depend on the past history of the process. The methods of the present paper apply to this problem as well. The principal modification required is that, in forming the A_{ij} of (3.4), (3.7), (3.10), and the other corresponding equations, one must take into account the probability that the process will terminate. In the stationary case one can derive an integral equation analogous to (4.6) and (5.19). Other problems are treated similarly.

Cornell University

Reprinted from
Econometrica **20** (1952), 187-222

THE INVENTORY PROBLEM: II. CASE OF UNKNOWN DISTRIBUTIONS OF DEMAND*

By A. Dvoretzky,[1] J. Kiefer, and J. Wolfowitz

1. INTRODUCTION

1.1 The inventory problem is the problem of stocking in anticipation of future demand. In the first part of this paper,[2] we treated the case when the demand was given by completely specified probability distribution functions. In Part II we deal with the case when the above distribution functions are specified only to the extent of their belonging to a given set of distribution functions. It is the study of this more general case that requires the methods of mathematical statistics, and we treat it in the spirit of Wald's decision theory.

Part II is so written that it can be read independently of Part I; although, occasionally, the reader is referred to I for motivation or a more detailed argument. After the introduction in Section 1, we present in Section 2 the relevant points of decision theory formulated in the framework of the inventory problem. Section 3 treats the case of a finite number of intervals (this corresponds to I, 2 and I, 3). The case of infinitely many time intervals and, in particular, the stationary case is studied in Section 4 (corresponding to I, 4). The last Section, 5, deals briefly with various generalizations such as time lag (I, 5), the simultaneous consideration of many commodities (I, 6), and related matters.

1.2. As in most of I we assume that orders are fulfilled instantaneously (see, however, Section 5). We denote by x_i and y_i $(0 \leqslant x_i \leqslant y_i < \infty)$[3] the initial and the starting stock respectively at the beginning of the ith time interval;[4] $y_i - x_i$ represents the amount ordered (and received) at the beginning of this time interval. The demand at the ith interval is denoted by D_i; it is a random variable which may depend also on x_i, y_i and the values of x_j, y_j, and D_j for $1 \leqslant j \leqslant i - 1$. For brevity we introduce the notations (see I, 3):

$$B_i = (x_1, \cdots, x_i, y_1, \cdots, y_i, D_1, \cdots, D_{i-1}),$$

$$\beta_i = (a_1, \cdots, a_i, b_1, \cdots, b_i, d_1, \cdots, d_{i-1})$$

so that the distribution of the demand D_i in the ith time interval may, inter alia, depend also on the particular value β_i assumed by B_i.

* Research under contract with the Office of Naval Research.

[1] On leave from the Hebrew University, Jerusalem, Israel.

[2] "The Inventory Problem: I. Case of Known Distributions of Demand," Econometrica, Vol. 20, April, 1952, pp. 187–222. To be referred to, hereafter, as I.

[3] This condition, allowing the ordering of nonnegative amounts only, can easily be modified to permit returning or dumping (see I, 1).

[4] $i = 1, 2, \cdots, N$ in Section 3, $i = 1, 2, \cdots$ ad inf. in Section 4.

By giving a distribution of demand F, we understand prescribing for all[5] β_i the probability distribution of the demand D_i in the ith interval under the condition $B_i = \beta_i$. We use the relation

$$F_i(z \mid B_i = \beta_i ; F) = Pr\,\{D_i \leqslant z \mid B_i = \beta_i ; F\}$$

thus exhibiting the distribution of demand F.

Very often the actual demands D_j ($j = 1, \cdots, i - 1$) are known only when $D_j < y_j$, whereas in the other case it is only known that $D_j \geqslant y_j$. Such cases are included in our model provided $F_i(z \mid B_i = \beta_i ; F)$ is the same for all β_i for which the quantities a_j, b_j, and min (b_j, d_j) are identical ($j < i$). (A similar remark applies to the case when D_j is known exactly whenever $D_j \leqslant y_j$, whereas in the remaining cases it is only known that $D_j > y_j$, etc.,). This, in particular, is so whenever F_i does not depend on the quantities d_j. Similar remarks apply to the functions W_i and G_i introduced below.

We shall also have occasion to use the notations

$$B_i' = (x_1, \cdots, x_i, y_1, \cdots, y_{i-1}, D_1, \cdots, D_{i-1}),$$
$$B_i'' = (x_1, \cdots, x_i, y_1, \cdots, y_i, D_1, \cdots, D_i),$$

as well as β_i' and β_i'' obtained from B_i' and B_i'' on replacing x, y, and D by a, b, and d respectively.

$W_i(x_i, y_i ; d_i \mid B_{i-1}'' = \beta_{i-1}'' ; F)$ represents the loss in the ith interval when F is the distribution of demand, B_{i-1}'' assumes the value β_{i-1}'', d_i is the actual demand and x_i and y_i are, respectively, the initial and starting stocks in the ith interval. It is assumed that W_i and F_i are Borel measurable (separately and jointly) functions of x_i, y_i, d_i as well as of the arguments of β_{i-1}''. We furthermore assume W_i to be nonnegative. The last assumption may be weakened to a very large extent (cf. footnote 5, I) without modifying our results. We make this assumption since it is a natural one in most applications.

Given x_i, y_i, β_{i-1}'', and F the expected loss in the ith interval is

$$\text{(1.1)}\quad \begin{aligned} &V_i(x_i, y_i \mid B_{i-1}'' = \beta_{i-1}''; F) \\ &\quad = \int_0^\infty W_i(x_i, y_i ; z \mid B_{i-1}'' = \beta_{i-1}''; F)\, dF_i(z \mid B_i = \beta_i; F). \end{aligned}$$

(Whenever the integral does not converge absolutely we put $V_i = \infty$.)

In I only a single F was considered; it was therefore unnecessary to allow for the dependence of W_i (and V_i) on F. In the more general situation considered in Part II, it is essential to take this dependence into

[5] Often the quantities y_i are restricted by a condition such as $y_i \leqslant H_i < \infty$; in such cases it is clearly sufficient to consider β_i with $x_i \leqslant y_i \leqslant H_i$. A similar remark applies to D_i. The domains of x_i, y_i, and D_i may be restricted in any other consistent way throughout this paper. The corresponding restrictions then apply to β_i.

consideration. Although in many applications W_i does not depend explicitly on F (see also the discussion following (3.3) below) this is by no means generally so. Moreover, even when W_i does not depend on F, the associated V_i generally does. Thus even in the simple case when for every i and F the expected loss is given by (2.9) of I it is clear that V_i (though not W_i) depends on F.

1.3. In I an ordering policy Y specified for every i exactly what amount to order in the ith interval when x_i and β''_{i-1} were given. As remarked (footnote 6, I) it was not necessary to consider randomized ordering policies. In the present, more general situation this is no longer so and we must take into account randomized ordering policies. A (randomized) *ordering policy* Y is a set of distribution functions $G_i(y \mid x_i, B''_{i-1} = \beta''_{i-1})$ defined for all[5] i, x_i, and β''_{i-1} and satisfying[3]

$$G_i(y \mid x_i, B''_{i-1} = \beta''_{i-1}) = 0 \quad \text{whenever } y < x_i. \tag{1.2}$$

Thus, given x_i and the past history β''_{i-1}, the starting stock y_i (and the amount $y_i - x_i$ to be ordered) have to be chosen by an independent chance mechanism for which $Pr\{y_i \leqslant y\} = G_i(y \mid x_i, B''_{i-1} = \beta''_{i-1})$.

If the ordering policy Y is such that, for all i, x_i, and β''_{i-1}, the distribution functions G_i are degenerate (i.e., can assume only the values 0 and 1) then we have a (nonrandomized) ordering policy in the sense of I. If the very general conditions of Theorem 5.2 of the paper by Dvoretzky, Wald, and Wolfowitz[6] are satisfied, then according to this theorem randomization is unnecessary.

We now make the same measurability assumptions for G_i as are made for nonrandomized ordering policies in the paragraph immediately following equation (3.3) of I.

The class of all available ordering policies will be denoted by Υ.

1.4. We introduce discount factors α_i $(0 \leqslant \alpha_i \leqslant 1)$ so that [cf. I(3.1)]

$$A_{ii}(y_i \mid B'_i = \beta'_i ; F) = \alpha_i V_i(x_i, y_i \mid B''_{i-1} = \beta''_{i-1} ; F) \tag{1.3}$$

is the present value of the expected loss V_i in the ith interval for given x_i, y_i, β''_{i-1}, and F. Similarly to I(3.2), for a given Y, we define recursively A_{ij} for $1 \leqslant j < i$ through[7]

$$\begin{aligned} &A_{ij}(y_j \mid B'_j = \beta'_j; F) \\ &\quad = \int_0^\infty A^*_{i(j+1)}(x_{j+1} \mid B''_j = \beta''_j; F)\, d_{d_j} Pr(D_j < d_j \mid B_j = \beta_j; F), \end{aligned} \tag{1.4}$$

[6] A. Dvoretzky, A. Wald, and J. Wolfowitz, "Elimination of Randomization in Certain Statistical Decision Procedures and Zero-Sum Two-Person Games," *Annals of Mathematical Statistics*, Vol. 22, 1951, pp. 1–21.

[7] Unless otherwise indicated all finite integration limits are included in the range of integration.

with

$$A^*_{i(j+1)}(x_{j+1} \mid B''_j = \beta''_j ; F)$$

$$(1.5) \qquad = \int_{x_{j+1}}^{\infty} A_{i(j+1)}(y_{j+1} \mid B'_{j+1} = \beta'_{j+1}; F)\, dG_{j+1}(y_{j+1} \mid x_{j+1}, B''_j = \beta''_j)$$

$$\text{for} \quad 0 \leqslant j \leqslant i - 1.$$

From (1.2) and (1.4) we have

$$A_{ij}(y_j \mid B'_j = \beta'_j ; F)$$

$$(1.6) \qquad = \int_0^{y_j} A^*_{i(j+1)}(y_j - z \mid B''_j = \beta''_j ; F)\, dF_j(z \mid B_j = \beta_j ; F)$$

$$+ \int_{y_j^+}^{\infty} A^*_{i(j+1)}(0 \mid B''_j = \beta''_j ; F)\, dF_j(z \mid B_j = \beta_j ; F).$$

Noting that $B'_1 = (x_1)$ we define [cf. I(3.4)],

$$(1.7) \qquad A(F, Y \mid x_1) = \sum_i A^*_{i1}(x_1 \mid F);$$

$\sum$ extending from $i = 1$ to $i = N$ in Section 3 and over all positive integers in Section 4.

It is easily seen that $A(F, Y \mid x_1)$ represents the present value of the total expected loss when x_1 is the initial stock at the beginning of the first interval, Y is the ordering policy used by the ordering agency, and F is the distribution of demand. We shall call $A(F, Y \mid x_1)$ the *risk function at F corresponding to x_1 and Y*.

As will be seen in the following section, we shall always judge the goodness of an ordering policy by considering the corresponding risk function.

2. DECISION-THEORETIC CONSIDERATIONS

2.1. In this section we shall formulate our problem in the spirit of Wald's decision theory.[8] To formulate the concepts of this theory it is necessary to know in advance the class of distributions of demand to which F must belong. We shall denote this class by Ω.

[8] For a comprehensive treatment see A. Wald, *Statistical Decision Functions*, New York: John Wiley and Sons, 1950. Strictly speaking the inventory problem as treated here is not a decision problem in the sense of Wald's book. However, it is easily seen that the methods and results of decision theory extend also to the problems treated here. See also J. Wolfowitz, "On ϵ-Complete Classes of Decision Functions," *Annals of Mathematical Statistics*, Vol. 22 (1951) pp. 461–465 for a discussion of some of the other notions used here, and for a method which may be used herein to construct *finite* ϵ-essentially complete classes.

For every $x_1 > 0$, a metric[9] is introduced into the space Ω through

$$\rho(F, F' \mid x_1) = \sup_{Y \in \Upsilon} | A(F, Y \mid x_1) - A(F', Y \mid x_1) | \tag{2.1}$$

for all $F \in \Omega$ and $F' \in \Omega$.

The Borel sets of Ω (for a given x_1) are the subsets of Ω which are included in the smallest Borel field containing the open subsets of Ω in the sense of the topology induced by (2.1).

A probability measure ξ over Ω is a countably additive nonnegative set function defined for all Borel sets of Ω and satisfying $\xi(\Omega) = 1$. The set of all probability measures ξ is denoted by Ξ. (The set Ξ may depend on x_1; however, since x_1 is always fixed throughout the discussion, we do not indicate this dependence. A similar observation applies to other concepts below whose definitions depend on the Borel sets of Ω.)

We must assume that the risk $A(F, Y \mid x_1)$ is a measurable function of F (i.e., for every real t, the set of F for which $A(F, Y \mid x_1) < t$ is a Borel set). This can be secured by assuming that the functions W_i and F_i of Section 1 are measurable in F as well as in their other arguments; i.e., they are measurable, in the cartesian product of Ω and the Euclidean space representing the other arguments, with respect to the smallest Borel field containing the cartesian products of the Borel sets of the two spaces. It may be remarked that in many practical situations the space Ω may be considered at most countable, while x_i, y_i, and D_i must be integral multiples of some unit; in all such cases all measurability assumptions are automatically fulfilled.

We now extend the definition of the risk function A to cover also the case when F is replaced by ξ (in the language of decision theory this corresponds to allowing Nature to use mixed strategies). For every $x_1 > 0$, $Y \in \Upsilon$, and $\xi \in \Xi$, we define

$$A(\xi, Y \mid x_1) = \int_\Omega A(F, Y \mid x_1)\, d\xi.$$

If ξ is the distribution function assigning probability 1 to the distribution F, then the right side of the above reduces to $A(F, Y \mid x_1)$.

2.2. The decision theory approach is to regard the inventory problem (for each fixed x_1) as a zero-sum two-person game with payoff function $A(F, Y \mid x_1)$. In this subsection we define the relevant concepts of the theory of games and decision functions.

The inventory problem is said to be *determined* for x_1 if

$$\inf_{Y \in \Upsilon} \sup_{\xi \in \Xi} A(\xi, Y \mid x_1) = \sup_{\xi \in \Xi} \inf_{Y \in \Upsilon} A(\xi, Y \mid x_1) \tag{2.2}$$

holds.

[9] This definition allows also infinite distances; these are eliminated by Assumption F below.

An ordering policy Y is said to be a ***minimax ordering policy*** for x_1 if it satisfies

$$\sup_{F\in\Omega} A(F, Y \mid x_1) \leqslant \sup_{F\in\Omega} A(F, Y' \mid x_1) \qquad \text{for all } Y' \in \Upsilon. \tag{2.3}$$

We refer to Wald's book[8] for a discussion of the significance of these and related concepts.

As in I, our main interest is in effective approximation procedures. For this purpose we need also the various ϵ-concepts introduced below. Given any $\epsilon > 0$, Y is said to be an ***ϵ-minimax ordering policy*** for x_1 if it satisfies

$$\sup_{F\in\Omega} A(F, Y \mid x_1) < \sup_{F\in\Omega} A(F, Y' \mid x_1) + \epsilon \qquad \text{for all } Y' \in \Upsilon. \tag{2.4}$$

An ordering policy Y is said to be ***as good as*** an ordering policy Y' for x_1 if

$$A(F, Y \mid x_1) \leqslant A(F, Y' \mid x_1) \qquad \text{for all } F \in \Omega \tag{2.5}$$

holds; i.e., if for all $F \in \Omega$ the value at F of the risk function corresponding to Y' and x_1 is not smaller than that corresponding to Y and x_1. If Y is as good as Y' for x_1 and, conversely, Y' is as good as Y for x_1 then the two ordering policies are said to be ***equivalent*** for x_1. In this case there is no preference for one over the other if x_1 is the initial stock at the beginning of the first interval. If Y is as good as Y' for x_1 and they are not equivalent for x_1, then Y is said to be ***better***[10] for x_1 than Y'. Thus Y is better than Y' for x_1 if (2.5) is satisfied and, moreover, the ***strict*** inequality holds for at least one $F \in \Omega$.

For any fixed $\epsilon > 0$ the policy Y is said to be ***ϵ-as good*** for x_1 as Y' if

$$A(F, Y \mid x_1) < A(F, Y' \mid x_1) + \epsilon \qquad \text{for all } F \in \Omega. \tag{2.6}$$

If (2.6) holds and, conversely, Y' is ϵ-as good as Y for x_1, then the two ordering policies are said to be ***ϵ-equivalent*** for x_1. If (2.6) holds but Y and Y' are ***not*** ϵ-equivalent for x_1 then Y is said to be ***ϵ-better than*** Y' for x_1.

A class C of ordering policies is said to be ***complete*** for x_1 if for every ordering policy $Y' \notin C$ there exists an ordering policy $Y \in C$ which is better than Y' for x_1. A class C of ordering policies is called ***essentially complete*** for x_1 if for every Y' there exists a $Y \in C$ which is as good as Y' for x_1. Given $\epsilon > 0$, a class C is said to be ***ϵ-essentially complete*** for x_1 if for every ordering policy Y' there exists a $Y \in C$ satisfying (2.6). (One could define ϵ-complete classes in a similar manner but we do not need this concept in what follows).

[10] *Uniformly better* in Wald's terminology.

A class C which is ϵ-essentially complete for x_1 for all $\epsilon > 0$ is called *0-essentially complete for* x_1. In the following sections we shall announce some of our results for ϵ-essentially complete classes only. However, if C_n $(n = 1, 2, \cdots)$ is ϵ_n essentially complete for x_1 and $\epsilon_n \to 0$ as $n \to \infty$, then the union of the C_n is 0-essentially complete for x_1. Thus all results about ϵ-essentially complete classes have immediate corollaries for 0-essentially complete classes.

An ordering policy Y is said to be a *Bayes policy relative to* ξ for x_1 if it satisfies

$$A(\xi, Y \mid x_1) \leqslant A(\xi, Y' \mid x_1) \qquad \text{for all } Y' \in \Upsilon. \tag{2.7}$$

If, for a given $\epsilon > 0$, it satisfies

$$A(\xi, Y \mid x_1) < A(\xi, Y' \mid x_1) + \epsilon \qquad \text{for all } Y' \in \Upsilon, \tag{2.8}$$

then it is called an ϵ-*Bayes policy relative to* ξ for x_1.

2.3. In order to be in a position to apply the results of the theory of decision functions, it is necessary to make some further assumptions about the risk A and the space Ω. We shall not strive for the most general setup[11] and postulate the following two assumptions which hold in most situations:

Assumption F: For every[12] x_1 the risk $A(F, Y \mid x_1)$ is a uniformly bounded function of F and Y; i.e.,

$$A(F, Y \mid x_1) < M_{x_1} < \infty \quad \text{for all } F \in \Omega \text{ and } Y \in \Upsilon. \tag{2.9}$$

Assumption G: For every[12] x_1 the space Ω is conditionally compact in the sense of the metric (2.1).

As is shown in the Wald theory of decision functions, Assumptions F and G entail the following results:

The inventory problem is determined for all x_1; i.e., (2.2) always holds.

Every ϵ-minimax policy for x_1 is also an ϵ'-Bayes policy relative to ξ for x_1, for every $\epsilon' > \epsilon$ and for some suitable $\xi \in \Xi$. [For every $\epsilon > 0$ and x_1 there exists at least one ϵ-minimax policy for x_1; i.e., there always exists a Y satisfying (2.4). For every $\epsilon > 0$, x_1, and ξ there exists at least one ϵ-Bayes policy relative to ξ; i.e., there always exists a Y satisfying (2.8).]

For every $\epsilon > 0$, x_1, and ξ let $\bar{C}(\epsilon, x_1, \xi)$ denote the set of all ϵ-Bayes policies relative to ξ for x_1; then the class $\bar{C}(\epsilon, x_1) = \bigcup_{\xi \in \Xi} \bar{C}(\epsilon, x_1, \xi)$ is complete for x_1.

[11] For a more general setup under which the following results remain valid see e.g. Wald's book mentioned in footnote 8.

[12] If Assumptions F and G are fulfilled only for some x_1 then our conclusions hold for those values of the initial stock.

2.4. We now replace Assumptions F and G by more simple Assumptions which entail them [Assumptions H, I, and (2.13)].

Assumption H. There exists a finite number M satisfying

$$V_i(x_i, y_i \mid B''_{i-1} = \beta''_{i-1}; F) < M \tag{2.10}$$

for all $i, x_i, y_i, \beta''_{i-1}$, and F.

From (2.10) and (1.3)–(1.6) we have

$$A^*_{i1}(x_1 \mid F) < \alpha_i M. \tag{2.11}$$

From (2.11) and (1.7) we then have

$$A(F, Y \mid x_1) < M \sum_i \alpha_i \tag{2.12}$$

for all $F \epsilon \Omega$, $Y \epsilon \Upsilon$, and x_1. Thus, Assumption H entails Assumption F whenever

$$\sum_i \alpha_i < \infty. \tag{2.13}$$

In Section 3 the series of (2.13) is a finite series and thus (2.13) is automatically fulfilled. In the case of Section 4 we postulate (2.13) explicitly.[13]

It should be remarked that Assumption H together with (2.13) imply (2.9) with $M_{x_1} = M \sum_i \alpha_i$ *independent* of x_1. Thus Assumption F is fulfilled uniformly in x_1.

Next we proceed to replace Assumption G. We shall show that Assumption G follows from Assumption H, (2.13) and the following:

Assumption I: Ω is conditionally compact in the sense of the metric

$$\delta(F, F') = \delta_1(F, F') + \delta_2(F, F'), \tag{2.14}$$

where

$$\delta_1(F, F') = \sum_{i=1}^{\infty} 2^{-i} \sup \int_0^{\infty} \mid d_z[F_i(z \mid B_i = \beta_i; F) - F_i(z \mid B_i = \beta_i; F')] \mid, \tag{2.15}$$

and

$$\delta_2(F, F') = \sum_{i=1}^{\infty} 2^{-i} \sup \mid V_i(x_i, y_i \mid B''_{i-1} = \beta''_{i-1}; F) - V_i(x_i, y_i \mid B''_{i-1} = \beta''_{i-1}; F') \mid, \tag{2.16}$$

the suprema in (2.15) and (2.16) being over all β_i.

[13] This is Assumption C of Part I.

We give a sketch of the proof. Because of Assumptions H and (2.13), a "truncation" argument (see I, 4.1) shows that we need only consider the case of a fixed finite number N of time intervals (even in the case of Section 4). Given any $\epsilon > 0$, let $\epsilon' = \epsilon/N(1 + MN)$, where M satisfies (2.10). Let $F_1, \cdots, F_m$ be $2^{-N}\epsilon'$–dense in Ω in the sense of the metric δ. For any given F in Ω, there is an $\bar{F} = F_{j_0}$ $(1 \leqslant j_0 \leqslant m)$ such that $\delta(F, \bar{F}) \leqslant 2^{-N}\epsilon'$. For any x_1, we shall show that $\rho(F, \bar{F} \mid x_1) < \epsilon$; this will complete the proof.

Let any policy $Y = \{G_i(y \mid x_i, B''_{i-1} = \beta''_{i-1})\}, i = 1, \cdots, N$ be given. Let $1 \leqslant i \leqslant N$. Since the V_i corresponding to $\bar{F}$ and F are within ϵ' uniformly in the other arguments, the same is true of the corresponding A_{ii}. Hence, for every x_i and β''_{i-1}, the corresponding functions A^*_{ii} also differ by at most ϵ'. Since 2^{i-1} times the $(i - 1)$th term in the sum representing $\delta_1(\bar{F}, F)$ is $< \epsilon'$, it follows that for any y_{i-1} and β'_{i-1} the functions $A_{i(i-1)}$ corresponding to $\bar{F}$ and F differ by at most $\epsilon' + M\epsilon'$. For every x_{i-1} and β''_{i-2}, the corresponding $A^*_{i(i-1)}$ therefore also differ by at most $\epsilon'(1 + M)$. Similarly, the A^*_{i1} differ by at most $\epsilon'[1 + M(i - 1)]$, so that $A(F, Y \mid x_1)$ and $A(\bar{F}, Y \mid x_1)$ differ by less than $N(1 + MN)\epsilon' = \epsilon$.

We remark that in the special case often encountered in applications where W_i does not depend on F, Assumption I may be simplified to the condition that Ω be conditionally compact in the sense of the metric δ_1. In fact, because of Assumption H, it is easy to verify in this case that if $\{F^{(i)}\}$ $(i = 0, 1, 2, \cdots$, ad inf.,) is a sequence of distribution functions such that $\lim_{i\to\infty} \delta_1[F^{(i)}, F^{(0)}] = 0$, then $\lim_{i\to\infty} \delta_2[F^{(i)}, F^{(0)}] = 0$.

2.5. The following Assumption J is made primarily in the interests of simplified writing. It is obviously not restrictive in any real situation and is, indeed, automatically fulfilled whenever x_i, y_i, and D_i are integral multiples of a fixed unit.

ASSUMPTION J: All probability measures $F_i(z \mid B_i = \beta_i; F)$ are dominated by a σ-finite[14] measure on the nonnegative real line; i.e., there exists a σ-finite measure μ defined for all Borel sets of the real nonnegative axis such that we have

$$(2.17) \qquad F_i(z \mid B_i = \beta_i; F) = \int_0^z f_i(t \mid B_i = \beta_i; F)\, d\mu(t)$$

for all i, β_i, F, and z with suitable "generalized densities" f_i.

We shall refer to the distributions ξ of Ξ as *a priori probability distributions on* Ω. For every i, β_i, ξ, and Borel set ω of Ω, we define through the equation

[14] One could equivalently assume domination by a probability measure.

$$\xi(\omega \mid \beta''_{i-1})$$

(2.18)
$$= \frac{\int_\omega f_1(d_1 \mid B_1 = \beta_1; F) \cdots f_{i-1}(d_{i-1} \mid B_{i-1} = \beta_{i-1}; F)\, d\xi}{\int_\Omega f_1(d_1 \mid B_1 = \beta_1; F) \cdots f_{i-1}(d_{i-1} \mid B_{i-1} = \beta_{i-1}; F)\, d\xi}$$

the *a posteriori probability relative to* ξ that $F \in \omega$ given the past history β''_{i-1}. The a posteriori probability measure defined[15] by (2.18) will be denoted by $\xi_{\beta''_{i-1}}$ (we put $\xi_{\beta''_0} = \xi$).

3. CASE OF FINITELY MANY TIME INTERVALS

3.1. As in Section 2 of I, we start with the case of a single time interval.

Let the initial stock x_1, $\epsilon > 0$, and the a priori distribution ξ be given. An ordering policy Y, which in this case is characterized by a distribution function $G(y_1 \mid x_1)$, is an ϵ-Bayes policy relative to ξ for x_1 if and only if

(3.1)
$$\int_\Omega \int_0^\infty V_1(x_1, y_1 \mid F)\, dG_1(y_1 \mid x_1)\, d\xi < \inf_{y_1 \geq x_1} \int_\Omega V_1(x_1, y_1 \mid F)\, d\xi + \epsilon$$

holds. Here, as well as later, we use the well-known fact that, once ξ is given, the inf operation yields the same value whether applied to all ordering policies or restricted to nonrandomized ones.

If we put

$$V_1(x_1, y_1 \mid \xi) = \int_\Omega V_1(x_1, y_1 \mid F)\, d\xi,$$

then (3.1) becomes

(3.2)
$$\int_0^\infty V_1(x_1, y_1 \mid \xi)\, dG_1(y_1 \mid x_1) < \inf_{y_1 \geq x_1} V_1(x_1, y_1 \mid \xi) + \epsilon.$$

This is very similar to (2.5) of I, especially if one remarks that when $G_1(y_1 \mid x_1)$ is the degenerate distribution assigning probability 1 to the value $\bar{y}$, then the integral in (3.2) reduces to $V(x_1, \bar{y})$. Thus all the methods discussed in detail in (I, 2) apply also here. Moreover, it is clear that there exist nonrandomized solutions, i.e., with degenerate $G_1(y_1 \mid x_1)$ of (3.1).

[15] There can be no trouble due to ambiguity of (2.18) since this ambiguity can occur only on a set of $d_1, \cdots, d_{i-1}$ whose probability is zero when ξ is the a priori distribution. This ambiguity has, therefore, no effect on the constructions in the following sections.

There is, however, one major difference which makes the study in the present case more elaborate, even in principle, than that of (I, 2). We know from 2.3 that there is an ϵ-minimax policy among the class of all ϵ-Bayes policies. But in order to be sure that a class C of ordering policies contains an ϵ-minimax policy (or is ϵ-essentially complete) it is not in general enough that C contain, for every ξ, *one* ϵ-Bayes policy relative to ϵ. It is, however, sufficient that C include *all*[16] ϵ-Bayes policies relative to ξ. Thus, having the above applications in view, it is not enough to find the degenerate solutions of (3.1) or (3.2) but it is necessary to consider all solutions $G_1(y_1 \mid x_1)$. This remark applies throughout Part II of the paper.

In many important cases the loss function

$$W_1(x_1, y_1; d_1 \mid F) = W_1(x_1, y_1; d_1) \tag{3.3}$$

is independent of the distribution of demand F. In this case (3.1) may be rewritten, in a form more suited to calculation, as follows: Putting

$$f_1(z \mid x_1, y_1; \xi) = \int_\Omega f_1[z \mid B_1 = (x_1, y_1); F]\, d\xi \tag{3.4}$$

(see Assumption J; the density (3.4) corresponds to a mixture of the demands according to the a priori probability distribution ξ) equation (3.1) becomes

$$\begin{aligned} &\int_0^\infty \int_0^\infty W_1(x_1, y_1; z) f_1(z \mid x_1, y_1; \xi)\, d\mu(z)\, dG_1(y_1 \mid x_1) \\ &\qquad < \inf_{y_1 \geq x_1} \int_0^\infty W_1(x_1, y_1; z) f_1(z \mid x_1, y_1; \xi)\, d\mu(z) + \epsilon. \end{aligned} \tag{3.5}$$

3.2. We now consider the case of N time intervals. We shall show how to construct ϵ-Bayes policies relative to a given a priori probability distribution ξ. The method used will be a recursive one similar to that of (I, 3). The initial stock x_1 and $\epsilon > 0$ are fixed throughout the discussion that follows.

We define for $j = 1, \cdots, N$

$$\bar{V}_j(y_j \mid B'_j = \beta'_j; G_{j+1}, \cdots, G_n; F) = \sum_{i=j}^N A_{ij} \tag{3.6}$$

and

$$\bar{V}^*_j(x_j \mid B''_{j-1} = \beta''_{j-1}; G_j, \cdots, G_n; F) = \sum_{i=j}^N A^*_{ij} \tag{3.7}$$

[16] This can be relaxed by taking a set of ϵ-Bayes policies which are sufficiently dense in the set of all ϵ-Bayes policies in the sense of the metric obtained from (2.1) by interchanging the roles of F and Y. (It is of course also permissible to replace ϵ by a smaller $\epsilon' > 0$.)

where we have omitted the arguments of A_{ij} and A^*_{ij}. (These functions are defined in (1.4)–(1.6) and depend on the ordering policy Y through G_ν for $\nu > j$ and $\nu \geqslant j$ respectively.) The functions obtained from (3.6) and (3.7) by integrating over Ω with respect to the a posteriori probability measure $\xi_{\beta''_{j-1}}$ of 2.5 will again be denoted by $\bar{V}$ and $\bar{V}^*$ respectively, but with F replaced by ξ. Thus, we have in particular

$$\bar{V}_1^*(x_1 \mid G_1, \cdots, G_N; \xi) = A(\xi, Y \mid x_1) \tag{3.8}$$

by (1.7).

As in 3.1 it is possible to construct a distribution function $G_N^* = G_N^*(y_N \mid x_N, B''_{N-1} = \beta''_{N-1})$ of y_N which satisfies

$$\bar{V}_N^*(x_N \mid B''_{N-1} = \beta''_{N-1}; G_N^*; \xi) < \inf_{y_N \geq x_N} \bar{V}_N(y_N \mid B'_N = \beta'_N; \xi) + \frac{\epsilon}{N}. \tag{3.9}$$

Having constructed $G_{j+1}^*, \cdots, G_N^*$ we then construct

$$G_j^* = G_j^*(y_j \mid x_j, B''_{j-1} = \beta''_{j-1})$$

so that it satisfies

$$\begin{aligned} &\bar{V}_j^*(x_j \mid B''_{j-1} = \beta''_{j-1}; G_j^*, \cdots, G_N^*; \xi) \\ &\quad < \inf_{y_j \geq x_j} \bar{V}_j(y_j \mid B'_j = \beta'_j; G_{j+1}^*, \cdots, G_N^*; \xi) + \frac{\epsilon}{N}. \end{aligned} \tag{3.10}$$

It is easily seen that (3.9) and (3.10) for $j = 1, \cdots, N - 1$ yield an ordering policy $Y^* = \{G_1^*, \cdots, G_N^*\}$ satisfying, by (3.8),

$$A(\xi, Y^* \mid x_1) < A(\xi, Y \mid x_1) + \epsilon \qquad \text{for all } Y \in \Upsilon; \tag{3.11}$$

i.e., Y^* is an ϵ-Bayes policy relative to ξ for x_1.

The above construction is similar to that employed in finding a Bayes solution for an N-stage decision problem.[17]

The class of all ϵ-Bayes solutions constructed in the above manner obviously comprises all ϵ/N-Bayes solutions and is, hence, complete for x_1.

4. CASE OF INFINITELY MANY TIME INTERVALS

4.1. In Section 4 we shall treat the case where an infinite number of time intervals is under consideration, corresponding to the case of (I, 4). In 4.1 we shall study briefly the general case, while 4.2 will be devoted to the important special case of stationarity (see also Section 5) and independence.

Because of Assumption H and (2.13) it is evident that for any given

[17] One may find such a construction on pp. 117–119 of Wald's book mentioned in footnote 8.

ξ, x_1, and $\epsilon > 0$, the following construction gives an ϵ-Bayes solution relative to ξ for x_1: Let $N = N(\epsilon)$ be such that $\sum_{i=N+1}^{\infty} \alpha_i < \epsilon/3M$, where M is as in Assumption H. For $i > N$, use any convenient sequence $\{G_i\}$ [e.g., $G_i(x_i \mid x_i, B''_{i-1} = \beta''_{i-1}) = 1$ for all x_i, β''_{i-1}, and $i > N$). This sequence may be taken to be the same for all x_i and ξ in this and the next paragraph. For $i \leqslant N$, use a set of G_i which constitute a $2\epsilon/3$-Bayes solution (constructed by the method of Section 3) if one considers only the first N intervals. We shall denote by $C(x_1, \epsilon)$ the class of all ordering policies of the above structure for a given x_1 and all ξ. [We recall that the G_i for $i > N$ may be taken to be fixed throughout $C(x_1, \epsilon)$.]

Since $\sum_{i=N+1}^{\infty} A_{i1}^* < \epsilon/3$ for all F, Y, and x_1, it follows that if $\bar{Y} = \{\bar{G}_i\}$ is an $\epsilon/3$-Bayes solution relative to ξ and $\bar{A}_{i1}^*$ is the value of A_{i1}^* when $\bar{Y}$ is used,

$$\begin{aligned}\sum_{i=1}^{N} \bar{A}_{i1}^*(x_1 \mid \xi) \leqslant A(\xi, \bar{Y} \mid x_1) < \frac{\epsilon}{3} + \inf_{Y \epsilon T} A(\xi, Y \mid x_1) \\ < \frac{2\epsilon}{3} + \inf_{Y \epsilon T} \sum_{i=1}^{N} \bar{A}_{i1}^*(x_1 \mid \xi),\end{aligned} \tag{4.1}$$

so that $\bar{G}_i = G_i^*$, $i = 1, \cdots, N$, for some $Y^* = \{G_i^*\}$ in $C(x_1, \epsilon)$. It follows that $A(Y^*, F \mid x_1) - A(\bar{Y}, F \mid x_1) < \epsilon$ for all F in Ω. Since the class of all $\epsilon/3$-Bayes solutions is complete, we conclude that $C(x_1, \epsilon)$ is ϵ-essentially complete for x_1.

4.2. The remainder of Section 4 will be devoted to the case of stationarity and independence corresponding to (I, 4.2–4.5). We assume that the generalized densities are independent of i and β_i; they may thus be written as $f(z \mid F)$. The loss function W_i is similarly assumed to be independent of i and β''_{i-1}; thus we denote it by $W(x, y; d \mid F)$, which exhibits all the arguments on which it may depend. We furthermore assume $\alpha_i = \alpha^{i-1}$ with $0 < \alpha < 1$. The a posteriori distributions $\xi_{\beta''_i}$ depend now only on ξ and $d_1, \cdots, d_i$ and will be denoted by $\xi_{d_1,\ldots,d_i}$.

Let

$$\ell(x; \xi) = \inf_{Y \epsilon T} A(\xi, Y \mid x). \tag{4.2}$$

If ξ assigns probability 1 to a single F then we write $\ell(x, F)$ instead of $\ell(x; \xi)$; this is obviously the function of (4.2) of I. We proceed to derive, in the manner of (I, 4), an integral equation satisfied by $\ell(x; \xi)$. In this derivation we assume that $\ell(x; F)$ is measurable in x and F and that $\ell(x; \xi_{d_1,\ldots,d_i})$ is measurable in $x, d_1, \cdots, d_i$. As in (I, 4.2), we note that because of the independence and stationarity assumption, the outlook of the ordering agency at the beginning of the second time interval is exactly the same as at the beginning of the first interval except that x_1 and ξ have been replaced by x_2 and ξ_{d_1}, respectively. It follows that the present expected value (before any ordering is done) of the sum of the losses in all intervals except the first, incurred by ordering to y_1 in the first time interval and by proceeding thereafter in an optimum manner, is

$$(4.3)\quad \alpha \int_0^{y_1} \ell(y_1 - z; \xi_z) f(z \mid \xi)\, d\mu(z) + \alpha \int_{y_1+}^{\infty} \ell(0; \xi_z) f(z \mid \xi)\, d\mu(z),$$

where $f(z \mid \xi)$ is given by (3.4) and where we have used the fact that $x_{j+1} = \max(0, y_j - d_j)$. We conclude that $\ell(x; \xi)$ must satisfy the integral equation

$$(4.4)\quad \begin{aligned} \ell(x; \xi) = \inf_{y \geq x} \Big\{ \int_\Omega V(x, y \mid F)\, d\xi + \alpha \int_0^{y} \ell(y - z; \xi_z) f(z \mid \xi)\, d\mu(z) \\ + \alpha \int_{y+}^{\infty} \ell(0; \xi_z) f(z \mid \xi)\, d\mu(z) \Big\}. \end{aligned}$$

We remark that if ξ gives probability one to a single F, (4.4) reduces to equation (4.6) of I. This fact may be helpful in obtaining a complete solution to (4.4) by the iterative method noted below. We also note that the essential point in the above derivation is the stationarity; a similar equation may be derived without assuming independence (see Section 5).

Denoting the expression in braces in (4.4) by $h(x, y; \xi)$, a Bayes solution relative to ξ for x_1 may be described as follows:[18] At the first interval, use a $G_1(y \mid x_1)$ for which

$$(4.5)\quad \int_0^{\infty} h(x_1, y; \xi)\, dG_1(y \mid x_1) = \min_{y \geq x_1} h(x_1, y; \xi).$$

At the jth interval ($j > 1$) use a $G_j(y \mid x_j)$ for which (4.5) holds with x_1, G_1, and ξ replaced by x_j, G_j, and $\xi_{d_1, \cdots, d_{j-1}}$, respectively. It is thus unnecessary for G_j to depend on the past history except through x_j and $\xi_{d_1, \cdots, d_{j-1}}$.

If $h(x, y; \xi)$ is not available but one employs an approximation $h_1(x, y; \xi)$ (obtained, for example, by the iterative process noted below) such that

$$(4.6)\quad | h_1(x, y; \xi) - h(x, y; \xi) | < \epsilon$$

[18] The existence of a Bayes solution is assumed only for ease of exposition in the present paragraph, and is unnecessary in the remainder of this section. If a Bayes solution does not exist but one uses at the jth stage a G_j for which the left side of (4.5) is within ϵ of the right side (with min, x_1, G_1, and ξ replaced by inf, x_j, G_j, and $\xi_{d_1, \cdots, d_{j-1}}$ respectively), the considerations of the next paragraph apply.

for all x, y, and ξ, then the total loss incurred by using h_1 instead of h will not exceed $\ell(x_1; \xi)$ by more than $\epsilon/(1 - \alpha)$. In particular, if one denotes by $\bar{G}(y, x_1, \xi)$ a G_1 for which (for any given ξ) the left member of (4.5) is within ϵ of the right member (with min replaced by inf), the policy resulting from using $\bar{G}(y \mid x_j, \xi_{d_1,\ldots,d_{j-1}})$ at the jth stage ($j > 1$) is an $\epsilon/(1 - \alpha)$-Bayes policy relative to ξ for x_1. The class $C_1(x_1, \epsilon)$ of all policies which may be constructed in this way contains an ϵ'-Bayes solution relative to ξ for x_1 for every ξ and every $\epsilon' > 0$. Let $\bar{C}_1(x_1, \epsilon)$ be the totality of all ordering policies obtained in the following manner: Let $Y^1, Y^2, \cdots, Y^n$ be any n policies in $C_1(x_1, \epsilon)$ ($n > 1$). Let $\lambda_1, \lambda_2, \cdots, \lambda_n$ be any positive numbers whose sum is one. The ordering agency performs a single chance experiment whose n possible outcomes are the integers $1, 2, \cdots, n$ with respective probabilities $\lambda_1, \lambda_2, \cdots, \lambda_n$. When the outcome of the experiment is i, the ordering agency proceeds in every interval according to the policy $Y^i (i = 1, 2, \cdots, n)$. Then it can be proved (for example, geometrically) that the class $\bar{C}_1(x_1, \epsilon)$ is 0-essentially complete for x_1 if Ω is finite, and hence also in the general case when Assumption G is satisfied.

Denote by $J\ell(x; \xi)$ [the result of the operator J applied to the function $\ell(x; \xi)$] the expression on the right side of (4.4). If $t(x; \xi)$ is any bounded measurable function of its arguments, it may be shown exactly as in (I, 4.2) that for any positive integer n,

$$\text{(4.7)} \qquad \sup_{x, \xi} | J^n t(x; \xi) - \ell(x; \xi) | \leq \alpha^n \sup_{x, \xi} | t(x; \xi) - \ell(x; \xi) |.$$

Thus, the $\ell(x; \xi)$ of (4.2) is the unique bounded solution of (4.4) and (4.7) indicates an iterative method for obtaining the function $\ell(x; \xi)$. Similar remarks apply to the function $h(x, y; \xi)$ [see I, (4.21)]. More precise bounds on the speed of convergence of $J^n t$ to ℓ may be obtained as in (I, 4.3).

The remarks of (I, 4.3) concerning the case when there exists a finite Z such that $y_i \leq Z$ for all i, also apply to the present situation. Finally, one may proceed as in (I, 4.5) to give an integral equation (and iterative solution thereof) for $A(F, Y \mid x)$, for any F [or for $A(\xi, Y \mid x)$ simultaneously for all ξ] and any Y for which G_i depends only on x_i and $\xi_{d_1,\ldots,d_{i-1}}$ but not on i, β''_{i-1}, or ξ in any other way [so that $\bar{G}_i(y \mid x_i) = \bar{G}(Y \mid x_i, \xi_{d_1,\ldots,d_{j-1}})$ may be written withou at subscripts on $\bar{G}$].

5. GENERALIZATIONS

5.1. The generalizations of the results of the preceding sections to the cases corresponding to those of Sections 5 and 6 of Part I may be carried out by combining the methods of the latter with those of the present

Part II. Without giving a detailed presentation, we shall for the sake of completeness state briefly some of these additional results. In 5.1 we shall mention the case of time-lag corresponding to (I, 5.1–5.3); in 5.2, the general case of stationarity without independence corresponding to (I, 5.3) will be noted; and in 5.3 we shall mention the case of several commodities corresponding to (I, 6).

In Sections 3 and 4 it was assumed that every order placed by the ordering agency was instantly filled. If instead there is a lag of T intervals (or, more generally, a variable lag: see (I, footnote 12), the functions A_{ij} and A^*_{ij} must be modified to functions analogous to those of equations (5.1) and (5.2) of I. The functions W_i, V_i, $\bar{V}_i$, and $\bar{V}^*_i$ are of course similarly modified. Bayes solutions, complete classes, etc., are defined with respect to the corresponding risk function. ϵ-Bayes solutions may be constructed by procedures like those of Section 3 and of 4.1, the corresponding results on complete classes remaining valid.

In the case of independence and stationarity corresponding to (I, 5.2) (i.e., 4.2 with lag), analogous results apply. In particular, one can again derive a fundamental integral equation, the unique bounded solution of which may be used as in 4.2 to construct Bayes solutions or ϵ-Bayes solution. As in the case of (I, 5.2), such a policy (corresponding to the $\bar{G}$ of 4.2) must now depend also on the quantities ordered in the previous $T - 1$ intervals. The other results of 4.2 on an iterative method of solution of the integral equation, complete classes, etc., may similarly be extended to the present case.

5.2. The results of Sections 3 and 4 and of 5.1 may also be generalized to the case where the ordered quantities may be delivered in several parts at several future times; these parts and times may also be chance variables. With any or all of the above modifications, if the loss function and the joint distribution function of all relevant chance variables at the beginning of an interval have a stationary structure for all elements of Ω (no independence being postulated), an integral equation analogous to that of equation (5.19) of I may be derived. However, it is important to notice that in the construction of a Bayes solution or ϵ-Bayes solution, even when no lag is present, the policy corresponding to the $\bar{G}$ of 4.2 must now depend on other quantities than x and the a posteriori distribution.[19]

For example, in the case of no lag if the distribution of demand as well as the expected loss in the ith interval depend only on d_{i-1}, x_i, y_i, and the element F of Ω, one obtains

[19] For an analogous discussion of the stationary case in problems of decision theory, see J. Kiefer, "Sequential Decision Problems in the Stationary Case," abstract to appear in the *Annals of Mathematical Statistics*.

$$
\begin{aligned}
\ell(x; d, \xi) = \inf_{y \geq x} \Big\{ \int_\Omega V(x, y \mid d; F)\, d\xi \\
(5.1) \qquad + \alpha \int_0^y \ell(y - z; z, \xi_z) f(z \mid x, y, d; \xi)\, d\mu(z) \\
+ \alpha \int_{y+}^\infty \ell(0; z, \xi_z) f(z \mid x, y, d; \xi)\, d\mu(z) \Big\},
\end{aligned}
$$

where $f(z \mid x, y, d; \xi)$ is given by integrating the generalized density defined in 2.5 over Ω with respect to ξ, and where $\ell(x; d, \xi)$ is the infimum of the present value of the total expected loss when x is the starting stock, d was the demand in the immediately preceding interval, and ξ is the a posteriori distribution at the beginning of the present interval.

5.3. All of the previous results of this paper may be extended to the case where several commodities, several warehouses, and several consumers are simultaneously considered. The principal modifications necessary for this treatment are that x_i and y_i must be replaced by matrices whose elements represent initial and starting stocks of various commodities at various warehouses, that D_i becomes the matrix of various demands by various consumers, and that besides the ordering policy one must now consider a distributing policy which, after the manifestation of demand in each interval, indicates (perhaps in a randomized manner) how the available stocks are to be allocated from the various warehouses to the various consumers. The reader is referred to (I, 6) for further details.

The theory of the present paper can be applied in an obvious way to various modifications of the inventory problem some of which were discussed in I, e.g., the inventory problem with speculation in commodity prices, the problem mentioned in (I, 6.2) where the production proceeds in several stages, and the case of (I, 6.2) where the number of intervals to be considered is a chance variable.

Cornell University

Reprinted from
Econometrica **20** (1952), 450-466

Reprinted from THE ANNALS OF MATHEMATICAL STATISTICS
Vol. 23, No. 3, September, 1952

STOCHASTIC ESTIMATION OF THE MAXIMUM OF A REGRESSION FUNCTION[1]

BY J. KIEFER AND J. WOLFOWITZ

Cornell University

1. Summary. Let $M(x)$ be a regression function which has a maximum at the unknown point θ. $M(x)$ is itself unknown to the statistician who, however, can take observations at any level x. This paper gives a scheme whereby, starting from an arbitrary point x_1, one obtains successively $x_2, x_3, \cdots$ such that x_n converges to θ in probability as $n \to \infty$.

2. Introduction. Let $H(y \mid x)$ be a family of distribution functions which depend on a parameter x, and let

$$M(x) = \int_{-\infty}^{\infty} y \, dH(y \mid x). \tag{2.1}$$

We suppose that

$$\int_{-\infty}^{\infty} (y - M(x))^2 \, dH(y \mid x) \leqq S < \infty, \tag{2.2}$$

and that $M(x)$ is strictly increasing for $x < \theta$, and $M(x)$ is strictly decreasing for $x > \theta$. Let $\{a_n\}$ and $\{c_n\}$ be infinite sequences of positive numbers such that

$$c_n \to 0, \tag{2.3}$$

$$\sum a_n = \infty, \tag{2.4}$$

$$\sum a_n c_n < \infty, \tag{2.5}$$

$$\sum a_n^2 c_n^{-2} < \infty. \tag{2.6}$$

(For example, $a_n = n^{-1}$, $c_n = n^{-1/3}$.)

We can now describe a recursive scheme as follows. Let z_1 be an arbitrary number. For all positive integral n we have

$$z_{n+1} = z_n + a_n \frac{(y_{2n} - y_{2n-1})}{c_n}, \tag{2.7}$$

where y_{2n-1} and y_{2n} are independent chance variables with respective distributions $H(y \mid z_n - c_n)$ and $H(y \mid z_n + c_n)$. Under regularity conditions on $M(x)$ which we shall state below we will prove that z_n converges stochastically to θ (as $n \to \infty$).

The statistical importance of this problem is obvious and need not be discussed. The stimulus for this paper came from the interesting paper by Robbins and Monro [1] (see also Wolfowitz [2]).

[1] Research under contract with the Office of Naval Research. Presented to the American Mathematical Society at New York on April 25, 1952.

While we have no need to postulate the existence of the derivative of $M(x)$ (indeed, $M(x)$ can be discontinuous), the spirit of our regularity assumptions postulated below is as follows. (a) If $M(x)$ did have a derivative it would be zero at $x = \theta$. Hence we would have expected the derivative not to be too large in a neighborhood of $x = \theta$. (b) If, at a distance from θ, $M(x)$ were very flat, then movement towards θ would be too slow. Hence outside of a neighborhood of $x = \theta$ we would have liked the absolute value of the derivative to be bounded below by a positive number. (c) If $M(x)$ rose too steeply in places we might through mischance get a movement of z_n which would throw us far out from θ. If there were many such steep places z_n could be made to approach $+\infty$ or $-\infty$ with positive probability. We would therefore have postulated a Lipschitz condition.

From the mathematical point of view it would be aesthetic to weaken the conditions. From the practical point of view it might be objected that these conditions prevent $M(x)$ from being a function which flattens out toward the x-axis, for example, $M(x) = e^{-x^2}$, or from being a function which drops off steadily faster to $-\infty$, for example, $M(x) = -x^2$. Now in any practical situation one can always give a priori an interval $[C_1, C_2]$ such that $C_1 \leqq \theta \leqq C_2$. It will be sufficient if our conditions are fulfilled in this interval.

Suppose, however, that some $z_n \pm c_n$ falls outside the interval $[C_1, C_2]$ and one cannot take an observation at that level. If one then moves z_n so that the offending $z_n \pm c_n$ is at C_1 or C_2, as the case may be, and proceeds as directed by (2.7), then our conclusion remains valid.

We postulate the following regularity conditions on $M(x)$.

CONDITION 1. There exist positive β and B such that

$$|x' - \theta| + |x'' - \theta| < \beta \text{ implies } |M(x') - M(x'')| < B|x' - x''|. \tag{2.8}$$

CONDITION 2. There exist positive ρ and R such that

$$|x' - x''| < \rho \text{ implies } |M(x') - M(x'')| < R. \tag{2.9}$$

CONDITION 3. For every $\delta > 0$ there exists a positive $\pi(\delta)$ such that

$$|z - \theta| > \delta \text{ implies } \inf_{\frac{1}{2}\delta > \epsilon > 0} \frac{|M(z+\epsilon) - M(z-\epsilon)|}{\epsilon} > \pi(\delta). \tag{2.10}$$

3. Proof that z_n converges stochastically to θ. Let

$$b_n = E(z_n - \theta)^2, \tag{3.1}$$

$$U_n(z) = (z - \theta)\, E\{y_{2n} - y_{2n-1} \mid z_n = z\}, \tag{3.2}$$

$$U_n^+(z) = \tfrac{1}{2}(U_n(z) + |U_n(z)|),\; U_n^-(z) = \tfrac{1}{2}(U_n(z) - |U_n(z)|), \tag{3.3}$$

$$P_n = E(U_n^+(z_n)),\; N_n = E(U_n^-(z_n)), \tag{3.4}$$

$$e_n = E(y_{2n} - y_{2n-1})^2. \tag{3.5}$$

From (2.7) we have

$$b_{n+1} = b_n + 2\frac{a_n}{c_n}(P_n + N_n) + \frac{a_n^2}{c_n^2}e_n. \tag{3.6}$$

Adding the expressions obtained from (3.6) for $b_{j+1} - b_j$ for $1 \leqq j \leqq n$, we obtain

$$b_{n+1} = b_1 + 2\sum_{j=1}^{n}\frac{a_j}{c_j}P_j + 2\sum_{j=1}^{n}\frac{a_j}{c_j}N_j + \sum_{j=1}^{n}\frac{a_j^2}{c_j^2}e_j. \tag{3.7}$$

Noting that $U_n^+(z) \geqq 0$ and that $U_n^+(z) > 0$ implies that $|z - \theta| < c_n$ because $M(x)$ is monotonic for $x < \theta$ and for $x > \theta$, it follows from (2.8) that, for all n for which $c_n < \frac{1}{2}\beta$, we have

$$0 \leqq U_n^+(z) < 2Bc_n^2 \tag{3.8}$$

It follows from (2.5) and (3.8) that the positive-term series

$$\sum_{n=1}^{\infty}\frac{a_n}{c_n}P_n \tag{3.9}$$

converges, say to α. From (2.9) we have

$$[M(z_n + c_n) - M(z_n - c_n)]^2 < R^2 \tag{3.10}$$

for n sufficiently large. Also for large enough n,

$$\begin{aligned} &E\{(y_{2n} - y_{2n-1})^2 \mid z_n\} \\ &= E\{(y_{2n} - M(z_n + c_n))^2 + (y_{2n-1} - M(z_n - c_n))^2 \mid z_n\} \\ &\quad + [M(z_n + c_n) - M(z_n - c_n)]^2 \leqq 2S + R^2 \end{aligned} \tag{3.11}$$

by (2.2) and (3.10). Hence for large enough n

$$E[y_{2n} - y_{2n-1}]^2 \leqq 2S + R^2. \tag{3.12}$$

Consequently from (2.6) we obtain that the positive-term series

$$\sum_{n=1}^{\infty}\frac{a_n^2}{c_n^2}e_n \tag{3.13}$$

converges, say to γ. Hence, since $b_{n+1} \geqq 0$, it follows from (3.7) that

$$2\sum_{j=1}^{n}\frac{a_j}{c_j}N_j \geqq -b_1 - 2\alpha - \gamma > -\infty, \tag{3.14}$$

so that the negative-term series

$$\sum_{n=1}^{\infty}\frac{a_n}{c_n}N_n \tag{3.15}$$

converges.

Let

$$K_n = \left| \frac{M(z_n + c_n) - M(z_n - c_n)}{c_n} \right|. \tag{3.16}$$

Then

$$E\{K_n \,|\, z_n - \theta \,|\,\} = \frac{P_n - N_n}{c_n}. \tag{3.17}$$

From the convergence of (3.9) and (3.15) and the divergence of $\sum a_n$, it follows that

$$\liminf_{n\to\infty} E\{K_n \,|\, z_n - \theta \,|\,\} = 0. \tag{3.18}$$

Let $n_1 < n_2 < \cdots$ be an infinite sequence of positive integers such that

$$\lim_{j\to\infty} E\{K_{n_j} \,|\, z_{n_j} - \theta \,|\,\} = 0. \tag{3.19}$$

We assert that $(z_{n_j} - \theta)$ converges stochastically to zero as $j \to \infty$. For if not, there would exist two positive numbers δ and ϵ and a subsequence $\{t_j\}$ of $\{n_j\}$ such that, for all j,

$$P\{\,|\, z_{t_j} - \theta \,|\, > \delta\} > \epsilon, \tag{3.20}$$

which implies that

$$E\{K_{t_j} \,|\, z_{t_j} - \theta \,|\,\} \geqq \delta\epsilon\pi\left(\frac{\delta}{2}\right) > 0 \tag{3.21}$$

for all j for which $c_{t_j} < \frac{1}{2}\delta$. But (3.21) contradicts (3.19) and the stochastic convergence to zero of $(z_{n_j} - \theta)$ is proved.

Let η and ϵ be arbitrary positive numbers. The proof of the theorem will be complete if we can show the existence of an integer $N(\eta, \epsilon)$ such that

$$P\{\,|\, z_n - \theta \,|\, > \eta\} \leqq \epsilon \text{ for } n > N(\eta, \epsilon). \tag{3.22}$$

Let s be a positive number such that

$$\frac{s^2 + s}{\eta^2} < \frac{\epsilon}{2}. \tag{3.23}$$

Because z_{n_j} converges stochastically to θ there exists an integer N_0 such that

$$P\{|\, z_{N_0} - \theta \,| \geqq s\} < \frac{\epsilon}{2}. \tag{3.24}$$

We may also choose N_0 so large that

$$c_n < \min\left(\frac{\rho}{2}, \frac{\beta}{2}\right) \text{ for all } n \geqq N_0, \tag{3.25}$$

and

$$\sum_{n=N_0}^{\infty} \frac{a_n^2}{c_n^2} < \frac{s}{2R^2 + 4S}, \tag{3.26}$$

and

$$\sum_{n=N_0}^{\infty} a_n c_n < \frac{s}{8B}. \tag{3.27}$$

Proceeding in a manner similar to that used to obtain (3.7), we have, for each $n > N_0$,

$$\begin{aligned} E\{(z_n - \theta)^2 \mid z_{N_0} = z\} &= (z - \theta)^2 + 2 \sum_{j=N_0}^{n-1} \frac{a_j}{c_j} E\{U_j \mid z_{N_0} = z\} \\ &\quad + \sum_{j=N_0}^{n-1} \frac{a_j^2}{c_j^2} E\{(y_{2j} - y_{2j-1})^2 \mid z_{N_0} = z\} \\ &\leqq (z - \theta)^2 + 2 \sum_{j=N_0}^{\infty} \frac{a_j}{c_j} E\{U_j^+ \mid z_{N_0} = z\} + (R^2 + 2S) \sum_{j=N_0}^{\infty} \frac{a_j^2}{c_j^2} < (z - \theta)^2 + s. \end{aligned} \tag{3.28}$$

Using (3.23), (3.28), and Tchebycheff's inequality, we have

$$P\{|z_n - \theta| > \eta \,\big|\, |z_{N_0} - \theta| < s\} < \frac{\epsilon}{2}. \tag{3.29}$$

The inequalities (3.24) and (3.29) show that (3.22) holds for $N(\eta, \epsilon) = N_0$, and the proof is complete.

4. Further problems. The following remarks about further problems apply also to [1].

A. An obvious problem is to determine sequences $\{c_n\}$ and $\{a_n\}$ which would be optimal in some reasonable sense.

B. An important problem is to determine a stopping rule, that is, a rule by which the statistician decides when he is sufficiently close to θ.

C. This problem is a combination of *B* and a generalization of *A*, that is, to determine an optimal procedure with its stopping rule.

REFERENCES

[1] H. Robbins and S. Monro, "A stochastic approximation method," *Annals of Math. Stat.*, Vol. 22 (1951), pp. 400–407.

[2] J. Wolfowitz, "On the stochastic approximation method of Robbins and Monro," *Annals of Math. Stat.*, Vol. 23 (1952), pp. 457–461.

Reprinted from The Annals of Mathematical Statistics
Vol. 23, No. 4, December, 1952

SEQUENTIAL MINIMAX ESTIMATION FOR THE RECTANGULAR DISTRIBUTION WITH UNKNOWN RANGE[1]

By J. Kiefer

Cornell University

1. Summary. This paper is concerned with sequential minimax estimation of the parameter θ $(0 < \theta < \infty)$ of the density function (2.1), when the observations are independently and identically distributed with this density, each observation costs the same amount $c > 0$, and the weight function is as given in Section 2. A procedure requiring a fixed sample size is shown to be a minimax solution for this problem.

2. Introduction. An important problem in the theory of statistical decision functions[2] is that of minimax sequential estimation of the parameter of an (unknown) member of a given family of distribution functions when the observations are taken on chance variables which are independently and identically distributed and when the cost of taking n observations is cn with $c > 0$, regardless of the way in which they are taken. This problem was solved for the case of point estimation of the mean of the rectangular distribution from $\theta - \frac{1}{2}$ to $\theta + \frac{1}{2}$ $(-\infty < \theta < \infty)$ for the weight function $W(\theta, d) = A(\theta - d)^2$ by Wald [1]; the minimax sequential estimation problem for the normal distribution was solved for a variety of normal decision spaces and weight functions by Wolfowitz [2] (see also [3]); certain computational simplifications in the course of both of these cases were given by Blyth [4].

The present paper is devoted to a problem of sequential minimax estimation for the case where the family of possible distribution functions consists of all distributions for which the successive observations are independently and identically distributed with rectangular density function from 0 to θ (equation (2.1)), for $\theta \in \Omega = \{\theta \mid 0 < \theta < \infty\}$, and where the cost of taking n observations is cn $(c > 0)$ regardless of the way in which the observations are taken. The object is to estimate θ, the terminal decision space being $D = \{d \mid 0 < d < \infty\}$. The weight function is $W(\theta, d) = A(\theta - d)^2/\theta^2$, i.e., the loss incurred by making decision d when θ is the true parameter is A times the square of the fractional error in estimating θ. Thus, the minimax problem considered in this paper is that of finding a sequential estimation procedure which minimizes $\sup_\theta [cE_\theta(n) + AE_\theta(\theta - d)^2/\theta^2]$. A word is in order concerning our choice of weight function. The reason we do not study the problem for such weight functions as $A(\theta - d)^2$ or $A(\theta - d)^2/\theta$ is that for such weight functions, the supremum of the risk over all θ is infinite for every decision function, so that

[1] Research under a contract with the Office of Naval Research.

[2] See [5] and [6] for an exposition of this theory and an explanation of the terminology used herein.

Reprinted from The Annals of Mathematical Statistics
Vol. 23, No. 4, December, 1952

SEQUENTIAL MINIMAX ESTIMATION FOR THE RECTANGULAR DISTRIBUTION WITH UNKNOWN RANGE[1]

By J. Kiefer
Cornell University

1. Summary. This paper is concerned with sequential minimax estimation of the parameter $\theta(0 < \theta < \infty)$ of the density function (3.1) when the observations are independently and identically distributed with this density, each observation costs the same amount $c > 0$, and the weight function is as given in Section 2. A procedure requiring a fixed sample size is shown to be a minimax solution for this problem.

2. Introduction. An important problem in the theory of statistical decision functions[2] is that of minimax sequential estimation of the parameter of an (unknown) member of a given family of distribution functions when the observations are taken on chance variables which are independently and identically distributed and when the cost of taking n observations is cn (with $c > 0$) regardless of the way in which they are taken. This problem was solved for the case of point estimation of the mean of the rectangular distribution from $\theta - \frac{1}{2}$ to $\theta + \frac{1}{2}$ $(-\infty < \theta < \infty)$, for weight function $W(\theta, d) = (\theta - d)^2$ by Wald [1]; the minimax sequential estimation problem for the normal distribution was solved for a variety of terminal decision spaces and weight functions by Wolfowitz [2] (see also [3]); certain extensions and modifications of the results of both of these cases were given by Blyth [4].

The present paper is devoted to a problem of sequential minimax estimation for the case where the family of possible distribution functions consists of all distributions for which the successive observations are independently and identically distributed with rectangular density function from 0 to θ (equation (3.1)) for $\theta \in \Omega = \{\theta \mid 0 < \theta < \infty\}$ and where the cost of taking n observations is $cn(c > 0)$ regardless of the way in which the observations are taken. The object is to estimate θ, the terminal decision space being $D = \{d \mid 0 \leqq d < \infty\}$. The weight function is $W(\theta, d) = [(\theta - d)/\theta]^2$; i.e., the loss incurred by making decision d when θ is the true parameter is the square of the fractional error in estimating θ. Thus, the minimax problem considered in this paper is that of finding a sequential estimation procedure which minimizes $\sup_\theta\{cE_\theta(n) + E_\theta[(\theta - d)/\theta]^2\}$. A word is in order concerning our choice of weight function. The reason we do not study the problem for such weight functions as $|\theta - d|$, $(\theta - d)^2$, or $[(\theta - d)^2/\theta]$ is that for such weight functions the supremum of the risk over all $\theta \in \Omega$ is infinite for every decision function, so that

[1] Research under a contract with the Office of Naval Research.

[2] See Wald [1] for an exposition of this theory and an explanation of the nomenclature used herein.

every decision function is minimax. In addition, weight functions which depend only on d/θ (such as $[(\theta - d)/\theta]^2$) have a structure which essentially simplifies matters when estimating a scale parameter. On the other hand, it does not seem convenient in the present case to consider simultaneously a large class of weight functions as was possible in the cases of symmetrical densities studied in [2] and [4]. We therefore treat only one typical weight function here, noting that the same method should be applicable to many others.

With Ω, D, and $W(\theta, d)$ as described above, we shall prove that there is a minimax solution for which a fixed number of observations is taken. Specifically, the function $r(m)$ of (3.20) (which is the constant risk corresponding to taking a sample of fixed sample size m and then estimating θ by the expression of (2.1) with m for m_0) has at most two minima (if there are two, they are for successive values of m; moreover, there is only one minimum for all but a denumerable set of values of c). A minimax decision function is given by taking m_0 observations $y_1, y_2, \cdots y_{m_0}$, where $r(m_0)$ is the minimum of $r(m)$ (if there are two minima, at m_0 and $m_0 + 1$, one may randomize in any way between the decisions to take m_0 or $m_0 + 1$ observations); and by then estimating θ by

$$\frac{m_0 + 2}{m_0 + 1} \max (y_1, \cdots, y_{m_0}) \tag{2.1}$$

if $m_0 > 0$ (we replace m_0 by $m_0 + 1$ throughout (2.1) if the latter number of observations is taken when there are two minima), and by 0 if $m_0 = 0$. The risk corresponding to this decision function is then $r(m_0)$ for all values of $\theta \varepsilon \Omega$. It follows, incidentally, that this decision function is uniformly best among all cogredient procedures (see [4]). It is also a minimax solution for some related problems discussed in Section 3 of [4].

The method of proof is to calculate a lower bound on the Bayes risk when the a priori density on Ω is given by (3.4). It follows from (3.24) that as the parameter a of (3.4) approaches zero, the corresponding Bayes risk approaches $r(m_0)$; hence, by an argument like that of [1], p. 167, the procedure described in the previous paragraph is a minimax solution. The lower bound (3.24) is calculated in detail, since the necessary steps in its calculation differ somewhat from those of [1], [2], and [4]. We also note that, in this case of estimating a scale parameter, the tool used in [1], [2], and [4] of attempting to attain a "uniform a priori distribution on the real line" in the location parameter case is replaced by trying to attain the "a priori density" $1/\theta$. The proof is somewhat shortened by restricting the positive range of $\lambda_a(\theta)$ to values $\theta < 1$. This asymmetry manifests itself in the fact that the estimator of (3.7) does not tend to a minimax solution as $a \to 0$.

The fact that $\lambda_a(\theta)$ is positive only for $\theta < 1$ also shows that the fixed sample procedure described above is minimax for the problem of estimating θ when the above setup is altered by making $\Omega = \{\theta \mid 0 < \theta < b\}$, where $0 < b < \infty$: the argument of Section 3 shows this for $b = 1$, and the result for general $b = b'$ follows immediately from the case $b = 1$ if one considers there the problem of estimating $b'\theta$ from the sequence $\{b'Y_i\}$ of chance variables. Similarly, by considering for each value of a in Section 3 the problem of estimating $ba\theta$ from the sequence $\{baY_i\}$, one sees that our fixed sample procedure is also minimax for the problem of estimating θ when our original setup is altered by making $\Omega = \{\theta \mid b < \theta < \infty\}$. However, the given procedure is obviously not admissible if $m_0 > 0$ (or $m_0 \geqq 0$ in the second case): for example, a trivially better procedure in the first case when $m_0 > 0$ is to estimate θ by b whenever the expression of (2.1) is $> b$.

Finally, we remark that the problem of estimating θ for the case where the $f(y; \theta)$ of (3.1) is replaced by $1/(2\theta)$ for $-\theta < y < \theta$, is obviously identical to the one we consider: one has only to note that after n observations a sufficient statistic is still given by (3.2) if only Y_i is replaced by $|Y_i|$ for $i = 1, \cdots, n$. It is also of interest to note that our problem may be translated (by considering $T_i = e^{-Y_i}$, $\phi = e^{-\theta}$) into that of sequential minimax estimation of the parameter ϕ of the density $e^{-(t-\phi)}$ for $t > \phi$, 0 otherwise $(-\infty < \phi < \infty)$, when the weight function is $W(\phi, d) = (1 - e^{-(d-\phi)})^2$.

3. Calculations. For brevity, we shall throughout this section state the values of density functions and discrete probability functions only over the domains where they are positive. Let $Y_1, Y_2, \cdots$ be a sequence of independently and identically distributed chance variables, each with density function

$$f(y; \theta) = 1/\theta \qquad 0 < y < \theta, \tag{3.1}$$

where $\theta \,\varepsilon\, \Omega = \{\theta \mid 0 < \theta < \infty\}$. Define

$$X_n = \max\{Y_1, \cdots, Y_n\}. \tag{3.2}$$

Clearly, if observations $y_1, \cdots, y_n$ on $Y_1, \cdots, Y_n$ are taken, then X_n is a sufficient statistic for θ; i.e., for any a priori probability distribution on Ω, the a posteriori distribution of θ depends on $y_1, \cdots, y_n$ only through the value x_n taken on by X_n. Thus, in constructing sequential Bayes solutions, we may restrict ourselves to decision functions for which the (perhaps randomized) rule for stopping and estimation depends, after n observations, only on x_n. The density function of X_n is given by

$$g_n(x; \theta) = \frac{nx^{n-1}}{\theta^n}, \qquad 0 < x < \theta. \tag{3.3}$$

For $0 < a < 1$, we define

$$\lambda_a(\theta) = \frac{1}{\log(1/a)} \frac{1}{\theta}, \qquad a < \theta < 1. \tag{3.4}$$

If $\lambda_a(\theta)$ is the a priori density function on Ω and $y_1, \cdots, y_n$ have been observed, the a posteriori density of θ given that $X_n = x$ is easily computed to be

$$h_n(\theta \mid X_n = x) = \frac{nz^n}{1 - z^n} \cdot \frac{1}{\theta^{n+1}}, \qquad z < \theta < 1, \tag{3.5}$$

where $z = \max(a, x)$ and we note that $P\{z < 1\} = 1$.

The a posteriori loss (excluding cost of experimentation) if one stops after n observations and uses d to estimate θ, is

$$(3.6)\qquad \begin{aligned} W_n^*(d, z) &= \int_0^1 \left(\frac{d-\theta}{\theta}\right)^2 h_n(\theta \mid X_n = x)\, d\theta \\ &= 1 + \frac{n}{z^2(1-z^n)}\left[z^{n+2}\left(\frac{2d}{n+1} - \frac{d^2}{n+2}\right) - \left(\frac{2dz}{n+1} - \frac{d^2}{n+2}\right)\right]. \end{aligned}$$

The unique minimum of W_n^* with respect to d is easily seen to occur for

$$(3.7)\qquad d = \frac{n+2}{n+1} \cdot \frac{1 - z^{n+1}}{1 - z^{n+2}} \cdot z,$$

the corresponding value of W_n^* being

$$(3.8)\qquad W_n^{**}(z) = 1 - \frac{n(n+2)}{(n+1)^2} \cdot \frac{(1 - z^{n+1})^2}{(1 - z^n)(1 - z^{n+2})}.$$

For $n = 0$, the integral in (3.6) must be altered by replacing h_n by λ_a; the final expression must be changed accordingly. Equation (3.7) then holds with $z = a$, and (3.8) becomes $1 - 2(1 - a)/[(1 + a) \log (1/a)]$.

Next we note that when $f(y; \theta)$ is the density of each Y_i, the conditional distribution function of X_n given that $X_{n-1} = u$ assigns probability mass u/θ at the point $x = u$ and density $1/\theta$ for $u < x < \theta$. For $n = 1$, the distribution of X_1 is of course given by the density $f(x; \theta)$. We conclude that if $\lambda_a(\theta)$ is the a priori density on Ω, the distribution of X_1 is given by the density

$$(3.9)\qquad p_1(x) = \int_0^1 f(x; \theta)\lambda_a(\theta)\, d\theta = \begin{cases} \dfrac{1}{\log (1/a)} \cdot \dfrac{(1-a)}{a} & \text{if } x \leqq a, \\[2ex] \dfrac{1}{\log (1/a)} \cdot \dfrac{(1-x)}{x} & \text{if } a < x < 1; \end{cases}$$

and that (using (3.5) with n replaced by $n - 1$), for $n > 1$, the conditional distribution of X_n given that $\lambda_a(\theta)$ is the a priori density and $X_{n-1} = u$, is given, if $u \leqq a$, by

$$(3.10)\qquad \begin{aligned} P_n\{X = u\} &= \frac{n-1}{n} \cdot \frac{1-a^n}{1-a^{n-1}} \cdot \frac{u}{a}, \\ p_n(x \mid u) &= \begin{cases} \dfrac{n-1}{n} \cdot \dfrac{1-a^n}{1-a^{n-1}} \cdot \dfrac{1}{a}, & u < x \leqq a, \\[2ex] \dfrac{n-1}{n} \cdot \dfrac{1-x^n}{1-a^{n-1}} \cdot \dfrac{a^{n-1}}{x^n}, & a < x < 1; \end{cases} \end{aligned}$$

and, if $u > a$, by

$$(3.11)\qquad \begin{aligned} P_n\{X = u\} &= \frac{n-1}{n} \cdot \frac{1-u^n}{1-u^{n-1}}, \\ p_n(x \mid u) &= \frac{n-1}{n} \cdot \frac{1-x^n}{1-u^{n-1}} \cdot \frac{u^{n-1}}{x^n}, \qquad u < x < 1; \end{aligned}$$

where in each case P_n is the probability mass at $x = u$ and $p_n(x \mid u)$ is the density elsewhere.

Equations (3.10) and (3.11) yield for the conditional distribution of $Z_n = \max(X_n, a)$ given that $\lambda_a(\theta)$ is the a priori density and that $Z_{n-1} = v$, for all $n > 1$,

$$(3.12)\qquad \begin{aligned} Q_n\{Z = v\} &= \frac{n-1}{n} \cdot \frac{1 - v^n}{1 - v^{n-1}}, \\ q_n(z \mid v) &= \frac{n-1}{n} \cdot \frac{1 - z^n}{1 - v^{n-1}} \cdot \frac{v^{n-1}}{z^n}, \qquad v < z < 1, \end{aligned}$$

where again q_n is a density and Q_n is the probability mass at $z = v$.

Let $\overline{W}_{n-1}(v)$ be the conditional expected value of $W_n^{**}(Z_n)$ given that $\lambda_a(\theta)$ is the a priori density and that $Z_{n-1} = v$ (where we define $Z_0 = a$). Using (3.8) and (3.9), we have

$$(3.13a)\qquad \begin{aligned} \overline{W}_0(a) &= E\{W_1^{**}(Z_1)\} \\ &= W_1^{**}(a) \int_0^a p_1(z)\, dz + \int_a^1 W_1^{**}(z) p_1(z)\, dz \\ &= 1 - \frac{3}{4 \log(1/a)} \left\{ \frac{(1 - a^2)^2}{(1 - a^3)} + \int_a^1 \frac{(1 - z^2)^2}{z(1 - z^3)}\, dz \right\} \\ &< 1 - \frac{3}{4 \log(1/a)} \int_a^1 \left[\frac{1}{z} - 1 + \frac{(1 - z)^2}{(1 - z^3)} \right] dz \\ &< 1 - \frac{3}{4 \log(1/a)} \left[\log \frac{1}{a} - (1 - a) \right] < \tfrac{1}{4} + \frac{1}{\log(1/a)}. \end{aligned}$$

For n > 1, we have from (3.8) and (3.12),

$$(3.13b)\qquad \begin{aligned} \overline{W}_{n-1}(v) &= E\{W_n^{**}(Z_n) \mid v\} \\ &= W_n^{**}(v) Q_n(Z = v) + \int_v^1 W_n^{**}(z) q_n(z \mid v)\, dz \\ &= 1 - \frac{(n-1)(n+2)}{(n+1)^2(1 - v^{n-1})} \\ &\qquad \cdot \left\{ \frac{(1 - v^{n+1})^2}{(1 - v^{n+2})} + v^{n-1} \int_v^1 \frac{(1 - z^{n+1})^2}{z^n(1 - z^{n+2})}\, dz \right\}. \end{aligned}$$

The term in the last set of braces in (3.13b) may be written as

$$(3.14)\qquad \begin{aligned} 1 - v^{n-1} &+ \frac{v^{n-1}(1 - v)[1 + v - v^2 - v^{n+2}]}{(1 - v^{n+2})} \\ &+ v^{n-1} \int_v^1 \left[\frac{1}{z^n} - \frac{z(2 - z - z^{n+1})}{(1 - z^{n+2})} \right] dz \\ &> (1 - v^{n-1}) + \frac{1}{n-1}(1 - v^{n-1}) - v^{n-1} \int_v^1 2z\, dz \\ &= \frac{n}{n-1}(1 - v^{n-1}) - v^{n-1}(1 - v^2). \end{aligned}$$

We conclude that, whenever $n > 1$,

$$\overline{W}_{n-1}(v) < 1 - \frac{n(n+2)}{(n+1)^2} + \frac{(n-1)(n+2)}{(n+1)^2} \cdot \frac{v^{n-1}(1-v^2)}{(1-v)^{n-1}} \tag{3.15}$$
$$< \frac{1}{(n+1)^2} + \frac{1}{\log(1/v)},$$

where in the last step we have used the fact that $(n-1)(n+2)(n+1)^{-2} < \frac{1}{2}$ if $n = 2$ and < 1 otherwise, that $(1-v^2)(1-v^{n-1})^{-1} < 2$ if $n = 2$ and $\leqq 1$ otherwise, and that if $n > 1$ we have $v^{n-1} \leqq v < (\log 1/v)^{-1}$. From (3.13a) and (3.15), we have for all $n > 0$,

$$\overline{W}_{n-1}(v) < \frac{1}{(n+1)^2} + \frac{1}{\log(1/v)}. \tag{3.16}$$

Similarly, we have from (3.8) for $n > 1$,

$$W^{**}_{n-1}(v) = 1 - \frac{(n-1)(n+1)}{n^2} \cdot \frac{(1-v^n)^2}{(1-v^{n-1})(1-v^{n+1})}$$
$$= 1 - \frac{(n-1)(n+1)}{n^2}\left[1 + \frac{v^{n-1}(1-v)^2}{(1-v^{n-1})(1-N^{n+1})}\right] \tag{3.17}$$
$$> \frac{1}{n^2} - v^{n-1} \geqq \frac{1}{n^2} - v > \frac{1}{n^2} - \frac{1}{\log(1/v)};$$

and, for $n = 1$ (putting $v = a$),

$$W^{**}_0(v) = 1 - \frac{2(1-a)}{(1+a)\log(1/a)} > 1 - \frac{2}{\log(1/v)}. \tag{3.18}$$

Combining (3.16), (3.17), and (3.18), we have for all $m \geqq 0$,

$$W^{**}_m(v) - \overline{W}_m(v) > \frac{2m+3}{(m+1)^2(m+2)^2} - \frac{3}{\log(1/v)}. \tag{3.19}$$

We now define, for all integers $m \geqq 0$,

$$r(m) = cm + \frac{1}{(m+1)^2}. \tag{3.20}$$

We note that $r(m+1) - r(m) = c - (2m+3)/((m+1)^2(m+2)^2)$. The function $r(m)$ evidently has at most two minima (if there are two, they are for consecutive values of m). Denote by m_0 the first integer for which $r(m_0)$ is a minimum. Let ϵ $(0 < \epsilon < 1)$ be such that $3\epsilon < r(m_0 - 1) - r(m_0)$ (if $m_0 = 0$, the last restriction is omitted). Let $d = e^{-1/\epsilon}$ and $a = e^{-1/\epsilon^2}$. Let m_1 be the smallest integer not less than $1/c$.

For any integer $K > 0$, if $\lambda_a(\theta)$ is the a priori density we have (noting that $d > a$)

$$P\{X_K \geqq d\} = \int_d^1 \int_x^1 g_K(x; \theta)\lambda_a(\theta)\, d\theta\, dx = \int_d^1 \frac{1 - x^K}{x \log(1/a)}\, dx \tag{3.21}$$

$$= \frac{\log(1/d)}{\log(1/a)} - \frac{1 - d^K}{K \log(1/a)} < \epsilon.$$

We note that, after m observations ($m = 0, 1, \cdots$, ad inf. and putting $v = a$ if $m = 0$), any Bayes solution will certainly prescribe taking another observation if $W_m^{**}(v) - \overline{W}_m(v) - c > 0$, since this quantity is the a posteriori expected saving over stopping after m observations if instead one takes one additional observation and then stops and makes the best terminal decision.

We also note that, since $(\log 1/a)^{-1} = \epsilon^2 < \epsilon$, it follows from (3.21) that, when $\lambda_a(\theta)$ is the a priori density,

$$P\left\{\frac{1}{\log(1/Z_i)} < \epsilon \text{ for } i = 1, 2, \cdots, m_0 + m_1\right\} = P\left\{\frac{1}{\log(1/Z_{m_0+m_1})} < \epsilon\right\}$$

$$= P\left\{\frac{1}{\log(1/X_{m_0+m_1})} < \epsilon\right\} = P\{X_{m_0+m_1} < d\} > 1 - \epsilon.$$

Since $r(m - 1) - r(m)$ is a decreasing function of $m(m > 0)$ and since $3\epsilon < r(m_0 - 1) - r(m_0)$, we conclude that, if $m_0 > 0$, the event

$$\frac{1}{\log(1/Z_{m_0+m_1})} < \epsilon \tag{3.23}$$

entails the event $(\log(1/Z_{m_0-1}))^{-1} < \epsilon$, which entails $3(\log(1/Z_{m_0-1}))^{-1} < r(m_0 - 1) - r(m_0)$; or, equivalently, $-3(\log(1/Z_i))^{-1} + r(i) - r(i + 1) > 0$ for $i = 0, 1, \cdots, m_0 - 1$. Finally, it follows from (3.19) that this entails the event $W_i^{**}(v) - \overline{W}_i(v) - c > 0$ for $i = 0, 1, \cdots, m_0 - 1$; and, for any Bayes solution relative to $\lambda_a(\theta)$, this entails the event that at least m_0 observations will be taken. Furthermore, the last statement is always true for $m_0 = 0$.

Similarly, we note from (3.17) and (3.18) that the event (3.23) certainly entails the event $W_i^*(v) > (1/(1 + i)^2) - 2\epsilon$ for $i = m_0, m_0 + 1, \cdots, m_0 + m_1$. That is, if a terminal decision is made after exactly i observations ($i = m_0, \cdots, m_0 + m_1$), the total a posteriori loss plus cost of experimentation will be $> ci + (1/(1 + i)^2) - 2\epsilon \geqq cm_0 + (1/(1 + m_0)^2) - 2\epsilon$. Moreover, it follows from the definition of m_1 that this last expression is less than the cost of experimentation alone if more than $m_0 + m_1$ observations are taken.

To summarize, then, the event (3.23) implies for any Bayes solution relative to $\lambda_a(\theta)$ that the experiment will terminate with a total a posteriori loss plus cost of experimentation exceeding $cm_0 + (1/(1 + m_0)^2) - 2\epsilon$. But it follows from (3.22) that (3.23) occurs with probability $>1 - \epsilon$. Since $m_0c + (1/(m_0 + 1)^2) \leqq 1$, it follows that the Bayes risk relative to $\lambda_a(\theta)$ exceeds

$$(1 - \epsilon)\left(m_0 c + \frac{1}{(m_0 + 1)^2} - 2\epsilon\right) > m_0 c + \frac{1}{(m_0 + 1)^2} - 3\epsilon. \tag{3.24}$$

Since ϵ may be taken to be arbitrarily small in magnitude, we conclude (see Section 2) that the fixed sample procedure described in Section 2 is indeed minimax.

REFERENCES

[1] ABRAHAM WALD, *Statistical Decision Functions*, John Wiley and Sons, 1950.

[2] J. WOLFOWITZ, "Minimax estimates of the mean of a normal distribution with known variance," *Annals of Math. Stat.*, Vol. 21 (1950), pp. 218–230.

[3] C. STEIN AND A. WALD, "Sequential confidence intervals for the mean of a normal distribution with known variance," *Annals of Math. Stat.*, Vol. 18 (1947), pp. 427–433.

[4] C. R. BLYTH, "On minimax statistical procedures and their admissibility," *Annals of Math. Stat.*, Vol. 22 (1951), pp. 22–42.

Reprinted from THE ANNALS OF MATHEMATICAL STATISTICS
Vol. 23, No. 4, December, 1952

ON MINIMUM VARIANCE ESTIMATORS[1]

BY J. KIEFER

Cornell University

Chapman and Robbins [1] have given a simple improvement on the Cramér-Rao inequality without postulating the regularity assumptions under which the latter is usually proved. The purpose of this note is to show by examples how a similarly derived stronger inequality (see equation (2)) may be used to verify that certain estimators are uniform minimum variance unbiased estimators. This stronger inequality is that which (under additional restrictions) was shown in [2] to be the best possible, but is in a more useful form for applications than the form given in [2]. For simplicity we consider only an inequality on the variance of unbiased estimators, but inequalities on other moments than the second (see [2]), or for biased estimators, may be found similarly. The two examples considered here are ones where the regularity conditions of [2] are not satisfied, where the method of [1] does not give the best bound, and where the method of this note is used to find the best bound and thus to verify that certain estimators are uniform minimum variance unbiased. (For the examples considered this also follows from completeness of the sufficient statistic; the method used here applies, of course, more generally.)

Let X be a chance variable with density $f(x; \theta)$ with respect to some fixed σ-finite measure μ. ($\theta \varepsilon \Omega$, $x \varepsilon \mathfrak{X}$). We suppose suitable Borel fields to be given and $f(x; \theta)$ to be measurable in its arguments. Ω is a subset of the real line. For each θ, let $\Omega_\theta = \{h \mid (\theta + h) \varepsilon \Omega\}$. For fixed θ, let λ_1 and λ_2 be any two probability measures on Ω_θ such that $E_i h = \int_{\Omega_\theta} h d\lambda_i(h)$ exists for $i = 1, 2$. Then, for any

[1] Research sponsored by the Office of Naval Research.

$t(x)$ for which $E_\theta t = \theta$, we have

$$\text{(1)} \quad \int_{\mathfrak{X}} (t - \theta)\sqrt{f(x;\theta)} \left\{ \frac{\int_{\Omega_\theta} f(x;\theta + h)\, d[\lambda_1(h) - \lambda_2(h)]}{f(x;\theta)} \right\} \sqrt{f(x;\theta)}\, d\mu = E_1 h - E_2 h.$$

Applying Schwarz's inequality, we have after some obvious manipulations,

$$\text{(2)} \qquad E_\theta(t - \theta)^2 \geqq \sup \left\{ \frac{(E_1 h - E_2 h)^2}{\int_{\mathfrak{X}} \frac{\left\{ \int_{\Omega_\theta} f(x;\theta + h)\, d[\lambda_1(h) - \lambda_2(h)] \right\}^2}{f(x;\theta)}\, d\mu} \right\},$$

where for each θ the supremum is taken over all λ_1 and λ_2 for which $\lambda_1 \neq \lambda_2$ and for which the integrand of the integral over $\mathfrak{X}$ is defined a.e. (μ).

We remark that the supremum of (2) is easily seen to be unimproved if λ_i and $E_i h$ are multiplied by real numbers c_i ($i = 1, 2$) with respect to which the supremum is also taken. From this fact it is easy to verify that the right side of (2) must coincide with the expression given in Theorem 4 of [2] (for $s = 2$ there), and which Barankin shows (under the assumption that $f(x;\theta + h)/f(x;\theta)$ is defined a.e. (μ) and (for our case) belongs to L_2 with respect to the measure $\nu(A) = \int_A f(x;\theta)\, d\mu$ for all $h \varepsilon \Omega_\theta$) to be the best possible bound. However, the form of equation (2) is more useful for applications, since one can sometimes find λ_i for which the bound is attained but where no discrete λ_i (essentially what are used in the form of [2]) actually give this bound.

It will often suffice in applications to let λ_2 give measure one to the point $h = 0$. This gives

$$\text{(3)} \qquad E_\theta(t - \theta)^2 \geqq \sup_{\lambda_1} \left\{ \frac{(E_1 h)^2}{\int_{\mathfrak{X}} \frac{\left[\int_{\Omega_\theta} f(x;\theta + h)\, d\,\lambda_1(h) \right]^2}{f(x;\theta)}\, d\mu - 1} \right\}.$$

If we consider only those λ_1 which give measure one to a single h, we obtain

$$\text{(4)} \qquad E_\theta(t - \theta)^2 \geqq \frac{1}{\inf_h \frac{1}{h^2} \left\{ \int_{\mathfrak{X}} \frac{[f(x;\theta + h)]^2}{f(x;\theta)}\, d\mu - 1 \right\}},$$

where the infimum is over all $h \neq 0$ for which $h \varepsilon \Omega_\theta$ and for which $f(x;\theta) = 0$ implies $f(x;\theta + h) = 0$ a.e. (μ). The latter is precisely the condition of equation (2) of [1], the result of which thus coincides with (4).

We now give two examples where the right side of (3) suffices to give the best bound, where the right side of (4) does not give the best bound, and where the previously mentioned restrictions of [2] are not satisfied. In both examples μ is Lebesgue measure on the real line.

EXAMPLE 1. We have n observations from a rectangular distribution from 0 to θ ($\Omega = \{\theta \mid \theta > 0\}$). It suffices to consider the maximum Y of the observations, whose density is ny^{n-1}/θ^n for $0 \leqq y \leqq \theta$, and 0 elsewhere. For $n = 1$, the denominator of the right side of (4) becomes $\inf_{-\theta<h<0}\{-1/[h(\theta + h])\}$, so that (4) gives the bound $\theta^2/4$. It would be too tedious to carry this calculation out for each n, but it can be shown that, as $n \to \infty$, (4) asymptotically gives the bound $.648\theta^2/n^2$. On the other hand, if we put $d\lambda_1(h) = [(n + 1)/\theta]\ (h/\theta + 1)^n\, dh$ for $-\theta < h < 0$, the term in braces on the right side of (3) becomes $\theta^2/[n(n + 2)]$, which is in fact attained as the variance of the unbiased estimator $[(n + 1)/n]Y$.

EXAMPLE 2. We have m observations from the distribution with density $e^{-(x-\theta)}$ for $x \geqq \theta$ and 0 elsewhere (Ω is the real line). Here the minimum Z of the observations is sufficient and has density $me^{-m(z-\theta)}$, $z \geqq \theta$. The denominator of (4) is $\inf_{h>0}\ ([e^{mh} - 1]/h^2)$. The infimum is attained for $mh = 1.5936$, and yields $.648/m^2$ as the bound given by (4). On the other hand, putting $d\lambda_1(h) = me^{-mh}\, dh$ for $0 < h < \infty$ and 0 otherwise, the expression in braces of (3) becomes $1/m^2$, which is actually attained as the variance of the unbiased estimator $Z - 1/m$.

REFERENCES

[1] D. C. CHAPMAN AND H. ROBBINS, "Minimum variance estimation without regularity assumptions," *Annals of Math. Stat.*, Vol. 22 (1951), pp. 581–586.

[2] E. BARANKIN, "Locally best unbiased estimates," *Annals of Math. Stat.*, Vol. 20 (1949), pp. 477-501.

Reprinted from THE ANNALS OF MATHEMATICAL STATISTICS
Vol. 24, No. 1, March, 1953

ON WALD'S COMPLETE CLASS THEOREMS[1]

BY J. KIEFER

Cornell University

1. Summary. The purpose of this paper is to prove certain results concerning complete classes of strategies, some of which were announced in an abstract in *Bull. Am. Math. Soc.*, Vol. 57 (1951), p. 372.

2. Introduction. Except where explicitly stated to the contrary, we shall use the nomenclature and notation of Chapter 2 of [1] concerning zero-sum two-person games. Our considerations here do not require, however, that the payoff function $K(a, b)$ be bounded (or finite), but merely that it be bounded (by zero, without loss of generality) from below (because if unbounded in both directions, expectation relative to a mixed strategy might be undefined). This generalization is of use in some games and statistical work, as will be seen below. We remark without proof that such results as weakened forms of Theorems 1 and 4 of [2] may be proved under this set up. For example, we shall later use the following:

THEOREM 1. *Suppose that* Ξ *and* H *are convex spaces of allowable strategies for players* 1 *and* 2, *respectively, that* $0 \leqq K(a, b) \leqq \infty$, *that* H *is weakly compact relative to* Ξ *in the sense of Wald* (i.e., *for any sequence* $\{\eta_i\}$ *in* H *there is a subsequence* $\{\eta_{i_j}\}$ *and an* η_0 *in* H *such that* $\liminf_{j\to\infty} K(\xi, \eta_{i_j}) \geqq K(\xi, \eta_0)$ *for all* ξ *in* Ξ), *and that there exists a sequence* $\{\xi_i\}$ *in* Ξ *such that for any* ξ *in* Ξ *and* η *in* H *there is a subsequence* $\{\xi_{i_j}\}$ *(all of whose elements may be the same) which may depend on* ξ *and* η *and is such that* $\lim_{j\to\infty} K(\xi_{i_j}, \eta) \geqq K(\xi, \eta)$. *Then the game is determined.*

The weak compactness assumption is enough to assure the existence of a minimax strategy for player 2. The above conditions may be weakened as in Theorem 4 of [2] or even further, and a generalization of Theorem 5 of [2] (which should be corrected there by assuming g_0 to be independent of ϵ) may similarly be proved.

3. Admissible strategies and complete classes. Wald considered two types of complete class theorems: those which give conditions under which the class of admissible strategies (e.g., of player 2) is complete, and those which give conditions under which the class of minimal strategies in the strict or wide sense is complete. The latter will occupy most of this paper. We remark, regarding the former, that the proof used by Wald in Theorem 2.22 of [1] actually suffices to prove the following:

THEOREM 2. *Let* Ξ *and* H *be arbitrary spaces of mixed strategies with the property*

Received 3/4/52.

[1] Research sponsored by the Office of Naval Research.

that there exists a denumerable subset Ξ^* *of* Ξ *such that, if* η' *and* η'' *are any members of* H *for which* $K(\xi, \eta') \leqq K(\xi, \eta'')$ *for all* ξ *with strict inequality for some* ξ, *then there is a* ξ' *in* Ξ^* *with* $K(\xi', \eta') < K(\xi', \eta'')$. *Suppose also that* H *is weakly compact relative to* Ξ *in the sense of Wald. Then the class of all admissible strategies of player* 2 *is minimal complete.*

Note that $K(\xi, \eta)$ is not assumed bounded. An application of this theorem which indicates the usefulness of the hypothesis as stated herein over the stronger condition stated in Theorem 2.22 of [1], will be given in the next paragraph. We remark here that the condition of Theorem 2 is not necessary; for example, let A as well as B consist of all integers, Ξ and H consist of all probability measures on A and B, and $K(a, b) = 0$ if $a = b > 0$ and $K(a, b) = 2^{-|b|}$ otherwise; the class of all admissible strategies (those giving probability 1 to a single element $b > 0$) is then minimal complete, but H is not even weakly compact for every sequence of strategies for which each strategy is better than its predecessor (as is evidenced by the sequence of pure strategies $b = 0, -1, -2, \cdots$). On the other hand, the theorem does not remain valid if only weak compactness (but not the condition on Ξ^*) is assumed. For example, let A as well as B consist of all ordinals less than the first uncountable ordinal, let Ξ and H consist of all discrete probability measures on A and B, and let $K(a, b) = -1$ or 1 according to whether $a < b$ or $a \geqq b$, respectively. Then the condition of weak compactness is satisfied, but no strategy is admissible. (This example also illustrates why weak compactness alone is not enough to insure the determinateness of the game.) The above theorem may be generalized in an obvious manner by replacing the condition of weak compactness by a similar one on all well-ordered subsets of H whose power does not exceed that of some infinite Ξ^* with the stated property. (It is enough to consider only subsets of H whose members become "better" with increasing index.) It follows that the bicompactness condition used in Theorem 3 of [2], which implies such a condition for *every* subset of H, also implies the conclusion of Theorem 2 above.

As an important statistical application of Theorem 2, which also illustrates the advantage of using the condition on Ξ^* stated therein rather than that of the separability of Ξ^* in the sense of intrinsic metric (2.4) of [1], we shall now prove the following:

THEOREM 3. *Under Assumptions* 3.1 *to* 3.6 *of* [1], *the class of all admissible decision functions is minimal complete.*

This theorem extends the result of Theorem 2.22 of [1] to the setup of Chapter 3 of [1]. To prove it we let $\Xi^* = \bigcup_{i=1}^{\infty} \Xi_i$, where Ξ_i is a denumerable set of a priori distributions which is dense in Ξ in the sense of the metric $\rho_i(\xi_1, \xi_2) = \sup | r(\xi_1, \delta) - r(\xi_2, \delta) |$, the supremum being taken over all decision functions δ requiring at most i stages of experimentation. The existence of such Ξ_i follows from Theorems 3.3 and 2.16 of [1]. We shall show that Ξ^* satisfies the assumption of Theorem 2. Let δ_1 and δ_2 be two decision functions and ϵ a positive number such that $r(\xi) \geqq 0$ and $\sup_\xi r(\xi) > 2\epsilon$, where (using the notation of [1]) $r(\xi) = r(\xi, \delta_1) - r(\xi. \delta_2)$ (with the definition $\infty - \infty = 0$). We need only show that

$r(\xi') > 0$ for some ξ' in Ξ^*. Let $r_{1,m}(\xi) = \sup r(\xi, \delta^m)$, the supremum being over all δ^m requiring not more than m stages of observation and such that $r(\xi, \delta^m) \leqq r(\xi, \delta_1) + \epsilon$; if no such δ^m exists, we define $r_{1,m}(\xi) = 0$. Similarly, let $r_{2,m}(\xi) = \inf r(\xi, \delta^m)$, the infimum being over all δ^m requiring not more than m stages of observation and such that $r(\xi, \delta^m) \geqq r(\xi, \delta_2)$; if no such δ^m exist, we define $r_{2,m}(\xi) = +\infty$. Clearly, for each ξ, $r_{1,m}(\xi)$ is nondecreasing with m (we assume without loss of generality that the weight function W is nonnegative) and $r_{2,m}(\xi)$ is nonincreasing with m. Moreover, noting Lemma 3.3 of [1], and defining $r_m(\xi) = r_{1,m}(\xi) - r_{2,m}(\xi)$, we see that $r(\xi) \leqq \lim_{m\to\infty} r_m(\xi)$for every ξ for which $r(\xi, \delta_2)$ is finite. Moreover, $r_m(\xi) \leqq r(\xi) + \epsilon$ is nondecreasing in m, so that

$$\begin{aligned}\epsilon + \sup_{\xi\epsilon\Xi^*} r(\xi) &\geqq \sup_{\xi\epsilon\Xi^*} \lim_{m\to\infty} r_m(\xi) = \sup_{\xi\epsilon\Xi^*} \sup_m r_m(\xi)\\ &= \sup_m \sup_{\xi\epsilon\Xi^*} r_m(\xi) = \sup_m \sup_{\xi\epsilon\Xi} r_m(\xi) = \sup_{\xi\epsilon\Xi} \sup_m r_m(\xi)\\ &\geqq \sup_{\xi\epsilon\Xi} r(\xi) > 2\epsilon,\end{aligned}$$

completing the proof. (It is essential here that $r_m(\xi)$ is increasing in m, so that the operations "lim" and "sup" may be interchanged.)

4. Minimal strategies and complete classes. We now turn to our main theorem, which generalizes Theorem 2.25 of [1]. The proof of the theorem is followed by two applications. The first of these is an essential strengthening of Theorems 3.17 and 3.20 of [1] regarding statistical decision functions. The second weakens the conditions of Theorem 2.25 of [1], even when $K(a, b)$ is bounded.

The idea of forming a new game with payoff function $K^*(a, b)$ is Wald's, and the proof of the first part of the conclusion of the theorem below is that of Theorem 2.25 of [1] if $K(a, b)$ is bounded. (The last part of the conclusion was proved under the stronger conditions that $K(a, b)$ is bounded and A and B are compact, so that minimality in the wide and strict senses are equivalent, in Theorem 3.10 of [4].) In the bounded case, any condition entailing the determinateness of the game and existence of a minimax strategy for player 2 and whose validity relative to K implies its validity relative to K^* (e.g., the condition of Theorem 2.25 of [1] or of Theorem 3 of [2]), also obviously entails the conclusion of the theorem below. When $K(a, b)$ is unbounded, one must be careful to use $K^*(\xi, \eta)$ only where $K(\xi, \eta)$ and $K(\xi, \eta_0)$ are not both infinite. Otherwise, K^* may not be properly defined. At the same time, it is useful to state the theorem in terms of the Ξ_N of the theorem rather than only in terms of A^*, since in many applications the Ξ_N may be chosen so that K^* is bounded from below on each Ξ_N (but not necessarily on A^*), so that in verifying condition (b) in applications one may use such results as that italicized in the first paragraph of this paper.

We recall (putting $\infty - \infty = 0$ in our case) that a strategy η' is minimal in the wide sense if

$$\text{(1)} \qquad \inf_{\xi\epsilon\Xi} [K(\xi, \eta') - \inf_{\eta\epsilon H} K(\xi, \eta)] = 0.$$

THEOREM 4. *Suppose* $0 \leqq K(a, b) \leqq \infty$, $\Xi \supset A$, $\mathrm{H} \supset B$, *and that for any* η_0 *for which* $\inf_a K(a, \eta_0) < \infty$ *and which is not a member of the class* C_W *of all minimal strategies in the wide sense, there exists a sequence* $\{\Xi_i\}(i = 1, 2, \cdots, \text{ad inf})$ *of subsets of* Ξ *such that*

(a) $\liminf_{N\to\infty} \Xi_N \supset A^* = \{a \mid K(a, \eta_0) < \infty\}$; *for every* N, $K(\xi, \eta_0) < \infty$ *for all* ξ *in* Ξ_N; *if* $\sup_a K(a, \eta_0) < \infty$, $\Xi_N \supset A$;

(b) *the game relative to* Ξ_N, H, *and* $K^*(\xi, \eta) = K(\xi, \eta) - K(\xi, \eta_0)$ *is determined and player 2 has a minimax strategy for this game.*

If $\sup_{a,b} K(a, b) = +\infty$, *suppose also that* H *is weakly compact relative to* A^* *for each* $\eta_0 \notin C_W$ *for which* $\inf_a K(a, \eta_0) < \sup_a K(a, \eta_0) = +\infty$. (*If* H *is weakly compact relative to* A, *this is automatically satisfied.*)

Then C_W *is complete. Moreover, for any* η_0 *not in* C_W *there is an* η_1 *in* C_W *and an* $\epsilon > 0$ *such that* $K(\xi, \eta_1) \leqq K(\xi, \eta_0) - \epsilon$ *for all* ξ *in* Ξ.

PROOF. We suppose $C_W \neq \mathrm{H}$, or the theorem is trivial; in particular, $\inf_{a,b} K(a, b) < \infty$, since otherwise $C_W = \mathrm{H}$. We now show that C_W is not empty. If there is an $\eta_0 \notin C_W$ with $\sup_a K(a, \eta_0) < R < \infty$, it follows from (b) that there is a minimax strategy η' relative to Ξ_N, H, and K^*. Since this game is determined and $\Xi \supset \Xi_N \supset A$ in this case, it is easy to verify that the game relative to Ξ, H, and K^* is determined, that η' is minimax for it, and hence that η' is minimal in the wide sense relative to Ξ, H, and K^* (since $0 \geqq K^*(a, \eta') \geqq -R$, the proof of Theorem 2.17 of [1] applies), and hence relative to Ξ, H, and K. On the other hand, if no such η_0 exists, the first sentence of the proof shows that there must exist an η_0 with non-empty A^* and (by the assumption following (b)) such that there exists a minimal strategy relative to any member of A^*. At any rate, C_W is not empty.

Let η_0 be any member of H which is not in C_W. If $K(a, \eta_0) = +\infty$ for all a, any η' in C_W (which is non-empty by the previous paragraph) is uniformly better than η_0 and is such that $K(\xi, \eta') \leqq K(\xi, \eta_0) - 1$ for all ξ. Hence, we may assume in what follows that $\inf_a K(a, \eta_0) < \infty$, and that the Ξ_N corresponding to this η_0 are non-empty for all N not less than some N_0. We now let $\Xi^* = \bigcup_{N=1}^{\infty} \Xi_N$, and define

$$\epsilon = \inf_{\xi \in \Xi^*} [K(\xi, \eta_0) - \inf_\eta K(\xi, \eta)]. \tag{2}$$

(It is clear that $\inf_\eta K(\xi, \eta) < \infty$ for all ξ. Otherwise, every η in H would be minimal and we would have $C_W = \mathrm{H}$.) Clearly, $\epsilon > 0$, or by (1) (with $\eta' = \eta_0$) η_0 would be minimal in the wide sense. Moreover, $\epsilon < \infty$, since Ξ_{N_0} is non-empty.

For any $N \geqq N_0$, let η_N be a minimax strategy for the game described in (b), so that

$$\sup_{\xi \in \Xi_N} \inf_\eta K^*(\xi, \eta) = \inf_\eta \sup_{\xi \in \Xi_N} K^*(\xi, \eta) = \sup_{\xi \in \Xi_N} K^*(\xi, \eta_N). \tag{3}$$

The common value of (3) is less than or equal to $-\epsilon$; for if, to the contrary, it were $-\epsilon + 2\rho$ for some $\rho > 0$, there would by (3) exist a ξ_0 in Ξ_N for which

$$-\epsilon + \rho \leqq \inf_\eta K^*(\xi_0, \eta) = \inf_\eta K(\xi_0, \eta) - K(\xi_0, \eta_0), \tag{4}$$

which would contradict (2). Hence, we must have

(5) $$K(\xi, \eta_N) \leqq K(\xi, \eta_0) - \epsilon \qquad \text{for all } \xi \text{ in } \Xi_N.$$

Let the subsequence $\{N_j\}$ $(j = 1, 2, \cdots, \text{ad inf})$ of the positive integers and the strategy $\eta^* \epsilon \mathrm{H}$ be (as guaranteed by weak compactness relative to A^* if $\sup_a K(a, \eta_0) = +\infty$, and putting $\eta^* = \eta_{N'}$ if $\sup_a K(a, \eta_0) < \infty$ and assuming without loss of generality that $\Xi_N = \Xi_{N'}$ for $N > N'$ in this case) such that

(6) $$\liminf_{j \to \infty} K(a, \eta_{N_j}) \geqq K(a, \eta^*) \qquad \text{for all } a \text{ in } A^*.$$

It follows from (5) that

(7) $$K(a, \eta^*) \leqq K(a, \eta_0) - \epsilon \qquad \text{for all } a \text{ in } A^*;$$

that is, $K(a, \eta^*) \leqq K(a, \eta_0) - \epsilon$ for all a for which $K(a, \eta_0) < \infty$. Since the latter set is nonempty, η^* is uniformly better than η_0, and in fact

(8) $$K(\xi, \eta^*) \leqq K(\xi, \eta_0) - \epsilon \qquad \text{for all } \xi \text{ in } \Xi.$$

The minimality in the wide sense of η^* (i.e., the verification of (1) for $\eta' = \eta^*$) is a direct consequence of (8), the fact that Ξ^* is nonempty, and (2). This completes the proof of the theorem.

APPLICATION I. In the terminology of Chapter 3 of [1], let $\mathfrak{D}$ be the class of all decision functions and $\mathfrak{D}_b$ the class of all decision functions with bounded risk functions. Let C_s be the class of all Bayes solutions in the strict sense and C_W the class of all Bayes solutions in the wide sense. Wald showed that, under Assumptions 3.1 to 3.6 of [1], C_W is complete relative to $\mathfrak{D}_b$ (Theorem 3.17 of [1]), and that, under Assumptions 3.1 to 3.7 of [1], C_s is complete relative to $\mathfrak{D}_b$ (Theorem 3.20 of [1]). (These theorems were also proved by Wald under stronger conditions in [3], [4], and [5], and were stated under stronger conditions in [6]. In [3] and [4] (by Condition 7 of the latter) the risk function is always bounded. Theorems 2.6, 2.7, 3.5, and 3.6 of [5] are stated correctly, relative to $\mathfrak{D}_b$. The proofs of Theorems 2.5 and 3.4 of [5] are correct only if the statement of these theorems is interpreted relative to $\mathfrak{D}_b$; otherwise, the statement following equation (2.72) of [5] is false, since the W^* defined there need not satisfy Condition 2.2 of [5]). If, using Wald's notation and in particular putting δ_0 for η_0 and $r(F, \delta_0)$ for $K(a, \eta_0)$, one defines the Ξ_N of our theorem to consist of all ξ for which $\xi(A_N) = 1$, where $A_N = \{a \mid K(a, \eta_0) \leqq N\}$, it is easy to verify that A_N, the terminal decision space D^t, and the weight function $W^*(F, d) = W(F, d) - r(F, \delta_0)$ (when restricted to A_N) satisfy Assumptions 3.1 to 3.6 (and 3.7) of [1] whenever Ω, D^t, and $W(F, d)$ satisfy the corresponding assumptions. Hence, Theorems 3.4, 3.7, and 3.2 of [1] imply (putting $r^*(F, \delta) = r(F, \delta) - r(F, \delta_0)$ for our K^*) that condition (b) and the condition which follows it in our theorem are satisfied, so that the conclusion of Theorem 4 holds. Hence, we have proved the following:

THEOREM 5. *In the statements of Theorem* 3.17 *and Theorem* 3.20 *of* [1], $\mathfrak{D}_b$ *may be replaced by* $\mathfrak{D}$.

The proof for Theorem 3.20 uses Theorem 3.15. The last part of the conclusion of the Theorem 4, when applied to the present case, yields a result not proved in [1] but proved under stronger conditions (e.g., all risk functions are bounded) in Theorem 4.11 of [4].

APPLICATION II. Suppose $0 \leq K(a, b) \leq \infty$, that Ξ and H are convex, that H is weakly compact relative to Ξ, and that there is a countable subset $\Xi^* = \{\xi_i\}$ of Ξ such that, given any ξ in Ξ, there is a subsequence $\{\xi_{i_j}\}$ of Ξ^* (whose elements are not necessarily different) such that $\lim_{j\to\infty} K(\xi_{i_j}, \eta) = K(\xi, \eta)$ for all η in H. We define $\Xi_N = \{\xi \mid K(\xi, \eta_0) < N; \xi \epsilon \Xi\}$, and we note that only (b) need be verified to assure the applicability of Theorem 4. It is easy to verify that H is weakly compact relative to Ξ_N and the payoff function K^*. Moreover, for any ξ in Ξ_N there is by assumption a subsequence $\{\xi_{i_j}\}$ of Ξ^* with $\lim_{j\to\infty} K(\xi_{i_j}, \eta) = K(\xi, \eta)$ for all η in H. In particular, this holds for $\eta = \eta_0$, so that $K(\xi_{i_j}, \eta_0) < N$ for sufficiently large j. We conclude that $\Xi^* \cap \Xi_N$ satisfies relative to Ξ_N, H, K, and hence relative to Ξ_N, H, K^*, the same relationship that Ξ^* did to Ξ, H, K. From Theorem 1 stated in the first paragraph of this paper, we conclude that (b) is satisfied.

Even when $K(a, b)$ is bounded, the above condition of weak sequential separability is weaker than the strong separability condition used in Theorem 2.25 of [1].

REFERENCES

[1] A. WALD, *Statistical Decision Functions*, John Wiley and Sons, New York, 1950.

[2] S. KARLIN, "Operator treatment of minimax principle," *Ann. Mathematics Studies, no.* 24, Princeton University Press, (1950), pp. 133–154.

[3] A. WALD, "An essentially complete class of admissible decision functions," *Ann. Math. Stat.*, Vol. 18 (1947), pp. 549–555.

[4] A. WALD, "Foundations of a general theory of sequential decision functions," *Econometrica*, Vol. 15 (1947), pp. 279–313.

[5] A. WALD, "Statistical decision functions," *Ann. Math. Stat.*, Vol. 20 (1949), pp. 165–205.

[6] A. WALD, "Basic ideas of a general theory of statistical decision rules," *Proceedings of the International Congress of mathematicians, 1950*, American Mathematical Society, 1952, pp. 231–243.

CORRECTION OF A PROOF*

BY J. KIEFER
Cornell University

In the proof of Theorem 3 of "On Wald's Complete Class Theorems" (*Ann. Math. Stat.*, Vol. 24 (1953), pp. 70–75), the inequality appearing in the definition of $r_{2,m}(\xi)$ should be altered to read $r(\xi, \delta^m) \geq r(\xi, \delta_2) - \epsilon/2$; the remainder of the proof is then easily altered to give the desired result. Without the $\epsilon/2$, one would still have to prove that the space $\mathfrak{D}$ is large enough to give $\lim_{m\to\infty} r_{2,m}(\xi) < \infty$. The author is indebted to Mr. Jerome Sacks for pointing out this fact.

* Received 7/11/53.

Reprinted from *Ann. Math. Statist.* **24** (1953).

SEQUENTIAL MINIMAX SEARCH FOR A MAXIMUM

J. KIEFER[1]

The problem formulated below was motivated by that of determining an interval containing the point at which a unimodal function on the unit interval possesses a maximum, without postulating regularity conditions involving continuity, derivatives, etc. Our solution is to give, for every $\epsilon>0$ and every specified number N of values of the argument at which the function may be observed, a procedure which is ϵ-minimax (see (1) below) among the class of all sequential nonrandomized procedures which terminate by giving an interval containing the required point, where the payoff of the computer to nature is the length of this final interval. (The same result holds if, e.g., we consider *all* nonrandomized procedures and let the payoff be length of interval plus c or 0 according to whether the interval fails to cover or covers the desired point, where $c \geqq 1/U_{N+1}$, the latter being defined below.) The analogous problem where errors are present in the observations was considered in [1], but no optimum results are yet known for that more difficult case.

Search for a maximum is a "second-order" search in the sense that information is given by *pairs* of observations. Thus, if $x_1<x_2$ and $f(x_1)\geqq f(x_2)$, and f is a member of the class $\mathcal{J}$ described below, the point $x^{(f)}$ defined below (the point where f attains its maximum if the latter exists) must lie to the left of x_2. If we postulate only that $f\in\mathcal{J}$, this is essentially the only information we have about $x^{(f)}$: i.e., that future observations should be taken to the left of x_2. Similarly, the problem of finding the point $x^{[f]}$ at which a strictly increasing function f on the unit interval attains a value α (weaker restrictions on f can be made) is essentially first-order in the sense that every single observation gives information about where to take the next. (See [2; 3] for the corresponding statistical problem.) The minimax procedure in that case is successively to split the interval in which $x^{[f]}$ is known to lie into equal parts at the point of next observation, in an obvious manner. It may be fruitful to consider higher-order search problems, as will be done in a future paper.

Let the class $\mathcal{J}$ consist of every function f from the closed unit interval I into the reals (R) and for which there is an $x^{(f)}\in I$ such that f is either strictly increasing for $x\leqq x^{(f)}$ and strictly decreasing for $x>x^{(f)}$, or else strictly increasing for $x<x^{(f)}$ and strictly decreasing

Received by the editors September 17, 1952.

[1] Research under a contract with the ONR.

for $x \geqq x^{(f)}$. Let $\mathcal{D}$ denote the space of all closed intervals D which are subsets of I. An element $D \in \mathcal{D}$ will be called a terminal decision. Let N be given integer $\geqq 2$. A (nonrandomized) strategy S is a set $S = \{x_1, g_2, \cdots, g_N, s, t\}$ consisting of a number $x_1 \in I$, functions g_k from $I^{k-2} \times R^{k-1}$ into I $(2 \leqq k \leqq N)$, and functions s and t from $I^{N-1} \times R^N$ into I and with $s \leqq t$. A strategy S is used as follows: one observes (or computes) in order

$$f(x_1), f(x_2), \cdots, f(x_N) \text{ with } x_k = g_k[x_2, \cdots, x_{k-1}, f(x_1), \cdots, f(x_{k-1})]$$

for $2 \leqq k \leqq N$ $(x_2 = g_2[f(x_1)])$, and then selects the closed interval

$$D(f, S) = [s(x_2, \cdots, x_N, f(x_1), \cdots, f(x_N)), \quad t(x_2, \cdots, x_N, f(x_1), \cdots, f(x_N))]$$

as terminal decision. Let $\mathfrak{S}_N$ be the class consisting of every strategy S requiring N observations and for which $x^{(f)} \in D(f, S)$ for all f in $\mathcal{F}$. Our problem is, given any $\epsilon > 0$ and $N \geqq 2$, to find an $S_N^* \in \mathfrak{S}_N$ such that

$$\sup_{f \in \mathcal{F}} L(D(f, S_N^*)) \leqq \inf_{S \in \mathfrak{S}_N} \sup_{f \in \mathcal{F}} L(D(f, S)) + \epsilon, \tag{1}$$

where $L(D)$ is the length of the interval D.

We shall now describe an S_N^* which, it will be shown, satisfies (1). Let U_n be the nth Fibonacci number $(U_0 = 0,\ U_1 = 1,\ U_n = U_{n-1} + U_{n-2}$ for $n \geqq 2)$. S_2^* is defined by $x_1 = 1/2$, $x_2 = 1/2 + \epsilon$, and $[s, t] = [0, x_2]$ or $[x_1, 1]$ according to whether $f(x_1) \geqq f(x_2)$ or $f(x_1) < f(x_2)$. Suppose S_{N-1}^* has been defined $(N \geqq 3)$. We then define S_N^* as follows: $x_1 = U_{N-1}/U_{N+1}$, $x_2 = 1 - x_1 = U_N/U_{N+1}$. According to whether $f(x_1) \geqq f(x_2)$ or $f(x_1) < f(x_2)$, let $h(x) = xU_{N+1}/U_N$ or $h(x) = (-U_{N-1} + xU_{N+1})/U_N$, let $y = h(x)$ and $f^*(y) = f(h^{-1}(y))$ for $y \in I$, and define $y_2 = h(x_1)$ or $y_1 = h(x_2)$ in the respective cases. Thus, $y_2 = U_{N-1}/U_N$ or $y_1 = U_{N-2}/U_N$. Use S_{N-1}^* on the variable y and function f^* (for $y \in I$), noting that either $f^*(y_2)$ or $f^*(y_1)$ has already been observed. Obviously, $S_N^* \in \mathfrak{S}_N$.

We shall now prove inductively that S_N^* satisfies (1). This is obvious for $N = 2$. Assuming it to be true for $N \leqq n$, we now prove (1) to hold for $N = n+1$. Suppose the latter to be false. Then, since $L(D(f, S_N^*)) \leqq \epsilon + 1/U_{N+1}$, there would exist a procedure $\bar{S} \in \mathfrak{S}_{n+1}$ for which

$$\sup_{f \in \mathcal{F}} L(D(f, \bar{S})) < 1/U_{n+2}. \tag{2}$$

We may suppose that, under $\bar{S}$, g_2 is a constant. For otherwise we could define a procedure $\hat{S}$ by using the x_1 of $\bar{S}$ and thereafter using $\bar{S}$ on the function $f(x)-f(x_1)$; the g_2 of this procedure is then constant (equal to the $g_2(0)$ of $\bar{S}$), and $\hat{S}$ clearly satisfies (2) (with $\bar{S}$ replaced by $\hat{S}$) if $\bar{S}$ does. We hereafter denote the x_1 and x_2 of $\bar{S}$ by b and $b+a=1-c$ (say) with $a\geqq 0$; the corresponding values for S^*_{n+1} will be denoted by $d=U_n/U_{n+2}$ and $d+e=1-d=U_{n+1}/U_{n+2}$.

We next show that, as a consequence of (2),

$$(3) \qquad a+b\leqq d+e \quad \text{and} \quad a+c\leqq d+e.$$

We prove the first inequality of (3), the proof of the second being similar. Suppose, to the contrary, that $a+b>d+e$. We shall construct a procedure $S'\in\mathfrak{S}_n$ for which

$$(4) \qquad \sup_{f\in\mathcal{J}} L(D(f,S'))<1/U_{n+1},$$

contradicting the induction hypothesis. For any $f\in\mathcal{J}$ define f' for $y\in I$ by

$$(5) \qquad f'(y)=\begin{cases}\exp\left\{f\left(\dfrac{y}{a+b}\right)\right\} & \text{if } 0\leqq y<a+b,\\ -y & \text{if } a+b\leqq y\leqq 1.\end{cases}$$

It is easy to verify that $f'\in\mathcal{J}$ and that $x^{(f')}=(a+b)x^{(f)}$. Under S' we observe $\exp f(b/(a+b))=f'(b)$, calculate $f'(a+b)=-a-b$, treat these as the first two observations on f' according to $\bar{S}$, and take the remaining $n-1$ observations on f under S' by using $\bar{S}$ on f' as follows: if the kth observation on f' under $\bar{S}$ is to be taken at a value $y_k\geqq a+b$ (for $k\geqq 3$), S' puts $x_{k-1}=0$ and ignores the value of $f(x_{k-1})$, $f'(y_k)$ is calculated from the last line of (5), and y_{k+1} is computed under $\bar{S}$; if the kth observation on f' under $\bar{S}$ is to be taken at a value $y_k<a+b$, the $(k-1)$th observation on f according to S' is taken at $x_{k-1}=y_k/(a+b)$, the value of $f'(y_k)$ is then computed from (5) from the observed $f(x_{k-1})$, and y_{k+1} is determined by $\bar{S}$. After n observations on f ($n+1$ on f'), we put $D(f,S')=[s'/(a+b),\min(1,t'/(a+b))]$, where $[s',t']=D(f',\bar{S})$. Clearly $x^{(f)}\in D(f,S')$, so that $S'\in\mathfrak{S}_N$. Moreover, since $L(D(f,S'))\leqq L(D(f',\bar{S}))/(a+b)$, if $a+b>d+e=U_{n+1}/U_{n+2}$, equation (2) would imply equation (4). This completes the proof of (3).

The second inequality of (3) and the fact that $a+b+c=2d+e=1$ show that $b\geqq d$ (similarly, $c\geqq d$). We shall use this and (2) to construct a procedure $S''\in\mathfrak{S}_{n-1}$ for which

$$\sup_{f \in \mathcal{J}} L(D(f, S'')) < 1/U_n; \tag{6}$$

this contradiction of our induction hypothesis will then imply that (2) is false, completing our proof. To this end, for any $f \in \mathcal{J}$ define f'' for $y \in I$ by

$$f''(y) = \begin{cases} \exp\left\{f\left(\frac{y}{b}\right)\right\} & \text{for } 0 \leqq y < b, \\ -y & \text{for } b \leqq y \leqq 1. \end{cases} \tag{7}$$

Clearly, $f'' \in \mathcal{J}$ and $x^{(f'')} = bx^{(f)}$. S'' is defined by using $\bar{S}$ on f'' in a manner similar to that used in the previous paragraph on f' to define S': the first two observations on f'' under $\bar{S}$ are $f''(b) = -b$ and $f''(a+b) = -a-b$; thereafter, the kth observation on f'' under $\bar{S}$ corresponds in an obvious manner to the $(k-2)$th on f under S'', and $D(f, S'') = [s''/b, \min(1, t''/b)]$, where $[s'', t''] = D(f'', \bar{S})$. Thus, $L(D(f, S'')) \leqq L(D(f'', \bar{S}))/b$; since $b \geqq d = U_n/U_{n+2}$, this and (2) imply (6), completing the proof of our assertion.

We remark that a minimax procedure (one satisfying (1) with $\epsilon = 0$) does not exist in the above problem, as is evident from the case $N = 2$. The procedure S_N^* defined above may be improved upon by noting that, whenever the two largest observations are equal, $x^{(f)}$ must lie between the two corresponding values of x for any $f \in \mathcal{J}$. An interesting procedure which is not minimax for any fixed N but will often be useful in applications is the strategy S^* defined as follows: let $x_2 = 1 - x_1 = -1/2 + 5^{1/2}/2 = .618 = \mu$ (say). If $f(x_1) \geqq f(x_2)$, define $v(x) = x/\mu$ and $y = v(x)$, and $f^*(y) = f(v^{-1}(y))$ for $y \in I$. Putting $y_2 = v(x_1) = \mu$ and $y_1 = 1 - \mu$, we then use S^* on the variable y and function f^*, where we already have observed $f^*(y_2)$. (A similar procedure applies if $f(x_1) < f(x_2)$.) Continuing in this manner, at every stage we have the same geometric configuration, unlike the case of S_N^*. The advantage of this is that if the number of observations is not specified in advance but is determined after several values have been observed (e.g., more observation might be taken if f appears to be sharply peaked near its maximum), the use of any S_N^* (or sequence of S_N^*'s) can lead to great inefficiency if one decides after N observations to take more. When N is large, if S^* is used for N observations, the length of the final interval is about 1.17 times that of S_N^* (with $\epsilon \to 0$).

References

1. J. Kiefer and J. Wolfowitz, *Stochastic estimation of the maximum of a regression function*, Ann. Math. Statist. vol. 23 (1952) pp. 462–466.

2. H. Robbins and S. Monro, *A stochastic approximation method*, Ann. Math. Statist. vol. 22 (1951) pp. 400–407.

3. J. Wolfowitz, *On the stochastic approximation method of Robbins and Monro*, Ann. Math. Statist. vol. 23 (1952) pp. 457–461.

CORNELL UNIVERSITY

Reprinted from
Proc. Amer. Math. Soc. **4** (3), 502–506 (1953)

Reprinted from THE ANNALS OF MATHEMATICAL STATISTICS
Vol. 24, No. 2, June, 1953

SEQUENTIAL DECISION PROBLEMS FOR PROCESSES WITH CONTINUOUS TIME PARAMETER. TESTING HYPOTHESES[1]

A. DVORETZKY, J. KIEFER AND J. WOLFOWITZ

Hebrew University, Jerusalem, Cornell University and University of California at Los Angeles

Summary. The purpose of the present paper is to contribute to the sequential theory of testing hypotheses about stochastic processes with a continuous parameter (say, t which one may think of as time). Sequential decision problems about such processes seem not to have been treated before. Subsequently we shall treat problems of point and interval estimation and general sequential decision problems for such processes. The results, in addition to their interest per se and their practical importance, also shed light on the corresponding results for discrete stochastic processes. The subjects of sequential analysis and the theory of decision functions were founded by Wald, and we treat our present subjects in the spirit of his approach. The general results of decision theory, such as the complete class theorem, carry over to sequential problems about stochastic processes with continuous time parameter. As specific examples we treat the Wiener and Poisson processes and obtain, for example, the exact power function. (For discrete processes the corresponding known results, due to Wald, are approximations).

1. Introduction. Let $\{x_1(t), t \geqq 0\}$ and $\{x_2(t), t \geqq 0\}$ be two different stochastic processes. The statistician observes continuously, beginning at $t = 0$, a process $\{x(t), t \geqq 0\}$ which is either $\{x_1(t)\}$ or $\{x_2(t)\}$, and wishes to decide, as soon as possible, whether $\{x(t)\}$ is $\{x_1(t)\}$ or $\{x_2(t)\}$. "As soon as possible" means the following here. Let T be the time when he reaches a decision (in general this may be a chance variable and need not be a constant). Let E_iT denote the expected value of T when $\{x(t)\} = \{x_i(t)\}$, $i = 1, 2$. Let α_1, α_2 be two positive constants, $\alpha_1 + \alpha_2 < 1$. Subject to the requirement that the probability of an incorrect decision when $\{x(t)\} = \{x_i(t)\}$ be at most α_i, the problem is to give a procedure for deciding between $\{x_1(t)\}$ and $\{x_2(t)\}$ such that $E_i(T)$ is a minimum for $i = 1, 2$. This is simply the same formulation for stochastic processes with continuous parameter as was originally given by Wald ([3], [4]) for stochastic processes with a discrete parameter.

In this paper we shall limit ourselves to stochastic processes which fulfill the following conditions. For every $t \geqq 0$, $x(t)$ is a sufficient statistic for the process, that is, the conditional distribution of the chance function $x(\tau), 0 \leqq \tau \leqq t$, given $x(t)$, is, with probability one for every t, the same for the processes $\{x_1(t)\}$

Received 11/4/52.

[1] This work was sponsored by the Office of Naval Research under a contract with Columbia University and under a National Bureau of Standards contract with the University of California at Los Angeles.

and $\{x_2(t)\}$. For every t and x, both $x_1(t)$ and $x_2(t)$ have frequency functions, say $f_1(x, t)$ and $f_2(x, t)$, respectively. Let

$$\lambda(t) = \log \frac{f_2(x(t), t)}{f_1(x(t), t)} \qquad (\lambda(0) = 0). \tag{1.1}$$

Finally we postulate that the $\lambda(t)$ process is one of stationary independent increments, that is, a) for every positive integral k, every $h > 0$, and every sequence $t_1 < t_2 < \cdots < t_k \leqq t$, $\lambda(t + h) - \lambda(t)$ is distributed independently of $\lambda(t_1), \cdots, \lambda(t_k)$; b) the distribution of $\lambda(t + h) - \lambda(t)$ depends only upon h and not upon t.

Thus our theory will include the following problems: 1) testing hypotheses about the parameter of a continuous Poisson process with stationary independent increments (to be discussed in detail below in Section 3); 2) testing hypotheses about the mean of a Wiener process (to be discussed in detail below in Section 4); 3) testing hypotheses about the value of $p(0 < p < 1)$ in the following process with stationary independent increments (called the negative binomial): the probability that $x(t) = k$ for every nonnegative integer k is

$$\Gamma(t + k)p^t(1 - p)^k/\Gamma(k + 1)\Gamma(t);$$

4) testing hypotheses about the value of $\theta(\theta > 0)$ in the following process with stationary independent increments (called the Gamma process): the probability density of $x(t)$ at $x(x \geqq 0)$ is given by $x^{t-1}e^{-x/\theta}/\Gamma(t)\theta^t$.

In practice it is, of course, impossible to observe without error a sample function of a continuous process such as the Poisson process or the Wiener process. Yet in many cases these processes do constitute an excellent approximation to physical reality. For example, the incidence of mesons on a Geiger counter is generally assumed to follow a Poisson process. If the recording lag and the dead time of the Geiger counter are very small, a physicist could use the present theory to decide between two possible values of meson density. In this case continuous observation means simply exact registration of incidence times. As another example, our method, or a modification of it, may be applied to problems of life testing.

Moreover, there are several distinct advantages of the continuous parameter procedure over the discrete one. These are as follows.

The expected duration of observing the process before reaching a decision about which hypothesis to adopt can obviously only be shortened by allowing continuous observation.

Moreover, there are many cases, notably the Poisson and Wiener processes, in which an exact determination of the optimal procedure is possible in the continuous case, while in the discrete case so far only approximations have been derived. Thus, even when treating the discrete case, the continuous case, which is easier to treat, may be used to derive approximations when the unit of time is small.

There may also be other advantages in special problems. Thus it is seen in

Section 3 that in the continuous Poisson process the solution does not depend, as in the discrete case, on the values of the two parameters λ_1 and λ_2, but only on their ratio λ_2/λ_1.

2. Application of the Wald sequential procedure. Optimum character of the test. A careful examination of the results of [5] and [6] shows that their conclusions in no way require that the processes be discrete in time, and under the assumptions about the processes made in the preceding section the following results hold.

i) Let a and b, $b < 0 < a$, be given numbers, and let us employ the Wald sequential probability ratio test as follows. As long as $\ell(t)$ lies between b and a, continue observing $\{x(t)\}$. As soon as $\ell(t) \leqq b$, stop observing $\{x(t)\}$ and decide $\{x(t)\} = \{x_1(t)\}$. As soon as $\ell(t) \geqq a$, stop observing $\{x(t)\}$ and decide $\{x(t)\} = \{x_2(t)\}$. Let $\alpha_i(a, b)$ be the probability of error and $E_i(T \mid a, b)$ be the expected value of T when $\{x(t)\} = \{x_i(t)\}$, $i = 1, 2$. For any other procedure with respective probabilities of error α_1^* and α_2^* and respective expected values E_1^*T an E_2^*T, we have that $\alpha_i^* \leqq \alpha_i$, $i = 1, 2$ implies $E_i^*T \geqq E_i(T \mid a, b)$, that is, the optimum character of the Wald sequential probability ratio test (with respect to all randomized as well as nonrandomized procedures).

ii) Let c, W_1 and W_2 be positive numbers, and let g_i be the a priori probability that $\{x(t)\} = \{x_i(t)\}$, $i = 1, 2$ (cf. remarks about a priori probability distributions in [5] and [6]). There exist two numbers $a(c, W_1, W_2, g_1, g_2)$ and $b(c, W_1, W_2, g_1, g_2)$ such that, if the statistician continues to observe $\{x(t)\}$ until either $\ell(t) \leqq b$ or $\ell(t) \geqq a$, and then decides respectively that $\{x(t)\} = \{x_1(t)\}$ or $\{x(t)\} = \{x_2(t)\}$, he will minimize $g_1(\alpha_1 W_1 + cE_1T) + g_2(\alpha_2 W_2 + cE_2T)$ with respect to all possible procedures for deciding between $\{x_1(t)\}$ and $\{x_2(t)\}$, where E_iT is the expected value of T when $\{x(t)\} = \{x_i(t)\}$, $i = 1, 2$. (It is of course assumed that $a \geqq b$, with the equality sign not excluded. Also $\ell(0) = 0$. Thus if $a = b$, or $a \leqq 0$ or $b \geqq 0$, the decision will always be made at time $t = 0$.)

It is to be understood that any procedure which the statistician will employ should be such that the quantities α_1, α_2, E_1T, and E_2T will be well defined. The consideration of questions of measurability is a little more involved for our problem than it is in [5] and [6], but because of the assumptions on the processes made in the preceding section it can be carried out without difficulty. We shall therefore omit consideration of such questions.

From the remarks at the end of Section 1 and well known results of sequential analysis (see Stein [2]), it follows that $E_iT^k < \infty$ for any sequential probability ratio test and any positive k.

Other important results of sequential analysis established for discrete processes apply also to the continuous parameter case. For example, let $\{z(t), t \geqq 0\}$ $(z(0) = 0)$, be a process with stationary independent increments. Assume that $Ez(1)$ exists and denote it by h. Suppose that one has any stopping rule, that is, there is defined a positive chance variable T such that the set of chance functions for which $T = t$ is defined only by conditions on $z(\tau)$, $0 \leqq \tau \leqq t$. Then Wald's equation ([3], [7])

$$(2.1) \qquad Ez(T+) = hE(T)$$

holds. Suppose also that $Ee^{uz(1)}$ exists for all real u, and denote it by $\phi(u)$. Then Wald's fundamental identity ([4], p. 159)

$$(2.2) \qquad Ee^{uz(T+)}(\phi(u))^{-T} = 1$$

holds for many stopping rules, including in particular the rule where $T = t$ if $z(t) \geqq a$ or $z(t) \leqq b$, while $b < z(\tau) < a$ for $\tau < t$. Here a and b are constants, $a > 0$, $b < 0$. The simplest way of proving these results is to derive them as immediate consequences of a theorem of J. L. Doob on martingales with a continuous parameter ([1], Chap. VII, Theorem 11.8). For (2.1) the martingale process is $\{z(t) - ht\}$, and for (2.2) the martingale process is $\{e^{uz(t)}(\phi(u))^{-t}\}$. Another, more laborious way, of proving these results is to consider the process $\{z(t)\}$ only at time intervals which are integral multiples of Δ, proceed as in [4] or [7], and then let Δ approach zero. This is, however, a laborious way of proving a special case of the martingale theorem.

3. The Wiener process. Let $\{x_1(t)\}$ and $\{x_2(t)\}$ be Wiener processes ($t \geqq 0$, $x_1(0) = x_2(0) = 0$) each with a variance which without loss of generality we take to be one per unit of time. Let m_1 and m_2 ($m_1 \neq m_2$) be the mean values per unit time of $\{x_1(t)\}$ and $\{x_2(t)\}$, respectively. Thus we have the following situation: $t \geqq 0$ is a continuous (time) parameter. For any a_1, $a_2(0 < a_1 < a_2)$, $x_i(a_2) - x_i(a_1)$ is normally distributed with mean $m_i(a_2 - a_1)(i = 1, 2)$ and variance $(a_2 - a_1)$. For any integral k and sequence $a_1^1 < a_2^1 \leqq a_1^2 < a_2^2 \leqq \cdots \leqq a_1^k < a_2^k$, the k chance variables $x_i(a_2^j) - x_i(a_1^j)$, $j = 1, \cdots, k$, $i = 1, 2$, are independently distributed. The statistician observes continuously, beginning at $t = 0$, a process $\{x(t)\}$ which is either $\{x_1(t)\}$ or $\{x_2(t)\}$, and wishes to decide whether $\{x(t)\} = \{x_1(t)\}$ or $\{x(t)\} = \{x_2(t)\}$.

At time t_0 the quantity $x(t_0)$ is sufficient for deciding between $\{x_1(t)\}$ and $\{x_2(t)\}$, that is, it is unnecessary to know the previous history of the process. The likelihood ratio $L(x(t), t))$ at time t is given by

$$L(x(t), t) = \frac{\dfrac{1}{\sqrt{2\pi t}} e^{-\frac{1}{2}((x(t)-m_2t)^2/t)}}{\dfrac{1}{\sqrt{2\pi t}} e^{-\frac{1}{2}((x(t)-m_1t)^2/t)}}.$$

Hence

$$\mathfrak{L}(t) = x(t)\,(m_2 - m_1) - \frac{t}{2}\,(m_2^2 - m_1^2).$$

The sample functions of the $\{x(t)\}$ process are continuous with probability one. We choose a and b, $b < 0 < a$, such that the statistician will continue to observe $\{x(t)\}$ only until $\mathfrak{L}(t) = a$ or $\mathfrak{L}(t) = b$. In the first case he will decide that $\{x(t)\} =$

$\{x_2(t)\}$, in the second case that $\{x(t)\} = \{x_1(t)\}$. We shall now find $\alpha_i(a, b)$, $E_i(T \mid a, b)$, and the distribution function of T. The same problem for the discrete stochastic process when one observes $\{x(t)\}$ only at $t = 1, 2, \cdots$ has been studied by Wald ([3], [4]) who gave, inter alia, approximations for these quantities. An examination of his argument shows that, in his problem, his results are approximate only because he neglects the excess of $\dotplus(T)$ over a or b. In our problem this excess is zero with probability one, and Wald's formulae cease to be mere approximations and become exact. Thus we have, for example, ([4], p. 50, equation (3.42))

$$\alpha_1(a, b) = \frac{1 - e^b}{e^a - e^b} \tag{3.1}$$

$$\alpha_2(a, b) = \frac{e^b(e^a - 1)}{e^a - e^b}. \tag{3.2}$$

For any Wiener process with variance one per unit of time, not necessarily either $\{x_1(t)\}$ or $\{x_2(t)\}$, the probability that $\dotplus(t)$ will reach b before reaching a is given *exactly* by [4], page 50, equation (3.43). Call this probability H. Then, for any Wiener process with variance one per unit of time, not necessarily $\{x_1(t)\}$ or $\{x_2(t)\}$, $ET = (Hb + (1 - H)a)/h$ ([4], page 53, equation (3.57)). These results can be derived from (2.1) and (2.2) by Wald's methods. Also the density function of T is given exactly by formula (A:194) on page 195 of [4].

In practice one has to find a and b to correspond to given values α_1 and α_2. Solving (3.1) and (3.2) we obtain

$$a = \log \frac{1 - \alpha_2}{\alpha_1}, \tag{3.3}$$

$$b = \log \frac{\alpha_2}{1 - \alpha_1}. \tag{3.4}$$

All of the above results are exact because the excess of $\dotplus(T)$ over the boundaries a and b is zero with probability one. For the same reason one may already infer the optimal character of the Wald sequential probability ratio test for testing hypotheses about the mean of a Wiener process from the approximations and heuristic arguments given by Wald on pages 196–199 of [4].

One may raise the question how to test hypotheses about the variance of a Wiener process. However, a scrutiny of the problem shows that from a knowledge of a sample function in any interval, no matter how small, one can, with probability one, determine the variance to any arbitrary accuracy, so that the problem is trivial. For suppose $\{x(t)\}$ is a Wiener process with mean value m and variance v, both per unit of time. Suppose the process has been observed from $t = 0$ to $t = \mathrm{H}_0$, where H_0 is any positive number. Let N be any integer which will later approach infinity, and write $t_i = iH_0/N$, $i = 0, 1, \cdots, N$. For any i from 1 to N we have

$$E(x(t_i) - x(t_{i-1}))^2 = v\frac{H_0}{N} + m^2\frac{H_0^2}{N^2}.$$

Now, for $i = 1, \cdots, N$, the chance variables

$$\left\{(x(t_i) - x(t_{i-1}))^2 - v\frac{H_0}{N} - m^2\frac{H_0^2}{N^2}\right\}$$

are identically and independently distributed, with variance of order N^{-2} and fourth moment of order N^{-4}. Hence the fourth moment of

$$\frac{\sum_{i=1}^{N}(x(t_i) - x(t_{i-1}))^2}{H_0} - v - \frac{m^2 H_0}{N}$$

is of order N^{-2}. Consequently, for any $\epsilon > 0$ we have that

$$P\left\{\left|\frac{\sum_{i=1}^{N}(x(t_i) - x(t_{i-1}))^2}{H_0} - v - \frac{m^2 H_0}{N}\right| > \epsilon\right\} < \frac{C}{\epsilon^4 N^2}$$

where C is a suitable constant. Since the series $\sum N^{-2}$ converges it follows immediately from the Borel-Cantelli lemma that $(\sum_{i=1}^{N}(x(t_i) - x(t_{i-1}))^2)/H_0$ converges to v with probability one as $N \to \infty$.

4. The Poisson process. In this section we treat the problem of deciding which of two values given in advance represents the correct mean occurrence time of a Poisson process with stationary independent increments.

The probability that a Poisson process with mean occurrence time λ will result in exactly k occurrences between times $t = 0$ and $t = T$ is

$$\frac{(\lambda T)^k}{k!}e^{-\lambda T} \qquad (k = 0, 1, 2, \cdots). \tag{4.1}$$

Let $H_i (i = 1, 2)$ denote the hypothesis that $\lambda = \lambda_i$, where λ_1 and λ_2 are any two different positive numbers. It is clear that the two corresponding processes satisfy the conditions imposed in the introduction. Hence, given two positive numbers α_1, α_2, $(\alpha_1 + \alpha_2 < 1)$, the optimal test procedure for deciding between H_1 and H_2 which satisfies the condition that the probability of a wrong decision when H_i is true does not exceed $\alpha_i (i = 1, 2)$ is given by a Wald sequential probability ratio test.

More specifically, in view of (4.1) we have

$$\ell(t) = x(t)\log\frac{\lambda_2}{\lambda_1} + (\lambda_2 - \lambda_1)t. \tag{4.2}$$

Thus, assuming $\lambda_2 > \lambda_1$, the best decision rule is specified by two numbers a, $b (b < 0 < a)$ in the manner described in the introduction.

Suppose now that α_1 and α_2 are the actual probabilities of error. According to Wald ([4], p. 196) we have

$$\frac{1 - \alpha_2}{\alpha_1} = \frac{P_2(H_2)}{P_1(H_2)}, \qquad \frac{\alpha_2}{1 - \alpha_1} = \frac{P_2(H_1)}{P_1(H_1)}, \tag{4.3}$$

where $P_i(H_j)$ is the probability that hypothesis H_j is accepted when hypothesis H_i is true. By the argument used by Wald we have

$$e^{\inf_i \text{ł}(T)} = \inf_i e^{\text{ł}(T)} \leqq \frac{P_2(H_i)}{P_1(H_i)} \leqq \sup_i e^{\text{ł}(T)} = e^{\sup_i \text{ł}(T)} \tag{4.4}$$

the $\sup_i$ and $\inf_i$ being taken over all values of $\text{ł}(T)$ where the observation is stopped at time T with the decision to adopt H_i. In our case we know that if the decision to accept H_2 is adopted at time T we must have $\text{ł}(T) \geqq a$, while $\text{ł}(t) < a$ for $t < T$. Since (see (4.2)) $\text{ł}(t + 0) - \text{ł}(t) \leqq \log \lambda_2/\lambda_1$ with probability 1 we have from (4.3) and (4.4)

$$e^a \leqq \frac{1 - \alpha_2}{\alpha_1} \leqq \frac{\lambda_2}{\lambda_1} e^a. \tag{4.5}$$

Similarly if at time T we decide to terminate observation and adopt H_1 we must have $\text{ł}(T) \leqq b$ and $\text{ł}(t) > b$ for $t < T$. Since with probability 1 we have $\text{ł}(t) \geqq \text{ł}(t - 0)$ we find that $\text{ł}(T) = b$ with probability 1. Therefore

$$\frac{\alpha_2}{1 - \alpha_1} = e^b. \tag{4.6}$$

We see here one of the advantages of continuous observation over observation at discrete times only. If we were treating the problem in the conventional manner we would have (4.6) replaced by an inequality, while only the first of the inequalities (4.5) could be derived in the above manner.

Thus we have

$$b = \log \frac{\alpha_2}{1 - \alpha_1} \tag{4.7}$$

and

$$\log \frac{\lambda_1}{\lambda_2} + \log \frac{1 - \alpha_2}{\alpha_1} \leqq a \leqq \log \frac{1 - \alpha_2}{\alpha_1}. \tag{4.8}$$

We now proceed to give a method for the exact computation of a. Without additional effort we shall also find the power function of the test.

We put

$$R(t) = \frac{\text{ł}(t)}{\log \dfrac{\lambda_2}{\lambda_1}} = x(t) - ct \tag{4.9}$$

where $c = (\lambda_2 - \lambda_1)/\log (\lambda_2/\lambda_1)$. Together with the process $\{x(t)\}$ we have to consider also processes differing from it by a constant; that is, we consider processes with arbitrary $x(0)$.

For given a and b, let $V_\lambda(r)$ be the probability that the procedure described above will terminate with the adoption of H_2 when the Poisson parameter is really λ and $R(0) = r$. We then have

$$R(\Delta t) = \begin{cases} r - c\Delta t \\ r + 1 - c\Delta t \text{ with probability} \\ \text{any other value} \end{cases} \begin{cases} 1 - \lambda\Delta t + o(\Delta t) \\ \lambda\Delta t + o(\Delta t) \\ o(\Delta t) \end{cases}$$

where the $o(\Delta t)$ terms are all smaller than $\lambda^2\Delta t^2$ for $0 < \Delta t < 1/\lambda$.

Putting

$$K = \frac{b}{\log \frac{\lambda_2}{\lambda_1}}, \qquad J = \frac{a}{\log \frac{\lambda_2}{\lambda_1}} \tag{4.10}$$

we have

$$V_\lambda(r) = 0 \qquad \text{for } r \leqq K$$

$$V_\lambda(r) = 1 \qquad \text{for } r \geqq J$$

while for $K < r < J$ we have

$$V_\lambda(r) = (1 - \lambda\Delta t)V_\lambda(r - c\Delta t) + \lambda\Delta t V_\lambda(r + 1 - c\Delta t) + o(\Delta t) \tag{4.11}$$

with $|\,o(\Delta t)\,| < \lambda^2(\Delta t)^2$ for $0 < \Delta t < 1/\lambda$. It follows at once that $V_\lambda(r)$ is continuous for $K \leqq r < J$. (It will be discontinuous at $r = J$.) Rewriting (4.11) as

$$\frac{V_\lambda(r) - V_\lambda(r - c\Delta t)}{\Delta t} = -\lambda V_\lambda(r - c\Delta t) + \lambda V_\lambda(r + 1 - c\Delta t) + \frac{o(\Delta t)}{\Delta t} \tag{4.12}$$

and letting $\Delta t \to 0$ we see that $V_\lambda(r)$ is differentiable in the interval $K < r < J$ with the exception of the point $r = J - 1$ (in case $K < J - 1$). Thus we have the difference-differential equation

$$cV_\lambda'(r) + \lambda V_\lambda(r) = \lambda V_\lambda(r + 1) \tag{4.13}$$

for $K < r < J$ and $r \neq J - 1$. The unique solution in $K < r < J$ is determined by the conditions: (i) $V_\lambda(r)$ continuous for $\lambda < J$, (ii) $V_\lambda(K) = 0$, (iii) $V_\lambda(r) = 1$ for $r \geqq J$.

Let $n(r)$ be the integer such that

$$J - r - 1 \leqq n(r) < J - r. \tag{4.14}$$

It is easy to verify that, for $K \leqq r < J$,

$$V_\lambda(r) = 1 + Ce^{-(\lambda/c)r} \sum_{i=0}^{n(r)} \frac{(-1)^i}{i!} \left[(J - r - i) \frac{\lambda}{c} e^{-\lambda/c} \right]^i \tag{4.15}$$

satisfies (4.13) and (i) for every choice of the constant of integration C. To satisfy also (ii) one has merely to choose

$$C = -e^{(\lambda/c)K} \Big/ \sum_{i=1}^{n(K)} \frac{(-1)^i}{i!} \left[(J - K - i) \frac{\lambda}{c} e^{-\lambda/c} \right]^i. \tag{4.16}$$

Putting $r = 0$ to represent the start of the actual probability ratio test as used in applications, we have from (4.15) and (4.16) that the "OC" function corresponding to the given values of λ_1, λ_2, a and b is given by, say,

$$g\left(\frac{\lambda}{c}\right) = 1 - e^{(\lambda/c)K} \frac{\displaystyle\sum_{i=0}^{n(0)} \frac{(-1)^i}{i!}\left[(J - i)\frac{\lambda}{c}e^{-\lambda/c}\right]^i}{\displaystyle\sum_{i=0}^{n(K)} \frac{(-1)^i}{i!}\left[(J - K - i)\frac{\lambda}{c}e^{-\lambda/c}\right]^i}. \tag{4.17}$$

(K is not displayed since it is given explicitly by (4.10).) Now J should be determined so that

$$g\left(\frac{\lambda_1}{c}\right) = g\left(\frac{\log \dfrac{\lambda_2}{\lambda_1}}{\dfrac{\lambda_2}{\lambda_1} - 1}\right) = \alpha_1;$$

(4.18)

$$g\left(\frac{\lambda_2}{c}\right) = g\left(\frac{\log \dfrac{\lambda_2}{\lambda_1}}{1 - \dfrac{\lambda_1}{\lambda_2}}\right) = 1 - \alpha_2.$$

Each of the equations (4.18) follows from the other and either one may be used to find J.

It should be noticed that the dependence of K and J on λ_1 and λ_2 is only through the ratio λ_2/λ_1. This follows from (4.10) and (4.17) and could also have been foreseen from the nature of the problem. This remark is useful in the numerical tabulation of the values of J and K or, equivalently, of a and b. (The fact that the λ_i are involved only through their ratio is due to the fact that they are not attached to a given time-unit. In the discrete parameter problem there is an absolute unit of time and hence the two λ_i enter as two parameters. The simplification mentioned above therefore does not occur.)

We now derive, in a manner similar to that used above, an expression for the moment generating function $M_\lambda(u; r) = Ee^{uT}$ of the observation time T necessary to reach a decision when $R(0) = r$ and the true Poisson parameter is λ. From a result of C. Stein [2] it follows that for given J, K and λ there is a positive number $u_0 = u_0(J, K, \lambda)$ such that $M_\lambda(u; r)$ is analytic and uniformly bounded in r for each complex u with real part smaller than u_0. By definition we have $M_\lambda(u; r) = 1$ for $r \leqq K$ or $r \geqq J$. In the same way as (4.11) was derived we obtain (for each u with real part smaller than u_0) for $K < r \leqq J - 1$

$$\begin{aligned} M_\lambda(u; r) = {} & (1 - \lambda\Delta t)E\{e^{ut} \mid R(0) = r, \quad R(\Delta t) = r - c\Delta t\} \\ & + \lambda\Delta t E\{e^{ut} \mid R(0) = r, \quad R(\Delta t) = r + 1 - c\Delta t\} + o(\Delta t) \\ = {} & (1 - \lambda\Delta t)e^{u\Delta t}E\{e^{u(t-\Delta t)} \mid R(\Delta t) = r - c\Delta t\} \\ & + \lambda\Delta t E\{e^{ut} \mid R(\Delta t) = r + 1 - c\Delta t\} + o(\Delta t), \end{aligned}$$

or

$$M_\lambda(u; r) = (1 - \lambda\Delta t)(1 + u\Delta t)M_\lambda(u; r - c\Delta t) + \lambda\Delta t M_\lambda(u; r + 1 - c\Delta t) + o(\Delta t). \tag{4.19}$$

This form is also valid for $J - 1 < r < J$ since $M_\lambda(u; r + 1 - c\Delta t) = 1 + o(1)$ for $r > J - 1$. Since the $o(\Delta t)$ term and $M_\lambda(u; r)$ are uniformly bounded in r we deduce, as in the case of $V_\lambda(r)$, that, considered as a function of r, $M_\lambda(u; r)$ is continuous for $r < J$, possesses a derivative for $K < r < J$ and $r \neq J - 1$ and satisfies in the last range the equation

$$c\frac{\partial}{\partial r} M_\lambda(u; r) + (\lambda - u)M_\lambda(u; r) = \lambda M_\lambda(u; r + 1). \tag{4.20}$$

It can be verified that the solution of (4.20) satisfying the required boundary conditions is given for $K \leqq r < J$ by

$$\begin{aligned} M_\lambda(u; r) = {} & \left(\frac{\lambda}{\lambda - u}\right)^{n(r)+1} \\ & + C(u)e^{-r(\lambda-u)/c} \sum_{i=0}^{n(r)} \frac{(-1)^i}{i!}\left[(J - r - i)\frac{\lambda}{c}e^{-(\lambda-u)/c}\right]^i - \frac{\lambda u}{(\lambda - u)^2} \\ & \cdot e^{(\lambda-u)(J-r-1)/c} \sum_{i=0}^{n(r)-1} \left(\frac{\lambda}{\lambda - u}e^{-(\lambda-u)/c}\right)^i \sum_{j=0}^{i} \frac{(-1)^j}{j!}\left[(J - r - i - 1)\frac{\lambda - u}{c}\right]^j \end{aligned} \tag{4.21}$$

with $C(u)$ determined so that $M_\lambda(u; K) \equiv 1$.

Let $Z_\lambda(r)$ be the expected length of time before a final decision is adopted. Then $Z_\lambda(r) = \partial/(\partial u)\, M_\lambda(u, r) \mid_{u=0}$. Since $C(0) = 0$ in (4.21) we obtain, on putting $C'(0) = C'$,

$$\begin{aligned} Z_\lambda(r) = {} & \frac{n(r) + 1}{\lambda} + C'e^{-(\lambda/c)r} \sum_{i=0}^{n(r)} \frac{(-1)^i}{i!}\left[(J - r - i)\frac{\lambda}{c}e^{-\lambda/c}\right]^i \\ & - \frac{1}{\lambda}e^{(\lambda/c)(J-r-1)} \sum_{i=0}^{n(r)-1} e^{-(\lambda/c)i} \sum_{j=0}^{i} \frac{(-1)^j}{j!}\left[(J - r - i - 1)\frac{\lambda}{c}\right]^j \end{aligned} \tag{4.22}$$

for $K \leqq r < J$ (of course $Z_\lambda(r) \equiv 0$ outside this range and C' is determined so that $Z_\lambda(K) = 0$).

(One could derive (4.22) without using the moment generating function by establishing the equation

$$cZ'_\lambda(r) + \lambda Z_\lambda(r) = 1 + \lambda Z_\lambda(r + 1)$$

for $K < r < J$, $r \neq J - 1$.)

If we write in a more explicit manner $Z_\lambda(r \mid \lambda_1, \lambda_2)$ for $Z_\lambda(r)$ with J and K determined as explained above, it is easily seen that

$$Z_{\alpha\lambda}(r \mid \alpha\lambda_1, \alpha\lambda_2) = \frac{1}{\alpha} Z_\lambda(r \mid \lambda_1, \lambda_2) \tag{4.23}$$

for every positive α.

It is possible to treat the negative binomial process in a manner essentially the same in which we have treated the Poisson process above. A complication is caused by the fact that the probability that the chance variable will exceed one in a small time interval is of the same order of magnitude as the probability that the chance variable will be one.

The authors are obliged to Professor J. L. Doob for several helpful remarks.

REFERENCES

[1] J. L. Doob, *Stochastic Processes*, John Wiley and Sons, 1953.
[2] C. Stein, "A note on cumulative sums," *Ann. Math. Stat.*, Vol. 17 (1946), pp. 498–499.
[3] A. Wald, "Sequential tests of statistical hypotheses, *Ann. Math. Stat.*, Vol. 16 (1945), pp. 117–186.
[4] A. Wald, *Sequential Analysis*, John Wiley and Sons, 1947.
[5] A. Wald and J. Wolfowitz, "Optimum character of the sequential probability ratio test," *Ann. Math. Stat.*, Vol. 19 (1948), pp. 326–329.
[6] A. Wald and J. Wolfowitz, "Bayes solutions of sequential decision problems," *Ann. Math. Stat.*, Vol. 21 (1950), pp. 82–99.
[7] J. Wolfowitz, "The efficiency of sequential estimates and Wald's equation for sequential processes," *Ann. Math. Stat.*, Vol. 18 (1947), pp. 215–230.

CORRECTIONS TO "SEQUENTIAL DECISION PROBLEMS FOR PROCESSES WITH CONTINUOUS TIME PARAMETER. TESTING HYPOTHESES"

By A. Dvoretzky, J. Kiefer, and J. Wolfowitz

The following corrections should be made on p. 259 of the above-titled paper (*Ann. Math. Stat.*, Vol. 24(1953), pp. 254–264): The mean occurrence time is $1/\lambda$, not λ, on line 4 of Section 4. In (4.2) the plus sign should be a minus sign.

Reprinted from
Ann. Math. Statist. **30** (1959), 1265

Reprinted from THE ANNALS OF MATHEMATICAL STATISTICS
Vol. 24, No. 3, September, 1953

SEQUENTIAL DECISION PROBLEMS FOR PROCESSES WITH CONTINUOUS TIME PARAMETER. PROBLEMS OF ESTIMATION[1]

BY A. DVORETZKY, J. KIEFER AND J. WOLFOWITZ

Hebrew University, Jerusalem, Cornell University, and University of California at Los Angeles

Summary. In a recent paper [1] the authors began the study of the theory of sequential decision functions for stochastic processes with a continuous time parameter. This paper treated the standard problem of testing hypotheses, and the advantage of being able to stop at an arbitrary time point (not necessarily a multiple of some unit given in advance) was demonstrated in several cases, notably in that of deciding between two Poisson processes. The optimal tests were Wald probability ratio tests and thus truly sequential. In the present paper we treat the problem of estimation, and study in detail the Poisson, Gamma, Normal and Negative Binomial processes. It turns out for these processes that, with a proper weight function, the minimax (sequential) rule reduces to a fixed-time rule. Though we confine ourselves to point-estimation it is clear that similar methods apply to interval estimation. It may also be remarked that the case when the time-parameter is discrete need not be treated separately. For example, as described in Section 6.1, the results of Sections 2 and 3 imply analogous results in the case of discrete time, which in turn imply certain results proved in [3] and (in the nonsequential case) in [2] by other methods. The treatment of some other problems in estimation is discussed in Section 6. This paper may be read independently of [1].

1. Preliminaries. Let $X(t \mid \omega)$, $t \geqq 0$, $\omega \,\varepsilon\, \Omega$, be a family of stochastic processes in time t which depend on a parameter ω. Let $c(t)$, $t \geqq 0$, be a given cost function which represents the cost to the statistician of observing the process up to time t. For every ω in Ω and $\bar{\omega}$ in the terminal decision space D[2] let $W(\omega, \bar{\omega})$ be the weight function, that is, the loss involved in giving the estimate $\bar{\omega}$ when ω is the correct value of the parameter. Let (T, δ) be a pair of functionals of the sample function $x(t)$ into $(0 \leqq T \leqq \infty, D)$, where δ depends on $x(t)$ only through its values for $0 \leqq t \leqq T$ if $T < \infty$ (if $T = \infty$, δ is undefined, but in accordance with our assumptions on $c(t)$ below we define the quantity of (1) to be ∞ if this event occurs with positive probability under ω). The decision rule corresponding to these functionals is: observe up to time T and then (in case T is finite) adopt the esti-

Received 11/28/52.

[1] This work was sponsored by the Office of Naval Research under a contract with Columbia University and under a National Bureau of Standards contract with the University of California at Los Angeles.

[2] In Sections 3 and 5, we take $\Omega = D$. In Section 2, $D = \Omega + \{\lambda = 0\}$. In Section 4, $D = \Omega + \{\omega = 0\} = \Omega + \{p = 1\}$.

mate δ. If T is a constant independent of the sample function $x(t)$ then the procedure is not truly sequential. It is called a fixed-size or fixed-time estimation procedure. Throughout this paper we shall by $x(t)$ mean $x(t+)$; that is, the sample functions are to be considered as continuous from the right.

For a given ω the risk associated with such a procedure is defined by

$$R_\omega(T, \delta) = E_\omega\{c(T) + W(\omega, \delta)\} \tag{1}$$

where E_ω denotes the expected value under the assumption that ω is the true value of the parameter, provided the expected value exists. Assuming the expected value to exist for every $\omega \varepsilon \Omega$ we define the maximum risk associated with T and δ by

$$R(T, \delta) = \sup R_\omega(T, \delta), \tag{2}$$

the supremum taken over all $\omega \varepsilon \Omega$.

An estimation procedure $(\hat{T}, \hat{\delta})$ is called *minimax* if

$$R(\hat{T}, \hat{\delta}) \leqq R(T, \delta) \tag{3}$$

for any functionals T and δ for which (2) is defined. If no minimax estimation rule exists, it is still possible to define a minimax sequence of decision rules $\hat{T}_n, \hat{\delta}_n (n = 1, 2, \cdots)$, that is, a sequence for which

$$\lim_{n=\infty} R(\hat{T}_n, \hat{\delta}_n) = \inf R(T, \delta). \tag{4}$$

In the cases we treat it will be shown that a minimax rule does exist. However, a slight relaxation of the assumptions (e.g., dropping the continuity assumption about the cost function $c(t)$) may affect the existence of minimax rules, and in such cases one has to modify the argument only slightly in order to find a minimax sequence (which, in the cases treated below, may be taken to consist of fixed-time rules).

Let ζ be a Borel field of subsets of Ω and $R_\omega(\delta, T)$ be a measurable function of ω with respect to ζ. Let $F(\omega)$ be a probability distribution on Ω. Then, assuming the integral to exist, we define

$$R_F(T, \delta) = \int_\Omega R_\omega(T, \delta) dF(\omega). \tag{5}$$

The estimation rule (T_F, δ_F) is called a Bayes rule for F if

$$R_F(T_F, \delta_F) = \inf R(T, \delta). \tag{6}$$

We shall denote by δ^T fixed time estimation rules with constant observation time T, and in this case we shall write δ^T instead of the pair (T, δ). We define

$$r_\omega(\delta^T) = E_\omega W(\omega, \delta^T) \tag{7}$$

and

$$r_F(\delta^T) = \int_\Omega r_\omega(\delta^T)\, dF(\omega). \tag{8}$$

δ_F^T is called a *T-Bayes estimation rule* for F if (8) assumes its minimum for $\delta^T = \delta_F^T$.

Let A be any set of sample curves $x(t)$ for which the probability $P\{x(t) \varepsilon A\}$ is defined and is a measurable function of ω. Let $F(\omega)$ be any distribution function over Ω. Then for every A for which $P(A) = \int_\Omega P\{x(t) \varepsilon A\} dF(\omega) > 0$ we define the a posteriori probability distribution $F(\omega \mid A)$ by assigning to every Borel set $S \varepsilon \zeta$ the probability $P(A)^{-1} \int_S P\{x(t) \varepsilon A\} dF(\omega)$. The *a posteriori T-risk* corresponding to F and δ^T is defined by

$$(9) \qquad r_F(\delta^T \mid A) = \int_\Omega r_\omega(\delta^T)\, dF(\omega \mid A).$$

If $r_F(\delta^T \mid A)$ is independent of A we say that *the a posteriori risk is independent of the sample* $x(t)$. (It is assumed in the sequel that "many" sets A with the above property exist. This is of course the case for the processes usually encountered in mathematical statistics and in particular with families of processes, like those treated in this paper, with which are associated sufficient statistics of a simple nature. Since our primary interest is in statistical applications there seems to be no point in inserting a lengthy technical discussion of the precise measurability properties required in order to insure that the class of sets A will be sufficiently rich.)

We shall make frequent use of the following obvious remark, which is a familiar tool in decision theory (see, e.g., [4]).

Suppose that, for every $T \geqq 0$, there exists a sequence of probability distributions $F_n (n = 1, 2, \cdots)$ for which there are corresponding T-Bayes solutions δ_n^T with the property that the a posteriori risk associated with F_n and δ_n^T is independent of the sample $x(t)$, and suppose that there exists $\hat{\delta}^T$ for which

$$(10) \qquad r(T) = \sup_\omega r_\omega(\hat{\delta}^T) = \lim_{n=\infty} r_{F_n}(\delta_n^T).$$

If there exists a T_0 $(0 \leqq T_0 < \infty)$ for which

$$(11) \qquad c(T_0) + r(T_0) = \min_{T \geqq 0} [c(T) + r(T)]$$

holds, then the fixed time rule $\hat{\delta}^{T_0}$ is a minimax estimation rule.

The proof of this assertion is evident. Indeed, the conclusion remains valid under weaker assumptions. As this is not needed for the sequel we just point out that we could have dropped the assumption of risk independent of the sample and replaced (10) by

$$\sup_\omega r_\omega(\hat{\delta}^T) = \lim_{n=\infty} \inf_A r_{F_n}(\delta_n^T \mid A).$$

It may also be worth while to remark that if no T_0 satisfying (11) exists, we still have a minimax sequence of estimation rules all of which are fixed-time rules. (These results clearly remain valid also if randomized rules are considered.)

In the examples treated in the sequel $r(T)$ is a nonnegative continuous function.

We assume that the cost function $c(T)$ is nonnegative, lower semicontinuous, and tends to infinity as $T \to \infty$. These assumptions guarantee the existence of a T_0 which satisfies (11).

We remark, finally, that in the examples of Sections 2 and 3, the minimum of (11) will always be achieved for $T > 0$, since $R(0, \delta) = \infty$ for all δ. In the examples of Sections 4 and 5, this need not be the case. Analogous remarks apply to the discussion of Section 6.

2. The Poisson process. This is defined for every $\lambda > 0$ as a process $X_\lambda(t)$ with independent stationary increments which satisfies

$$P\{X_\lambda(t) = x\} = \frac{(\lambda t)^x}{x!} e^{-\lambda t} \qquad (x = 0, 1, 2, \cdots) \tag{12}$$

for all $t \geqq 0$.

We let Ω be the half-line $0 < \lambda < \infty$ and ζ consist of the usual Borel sets.[2] Our problem is to estimate the mean λ. It is well known that $x(T)$ is a sufficient statistic for λ when the sample curve $x(t)$ is observed for $0 \leqq t \leqq T$.

As weight function we take, following Hodges and Lehmann [3] and Girshick and Savage [2],

$$W(\lambda, \delta) = \frac{1}{\lambda} (\delta - \lambda)^2. \tag{13}$$

This is the squared error measured in terms of the variance. As these authors point out, the classical squared error $(\delta - \lambda)^2$ gives, for every finite time, infinite minimax risk, and is thus of no interest unless some additional information about λ is known.

Let $F_n(\lambda)$, $n = 1, 2, \cdots$ be the probability distribution on the half-line $\lambda > 0$ with density

$$f_n(\lambda) = \frac{1}{n} e^{-\lambda/n} \qquad (0 < \lambda < \infty). \tag{14}$$

Let the process be observed during the time $0 \leqq t \leqq T$. The a posteriori probability distribution when $x(T) = x$ is well defined and its density is given by

$$f_n(\lambda \mid x) = \frac{\lambda^x}{x!} \left(T + \frac{1}{n}\right)^{x+1} e^{-\lambda(T+1/n)} \qquad (0 < \lambda < \infty).$$

The a posteriori T-risk (see (9)) is given by

$$r_n(\delta^T \mid x) = \int_0^\infty \frac{1}{\lambda} (\delta^T - \lambda)^2 f_n(\lambda \mid x)\, d\lambda.$$

It is easily seen that this is minimized by taking

$$\delta^T(x(T) = x) = 1 \Big/ \int_0^\infty \frac{1}{\lambda} f_n(\lambda \mid x)\, d\lambda = \frac{x}{T + 1/n}.$$

Therefore the T-Bayes solution corresponding to $F_n(\lambda)$ is given by

$$\hat{\delta}_n^T = \frac{x(T)}{T + 1/n}. \tag{15}$$

The corresponding a posteriori risk is

$$r_n(\hat{\delta}_n^T \mid x) = \frac{1}{T + 1/n}, \tag{16}$$

which is independent of $x(T)$ (hence, $x(T)$ being a sufficient statistic, of the sample).

On the other hand, taking

$$\hat{\delta}^T = \frac{x(T)}{T} \tag{17}$$

we see that

$$r_\lambda(\hat{\delta}^T) = e^{-\lambda T} \sum_{x=0}^{\infty} x \frac{(\lambda T)^x}{x!} = \frac{1}{T}$$

for all $\lambda > 0$. Thus we have the following.

For the Poisson process (12) *with* $0 < \lambda < \infty$ *and weight function* (13) *the fixed-time estimate* (17) *with* $T = T_0$ *given by*

$$c(T_0) + \frac{1}{T_0} = \min_{T>0} \left[c(T) + \frac{1}{T} \right] \tag{18}$$

is minimax.

3. The Gamma process. This is defined for every pair of positive numbers r and θ as a process $X_{r,\theta}(t)$ with independent stationary increments such that $X_{r,\theta}(0) = 0$ and for every $t > 0$ and $x > 0$

$$P\{X_{r,\theta}(t) < x\} = \int_0^x \frac{x^{rt-1}}{\Gamma(rt)\theta^{rt}} e^{-x/\theta} \, dx. \tag{19}$$

The parameter r will be assumed known, and the space Ω will consist of the half-line $0 < \theta < \infty$, the Borel sets being the ordinary ones.[2] Here again it is well known that if the sample curve $x(t)$ is given only for $t \leqq T$ then $x(T)$ is a sufficient statistic for θ.

As weight function we take

$$W(\theta, \delta) = \left[\left(\frac{\delta}{\theta} \right)^\gamma - 1 \right]^2, \tag{20}$$

γ being an arbitrary positive number. For $\gamma = 1$ this weight function, like (13), is proportional to the square error of the mean measured in terms of the variance and occurs in Hodges and Lehmann [3] and Girshick and Savage [2].

Let $F_n(\theta)$, $n = 1, 2, \cdots$ be the probability distribution over $\theta > 0$ with density

$$f_n(\theta) = \frac{\theta^{-1-1/n} e^{-1/\theta}}{\Gamma(1/n)} \qquad (0 < \theta < \infty). \tag{21}$$

The a posteriori probability distribution when $x(T) = x$ has the density

$$f_n(\theta \mid x) = \frac{(x+1)^{rT+1/n} e^{-(x+1)/\theta}}{\Gamma(rT + 1/n)\theta^{rT+1+1/n}} \qquad (0 < \theta < \omega). \tag{22}$$

The a posteriori T-risk is given by

$$r_n(\delta^T \mid x) = \int_0^\infty \left[\left(\frac{\delta}{\theta}\right)^\gamma - 1\right]^2 f_n(\theta \mid x)\, d\theta. \tag{23}$$

It is minimized by taking

$$\begin{aligned} \hat{\delta}_n^T(x(T) = x) &= \left[\int_0^\infty \theta^{-\gamma} f_n(\theta \mid x)\, d\theta \Big/ \int_0^\infty \theta^{-2\gamma} f_n(\theta \mid x)\, d\theta\right]^{1/\gamma} \\ &= (x+1)\left[\frac{\Gamma(rT + \gamma + 1/n)}{\Gamma(rT + 2\gamma + 1/n)}\right]^{1/\gamma}. \end{aligned} \tag{24}$$

Substituting this value in (23) we obtain

$$r_n(\hat{\delta}_n^T \mid x) = 1 - \frac{\Gamma^2(rT + \gamma + 1/n)}{\Gamma(rT + 1/n)\Gamma(rT + 2\gamma + 1/n)}, \tag{25}$$

which is independent of $x(T)$.

On the other hand the estimator

$$\hat{\delta}^T = \left[\frac{\Gamma(rT + \gamma)}{\Gamma(rT + 2\gamma)}\right]^{1/\gamma} x(T) \tag{26}$$

gives

$$r_\theta(\hat{\delta}^T) = 1 - \frac{\Gamma^2(rT + \gamma)}{\Gamma(rT)\Gamma(rT + 2\gamma)} \tag{27}$$

for all $0 < \theta < \infty$. Since (27) is independent of θ and is the limit of (25) as $n \to \infty$ we have:

For the Gamma process (14) *with fixed r, unknown θ* $(0 < \theta < \infty)$ *and weight function* (20), *the fixed time estimate* (26) *with $T = T_0$ given by*

$$c(T_0) - \frac{\Gamma^2(rT_0 + \gamma)}{\Gamma(rT_0)\Gamma(rT_0 + 2\gamma)} = \min_{T>0}\left[c(T) - \frac{\Gamma^2(rT + \gamma)}{\Gamma(rT)\Gamma(rT + 2\gamma)}\right] \tag{28}$$

is minimax.

If instead of using the weight function (20) we use, following Girshick and Savage [2],

$$W_1(\theta, \delta) = \log^2 \frac{\delta}{\theta}, \tag{29}$$

that is, the squared error when $\log\theta$ is considered as the parameter, then we find that for the distributions (21) the a posteriori T-risk is minimized by taking

$$\hat{\delta}_n^T(x(T) = x) = \exp\int_0^\infty \log\theta f_n(\theta \mid x)\, d\theta = (x+1)\exp\left\{-\frac{\Gamma'(rT+1/n)}{\Gamma(rT+1/n)}\right\}$$

and that the corresponding value of the a posteriori risk is

$$\frac{\Gamma''(rT+1/n)}{\Gamma(rT+1/n)} - \left[\frac{\Gamma'(rT+1/n)}{\Gamma(rT+1/n)}\right]^2,$$

which again is independent of x. Since

$$\delta^T = x(T)e^{-\Gamma'(rT)/\Gamma(rT)} \tag{30}$$

has the constant risk function

$$\frac{\Gamma''(rT)}{\Gamma(rT)} - \left[\frac{\Gamma'(rT)}{\Gamma(rT)}\right]^2,$$

we have as before:

If, instead of (20), *the weight function* (29) *is used, then the fixed time estimate* (30) *with* $T = T_0$ *given by*

$$c(T_0) + \frac{\Gamma''(rT_0)}{\Gamma(rT_0)} - \left[\frac{\Gamma'(rT_0)}{\Gamma(rT_0)}\right]^2 = \min_{T>0}\left\{c(T) + \frac{\Gamma''(rT)}{\Gamma(rT)} - \left[\frac{\Gamma'(rT)}{\Gamma(rT)}\right]^2\right\} \tag{31}$$

is minimax.

4. The Negative Binomial process. This is defined for every $\omega > 0$ as a process $X_\omega(t)$ with independent stationary increments satisfying $X_\omega(0) = 0$ and

$$P\{X_\omega(t) = x\} = \frac{\Gamma(t+x)}{\Gamma(t)\Gamma(x+1)}\frac{\omega^x}{(1+\omega)^{x+t}} \quad (x = 0, 1, 2, \cdots) \tag{32}$$

for every $t > 0$. It is customary to put

$$p = \frac{1}{1+\omega}, \qquad q = \frac{\omega}{1+\omega}; \tag{33}$$

then (32) becomes

$$P\{X_\omega(t) = x\} = \frac{\Gamma(t+x)}{\Gamma(t)\Gamma(x+1)}p^t q^x \qquad (x = 0, 1, 2, \cdots).$$

As Ω we take the half line $0 < \omega < \infty$, the Borel sets being the usual ones.[2] It is easy to see that $x(T)$ is a sufficient statistic for ω when $x(t)$ is observed for $0 \leqq t \leqq T$.

$X_\omega(t)$ has mean ωt and variance $\omega(1+\omega)t$. It is easily seen that the square error $(\delta - \omega)^2$ would give an infinite minimax risk. We therefore use as weight function

$$W(\delta, \omega) = \frac{(\delta - \omega)^2}{\omega(1 + \omega)} = \frac{p^2}{q}\left(\delta - \frac{q}{p}\right)^2 \tag{34}$$

which is proportional to the square error measured in units equal to the variance.

Let $F_n(p)$, $n = 1, 2, \cdots$ be the probability distribution over $0 < p < 1$ with density

$$f_n(p) = \frac{\Gamma(2 + 1/n)}{\Gamma(2)\Gamma(1/n)} p^{-1+1/n} q \qquad (0 < p < 1). \tag{35}$$

If the process is observed for the period $0 \leqq t \leqq T$ and $x(T) = x$ then the a posteriori probability distribution has density

$$f_n(p \mid x) = \frac{\Gamma(x + 2 + T + 1/n)}{\Gamma(x + 2)\Gamma(T + 1/n)} p^{T-1+1/n} q^{x+1} \qquad (0 < p < 1).$$

The a posteriori T-risk is

$$r_n(\delta^T \mid x) = \int_0^1 \frac{p^2}{q}\left(\delta^T - \frac{q}{p}\right)^2 f_n(p \mid x)\,dp,$$

and is minimized by taking

$$\delta^T(x(T) = x) = \frac{\int_0^1 p f_n(p \mid x)\,dp}{\int_0^1 \frac{p^2}{q} f_n(p \mid x)\,dp} = \frac{x + 1}{T + 1 + 1/n}.$$

Therefore the T-Bayes estimate corresponding to the a priori distribution $F_n(p)$ with density (35) is given by $\hat{\delta}_n^T = (x + 1)/(T + 1 + 1/n)$. The corresponding a posteriori risk is

$$r_n(\hat{\delta}_n^T \mid x) = \frac{1}{T + 1 + 1/n}, \tag{36}$$

which is independent of $x(T)$ and, therefore, of the sample.

For every given ω, $0 < \omega < \infty$, the estimator

$$\hat{\delta}^T = \frac{x(T)}{T + 1} \tag{37}$$

gives the risk

$$\frac{p^2}{q} \sum_{x=0}^{\infty} \left(\frac{x}{T + 1} - \frac{q}{p}\right)^2 \frac{\Gamma(T + x)}{\Gamma(T)\Gamma(x + 1)} p^T q^x = \frac{T + q}{(T + 1)^2}.$$

The supremum of this expression for $0 \leqq \omega < \infty$ is $1/(T + 1)$ by (33); that is, equal to the limit of (36) as $n \to \infty$. Hence (10) holds and the remark of Section 2 may be applied to give the following.

For the Negative Binomial process (32) *with* $0 < \omega < \infty$ *and weight function* (34) *the fixed time estimate* (37) *with* $T = T_0$ *given by*

$$c(T_0) + \frac{1}{T_0 + 1} = \min_{T \geq 0} \left[c(T) + \frac{1}{T + 1} \right] \tag{38}$$

is a minimax estimate.

It may be worth while to remark that (37) is a biased estimate. Indeed the expected value of it for given ω is $T\omega/(T + 1)$. The unbiased estimate x/T gives a constant risk $1/T$.

5. The Normal (Wiener) process. This is defined for every real μ and positive σ as a process with stationary independent increments such that $X_{\mu,\sigma}(0) = 0$ and

$$P\{X_{\mu,\sigma}(t) < x\} = \frac{1}{\sqrt{2\pi t}\sigma} \int_{-\infty}^{x} e^{-(x-\mu t)^2/2\sigma^2 t}\, dx$$

for every real x and $t > 0$.

The parameter σ will be assumed known and the space Ω will consist of the real line $-\infty < \mu < \infty$, the Borel sets being the ordinary ones.[2]

As weight function we take any function of the form

$$W(\mu, \delta) = w(\,|\, \mu - \delta \,|\,) \tag{39}$$

with $w(x)$ nonnegative and nondecreasing for $x \geqq 0$.

In the present case it is not necessary to perform computations similar to those of the preceding sections, since it is easily seen that the arguments of Wolfowitz [4], where a discrete time parameter was considered, carry over to the present case.

The fixed time estimator

$$\hat{\delta}^T = \frac{x(T)}{T} \tag{40}$$

gives for $T > 0$ and any real μ the T-risk

$$r_\mu(\hat{\delta}^T) = \frac{\sqrt{2T}}{\sqrt{\pi}\sigma} \int_0^\infty w(x) e^{-x^2 T/2\sigma^2}\, dx. \tag{41}$$

Moreover it is easily seen that $r_\mu(\hat{\delta}^T)$ is, as a function of T, for $T > 0$, nonincreasing, continuous from the right and that

$$\lim_{T \downarrow 0} r_\mu(\hat{\delta}^T) = \lim_{x \to \infty} w(x).$$

Disregarding the trivial cases 1) when

$$\int_0^\infty w(x) e^{-x^2 h}\, dx \tag{42}$$

is divergent for all h, that is, when the risk is always infinite, and 2) $w(x) \equiv 0$ when the value of the estimator is of no consequence, we have:

For the Normal process with fixed variance σ^2, *unknown mean* μ *and weight function* (39) *with* $w(x)$, $x \geqq 0$, *not identically zero, nondecreasing, and such that* (42) *converges for at least one value of h, the fixed time estimate* (40) *with* $T = T_0$ *given by*

$$(43)\quad c(T_0) + \frac{\sqrt{2T_0}}{\sqrt{\pi}\sigma}\int_0^\infty w(x)e^{-x^2T_0/2\sigma^2}\,dx = \min_{T\geqq 0}\left[c(T) + \frac{\sqrt{2T}}{\sqrt{\pi}\sigma}\int_0^\infty w(x)e^{-x^2T/2\sigma^2}\,dx\right]$$

is minimax, where the term following $c(T)$ *and* $c(T_0)$ *in* (43) *is replaced by* $\sup_x w(x)$ *for* $T = 0$ *or* $T_0 = 0$, *and where* (40) *may be replaced by any estimator if the minimum of* (43) *is at* $T = 0$ *(which can only occur if w is bounded).*

For further remarks on the normal process see the next section, especially 6.4 and 6.5.

6. Generalizations and other remarks.

6.1. It is by no means impossible to have practically continuous observation of a stochastic process. However, our results apply without any modification to stochastic processes with a discrete time parameter or, more generally, to the case when the observations can be made only at times belonging to some set given in advance (there is, of course, no loss of generality in assuming the time-parameter continuous). Indeed, if I is any closed subset of the reals and an observation at time T may be made only if $T \varepsilon I$, all our results remain valid provided T_0 and T in (11) [respectively (18), (28), (31), (38) and (43)] are restricted by the condition that they belong to I. (The same end could be achieved by having $c(t) = \infty$ for $t \not\varepsilon I$ and dropping or suitably modifying the assumption that $c(t)$ is continuous.)

For the special case $I = \{0, 1, 2, \cdots\}$ we have the usual discrete time case; if, furthermore, $c(t)$ is a linear function of t we have the classical sequential case. In this classical case, Hodges and Lehmann [3] obtained by a different method the fact that the fixed-sample estimator of Section 2 on the Poisson process as well as the first estimator of Section 3 on the Gamma process (for the weight function (20) with $\gamma = 1$), both with $T = n$, minimize $\sup_\omega E_\omega W(\omega, \delta)$ subject to $\sup_\omega E_\omega N \leqq n$, where N is the number of observations required (a chance variable) and n is a given positive integer. These results are implied by, but do not imply, ours (see [5], Lemma 5); as remarked in [3], their method does not seem applicable to our problem. The Gamma process with the weight function (29) was considered nonsequentially by Girshick and Savage [2] who, using a different method, established that (30) is a minimax estimator for the fixed sample size problem. As far as we know the Negative Binomial process has never been treated before even nonsequentially.

6.2. The impossibility, in practice, of observing a process for a continuous range of time may be taken care of in the following manner. We replace $c(t)$ by $c(n; t_1, t_2, \cdots, t_n)$ which represents the cost of taking n observations at times $t_1 < t_2 < \cdots < t_n$. The function $c(n; t_1, \cdots, t_n)$ is assumed to satisfy appropriate conditions such as being nonnegative and satisfying $c(n+1; t_1, \cdots, t_{n+1}) \geqq c(n; t_1, \cdots, t_{i-1}, t_{i+1}, \cdots, t_n)$ for $n = 0, 1, \cdots$ and $i = 1, 2, \cdots, n+1$. Our results easily carry over to this case. Thus, for example, for the Poisson process with the weight function considered in Section 2 a minimax estimation procedure is to take a single observation at time $T = T_0$ for which $c(1; T) + 1/T$ becomes a minimum, and to estimate λ by $x(T_0)/T_0$.

This modification of the problem may be combined with that of Section 6.1 by considering only times belonging to a given set I.

6.3. Another modification of the sequential estimation problem is the following: The statistician is required to estimate ω continuously by a function $\delta(t)$ which is a functional of the observed process up to time t, and the loss function is $\int_0^\infty W(\delta(t), \omega)dG(t)$ where $G(t)$ is a monotone nondecreasing function. Our methods apply also to this modified problem. For example in the case of the Poisson process with the weight function used in Section 2 a minimax procedure is obtained by taking $\delta(t) = x(t)/t$.

This formulation may be combined with that of Section 6.2 by having a cost function $c(n; t_1, \cdots, t_n; v, \tau_1, \cdots, \tau_v)$ which expresses the cost of observing the process at times $t_1 < \cdots < t_n$ and changing the estimator $\delta(t)$ at times $\tau_1 < \cdots < \tau_v$. Here again if for every T we can find a sequence of probability distributions with T-Bayes solutions for which the a posteriori risk is independent of the sample and an estimator $\hat{\delta}_T$ satisfying (10) we deduce that a minimax procedure is obtained as follows: choose $n, t_1, \cdots, t_n, v, \tau_1, \cdots, \tau_v$ so as to minimize

$$c(n; t_1, \cdots, t_n; v, \tau_1, \cdots, \tau_n) + \int_0^\infty r(\bar{t})\, dG(t)$$

where $\bar{t} = \max_{t_i \leqq \tau_j} t_i$ for $\tau_j \leqq t < \tau_{j+1}$ (with $\tau_{v+1} = \infty$), and estimate by $\delta(t) = \hat{\delta}^{\bar{t}}$. It is easily seen that if c reduces to a function of n and v only which is monotone in both arguments, then one can choose the τ_v from among the t_i.

This modification may also be considered together with that of Section 6.1. We may further combine it with a weight function which is dependent on the time, etc.

6.4. Throughout the paper we dealt with the problem of point estimation, but it is possible to treat similarly the problem of sequential interval estimation (including that of one-sided estimation). In particular, for the case of the Normal process the results of Wolfowitz [4] carry over to the case of a continuous time parameter.

6.5. We would like now to make some remarks about a class of estimation problems best exemplified by the problem of estimating the variance of a Normal Process.

Let T_1 and $T_2 > T_1$ be any two nonnegative numbers and put $t_{m,n} = T_1 + (n/2^m)(T_2 - T_1)$ for $m = 1, 2, \cdots; n = 0, 1, \cdots, 2^m$. It is well known (see [1]) that if $x(t)$ is a sample function of the Normal process then

$$\lim_{m=\infty} \sum_{n=1}^{2^m} [x(t_{m,n}) - x(t_{m,n-1})]^2 = \sigma^2(T_2 - T_1)$$

with probability 1. Therefore if one could observe a Normal process *without error* for an arbitrarily short period of time one would know the correct value of σ with probability 1. To make the problem practical it is necessary to modify the problem somewhat, for example, in the manner suggested in Section 6.2. We observe that if $X(t)$ is the Normal process with mean μ and variance σ^2 then, for every positive T and t, the random variable $(1/2t)[X(T + t) - X(T) - \mu t]^2$ has the Gamma distribution given by (19) with $r = \frac{1}{2}$, $\theta = \sigma^2$ and $t = 1$. Therefore, if the mean is known and the problem is to estimate the variance we could apply the results of Section 3 (with the modifications suggested in Section 6.2). If both the mean and the variance are unknown again only a slight change, corresponding to the loss of one degree of freedom in the chi square distribution, is necessary.

The situation encountered in this last subsection does not occur if the process is observed continuously but not exactly. This may be done in various ways, a suggestive one for estimating the variance being the following. The process is observed continuously but only deviations exceeding a prescribed size Δ are recorded; that is, we are given a sequence of real numbers $0 = t_0 < t_1 < t_2 < \cdots$ having the property that $| x(t_{n+1}) - x(t_n) | \geqq \Delta$ while $| x(t) - x(t_n) | < \Delta$ for $t_n < t < t_{n+1}$, $(n = 0, 1, 2, \cdots)$.

6.6. One may also consider for continuous (in time) processes such problems as those of sequential unbiased estimation (see [6]) and of unbiased estimation at the conclusion of sequential hypothesis testing (see [7], [8]). For example, for the first of these, one can prove an analogue of the extension of the Cramér-Rao inequality proved in [6], where for our setup En is replaced by ET and $f(x, \theta)$ is replaced by the probability function or density function of $x(1)$ in equation (4.5) of [6]. Under regularity conditions analogous to those of [6], and which are satisfied for the four processes considered herein, the proof (also valid for biased estimators) may be carried out by dividing the time axis into intervals of equal length and allowing the length to approach zero. In particular, the fixed duration estimator of duration T and with estimator $x(T)/T$ is an unbiased estimator of λ, $r\theta$, $(1/p) - 1$, and μ (for the cases considered in Sections 2, 3, 4, and 5, respectively) for which equality holds in this extended Cramér-Rao inequality. This inequality could also be used to apply the technique of [3] to our problems. The two problems described above will both be considered in detail in a future paper.

Finally, we remark that many of Wald's general results on decision functions (complete class theorems, etc.) carry over to the present case of continuous time processes under suitable assumptions. As in [1], the main difficulties in the general theory are ones of measurability, and we shall not bother with them here. We shall return to these problems in a future publication.

REFERENCES

[1] A. Dvoretzky, J. Kiefer, and J. Wolfowitz, "Sequential decision problems for processes with continuous time parameter. Testing hypotheses," *Ann. Math. Stat.*, Vol. 24 (1953), pp. 254–264.

[2] M. A. Girshick and L. J. Savage, "Bayes and minimax estimates for quadratic loss functions," *Proceedings of the Second Berkeley Symposium on Mathematical Statistics and Probability*, University of California Press, 1951, pp. 53–74.

[3] J. L. Hodges, Jr., and E. L. Lehmann, "Some applications of the Cramér-Rao inequality," *Proceedings of the Second Berkeley Symposium on Mathematical Statistics and Probability*, University of California Press, 1951, pp. 13–22.

[4] J. Wolfowitz, "Minimax estimates of the mean of a normal distribution with known variance," *Ann. Math. Stat.*, Vol. 21 (1950), pp. 218–230.

[5] C. R. Blyth, "On minimax statistical decision procedures and their admissibility," *Ann. Math. Stat.*, Vol. 22 (1951), pp. 22–42.

[6] J. Wolfowitz, "The efficiency of sequential estimates and Wald's equation for sequential processes," *Ann. Math. Stat.*, Vol. 18 (1947), pp. 215–230.

[7] M. A. Girshick, F. Mosteller, and L. J. Savage, "Unbiased estimates for certain binomial sampling problems," *Ann. Math. Stat.*, Vol. 17 (1946), pp. 13–23.

[8] D. Blackwell, "Conditional expectation and unbiased sequential estimation," *Ann. Math. Stat.*, Vol. 18 (1947), pp. 105–110.

ON THE OPTIMAL CHARACTER OF THE (*s*, *S*) POLICY IN INVENTORY THEORY[1]

By A. Dvoretzky, J. Kiefer and J. Wolfowitz

1. Introduction

It is the practice among many economists and practical people to assume that the optimal method of controlling inventories (from the point of view of minimizing expected loss or, equivalently, maximizing expected profit) is to specify two appropriate numbers, s and S say, with $0 \leqslant s \leqslant S$, to order goods when and only when the stock at hand, say x, is smaller than s, and then to order the quantity $S - x$ so as to bring the stock up to S. We shall henceforth refer to such a policy as an (s, S) policy. The authors showed in a previous paper ([**2**], pp. 194–196) that the optimal inventory policy need not be an (s, S) policy even in very simple cases. In the present paper we study in detail the problem of when an (s, S) policy is optimal for the simplest and probably the most important case of the inventory problem (see also [**1**]).

An inventory problem will be said to satisfy the (s, S) *assumption* if there exists an (s, S) policy which is an optimal ordering policy for this problem.[2] Our purpose is to obtain necessary and sufficient conditions for the validity of the (s, S) assumption for some classes of inventory problems. These conditions will be stated in terms of the distribution function of demand.

2. Statement of the Problems

As stated in Section 1 we confine our study to a simple but important case. We assume that the expected loss $V(x, y)$ when x is the initial stock and y is the starting stock[3] ($0 \leqslant x \leqslant y < \infty$) is given by

$$V(x, y) = cy + A[1 - F(y)] + \begin{cases} 0 \text{ if } y = x \\ K \text{ if } y > x. \end{cases} \tag{2.1}$$

[1] This work was sponsored by the Office of Naval Research under a contract with Columbia University and under a National Bureau of Standards contract with the University of California at Los Angeles.

[2] There may also be equally good ordering policies that are not (s, S). There are, of course, always such policies that differ from the optimal policy on sets of measure zero. However, when the (s, S) assumption is satisfied but the optimal (s, S) policy is not unique, there will be non-(s, S) optimal policies of another kind (falling, so to speak, "between" the different (s, S) policies). See the relevant remarks in Sections 3 and 4.

[3] Following the terminology of [**2**] we refer to the stock before ordering as the "initial" stock, the starting stock being obtained from the initial one by adding to it the amount ordered.

Here c, A, K are positive constants which denote, respectively, the carrying cost per unit of goods, the loss involved if it is impossible to satisfy the demand for goods, and the ordering cost. $F(z)$ is the distribution function of demand; it is the probability that the random demand D does not exceed z, i.e.,

$$F(z) = P\{D \leqslant z\}. \tag{2.2}$$

For a given initial stock x it is obviously best to order an amount $Y - x$ with $Y = Y(x)$ satisfying

$$V(x, Y) = \min_{y \geqslant x} V(x, y). \tag{2.3}$$

Since, by (2.2), $F(z)$ is nondecreasing and continuous to the right, it follows from the definition (2.1) that $V(x, y)$ is also continuous to the right and that the minimum in (2.3) always exists.

Our purpose is to characterize the distribution functions F of demand for which the above inventory problem satisfies the (s, S) assumption. There naturally arise several problems of this kind; e.g., we may require that the (s, S) assumption be satisfied for all choices of c, A and K, or we may fix c and A and require that the (s, S) assumption hold for all K, etc. Since the problem is not changed if all three parameters c, A, K are multiplied by the same positive constant, the fixing of only one parameter amounts to no restriction at all. We are thus led to the problem of characterizing the following five classes of distribution functions F:

(a) The class $C(c_0, A_0, K_0)$ of all F for which the (s, S) assumption holds when $c = c_0$, $A = A_0$, $K = K_0$.

(b) The class $C(c_0, A_0, —) = \bigcap_K C(c_0, A_0, K)$ of all F for which the (s, S) assumption holds for given $c = c_0$, $A = A_0$ and all choices of the ordering cost K.

(c) The class $C(—, A_0, K_0) = \bigcap_c C(c, A_0, K_0)$ of all F for which the (s, S) assumption holds for given $A = A_0$, $K = K_0$ and all choices of the carrying cost c.

(d) The class $C(c_0, —, K_0) = \bigcap_A C(c_0, A, K_0)$ of all F for which the (s, S) assumption holds for given $c = c_0$, $K = K_0$ and all choices of the penalty A for not satisfying demand.

(e) The class $C = \bigcap_{c,A,K} C(c, A, K) = \bigcap_{c,A} C(c, A, —) = \bigcap_{A,K} C(—, A, K) = \bigcap_{c,K} C(c, —, K)$ of all F for which the (s, S) assumption holds for all choices of c, A and K.

It is quite easy to answer problems (a) and (b). A worthwhile solution of the other problems is less obvious. The next five sections contain the solution of the five problems mentioned above. In Section 8 the solutions to problems (d) and (e) are reformulated in a form that makes their graphical verification extremely easy. Finally Section 9 contains some remarks on the extension of the above results. In particular, the important case when D can assume only integral values is considered.

3. PROBLEM (a) AND PRELIMINARY CONSIDERATIONS

Consider the function

$$g(x) = V(x, x) = c_0 x + A_0[1 - F(x)], \qquad (x \geqslant 0). \tag{3.1}$$

Let y_0 be a nonnegative number which satisfies

$$g(y_0) = \min_{x \geqslant 0} g(x). \tag{3.2}$$

(As was remarked after (2.3) such a number always exists.) Let x_0, $(0 \leqslant x_0 \leqslant y_0)$ be defined through[4]

$$x_0 = \max_{\substack{x \in A \\ 0 \leqslant x \leqslant y_0}} x \tag{3.3}$$

where

$$A = \{x \mid x \leqslant x' \leqslant y_0 \text{ implies } g(x') \leqslant g(y_0) + K_0\}.$$

These definitions become clear from Figure 1. If we name "graph" of $g(x)$ the continuous curve obtained from the set (x, z) with $z = g(x)$ by adjoining to it, at all points of discontinuity, the vertical segment $g(x) < z \leqslant g(x - 0)$, then x_0 is the abscissa of the first point to the left of y_0 where the horizontal line $z = g(y_0) + K_0$ cuts the graph of $g(x)$.

It is clear that (s, S) with $s = x_0$, $S = y_0$ is an optimal ordering policy if and only if the following conditions hold:

(i) *For every* $x_1 > y_0$ *we have* $\min_{x \geqslant x_1} g(x) \geqslant g(x_1) - K_0$;

(ii) $g(x) \geqslant g(y_0) + K_0$ *for* $0 \leqslant x < x_0$.

Condition (i) implies that one cannot do better by ordering when the initial stock x is greater than y_0 (hence also not when $x \geqslant x_0$, since by definition of x_0 one cannot do better by ordering when $x_0 \leqslant x \leqslant y_0$). Condition (ii) implies that one cannot do better than order $y_0 - x$ when the initial stock is smaller than x_0. Conversely, if it pays to order when the initial stock is some $x_1 > y_0$, then Condition (i) cannot hold, and a similar remark applies to Condition (ii).

[4] $\{x \mid \cdots\}$ indicates the set of x having the properties $\cdots$. Usually only one x has the particular properties indicated, but there may also be an interval in which case the right end point of this interval is taken as x_0.

Remark: y_0 need not be uniquely determined by (3.2). Let y_0' be another value which satisfies (3.2) and define x_0' through (3.3), in it replacing y_0 by y_0'. Then Conditions (i) and (ii) hold if and only if the analogous conditions obtained on replacing x_0, y_0 by x_0', y_0' also hold. Hence, if Conditions (i) and (ii) are satisfied for some choice of y_0, the optimal (s, S) policy is determined up to the following: As S we may take any solution y_0 of (3.2), and then $s = x_0$ is uniquely determined through (3.3), except when there exists an interval $x_1 \leqslant x < x_0$ such that $g(x) = g(y_0) + K$ throughout this interval. If this is the case then we may take $s = x$ where x is any number satisfying $x_1 \leqslant x \leqslant x_0$.

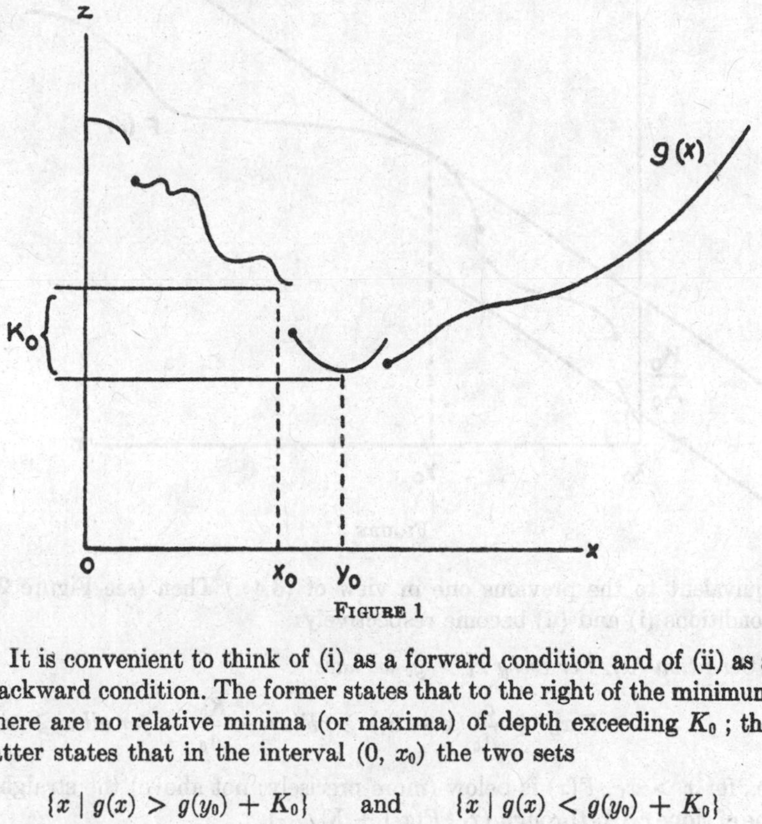

FIGURE 1

It is convenient to think of (i) as a forward condition and of (ii) as a backward condition. The former states that to the right of the minimum there are no relative minima (or maxima) of depth exceeding K_0; the latter states that in the interval $(0, x_0)$ the two sets

$$\{x \mid g(x) > g(y_0) + K_0\} \quad \text{and} \quad \{x \mid g(x) < g(y_0) + K_0\}$$

are separated, the first lying to the left of the second.

It is useful for the sequel to restate Conditions (i) and (ii) in a different form. To this end we note that

$$1 - \frac{g(x)}{A_0} = F(x) - \frac{c_0}{A_0}\,x. \tag{3.4}$$

Consider now the points (x, z) with $z = F(x)$, $(x \geqslant 0)$ and a straight line of slope c_0/A_0 completely above this set of points. As the line is moved downward parallel to itself it will pass through a position where it has, for the first time, a point of contact with the set (x, z).[5] Let $[y_0, F(y_0)]$ be any such point of contact. (This way of defining y_0 is

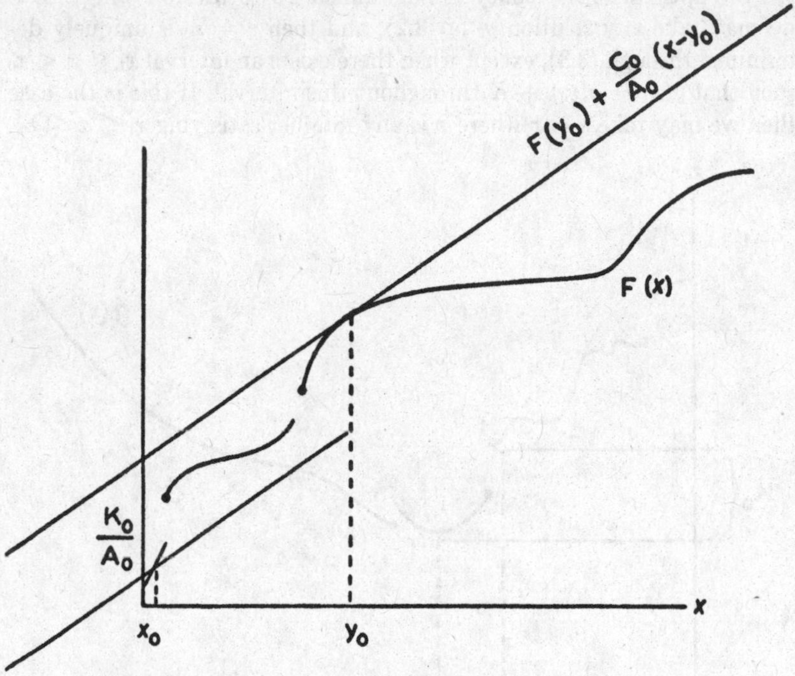

FIGURE 2

equivalent to the previous one in view of (3.4).) Then (see Figure 2) Conditions (i) and (ii) become respectively:

Condition (α). *For every $x_1 > y_0$ we have*

$$F(x_1) + \frac{c_0}{A_0}(x - x_1) \geqslant F(x) - \frac{K_0}{A_0} \qquad \text{for all } x \geqslant x_1;$$

i.e., for $x > x_1$, $F(x)$ is below (more precisely: not above) the straight line of slope c_0/A_0 through $(x_1, F(x_1) + K_0/A_0)$.

Condition (β). *There exists a number x_0, $0 \leqslant x_0 \leqslant y_0$, such that*

$$F(x) \begin{Bmatrix} \leqslant \\ \geqslant \end{Bmatrix} F(y_0) + \frac{c_0}{A_0}(x - y_0) - \frac{K_0}{A_0} \text{ for } \begin{cases} 0 \leqslant x < x_0 \\ x_0 \leqslant x \leqslant y_0 \end{cases};$$

[5] This follows from the fact that F is bounded, nondecreasing and continuous to the right.

i.e. the graph of $F(x)$ to the left of y_0 cannot be cut more than once by the straight line of slope c_0/A_0 through y_0, $(F(y_0) - K_0/A_0)$.

4. PROBLEM (b)

From Conditions (i) and (ii) it is easy to obtain a characterization of $C(c_0, A_0, —)$. All that is required is that (i) and (ii) hold for all $K_0 > 0$.

Condition (i) for all $K_0 > 0$ is obviously equivalent to: $g(x)$ is monotone nondecreasing for $x > y_0$. Similarly, Condition (ii) becomes: $g(x)$ is monotone nonincreasing for $x < y_0$.

Summing up we have: *F belongs to the class $C(c_0, A_0, —)$ if and only if $g(x)$ is U-shaped.*[6]

It follows in particular that $F(x)$ is continuous for $x \geqslant y_0$. It is also obvious that whereas S in the optimal (s, S) policy may be taken independently of K_0, this is not the case for s (except when $g(x)$ degenerates to a monotonic function in which event we may always take $s = S = 0$).

We also rephrase Conditions (α) and (β) for this case. They become respectively:

Condition (α'). For every $x_1 > y_0$ we have

$$F(x_1) + \frac{c_0}{A_0}(x - x_1) \geqslant F(x) \qquad \textit{for all } x > x_1;$$

i.e., for $x > x_1$, $F(x)$ cannot be above the straight line of slope c_0/A_0 through $[x_1, F(x_1)]$.

Condition (β'). Every straight line of slope c_0/A_0 can cut the graph of $F(x)$, $0 \leqslant x \leqslant y_0$ at most once.[7]

It should, perhaps, be remarked that Condition (α') implies that the probability density[8] of D cannot exceed c_0/A_0 for $x \geqslant y_0$.

5. PROBLEM (c)

In this section we characterize, in an easily verifiable manner, the class $C(—, A_0, K_0)$. For this purpose, as well as for the following sections, it is necessary to introduce some notions and notations.

Given any distribution function $F(y)$ $(0 \leqslant y < \infty)$, we denote by

[6] A function $g(x)$ is U-shaped if it is monotonically nonincreasing up to some value of x, and thereafter is monotonically nondecreasing ($g(x)$ may degenerate to a monotonically nondecreasing function).

[7] The possibility of there being more than one point of the graph on the straight line is not excluded, but in such a case all the points with intermediary abscissae must also be on the same straight line.

[8] Or the upper right and upper left probability densities wherever the probability density does not exist.

$F^*(y)$, $0 \leqslant y < \infty$, the *minimal concave* function $\geqslant F(y)$. $F^*(y)$ is the function defined for $y \geqslant 0$ through

$$(5.1) \quad F^*(y) = \sup_{0 \leqslant y' \leqslant y < y''} \left[\frac{y'' - y}{y'' - y'} F(y') + \frac{y - y'}{y'' - y'} F(y'') \right].$$

It is obvious from (5.1) that $F^*(y)$ is concave and monotonically nondecreasing. For all $y \geqslant 0$ we have $F^*(y) \geqslant F(y)$ and there are arbitrarily large y for which $F^*(y) = F(y)$. Furthermore, if for some value of y we have $F^*(y) > F(y)$, then $F^*(y)$ is linear in some neighborhood of y.

A *line of support* of a concave function $G(y)$, $0 \leqslant y < \infty$, at y_1 is a straight line $z = l(y)$ satisfying $l(y) \geqslant G(y)$ for $y \geqslant 0$ with equality holding for $y = y_1$.

We can now state the following result:

THEOREM 1. *F belongs to the class $C(-, A_0, K_0)$ if and only if, for every y_1, $(0 \leqslant y_1 < \infty)$ and every line of support $l(y)$ of $F^*(y)$ at y_1, the set of points $T^- = \{y \mid F(y) < l(y) - K_0/A_0, 0 \leqslant y < y_1\}$ is to the left of the set $T^+ = \{y \mid F(y) > l(y) - K_0/A_0, 0 \leqslant y < y_1\}$.*

PROOF. *Necessity.* Assume the condition is violated for y_1. Let $l(y)$ be a line of support of $F^*(y)$ at y_1. By the remarks following (5.1) there exists a number y_0, $y_0 \geqslant y_1$, such that $F(y_0) = F^*(y_0) = l(y_0)$. (If $F(y_1) = F^*(y_1)$ we take $y_0 = y_1$, otherwise $F^*(y) = l(y)$ for y sufficiently near to y_1 and there exists a largest value y for which the last equality holds; we take y_0 as this largest y.) If we define c_0 by $c_0/A_0 =$ slope of $l(y)$, then the above y_0 has the properties which define y_0 in Section 3 (see discussion preceding Condition (α) in Section 3).

If the condition of the theorem is violated for y_1 and $l(y)$, it is of course also violated for y_0 and $l(y)$. But this then means that Condition (β) of Section 3 is not satisfied for $c = c_0$ defined above, and thus no (s, S) policy is optimal for this choice of c.

Sufficiency. The proof of necessity shows that the condition of Theorem 1 is equivalent to the statement that Condition (β) holds for the given values A_0 and K_0 and every choice of c_0. To complete the proof it remains to show that Condition α is also satisfied.

First we prove that the condition of Theorem 1 implies

$$(5.2) \qquad F(x) \geqslant F^*(x) - K_0/A_0 \qquad (x \geqslant 0).$$

Indeed, let (5.2) be false for $x = x_1$. Then $F^*(x)$ is linear in the neighborhood of x_1. Let (y', y_1) be the largest segment containing x_1 in which $F^*(x)$ is linear, and denote by $l(y)$ the straight line which coincides with $F^*(x)$ on this segment. Then $l(y)$ is a line of support of $F^*(x)$ at $x = y_1$, and in the notation of Theorem 1 we have $y' \in T^+$, $x_1 \in T^-$.

Since $0 \leqslant y' < x_1 < y_1$ this contradicts the condition of the theorem and (5.2) is thus established.

To complete the proof it remains to show that (5.2) implies Condition (α) of Section 3 for every choice of c_0. As remarked above, the straight line of slope c_0/A_0 through $[y_0, F(y_0)]$ is a line of support of $F^*(y)$ at y_0. Let $x_1 > y_0$ and denote by $l_1(y)$ the line of support of $F^*(y)$ at $y = x_1$. Obviously $F(x) \leqslant l_1(x)$ for all x, hence by (5.2) $F(x) \leqslant F(x_1) + K_0/A_0 + (c_1/A_0)(x - x_1)$ where c_1/A_0 is the slope of $l_1(x)$. But F^* is concave and $x_1 > y_0$, hence $c_1 \leqslant c_0$ and therefore we have for $x > x_1$

$$F(x) \leqslant F(x_1) + \frac{K_0}{A_0} + \frac{c_0}{A_0}(x - x_1).$$

This is precisely Condition (α).

Remark: It is easy to see that (5.2) is equivalent to Condition (α), but this is not required for our proof. It is very easy to verify whether (5.2) holds; since (5.2) is a necessary condition for $F \in C(—, A_0, K_0)$, it may be worthwhile to check whether it is fulfilled before trying to see if the condition of Theorem 1 is satisfied.

6. PROBLEM (d)

In this section we characterize the class $C(c_0, — K_0)$.

THEOREM 2. *F belongs to the class $C(c_0, - K_0)$ if and only if for every $y_1(0 \leqslant y_1 < \infty)$ and every line of support $l(y)$ of $F^*(y)$ at y_1, the set of points $\{y \mid F(y) < l(y - K_0/c_0), 0 \leqslant y < y_1\}$ is to the left of the set $\{y \mid F(y) > l(y - K_0/c_0), 0 \leqslant y < y_1\}$.*

PROOF. *Necessity.* Given any $A = A_0 > 0$ let $l(y)$ be a line of support of $F^*(y)$ with slope c_0/A_0 and let y_1, $0 \leqslant y_1 < \infty$, be such that $F^*(y_1) = F(y_1) = l(y_1)$. Then

$$l\left(y - \frac{K_0}{c_0}\right) = F(y_1) + \left(y - \frac{K_0}{c_0} - y_1\right)\frac{c_0}{A_0} = F(y_1) + \frac{c_0}{A_0}(y - y_1) - \frac{K_0}{A_0}.$$

Thus the condition of Theorem 2 is equivalent to Condition (β) holding for all A_0 and is therefore necessary.

Sufficiency. We need only to show that Condition (α) holds for all A_0.

It is easily seen from the derivation of (5.2) in the preceding section, that the condition of the theorem implies

$$F(y) \geqslant F^*\left(y - \frac{K_0}{c_0}\right) \quad \text{for} \quad y \geqslant \frac{K_0}{c_0}. \tag{6.1}$$

From (6.1), the monotonicity and concavity of F^*, and the fact that $F(y_0) = F^*(y_0)$, we have, for $x_1 \geqslant y_0$,

$$F^*(x_1) - F(x_1) \leqslant F^*(x_1) - \max\,[F^*(y_0), F^*(x_1 - K_0/c_0)]$$

$$= F^*(x_1) - F^*[\max(y_0, x_1 - K_0/c_0)]$$
$$\leqslant F^*(y_0 + K_0/c_0) - F^*(y_0) \leqslant (K_0/c_0)\cdot(c_0/A_0) = K_0/A_0 .$$

Hence, again using the concavity of F^*, we have, for $x \geqslant x_1 \geqslant y_0$,

$$F(x) - F(x_1) \leqslant F^*(x) - F^*(x_1) + F^*(x_1) - F(x_1)$$
$$\leqslant (c_0/A_0)(x - x_1) + K_0/A_0 .$$

Thus Condition (α) is satisfied and the proof is complete.

7. PROBLEM (e)

From either of Theorems 1 and 2 we deduce immediately the following

THEOREM 3. *F belongs to the class C if and only if it is concave for* $y \geqslant 0$.

Thus $F \in C$ if, and only if, left and right probability densities exist for $y > 0$ and each of these is monotonically nonincreasing (the two densities are, of course, equal except at an at most countable set of values of y).

The condition of Theorem 3 is never satisfied when D is a discrete random variable, e.g., in the important case when D is always an integral multiple of some unit. In this case it is reasonable, however, also to restrict the stock and the amounts ordered and not to allow them to assume arbitrary positive values. For a further discussion we refer to Section 9 below.

8. GEOMETRIC FORMULATIONS

A straight line $l(y)$ *crosses* $F(y)$ *at most once to* the left of $y_1(>0)$[9] if the interval $(0, y_1)$ can be divided into two intervals[10] such that $F(y) \leqslant l(y)$ in one of them and $F(y) \geqslant l(y)$ in the other. Using this terminology it is very easy to give a graphical interpretation of Theorems 1 and 2.

Theorem 1 becomes:

$F \in C(—, A_0, K_0)$ *if and only if every support line of* $F^*(y) - K_0/A_0$ *crosses* $F(y)$ *at most once to the left of the point of support.*

Similarly Theorem 2 may be reformulated:

$F \in C(c_0, —, K_0)$ *if and only if every support line of* $F^*(y - K_0/c_0)$ *crosses* $F(y)$ *at most once to the left of the point of support.*

The reformulation of Theorem 2 needs some justification. Let $l(y)$ be a line of support of $F^*(y - K_0/c_0)$ at y_1. Then the condition of Theorem 2 is equivalent to the statement that $l(y)$ crosses $F(y)$ at most once to the left of $y_1 - K_0/c_0$. We must prove that no trouble can be caused

[9] Or to the left of the point $[y_1, l(y_1)]$.

[10] One of the two may be degenerate.

by adjoining the interval $(y_1 - K_0/c_0, y_1)$. It is obviously sufficient to show this for y_1 with $F^*(y - K_0/c_0) < l(y)$ for $y > y_1$. But then $F^*(y)$ is not linear throughout any neighborhood of $y_1 - K_0/c_0$. Hence $F(y_1 - K_0/c_0) = F^*(y_1 - K_0/c_0) = l(y_1)$, and therefore $F(y) \geqslant l(y)$ for $y_1 - K_0/c_0 \leqslant y \leqslant y_1$ as required.

It is very easy to check the above conditions graphically by means of a simple linkage mechanism such as, for example, is shown in the following figure:

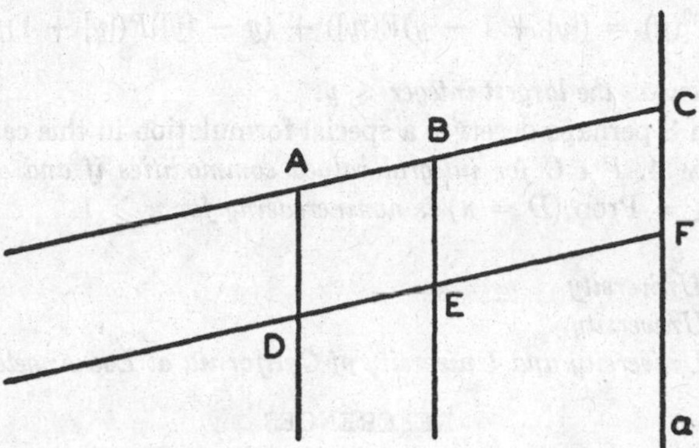

FIGURE 3

Here $\overline{AB} = \overline{DE}$, $\overline{BC} = \overline{EF}$; the lines ABC and DEF pivot around A, B, D, E, and C, F slide freely on the bar a. To check the condition of Theorem 1, for example, one adjusts D and E so that $\overline{AD} = \overline{BE} = K_0/A_0$ and sets the line a parallel to the ordinate axis. Then one moves the line ABC through the positions where it is a line of support of $F^*(y)$ and observes where the parallel line DEF cuts $F(y)$. (It is not necessary for this purpose to draw $F^*(y)$.)

9. THE DISCRETE CASE

The method of Sections 3–6 can be applied to characterize many other classes of demand functions for which the (s, S) assumption holds. Usually, however, the conditions obtained are clumsy and not easily verifiable.

There is, however, one remark that should be made. Let us restrict ourselves to the most important case when the commodity considered comes in integral multiples of some unit. We may then assume that the demand D can take only integral values, and similarly the stock at hand and the amount ordered are integers. In this case all our conditions must be satisfied only for integral values of the arguments. Thus the condition of Section 4 becomes: $F \in C(c_0, A_0, —)$ *if and only* *if the sequence $g(n)$, $n = 0, 1, 2, \cdots$, is decreasing up to some n_0 and nondecreasing for $n \geqslant n_0$.*

Similarly we must worry about Conditions (α) and (β) only for integral values of the arguments. Thus in Theorems 1–3 it is necessary to consider $F(y)$ only for integral values of y. An equivalent method of doing this is the following:

Theorems 1–3 become valid for integral valued commodities if $F(y)$ is replaced by[11]

$$F^0(y) = ([y] + 1 - y)F([y]) + (y - [y])F([y] + 1),$$

where $[y]$ denotes the largest integer $\leqslant y$.

Theorem 3 perhaps deserves a special formulation in this case:

THEOREM 4. *$F \in C$ for integral valued commodities if and only if the sequence $\beta_n = \text{Prob}\ (D = n)$ is nonincreasing for $n \geqslant 1$.*

Hebrew University
Cornell University
Cornell University and University of California at Los Angeles

REFERENCES

[1] ARROW, K. J., T. HARRIS, AND J. MARSCHAK, "Optimal Inventory Policy," ECONOMETRICA, Vol. 19, July, 1951, pp. 250–272.

[2] DVORETZKY, A., J. KIEFER, AND J. WOLFOWITZ, "The Inventory Problem: I, Case of Known Distributions of Demand," ECONOMETRICA, Vol. 20, April, 1952, pp. 187–222.

[11] This amounts to redefining $F(y)$ for nonintegral y by linear interpolation between the two integers nearest y.

Reprinted from
Econometrica **21** (4), 586–596 (1953)

ON THE THEORY OF QUEUES WITH MANY SERVERS(1)

BY

J. KIEFER AND J. WOLFOWITZ

1. Introduction. The physical original of the mathematical problem to which this paper is devoted is a system of s "servers," who can be machines in a factory, ticket windows at a railroad station, salespeople in a store, or the like. Individuals (clients) who are to be served by these servers arrive at random and the duration of anyone's service (e.g., stay at the ticket window) is a chance variable whose distribution function may be arbitrary. The phrase "at random" used above is not to be interpreted to mean that the interval between successive arrivals is to have an exponential distribution. The assumption of an exponential or other special distribution for either the interval between arrivals or the service time of an individual or both usually makes the problem much easier. We also allow the distribution of the interval between arrivals to be arbitrary. The queue discipline is "first come, first served." The system is described precisely in §2.

In this system the waiting time of the individual who is ith in order of arrival, i.e., the time which elapses between his arrival and the beginning of his service, is a chance variable whose distribution function depends upon i. In §3 we prove that this distribution function approaches a limit as $i \rightarrow \infty$. This limit may not be a distribution function because its variation may be less than one. We assume that the expected value of the time interval between the arrivals of successive clients and the expected value of the service time of an individual both exist. In terms of these one defines a quantity ρ in §6. The situation may then be classified according as $\rho < 1$ or $\rho \geqq 1$. In the former and interesting case the limiting function is a distribution function (§6), in the latter case it is not a distribution function (§7). The limiting function is (a marginal function) obtained from a function which satisfies an integral equation derived in §3. This integral equation is satisfied by a unique distribution function on s-space when $\rho < 1$, and by no distribution function when $\rho \geqq 1$ (§8). These results for the case of one server were obtained by Lindley [1]. The problem when there are many servers offers many difficulties not present when there is only one server. The methods of the present paper are different from those of [1]. The proof of the result of §7, that the limit is not a distribution function when $\rho \geqq 1$, is obtained by reducing the problem to the case $s=1$ by using our lemma of §4, and then employing the corresponding result of [1]; except for this argument our paper is self-contained. For special distributions of the time between successive arrivals

Presented to the Society, October 24, 1953; received by the editors August 13, 1953.

(1) Research under contract with the Office of Naval Research.

1

and of the service time the results of the present paper have been obtained by various authors (we refer the reader to [5] and [6] which contain extensive bibliographies). The methods of these authors make use of their special assumptions in an essential way. The novelty of the results of the present paper lies in the fact that no restrictions are imposed on the distributions, with the exception of the assumption of finite first moment(2). Thus the results of the present paper include the corresponding ones of previous papers as special cases(3).

Mathematically speaking, our study is one of the ergodic character of the waiting time in our system, and the conditions under which the distribution of the latter approaches stability. Our problem can be reduced, and actually is so reduced by us, to studying a random walk in s-space with certain impassable but not absorbing barriers. We actually show that, when $\rho<1$, the distribution function of the particle engaged in the random walk approaches a limiting distribution which is the same no matter what the original starting point of the particle (§8).

Perhaps our principal device is to dominate the stochastic process to be studied by a lattice process to which we then apply available theorems from the theory of Markoff processes with discrete time parameter and denumerably many states. This device makes possible the argument of §6 and is also employed in §8. We are of the opinion that this device could be applied to other ergodic problems connected with random walks.

When the original process is a lattice process, i.e., when the chance variables R_1 and g_1 (defined in §2) take, with probability one, only values which are integral multiples of some positive number c, and when $\rho<1$, the limiting probabilities (which are shown to exist in §6) are reciprocals of certain mean recurrence times (this follows from the application of Theorem 2 of Chapter 15 of [3] to the argument of §6). Monte Carlo methods (see, e.g., [4]) may perhaps then be profitably employed to solve the integral equation (3.8).

It would be very desirable and interesting to solve the integral equation (3.8), at least for interesting or important functions G and H (see §2). This, however, is likely to be very difficult. Even in the simplest case, when $s=1$, the equation becomes the Wiener-Hopf equation, which has been of considerable interest to physicists but has been solved only in special cases. Some special cases of the equation (3.8) are discussed in [5], [6], and [1]. It may also interest the pure analyst that one can, by probabilistic methods

(2) Under stronger assumptions (e.g., existence of all moments), F. Pollaczek in recent notes (C. R. Acad. Sci. Paris vol. 236 (1953) pp. 578–580, 1469–1470) gives formally an integral equation for the Laplace transform of F (to be defined below), but does not consider the questions of the present paper.

(3) In a paper to be published elsewhere which makes extensive use of the present paper, the authors obtain, under minimal conditions, theorems on convergence of the mean of various variables connected with the queueing process.

like ours, prove the existence or non-existence of distribution function solutions of (3.8).

Finally, in §9 we discuss the limiting distribution (as $i\to\infty$) of the queue size, i.e., of the number waiting to be served when the service of the ith customer begins.

We are obliged to Professor J. L. Doob for helpful discussions.

2. Description of the system. The system consists of s ($\geqq 1$) machines, $M_1, \cdots, M_s$. The ith individual arrives at time t_i ($\geqq 0$), with, of course, $t_i \leqq t_{i+1}$. If all machines are in service at his arrival he takes his place in the queue. His service begins as soon as at least one machine is unoccupied, and all individuals with smaller indices have been or are being served. If more than one machine becomes unoccupied at the time when it is the ith individual's turn to be served, we shall assume, for definiteness, that he takes his place at the unoccupied machine with smallest index.

Let $t_0=0$, $g_i=t_i-t_{i-1}$ for all $i\geqq 1$. We assume that the g_i are independently and identically distributed chance variables; let $G(z)=P\{g_1\leqq z\}$, where $P\{\ \}$ is the probability of the relation in braces. Throughout the paper we assume that $G(0)<1$; the case $G(0)=1$ is too trivial to discuss. We assume $Eg_1<\infty$.

Let R_i be the length of time the ith person spends being serviced by a machine. We assume that the R_i are independently and identically distributed chance variables, distributed independently of the g_i; let $H(z)=P\{R_1\leqq z\}$. We assume $ER_1<\infty$. We also assume $H(0)<1$, the case $H(0)=1$ being trivial.

Let $w_{i1}+t_i$ be the time at which service of the ith individual begins; w_{i1} is his waiting time. Then the ith individual leaves his machine at the time $w_{i1}+t_i+R_i$.

Let $u_{ij}+t_i$ be the time at which the jth machine finishes serving the last of those among the first $(i-1)$ individuals which it serves. Let $u'_{ij}=\max(0, u_{ij})$. Let $w_{i1}, \cdots, w_{is}$ be the quantities $u'_{i1}, \cdots, u'_{is}$ arranged in order of increasing size. It is easy to see that this definition of w_{i1} coincides with the former.

Let

$$(2.1)\qquad F_i(x_1, \cdots, x_s) = P\{w_{i1} \leqq x_1, \cdots, w_{is} \leqq x_s\}.$$

If ever $x_j>x_{j+1}$ we may, since $w_{ij}\leqq w_{i(j+1)}$, replace x_j by x_{j+1} in both members of (2.1) without changing the value of either.

Write $w_i=(w_{i1}, \cdots, w_{is})$. The earliest times at which the various machines could attend to the $(i+1)$st individual are $t_i+w_{i1}+R_i$, $t_i+w_{i2}, \cdots$, t_i+w_{is}. If t_{i+1} is greater than or equal to any of these quantities the $(i+1)$st individual finds at least one machine unoccupied at his arrival and does not have to wait at all. If t_{i+1} is less than all these quantities the $(i+1)$st individual has to wait for the first machine to be unoccupied. Since $t_{i+1}=t_i+g_{i+1}$, w_{i+1} is obtained from w_i as follows: Subtract g_{i+1} from *every* component of $(w_{i1}+R_i, w_{i2}, w_{i3}, \cdots, w_{is})$. Rearrange the resulting quantities in ascending order and replace all negative quantities by zero. The ensuing result is w_{i+1}.

3. **Recursion formula for F_i. Existence of the limit of F_i as $i\to\infty$.** Let $\phi_j(a, b, c)$, $j=1, \cdots, s$, be the value of $w_{(i+1),j}$ when $w_i=a$, $R_i=b$, $g_{i+1}=c$. If d is a point in s-space we shall say that $a\leqq d$ if every coordinate of a is not greater than the corresponding coordinate of d. If now $a\leqq d$ then obviously

$$\phi_j(a, b, c) \leqq \phi_j(d, b, c)$$

for $1\leqq j\leqq s$. Applying this argument k times we obtain the following result: Let $R_{i+j-1}=b_{i+j-1}$, $g_{i+j}=c_{i+j}$, $j=1, \cdots, k$. Let $w_{i+k}=e_1$ when $w_i=a_1$, and let $w_{i+k}=e_2$ when $w_i=a_2$. Then $a_1\leqq a_2$ implies $e_1\leqq e_2$.

Let S be the totality of points $(x_1, x_2, \cdots, x_s)$ of Euclidean s-space such that $0\leqq x_1\leqq x_2\leqq \cdots \leqq x_s$. Let x and y be generic points of S. For $i\geqq 1$, let

$$F_i(x \mid y) = P\{w_i \leqq x \mid w_1 = y\}.$$

Let 0 be the origin in s-space. Then

$$F_i(x \mid 0) = F_i(x)$$

and

$$F_{i+1}(x) = \int F_i(x \mid y)dF_2(y). \tag{3.1}$$

The conclusion of the preceding paragraph enables us to conclude that $y_1\in S$, $y_2\in S$, $y_1\leqq y_2$, imply

$$F_i(x \mid y_1) \geqq F_i(x \mid y_2)$$

for every x and every i. Now

$$F_{i+1}(x) - F_i(x) = \int [F_i(x \mid y) - F_i(x \mid 0)]dF_2(y). \tag{3.2}$$

Since the integrand is never positive we have that

$$F_{i+1}(x) \leqq F_i(x) \tag{3.3}$$

for all x and i. From (3.3) it follows that $F_i(x)$ approaches a limit, say $F(x)$, which is nondecreasing in every component of x, continuous to the right, and assigns non-negative measure to all rectangles. It need not, however, be a distribution function, i.e., its variation over S (hence over all of s-space) may be less than one.

Write

$$\phi(a, b, c) = (\phi_1(a, b, c), \cdots, \phi_s(a, b, c)). \tag{3.4}$$

For given $x\in S$, b, c, let $\psi(x, b, c)$ be the totality of points $y\in S$ such that

$\phi(y, b, c) \leqq x$ and $\in S$. Obviously

$$F_{(i+1)}(x) = \int P_i\{\psi(x, b, c)\} dH(b) dG(c) \tag{3.5}$$

where P_i is the measure according to F_i. This equation determines each F_i uniquely by recursion, since of course $F_1(0) = 1$. For $s = 1$ and $x \geqq 0$, $\psi(x, b, c)$ is $\{y \mid 0 \leqq y \leqq x - b + c\}$. Hence (3.5) becomes, for $x \geqq 0$,

$$F_{(i+1)}(x) = \int F_i(x - b + c) dH(b) dG(c), \tag{3.6}$$

an equation due to Lindley [1]. (In (3.6) it is understood that $F_i(x - b + c) = 0$ whenever $x - b + c < 0$.) For $s = 2$ and $x = (x_1, x_2) \in S$ we have that $\psi(x, b, c)$ is the set of points $y \in S$ such that $y \leqq (x_1 - b + c, x_2 + c)$, together with the set of points $y \in S$ such that $y \leqq ([\min (x_2 - b + c, x_1 + c)], x_1 + c)$. We extend the definition of $F_i(y_1, y_2)$ to all of s-space in the natural way as follows: $F_i(y_1, y_2) = 0$ if either y_1 or $y_2 < 0$, $F_i(y_1, y_2) = F_i(y_2, y_2)$ if $y_1 > y_2$. Then for all (x_1, x_2) in S and $s = 2$, (3.5) becomes

$$\begin{aligned} F_{(i+1)}(x_1, x_2) = \int & [F_i(x_1 - b + c, x_2 + c) + F_i(x_2 - b + c, x_1 + c) \\ & - F_i(x_1 - b + c, x_1 + c)] dH(b) dG(c). \end{aligned} \tag{3.7}$$

In general, when (3.5) is written in the form of (3.6) and (3.7) the integrand contains $(2^s - 1)$ terms. With the integrand in this form let $i \to \infty$ in (3.5). By Lebesgue's bounded convergence theorem we obtain for $x \in S$,

$$F(x) = \int P_\infty\{\psi(x, b, c)\} dH(b) dG(c) \tag{3.8}$$

where P_∞ is the measure according to $F(x)$. (When $s = 1$ or 2 equation (3.8) becomes (3.6) and (3.7) with the subscripts of F deleted.) This is an integral equation satisfied by $F(x)$. We shall later prove that, when $\rho < 1$, $F(x)$ is a distribution function (d.f.), and the only d.f. over S which satisfies (3.8). Moreover, we shall prove that, when $\rho \geqq 1$, $F(x)$ is not a d.f., and (3.8) has no solution which is a d.f. over S.

We remark that (3.8) implies that if $F(x)$ is a d.f., the latter defines a stationary absolute probability distribution for our (Markoff) stochastic process, i.e., if w_1 is distributed according to $F(x)$ then w_i has this distribution for every value of i.

Write $\bar{x}_1 = (x_1, \infty, \cdots, \infty)$, $F_i^*(x_1) = F_i(\bar{x}_1)$. Then, from (3.5), the Lebesgue bounded convergence theorem, and the structure of ψ, it follows that

$$F^*_{(i+1)}(x_1) = \int P_i\{\psi(\bar{x}_1, b, c)\} dH(b) dG(c). \tag{3.9}$$

We proved earlier that

$$F_i(x) \geqq F_{(i+1)}(x) \tag{3.10}$$

for every i and x. In (3.10) let the last $(s-1)$ coordinates of x approach infinity. We obtain

$$F_i^*(x_1) \geqq F_{(i+1)}^*(x_1). \tag{3.11}$$

We conclude that $\lim_{i\to\infty} F_i^*(x_1)$ exists; call it $F^*(x_1)$, say. Clearly we have

$$F^*(x_1) \geqq F(\bar{x}_1). \tag{3.12}$$

We shall prove in §5 that equality holds in (3.12). It will then follow from (3.8) that

$$F^*(x_1) = \int P_\infty\{\psi(\bar{x}_1, b, c)\} dH(b) dG(c). \tag{3.13}$$

4. **An essential lemma.** In this section we shall prove the following

LEMMA.

$$\lim_{y'\to\infty} \liminf_{i\to\infty} P\{w_{i,s} - w_{i,1} \leqq y'\} = 1$$

for $s>1$.

Proof. Let

$$B_i = (s-1)w_{i,s} - \sum_{j=1}^{s-1} w_{i,j} \tag{4.1}$$

for $i \geqq 1$. It follows easily from the way in which $w_{(i+1)}$ is obtained from w_i that

$$B_{(i+1)} \leqq \begin{cases} B_i - R_i & \text{when } R_i \leqq w_{is} - w_{i1}, \\ B_i - s(w_{is} - w_{i1}) + (s-1)R_i \leqq (s-1)R_i & \\ & \text{when } R_i \geqq w_{is} - w_{i1}. \end{cases} \tag{4.2}$$

In either case we have

$$B_{(i+1)} \leqq \max(B_i - R_i, (s-1)R_i). \tag{4.3}$$

Applying (4.3) to B_i we obtain

$$B_{(i+1)} \leqq \max(B_{i-1} - R_{i-1} - R_i, (s-1)R_{i-1} - R_i, (s-1)R_i). \tag{4.4}$$

Continuing in this manner and noting that $B_1 = 0$ we obtain

$$\begin{aligned} B_{i+1} \leqq \max\,[(s-1)R_i, (s-1)R_{i-1} - R_i, (s-1)R_{i-2} - R_{i-1} - R_i, \\ \cdots, (s-1)R_1 - R_2 - \cdots - R_i] = Y_i \text{ (say)}. \end{aligned} \tag{4.5}$$

Since the R_j are independently and identically distributed, we may interchange indices j and $i-j+1$, $j=1, \cdots, i$, in the middle member of (4.5) without altering its distribution. Hence, setting $h=(s-1)^{-1}$, we have

$$P\{Y_n \leqq y'\} = P\{R_1 \leqq hy', R_2 \leqq h(R_1 + y'), \cdots, R_n \leqq h(R_1 + \cdots + R_{n-1} + y')\} \tag{4.6}$$

(where, for $n=1$, we replace $R_1+ \cdots +R_{n-1}$ by 0). Since $B_i \geqq w_{is}-w_{i1}$, our proof will be complete if we show that

$$\lim_{y'\to\infty} \liminf_{n\to\infty} P\{Y_n \leqq y'\} = 1. \tag{4.7}$$

From the strong law of large numbers we have

$$P\left\{\lim_{n\to\infty} \frac{1}{n}\sum_{i=1}^{n} R_i = ER_1,\ \lim_{n\to\infty} \frac{R_n}{n} = 0\right\} = 1. \tag{4.8}$$

Now, for $y' \geqq 0$,

$$\begin{aligned} &P\{Y_n \leqq y'\} \\ &\geqq P\{R_n \leqq h(R_1 + \cdots + R_{n-1} + y') \text{ for } n = 1, 2, \cdots, \text{ad inf.}\} \\ &= P\left\{\frac{R_n}{hn} \leqq \frac{R_1 + \cdots + R_{n-1}}{n} + \frac{y'}{n} \text{ for } n = 1, 2, \cdots, \text{ad inf.}\right\}. \end{aligned} \tag{4.9}$$

Because of (4.8), for any $\epsilon>0$ there exists an integer N such that

$$P\left\{\frac{R_n}{hn} \leqq \frac{R_1 + \cdots + R_{n-1}}{n} \text{ for } n > N\right\} > 1 - \epsilon.$$

Clearly, there is a value y_0' such that for $y'>y_0'$

$$P\left\{\frac{R_n}{hn} \leqq \frac{R_1 + \cdots + R_{n-1}}{n} + \frac{y'}{n} \text{ for } n \leqq N\right\} > 1 - \epsilon.$$

Equation (4.7) is an immediate consequence.

5. **Certain immediate consequences of the lemma.**

(A) $F^*(x_1) = F(\bar{x}_1)$.

Proof. Let $x(x_1, y')$ be the point $x_1, y', \cdots, y'$. From the lemma it follows at once that for any $\epsilon>0$ and i and y' sufficiently large we have

$$| P\{w_{i1} \leqq x_1\} - P\{w_{i1} \leqq x_1, w_{i2} \leqq y', \cdots, w_{is} \leqq y'\} | < \epsilon. \tag{5.1}$$

Let $i\to\infty$. We obtain

$$| F^*(x_1) - F(x(x_1, y')) | \leqq \epsilon. \tag{5.2}$$

Let $y'\to\infty$. We obtain

$$| F^*(x_1) - F(\bar{x}_1) | \leqq \epsilon. \tag{5.3}$$

Since ϵ was arbitrary the desired result follows.

(B) Either F and F^* are both distribution functions or neither is a distribution function.

This follows from the fact that (A) above implies that $\lim F(x)$ as all coordinates of x approach infinity is the same as $\lim F^*(x_1)$ as x_1 approaches infinity.

6. Proof that F is a distribution function when $\rho<1$. We define

$$\rho = (ER_1)(sEg_1)^{-1}. \tag{6.1}$$

We shall now prove that, if $\rho<1$, $F(x)\to 1$ as all coordinates of x approach infinity. Then, by (3.12), we have $\lim F^*(x_1)=1$ as $x_1\to\infty$.

I. We show that it is sufficient to prove this result in a "dominating" case where, for some $c>0$,

$$\sum_{i=0}^{\infty} P\{R_1 = ci\} = \sum_{i=0}^{\infty} P\{g_1 = ci\} = 1. \tag{6.2}$$

Let $[a]$ be the largest integer $\leqq a$ and for some one $c>0$ define, for all i,

$$g_i' = c\left[\frac{g_i}{c}\right], \qquad R_i' = c\left[\frac{R_i}{c}\right] + c.$$

Then $g_i' \leqq g_i$ and $R_i' \geqq R_i$. Let w_i' be the same function of $\{g_j'\}$ and $\{R_j'\}$ that w_i is of $\{g_j\}$ and $\{R_j\}$. It follows from an argument like that of §3 that $w_i' \geqq w_i$ for all i. Hence if we can show that

$$\lim_{y'\to\infty} \liminf_{i\to\infty} P\{w_{i1}' \leqq y', \cdots, w_{is}' \leqq y'\} = 1 \tag{6.3}$$

it will follow that

$$\lim_{y'\to\infty} F(y', y', \cdots, y') = 1 \tag{6.4}$$

which is the desired result.

We have

$$Eg_j' \geqq Eg_j - c, \qquad ER_j' \leqq ER_j + c.$$

Hence, if c is sufficiently small, $(ER_1')(sEg_1')^{-1}<1$.

In the remainder of this section we assume that (6.2) is satisfied with $\rho<1$, and we shall prove (6.3) for this process.

II. We show that (6.3) is valid if $P\{R_1=0\}>0$. We recall that $P\{g_1=0\}<1$. Hence, for any i and integral $a_1, \cdots, a_s$, with $0\leqq a_1 \leqq a_2 \leqq \cdots \leqq a_s$,

$$P\{w_{(i+j)} = 0 \text{ for some } j \mid w_{i,1} = a_1c,\ w_{i,2} = a_2c, \cdots, w_{i,s} = a_sc\}$$

is positive. Let Z be the totality of all points $(a_1c, a_2c, \cdots, a_sc)$ with integral

a's such that $0 \leqq a_1 \leqq a_2 \leqq \cdots \leqq a_s$. Let z be a generic point of Z. The points z are the states of our Markoff process $\{w_i\}$. The preceding argument shows that the origin 0 and all the points z which can be reached by the process $\{w_i\}$ with positive probability form a chain C which is irreducible. C is aperiodic, since

$$P\{w_{i+1} = 0 \mid w_i = 0\} \geqq P\{R_1 = 0\} > 0. \tag{6.5}$$

The desired result then follows by III below.

III. We shall show that, if C is aperiodic and irreducible, (6.3) holds. From §3 or [3, Chap. 15, Theorems 1 and 2], it follows that

$$\lim_{i\to\infty} P\{w_i = z\}$$

exists for all z in C; call it $f(z)$. From the theorems of [3] cited above it follows that

$$\sum_{z\in C} f(z) = 0 \quad \text{or} \quad 1. \tag{6.6}$$

Our result is proved if we show that the sum in (6.6) is 1. Suppose it were 0; every $f(z)$ is then zero. We would then have

$$F(y', y', \cdots, y') = 0 \tag{6.7}$$

for every y'. Hence from (A), §5, we obtain, using (6.7),

$$F^*(x_1) = 0 \tag{6.8}$$

for every x_1. From the definition of w_i and the fact that $\rho < 1$ we obtain that there exists an $M > 0$ such that, whenever

$$M \leqq a_1 \leqq a_2 \leqq \cdots \leqq a_s, \tag{6.9}$$

we have

$$E\left\{\sum_{j=1}^{s} w_{(i+1),j} \mid w_{ij} = a_j,\ j = 1, \cdots, s\right\} < \sum_{j=1}^{s} a_j - \delta \tag{6.10}$$

for some $\delta > 0$. It is to be noted that, whether (6.9) holds or not, the left member of (6.10) is never greater than

$$\sum_{j=1}^{s} a_j + ER_1. \tag{6.11}$$

Since $F^*(x_1) \equiv 0$ we can find an $N > 0$ such that for $i \geqq N$ we have

$$P\{w_{i,1} < M\} < \frac{\delta}{\delta + ER_1}. \tag{6.12}$$

Then, for $i \geqq N$, we have

$$E\left\{\sum_{j=1}^{s} w_{i+1,j} - \sum_{j=1}^{s} w_{i,j}\right\}$$

$$(6.13)\qquad = E\left\{E\left\{\sum_{j=1}^{s} w_{i+1,j} - \sum_{j=1}^{s} w_{i,j} \,\middle|\, w_{ij,}j = 1, \cdots, s\right\}\right\}$$

$$< \frac{\delta}{\delta + ER_1}(ER_1) + \frac{ER_1}{\delta + ER_1}(-\delta) = 0.$$

Hence, for $i>0$,

$$(6.14)\qquad E\sum_{j=1}^{s} w_{i+N,j} < E\sum_{j=1}^{s} w_{N,j}$$

so that $E\sum_{j=1}^{s} w_{i,j}$ is bounded uniformly in i. This contradicts (6.7) and proves III.

IV. We now suppose $P\{R_1=0\}=0$, and we construct a suitable "dominating" process for which we can prove results analogous to II and III.

Let k be a positive integer such that $(sk-1)>0$,

$$(6.15)\qquad P\{R_1 \geqq (sk-1)c\} = 1,$$

and

$$(6.16)\qquad P\{R_1 = (sk-1)c\} > 0.$$

If necessary, we can decrease the c of (6.2) so that such a k can always be found. If now

$$(6.17)\qquad P\{g_1 \geqq skc\} > 0,$$

then it is clear that

$$(6.18)\qquad P\{w_{i+j} = 0 \text{ for some } j > 0 \mid w_i = z\} > 0$$

for every z in C. Hence C is irreducible. It is also aperiodic, because

$$(6.19)\qquad P\{w_{i+1} = 0 \mid w_i = 0\} > 0.$$

Hence the desired result follows by III. We therefore assume that (6.17) does not hold, i.e., that

$$(6.20)\qquad P\{g_1 \leqq (sk-1)c\} = 1.$$

Let m be the largest integer for which

$$P\{g_1 = mc\} > 0.$$

Let A_1 be the set of α (say) non-negative integers $j<m$ such that

$$P\{g_1 = jc\} = 0, \qquad j \in A_1.$$

Let $\{g_i'\}$ be independently and identically distributed chance variables with the following distribution:

$$P\{g_1' = jc\} = \lambda, \qquad j \in A_1,$$
$$P\{g_1' = mc\} = P\{g_1 = mc\} - \alpha\lambda,$$
$$P\{g_1' = jc\} = P\{g_1 = jc\}, \qquad j \neq m, j \notin A_1.$$

Here λ is a small positive number, whose choice will be more fully described shortly, but which should in any case be such that $P\{g_1 = mc\} - \alpha\lambda > 0$ and $\lambda < P\{R_1 = (sk-1)c\}$.

Let A_2 be the totality of integers $j > (sk-1)$ such that

$$P\{R_1 = jc\} = 0, \qquad j \in A_2.$$

Let $\{R_i'\}$ be independently and identically distributed chance variables, independent of $\{g_i'\}$ and with the following distribution:

$$P\{R_1' = jc\} = \frac{\lambda}{2^j}, \qquad j \in A_2,$$

$$P\{R_1' = jc\} = P\{R_1 = jc\}, \qquad j \neq (sk-1), j \notin A_2,$$
$$P\{R_1' \geqq (sk-1)c\} = 1.$$

We choose $\lambda > 0$ so small that

$$(ER_1')(sEg_1')^{-1} < 1.$$

Any such λ will suffice.

Let $\{w_i'\}$ be the same functions of $\{R_i', g_i'\}$ as w_i are of $\{R_i, g_i\}$. Let F_i' and F' be the corresponding functions for the primed w_i. Comparing corresponding sequences in the manner of §3 we obtain that

$$F_i'(x) \leqq F_i(x) \qquad \text{for every } x.$$

Hence

$$F'(x) = \lim F_i'(x) \leqq F(x) \qquad \text{for every } x.$$

If, therefore, we prove the desired result for the system $\{w_i'\}$ we have a fortiori proved the desired result for the system $\{w_i\}$. We may therefore drop the accents and henceforth assume that

$$P\{g_1 = jc\} > 0, \qquad j = 0, \cdots, (sk-1), \tag{6.21}$$

$$P\{g_1 \leqq (sk-1)c\} = 1, \tag{6.22}$$

$$P\{R_1 = jc\} > 0, j = (sk-1), (sk), (sk+1), \cdots, \tag{6.23}$$

$$P\{R_1 \geqq (sk-1)c\} = 1. \tag{6.24}$$

(We note that these imply that we are in the case $s > 1$, for $s = 1$ would violate the requirement that $\rho<1$.)

Let

$$z^* = (skc, skc, \cdots, skc).$$

Let

$$z = (a_1c, a_2c, \cdots, a_sc)$$

with $0\leqq a_1\leqq a_2\leqq \cdots \leqq a_s$ be any point in Z. Let

$$L = w_{is} + skc.$$

If $R_{i+j-1}=L-w_{ij}$ and $g_{i+j}=0, j=1, \cdots, s$, an event of positive probability, we have

$$w_{i+s} = (L, L, \cdots, L).$$

If $R_{i+s+j-1}=(sk-1)c$ and $g_{i+s+j}=kc, j=1, \cdots, s$, again an event of positive probability, we have, since $L\geqq skc$, that

$$w_{i+2s} = (L - c, L - c, \cdots, L - c).$$

Applying the above argument a_s times we conclude that, for any z and i, $P\{w_{i+j}=z^* \text{ for some } j\geqq 0 | w_i=z\}>0$. Let D be the set of all points in Z which can be reached from z^* with positive probability. The above argument shows that the states of D form an irreducible Markoff chain. This chain is aperiodic, because a modification of the above argument shows (using (6.23)) that there exists a number N such that, whatever be $n\geqq N$, there is a positive probability of moving from z^* back to z^* in exactly n steps.

If now, with probability one, $w_i\in D$ for some i, an argument similar to that of III applies and the desired result is proved.

V. We now prove that, with probability one, $w_i\in D$ for some i.

From (6.23) and the fact that $P\{g_1=0\}>0$ it follows that any point $z=(a_1c, \cdots, a_sc)\in Z$ such that $(2sk-1)\leqq a_1$ is a member of D. We now note that the probability of entering D in at most s steps from any point z not in D is bounded below by a number (say) $\mu>0$, independently of z (not in D). To see this, we note that this can be accomplished in at most s steps where each $R=2skc$ and each $g=0$. From this it follows that the probability of entering D for some i is one.

The proof of the result of this section is now complete.

7. Proof that F is not a distribution function when $\rho\geqq 1$. To prove this result we must in addition assume that, when $\rho=1$,

$$P\{R_i - sg_i = 0\} < 1. \tag{7.1}$$

For if (7.1) does not hold we have, for some $e>0$,

$$P\{g_1 = e\} = P\{R_1 = se\} = 1. \tag{7.2}$$

(Hence $\rho=1$ here.) Therefore, with probability one,

$$
\begin{aligned}
w_1 &= (0, 0, \cdots, 0),\\
w_2 &= (0, 0, \cdots, 0, (s-1)e),\\
w_3 &= (0, 0, \cdots, 0, (s-2)e, (s-1)e),\\
&\vdots\\
w_s &= (0, e, 2e, \cdots, (s-1)e),\\
w_i &= w_s, \qquad i > s.
\end{aligned}
$$

Hence a limiting distribution function F does exist.

We therefore assume that $\rho \geqq 1$ and (7.1) holds. We shall show that $F(x) \equiv 0$, and hence (see §5, A) that $F^*(x_1) \equiv 0$.

Let $\{L_i\}$ be a sequence of chance variables defined as follows: $L_1 = 0$ with probability one. For $i \geqq 1$

$$L_{i+1} = \max(0, L_i + R_i - sg_{i+1}).$$

Thus L_i is the waiting time of the ith individual in a system such as described in §2 where $s=1$, the service time of the ith individual is R_i, and the interval between the ith arrival and $(i+1)$st arrival is sg_{i+1}. In this system $\rho \geqq 1$, so that the theorem of Lindley (which treats the case $s=1$) is applicable, i.e.,

$$\lim_{x'\to\infty} \lim_{i\to\infty} P\{L_i \leqq x'\} = 0. \tag{7.3}$$

Now, if $0 \leqq a_1 \leqq a_2 \leqq \cdots \leqq a_s$, and b, c, and d are non-negative numbers with $b \leqq \sum_{j=1}^{s} a_j$, we clearly have $\max(0, a_1+c-d) + \sum_{j=2}^{s} \max(0, a_j - d) \geqq \max(0, b+c-sd)$. We conclude, using induction, that for all $i \geqq 1$,

$$L_i \leqq \sum_{j=1}^{s} w_{i,j} \tag{7.4}$$

with probability one. It follows from (7.3) and (7.4) that

$$
\begin{aligned}
\lim_{x'\to\infty} F(x', x', \cdots, x') &= \lim_{x'\to\infty} \lim_{i\to\infty} P\{w_{i,j} \leqq x', j = 1, \cdots, s\}\\
&\leqq \lim_{x'\to\infty} \lim_{i\to\infty} P\left\{\sum_{j=1}^{s} w_{i,j} \leqq sx'\right\} \qquad (7.5)\\
&\leqq \lim_{x'\to\infty} \lim_{i\to\infty} P\{L_i \leqq sx'\} = 0,
\end{aligned}
$$

which proves the desired result.

8. **Proof that $\lim_{i\to\infty} F_i(x \mid y)$ exists and is independent of y. Uniqueness of the solution of the integral equation (3.8).** Suppose first that $\rho \geqq 1$. Since $F_i(x \mid y) \leqq F_i(x)$ for every i and every x and $y \in S$, it follows from the results

of §7 that

$$\lim_{i\to\infty} F_i(x \mid y) = 0$$

when $\rho \geqq 1$. We shall shortly show that, when $\rho \geqq 1$, (3.8) has no solution which is a distribution function over S.

Assume, therefore, that $\rho < 1$, which is the interesting case. We shall show that, for all x and y in S, the ergodic property

$$\lim_{i\to\infty} F_i(x \mid y) = F(x) \tag{8.1}$$

holds. From this it follows easily that (3.8) has at most one solution which is a distribution function over S (thus, by §6, it has exactly one such solution). For, suppose, to the contrary, that there were another such distribution function, say $V(x)$. It is clear then that, if w_1 is distributed according to $V(x)$, so are $w_2, w_3, \cdots$, so that $V(x)$ is the limiting distribution. On the other hand, it follows from (8.1) and the Lebesgue bounded convergence theorem applied to

$$V(x) = \int F_i(x \mid y) dV(y)$$

that

$$V(x) = \int F(x) dV(y) = F(x)$$

which is the desired result. (Thus we have proved that, when $\rho < 1$, $F(x)$ is the unique stationary absolute probability distribution; see the paragraph following equation (3.8).)

Conversely, if (3.8) has a solution V which is a distribution function over S, then

$$F(x) = \lim_{i\to\infty} F_i(x) \geqq \lim_{i\to\infty} \int F_i(x \mid y) dV(y) = V(x),$$

so that (from the result of §7) $\rho < 1$ and hence $V(x) = \lim_{i\to\infty} F_i(x)$ is the unique solution of (3.8).

Denote by $[a, b]$ and $[c, d]$ the smallest closed intervals for which

$$P\{a \leqq R_1 \leqq b\} = P\{c \leqq g_1 \leqq d\} = 1.$$

We shall conduct the proof separately for several cases.

Case 1: $b > sc$. Let $b - sc = 2\nu > 0$. Then, for any positive integer n,

$$P\{w_{sn,1} > \nu n\} = q_n > 0.$$

Fix n. For any x and $\delta > 0$ there exists an integer M such that, for all $j \geqq M$

and $k>0$, we have

$$|F_j(x) - F_{j+k}(x)| < q_n\delta.$$

We recall that, if $y_1 \leqq y_2$, $y_1 \in S$, $y_2 \in S$, we have, for all i,

$$F_i(x \mid y_1) \geqq F_i(x \mid y_2).$$

Hence, for $j \geqq M$, we have

$$\begin{aligned} 0 &\leqq q_n[F_j(x) - F_j(x \mid (nv, nv, \cdots, nv))] \\ &\leqq \int [F_j(x) - F_j(x \mid y)] dF_{sn}(y) \qquad (8.2) \\ &< q_n\delta + \int [F_{j+sn-1}(x) - F_j(x \mid y)] dF_{sn}(y) = q_n\delta. \end{aligned}$$

Therefore

$$0 \leqq [F_j(x) - F_j(x \mid y)] < \delta \qquad (8.3)$$

for all $y \leqq (nv, nv, \cdots, nv)$, $y \in S$, and all $j \geqq M$. Since x, n, and δ were arbitrary this proves (8.1) for Case 1.

Case 2: $a<d$. Let y be any point in S, and

$$p_n(y) = P\{w_n = 0;\ w_i \neq 0,\ i < n \mid w_1 = y\}.$$

We shall show that, for all y in S,

$$\sum_{n=1}^{\infty} p_n(y) = 1. \qquad (8.4)$$

This is sufficient to prove the desired result, because

$$F_i(x \mid y) - \sum_{n \leqq i} p_n(y) F_{i-n}(x) \qquad (8.5)$$

then approaches zero as $i \to \infty$.

To prove (8.4) we proceed as in §6 to construct a "dominating" random walk on a lattice. The walk begins at a point on the lattice all of whose coordinates are no less than the corresponding ones of y. As in §6 one proves that with probability one the walk enters an irreducible aperiodic chain. Since $a<d$ this chain contains the origin. Since $\rho<1$ and $F(x)$ is a distribution function this chain constitutes a positive recurrent class. For an irreducible, recurrent class (8.4) must hold for all y in the class. Since the walk enters the class with probability one, (8.4) holds for all y in S.

Case 3: $c=d \leqq a=b \leqq sc$. In this case we have $P\{R_1=b\}=P\{g_1=c\}=1$. Since $\rho<1$, we also have $b<sc$.

Let $y^*=(y_1, \cdots, y_s)$ be any point in S, and let $y''=(y_s, \cdots, y_s)$. Given the process $\{R_i, g_i\}$, let w_i^* be the position of w_i if $w_1=y^*$, let w_i' be the position of w_i if $w_1=0$, and let w_i'' be the position of w_i if $w_1=y''$. Clearly, $w_i \leqq w_i^* \leqq w_i''$ for all i with probability one, and for $i \geqq s$ we have $w_i' = \bar{w}$, where $\bar{w}$ is defined by

$$\bar{w} = (0, u_{s-1}, \cdots, u_2, u_1), \tag{8.6}$$

where

$$u_j = \max (b - jc, 0).$$

We shall show that $w_i'' = \bar{w}$ with probability one for i sufficiently large, which implies that for sufficiently large i with probability one, $w_i^* = \bar{w}$, and proves the desired result.

It is clear that, for all i,

$$\begin{aligned} w''_{i+1,j} &= \max (0, w''_{i,j+1} - c) \qquad \text{for } 1 \leqq j < s, \\ w''_{i+1,s} &= w''_{i,1} + b - c. \end{aligned} \tag{8.7}$$

For $n \geqq 0$ and $0 \leqq i \leqq s-1$, we evidently have

$$w''_{1+ns+i,1} \leqq w''_{ns+1,1}. \tag{8.8}$$

Let N be a positive integer such that $y_s - N(sc-b) \leqq 0$. Then $w''_{ns+1,1}=0$ for $n \geqq N$, and hence from (8.8) we have $w''_{i,1}=0$ for $i \geqq Ns+1$. It follows from (8.7) that $w_i'' = \bar{w}$ for $i \geqq (N+1)s$.

Case 4: *$d \leqq a$, $b \leqq sc$, and either $a<b$ or $c<d$.* Let u_j be as in (8.6) and for $\epsilon>0$ define $\bar{w}^\epsilon = (0, u^\epsilon_{s-1}, \cdots, u^\epsilon_2, u^\epsilon_1)$, where $u^\epsilon_j = \max (0, u_j - \epsilon)$. From the definition of b and c we have that, for every $\epsilon>0$,

$$P\{ w_{s,j} \geqq u^\epsilon_{s-j+1}, j = 1, \cdots, s\} = \gamma > 0.$$

An argument like that of Case 1 (with γ for q_n and F_s for F_{sn}) then shows that

$$\lim_{i \to \infty} F_i(x \mid y) = F(x)$$

for all $y \in \Gamma^\epsilon = \{y \mid y \in S, y \leqq \bar{w}^\epsilon\}$ and all x.

Let y be any point in S, and

$$p^\epsilon_n(y) = P\{ w_n \in \Gamma^\epsilon, w_i \notin \Gamma^\epsilon, i < n \mid w_1 = y\}.$$

We shall show that, for some $\epsilon>0$ and all $y \in S$, we have

$$\sum_{n=1}^{\infty} p^\epsilon_n(y) = 1; \tag{8.9}$$

from this and the result of the previous paragraph, the desired result is proved in the manner of the first paragraph of Case 2.

Let

(8.10) $$E = \{y \mid y = (y_1, \cdots, y_s) \in S;\ y_1 = 0;\ y_s \leqq (s-1)b\}.$$

In order to prove (8.9) for some $\epsilon > 0$ and for all $y \in S$, it clearly suffices to show that

(8.11) $$P\{w_i \in E \text{ for infinitely many } i \mid w_1 = y\} = 1$$

for all $y \in S$, and that there exists a positive integer M and positive numbers α and ϵ such that

(8.12) $$P\{w_M \in \Gamma^\epsilon \mid w_1 = y\} > \alpha$$

for all $y \in E$.

We first prove (8.11). To this end, let $y = (y_1, \cdots, y_s)$ be any fixed point in S. Since we have always assumed $d > 0$, we have in Case 4 that $a > 0$. It follows from equation (4.3) that for $n > (s-1)y_s/a$ we have

(8.13) $$P\{B_n \leqq (s-1)b \mid w_1 = y\} = 1.$$

Let $\{e_i\}$, $\{f_i\}$ be any sequences of non-negative numbers, and let $\{v_i\}$ be the corresponding sequence of values of $\{w_i\}$ when $w_1 = y^*$, $R_i = e_i$, and $g_i = f_i$. Then, if $v_{i1} = 0$ for only finitely many i, it would follow that $\liminf_{n\to\infty} (1/n)\sum_{i=1}^{n} (e_i - sf_{i+1}) \geqq 0$. However, since $\rho < 1$, the strong law of large numbers implies that

$$P\left\{\lim_{n\to\infty} \frac{1}{n} \sum_{i=1}^{n} (R_i - sg_{i+1}) = ER_1 - sEg_1 < 0\right\} = 1.$$

Hence

(8.14) $$P\{w_{i,1} = 0 \text{ for infinitely many } i \mid w_1 = y^*\} = 1$$

for all $y^* \in S$. Equation (8.14) is a fortiori true for the original process, and (8.11) is an immediate consequence of (4.1), (8.13), and (8.14).

It remains to prove (8.12). We recall that in Case 4 we have $c < b \leqq sc$ and that there are numbers b', c' such that $P\{R_1 \leqq b'\} = p > 0$, $P\{g_1 \geqq c'\} = q > 0$, and $b' - c' = b - c - \epsilon$ for some $\epsilon > 0$. An obvious modification of the argument of Case 3 (put b', c' for b, c) shows that if $w_1 = y \in E$ and if $R_j = b'$ and $g_{j+1} = c'$ for $1 \leqq j < M$ where M/s is the greatest integer contained in $2 + (s-1)b/(sc' - b')$, then

(8.15) $$w_{M,i} \leqq \max(0, b' - (s+1-i)c') \leqq u^\epsilon_{s-i+1}.$$

Equation (8.15) is a fortiori true if $R_j \leqq b'$ and $g_{j+1} \geqq c'$ for $1 \leqq j < M$ (the argument being similar to that of §3). We conclude that (8.12) is satisfied for ϵ and M as defined here and for $\alpha = (pq)^{M-1}$.

9. **Distribution of the number of individuals waiting in the queue.** In order to avoid trivial cases and the circumlocutions required to dispose of them, we shall assume in this section that $G(0) = 0$. This means that the prob-

ability is zero that two or more individuals arrive simultaneously.

Let Q_i be the number of arrivals in the open time interval (t_i, t_i+w_{i1}); i.e., Q_i is the number of individuals in the queue waiting to be served, just before the service of the ith individual begins.

Since $g_{i+1}, g_{i+2}, \cdots$ are independent of t_i, we have

$$\begin{aligned} P\{Q_i \geqq n\} &= \int P\{g_1 + g_2 + \cdots + g_n < a\} dF_i^*(a) \\ &= \int G^{n*}(a-) dF_i^*(a), \end{aligned} \tag{9.1}$$

where $G^{n*}(a)$ denotes the n-fold convolution of $G(a)$ with itself. Since $F_i^*(a)$ tends nonincreasingly to $F^*(a)$ as $i\to\infty$ for all a, and since $G^{n*}(a-)$ is continuous from the left, we obtain, in the case $\rho<1$,

$$\lim_{i\to\infty} P\{Q_i \geqq n\} = \int G^{n*}(a-) dF^*(a). \tag{9.2}$$

If $\rho \geqq 1$, equation (9.1) shows that $\lim_{i\to\infty} P\{Q_i \geqq n\} = 1$ for all n, except in the trivial case where $P\{R_1 - sg_1 = 0\} = 1$.

References

1. D. V. Lindley, *The theory of queues with a single server*, Proc. Cambridge Philos. Soc. vol. 48 (1952), part 2, pp. 277–289.

2. J. V. Uspensky, *Introduction to mathematical probability*, New York, McGraw-Hill, 1937.

3. W. Feller, *An introduction to probability theory and its applications*, New York, Wiley, 1950.

4. N. Metropolis and S. Ulam, *The Monte Carlo method*, Journal of the American Statistical Association vol. 44 (1949) pp. 335–341.

5. D. G. Kendall, *Stochastic processes occurring in the theory of queues and their analysis by the method of the imbedded Markov chain*, Ann. Math. Statist. vol. 24 (1953) pp. 338–354.

6. ———, *Some problems in the theory of queues*, Journal Royal Statistical Society (B) vol. 13 (1951) pp. 151–173 and 184–185.

Cornell University,
Ithaca, N. Y.

Reprinted from
Trans. Am. Math. Soc. **78** (1), 1–18 (1955)

Reprinted from THE ANNALS OF MATHEMATICAL STATISTICS
Vol. 26, No. 2, June, 1955

ON TESTS OF NORMALITY AND OTHER TESTS OF GOODNESS OF FIT BASED ON DISTANCE METHODS

BY M. KAC, J. KIEFER,[1] AND J. WOLFOWITZ[2]

Cornell University

Summary. The authors study the problem of testing whether the distribution function (d.f.) of the observed independent chance variables $x_1, \cdots, x_n$ is a member of a given class. A classical problem is concerned with the case where this class is the class of all normal d.f.'s. For any two d.f.'s $F(y)$ and $G(y)$, let $\delta(F, G) = \sup_y |F(y) - G(y)|$. Let $N(y \mid \mu, \sigma^2)$ be the normal d.f. with mean μ and variance σ^2. Let $G_n^*(y)$ be the empiric d.f. of $x_1, \cdots, x_n$. The authors consider, inter alia, tests of normality based on $v_n = \delta(G_n^*(y), \quad N(y \mid \bar{x}, s^2))$ and on $w_n = \int (G_n^*(y) - N(y \mid \bar{x}, s^2))^2 \, d_y N(y \mid \bar{x}, s^2)$. It is shown that the asymptotic power of these tests is considerably greater than that of the optimum χ^2 test. The covariance function of a certain Gaussian process $Z(t)$, $0 \leqq t \leqq 1$, is found. It is shown that the sample functions of $Z(t)$ are continuous with probability one, and that

$$\lim_{n \to \infty} P\{nw_n < a\} = P\{W < a\}, \quad \text{where} \quad W = \int_0^1 [Z(t)]^2 \, dt.$$

Tables of the distribution of W and of the limiting distribution of $\sqrt{n}\, v_n$ are given. The role of various metrics is discussed.

1. Introduction. Let $x_1, \cdots, x_n$ be n independent chance variables with the same cumulative distribution function $G(x)$ (i.e., $G(x) = P\{x_1 < x\}$) which is unknown to the statistician. It is desired to test the hypothesis that $G(x)$ is a normal distribution. This is an old problem of considerable interest which has received a fair share of attention in the literature.

A commonly used test consists essentially in testing whether the third moment of $G(x)$ about its mean is zero and whether the ratio of the fourth moment about the mean to the square of the second moment about the mean is three. It is obvious that this is not really a test of normality, because there are many non-normal distributions which satisfy these conditions on the moments.

Perhaps the best of the commonly used large sample tests of normality is the χ^2 test due to Karl Pearson; see for example Cramér [1], Sections 30.1 and 30.3, and the recent results of Chernoff and Lehmann [2]. The asymptotic power of the χ^2 test was studied by Mann and Wald [3]. (It is true that these authors

Received August 30, 1954, revised January 10, 1955.

[1] Research under contract with the Office of Naval Research. The computations of the tables of the present paper were performed under the auspices of the Office of Naval Research to whom the authors are greatly obliged.

[2] The research of this author was supported in part by the United States Air Force under Contract No. AF18(600)-685, monitored by the Office of Scientific Research.

studied the problem of goodness of fit for a simple hypothesis, which in our problem would correspond to knowing the mean and variance of $G(x)$. However, it is plausible that the comparison in power which we will make below will be true a fortiori for our problem.) One of their principal results is the following. Define the distance $\delta(H_1, H_2)$ between any two distribution functions H_1 and H_2 by

$$\delta(H_1, H_2) = \sup_x |H_1(x) - H_2(x)|.$$

Suppose one tests the null hypothesis that $G(x) \equiv R(x)$, where $R(x)$ is a given continuous distribution, at some fixed level of significance. In [3] it is shown that the χ^2 test based on n observations and k_n intervals of equal probability under R, where k_n is chosen for each n so as to minimize the value Δ_n for which the minimum of the power function among alternatives R^* with $\delta(R^*, R) \geqq \Delta_n$ is $\frac{1}{2}$, gives $\Delta_n \doteq cn^{-2/5}$ when n is large. In Section 5 we shall show that the result of Mann and Wald just stated also holds if δ is replaced by the measure of discrepancy γ introduced in Section 5.

Let

$$\psi_x(a) = \begin{cases} 0, & x \leqq a; \\ 1, & x > a. \end{cases}$$

Let

$$G_n^*(x) = \frac{1}{n} \sum_{i=1}^{n} \psi_x(x_i)$$

be the empiric d.f. of $x_1, \cdots, x_n$, that is, $G_n^*(x)$ is the proportion of x_i's less than x. The asymptotic distribution of $\delta(G, G_n^*)$ was found by Kolmogoroff [4] and it is known that $\delta(G, G_n^*)$ is of the order $n^{-1/2}$ in probability (that is, with probability arbitrarily close to one). If now one bases the test of the hypothesis that $G(x) \equiv R(x)$ on $\delta(R, G_n^*)$, with large values of δ significant, it follows that, for large n, it is sufficient that $\delta(R, R^*)$ be of the order $n^{-1/2}$ for the probability of rejecting the null hypothesis to be appreciable ($\geqq \frac{1}{2}$). To put matters a little differently: Let $\delta(R, R^*) = h$ be small (so that n has to be large in order to distinguish between R and R^*). Then, if n has to be equal to N in order to guarantee that the power at R^* of the χ^2 test of goodness of fit be at least $\frac{1}{2}$, when one uses the test based on δ it is enough that n be of the order of $N^{4/5}$. This is a considerable improvement (for large n). The result just stated holds also for the classical "ω^2" test if δ is replaced in the above by γ; this follows from the results of Section 5.

Let us return to the problem of testing the composite null hypothesis whether $G(x)$ is normal (its mean and variance being unknown). One of the present authors has been developing the minimum distance method for estimation and testing hypotheses in a number of papers (Wolfowitz [5], [6], [7], [8]). In accord with this method it is proposed in [5] (page 149) that this test of normality be based on $\delta(G_n^*, N^{**})$ with the large values critical. Here N^{**} is the class of all normal distributions, and the distance of any d.f. $H(x)$ from the class N^{**} is given by

$$\delta(H, N^{**}) = \inf_{N \varepsilon N^{**}} \delta(H, N).$$

In Sections 2, 3, and 4 we investigate tests based on various "distance" criteria, constructed in the spirit of the above discussion, and in Section 5 we discuss the asymptotic power of these tests. As stated in [5], the minimum distance method is not limited merely to testing normality, and in Section 4 we discuss its application to other tests of goodness of fit.

Throughout this paper we discuss tests, computations, and power considerations in terms of particular (normal or rectangular) examples, but it will be obvious that the results of Sections 2, 3, and 5 may in general be carried over to testing composite hypotheses involving parametric families. The minimum distance method applies in principle to even more complicated families of distribution functions about which one desires to test a hypothesis. For hypotheses of a more complicated nature than our examples, there will often exist the additional complication that the test criterion may not be distribution-free under the null hypothesis.

Tests may also be constructed using other "distances" than those mentioned in Section 2 and above (see also [5], pp. 148–149, and [6], p. 10 in this connection), involving other "estimators" than those of Sections 2 and 3, and involving other modifications of the notion of "distance methods" as motivated in this section.

2. Testing normality. For convenience we divide this section into several subsections.

2.1. The computation of $\delta(G_n^*, N^{**})$ offers considerable difficulty; unpublished work on this subject has been done by Blackman. The distribution of $\delta(G_n^*, N^{**})$ under the null hypothesis is still unknown. (It is easy to verify that, when $G(x)$ is a member of N^{**}, the distribution of $\delta(G_n^*, N^{**})$ does not depend on $G(x)$. Thus the composite null hypothesis determines uniquely the distribution of the test criterion.) In Section 4 we shall give an example of another problem of testing hypotheses where the limiting distribution of the minimum distance criterion is explicitly calculated. In the present Section 2 we shall consider some other "distance" tests. (For the case where either the mean or variance is assumed known, the test corresponding to that discussed in this section is being studied by Darling [18]; the suggestion of using such tests is apparently due to Cramér.) Let

$$\bar{x} = \frac{1}{n}\sum_1^n x_i, \qquad s^2 = \frac{1}{n}\sum_1^n x_i^2 - \bar{x}^2.$$

Let $N(x \mid \bar{x}, s^2)$ be the normal distribution function with mean $\bar{x}$ and variance s^2. One can base the test of normality on $v_n = \delta(G_n^*(x), N(x \mid \bar{x}, s^2))$, with the large values critical. It is easy to see that, when $G(x)$ is actually a member of

N^{**}, the distribution of v_n does not depend on $G(x)$. The distribution of v_n does not seem easy to obtain, except, for example, by Monte Carlo methods. Another test criterion, similar to the above but of the "ω^2" type, is to base the test on

$$w_n = \frac{1}{n}\int_0^1 Z_{r,n}^2\,dr$$

where $Z_{r,n}$ is defined in 2.2.[3] (The idea of the "ω^2" test, which is defined precisely in Section 6, is due to von Mises, with a modification by Smirnov.) Sections 2.2 to 2.6 are concerned with the limiting distribution of nw_n as $n \to \infty$ when G is normal. The power of the tests based on v_n and w_n is discussed in Section 5. In Section 3 we treat briefly an example which illustrates the construction of test criteria which are similar to v_n and w_n but which may use "inefficient" statistics to estimate which member of N^{**} G is (if it is); such techniques may have obvious practical importance.

2.2. Let $x_1, x_2, \cdots$ be independent, identically distributed Gaussian random variables with zero mean and unit variance. Let $G_n^*(x)$ be as defined in Section 1. We shall use the following notation:

$$\phi(x) = \phi(-x) = \frac{1}{\sqrt{2\pi}}e^{-x^2/2}, \qquad I(x) = \int_x^\infty \phi(y)\,dy,$$

$$J(y) = I^{-1}(1-y) = \{x \mid y = 1 - I(x)\}, \qquad g(x) = \phi(x)/I(x).$$

We also let $[z]$ = smallest integer $\geqq z$. For $0 \leqq r \leqq 1$, we put $U_{r,n} = [rn]$th from the bottom among the ordered $x_1, \cdots, x_n$. Finally, let[4]

$$Z_{r,n} = \sqrt{n}[G_n^*(\sqrt{S_n'}J(r) + \bar{x}_n) - r]$$

where

$$\bar{x}_n = \frac{1}{n}\sum_{i=1}^n x_i, \qquad S_n = \frac{1}{n}\sum_{i=1}^n x_i^2, \qquad S_n' = S_n - \bar{x}_n^2.$$

(Remark: S_n and S_n' may be used equally well in the definition of $Z_{r,n}$ for the purpose of obtaining the limiting distribution; in applications S_n' should probably be used.)

2.3. We shall show that for $0 \leqq r_1 < r_2 < \cdots < r_k \leqq 1$, the quantities $Z_{r_i,n}$, where $1 \leqq i \leqq k$, are asymptotically jointly normally distributed with zero means and covariance function (for $Z_{s,n}$ and $Z_{t,n}$ as $n \to \infty$)

$$K(s, t) = \min(s, t) - st - (2\pi)^{-1}(1 + J(s)J(t)/2)e^{-[J(s)^2+J(t)^2]/2}. \tag{2.1}$$

We shall show here that (2.1) follows from the fact proved in Section 2.4, that $\sqrt{n}x_n$, $\sqrt{n}(S_n - 1)$, and $\sqrt{n}(U_{r_i,n} - J(r_i))$, where $1 \leqq i \leqq k$, are jointly

[3] This definition is equivalent to that given in the summary at the beginning of the paper.

[4] Elsewhere in this paper, in the summary, for example, the conventional symbol s^2 is sometimes used instead of the typographically easier (here) symbol S_n'. Both represent the same thing. Also $\bar{x}_n$ and $\bar{x}$ are used interchangeably in a manner to cause no confusion.

asymptotically normal with means 0 and covariance matrix (presented in the same order) given by

$$
(2.2)\qquad \begin{pmatrix} 1 & 0 & 1 & 1 & \cdots & 1 \\ 0 & 2 & J(r_1) & J(r_2) & \cdots & J(r_k) \\ 1 & J(r_1) & \lambda_{11} & \lambda_{12} & \cdots & \lambda_{1k} \\ \cdots & \cdots & \cdots & \cdots & \cdots & \cdots \\ 1 & J(r_k) & \lambda_{k1} & \lambda_{k2} & \cdots & \lambda_{kk} \end{pmatrix},
$$

$$
\lambda_{ij} = \lambda_{ji} = \frac{r_i(1 - r_j)}{\phi(J(r_i))\phi(J(r_j))}, \qquad r_i \leqq r_j .
$$

In fact, all these results are well known ([1], p. 364, 369) except for the joint normality and the last k entries in the first two rows (and columns).

Assume then for the moment that the above is proved. The event $\{Z_{r,n} \leqq z\}$ may be written as $\{[\text{number of } x_i \text{ which are} < \sqrt{S'_n}J(r) + \bar{x}_n] \leqq nr + \sqrt{n}z\}$. This in turn is the same as $\{U_{r+z/\sqrt{n},\, n} \geqq \sqrt{S'_n}J(r) + \bar{x}_n\}$, or

$$
\{\sqrt{n}\bar{x}_n + \sqrt{n}J(r)(\sqrt{S'_n} - 1) - \sqrt{n}(U_{r+z/\sqrt{n},\,n} - J(r + z/\sqrt{n}))
$$
$$
\leqq \sqrt{n}[J(r + z/\sqrt{n}) - J(r)]\}.
$$

As $n \to \infty$, neglecting terms of higher order in probability, we may replace $\sqrt{n}[J(r + z/\sqrt{n}) - J(r)]$ by $zJ'(r)$ and $\sqrt{n}(\sqrt{S'_n} - 1)$ by $\frac{1}{2}\sqrt{n}(S_n - 1)$. We conclude that

$$
\lim_{n\to\infty} P\{Z_{r_1,n} \leqq z_1, \cdots, Z_{r_k,n} \leqq z_k\}
$$

$$
(2.3)\qquad = \lim_{n\to\infty} P\left\{\frac{1}{J'(r_i)}[\sqrt{n}\bar{x}_n + \tfrac{1}{2}\sqrt{n}J(r_i)(S_n - 1) - \sqrt{n}(U_{r_i+z_i/\sqrt{n},\,n} - J(r_i + z_i/\sqrt{n}))] \leqq z_i, \quad i = 1, \cdots, k\right\}.
$$

Thus, $Z_{r_1,n}, Z_{r_2,n}, \cdots, Z_{r_k,n}$ are jointly asymptotically normal, with the same limiting distribution as the quantities

$$
[\sqrt{n}\bar{x}_n + \tfrac{1}{2}\sqrt{n}J(r_i)(S_n - 1) - \sqrt{n}(U_{r_i,n} - J(r_i))]/J'(r_i), \quad i = 1, \cdots, k.
$$

From this and (2.2), the result (2.1) follows by direct computation and the fact that $J'(r) = 1/\phi(J(r))$.

2.4. It remains to prove the joint asymptotic normality of the quantities mentioned in the paragraph following (2.1), and to verify the last k entries of the first two rows (and columns) of (2.2). We shall in fact compute here only the limiting distribution of $\sqrt{n}\bar{x}_n$, $\sqrt{n}(S_n - 1)$, and $\sqrt{n}(U_{r,n} - J(r))$ where $0 < r < 1$; from this will follow the desired result of (2.2), and the method of proof of joint asymptotic normality for the $k + 2$ random variables previously mentioned will be evident from that for the case $k = 1$ considered here (note especially p. 369 of [1] in this connection).

We begin by introducing a process which will enable us to simplify this computation. For fixed n, let $Y_{[rn]}$ be a chance variable whose distribution function is that of $U_{r,n}$ above. Let Y_i, where $1 \leqq i \leqq n$ and $i \neq [rn]$, be random variables whose joint conditional distribution, given that $Y_{[rn]} = y$, is such that these $n - 1$ random variables are (conditionally) independent with Y_i having a (conditional) truncated normal distribution given by

$$P\{Y_i < z \mid Y_{[rn]} = y\} = \begin{cases} \min[1, \quad (1 - \mathrm{I}(z))/(1 - \mathrm{I}(y))], & 1 \leqq i < [rn] \\ \max[0, \quad 1 - \mathrm{I}(z)/\mathrm{I}(y)], & [rn] < i \leqq n. \end{cases}$$

If now $Y_1, \cdots, Y_n$ are reordered with probability $1/n!$ for each possible reordering, and the resulting reordered variables are labeled $X_1^1, \cdots, X_n^1$, it is easily verified that $X_1^1, \cdots, X_n^1$ have the same joint distribution as the $x_1, \cdots, x_n$ considered in Section 1. We shall use the process $Y_1, \cdots, Y_n$ to compute the conditional distribution of S_n and $\bar{x}_n$, given that $U_{r,n} = y = J(r) + w/\sqrt{n}$, say. Let

$$([nr] - 1)\bar{X}_n^1 = \sum_1^{[rn]-1} Y_i, \qquad (n - [nr])\bar{X}_n^2 = \sum_{[rn]+1}^{n} Y_i,$$

$$([nr] - 1)S_n^1 = \sum_1^{[rn]-1} Y_i^2, \qquad (n - [nr])S_n^2 = \sum_{[rn]+1}^{n} Y_i^2.$$

It is easy to compute that for the truncated normal distribution from y to ∞ mentioned above, the first four moments about the origin are

$$\mu_1(y) = g(y), \qquad \mu_2(y) = 1 + y\,g(y),$$

$$\mu_3(y) = (y^2 + 2)\,g(y), \qquad \mu_4(y) = 3 + (y^3 + 3y)\,g(y).$$

The corresponding moments for the truncated distribution from $-\infty$ to y are clearly $-\mu_1(-y)$, $\mu_2(-y)$, $-\mu_3(-y)$, and $\mu_4(-y)$. From these and [1], p. 364, we obtain that

$$A_n = \sqrt{[nr] - 1}\,(\bar{X}_n^1 + g(-y)), \qquad B_n = \sqrt{[nr] - 1}(S_n^1 - 1 + y\,g(-y))$$

are asymptotically conditionally normal with means 0 and covariance matrix

$$\begin{pmatrix} 1 - y\,g(-y) - [g(-y)]^2 & -(y^2 + 1)\,g(-y) - y[g(-y)]^2 \\ -(y^2 + 1)\,g(-y) - y[g(-y)]^2 & 2 - (y^3 + y)\,g(-y) - y^2[g(-y)]^2 \end{pmatrix},$$

and that $C_n = \sqrt{n - [nr]}\,(\bar{X}_n^2 - g(y))$ and $D_n = \sqrt{n - [nr]}\,(S_n^2 - 1 - yg(y))$ are asymptotically conditionally normal with means 0 and covariance matrix

$$\begin{pmatrix} 1 + y\,g(y) - [g(y)]^2 & (y^2 + 1)\,g(y) - y[g(y)]^2 \\ (y^2 + 1)\,g(y) - y[g(y)]^2 & 2 + (y^3 + y)\,g(y) - y^2[g(y)]^2 \end{pmatrix},$$

with (A_n, B_n) conditionally independent of (C_n, D_n). By using the Liapounoff condition one proves that the approach to normality is uniform for any finite w-interval. Let $O_p(\ )$ denote "order of ( ) in probability" (e.g., [9]). Noting that $\sqrt{[nr]} - \sqrt{nr} = O(1/\sqrt{n})$ and that $\sqrt{n}(y/n) = O_p(1/\sqrt{n})$, we obtain as an expression for $\sqrt{n}\bar{x}_n$, given that $U_{r,n} = y$,

$$\sqrt{n}\bar{x}_n = \sqrt{r}A_n + \sqrt{1-r}\,C_n + \sqrt{n}[-r\,g(-y) + (1-r)\,g(y)] + O_p(1/\sqrt{n}).$$

But

$$-r\,g\,(-J(r) - w/\sqrt{n}) + (1-r)\,g\,(J(r) + w/\sqrt{n}) = \frac{w}{\sqrt{n}} \cdot \frac{[\phi(J(r))]^2}{r(1-r)} + O_p\left(\frac{1}{n}\right),$$

so that

$$\sqrt{n}\bar{x}_n = \sqrt{r}A_n + \sqrt{1-r}\,C_n + w\,\frac{[\phi(J(r))]^2}{r(1-r)} + O_p\left(\frac{1}{\sqrt{n}}\right).$$

In similar fashion, we obtain

$$\sqrt{n}(S_n - 1) = \sqrt{r}B_n + \sqrt{1-r}D_n + wJ(r)\,\frac{[\phi(J(r))]^2}{r(1-r)} + O_p\left(\frac{1}{\sqrt{n}}\right).$$

In both cases the $O_p(1/\sqrt{n})$ term is uniform in every bounded interval of w. Since $W_n = \sqrt{n}(U_{r,n} - J(r))$ is asymptotically normal with mean 0 and variance $r(1-r)/[\phi(J(r))]^2$, the desired asymptotic joint normality follows easily from the last two displayed expressions. The covariance of $\sqrt{n}\bar{x}_n$ and W_n is most easily computed as $E\{W_n \cdot E\{\sqrt{n}\bar{x}_n \mid W_n\}\}$, that of $\sqrt{n}(S_n - 1)$ and W_n being computed similarly; these give the desired results for the last k entries of the first two rows (and columns) of (2.2).

2.5. We now show that there exists a representation of any Gaussian process with mean identically zero and a covariance function like that of (2.1), whose sample functions are continuous with probability one (w.p. 1).

Let $W(t)$, $0 \leqq t \leqq 1$, be the Kac-Siegert representation ([12]) of a Gaussian process with continuous covariance function $K'(s, t)$. Let $\{\lambda_k\}$ be the eigenvalues and $\{\varphi_k(t)\}$ the corresponding normalized eigenfunctions of $K'(s, t)$. Suppose furthermore that $g(t)$ in L^2 is such that

$$K'(s, t) - g(s)g(t) = K''(s, t)$$

is a covariance function. We shall show how to construct explicitly a process $Z(t)$ with covariance function $K''(s, t)$ in terms of the process $W(t)$. (All processes studied in this section are to have mean zero.)

We first prove two lemmas.

LEMMA 1. *A necessary and sufficient condition that $K''(s, t)$ be positive definite is that*

$$\beta^2 = \sum_{k=1}^{\infty} \frac{g_k^2}{\lambda_k} \leqq 1, \quad g_k = \int_0^1 g(t)\,\varphi_k(t)\,dt.$$

(If a λ_k is 0, it follows immediately from the positivity of $K''(s, t)$ that $g(t)$ is orthogonal to all the eigenfunctions belonging to λ_k; thus the corresponding g_k vanish and we interpret g_k^2/λ_k as zero.)

Only the necessity of the condition is used in the application of the lemma. The proof of sufficiency is included because it is so brief and sufficiency seems to be of interest.

1. *Necessity*. Set $\psi(t) = \sum_1^n v_k \varphi_k(t)$ and note that by definition

$$\int_0^1 \int_0^1 (K'(s, t) - g(s)\, g(t))\, \psi(s)\, \psi(t)\, ds\, dt \geqq 0.$$

Evaluating the double integral we obtain

$$\sum_1^n v_k^2 \lambda_k \geqq \left(\sum_1^n v_k g_k\right)^2,$$

and setting $v_k = g_k/\lambda_k$

$$\sum_1^n \frac{g_k^2}{\lambda_k} \geqq \left(\sum_1^n \frac{g_k^2}{\lambda_k}\right)^2.$$

Thus $\sum_1^n g_k^2/\lambda_k \leqq 1$ for *every* n and the theorem follows.

2. *Sufficiency*. If $\sum_1^\infty g_k^2/\lambda_k \leqq 1$ we have

$$\left(\sum_1^\infty v_k g_k\right)^2 = \left(\sum_1^n v_k \sqrt{\lambda_k} \frac{g_k}{\sqrt{\lambda_k}}\right)^2 \leqq \left(\sum_1^\infty v_k^2 \lambda_k\right)\left(\sum_1^\infty \frac{g_k^2}{\lambda_k}\right) \leqq \sum_1^\infty v_k^2 \lambda_k,$$

and positiveness of $K'(s, t) - g(s)\, g(t)$ follows.

LEMMA 2. *The series* $\sum_1^\infty g_k \varphi_k(t)$ *converges uniformly (to* $g(t)$*).*

We have $| \sum_m^n g_k \varphi_k(t)| \leqq \sqrt{\sum_m^n g_k^2/\lambda_k}\, \sqrt{\sum_m^n \lambda_k \varphi_k^2(t)}$. By Mercer's theorem, $\sum_1^\infty \lambda_k \varphi_k^2(t) = K(t, t)$. Hence $| \sum_m^n g_k \varphi_k(t)| \leqq \sqrt{K(t, t)}\, \sqrt{\sum_m^n g_k^2/\lambda_k}$. Since $\sum g_k^2/\lambda_k$ converges (by Lemma 1), Lemma 2 follows.

We are now ready to prove that

$$Z(t) = W(t) - \frac{(1 - \sqrt{1 - \beta^2})}{\beta^2}\, g(t) \sum_{k=1}^\infty \frac{g_k}{\lambda_k} \int_0^1 W(t)\, \varphi_k(t)\, dt \tag{2.4}$$

has the covariance function $K''(s, t)$.

Note first that the chance variables

$$\left\{\frac{1}{\sqrt{\lambda_k}} \int_0^1 W(t) \varphi_k(t)\, dt\right\}$$

are independently and normally distributed, with mean 0 and variance 1. Since $\sum_1^\infty g_k^2/\lambda_k < \infty$, the series

$$\sum_1^\infty \frac{g_k}{\sqrt{\lambda_k}} \cdot \frac{1}{\sqrt{\lambda_k}} \int_0^1 W(t) \varphi_k(t)\, dt$$

converges in the mean (and even w.p. 1) and thus defines a random variable.

To calculate the covariance of $Z(t)$ we need

$$E\left\{W(t)\sum_{k=1}^{\infty}\frac{g_k}{\lambda_k}\int_0^1 W(t)\varphi_k(t)\,dt\right\}.$$

But $E\{W(t)\int_0^1 W(t)\varphi_k(t)\,dt\} = \int_0^1 E\{W(t)W(t)\}\varphi_k(t)\,dt = \int_0^1 K'(t,\ t)\varphi_k(t)\,dt = \lambda_k\varphi_k(t)$, and hence

$$E\left\{W(t)\sum_{k=1}^{\infty}\frac{g_k}{\lambda_k}\int_0^1 W(t)\varphi_k(t)\,dt\right\} = \sum_{k=1}^{\infty} g_k\,\varphi_k(t) = g(t).$$

The covariance function of $Z(t)$ now comes out to be

$$K'(s,t) - 2\beta^{-2}(1-\sqrt{1-\beta^2})\,g(s)g(t) + \beta^{-2}(1-\sqrt{1-\beta^2})^2 g(s)g(t)$$
$$= K'(s,t) - g(s)g(t) = K''(s,t).$$

Since $g(t)$ is continuous (by Lemma 2) it follows, in particular, that the sample functions of $Z(t)$ are continuous w.p. 1 if the sample functions of $W(t)$ are continuous w.p. 1. Now let $K'(s,t) = \min(s,t) - st$. Then the sample functions of $W(t)$ are continuous w.p. 1. The application of the above result twice (once for each remaining negative product in (2.1)) then proves that the representation $Z(t)$, which is Gaussian with mean zero and covariance function (2.1), has continuous sample functions, w.p. 1.

2.6. From the results of Sections 2.2 to 2.5, it is easily verified that the demonstration given by Kac on pp. 197–198 of [10] for the case $K(s,t) = \min(s,t) - st$ carries over with only slight modifications to the case now under discussion where $K(s,t)$ is given by (2.1). We conclude that

$$\lim_{n\to\infty} P\{nw_n < a\} = P\{W < a\}, \tag{2.5}$$

where $W = \int_0^1 [Z(t)]^2\,dt$ and $Z(t)$ is Gaussian with covariance function given by (2.1) and sample functions continuous w.p. 1. Modifying slightly the technique of Kac and Siegert [11], [12] as applied on pp. 199–200 of [10], we now study the distribution of Z. We write

$$K(s,t) = K^*(s,t) - h_1(s)\,h_1(t) - h_2(s)\,h_2(t) \qquad 0 \leqq s, t \leqq 1, \tag{2.6}$$

where $K^*(s,t) = \min(s,t) - st$ and $h_k(s) = (2\pi)^{-1/2}[J(s)/\sqrt{2}]^{k-1}e^{-[J(s)]^2/2}$ for $k = 1, 2$. Let $\lambda_1, \lambda_2, \cdots$ be the eigenvalues (all positive, since K is positive definite) of the integral equation

$$\int_0^1 K(x,y)\,\varphi(y)\,dy = \lambda\,\varphi(x). \tag{2.7}$$

Following the demonstration of [10], we conclude that the characteristic function of W is

$$Ee^{iW\xi} = \prod_{j=1}^{\infty}(1-2i\xi\lambda_j)^{-1/2}. \tag{2.8}$$

Thus, we may express W as

$$W = \sum_{j=1}^{\infty} R_j, \tag{2.9}$$

where the R_j are independent and R_j has density function

$$\frac{1}{\sqrt{2\pi r \lambda_j}} e^{-r/2\lambda_j}, \qquad r > 0. \tag{2.10}$$

(We remark that it may of course be easier to obtain the convolution with itself of the distribution of W rather than the latter itself. This could be used if one based a test of normality on the sum of two W_m's, each computed from half the sample of size $n = 2m$.)

We now give a procedure for finding the λ_j. For any eigenvalue λ and corresponding eigenfunction φ (not necessarily normalized) of (2.7), write

$$C_i = \int_0^1 h_i(y)\, \phi(y)\, dy, \qquad i = 1, 2. \tag{2.11}$$

We can rewrite (2.7) as

$$\int_0^1 K^*(x, y)\, \varphi(y)\, dy - C_1 h_1(x) - C_2 h_2(x) = \lambda\, \varphi(x). \tag{2.12}$$

Differentiating twice with respect to x and writing $\mu^2 = 1/\lambda$ (we may consider $\mu > 0$ in the sequel), we obtain

$$\varphi''(x) + \mu^2 \varphi(x) = -C_1 \mu^2 h_1''(x) - C_2 \mu^2 h_2''(x). \tag{2.13}$$

Any eigenvalue λ and eigenfunction $\phi(x)$ of (2.7) satisfy (2.13), (2.11), and

$$\varphi(0) = \phi(1) = 0. \tag{2.14}$$

Conversely if λ and $\phi(x)$ satisfy these conditions they are an eigenvalue and eigenfunction of (2.7). For let

$$\int_0^1 K(x, y)\, \phi(y)\, dy = \lambda\, \theta(x).$$

As we obtained (2.13) we get

$$\theta''(x) + \mu^2 \phi(x) = -C_1 \mu^2 h_1''(x) - C_2 \mu^2 h_2''(x).$$

Hence $\theta''(x) = \phi''(x)$, and since also $\theta(0) = \theta(1) = 0$ we have $\theta(x) = \phi(x)$.

Our problem now is to find a value μ for which there exist a function $\phi(x)$ and constants C_1 and C_2 satisfying (2.13), (2.14), and (2.11). For given C_1 and C_2 the general solution of (2.13) can be written

$$\phi(x) = A \sin \mu x + B \cos \mu x - \mu C_1 g_1(x) - \mu C_2 g_2(x), \tag{2.15}$$

where

$$g_i(x) = \int_x^{1/2} h_i''(t) \sin \mu (t - x)\, dt. \tag{2.16}$$

Applying conditions (2.14) and (2.11) gives

$$0 = B - \mu C_1 g_1(0) - \mu C_2 g_2(0), \qquad 0 = A \sin \mu + B \cos \mu - \mu C_1 g_1(1) - \mu C_2 g_2(1),$$

$$\tag{2.17} \begin{aligned} C_i = A \int_0^1 h_i(x) \sin \mu x \, dx + B \int_0^1 h_i(x) \cos \mu x \, dx \\ - \mu C_1 \int_0^1 h_i(x)\, g_1(x) \, dx - \mu C_2 \int_0^1 h_i(x)\, g_2(x) \, dx. \end{aligned}$$

These equations have a nontrivial solution for A, B, C_1, and C_2 if and only if the determinant $D(\mu)$ of the coefficient of these four quantities is zero. Hence the eigenvalues of (2.7) are determined by the roots of $D(\mu) = 0$.

The following method of computing $D(\mu)$ is due to R. J. Walker, to whom the authors are greatly obliged.

We first note some pertinent properties of $h_i(x)$ and $g_i(x)$.

$$h_1(1 - x) = h_1(x), \qquad h_1(0) = 0, \qquad h_1(\tfrac{1}{2}) = 1/\sqrt{2\pi}, \qquad h_1'(\tfrac{1}{2}) = 0;$$

$$h_2(1 - x) = -h_2(x), \qquad h_2(0) = 0, \qquad h_2(\tfrac{1}{2}) = 0, \qquad h_2'(\tfrac{1}{2}) = 1/\sqrt{2};$$

$$\begin{aligned} g_1(1 - x) &= \int_{1-x}^{1/2} h_1''(t) \sin \mu(t - 1 + x) \, dt = \int_x^{1/2} h_1''(1 - s) \sin \mu(s - x) \, ds \\ &= g_1(x); \end{aligned}$$

and similarly

$$g_2(1 - x) = -g_2(x).$$

It follows that $g_1(1) = g_1(0)$ and $g_2(1) = -g_2(0)$, and

$$\int_0^1 h_i(x)\, g_j(x) \, dx = \begin{cases} 2 \int_0^{1/2} h_i(x) g_i(x), & j = i; \\ 0, & j \neq i. \end{cases}$$

Also, using $\sin \mu x = \sin \tfrac{1}{2}\mu \cos \mu(\tfrac{1}{2} - x) - \cos \tfrac{1}{2}\mu \sin \mu(\tfrac{1}{2} - x)$, we get

$$\int_0^1 h_1(x) \sin \mu x \, dx = 2 \sin \tfrac{1}{2}\mu \int_0^{1/2} h_1(x) \cos \mu(\tfrac{1}{2} - x) \, dx,$$

$$\int_0^1 h_2(x) \sin \mu x \, dx = -2 \cos \tfrac{1}{2}\mu \int_0^{1/2} h_2(x) \sin \mu(\tfrac{1}{2} - x) \, dx,$$

with similar reductions for the coefficients of B in (2.17).

Introducing these simplifications we get by direct computation $D(\mu) = -2 D_1(\mu)\, D_2(\mu)$, where

$$\tag{2.18} \begin{aligned} D_1(\mu) = \cos \tfrac{1}{2}\mu \left[1 + 2\mu \int_0^{1/2} h_1(x)\, g_1(x) \, dx \right] \\ - 2\mu g_1(0) \int_0^{1/2} h_1(x) \cos \mu(\tfrac{1}{2} - x) \, dx, \end{aligned}$$

$$
\begin{aligned}
D_2(\mu) &= \sin \tfrac{1}{2}\mu \left[1 + 2\mu \int_0^{1/2} h_2(x)\, g_2(x)\, dx \right] \\
&\qquad - 2\mu g_2(0) \int_0^{1/2} h_2(x) \sin \mu(\tfrac{1}{2} - x)\, dx.
\end{aligned} \tag{2.19}
$$

These equations can be put in a form more suitable for computation. Integrating (2.16) by parts gives, for $i = 1$,

$$
g_1(x) = -\frac{\mu}{\sqrt{2\pi}} \cos \mu(\tfrac{1}{2} - x) + \mu h_1(x) - \mu^2 \int_x^{1/2} h_1(t) \sin \mu(t - x)\, dt
$$

$$
g_1(0) = -\frac{\mu}{\sqrt{2\pi}} \cos \tfrac{1}{2}\mu - \mu^2 \int_0^{1/2} h_1(t) \sin \mu t\, dt.
$$

Putting these in (2.18) gives

$$
\begin{aligned}
D_1(\mu) = \cos \tfrac{1}{2}\mu &\left[1 + 2\mu^2 \int_0^{1/2} h_1^2(x)\, dx \right] \\
&+ 2\mu^3 \left[\iint_{S+T} h_1(x)\, h_1(t) \cos \mu(\tfrac{1}{2} - x) \sin \mu t\, dA \right. \\
&\qquad \left. - \iint_T h_1(x)\, h_1(t) \sin \mu(t - x) \cos \tfrac{1}{2}\mu\, dA \right],
\end{aligned}
$$

where S is the triangle bounded by the lines $t = 0$, $t = x$, and $x = \frac{1}{2}$, and T the triangle bounded by $x = 0$, $x = t$, and $t = \frac{1}{2}$. The sum of the integrals over T reduces to

$$
\iint_T h_1(x)\, h_1(t) \cos \mu(\tfrac{1}{2} - t) \sin \mu x\, dA,
$$

which equals the integral over S. Hence, finally

$$
\begin{aligned}
D_1(\mu) &= \cos \tfrac{1}{2}\mu \left[1 + 2\mu^2 \int_0^{1/2} h_1^2(x)\, dx \right] \\
&\qquad + 4\mu^3 \int_0^{1/2} h_1(x) \sin \mu x\, dx \int_x^{1/2} h_1(t) \cos \mu(\tfrac{1}{2} - t)\, dt.
\end{aligned} \tag{2.20}
$$

Similarly, (2.19) becomes

$$
\begin{aligned}
D_2(\mu) &= \sin \tfrac{1}{2}\mu \left[1 + 2\mu^2 \int_0^{1/2} h_2^2(x)\, dx \right] \\
&\qquad + 4\mu^3 \int_0^{1/2} h_2(x) \sin \mu x\, dx \int_x^{1/2} h_2(t) \sin \mu(\tfrac{1}{2} - t)\, dt.
\end{aligned} \tag{2.21}
$$

The method just developed for obtaining the eigenvalues of (2.7) seems more accurate and computationally simpler than other methods, such as those employing trial functions. In Section 6 the smallest few zeros of the functions $D_1(\mu)$ and $D_2(\mu)$ of (2.20) and (2.21) are tabulated, and an approximation for the distribution of W is thereby obtained. This approximation is compared with empirical distribution functions of nw_n obtained by sampling. Empirical distribution functions of nw_n and $\sqrt{n}v_n$ are tabulated, and certain other interesting results of the sampling experiments are noted (for example, the joint distribution of the classical Kolmogoroff and von Mises statistics D_n and ω_n^2, which are defined precisely in Section 6).

3. Tests using quantiles. The relative ease of computing the limiting distributions of various possible test criteria of the type considered in this paper will of course depend on the particular problem. Thus, the use of the sample mean and variance in non-normal cases may lead to more complicated results than those of Section 2. A tool which may be used in all cases (where the hypothesized family has finitely many natural parameters, is normal or not, where a simple sufficient statistic does or does not exist, etc.) of density functions, with about equal complexity in all cases, is the use of sample quantiles (as many as necessary) to estimate the "true" d.f. if the null hypothesis is true. (Of course, the more unknown parameters and hence quantiles which must be used, the messier will be the result.) For computational reasons, tests constructed in this manner will sometimes be more practical to use than those involving a sufficient statistic. The results on power in Section 5 apply also here.

As an example, suppose the pth sample quantile $U_{p,n}$, $0 < p < 1$, is to be used to estimate the corresponding population parameter of a family of d.f.'s $F(x - \theta)$ for $-\infty < \theta < \infty$, in testing whether or not the "true" d.f. is a member of this family. We suppose without loss of generality that $F(0) = p$, and we denote by f the density function of F. We assume f to be continuous and positive in a neighborhood of 0.

Then letting $\psi = F^{-1}$ and $Z_{r,n} = \sqrt{n}[F_n(\psi(r) + U_{p,n}) - r]$, an argument like that of Sections 2.3 and 2.4 leads easily to the conclusion that, for $0 \leqq r_1 \leqq \cdots \leqq r_k \leqq 1$, the limiting distribution of $Z_{r_i,n}$, $1 \leqq i \leqq k$, is the same as that of

$$\sqrt{n}[U_{p,n} - [U_{r_i,n} - \psi(r_i)]] \,/\, \psi'(r_i), \qquad 1 \leqq i \leqq k,$$

and, in particular, is Gaussian. Putting $\gamma(r) = f(\psi(r))$ and

$$g(r) = \sqrt{p(1-p)}\,\frac{\gamma(r)}{\gamma(0)} - \frac{\min[p, r] - pr}{\sqrt{p(1-p)}},$$

we obtain for the limiting covariance function, for $0 \leqq s,t \leqq 1$,

$$\begin{aligned} K(s, t) &= \gamma(s)\,\gamma(t)\left[\frac{p(1-p)}{[\gamma(0)]^2} - \frac{\min(p, s) - ps}{\gamma(0)\,\gamma(s)} \right.\\ &\qquad\left. - \frac{\min(p, t) - pt}{\gamma(0)\,\gamma(t)} + \frac{\min(s, t) - st}{\gamma(s)\,\gamma(t)}\right] \\ &= g(s)\,g(t) - \frac{[\min(p, s) - ps][\min(p, t) - pt]}{p(1-p)} + \min(s, t) - st \end{aligned}$$

$$= g(s)\,g(t) + \begin{cases} \min(s, t) - st/p & s, t \leqq p, \\ \min(s - p, t - p) - \dfrac{(s-p)(t-p)}{1-p} & s, t \geqq p, \\ 0 & \text{otherwise.} \end{cases}$$

For computing purposes (e.g., by Monte Carlo methods), the last form of the covariance function is useful. Let $X_1(t)$ and $X_2(t)$ be Gaussian, each with the familiar covariance function $\min(s, t) - st$, and let X be a normal random variable with mean zero and variance 1, where X, X_1, and X_2 are independent. Defining $X_1(t) = X_2(t) = 0$ if $t < 0$ or $t > 1$, let

$$Z(t) = g(t)X + \sqrt{p}\,X_1\left(\frac{t}{p}\right) + \sqrt{1-p}\,X_2\left(\frac{t-p}{1-p}\right), \quad 0 \leqq t \leqq 1.$$

Then Z has the covariance function $K(s, t)$. As an example, if F is normal with unit variance and $p = \frac{1}{2}$, we have

$$(Zt) = [\tfrac{1}{2}e^{-\psi(t)^2/2} = \min(t, 1-t)]X + \frac{1}{\sqrt{2}}X_1(2t) + \frac{1}{\sqrt{2}}X_2(2t-1).$$

4. The rectangular distribution. In Section 4.1 we give an example where the minimum distance method statistic can be computed explicitly; in Section 4.2 we comment on the limiting distribution of test criteria of the type treated in Section 2 in the present case.

4.1. Let

$$F(x; \theta) = \begin{cases} 0 & x < \theta - \frac{1}{2}, \\ x - \theta + \frac{1}{2} & \theta - \frac{1}{2} \leqq x \leqq \theta + \frac{1}{2}, \\ 1 & \theta + \frac{1}{2} < x, \end{cases}$$

and let R denote the family of all such distribution functions for $-\infty < \theta < \infty$. It is desired to test the hypothesis that $x_1, x_2, \cdots, x_n$ are independently and identically distributed according to some member of R. We will be concerned with computations when this hypothesis is true (see Section 5 for remarks on power which apply also here), and denote by G the true member of R (i.e., the distribution function of X_1). G_n^* is as defined in Section 1. Let

$$D_n^+ = \sup_x (G_n^*(x) - G(x)), \qquad D_n^- = \sup_x (G(x) - G_n^*(x)).$$

In the present example the minimum distance criterion is easily seen to be

$$\delta(G_n^*, R) = \inf_\theta \sup_x |G_n^*(x) - F(x; \theta)| = \tfrac{1}{2}(D_n^+ + D_n^-). \tag{4.1}$$

The joint limiting distribution of $\sqrt{n}D_n^+$ and $\sqrt{n}D_n^-$ (as $n \to \infty$) is given by Doob ([13], p. 403) for $x,y > 0$ as

$$\lim_{n\to\infty} P\{\sqrt{n}D_n^- \leqq x, \sqrt{n}D_n^+ \leqq y\}$$

$$= 1 - \sum_{m=1}^{\infty} \{e^{-2(mx+(m-1)y)^2} + e^{-2(my+(m-1)x)^2} - 2e^{-2m^2(x+y)^2}\} \tag{4.2}$$

$$= G(x, y), \quad \text{say.}$$

Let $g(x, y) = \partial^2 G(x, y)/\partial x \partial y$. The series in (4.2) is uniformly and absolutely convergent and differentiable with respect to x and y outside of any neighborhood of the origin. Using the fact that the mixed derivative of the expression $e^{-2(ax+by)^2}$ is $4ab(u^2 - 1)e^{-u^2/2}$ where $u = (2ax + 2by)^2$, we obtain for the mth term of the series for $\int_0^z g(y, z - y)\, dy$ the expression

$$4m(m-1)[H(2mz) - H(2(m-1)z)] + 4m(4m^2z^2 - 1)H(2mz),$$

$$H(x) = \int_x^\infty (u^2 - 1)e^{-u^2/2}\, du = xe^{-x^2/2}. \tag{4.3}$$

The density function corresponding to the limiting distribution function of $B_n = \sqrt{n}(D_n^+ + D_n^-)$ is a sum of expressions given in (4.3). For $\sqrt{n}\delta(G_n^*, R) = \frac{1}{2}B_n = U$, say, the expression corresponding to (4.3) is

$$8m(m-1)[H(4mu) - H(4(m-1)u)] + 8m(16m^2u^2 - 1)H(4mu). \tag{4.4}$$

The series is absolutely convergent and in the sum of (4.4) from 1 to ∞, the coefficient of the expression $H(4ku)$ for $k \geqq 0$ is $8k(16k^2u^2 - 3)$, so that the density function of the limiting cumulative distribution function of U is

$$32u \sum_{m=1}^{\infty} m^2(16m^2u^2 - 3)e^{-8m^2u^2}. \tag{4.5}$$

Outside any given neighborhood of the origin, all terms of (4.5) are positive except for a finite number. Thus, integrating (4.5) we obtain for $u > 0$

$$\lim_{n\to\infty} P\{\sqrt{n}\delta(G_n^*, R) \leqq u\} = 1 - \sum_{m=1}^{\infty} (32m^2u^2 - 2)e^{-8m^2u^2}. \tag{4.6}$$

It is also of interest that in this example we can compute the limiting distribution of the minimum distance *estimator* of θ, namely, the random variable $T_n = T_n(x_1, \cdots, x_n)$ defined by

$$\delta(G_n^*(y), F(y; T_n)) = \delta(G_n^*, R), \tag{4.7}$$

which is satisfied by $T_n - \theta = \frac{1}{2}(D_n^- - D_n^+)$ when θ is the true parameter value. An analysis similar to that given above shows that, for $t > 0$,

$$\lim_{n\to\infty} P\{\sqrt{n}\,|T_n - \theta| \leqq t\} = 1 - 2\sum_{m=1}^{\infty} \frac{1}{4m^2 - 1} e^{-8m^2t^2}. \tag{4.8}$$

Of course, in this simple parametric example there are estimators of order $1/n$ in probability; in *estimation* problems, the minimum distance method is most useful in examples of a more nonparametric nature, where it often yields consistent estimators when other methods do not.

4.2. It is interesting to note that in the example of Section 4.1 and other similar cases it is simple to design along the lines of Section 2 a test of whether or not the unknown d.f. belongs to the specified class, where the limiting distribution of the test criterion when the null hypothesis is true is already known. These are the so-called "irregular" cases of estimation where an estimator of the unknown parameter(s) indexing the class exists whose deviation from the true parameter value is of lower order in probability than the usual $1/\sqrt{n}$ encountered in "regular" cases; for example, of order $1/n$ for the rectangular distribution with unknown location and/or range, or for the exponential distribution with known scale but unknown location parameter. For the sake of definiteness, we fix our attention on a Kolmogoroff-type criterion for the latter example, although the result applies equally to other distributions and other criteria (ω^2-type tests with different "weight"-functions, etc). Thus, the problem is to test, on the basis of n observations, whether or not the true d.f. is of the form

$$0, \quad x < \theta, \qquad 1 - e^{-(x-\theta)}, \quad x \geqq \theta,$$

for some real θ. Let $T_n = \min(x_1, \cdots, x_n)$. Then $P_\theta\{\lim_{n\to\infty}\sqrt{n}(T_n - \theta) = 0\} = 1$. Hence, if we compare the sample d.f. with the exponential c.d.f. as estimated by using T_n, by computing

$$B_n = \sqrt{n} \sup_{0<r<1} \left| F_n\left(T_n + \log \frac{1}{1-r}\right) - r \right|,$$

we may conclude that, when the null hypothesis is true, B_n has the same limiting distribution as the Kolmogoroff statistic.

Remarks on power like those of Section 5 apply also to the present case. The present remarks may also be modified to apply to situations where one but not all parameters have irregular estimators, for example, for the case of the exponential distribution with unknown location *and* scale. For the case of the rectangular distribution with unknown range studied in Section 4.1, it seems intuitively reasonable that a test constructed in the manner of the present section may be more powerful than the one considered there.

5. Asymptotic power of the tests of normality. The results of this section are carried out for the tests of normality mentioned in Section 2, but the remarks below concerning v_n may also be carried through for the minimum distance test, the test of Section 3, and in many other examples, and the remarks concerning v_n and w_n may be extended to many other "distance" criteria.

First we consider the test of size α based on v_n. The critical region is of the form $\{v_n > b(\alpha)/\sqrt{n}\}$, where $b(\alpha)$ is a constant, except for terms of lower order in n. Suppose $G(x) \equiv R_0(x)$, and that $\delta(R_0, N^{**}) = d/\sqrt{n}$. From the theorem of Kolmogoroff [4] we have that $\delta(R_0, G_n^*)$ is of the order $1/\sqrt{n}$ in probability (uniformly in R_0). Hence, for $0 < \beta < 1$ there is a number $d^* = d^*(\alpha, \beta)$ such that $d > d^*$ implies that $\delta(G_n^*, N^{**}) > b(\alpha)/\sqrt{n}$ with probability $\geqq 1 - \beta$. From the definition of $\delta(G_n^*, N^{**})$ we have

$$\delta(G_n^*, N(y \mid \bar{x}, s^2)) \geqq \delta(G_n^*, N^{**}).$$

Thus, if we are using the test of size α, the power is at least $1 - \beta$ for any alternative R_0 whose distance from N^{**} is $\geqq d^*(\alpha, \beta)/\sqrt{n}$, and the power of the test at R_0 approaches one as $\sqrt{n}\delta(R_0, N^{**})$ increases indefinitely. This is a result of the same order as obtains for testing goodness of fit of a simple hypothesis by use of Kolmogoroff's distribution (Section 1).

We now consider the "ω^2-type" test criterion w_n.[5] We consider the function

$$\gamma(F, H) = \left\{\int_{-\infty}^{\infty} [F(x) - H(x)]^2 \, d\left(\frac{F(x) + H(x)}{2}\right)\right\}^{1/2} \tag{5.1}$$

as a possible measure of discrepancy between two d.f.'s. This measure has been used by Lehmann [14] and others. (We remark that γ is not a metric, since it does not satisfy the triangle inequality. Another undesirable property of γ is that the discrepancy between the d.f.'s of two random variables X and Y may not be the same as that between the d.f.'s of $-X$ and $-Y$. Also, the failure of the formula for integration by parts in expressions like (5.1) necessitates slight complications, for example, in the second following paragraph. Neither of the last two difficulties is present if both d.f.'s are continuous. If the d.f.'s have jumps which are nowhere dense, one could eliminate these last difficulties by redefining γ, for example, by replacing each jump by a constant density over an interval of width ϵ about the jump and letting $\epsilon \to 0$ after integrating. The development which follows would not be materially altered by such a change in the definition of γ.)

If F and H are continuous,

$$\int_{-\infty}^{\infty} [F(x) - H(x)]^2 \, d[F(x) - H(x)] = 0,$$

and the integration in (5.1) may be carried out with respect to either F or H instead of $\frac{1}{2}(F + H)$. Hence, if $G(x) = R_0(x)$ is continuous,

$$\begin{aligned} w_n^{1/2} &= \left\{\int [G_n^*(x) - N(x \mid \bar{x}, s^2)]^2 \, d_x N(x \mid \bar{x}, s^2)\right\}^{1/2} \\ &\geqq \left\{\int [R_0(x) - N(x \mid \bar{x}, s^2)]^2 \, d_x N(x \mid \bar{x}, s^2)\right\}^{1/2} \\ &\quad - \left\{\int [R_0(x) - G_n^*(x)]^2 \, d_x N(x \mid \bar{x}, s^2)\right\}^{1/2} \\ &\geqq \gamma(R_0, N^{**}) - \delta(R_0, G_n^*), \end{aligned} \tag{5.2}$$

where

$$\gamma(R_0, N^{**}) = \inf_{N \varepsilon N^{**}} \gamma(R_0, N^{**}).$$

Now, $\delta(R_0, G_n^*)$ is of order $1/\sqrt{n}$ in probability (uniformly in R_0) and the critical region based on w_n is of the form $\{\sqrt{w_n} > c(\alpha)/\sqrt{n}\}$, where $c(\alpha)$ is constant except for terms of lower order in n. Hence, using (5.2), an argument like that

[5] In an unpublished manuscript, T. W. Anderson considers similar criteria for testing a simple hypothesis, and obtains similar results on asymptotic power. See also his abstract (*Ann. Math. Stat.*, Vol. 25 (1954), p. 174).

of the previous paragraph shows that there is a value $d'(\alpha, \beta)$ such that the test of size α has power $\geqq 1 - \beta$ for any continuous alternative R_0 for which $\gamma(R_0, N^{**}) \geqq d'(\alpha, \beta)/\sqrt{n}$.

We now consider what happens if R_0 is not continuous. It is easy to show, by consideration of the contribution to the integral of (5.1) at discontinuities of H, that, if F is continuous,

$$\int (F - H)^2 \, dF \geqq \tfrac{1}{4} \int (F - H)^2 \, dH,$$

so that

$$\int (F - H)^2 \, dF \geqq \tfrac{2}{5} \int (F - H)^2 \, d\left(\frac{F + H}{2}\right).$$

Hence, if R_0 is not continuous, the argument of the previous paragraph need only be altered by inserting the factor $\sqrt{2/5}$ before γ in the last expression of (5.2). The ensuing discussion of power then proceeds as before.

It is interesting to compare the measurements of distance δ and γ. Clearly, $\gamma(F, G) \leqq \delta(F, G)$ for all F,G. On the other hand, if there is a value x_0 for which $F(x_0) - G(x_0) = \delta$, it is clear from the monotonicity of F and G, using (5.1), that $[\gamma(F, G)]^2 \geqq \frac{1}{6}[\delta(F, G)]^3$. Thus, we have

$$\sqrt{1/6}\, \delta^{3/2} \leqq \gamma \leqq \delta, \tag{5.3}$$

where both equalities are attainable. We conclude that, whereas for any given $\beta > 0$, the power exceeds $1 - \beta$ for alternatives whose distance from N^{**} is of order $1/\sqrt{n}$ either according to δ for the test based on v_n or according to γ for the test based on w_n, the distance in terms of δ must be of order $1/\sqrt[3]{n}$ (in the worst case) to insure this for the latter test. For the former test, γ-distance of order $1/\sqrt{n}$ suffices.

We next verify the property of the χ^2-test relative to γ which was stated at the end of the third paragraph of the introduction. For brevity we shall use the notation of [3] without redefining symbols here; the reader may also refer to [3] for details of the argument which we omit. Suppose then that we are testing the hypothesis $G(x) = R(x) \equiv x$ for $0 \leqq x \leqq 1$ by means of the χ^2-test based on N observations and k_N intervals of equal length on the unit interval. If now

$$G(x) = F_{k_N}(x) \equiv \sum_{i=1}^{k_N} \frac{1}{k_N}\, \psi_x\left(\frac{2i - 1}{2k_N}\right)$$

(see Section 1 for the definition of ψ_x), then $\gamma(F_{k_N}, R) = 1/k_N\sqrt{6}$. Since F_{k_N} assigns the same probability as R to each of the k_N intervals, the power of the test against the alternative F_{k_N} is just the size of the test (assumed to be $< \frac{1}{2}$). We conclude that if the test gives power $\geqq \frac{1}{2}$ for all alternatives R^* satisfying $\gamma(R^*, R) \geqq \Gamma_N$, then we must have

$$k_N > 1/\Gamma_N\sqrt{6}. \tag{5.4}$$

(This is not the best possible inequality, but it suffices for our proof.) Consider now the distribution function

$$(5.5)\qquad H_a(x) = \begin{cases} (1+2a)x, & 0 \leqq x \leqq \frac{1}{2}, \\ 2a + (1-2a)x, & \frac{1}{2} \leqq x \leqq 1, \end{cases}$$

where $0 < a < \frac{1}{2}$.

A simple computation shows that $\gamma(H_a, R) = a/\sqrt{3} = \gamma_a$, say, and that (assuming for simplicity that k_N is even) when $G = H_a$ we have $\sum p_i^2 = (1+4a^2)/k_N = (1+12\gamma_a^2)/k_N$. Hence, the function ψ of [3] is given, when $G = H_a$, by

$$(5.6)\qquad \sigma'\psi(k_N) = 12(N-1)\,\gamma_a^2 - C\sqrt{2(k_N-1)}.$$

In order that the χ^2-test based on k_N intervals have power $\geqq \frac{1}{2}$ for all alternatives R^* with $\gamma(R^*, R) \geqq \Gamma_N$, it is necessary that the expression (5.6) be $\geqq 0$ asymptotically when we put $\gamma_a = \Gamma_N$, and that (5.4) be satisfied. We thus obtain, when N is large,

$$(5.7)\qquad \Gamma_N \geqq C'N^{-2/5}$$

where C' is a positive constant. From the result of [3] and the fact that $\gamma \leqq \delta$, we see that the reverse inequality to (5.7) is (for a different C') also true. Thus, $N^{-2/5}$ is indeed the smallest order of Γ_N which will give appreciable power for all alternatives R^* with $\delta(R, R^*) \geqq \Gamma_N$.

We shall now summarize the results proved thus far in this section. It is not known how the power function of the χ^2-test for composite hypotheses behaves, but it is plausible that the power function when testing a composite hypothesis by means of the χ^2-test (in any of its variations) is no better (in the sense we have used in measuring the goodness of a power function) than when testing a simple hypothesis. We have shown that *if* Γ *and* Δ *are small and the* χ^2*-test of size* $< \frac{1}{2}$ *of a simple hypothesis* $G = R$ *requires* N *observations to insure power* $\geqq \frac{1}{2}$ *at alternatives* R^* *for which* $\gamma(R^*, R) = \Gamma$ *(or* $\delta(R^*, R) = \Delta$*), so that* $N = C_1\Gamma^{-5/2}$ *(or* $N = C_2\Delta^{-5/2}$*), then the numbers of observations required by the Kolmogoroff and* ω^2 *tests to achieve the same minimum power at* Γ *(or* Δ*) are at most*

Kolmogoroff: $\quad n = C_3N^{4/5} = C_4\Gamma^{-2} \qquad (n = C_5N^{4/5} = C_6\Delta^{-2})$

ω^2: $\quad n = C_7N^{4/5} = C_8\Gamma^{-2} \qquad (n = C_9N^{6/5} = C_{10}\Delta^{-3}).$

The numbers of observations n *required by the tests based on* v_n *and* w_n *in testing composite hypotheses about parametric families are the same functions of* Γ *or* Δ *as for the Kolmogoroff and* ω^2 *tests of simple hypotheses.* The test based on v_n may thus be expected to be superior to the χ^2-test in the sense of both γ and δ, and that based on w_n may be expected to be superior in the sense of γ, at least for large N.

It might be supposed that the χ^2-test would show up to better advantage relative to the test based on v_n in terms of a metric like

$$\eta(R_0, N^{**}) = \inf_{N \in N^{**}} \int d\,|R_0 - N|.$$

However, this is not so; for fixed n, even if $\eta(R_0, N^{**})$ is near its maximum of 2, neither the best χ^2-test of [3] nor any of the other tests we have mentioned need have appreciable minimum power ($\geqq \frac{1}{2}$), and $\delta(R_0, N^{**})$ and $\gamma(R_0, N^{**})$ can be arbitrarily small. In fact, it is easy to see that no test can have the infimum of its power function over all alternatives R_0 with $\eta(R_0, N^{**}) = C > 0$ greater than the size of the test. (If N^{**} were a simple hypothesis, this would still be true.) In order better to compare the behavior of tests in terms of the metric η, we might therefore restrict our consideration to alternatives R_0 belonging to some regular class, for example, the class of d.f.'s with densities which cross that of each member of N^{**} at most M times. Under such a comparison the test based on v_n may be shown to be superior to that based on the χ^2-test, in the same sense as under our previous comparison.

The discussion of this section suggests very strongly that there is a "natural" distance with respect to which the power characteristics of a particular "distance" test criterion should be measured, and that a comparison of the power of such tests in terms of their own and other distances indicates that tests corresponding to strong metrics have the best global power characteristics. It is hoped to investigate this idea further.

6. Numerical results. In this section we list some pertinent experimental and computational results. The main purpose of the sampling experiments was to obtain estimates of the distributions of nw_n and $\sqrt{n}v_n$ which may be used in applications to test normality. As a check on the sampling experiments the same data were used to compute experimentally the d.f. of $\bar{x}$ and the d.f. of $\sqrt{n}D_n$ with n large, D_n being defined by

$$D_n = \delta(L(x), L_n^*(x))$$

where $L_n^*(x)$ is the empiric d.f. of n independent chance variables with the d.f. $L(x)$. Also, as another check on the experimentally obtained d.f. of nw_n, an approximation to the d.f. of W was computed, using (2.9).

Define

$$\omega_n^2 = \int (L(x) - L_n^*(x))^2\, dL(x).$$

The sampling experiments were conducted using 400 samples of size $n = 100$ and 400 samples of size $n = 25$, of random standard (mean zero, variance one) normal deviates from the well-known Rand Corporation series. From each sample the values of the sample mean ($\bar{x}$), sample variance (s^2), v_n, w_n, D_n, and ω_n^2, were computed. Thus, there were obtained 400 "observations" on $\sqrt{n}v_n$ and nw_n for $n = 25$ and $n = 100$, and the sample d.f.'s based on these observations serve as estimates of the d.f.'s of $\sqrt{n}v_n$ and nw_n. The known d.f. of $\bar{x}$

and the known d.f. of $\sqrt{n}D_n$ were compared with the experimentally obtained d.f.'s of $\bar{x}$ and $\sqrt{n}D_n$ as a check on the experimentally obtained d.f.'s of $\sqrt{n}v_n$ and nw_n.

It was found that the experimentally obtained d.f. of $\bar{x}$ agreed well with the known d.f., and the experimentally obtained d.f. of $\sqrt{n}D_n$ agreed well with the d.f. of $\sqrt{n}D_n$ as tabulated in [15]. In each case, the maximum difference between the experimentally obtained and the known actual d.f. was found to be small according to the tables [15]; this is the basis for saying the agreement was good. The agreement between the limiting d.f. of $n\omega_n^2$ (tabulated in [16]) and the experimentally obtained d.f. was found to be fairly close for $n = 100$ but not close for $n = 25$; the upper tails of the distributions seem to be the parts which come into agreement most rapidly with n. The distributions of $\sqrt{n}v_n$ and nw_n are concentrated closer to the origin than those of $\sqrt{n}D_n$ and $n\omega_n^2$, respectively. This is not entirely surprising since the covariance function (2.1) is smaller than that for the process corresponding to D_n and ω_n^2 (namely, min $(s, t) - st$). Scatter diagrams of the sample values of $(\sqrt{n}D_n, n\omega_n^2)$ indicate that the regression of $n\omega_n^2$ on $\sqrt{n}D_n$ is roughly parabolic with the conditional variance of $n\omega_n^2$ increasing with the value of $\sqrt{n}D_n$; this is not entirely surprising in view of the way in which ω_n^2 and D_n are computed from a sample. Although v_n and w_n are correlated with D_n and ω_n^2, they seem to be more strongly related to $\bar{x}$.

In Tables I and II, respectively, are given the estimates of the d.f.'s of $\sqrt{n}v_n$ and nw_n; commonly used percentage points are listed for convenience in Table III.

We now turn to the limiting distribution of nw_n. The first eight zeros of $D(\mu)$ are alternately zeros of $D_1(\mu)$ and $D_2(\mu)$. Their values, obtained by numerical computation of these functions for various values of μ, are 7.38, 8.62, 13.66, 15.14, 19.91, 21.52, 26.16, 27.87. The corresponding values of λ_j for $1 \leqq j \leqq 8$ are .01836, .01346, .00536, 00436, .00252, .00216, .00146, .00129. It is to be noted that λ_{2j-1} and λ_{2j} are both approximately c/j^2 (the corresponding property for the eigenvalues arising in the computation of $n\omega_n^2$ is that $\lambda_j = 1/\pi^2 j^2$). This (inferred) speed of convergence of the λ_j to 0 implies that the distribution of $W^* = \sum_1^4 \lambda_j \gamma_j$ should be a fairly good approximation to that of $W = \sum_1^\infty \lambda_j \gamma_j$,

TABLE I
Estimate of $Q_n(x) = P\{\sqrt{n}v_n \leqq x\}$.

x	$Q_{25}(x)$	$Q_{100}(x)$	x	$Q_{25}(x)$	$Q_{100}(x)$
.30	0	0	.75	.8100	.8425
.35	.0125	.0025	.80	.8600	.9025
.40	.0550	.0250	.85	.9125	.9350
.45	.1500	.0800	.90	.9500	.9600
.50	.2525	.1975	.95	.9725	.9775
.55	.3675	.3300	1.00	.9850	.9900
.60	.5025	.4775	1.05	.9950	.9925
.65	.6400	.6750	1.10	1.0000	.9975
.70	.7225	.7325	1.15	1.0000	1.0000

TABLE II

Estimate of $R_n(x) = P\{nw_n \leqq x\}$ and distribution function $H(x) = P\{W^* \leqq x\}$.

x	$R_{25}(x)$	$R_{100}(x)$	$H(x)$
.01	0	0	.108
.02	.0375	.0225	.297
.03	.1325	.1400	.471
.04	.2125	.3175	.609
.05	.4300	.4775	.712
.06	.5825	.6250	.787
.07	.6625	.7175	.843
.08	.7250	.7850	.883
.09	.7925	.8600	.914
.10	.8400	.8900	.937
.11	.8975	.9175	.953
.12	.9150	.9350	.966
.13	.9425	.9700	.975
.14	.9525	.9825	.982
.15	.9650	.9850	.987
.16	.9700	.9850	.997
.17	.9800	.9875	
.18	.9875	.9875	
.19	.9875	.9900	
.20	.9900	.9950	
.21	.9950	.9950	
.22	.9975	.9975	
.23	1.0000	1.0000	

TABLE III

Estimates of common percentage points of Q_n and R_n.

p	$Q_{25}^{-1}(p)$	$Q_{100}^{-1}(p)$	$R_{25}^{-1}(p)$	$R_{100}^{-1}(p)$
.20	.7435	.729	.0909	.0824
.10	.8225	.797	.1145	.1019
.05	.8980	.878	.1352	.1240
.02	.9685	.954	.1671	.1386
.01	1.0145	.989	.1957	.1859
.005	1.0465	1.062	.2053	.1957

and the distribution of W^* was therefore computed, using the method of [17] (here $\lambda_j\gamma_j$ is the R_j of (2.9)). The results are given in the last column of Table II.

From the fact noted above regarding the speed of convergence of the d.f. of $n\omega_n^2$, and the fact that Table II indicates (in the difference $R_{100}(x) - R_{25}(x)$) a much slower approach to the limiting distribution for $R_n(x)$ than that noted in our experiment for the ω_n^2-distribution, we would expect that the d.f. $H(x)$ of W^* should lie above the estimates $R_n(x)$ of Table II, being close to $R_{100}(x)$ only in the upper tail. This is what the last column of Table II actually shows, and an idea of how good the agreement is in the tail may be obtained by computing $M(x) = 20[H(x)]^{-1/2}[H(x) - R_{100}(x)]$; for $x = .13, .14, .15, .20$, one obtains $M(x) < 1$, which indicates very good agreement. Thus, in applications where n is large, it seems reasonable to use the last column of Table II, especially in the upper tail (which is the region that matters for statistical tests).

We are indebted to Prof. R. J. Walker and to Mr. R. C. Lesser of the Cornell Computing Center for carrying out the sampling and the computation of the λ_j's. The experimental data are available in the files of the Cornell Computing Center.

REFERENCES

[1] H. Cramér, *Mathematical Methods of Statistics*, Princeton University Press, 1946.
[2] H. Chernoff and E. L. Lehmann, "The use of maximum likelihood estimates in χ^2 tests for goodness of fit," *Ann. Math. Stat.*, Vol. 25 (1954), pp. 579–586.
[3] H. B. Mann and A. Wald, "On the choice of the number of class intervals in the application of the chi square test," *Ann. Math. Stat.*, Vol. 13 (1942), pp. 306–317.
[4] A. N. Kolmogoroff, "Sulla determinazione empirica di una leggi di distribuzione," *Giorn. Ist. Ital. Attuari*, Vol. 4 (1933), pp. 83–91.
[5] J. Wolfowitz, "Consistent estimators of the parameters of a linear structural relation," *Skand. Aktuarietids.*, Vol. 35 (1952), pp. 132–151.
[6] J. Wolfowitz, "Estimation by the minimum distance method," *Ann. Inst. Stat. Math., Tokyo*, Vol. 5 (1953), pp. 9–23.
[7] J. Wolfowitz, "Estimation by the minimum distance method in nonparametric stochastic difference equations," *Ann. Math. Stat.*, Vol. 25 (1954), pp. 203–217.
[8] J. Wolfowitz,[6] "Estimation of the components of stochastic structures," *Proc. Nat. Acad. Sci., U.S.A.*, Vol. 40 (1954), pp. 602–606.
[9] H. B. Mann and A. Wald, "On stochastic limit and order relationships," *Ann. Math. Stat.*, Vol. 14 (1943), pp. 217–226.
[10] M. Kac, "On some connections between probability theory and differential and integral equations," *Proceedings of the Second Berkeley Symposium on Mathematical Statistics and Probability*, University of California Press, 1951, pp. 180–215.
[11] M. Kac and A. J. F. Siegert, "On the theory of noise in radio receivers with square law detectors," *J. Applied Physics*, Vol. 18 (1947), pp. 383–397.
[12] M. Kac and A. J. F. Siegert, "An explicit representation of stationary Gaussian processes," *Ann. Math. Stat.*, Vol. 18 (1947), pp. 438–442.
[13] J. L. Doob, "Heuristic approach to the Kolmogorov-Smirnoff theorems," *Ann. Math. Stat.*, Vol. 20 (1949), pp. 393–403.
[14] E. L. Lehmann, "Consistency and unbiasedness of certain nonparametric tests," *Ann. Math. Stat.*, Vol. 22 (1951), pp. 165–179.
[15] Z. W. Birnbaum, "Numerical tabulation of the distribution of Kolmogoroff's statistic for finite sample size," *J. Amer. Stat. Assn.*, Vol. 47 (1952), pp. 425–441.
[16] T. W. Anderson and D. A. Darling, "Asymptotic theory of certain 'goodness of fit' criteria based on stochastic processes," *Ann. Math. Stat.*, Vol. 23 (1952), pp. 193–212.
[17] H. Robbins and E. J. G. Pitman, "Application of the method of mixtures of quadratic forms in normal variates," *Ann. Math. Stat.*, Vol. 20 (1949), pp. 552–560.
[18] D. A. Darling, "The Cramér-Smirnov test in the parametric case" (abstract), *Ann. Math. Stat.*, Vol. 24 (1953), p. 493.

---

[6] The sentence at the bottom of page 602 of [8] which reads, "It is well known that $F(z)$ determines α . . ." should read, "It is well known that the distribution of (x_i, x_{i+1}) determines α. . . ."

Reprinted from THE ANNALS OF MATHEMATICAL STATISTICS
Vol. 27, No. 1, March, 1956

ON THE CHARACTERISTICS OF THE GENERAL QUEUEING PROCESS, WITH APPLICATIONS TO RANDOM WALK[1]

BY J. KIEFER AND J. WOLFOWITZ

Cornell University

Summary. The authors continue the study (initiated in [1]) of the general queueing process (arbitrary distributions of service time and time between successive arrivals, many servers) for the case ($\rho < 1$) where a limiting distribution exists. They discuss convergence with probability one of the mean waiting time, mean queue length, mean busy period, etc. Necessary and sufficient conditions for the finiteness of various moments are given. These results have consequences for the theory of random walk, some of which are pointed out.

This paper is self-contained and may be read independently of [1]; the necessary results of [1] are quoted. No previous knowledge of the theory of queues is required for reading either [1] or the present paper.

Introduction. We recapitulate very briefly some of the results obtained in [1] in the notation of [1] to which we shall adhere without further mention.[2]

Let S be the totality of points $(x_1, x_2, \cdots, x_s)$ of Euclidean s-space such that $0 \leqq x_1 \leqq x_2 \leqq \cdots \leqq x_s$. Let x and y be generic points of S. Occasionally another letter will represent a point in S; it will always be clear from the context when this is so; for example, O will frequently denote the origin in s-space.

For $i \geqq 1$, let $t_i \geqq t_0 = 0$ be the time of arrival of the ith person at a system of $s \geqq 1$ machines, where he waits his turn until a machine is available to serve him, say at time $t_i + w_{i1} \geqq t_i$. This machine is then occupied by him for time $R_i \geqq 0$. Let $g_i = t_i - t_{i-1}$. $\{R_i\}$ and $\{g_i\}$ are independent sequences of identically distributed and independent chance variables. An s-dimensional random walk $\{w_i\}$, with w_{i1} its first component, is useful for the study of the theory of queues. The random walk $\{w_i\}$ is constructed as follows: $w_i = (w_{i1}, \cdots, w_{is})$. Unless the contrary is explicitly stated we have $w_1 = O$. To obtain w_{i+1} from w_i, reorder in ascending size the quantities

$$(w_{i1} + R_i - g_{i+1})^+, \qquad (w_{i2} - g_{i+1})^+, \qquad (w_{i3} - g_{i+1})^+, \cdots, (w_{is} - g_{i+1})^+.$$

The resulting sequence is w_{i+1}. We have $w_{i1} \leqq w_{i2} \leqq \cdots \leqq w_{is}$ for all i. As usual, $a^+ = (a + |a|)/2$. The times $t_i + w_{ij}$ $(1 \leqq j \leqq s)$ are easily seen to be the earliest times after (or at) t_i at which the s machines have finished serving those of the first $s - 1$ arrivals which they serve.

Let $F_i(F_i^*)$ be the d.f. (distribution function) of $w_i(w_{i1})$. It was shown in [1] that $F(x) = \lim_{i\to\infty} F_i(x)$ exists and satisfies a certain integral equation (I.E.);

Received Dec. 21, 1954.

[1] Research under contract with the Office of Naval Research.

[2] The definition of ν on p. 14 of [1] should be modified trivially to read $\nu = 1$ in the case $b = \infty$.

$F^*(z) = \lim_{i\to\infty} F_i^*(z)$ also exists. Assume $\rho = ER_i / sEg_i$ exists. F and F^* are d.f.'s if $\rho < 1$, and F is then the unique d.f. solution to the I.E. Except in the trivial case where $P\{R_i = sg_i\} = 1$, if $\rho \geqq 1$ then $F \equiv 0 \equiv F^*$, and the I.E. has no d.f. solution. Always $F^*(z) = F(z, \infty, \cdots, \infty)$. Results on the limiting length of the line are also proved in [1].

Let $F_i(x \mid y)$ be the d.f. of w_i, given that $w_1 = y$; i.e.,

$$F_i(x \mid y) = P\{w_i \leqq x \mid w_1 = y\}.$$

It was proved in [1] that, for all $y \in S$,

$$\lim_{i\to\infty} F_i(x \mid y) = F(x).$$

Throughout this paper we shall assume that $\rho < 1$. The case $\rho \geqq 1$ has little interest and was essentially disposed of in [1]; results proved in the present paper are trivial when $\rho \geqq 1$. Throughout this paper we shall assume that $Eg_1 < \infty$. However, it can be shown, always easily and sometimes trivially, that all the results of [1] and all the queueing results of the present paper except Theorem 3 are valid also when $Eg_1 = \infty$. In order to eliminate the completely trivial we also assume, as was done in [1], that $ER_1 > 0$, $Eg_1 > 0$. Since $\rho < 1$ we have then $0 < ER_1 < \infty$, $0 < Eg_1 < \infty$.

In two or three places below we shall cite the first paragraph of Section 3 of [1]. To ease the reader's task we now quote this paragraph in full:

Let $\varphi_j(a, b, c)$, $j = 1, \cdots, s$ be the value of $w_{(i+1),j}$ when $w_i = a$, $R_i = b$, $g_{i+1} = c$. If d is a point in s-space, we shall say that $a \leqq d$ if every coordinate of a is not greater than the corresponding coordinate of d. If now $a \leqq d$, then obviously

$$\varphi_j(a, b, c) \leqq \varphi_j(d, b, c)$$

for $1 \leqq j \leqq s$. Applying this argument k times we obtain the following result: Let $R_{i+j-1} = b_{i+j-1}$, $g_{i+j} = c_{i+j}$, $j = 1, \cdots, k$. Let $w_{i+k} = e_1$ when $w_i = a_1$, and let $w_{i+k} = e_2$ when $w_i = a_2$. Then $a_1 \leqq a_2$ implies $e_1 \leqq e_2$.

The results of [1] also imply that $F(x)$ determines a stationary and metrically transitive flow; this is the process $\{w_n^0\}$ defined in Section 1, below, where the relevant references to [1] are given.

1. Convergence of the mean waiting time. Let k be any positive number. Define $W_n = \sum_{i=1}^{s} w_{ni}$. Since w_{ni} is a nonnegative chance variable and $F_n(x) \to F(x)$, we easily have that

$$(1.1)\qquad \begin{aligned} \liminf_n\, (Ew_{ni})^k &\geqq \int (x_i)^k \, dF(x), \\ \liminf_n\, E(W_n)^k &\geqq \int (x_1 + \cdots + x_s)^k \, dF(x), \end{aligned}$$

where, of course, the right members may be infinite. From the fact (proved in

[1]) that $F_n(x)$ approaches $F(x)$ from above for every x, we have that

$$E(w_{ni})^k \leqq \int (x_i)^k \, dF(x).$$

Hence

$$\lim_n E(w_{ni})^k = \int (x_i)^k \, dF(x). \tag{1.2}$$

Let $F_n^W(z \mid y)$ be the d.f. of W_n, given that $w_1 = y(\varepsilon S)$. Hence $F_n^W(z \mid 0)$ is the d.f. of W_n. Then

$$F_{n+1}^W(z \mid 0) - F_n^W(z \mid 0) = \int [F_n^W(z \mid y) - F_n^W(z \mid 0)] \, dF_2(y).$$

It follows from the first paragraph of Section 3 of [1] that, if $y \varepsilon S$, the integrand in the last integral is never positive for any z. Hence the left member in the last equation is never positive for any z. Hence $F_n^W(z \mid 0)$ approaches its limit (which is a distribution function obtainable from $F(x)$ in an obvious way) from above. Consequently, as before,

$$E(W_n)^k \leqq \int \left(\sum_{i=1}^{s} x_i \right)^k dF(x).$$

From this and (1.1) we obtain

$$\lim_n E(W_n)^k = \int \left(\sum_{i=1}^{s} x_i \right)^k dF(x) = m_k' \text{ (say)}.$$

The question as to when $m_k' < \infty$ will be discussed in a later section. We define

$$m_k = \int (x_1)^k \, dF(x),$$

and

$$V_{nk} = \frac{1}{n} \sum_{i=1}^{n} (w_{i1})^k.$$

We now prove

Theorem 1. *We have, for any positive k,*

$$P\{\lim_{n \to \infty} V_{nk} = m_k\} = 1. \tag{1.3}$$

Proof. Let w_1^0 be an s-dimensional chance variable with the d.f. $F(x)$, and let w_{n+1}^0 be obtained from w_n^0 by using R_n and g_{n+1} in exactly the same manner as one obtains w_{n+1} from w_n. Thus w_n^0 pertains at time t_n. Then the process $\{w_n^0, n = 1, 2, \cdots\}$ is easily seen to be stationary, because $F(x)$ satisfies the integral equation derived in [1] (see Section 3 of [1] for details). It is proved in Section 8 of [1] that $F(x)$ is the only d.f. which satisfies the integral equation.

We shall show that this implies easily that there cannot be a Borel set B in s-dimensional Euclidean space such that

$$0 < \int_B dF < 1,$$

and $w_1^0 \varepsilon B$ implies with probability one that $w_n^0 \varepsilon B$, $n \geqq 2$. For let $\bar{B}$ be the complement of B, and $F(x \mid B)$ and $F(x \mid \bar{B})$ be, respectively, the conditional distribution functions on B and $\bar{B}$ implied by $F(x)$. Then $F(x \mid B)$ satisfies the integral equation. On a set of w_1^0 of probability one according to $F(x \mid \bar{B})$, $w_n^0 \varepsilon \bar{B}$ for $n \geqq 2$ with probability one, since otherwise $P\{w_n^0 \varepsilon B\}$ (when F is the distribution function of w_1^0) would not be independent of n, contradicting the stationarity of $\{w_n^0\}$. Hence $F(x \mid \bar{B})$ must also satisfy the integral equation. Clearly, $F(x \mid B)$ and $F(x \mid \bar{B})$ are not identical, in contradiction to the fact that $F(x)$ is the only d.f. that satisfies the integral equation. From the fact that there is no invariant set B such that $0 < \int_B dF < 1$, the fact that w_n^0 is a Markoff process, and Theorem 1.1, page 460 of [6] (which asserts that any set in the space of the Markoffian chance variables $w_1^0, w_2^0, \cdots$ that is invariant under a shift transformation differs from a set B by a set of probability zero), we conclude that the process w_n^0 is metrically transitive. Hence, by the ergodic theorem,

$$P\{\lim_{n\to\infty} V_{nk}^0 = m_k\} = 1, \tag{1.4}$$

where

$$V_{nk}^0 = \frac{1}{n}\sum_{i=1}^{n} (w_{i1}^0)^k,$$

and of course w_{i1}^0 is the first component of the vector w_i^0.

From the argument in the first paragraph of Section 3 of [1], it follows that always

$$V_{nk} \leqq V_{nk}^0. \tag{1.5}$$

Hence

$$P\{\limsup_{n\to\infty} V_{nk} \leqq m_k\} = 1. \tag{1.6}$$

We shall prove that also

$$P\{\liminf_{n\to\infty} V_{nk} \geqq m_k\} = 1. \tag{1.7}$$

This will prove the theorem.

We shall now deduce (1.7) from (1.4), and for this purpose divide the argument into consideration of the four cases of Section 8 of [1]. As there defined, denote by $[a, b]$ and $[c, d]$ the smallest closed intervals for which

$$P\{a \leqq R_1 \leqq b\} = P\{c \leqq g_1 \leqq d\} = 1.$$

Of course, b or d or both may be $+\infty$.

CASE 1: $b > sc$. Let t be so large that the point $T = (t, t, \cdots, t)$ of S is such that

$$\int_{x<T} dF(x) > 0.$$

It follows from (1.4) that there exists in S a point $x < T$ such that

$$P\{\lim_{n\to\infty} V^0_{nk} = m_k \mid w^0_1 = x\} = 1. \tag{1.8}$$

It is proved in [1] that there exists an integer r such that $P\{w_{(sr)} > T\} > 0$, say $=\alpha$. From this it follows that

$$P\{w_n > T \quad \text{for at least one } n\} = 1. \tag{1.9}$$

Let h be the smallest index n for which $w_n > T$; $h < \infty$ with probability one. Obviously $R_h, R_{h+1}, \cdots$ and $g_{h+1}, g_{h+2}, \cdots$ are distributed independently of h and w_h, and have the same distribution as $R_1, R_2, \cdots$ and $g_2, g_3, \cdots$. Consequently, if we define, for $n > h$,

$$V_{nk}(h) = \frac{(w_{h1})^k + (w_{(h+1),1})^k + \cdots + (w_{n,1})^k}{n},$$

we have, using (1.8) and the argument in the first paragraph of Section 3 of [1], that

$$P\{\liminf_{n\to\infty} V_{nk}(h) \geqq m_k\} = 1. \tag{1.10}$$

Obviously from the definition of $V_{nk}(h)$ it follows that

$$P\{\lim_{n\to\infty} (V_{nk}(h) - V_{nk}) = 0\} = 1. \tag{1.11}$$

The desired result (1.7) follows from (1.10) and (1.11).

CASE 2: $a < d$. It is proved in Section 8 of [1] that, in this case,

$$P\{w^0_n = 0 \quad \text{for some } n \geqq 1\} = 1. \tag{1.12}$$

The desired result (1.3) follows from (1.4) and (1.12) by means of an argument like that in Case 1.

CASE 3: $c = d \leqq a = b < sc$. It is proved in [1] that in this case there is a point in S, there called $\bar{w}$, such that

$$P\{w_n = w_{n+1} = \cdots = \bar{w} \quad \text{for some } n \geqq 1\} = 1. \tag{1.13}$$

The desired result (1.3) follows at once.

CASE 4: *$d \leqq a$, $b \leqq sc$, and either $a < b$ or $c < d$.* It is proved in [1] that, in this case, there exists an $\epsilon > 0$ such that the set

$$\Gamma^\epsilon = \{y \mid y \in S, y \leqq \bar{w}^\epsilon\},$$

where

$$\bar{w}^\epsilon = (0, u^\epsilon_{s-1}, u^\epsilon_{s-2}, \cdots, u^\epsilon_1)$$

and

$$u_j^\epsilon = \max(0, b - jc - \epsilon),$$

has the following properties:

(a) $$P\{w_n^0 \,\varepsilon\, \Gamma^\epsilon \quad \text{for some } n \geqq 1\} = 1.$$

(This implies at once that

(1.14) $$\int_{\Gamma^\epsilon} dF(x) > 0.)$$

(b) $$P\{w_s > \bar{w}^\epsilon\} > 0.$$

(This implies, using the argument in the first paragraph of Section 3 of [1], that

(1.15) $$P\{w_n > \bar{w}^\epsilon \quad \text{for at least one } n > 1\} = 1.)$$

The desired result now follows exactly as in Case 1, the place of T being taken by $\bar{w}^\epsilon$.

In exactly the same manner as that employed in this section we could have proved that

(1.16) $$P\left\{\lim_{n\to\infty} \frac{1}{n} \sum_{i=1}^{n} (W_i)^k = m_k'\right\} = 1$$

and similar theorems about other moments.

2. Generalization of the lemma of Section 4 of [1]. We shall prove the following essential generalization of the fundamental lemma of Section 4 of [1] both for its use as a tool in a subsequent section and for its intrinsic interest:

LEMMA. *If, for any positive $k > 0$,*

(2.1) $$ER_1^{k+1} < \infty,$$

then

(2.2) $$\sup_n E(w_{ns} - w_{n1})^k < \infty;$$

or, what is equivalent,

(2.3) $$\sup_n E\left((s-1)w_{ns} - \sum_{j=1}^{s-1} w_{nj}\right)^k < \infty.$$

PROOF. Define Y_i exactly as in (4.5) of [1], i.e.,

$$Y_i = \max[(s-1)R_i, (s-1)R_{i-1} - R_i, (s-1)R_{i-2} - R_{i-1} - R_i, \cdots, (s-1)R_1 - R_2 - \cdots - R_i].$$

Then (4.6) of [1] is

(2.4) $$L(y', n) = P\{Y_n \leqq y'\} = P\{R_1 \leqq hy', R_2 \leqq h(R_1 + y'), \cdots, R_n \leqq h(R_1 + \cdots + R_{n-1} + y')\},$$

where $h = (s-1)^{-1}$. Let $H(z)$ be the d.f. of R_1.

Define $L(y', 0) = 1$. Obviously $L(y', n)$ is nonincreasing in n and, for $n \geqq 0$,

$$\begin{aligned} L(y', n) - L(y', n+1) &= P\{Y_n \leqq y', R_{n+1} > h(R_1 + \cdots + R_n + y')\} \\ &\leqq P\{R_{n+1} > h(R_1 + \cdots + R_n + y')\} \\ &\leqq E\{1 - H(h[R_1 + \cdots + R_n + y'])\}. \end{aligned} \tag{2.5}$$

Hence

$$\begin{aligned} 1 - L(y', n) &= \sum_{i=1}^{n} [(Ly', i-1) - L(y', i)] \\ &\leqq \sum_{i=0}^{\infty} E\{1 - H(h[R_1 + \cdots + R_i + y'])\}. \end{aligned} \tag{2.6}$$

Let d be a small positive number and define

$$D_i = d \text{ when } R_i \geqq \frac{d}{h}$$

$$D_i = 0 \text{ otherwise.}$$

We choose d so small that $d < 1$ and

$$p = P\{D_1 = d\} > 0.$$

(We have earlier excluded the trivial case where $R_i = 0$ with probability one.) Since $R_i \geqq D_i/h$, if we replace the former by the latter in the right member of (2.6) we do not diminish any term of this member. It is well known (e.g., [2], p. 101) from approximations to the binomial distribution that, for suitable positive c_1, c_2, we have

$$P\left\{D_1 + \cdots + D_n \leqq \frac{npd}{2}\right\} < c_1 e^{-c_2 n} \tag{2.7}$$

When $k \geqq 1$ we have, from (2.6),

$$\begin{aligned} E(Y_n)^k &\leqq k \sum_{j=0}^{\infty} (j+1)^{k-1} P\{Y_n > j\} \\ &\leqq k \sum_{j=0}^{\infty} \sum_{i=0}^{\infty} (j+1)^{k-1} E\{1 - H(h[R_1 + \cdots R_i + j])\} \\ &\leqq k \sum_{j=0}^{\infty} \sum_{i=0}^{\infty} (j+1)^{k-1} E\{1 - H(D_1 + \cdots D_i + j)\} \\ &\leqq k \sum_{j=0}^{\infty} \sum_{i=0}^{\infty} (j+1)^{k-1} E\{1 - H(D_1 + \cdots + D_{i+j})\} \\ &\leqq \sum_{j=0}^{\infty} (j+2)^{k} E\{1 - H(D_1 + \cdots + D_j)\}. \end{aligned} \tag{2.8}$$

We have now, applying (2.7) to the right member of (2.8),

$$E(Y_n)^k \leqq c_1 \sum_{j=0}^{\infty} (j+2)^k e^{-c_2 j} + \sum_{j=0}^{\infty} (j+2)^k \left(1 - H\left(\frac{jpd}{2}\right)\right). \tag{2.9}$$

The first series on the right of (2.9) obviously converges. Now consider the second. We have

$$\begin{aligned} \sum_{j=0}^{\infty} (j+2)^k \left(1 - H\left(\frac{jpd}{2}\right)\right) &= \sum_{j=0}^{\infty} (j+2)^k P\left\{R_1 > \frac{jpd}{2}\right\} \\ &\leqq \left(\frac{2}{pd}\right)^{k+1} E(R_1+2)^{k+1}. \end{aligned} \tag{2.10}$$

In [1] (relation (4.5)) it is shown that

$$\left((s-1)w_{ns} - \sum_{j=1}^{s-1} w_{nj}\right) \leqq Y_{n-1}$$

Hence (2.3) and the lemma follow for $k \geqq 1$. The proof for $0 < k < 1$ is almost the same; only a few obvious changes are needed in (2.8), (2.9), and (2.10).

3. Finiteness of m_k'. Of great interest is the question of when m_k is finite. In this section we shall give a sufficient condition for m_k' to be finite (and hence a fortiori for $m_k \leqq m_k'$ to be finite). We shall later see that this condition is essentially necessary for m_k' to be finite.

THEOREM 2. *If* $k > 0$, *and*

$$ER_1^{k+1} < \infty, \tag{3.1}$$

then

$$m_k' < \infty, \tag{3.2}$$

and

$$m_k < \infty. \tag{3.3}$$

PROOF. We assume that there exists a number $T > 0$ such that $g_1 < T$ with probability one. When we bear in mind how w_{n+1} is related to w_n, it follows immediately that, if Theorem 2 holds in this case, it a fortiori holds in general.

In order to carry out the proof we shall assume that $m_k' = \infty$ and obtain a contradiction. Let A be the set $\{x \mid x_1 < T\}$. Then from (2.2) we obtain that

$$\sup_n \int_A (x_s)^k \, dF_n(x) < \infty, \tag{3.4}$$

and hence

$$\sup_n \int_A (x_1 + \cdots x_s)^k \, dF_n(x) < \infty. \tag{3.5}$$

From the manner in which we obtain w_{n+1} from w_n we have that

$$W_{n+1} = W_n + R_n - sg_{n+1} \tag{3.6}$$

if $w_{n1} \geqq T$, and always we have

$$W_{n+1} \leqq W_n + R_n . \tag{3.7}$$

We now note the inequality (2.15.1) on page 39 of [7], which states that $r > 1$, $x \geqq 0$, $y \geqq 0$ imply that

$$x^r - y^r \leqq rx^{r-1}(x - y). \tag{3.8}$$

Putting $r = k + 1$, $x = W_{n+1}$, $y = W_n$, we have, from (3.6),

$$\begin{aligned} W_{n+1}^{k+1} - W_n^{k+1} &\leqq (k+1)(W_n + R_n - sg_{n+1})^k(R_n - sg_{n+1}) \\ &= (k+1)W_n^k\left\{\left(1 + \frac{R_n - sg_{n+1}}{W_n}\right)^k (R_n - sg_{n+1})\right\}. \end{aligned} \tag{3.9}$$

Consider the expression in brackets in the last expression of (3.9). By (3.1), the boundedness of g_{n+1}, and the independence of W_n from g_{n+1} and R_n, the conditional expected value of this bracketed expression, given W_n, tends to $E(R_n - sg_{n+1}) < 0$ as $W_n \to \infty$. Hence, if $EW_n^k \to \infty (= m_k')$ as $n \to \infty$, (3.9) implies that

$$\lim_{n\to\infty} E\{W_{n+1}^{k+1} - W_n^{k+1} \mid w_{n1} \geqq T\} = -\infty. \tag{3.10}$$

Similarly, putting $x = W_n + R_n$, $y = W_n$, and noting that $(a + b)^k \leqq 2^k(a^k + b^k)$ if $a, b, k \geqq 0$, (3.7) yields

$$\begin{aligned} W_{n+1}^{k+1} - W_n^{k+1} \leqq (W_n + R_n)^{k+1} - W_n^{k+1} &\leqq (k+1)(W_n + R_n)^k R_n \\ &\leqq (k+1)2^k(W_n^k + R_n^k)R_n . \end{aligned} \tag{3.11}$$

From (3.1), (3.5) and the independence of R_n and W_n, we conclude that there is a number $c < \infty$ such that

$$\sup_n E\{W_{n+1}^{k+1} - W_n^{k+1} \mid w_{n1} < T\} < c. \tag{3.12}$$

From (3.10), (3.12), and the fact that (3.5) and $m_k' = \infty$ imply that $\bar{A}$ has probability $> \epsilon > 0$ according to F_n for all sufficiently large n, we conclude that there is an integer N_0 such that $EW_{n+1}^{k+1} \leqq EW_n^{k+1}$ for $n \geqq N_0$. Since, for $n \leqq N_0$, $EW_n^{k+1} \leqq E(R_1 + \cdots R_{N_0})^{k+1} < \infty$, we conclude that $\sup_n EW_n^{k+1} < \infty$, contradicting the assumption that $m_k' = \infty$. This completes the proof.

4. Necessity of the condition (3.1). The present section is devoted to the proof of

THEOREM 3. *If, for any positive k,*

$$ER_1^{k+1} = \infty \tag{4.1}$$

and $Eg_1 < \infty$, *then*

$$m_k' = \infty . \tag{4.2}$$

It will easily be seen from our proof that Theorem 3 is a fortiori true if $\rho \geqq 1$. Only the case $\rho < 1$ requires proof and this is the case we shall consider.

PROOF. Let m be so large that

$$\int_M dF(x) = \alpha > 0$$

where M is the set of all points $(x_1, x_2, \cdots, x_s)$ in S such that $x_s \leqq m$. We have already remarked in Section 1 that the process $\{w_n^0\}$ there defined is stationary and metrically transitive. Let $\nu_1^0, \nu_2^0, \cdots$ be the indices n for which $w_n^0 \varepsilon M$, and define

$$\mu_i^0 = \nu_{i+1}^0 - \nu_i^0 .$$

It follows from the ergodic theorem that

$$E\mu_i^0 = \frac{1}{\alpha} < \infty .$$

Let $\{w_n'\}$ be the process obtained from $\{w_n\}$ as follows: $w_1' = w_1 = 0$. Thereafter $w_n' = w_n$ until the first index n, say ν_1', such that $w_{\nu_1'} \varepsilon M$; define $w_{\nu_1'}' = 0$. We now obtain each successive w_{n+1}' from its predecessor w_n' by using R_n and g_{n+1} in exactly the same manner as w_{n+1} is obtained from w_n, until the next index, say ν_2', for which $w_{\nu_2'}'$ would be in M; instead set $w_{\nu_2'}' = 0$. Continue in this manner to define $\{w_n'\}$. Define $\mu_i' = \nu_{i+1}' - \nu_i'$. Then $\mu_1', \mu_2', \cdots$ are independent, identically distributed chance variables. It follows from the construction of the process $\{w_n'\}$ and the first paragraph of Section 3 of [1] that $E\mu_i' \leqq E\mu_i^0$. Hence $E\mu_i'$ is finite. It follows from the strong law of large numbers that

$$P\left\{\lim_{n\to\infty} \frac{\nu_n'}{n} = E\mu_1'\right\} = 1. \tag{4.3}$$

We shall later show that

$$P\left\{\lim_{n\to\infty} n^{-1} \sum_{i=1}^{n} (w_{is}')^k = \infty\right\} = 1. \tag{4.4}$$

Since $w_n' \leqq w_n$ it follows at once that

$$P\left\{\lim_{n\to\infty} n^{-1} \sum_{i=1}^{n} (w_{is})^k = \infty\right\} = 1. \tag{4.5}$$

Hence

$$P\left\{\lim_{n\to\infty} n^{-1} \sum_{i=1}^{n} (W_i)^k = \infty\right\} = 1. \tag{4.6}$$

The desired result (4.2) follows from (1.16) and (4.6).

It remains to prove (4.4). Let $j(n)$ be defined for all integral n by

$$\nu_{j(n)}' \leqq n < \nu_{j(n)+1}' .$$

We shall later prove that

(4.7) $$E\{(w'_{1s})^k + (w'_{2s})^k + \cdots + (w'_{\mu_1,s})^k\} = \infty.$$

From this and the strong law of large numbers it follows that

(4.8) $$P\left\{\lim_{n\to\infty} (j(n))^{-1} \sum_{i=1}^{\nu'_j(n)} (w'_{is})^k = \infty\right\} = 1.$$

From (4.3) and (4.8) we obtain that

(4.9) $$P\left\{\lim_{n\to\infty} (\nu'_{j(n)})^{-1} \sum_{i=1}^{\nu'_j(n)} (w'_{is})^k = \infty\right\} = 1.$$

From (4.9) we have at once that

(4.10) $$P\left\{\lim_{n\to\infty} (\nu'_{j(n)})^{-1} \sum_{i=1}^{n} (w'_{is})^k = \infty\right\} = 1.$$

Also

(4.11) $$P\left\{\lim_{n\to\infty} \frac{n}{\nu'_{j(n)}} = 1\right\} = 1.$$

From (4.10) and (4.11) we have the desired result (4.4).

It remains to prove (4.7). Let N be an integer so large that

(4.12) $$P\left\{\sum_{i=1}^{n} g_i < 2nEg_1 \text{ for all } n \geqq N\right\} > \tau > 0$$

The existence of such an N follows from the strong law of large numbers. We may also assume N so large that $2NEg_1 > m$. Let $T = 4NEg_1$. Suppose that $t \geqq T$ and the largest integer contained in $(t/4Eg_1)$ is t'. Then $t' \geqq N$, and (4.12) implies that the conditional probability of the event A_1,

(4.13) $$A_1 = \{\mu'_1 > t' \text{ and } w'_{ns} > 2t'Eg_1 \text{ for } 2 \leqq n \leqq t'\},$$

given that $w'_{2s} = t$, is greater than τ. ($\mu'_1 > t'$ is implied by the other events in (4.13).) When the event A_1 occurs, we have

(4.14) $$\sum_{n=1}^{\mu'_1} (w'_{ns})^k \geqq \sum_{n=1}^{t'} (w'_{ns})^k > t'(2t'Eg_1)^k \geqq ct^{k+1}$$

with $c > 0$. From (4.1) and the construction of the process $\{w'_n\}$ we have (by considering $((R_1 - g_2)^+)^{k+1}$ on the set where $g_2 < c$ where $c < \infty$ is chosen so that $P\{g_2 < c\} > 0$) that

(4.15) $$E(w'_{2s})^{k+1} = \infty.$$

The desired result (4.7) follows from (4.14) and (4.15). This completes the proof of Theorem 3.

The following theorem can be proved in essentially the same manner as Theorem 3:

THEOREM 4. *If, for a positive integer N, an integer j ($1 \leqq j \leqq s$), and a positive k*

$$E(w_{Nj})^{k+1} = \infty, \tag{4.16}$$

then

$$\int (x_j)^k \, dF(x) = \infty. \tag{4.17}$$

Theorem 3 is a special case of Theorem 4 for the case $N = 2, j = s$. For then (4.1) implies (4.16), and (4.17) implies (4.2). Let M_i denote the ith smallest of $R_1, \cdots, R_s$, and suppose

$$E(M_i)^{k+1} = \infty. \tag{4.18}$$

Then (4.16) holds with $N = s, j = i$. This also implies Theorem 3, for (4.1) implies (4.18) for $i = s$. Finally we remark that (4.18) with $i = 1$ implies

$$m_k = \infty. \tag{4.19}$$

5. Implications for the one-dimensional random walk. The results of the preceding sections imply not only results on the behavior of queues in general, but also results on the random walk in s-dimensional space. We shall content ourselves with pointing out two of these implications for the one-dimensional random walk, although the results for the s-dimensional walk obtained in earlier sections are more general and usually more difficult to prove. Without further remark all problems treated in this section are to be assumed to be one-dimensional.

THEOREM 5. *Let $u_1, u_2, \cdots$ be independent, identically distributed chance variables. Let $S_n = \sum_{i=1}^{n} u_i$, and define*

$$v = \sup(0, S_1, S_2, S_3, \cdots).$$

If

$$-\infty \leqq Eu_1 < 0, \tag{5.1}$$

and, for $k > 0$,

$$E(u_1^+)^{k+1} < \infty, \tag{5.2}$$

then

$$Ev^k < \infty. \tag{5.3}$$

THEOREM 6. *With the definitions of Theorem 5, if*

$$-\infty < Eu_1 < 0, \tag{5.4}$$

and, for $k > 0$,

$$E(u_1^+)^{k+1} = \infty, \tag{5.5}$$

then

$$Ev^k = \infty. \tag{5.6}$$

PROOF. Consider the process: $w_1^* = u_1^+$, $w_{n+1}^* = (w_n^* + u_{n+1})^+$, $n \geqq 1$. Let $F_n^*(z)$ be the d.f. of w_n^*, and let

$$F^*(z) = \lim_{n\to\infty} F_n^*(z)$$

when the latter exists. It was shown in [3] and follows from the results of [1] for the case $s = 1$ that, when $u_n = R_n - g_{n+1}$, $F^*(z)$ exists, is a distribution function, and equals the limiting d.f. $F(z)$ of w_n. It was also shown in [3] that the distribution function of v is then $F^*(z)$. An examination of the proofs of these statements shows that they are valid for the process $\{w_n^*\}$ even when u_n is not of the form $R_n - g_{n+1}$, provided only that (5.1) is satisfied. An examination of the proofs of Section 1 and Theorem 3 of the present paper shows that they too hold even if u_n is not of the form $R_n - g_{n+1}$. But then Theorem 6 is simply a restatement of Theorem 3.

It is sufficient to prove Theorem 5 for chance variables $\{u_n^*\}$, where $u_n^* = \max(u_n, -T)$ and $T > 0$ is so large that $Eu_n^* < 0$. But $u_n^* = (u_n^* + T) - T$ and is therefore of the form $R_n - g_{n+1}$, with $R_n = (u_n^* + T)$, $g_{n+1} \equiv T$. Theorem 5 is then simply a restatement of Theorem 2.

While the results of the present paper on the queueing process and the corresponding s-dimensional random walk are new, Theorems 5 and 6 on the one-dimensional random walk were also obtained by Darling, Erdös, and Kakutani, to whom the problem was communicated by us. These writers also obtained other related results, and they have informed us that many of these results are implicit in [4]. In the course of the present work we have had interesting discussions with Professor Shizuo Kakutani.

6. The mean queue length. As in [1], Section 9, let Q_i be the number of individuals in the queue waiting to be served, just before the service of the ith individual begins. To avoid trivial circumlocutions we assume $G(0) = 0$ ($G(x)$ is the d.f. of g_i). In [1] the limit $D(x)$ of $D_n(x)$, the d.f. of Q_n, is shown to exist and $D(x)$ is explicitly given. We shall now be concerned with

$$\bar{Q}_n = n^{-1} \sum_{i=1}^{n} Q_i .$$

Let $\{w_n^0\}$ be the process defined in Section 1. We now construct a process $\{w_n^0, Q_n^0\}$, where $Q_1^0, Q_2^0, \cdots$ remain to be defined. Let $t_n = \sum_{i=1}^n g_i$. We define

Q_n^0 to be equal to the number of indices i which satisfy

$$t_n < t_i \leqq t_n + w_{n1}^0. \tag{6.1}$$

It follows that the process $\{w_n^0, Q_n^0\}$ is stationary and metrically transitive, so that, by the ergodic theorem, $\bar{Q}_n^0 = n^{-1} \sum_{i=1}^n Q_i^0$ approaches a constant limit c,

$$c = \int x\, dD(x),$$

with probability one. (It is easy to prove that c is contained between $Ew_{n1}^0/Eg_1 - 1$ and Ew_{n1}^0/Eg_1.) Since $w_{n1} \leqq w_{n1}^0$ it follows from (6.1) that $Q_n \leqq Q_n^0$. Hence

$$P\{\limsup_n \bar{Q}_n \leqq c\} = 1. \tag{6.2}$$

Just as in Section 1, one proves that

$$P\{\liminf_n \bar{Q}_n \geqq c\} = 1. \tag{6.3}$$

Hence

$$P\{\lim_n \bar{Q}_n = c\} = 1. \tag{6.4}$$

7. The duration of busy periods. A busy period is a closed time interval, say $t' \leqq t \leqq t''$, such that all s servers are occupied throughout this interval, $t'' - t' > 0$, and the interval is maximal, i.e., if $\tau' \leqq t' < t'' \leqq \tau''$, $\tau'' - \tau' > t'' - t'$, then all s servers are not occupied for some time point in the interval (τ', τ''). The length of the busy period is $t'' - t'$, t' is its beginning, and t'' is its end. Let B_i be the sum of the lengths of all busy periods at or before t_i; if t_i is in the interior of a busy period, we count into B_i the length of the interval from the beginning of the period until t_i.

It is easy to verify that whether or not any time point t with $t_i < t < t_{i+1}$ is in a busy interval depends only on w_i, R_i, and g_{i+1}. Since the value of B_n is unaffected by removing from busy periods any of the points t_i $(1 \leqq i \leqq n)$ contained in them, it follows that the process

$$\{B_n, w_n\}, n = 1, 2, \cdots$$

is Markoffian.

Let $\{w_n^0\}$ be the process defined in Section 1. Define $B_1^0 = 0$. Define B_n^0, $n \geqq 2$, to be the same function of the process $\{w_n^0\}$ as B_n is of the process $\{w_n\}$. Since $w_n^0 \geqq w_n$ with probability one, it follows that $B_n^0 \geqq B_n$ with probability one.

Since the process $\{w_n^0\}$ is stationary and metrically transitive, so is the process

$$\{B_{n+1}^0 - B_n^0\}, n = 1, 2, \cdots.$$

Hence

$$P\left\{\lim \frac{B_n^0}{n} = E(B_2^0)\right\} = 1.$$

In essentially the same manner as in Section 1 one proves easily that

$$P\left\{\lim \frac{B_n}{n} = E(B_2^0)\right\} = 1.$$

From this we obtain immediately that

$$P\left\{\lim \frac{B_n}{t_n} = \frac{E(B_2^0)}{Eg_1}\right\} = 1.$$

This gives the long-term average time spent in busy periods.

The limiting distribution of the length of a busy period can be obtained in a very tedious but straightforward manner from the marginal distributions of the process $\{w_n^0\}$.

REFERENCES

[1] J. Kiefer, and J. Wolfowitz, "On the theory of queues with many servers," *Trans. Amer. Math. Soc.*, Vol. 78, 1, January 1955, pp. 1–18.

[2] J. V. Uspensky, "Introduction to mathematical probability," McGraw-Hill Book Company, Inc., New York, 1937.

[3] D. V. Lindley, "The theory of queues with a single server," *Proc. Cambridge Philos. Soc.*, Vol. 48 (1952), Part 2, pp. 277–89.

[4] P. Erdös, "On a theorem of Hsu and Robbins," *Ann. Math. Stat.*, Vol. 20 (1949), pp. 286–291.

[5] J. Wolfowitz, "The efficiency of sequential estimates etc.," *Ann. Math. Stat.*, Vol. 18 (1947), pp. 215–230.

[6] J. L. Doob, "Stochastic processes," John Wiley & Sons, Inc., New York, 1953.

[7] G. H. Hardy, J. E. Littlewood, and G. Polya, "Inequalities," Cambridge University Press, Cambridge, 1934.

Reprinted from THE ANNALS OF MATHEMATICAL STATISTICS
Vol. 27, No. 3, September, 1956

ASYMPTOTIC MINIMAX CHARACTER OF THE SAMPLE DISTRIBUTION FUNCTION AND OF THE CLASSICAL MULTINOMIAL ESTIMATOR

BY A. DVORETZKY,[1] J. KIEFER,[1] AND J. WOLFOWITZ[2]

Cornell University

0. Summary. This paper is devoted, in the main, to proving the asymptotic minimax character of the sample distribution function (d.f.) for estimating an unknown d.f. in $\mathfrak{F}$ or $\mathfrak{F}_c$ (defined in Section 1) for a wide variety of weight functions. Section 1 contains definitions and a discussion of measurability considerations. Lemma 2 of Section 2 is an essential tool in our proofs and seems to be of interest per se; for example, it implies the convergence of the moment generating function of G_n to that of G (definitions in (2.1)). In Section 3 the asymptotic minimax character is proved for a fundamental class of weight functions which are functions of the maximum deviation between estimating and true d.f. In Section 4 a device (of more general applicability in decision theory) is employed which yields the asymptotic minimax result for a wide class of weight functions of this character as a consequence of the results of Section 3 for weight functions of the fundamental class. In Section 5 the asymptotic minimax character is proved for a class of integrated weight functions. A more general class of weight functions for which the asymptotic minimax character holds is discussed in Section 6. This includes weight functions for which the risk function of the sample d.f. is not a constant over $\mathfrak{F}_c$. Most weight functions of practical interest are included in the considerations of Sections 3 to 6. Section 6 also includes a discussion of multinomial estimation problems for which the asymptotic minimax character of the classical estimator is contained in our results. Finally, Section 7 includes a general discussion of minimization of symmetric convex or monotone functionals of symmetric random elements, with special consideration of the "tied-down" Wiener process, and with a heuristic proof of the results of Sections 3, 4, 5, and much of Section 6.

1. Introduction and Preliminaries. Throughout this paper we shall denote by $\mathfrak{F}$ the class of all univariate d.f.'s and by $\mathfrak{F}_c$ the subclass of continuous members of $\mathfrak{F}$ (for the sake of definiteness, members of $\mathfrak{F}$ will be considered continuous on the right). Let R^n denote n-dimensional Euclidean space, and let G be any subspace of the space of all real-valued functions on R^1. For simplicity we assume $\mathfrak{F} \subset G$, although it is really only necessary that G contain the function S_n, defined below, for every $x^{(n)}$. Let B be the smallest Borel field on G such that every element of $\mathfrak{F}$ is an element of B and such that, for every positive integer k and all sets of real numbers $\{t_1, \cdots, t_k\}$ and $\{a_1, \cdots, a_k\}$ with $t_1 < t_2 < \cdots <$

Received May 31, 1955. Revised October 5, 1955.

[1] Research sponsored by the Office of Naval Research.

[2] The research of this author was supported in part by the United States Air Force under Contract No. AF18(600)-685 monitored by the Office of Scientific Research.

t_k, the set $\{g \mid g \varepsilon G; g(t_1) < a_1, \cdots, g(t_k) < a_k\}$ is in B. (Thus, we might have $G = \mathfrak{F}$ and B the Borel field generated by open sets in the common metric topology.) Let D_n be the class of all real-valued functions ϕ_n on $B \times R^n$ with the following properties: for each $x^{(n)} \varepsilon R^n$, $\phi_n(\cdot; x^{(n)})$ is a probability measure (B) on G; and for each $\Delta \varepsilon B$, $\phi_n(\Delta; \cdot)$ is Borel-measurable on R^n.

The problem which confronts the statistician may now be described. Let $X_1, \cdots, X_n$ be independently and identically distributed according to some d.f. F about which it is known only that $F \varepsilon \mathfrak{F}_c$ (or even $F \varepsilon \mathfrak{F}$). The statistician is to estimate F. Write $X^{(n)} = (X_1, \cdots, X_n)$. Having observed $X^{(n)} = x^{(n)} = (x_1, \cdots, x_n)$, the statistician uses the decision function ϕ_n as follows: a function $g \varepsilon G$ is selected by means of a randomization according to the probability measure $\phi_n(\cdot; x^{(n)})$ on G; the function g so selected (which need not be a member of $\mathfrak{F}$) is then the statistician's estimate of the unknown F. It is desirable to select a procedure ϕ_n which may be expected to yield a g which will lie close to the true F, whatever the latter may be; the term "close" will be made precise in succeeding sections. We note that the decision procedure ϕ_n^* which for each $x^{(n)}$ assigns probability one to the "sample d.f." S_n defined by

$$S_n(x) = (\text{number of } x_i \leqq x)/n$$

is a member of D_n.

We now turn (in this and the four succeeding paragraphs) to measure-theoretic considerations which are relevant to this paper. Our point of view is to waste as little space as possible on these considerations, since our results hold under any measurability assumptions which imply the meaningfulness of certain probabilities and integrals involving elements ϕ of D_n, and, in fact, our results hold even if these are interpreted as inner measures and integrals (which will be proper ones when $\phi = \phi_n^*$), as we shall now see.

In Sections 3, 4, and 6 we shall be concerned, for a given n, $\phi \varepsilon D_n$, $r > 0$, and $F \varepsilon \mathfrak{F}$, with the probability that, when the procedure ϕ_n is used and the X_i have d.f. F, the selected estimate g of F will satisfy the inequality

$$\sup_x | g(x) - F(x) | > r.$$

We shall denote this probability by

$$P_{F,\phi}\{\sup_x | g(x) - F(x) | > r\}. \tag{1.1}$$

It is clear when $\phi = \phi_n^*$ that this probability is well defined. This probability will also be meaningful if G is sufficiently regular; for example, if G consists of functions continuous on the right, the supremum in the displayed expression is unchanged if it is taken over rational x, and the probability in question is well defined. For our considerations it is not even necessary to restrict G in this way; we need not concern ourselves with questions of measurability of

$$\sup_x | g(x) - F(x) |,$$

since the optimal properties proved for ϕ_n^* hold if the supremum is taken only over the rationals (this last supremum is never greater than the supremum over all x and is equal to the latter when $g = S_n$). Thus, for arbitrary G and ϕ, the "probability" expression displayed above may be interpreted with the supremum taken over the rationals (or, alternately, as an inner measure, or as the infimum over all positive integers k and sets of real numbers $t_1, \cdots, t_k$ of

$$P_{F,\phi}\{\max_{1 \leqq i \leqq k} | g(t_i) - F(t_i) | > r\}).$$

In Sections 4 and 6 expressions such as

$$\int W(r)\, d_r P_{F,\phi} \left\{\sup_x | g(x) - F(x) | \leqq r\right\} \tag{1.2}$$

appear, the integral being taken over the nonnegative reals with $W \geqq 0$ and nondecreasing. The probability appearing here is to be interpreted as unity minus the probability previously displayed in (1.1), but the integral is to be interpreted as including a term $\gamma \lim_{r\to\infty} W(r)$ if $\gamma > 0$, where

$$\gamma = \lim_{r\to\infty} P_{F,\phi}\left\{\sup_x | g(x) - F(x) | > r\right\}.$$

In Sections 5 and 6 we will encounter such expressions as

$$r(F, \phi) = E_{F,\phi} \int W(g(t) - F(t), F(t))\, dF(t), \tag{1.3}$$

or such an expression with the first two symbols (operations) interchanged. Here $W(x, t)$ is defined for x real and $0 \leqq t \leqq 1$, is measurable (in the Borel sense on R^2), is nonnegative, and for each t is even in x and nondecreasing in x for $x \geqq 0$. $E_{F,\phi}$ is the operation of expectation when the procedure ϕ is used and the X_i have d.f. F. If $\phi = \phi_n^*$, $r(F, \phi)$ is clearly well defined. For other ϕ, any of a number of general assumptions on W and G will suffice to make the integral meaningful; for example, if W is continuous, $F \,\varepsilon\, \mathfrak{F}_c$, and G consists of functions continuous on the right, then the integral is determined by the values of g on the rationals, and $r(F, \phi)$ is meaningful. Weaker assumptions may be made, and, in fact, one could treat $r(F, \phi)$ as an inner integral (which is a proper integral when $\phi = \phi_n^*$) and still obtain the optimum properties for ϕ_n^* which are derived in this paper.

Finally, in Sections 3, 4, 5, and 6, the method of proof used involves integration of expressions such as (1.1), (1.2), and (1.3) with respect to probability measures ξ_{kn} on $\mathfrak{F}_c$. These ξ_{kn} will always be measures (B) and, in fact, will be of a very simple form. Sometimes the order of integration will be interchanged in these sections. If $\phi = \phi_n^*$, the above operations are all easily justified. For other ϕ these operations may be justified, as in the previous three paragraphs, by suitable regularity assumptions on G and W; or, again, the integrals in question may be considered as inner integrals.

2. Two Lemmas. In this section we shall state two lemmas (and a corollary to the second) which will be used to prove the results of Sections 3 and 4, respectively. Lemma 1 is due to Anderson [8], while Lemma 2 is derived from results of Smirnoff [9].

For any set S in R^n and any n-vector ρ, we write $S + \rho = \{x \mid x - \rho \in S\}$. Denote m-dimensional Lebesgue measure by μ_m. The case of Anderson's result which will be of use to us is the following:

LEMMA 1. *Let P be a (possibly degenerate)*[3] *normal probability measure on R^n with means zero, and let T be any convex body in R^n which is symmetric about the origin. Then $P(T) \geqq P(T + \rho)$ for all ρ.*

We shall also use (in Section 5) the trivial fact that the result of Lemma 1 holds for $n = 1$ when P is a normal probability measure truncated at $(-\beta, \beta)$ for $\beta > 0$. In Section 7 we shall mention briefly an application of the more general form of Lemma 1 given in [8].

Before stating Lemma 2, we shall introduce some notation. Let U denote the uniform d.f. (i.e., the d.f. whose density with respect to μ_1 is unity) on [0, 1], and write, for $r \geqq 0$,

$$\begin{aligned} G_n(r) &= P_U\left\{\sup_{0\leqq x\leqq 1} \mid S_n(x) - x \mid \leqq r/\sqrt{n}\right\}, \\ G_{k,n}(r) &= P_U\left\{\max_{1\leqq i\leqq k} \mid S_n(i/(k+1)) - i/(k+1) \mid \leqq r/\sqrt{n}\right\}, \\ G(r) &= 1 - 2\sum_{m=1}^{\infty} (-1)^{m+1} e^{-2m^2r^2}, \qquad (2.1) \\ H_n(r) &= P_U\left\{\sup_{0\leqq x\leqq 1} [S_n(x) - x] \leqq r/\sqrt{n}\right\}, \\ H_{k,n}(r) &= P_U\left\{\max_{1\leqq i\leqq k} [S_n(i/(k+1)) - i/(k+1)] \leqq r/\sqrt{n}\right\}, \\ H(r) &= 1 - e^{-2r^2}. \end{aligned}$$

Then

$$G_{k,n}(r) \geqq G_n(r) \quad \text{and} \quad H_{k,n}(r) \geqq H_n(r) \tag{2.2}$$

for all k, n, r. Moreover,

$$\begin{aligned} \lim_{k\to\infty} G_{k,n}(r) &= G_n(r), \\ \lim_{k\to\infty} H_{k,n}(r) &= H_n(r), \end{aligned} \tag{2.3}$$

and ([1], [2], [3])

$$\begin{aligned} \lim_{k\to\infty}\lim_{n\to\infty} G_{k,n}(r) &= \lim_{n\to\infty} G_n(r) = G(r), \\ \lim_{k\to\infty}\lim_{n\to\infty} H_{k,n}(r) &= \lim_{n\to\infty} H_n(r) = H(r). \end{aligned} \tag{2.4}$$

[3] The fact that the measure need not be n-dimensional necessitates only trivial modifications of the argument in [8].

We shall now prove the following:

LEMMA 2. *There exists a finite positive constant c such that*

$$1 - H_n(r) < ce^{-2r^2} \tag{2.5}$$

and

$$1 - G_n(r) < ce^{-2r^2} \tag{2.6}$$

hold for all $r \geqq 0$ *and all positive integers n.*

An immediate consequence is

COROLLARY 2. *If* $W(r)$ *is any nondecreasing nonnegative function defined for* $r > 0$, *then*

$$\lim_{n\to\infty} \int_0^\infty W(r)\, dH_n(r) = \int_0^\infty W(r)\, dH(r) \tag{2.7}$$

and

$$\lim_{n\to\infty} \int_0^\infty W(r)\, dG_n(r) = \int_0^\infty W(r)\, dG(r). \tag{2.8}$$

Indeed, the lim inf of the integral on the left side of (2.7) or (2.8) is always $\geqq$ the respective integral on the right side. Now, if $\int_0^\infty W(r)re^{-2r^2}\, dr = \infty$, then by (2.1), the integrals on the right side of (2.7) and (2.8) are both infinite and thus (2.7) and (2.8) hold in this case. If, on the other hand,

$$\int_0^\infty W(r)re^{-2r^2}\, dr < \infty,$$

then Corollary 2 follows from (2.4), (2.5), and (2.6), and in this case both sides of (2.7) and (2.8) are finite.

PROOF OF LEMMA 2. Since $1 - G_n(r) \leqq 2\,(1 - H_n(r))$, it suffices to prove (2.5). We shall deduce (2.5) from the explicit expression for $1 - H_n(r)$ given by Smirnoff [9]. Obviously, $1 - H_n(r) = 0$ for $r \geqq \sqrt{n}$, while for $0 < r < \sqrt{n}$, equation (50) of [9] asserts

$$1 - H_n(r) = (1 - r/\sqrt{n})^n + r\sqrt{n} \sum_{j=[r\sqrt{n}]+1}^{n-1} Q_n(j, r), \tag{2.9}$$

where $[x]$ denotes the greatest integer $\leqq x$ and

$$Q_n(j, r) = \binom{n}{j} (j - r\sqrt{n})^j (n - j + r\sqrt{n})^{n-j-1} n^{-n}. \tag{2.10}$$

In what follows we may, and do, restrict ourselves to $0 < r < \sqrt{n}$.

Taking logarithms and differentiating, it is seen that the maximum of $(1 - r/\sqrt{n})^n e^{2r^2}$ occurs at $r = 0$; hence,

$$\left(1 - \frac{r}{\sqrt{n}}\right)^n e^{2r^2} < 1. \tag{2.11}$$

A simple computation yields for all j with $r\sqrt{n} < j < n$,

$$\frac{d}{dr} \log Q_n(j, r) = \frac{-rn^2}{(j - r\sqrt{n})(n - j + r\sqrt{n})} - \frac{\sqrt{n}}{n - j + r\sqrt{n}}$$

$$< \frac{-4r}{1 - \frac{4}{n^2}\left(\frac{n}{2} - j + r\sqrt{n}\right)^2}$$

$$< -4r - \frac{16r}{n^2}\left(\frac{n}{2} - j + r\sqrt{n}\right)^2,$$

which on integrating gives

$$\text{(2.12)} \quad Q_n(j, r) < Q_n(j, 0) \exp\left[-2r^2 - \frac{8r^2}{n^2}\left(\frac{n}{2} - j + \frac{2r\sqrt{n}}{3}\right)^2 - \frac{4r^4}{9n}\right],$$

as well as

$$\text{(2.13)} \quad Q_n(j, r) < c_1 Q_n(j, 1) \exp\left[-2r^2 - \frac{8r^2}{n^2}\left(\frac{n}{2} - j + \frac{2r\sqrt{n}}{3}\right)^2 - \frac{4r^4}{9n}\right]$$

for $r \geqq 1$; here c_1 denotes a universal finite constant (and similarly, c_2, c_3, c_4, c_5 in the sequel).

We divide the sum of (2.9) into two parts: $\sum'$ will denote summation over those j for which

$$\text{(2.14)} \quad \left| j - \frac{n}{2} \right| \leqq \frac{n}{4}$$

and $\sum''$ will denote summation over the remaining values. It follows immediately from Stirling's formula that

$$Q_n(j, 0) < c_2 n^{-3/2}$$

for j satisfying (2.14). Hence we have from (2.12),

$$\sum' Q_n(j, r) < \frac{c_2}{n^{3/2}} e^{-2r^2} \sum' \exp\left[-8r^2\left(\frac{1}{2} + \frac{2r}{3\sqrt{n}} - \frac{j}{n}\right)^2\right]$$

$$< \frac{2c_2}{n^{3/2}} e^{-2r^2} \sum_{j=0}^{\infty} e^{-8r^2 j^2/n^2}$$

$$< \frac{2c_2}{\sqrt{n}} e^{-2r^2}\left(\frac{1}{n} + \int_0^\infty e^{-8r^2t^2}\, dt\right)$$

$$< \frac{c_3}{r\sqrt{n}} e^{-2r^2}.$$

Hence,

$$\text{(2.15)} \quad r\sqrt{n} \sum' Q_n(j, r) < c_3 e^{-2r^2}.$$

Let us now deal with the j occurring in $\sum''$, i.e., those for which (2.14) does not hold. If $2r\sqrt{n}/3 \leqq n/8$, then the second term in the exponent in (2.13) is $\leqq -(r^2/8)$ while otherwise $r > 3\sqrt{n}/16$ and the last term in the exponent in (2.13) is $< -(4/9)(3/16)^2r^2$. Thus, in both cases we have for $r > 1$,

$$Q_n(j, r) < c_1 Q_n(j, 1)e^{-2r^2}e^{-c_4r^2} < \frac{c_5}{r} Q_n(j, 1)e^{-2r^2}.$$

Hence we have from (2.9),

$$r\sqrt{n}\sum'' Q_n(j, r) < c_5 e^{-2r^2}\sqrt{n}\sum'' Q_n(j, 1) < c_6 e^{-2r^2}. \tag{2.16}$$

(2.11), (2.15), and (2.16) imply (2.5) for $1 < r < \sqrt{n}$ and thus obviously for all r.

3. Asymptotic minimax character of ϕ_n^* for a fundamental class of weight functions. In this section we shall prove the asymptotic minimax character of ϕ_n^* (as $n \to \infty$) in a sense which is fundamental in that the minimax character relative to all reasonable weight functions of a certain type will follow (in Section 4) from the results of the present section. We shall now prove the following strong property of ϕ_n^*:

THEOREM 3. *For every value $r > 0$,*

$$\lim_{n\to\infty} \frac{\sup_{F\varepsilon\mathfrak{F}_c} P_F\{\sup_x |S_n(x) - F(x)| > r/\sqrt{n}\}}{\inf_{\phi\varepsilon D_n} \sup_{F\varepsilon\mathfrak{F}_c} P_{F,\phi}\{\sup_x |g(x) - F(x)| > r\sqrt{n}\}} = 1. \tag{3.1}$$

In fact, the probability in the numerator of (3.1) is independent of F for $F \varepsilon \mathfrak{F}_c$ and is no greater for any $F \varepsilon \mathfrak{F} - \mathfrak{F}_c$ than for $F \varepsilon \mathfrak{F}_c$ (see [1]); as an immediate consequence of Theorem 3, we thus have

COROLLARY 3. *The result of Theorem 3 holds if $\mathfrak{F}_c$ is replaced by $\mathfrak{F}$ in its statement.*

We also remark that (3.9) and (3.20) below may be used to give an explicit bound on the departure of ϕ_n^* from minimax character; the integer N of (3.9) may be computed explicitly by merely keeping track of the constants which go into various error orders in the proof which follows; an explicit estimate of departure for $n \leqq N$ could be given similarly. With slightly more difficulty such a bound could also be computed in the cases treated in Sections 4, 5, and 6.

In order to prove (3.1), we shall exhibit a sequence $\{\xi_{kn}\}$ of a priori probability measures on $\mathfrak{F}_c$ such that, letting A_k (k a positive integer) denote the set consisting of the k points $i/(k + 1)$ (for $1 \leqq i \leqq k$), we have

$$\begin{aligned}
&\lim_{k\to\infty} \lim_{n\to\infty} \inf_{\phi\varepsilon D_n} \int P_{F,\phi}\{\sup_{a\varepsilon A_k} |g(a) - F(a)| > r/\sqrt{n}\}\, d\xi_{kn} \\
&\qquad = \lim_{k\to\infty} \lim_{n\to\infty} \int P_F\{\sup_{a\varepsilon A_k} |S_n(a) - F(a)| > r/\sqrt{n}\}\, d\xi_{kn} \\
&\qquad = \lim_{k\to\infty} \lim_{n\to\infty} P_U\{\sup_{a\varepsilon A_k} |S_n(a) - a| > r/\sqrt{n}\},
\end{aligned} \tag{3.2}$$

where U is the uniform distribution on [0, 1]. Now, the expression under the limit operations on the left side of (3.2) is, for each n and k, obviously no greater than the denominator of (3.1) for the same n. On the other hand, the right side of (3.2) is equal to the (positive) limit as $n \to \infty$ of the numerator of (3.1), by (2.4). Hence, (3.2) implies (3.1).

In order to prove (3.2), we shall for each k limit ourselves to measures ξ_{kn} which assign probability one to distribution functions in $\mathfrak{F}_c$ of the form

$$(3.3) \qquad F_k(x) = \sum_{i=1}^{k+1} p_i \, U_{ik}(x), \qquad p_i > 0, \sum p_i = 1,$$

where $U_{ik}(x)$ is the uniform probability distribution on the interval $[(i-1)/(k+1), i/(k+1)]$. For fixed k and n, it is easily seen that a sufficient statistic for the vector $\{p_i\}$ (and thus, for the family of F_k's of the form (3.3)) is given by the vector $T_k^{(n)} = \{T_{k1}^{(n)}, T_{k2}^{(n)}, \cdots, T_{k,k+1}^{(n)}\}$, where $T_{ki}^{(n)}$ is equal to the number of components of $X^{(n)}$ which lie in the interval $[(i-1)/(k+1), i/(k+1)]$. Hence, the validity of (3.2) will be implied by the following stronger result: Let B_k be the family of vectors $\pi = \{p_i, 1 \leqq i \leqq k+1\}$ satisfying $p_i \geqq 0, \sum p_i = 1$; $T_k^{(n)}$ has the multinomial distribution arising from n observations on $k+1$ types of objects, according to some $\pi \, \varepsilon \, B_k$, i.e., for integers $x_i \geqq 0$ with $\sum_1^{k+1} x_i = n$,

$$(3.4) \qquad P_\pi\{T_{ki}^{(n)} = x_i, 1 \leqq i \leqq k+1\} = \frac{n!}{x_1! \cdots x_{k+1}!} p_1^{x_1} \cdots p_{k+1}^{x_{k+1}};$$

$\mathcal{E}_n$ is the class of all (possibly randomized) vector estimators

$$\psi_n = \{\psi_{n1}, \cdots, \psi_{n,k+1}\}$$

of $\pi = \{p_i\}$ based on $T_k^{(n)}$ (ψ_n need not take on values in B_k); the ξ_{kn} are probability measures on B_k, which will be chosen so that

$$\begin{aligned}
\lim_{n\to\infty} \inf_{\psi_n \varepsilon \mathcal{E}_n} \int P_\pi\left\{\sup_i \left| \sum_{j=1}^{i} (\psi_{nj} - p_j) \right| > r/\sqrt{n}\right\} d\xi_{kn} \\
(3.5) \qquad = \lim_{n\to\infty} \int P_\pi\left\{\sup_i \left| \sum_{j=1}^{i} (T_{kj}^{(n)}/n - p_j) \right| > r/\sqrt{n}\right\} d\xi_{kn} \\
= \lim_{n\to\infty} P_{V_k}\left\{\sup_i \left| \sum_{j=1}^{i} (T_{kj}^{(n)}/n - 1/(k+1)) \right| > r/\sqrt{n}\right\},
\end{aligned}$$

where $V_k = \{1/(k+1), \cdots, 1/(k+1)\} \, \varepsilon \, B_k$. Taking limits as $k \to \infty$ (we have seen that this limit exists for the last expression of (3.5)), we see that the demonstration of (3.5) will imply that of (3.2). If we prove (3.5) with $\mathcal{E}_n$ replaced by the class of *non*randomized ψ_n, then (3.5) will a fortiori be true in the form stated above. Hence, in what follows, all ψ_n will be nonrandomized.

Some intuitive remarks are in order regarding the choice of ξ_{kn} (and the m_{kn} defining it) in the next paragraph. For simplicity, let us consider the case $k = 1$. We are then faced with a binomial estimation problem. The classical estimator of the parameter p_1 is asymptotically normal with maximum variance at $p_1 = \frac{1}{2}$ (this is V_1; in general, the corresponding phenomenon which concerns us occurs at $\pi = V_k$). In order to obtain our asymptotic Bayes result (3.5), we want ξ_{1n} to approximate a uniform measure on an interval of p_1 which has the following properties: on the one hand, the width e_n of this interval, when multiplied by $\sqrt{n}$, must tend to infinity with n; on the other hand, the width itself must tend to zero. In terms of the parameter $\sqrt{n}(p_1 - \frac{1}{2})$ and random variable $(T_{11}^{(n)} - n/2)/\sqrt{n}$, we will then be faced, asymptotically, with the problem of estimating the mean of a normal distribution (where, asymptotically, all real values are possible for the mean, with a uniform a priori distribution over a region whose width $\sqrt{n}e_n$ tends to ∞) with *almost constant variance*. The classical estimator will then be asymptotically Bayes for our weight function. Since a uniform a priori distribution would be slightly less simple to use (in keeping track of limits), we use instead one of the form (3.6) below; but the choice of the parameter m_{kn} therein is motivated by the remarks above.

Let $m = m_{k,n}$ = (greatest integer $\leqq n^{1/4}/k^2$), let $\epsilon = \epsilon_{k,n} = m/n$, and let $\xi_{k,n}$ be the probability measure on B_k which is given rise to by the probability density function

$$h_{k,n}(p_1, \cdots, p_k) = C_{k,n}\left[\left(1 - \sum_1^k p_i\right)\prod_{i=1}^k p_i\right]^m \tag{3.6}$$

with respect to Lebesgue measure on the k-simplex $\{0 \leqq \sum_1^k p_i \leqq 1,\ p_i \geqq 0\ (1 \leqq i \leqq k)\}$ and is zero elsewhere. Here

$$C_{k,n} = \Gamma([m+1][k+1])/[\Gamma(m+1)]^{k+1}.$$

Let $Y_{ki}^{(n)} = T_{ki}^{(n)}/n$. Let $\bar{\delta}_i = p_i - 1/(k+1)$. The a posteriori density of $\bar{\delta}_1, \cdots, \bar{\delta}_k$, given that $Y_{ki}^{(n)} = y_i\ (1 \leqq i \leqq k)$ (for possible values of the set $\{y_i\}$) when $\xi_{k,n}$ is the a priori probability measure on B_k is (the domain being obvious)

$$f_{k,n}(\delta_1, \cdots, \delta_k \mid y_1, \cdots, y_k) = \left[C_1 \prod_{i=1}^{k+1}\left(\delta_i + \frac{1}{k+1}\right)^{y_i+\epsilon}\right]^n, \tag{3.7}$$

where we have written $\delta_{k+1} = 1 - \sum_1^k \delta_i$ and $y_{k+1} = 1 - \sum_1^k y_i$ for typographical simplicity; here $(C_1)^n = \Gamma([m+1][k+1]+n)/\prod_1^{k+1}\Gamma(m+1+ny_i)$. Let $\bar{\eta}_i = \bar{\delta}_i - Y_{ki}^{(n)} + 1/(k+1)$. Then the a posteriori density of $\bar{\eta}_1, \cdots, \bar{\eta}_k$ under the same conditions is (the domain again being obvious)

$$\begin{aligned} f^*_{k,n}(\eta_1, \cdots, \eta_k \mid y_1, \cdots, y_k) &= [g_{k,n}(\eta_1, \cdots, \eta_k \mid y_1, \cdots, y_k)]^n \\ &= \left[C_1 \prod_{i=1}^{k+1}(y_i + \eta_i)^{y_i+\epsilon}\right]^n, \end{aligned} \tag{3.8}$$

where $\eta_{k+1} = -\sum_1^k \eta_i$.

We shall now prove that, for each k and each r^* with $0 < r^* < \infty$, we have for $n > N(k, r^*)$ (the latter will be defined below)

$$E_t P_a^* \left\{ \sup_i \left| \sum_{j=1}^{i} (p_j - Y_{kj}^{(n)}) \right| \leqq \frac{r}{\sqrt{n}} \right\}$$
$$\geqq E_t P_a^* \left\{ \sup_i \left| \sum_{j=1}^{i} (p_j - \psi_{nj}) \right| \leqq \frac{r}{\sqrt{n}} \right\} - n^{-1/9} \tag{3.9}$$

for all r with $0 \leqq r \leqq r^*$ and all ψ_{ni} (not necessarily positive or summing to unity); here P_a^* denotes a posteriori probability of π (i.e., of $\{p_j\}$) when (3.6) is the a priori distribution, while E_t denotes expectation with respect to the measure on $B_k \times R^{k+1}$ given by (3.6) and (3.4). Noting that the second integral in (3.5) is unity minus the left side of (3.9) and that for each k the left side of (3.9) tends to a limit as $n \to \infty$ (this will follow from (3.20) below), we see that (3.9) actually implies that the first and second expressions of (3.5) are equal for each k. On the other hand, the limiting joint distribution function of the set of random variables $\{\sqrt{n}[Y_{ki}^{(n)} - 1/(k+1)], 1 \leqq i \leqq k\}$ under V_k is well known to be that whose density is given in (3.20), below, if we set all $y_i = 1/(k+1)$ and let $n \to \infty$ in the latter; since (3.20), which is the asymptotic a posteriori joint density of the $(p_i - T_{ki}^{(n)}/n)$, is continuous in the y_i, and since the $Y_{ki}^{(n)}$ tend in probability (according to (3.6) and (3.4)) to $1/(k+1)$ as $n \to \infty$, it follows that the second and third expressions of (3.5) are equal. (This last follows also from the continuity in π of $\lim_{n\to\infty} P_\pi\{\ \}$ in the second expression of (3.5) and the fact that $\lim_{n\to\infty} \xi_{kn}(J) = 1$ for any neighborhood J of V_k.) Thus, our theorem will be proved if we prove (3.9), and we now turn to this proof.

In this demonstration our calculations will be performed under the conditions

$$\begin{aligned} &| y_i - 1/(k+1) | < 1/2(k+1) \qquad (1 \leqq i \leqq k+1), \\ &| \eta_i | < n^{-3/8}/4k(k+1) \qquad (1 \leqq i \leqq k+1), \\ &n > k^{40}. \end{aligned} \tag{3.10}$$

All orders $O(\cdot)$ will be uniform in the variables not appearing in the arguments. By (3.8),

$$\log g_{k,n} = \log C_1 + \sum_1^{k+1} (y_i + \epsilon) \log y_i + \sum_1^{k+1} (y_i + \epsilon) \log \left(1 + \frac{\eta_i}{y_i}\right). \tag{3.11}$$

From (3.10), we have

$$\left| \frac{\eta_i}{y_i} \right| < \frac{1}{2kn^{3/8}} \leqq \frac{1}{2}, \tag{3.12}$$

and hence

$$\log \left(1 + \frac{\eta_i}{y_i}\right) = \frac{\eta_i}{y_i} - \frac{\eta_i^2}{2y_i^2} + \theta_i \frac{\eta_i^3}{y_i^3},$$

with

$$|\theta_i| < 1, \qquad (1 \leqq i \leqq k+1). \tag{3.13}$$

Now, writing

$$(y_i + \epsilon) \log\left(1 + \frac{\eta_i}{y_i}\right) = \eta_i - \frac{\eta_i^2}{2y_i} + \theta_i \frac{\eta_i^3}{y_i^2} + \epsilon \frac{\eta_i}{y_i}\left(1 - \frac{\eta_i}{2y_i} + \theta_i \frac{\eta_i^2}{y_i^2}\right)$$

and remarking that $\sum_1^{k+1} \eta_i = 0$, that by (3.10) and (3.13)

$$\left|\sum_1^{k+1} \theta_i \frac{\eta_i^3}{y_i^2}\right| < (k+1)\frac{4(k+1)^2}{64k^3(k+1)^3 n^{9/8}} < \frac{1}{n^{9/8}},$$

and that by (3.10), (3.12), and the definition of ϵ

$$\epsilon\left|\sum_1^{k+1} \frac{\eta_i}{y_i}\left(1 - \frac{\eta_i}{2y_i} + \theta_i \frac{\eta_i^2}{y_i^2}\right)\right| < 2\epsilon \sum_1^{k+1}\left|\frac{\eta_i}{y_i}\right| < \frac{2}{k^2 n^{3/4}} \cdot \frac{k+1}{2kn^{3/8}} \leq \frac{2}{n^{9/8}},$$

we obtain

$$\sum_1^{k+1} (y_i + \epsilon) \log\left(1 + \frac{\eta_i}{y_i}\right) = -\frac{1}{2}\sum_1^{k+1} \frac{\eta_i^2}{y_i} + \frac{3\theta}{n^{9/8}}, \tag{3.14}$$

with $|\theta| < 1$. Combining (3.14) and (3.11), we have

$$\log g_{k,n} = \log C_1 + \sum_1^{k+1} (y_i + \epsilon) \log y_i - \frac{1}{2}\sum_1^{k+1} \frac{\eta_i^2}{y_i} + O(n^{-9/8}) \tag{3.15}$$

Next, we note that

$$(C_1)^n \prod_1^{k+1} y_i^{ny_i+m} = p_{n+(k+1)m}^{(k)}(ny_1 + m, \cdots, ny_{k+1} + m; y_1, \cdots, y_{k+1}), \tag{3.16}$$

where $p_N^{(k)}(w_1, \cdots, w_{k+1}; q_1, \cdots, q_{k+1})$ is the (multinomial) probability that among N independent, identically distributed random variables taking on the value i with probability $q_i(\sum_1^{k+1} q_i = 1, q_i \geqq 0)$, there will be w_i taking on the value $i(\sum w_i = N)$. Using the familiar representation of this probability in terms of binomial probabilities, the definition of m, the inequalities (3.10), and the estimate for binomial probabilities

$$\begin{aligned} &p_N^{(1)}(Np + t\sqrt{Np(1-p)}, N(1-p) - t\sqrt{Np(1-p)}; p, 1-p) \\ &\qquad = [2\pi Np(1-p)]^{-1/2} e^{-t^2/2}[1 + O(N^{-1/2})] \end{aligned} \tag{3.17}$$

for $|t| < C_6$ and $|p - \frac{1}{2}| < C_7 < \frac{1}{2}$ (given in [5], p. 135), we obtain (with a conservative estimate of error)

$$(C_1)^n \prod_1^{k+1} y_i^{ny_i+m} = (1 + O(n^{-1/8}))(2\pi n)^{-k/2} \prod_1^{k+1} y_i^{-1/2}. \tag{3.18}$$

Hence, in the region (3.10) we obtain from (3.15) and (3.18), writing again $\eta_{k+1} = -\sum_1^k \eta_i$ and $y_{k+1} = 1 - \sum_1^k y_i$,

$$\begin{aligned} &f_{k,n}^*(\eta_1, \cdots, \eta_k \mid y_1, \cdots, y_k) \\ &\qquad = (1 + O(n^{-1/8}))(2\pi n)^{-k/2}\left(\prod_1^{k+1} y_i\right)^{-1/2} \exp\left(-\frac{n}{2}\sum_1^{k+1} \frac{\eta_i^2}{y_i}\right). \end{aligned} \tag{3.19}$$

For the corresponding a posteriori joint density of $\bar{\gamma}_i = \sqrt{n}\bar{\eta}_i$, $i = 1, \cdots, k$, in the region (3.10), we thus obtain (writing $\gamma_{k+1} = -\sum_1^k \gamma_i$)

$$(3.20)\qquad (1 + O(n^{-1/8}))(2\pi)^{-k/2}\left(\prod_1^{k+1} y_i\right)^{-1/2} \exp\left(-\frac{1}{2}\sum_1^{k+1} \frac{\gamma_i^2}{y_i}\right).$$

Except for the first factor, this is a k-dimensional normal distribution centered at the origin. Note also that the probability assigned by this density to the complement of the region $|\bar{\eta}_i| < n^{-3/8}/4k(k+1)$ of (3.10) (for a single i) is (by Chebychev's inequality) $\leq [1 + O(n^{-1/8})]O(k^4 n^{-1/4})$, so that the probability of the above inequality on the $\bar{\eta}_i$ for all i according to (3.20) (using $k < n^{1/40}$) is at least $1 - O(n^{-1/8})$. Also, the p_i or (3.6) have means $1/(k+1)$ and variances $O(m^{-1}k^{-2}) = O(n^{-1/4})$, while $Y_{ki}^{(n)}$ (given the p_i) has mean p_i and variance $O(n^{-1})$, whatever the p_i may be. Hence, for a single i, the probability (according to (3.6) and (3.4)) that $|Y_{ki}^{(n)} - 1/(k+1)| < \frac{1}{2}(k+1)$ is

$$\geq P\{|p_i - 1/(k+1)| < \tfrac{1}{4}(k+1)\} \times P\{|Y_{ki}^{(n)} - p_i| < \tfrac{1}{4}(k+1)\,|p_i\}$$
$$\geq 1 - k^2\{O(n^{-1/4}) + O(n^{-1})\}.$$

The probability that $|Y_{ki}^{(n)} - 1/(k+1)| < 1/2(k+1)$ for all i is thus

$$\geq 1 - k^3 O(n^{-1/4}) \geq 1 - O(n^{-1/8}).$$

We conclude, then, that the region of $Y_{ki}^{(n)}, \bar{\eta}_i$ $(1 \leq i \leq k+1)$ specified in (3.10) (putting $Y_{ki}^{(n)}$ for y_i and $\bar{\eta}_i$ for η_i), and hence where (3.20) holds, has probability $1 - O(n^{-1/8})$ according to (3.6) and (3.4).

Now, for fixed $r^* > 0$, let $N_1(k, r^*)$ be such that if $n > N_1(k, r^*)$, then

$$8r^* n^{-1/2} < n^{-3/8}/4k(k+1)$$

and the probability under (3.20) that all $|\bar{\eta}_i|$ are $< n^{-3/8}/16k(k+1)$ is $\geq 1 - n^{-1/9}/2$; clearly, such a number $N_1(k, r^*)$ exists. For $0 < r \leq r^*$, let T_r be the region where $|\sum_{j=1}^{i} \gamma_j| \leq r$, $i = 1, \cdots, k+1$. Note that T_r is contained in the region where $|\gamma_j| \leq 2r^*$ for all j. If ρ is any vector all of whose $(k+1)$ components are $\leq n^{1/8}/8k(k+1)$ and if $n > N_1(k, r^*)$, then T_r and $T_r + \rho$ both lie entirely in the region of (3.10) (where (3.20) holds), whose probability according to (3.6) and (3.4) is $1 - O(n^{-1/8})$. Write C_r and D_r for the events in brackets on the left and right sides of (3.9), and define $L = L(X_1, \cdots, X_n, \psi_n)$ to be 1 or 0 according to whether or not

$$\max_j \sqrt{n}\,|Y_{kj}^{(n)} - \psi_{nj}| \leq n^{1/8}/8k(k+1).$$

From the previous remarks of this paragraph and Lemma 1 we conclude that

$$(3.21)\qquad E_i[L \cdot P_a^*\{C_r\}] \geq E_i[L \cdot P_a^*\{D_r\}] - n^{-1/9}/3$$

for $0 < r \leq r^*$, $n > N(k, r^*)$, and all ψ_n, where $N(k, r^*)$ is chosen (as it clearly may be because $n^{-1/8} = o(n^{-1/9})$) to be enough larger than $N_1(k, r^*)$ to give the term $n^{-1/9}/3$ in (3.21). On the other hand, if any component of ρ has magnitude $> n^{1/8}/8k(k+1)$, then with probability $1 - O(n^{-1/8})$ according to

(3.4) and (3.6), $T_r + \rho$ has a posteriori probability $<n^{-1/9}/2$. Hence, the $N(k, r^*)$ above may clearly also be chosen so large that

$$E_t[(1 - L)P_a^*\{D_r\}] - 2n^{-1/9}/3 \leqq 0 \tag{3.22}$$

for $0 < r \leqq r^*$, $n > N(k, r^*)$, and all ψ_n. Equation (3.9) follows from (3.21) and (3.22), completing the proof of Theorem 3.

We remark that ϕ_n^* will not be minimax in the sense of Theorem 3 for all r and fixed finite n. The first nontrivial case is that of $n = 3$. A tiresome but straightforward computation in this case shows that, among the procedures ϕ_c which for a given number c ($0 \leqq c \leqq \frac{1}{2}$) assign probability one to

$$g_c(x) = \begin{cases} 0 & \text{if } x > Z_1, \\ c & \text{if } Z_1 \leqq x < Z_2, \\ 1 - c & \text{if } Z_2 \leqq x < Z_3, \\ 1 & \text{if } Z_3 \leqq x, \end{cases}$$

where the Z_i are the ordered $X_i^{(3)}$, the expression $P_U\{\sup_x | g_c(x) - x | \leqq z\}$ is maximized for $\frac{1}{6} \leqq z \leqq \frac{1}{3}$ at $c = \frac{1}{3}$ (i.e., by ϕ_3^*), for $\frac{1}{3} \leqq z \leqq \frac{1}{2}$ by $c = z$, and for $\frac{1}{2} < z \leqq 1$ by any $c \geqq 1 - z$ (for $z \leqq \frac{1}{6}$, all values of c give probability zero). Similar remarks apply to the problems considered in the next three sections. For example, $E_U\{\sup_x | g_c(x) - x |\}$ in the above example is minimized by $c = [33 - 3(17)^{1/2}]/52 = 0.397$. Similar calculations are more easily made in the case studied in Section 4 (where the distribution of the maximum deviation need not be calculated), and such calculations may be found in the reference cited at the end of that section.

4. Other loss functions which are functions of distance. In this section we show that the asymptotic minimax character of ϕ_n^* proved in Section 3 may be extended to a broad class of weight functions. It turns out that it is unnecessary to start anew in order to prove this; the class of weight functions considered in Section 3 (see below) is the basic class in the sense that the minimax character relative to many other weight functions may be concluded from the results of Section 3 and the integrability result given in Corollary 2. It is clear that the method of attack used here, i.e., of carrying out the detailed proof of the minimax character for the basic class of weight functions and then extending to other weight functions, can be stated as a general theorem to apply to other statistical problems; we shall not bother to state this obvious extension in a general setting.

Throughout this section W will represent any nonnegative function defined on the nonnegative reals which is nondecreasing in its argument, not identically zero (the case $W \equiv 0$ is trivial), and which satisfies

$$\int_0^\infty W(r) r e^{-2r^2} \, dr < \tag{4.1}$$

The main result of this section is the following:

THEOREM 4. *Under the above assumptions on* W,

$$\text{(4.2)}\quad \lim_{n\to\infty} \frac{\sup_{F\varepsilon\mathfrak{F}_c} \int W(r)\, d_r P_F\{\sup_x \mid S_n(x) - F(x) \mid < r/\sqrt{n}\}}{\inf_{\phi\varepsilon D_n} \sup_{F\varepsilon\mathfrak{F}_c} \int W(r)\, d_r P_{F,\phi}\,\{\sup_x \mid g(x) - F(x) \mid < r/\sqrt{n}\}} = 1.$$

As in Section 3 (and for the same reason), an immediate corollary is

COROLLARY 4. *The result of Theorem 4 holds if* $\mathfrak{F}_c$ *is replaced by* $\mathfrak{F}$ *in its statement.*

PROOF OF THEOREM 4. By a reduction like that of Section 3, it is seen that (4.2) will be proved if, for the sequence $\{\xi_{kn}\}$ of Section 3, we can prove the following three statements, (4.3), (4.4), and (4.5):

$$\text{(4.3)}\quad \begin{aligned} &\lim_{n\to\infty} \inf_{\phi\varepsilon D_n} \int W(r)\, d_r P_{F,\phi}\,\{\max_{a\varepsilon A_k} \mid g(a) - F(a) \mid < r/\sqrt{n}\}\, d\xi_{kn} \\ &\quad = \lim_{n\to\infty} \int' W(r)\, d_r P_F\,\{\max_{a\varepsilon A_k} \mid S_n(a) - F(a) \mid < r/\sqrt{n}\}\, d\xi_{kn} \end{aligned}$$

for each positive integer k;

$$\text{(4.4)}\quad \begin{aligned} &\lim_{n\to\infty} \int W(r)\, d_r P_F\,\{\max_{a\varepsilon A_k} \mid S_n(a) - F(a) \mid < r/\sqrt{n}\}\, d\xi_{kn} \\ &\quad = \lim_{n\to\infty} \int W(r)\, d_r P_U\,\{\max_{a\varepsilon A_k} \mid S_n(a) - a \mid < r/\sqrt{n}\} \end{aligned}$$

for each positive integer k;

$$\text{(4.5)}\quad \begin{aligned} 0 &< \lim_{k\to\infty} \lim_{n\to\infty} \int W(r)\, d_r P_U\,\{\sup_{a\varepsilon A_k} \mid S_n(a) - a \mid < r/\sqrt{n}\} \\ &= \lim_{n\to\infty} \int W(r)\, d_r P_U\,\{\sup_{0\leq x\leq 1} \mid S_n(x) - x \mid < r/\sqrt{n}\} < \infty. \end{aligned}$$

(This includes, of course, proving the existence of the indicated limits.)

Firstly, (4.5) is an immediate consequence of (4.1), (2.4), (2.2), the continuity of G and of the d.f. $\lim_{n\to\infty} G_{k,n}$, and of Corollary 2.

In order to prove (4.4), we note first that, for fixed k and any $F\ \varepsilon\ \mathfrak{F}_c$, we have (similarly to (2.2)) the inequality $P_F\{\max_{a\varepsilon A_k} \mid S_n(a) - F(a) \mid \leqq r/\sqrt{n}\} \geqq G_n(r)$. Hence, by Corollary 2, the integral with respect to r on the left side of (4.4) is bounded uniformly in n and F. On the other hand, given any $\epsilon > 0$, there exists an integer N_0 such that, for $n > N_0$, ξ_{kn} assigns probability at least $1 - \epsilon$ to a set of F for which the expressions $P_F\{\quad\}$ and $P_U\{\quad\}$ of (4.4) differ by less than ϵ for all r (this rests on the continuity in π, for π in a neighborhood of V_k, of the normal approximation (for large n) to the joint distribution of the random variables $\sqrt{n}(Y_{ki}^{(n)} - p_i)$, $1 \leqq i \leqq k$). Since $P_U\{\quad\}$ is continuous in r, (4.4) follows.

Finally, we must prove (4.3). Consider any fixed k. Write $P_n^*(r; x^{(n)}, \phi)$ for the probability, calculated according to the a posteriori probability distribution of π (given that $X^{(n)} = x^{(n)}$ and when ξ_{kn} is the a priori probability measure on B_k) and the probability measure $\phi(\cdot\,; x^{(n)})$ on G (where $\phi \in D_n$ and perhaps $\phi = \phi_n^*$) of the set of (g, π) in $G \times B_k$ for which $\max_{a \in A_k}|g(a) - F(a)| < r/\sqrt{n}$. If (4.3) is false, there exists a value $\epsilon > 0$ such that, for every positive N, there is an $n > N$ and a $\phi_n \in D_n$ for which (the operation E_t being as defined in Section 3)

$$(4.6)\quad E_t \int W(r)\, d_r P_n^*(r; X^{(n)}, \phi_n) < E_t \int W(r)\, d_r P_n^*(r; X^{(n)}, \phi_n^*) - 2\epsilon.$$

It is clear from the preceding paragraphs that there is a real number $q > 0$ such that $W(q) > 0$ and

$$(4.7)\qquad E_t \int_q^\infty W(r)\, d_r P_n^*(r; X^{(n)}, \phi_n^*) < \epsilon$$

for all n. Write $W_q(r) = \min(W(r), W(q))$. Then (4.6) and (4.7) imply

$$(4.8)\qquad E_t \int_0^\infty W_q(r)\, d_r \{P_n^*(r; X^{(n)}, \phi_n) - P_n^*(r; X^{(n)}, \phi_n^*)\} < -\epsilon.$$

Since $W_q(r) \leqq W(q)$, the integral on the left side of (4.8) is $\geqq -W(q)$. Hence, (4.8) implies that, with probability at least $\epsilon/W(q)$ (under (3.6) and (3.4)), $X^{(n)}$ will be such that

$$(4.9)\qquad \int_0^\infty W_q(r)\, d_r \{P_n^*(r; X^{(n)}, \phi_n) - P_n^*(r; X^{(n)}, \phi_n^*)\} < -\epsilon.$$

Let $\epsilon' = \epsilon/2W(q)$. The discussion of the previous paragraph shows that we can find an R^* and M such that, for $n > M$, the probability (under (3.6) and (3.4)) will be $>1 - \epsilon'$ that $X^{(n)}$ will be such that

$$(4.10)\qquad P_n^*(R^*; X^{(n)}; \phi_n^*) > 1 - \epsilon'.$$

Let $\gamma_n = \sup_r\{P_n^*(r; X^{(n)}, \phi_n) - P_n^*(r; X^{(n)}, \phi_n^*)\}$. We shall show below that (4.9) implies

$$(4.11)\qquad \gamma_n > \epsilon'.$$

Then (4.10) and (4.11) (the latter of which is an event of probability at least $2\epsilon'$ according to (3.6) and (3.4)) will imply that for each $N > M$ there is an $n > N$ and a $\phi_n \in D_n$ for which, with probability $>\epsilon'$ according to (3.6) and (3.4), $X^{(n)}$ will be such that

$$(4.12)\qquad \{P_n^*(r; X^{(n)}, \phi_n) - P_n^*(r; X^{(n)}, \phi_n^*)\} > \epsilon'$$

for some r with $0 \leqq r < R^*$ (here r depends on n, $\phi^{(n)}$, $X^{(n)}$). This contradicts the fact that, with probability $1 - O(n^{-1/8})$ according to (3.6) and (3.4), the region T_r of the last paragraph of Section 3 was seen to maximize with respect to ρ (uniformly in $0 \leqq r \leqq R^*$), to within an (added) error of $O(n^{-1/9})$, the a posteriori probability of $T_r + \rho$. Thus, it remains only to prove that (4.9) implies (4.11). For fixed n, ϕ_n, $X^{(n)}$, abbreviate the bracketed expression in (4.9) as $B(r) - C(r)$. Let

$$B^*(r) = \begin{cases} 0 & \text{if } r \leqq 0. \\ \min (C(r) + \gamma_n, 1) & \text{if } r > 0. \end{cases} \tag{4.13}$$

Clearly, $B(r) \leqq B^*(r)$. Hence, since $W_q(r)$ is nondecreasing in r, we have

$$\int W_q(r)\, dB(r) \geqq \int W_q(r)\, dB^*(r). \tag{4.14}$$

Let α be the infimum of values r for which $B^*(r) = 1$. From (4.9), (4.14), and the fact that $B^*(r) - C(r)$ is constant for $0 < r < \alpha$, we obtain

$$\begin{aligned} \epsilon &< \int_{0-}^{\infty} W_q(r)\, d(C(r) - B(r)) \\ &\leqq \int_{0-}^{\infty} W_q(r)\, d(C(r) - B^*(r)) \\ &\leqq \int_{0+}^{\alpha-} W_q(r)\, d(C(r) - B^*(r)) + W(0)\gamma_n + W(q)\gamma_n \\ &\leqq 2W(q)\gamma_n , \end{aligned} \tag{4.15}$$

which proves (4.11) and thus completes the proof of Theorem 4.

5. Integral weight functions. In this section we consider weight functions W_n^* arising from integration of a function W in the following manner:

$$W_n^*(F, g) = \int_{-\infty}^{\infty} W(\sqrt{n}[g(x) - F(x)], F(x))\, dF(x). \tag{5.1}$$

Here $W(y, z)$, which is defined for y real and $0 \leqq z \leqq 1$, is nonnegative and is symmetric in y and nondecreasing in y for $y \geqq 0$; it may be thought of as a measure of the contribution to W_n^* arising from a deviation of $y\sqrt{n}$ of the estimator g from the true F at an argument x for which $F(x) = z$. Typical W's which might be of interest are $W(y, z) = |y|^p$ or 0 according to whether or not $a \leqq z \leqq b$ (here $p > 0$ and $0 \leqq a < b \leqq 1$), $W(y, z) = 0$ or 1 according to whether $|y| \leqq a$ or $|y| > a$ where a is a suitably chosen constant, $W(y, z) = y^2/z(1 - z)$, etc.

We now turn to considerations of the asymptotic minimax character of ϕ_n^* with respect to a sequence of risk functions $r_n(F, \phi) = E_{F,\phi}W_n^*(F, g)$, where $\phi \varepsilon D_n$. (The remainder of the present paragraph will be somewhat heuristic in order to compare the present problem with those of Sections 3 and 4; the statement and proof of Theorem 5 begin in the next paragraph.) These considerations are much easier than those of the previous two sections, since in obtaining a

Bayes solution with respect to the a priori probability measure ξ_{kn} of Section 3 it will suffice (as will be seen below) to minimize with respect to ϕ, *for each fixed* x (more precisely, for each irrational x),

$$(5.2)\quad r_{kn}(x, \phi, t_k^{(n)}) = \int_{B_k} E_\phi W(\sqrt{n}[g(x) - F(x, \pi)], F(x, \pi))\, d_\pi \xi_{kn}^*(\pi; x, t_k^{(n)});$$

here B_k is as in Section 3, $F(x, \pi)$ denotes the distribution function of (3.3) for a given value of $\pi = (p_1, \cdots, p_{k+1})$, and for any measurable subset B of B_k we set

$$(5.3)\quad \xi_{kn}^*(B, x, t_k^{(n)}) = \frac{\int_B f(x, \pi) P_\pi\{t_k^{(n)}\}\, d\xi_{kn}(\pi)}{\int_{B_k} f(x, \pi) P_\pi\{t_k^{(n)}\}\, d\xi_{kn}(\pi)},$$

where ξ_{kn} is given by (3.6) of Section 3, $f(x, \pi) = dF(x, \pi) / dx$ (this derivative exists for x irrational), and $P_\pi\{t_k^{(n)}\} = P_\pi\{T_{ki}^{(n)} = t_{ki}^{(n)}, 1 \leq i \leq k + 1\}$ is the probability function defined in (3.4). (Of course, ϕ in (5.2) may randomize over many g, which accounts for the presence of the E_ϕ operation.) Thus, present considerations will involve only the obtaining of a (univariate) normal approximation to the a posteriori distribution (more precisely, to a slight modification (5.3) of it) of $F(x, \pi)$ *for fixed irrational* x, which is much easier than the multivariate approximation (3.20) which it was necessary to obtain in Section 3. (We shall actually use (3.20), which implies easily the needed univariate approximation; however, the latter could have been obtained more easily directly.) The above remarks will be made precise in what follows. We hereafter denote the infimum of $r_{kn}(x, \phi, t_k^{(n)})$ over all ϕ in D_n by $r_{kn}^*(x, t_k^{(n)})$. The set of reals

$$\{z \mid 0 < z < \epsilon \text{ or } 1 - \epsilon < z < 1\}$$

will be denoted by I_ϵ for $0 < \epsilon < \frac{1}{2}$.

We now state Theorem 5. Our statement of this theorem is not the most general possible. (The set I_ϵ may be replaced by other sets where $W(y, z)$ is large, the continuity conditions on W may be weakened by considering continuous approximations to (a more general) measurable W, the integrability condition may be weakened, and W may be replaced by a distribution (rather than a density) in z so as to obtain results, e.g., on the estimation of F at a finite number of quantiles.) Rather, it is stated in a form which allows W to be any of the functions which would usually be of interest in applications, e.g., any of those functions given at the end of the first paragraph of this section, etc. (It should be noted that if the assumptions of Theorem 5 below were altered by deleting (5.5) and putting $\epsilon = 0$ in (5.4), then such weight functions as $y^2/z(1 - z)$ would be excluded. The circumlocution of including the condition (5.5) could be avoided in such cases if one could obtain a sufficiently strong bound on

$$P_U\{\sqrt{n}[S_n(x) - x] > r\sqrt{x(1 - x)}\}$$

which is independent of x. The difficulty of obtaining such an approximation is discussed in [4], p. 285.)

THEOREM 5. *Let* $W(y, z) \geqq 0$ *be defined for* $0 \leqq y < \infty, 0 < z < 1$ *and assume that* $W(y, z)$ *is monotone nondecreasing in* y *and (to avoid trivialities) that* $W(y, z)$ *is not almost everywhere zero (in the two-dimensional Lebesgue sense). Suppose further that* (a) *to every* z', $0 < z' < 1$, *not belonging to an exceptional set of linear measure zero, and every* $\delta > 0$ *there corresponds* $\epsilon(\delta, z') > 0$ *with the property that the set of* y *for which* $W(y, z)$ *is discontinuous for at least one* z *satisfying* $|z - z'| < \epsilon(\delta, z')$ *has exterior (linear Lebesgue) measure smaller than* δ. *Suppose also that* (b) *for each* ϵ *with* $0 < \epsilon < \frac{1}{2}$ *there is a function* $V(y, \epsilon)$ *such that* $W(y, z) \leqq V(y, \epsilon)$ *for* $\epsilon < z < 1 - \epsilon$ *and* $0 \leqq y < \infty$ *and such that*

$$\int_0^\infty V(y, \varepsilon) y e^{-2y^2}\, dy < \infty. \tag{5.4}$$

Suppose, finally, that (c)

$$\limsup_{\epsilon \to 0} \sup_n \int_{I_\epsilon} E_U W(\sqrt{n}[S_n(x) - x], x)\, dx = 0. \tag{5.5}$$

Then

$$\lim_{n \to \infty} \frac{\sup_{F \epsilon \mathfrak{F}_c} r_n(F, \phi_n^*)}{\inf_{\phi \epsilon D_n} \sup_{F \epsilon \mathfrak{F}_c} r_n(F, \phi)} = 1. \tag{5.6}$$

PROOF. $r_n(F, \phi_n^*)$ is, of course, independent of F for F in $\mathfrak{F}_c$. Because of Corollary 2 and the assumptions of Theorem 5, the numerator of (5.6) approaches a finite positive limit, say L, as $n \to \infty$. For any δ with $0 < \delta < L$ we may choose ϵ so small that $r_{n,\epsilon}(F, \phi_n^*)$ tends to a limit $> L - \delta$ when $n \to \infty$, where $r_{n,\epsilon}$ is the risk function corresponding to loss function $W_\epsilon(y, z)$ defined by

$$W_\epsilon(y, z) = \begin{cases} W(y, z) & \text{if } z \not\in I_\epsilon, \\ 0 & \text{if } z \in I_\epsilon. \end{cases} \tag{5.7}$$

It clearly suffices to prove (5.6) with r_n replaced by $r_{n,\epsilon}$. We hereafter drop the subscript ϵ on W_ϵ and $r_{n,\epsilon}$ and (because of (5.7)) may restate what is to be proved as (5.6) under the continuity assumption (a) on W and (replacing (5.4) and (5.5)) the assumption that $W(y, z) \leqq V(y)$ for $0 \leqq y < \infty$ and $0 < z < 1$, where

$$\int_0^\infty V(y) y e^{-2y^2}\, dy < \infty. \tag{5.8}$$

In what follows we denote (for fixed k, n, irrational x) by $P_x^*\{A\}$ the probability of any event A which is expressed in terms of $T_k^{(n)}$ when the probability function of $T_k^{(n)}$ is given by

$$
\begin{aligned}
&P\{T_{ki}^{(n)} = t_i, 1 \leqq i \leqq k+1\} \\
(5.9) \qquad &= \frac{1}{d(k, n, x)} \int_{B_k} f(x, \pi) P_\pi\{T_{ki}^{(n)} = t_i, 1 \leqq i \leqq k+1\}\, d\xi_{kn}(\pi),
\end{aligned}
$$

where P_π is given by (3.4) and $d(k, n, x)$ is the sum over all $(t_1, \cdots, t_{k+1})$ of the integral on the right side of (5.9). Expectation with respect to the probability function (5.9) will be denoted by E_x^*. We now have

$$
\begin{aligned}
&\int r_n(F, \phi)\, d\xi_{kn} = \int_0^1 \int_{B_k} E_{\pi,\phi} W(\sqrt{n}[g(x) - F(x, \pi)], F(x, \pi))\, d\xi_{kn}(\pi)\, dx \\
(5.10) \qquad &= \int_0^1 E_x^* r_{kn}(x, \phi, T_k^{(n)})\, d(k, n, x)\, dx,
\end{aligned}
$$

where the last integration (and each integration which follows) is over irrational x. Hence, in order to prove (5.6), it suffices to show that (5.8) and our continuity assumption on W imply that

$$
\begin{aligned}
&\lim_{k\to\infty} \lim_{n\to\infty} \int_0^1 E_x^* r_{kn}^*(x, T_k^{(n)})\, d(k, n, x)\, dx \\
(5.11) \qquad &= \lim_{n\to\infty} \int_0^1 E_U W(\sqrt{n}[S_n(x) - x], x)\, dx,
\end{aligned}
$$

since the right side of (5.11) is the limit of the finite positive numerator of (5.6).

Let x be an irrational number, $0 < x < 1$, which is a *non*exceptional z' of our continuity assumption (a). For fixed k with $1/(k+1) < \min(x, 1-x)$, we may write $x = (i_0 + t)/(k+1)$ with $1 \leqq i_0 \leqq k-1$ and $0 \leqq t < 1$. Write $q(r, \sigma^2) = (2\pi\sigma)^{-1/2} \exp(-r^2/2\sigma^2)$. We shall show that, given any x and k as above and any $\epsilon' > 0$, there is an integer $N = N(\epsilon', x, k)$ such that for $n > N$ we have $|d(k, n, x) - 1| < \epsilon'$ and such that, for $n > N$, P_x^* assigns probability at least $1 - \epsilon'$ to a set of $T_k^{(n)}$ values for which

$$
(5.12) \qquad r_{kn}^*(x, T_k^{(n)}) + \epsilon' > \int_{-\infty}^{\infty} W(y, x) q(y, x(1-x) + h)\, dy,
$$

where $h = (t^2 - t)/(k+1)$. But, for fixed irrational and nonexceptional x, the right side of (5.12) tends, as $k \to \infty$ (and thus, $h \to 0$), to the limit as $n \to \infty$ of the integrand in the right-hand member of (5.11). The integral of this limit is, by (5.8), the same as the right-hand member of (5.11). Thus, using (5.12) and applying Fatou's lemma to the left side of (5.11), we conclude that (5.11) will be proved if we demonstrate the statement of the sentence containing (5.12).

For fixed x and k as above and for any $\epsilon > 0$, ξ_{kn} assigns to the set of π for which $|f(x, \pi) - 1| < \epsilon$ a probability which tends to unity as $n \to \infty$. It follows that $d(k, n, x) \to 1$ as $n \to \infty$ and that (noting the relationship between ξ_{kn}^* and the f_{kn}^* of Section 3), for any $\epsilon > 0$ and for n sufficiently large, P_x^* assigns probability at least $1 - \epsilon$ to a set of values $t_k^{(n)}$ of $T_k^{(n)}$ for which, writing $y_i = t_{ki}^{(n)}/n$, the joint density function of the $\bar{\gamma}_i = \sqrt{n}(p_i - y_i)$ $(1 \leqq i \leqq k)$ according to $\xi_{kn}^*(\cdot, x, t_k^{(n)})$ in a spherical region centered at 0 in the space of the $\bar{\gamma}_i$ and of probability at least $1 - \epsilon$ according to ξ_{kn}^* is at least

$$(5.13) \qquad (1 - \epsilon)(2\pi)^{-k/2} \left(\prod_1^{k+1} y_i\right)^{-1/2} \exp\left(-\frac{1}{2}\sum_1^{k+1} \gamma_i^2/y_i\right).$$

Now, in the notation of Section 3, for $1 \leqq i \leqq k$,

$$F(i/(k + 1), \pi) = p_1 + \cdots + p_i .$$

Hence, if $T_{ki}^{(n)} = ny_i$ $(1 \leqq i \leqq k)$, we have (because of the form of (3.3))

$$F(x, \pi) = p_1 + \cdots + p_{i_0} + tp_{i_0+1} = (y_1 + \cdots + y_{i_0} + ty_{i_0+1}) + (\bar{\gamma}_1 + \cdots + \bar{\gamma}_{i_0} + t\bar{\gamma}_{i_0+1})/\sqrt{n}.$$

Now, $T_{ki}^{(n)}/n$ tends in probability (according to P_x^*) to $1/(k + 1)$, and expression (5.13) with $\epsilon = 0$ is continuous in the y_i (in the region where all $y_i > 0$). Moreover, if we had $\epsilon = 0$ in (5.13) and assumed the validity of this expression for all values of the γ_i and put all $y_i = 1/(k + 1)$, then $(\bar{\gamma}_1 + \cdots + \bar{\gamma}_{i_0})$ and $\bar{\gamma}_{i_0+1}$ would, according to (5.13), have a bivariate normal density function with means zero and covariance matrix

$$(5.14) \qquad \frac{1}{(k + 1)^2}\begin{pmatrix} i_0(k + 1 - i_0) & -i_0 \\ -i_0 & k \end{pmatrix}.$$

The corresponding density function of $\bar{\gamma}_i + \cdots + \bar{\gamma}_{i_0} + t\bar{\gamma}_{i_0+1}$ would then be normal with mean zero and variance

$$(5.15) \qquad [i_0(k + 1 - i_0) - 2ti_0 + t^2k]/(k + 1)^2 = x(1 - x) + h.$$

Hence, if we carry through this last argument with the actual form of (5.13) and its region of validity, we conclude that, for any $\epsilon'' > 0$ and for n sufficiently large, P_x^* assigns probability at least $1 - \epsilon''$ to a set of values $t_k^{(n)}$ of $T_k^{(n)}$ for which, on a real interval centered at 0 and of probability at least $1 - \epsilon''$ according to $\xi_{kn}^*(\cdot, x, t_k^{(n)})$, this last measure induces a distribution function J for

$$\sqrt{n}[F(x, \pi) - (y_1 + \cdots + y_{i_0} + ty_{i_0+1})] = \Lambda_x \text{ (say)}$$

whose absolutely continuous component has a corresponding density (the derivative of J) whose magnitude is at least

$$(5.16) \qquad (1 - \epsilon'')q(\lambda, x(1 - x) + h)$$

almost everywhere on this interval of λ-values.

Next, we note that

(5.17) $$W(\sqrt{n}[g(x) - F(x, \pi)], F(x, \pi)) = W(\rho - \Lambda_x, x + \mu + \Lambda_x/\sqrt{n}),$$

where $\rho = \sqrt{n}[g(x) - (y_1 + \cdots + y_{i_0} + ty_{i_0+1})]$ and

$$\mu = -x + (y_1 + \cdots + y_{i_0} + ty_{i_0+1}).$$

For fixed x and k as above, denote by α the right side of (5.12). Let β be such that the right side of (5.12) is at least $\alpha - \epsilon'/4$ if the limits of integration are changed to $(-\beta, \beta)$. Let $c = W(\beta, x)$. Let the δ of our assumption (a) be

$$\epsilon'/8cq(0, x(1 - x) + h),$$

and let $z' = x$ where x is nonexceptional. The set $0 \leqq y \leqq \beta$, $|z - x| \leqq \epsilon(\delta, x)$ minus a suitable countable set of open intervals of total length $<\delta$ covering the points of discontinuity is closed and bounded. Hence, W is uniformly continuous on this set. Hence, there is a value $\epsilon_1 > 0$ such that $W(y, z) \geqq W(y, x) - \epsilon'/4$ for $|x - z| \leqq \epsilon_1$ and $0 \leqq y \leqq \beta$ but y not in the excluded set. If $0 \leqq y \leqq \beta$ and y is in the exceptional set, y is in a maximal subinterval of the exceptional set of either the form $a < y < b$ with $a > 0$ or else of the form $0 \leqq y < b$. Define $\tilde{W}(y, x) = W(a, x)$ in the former case and $\tilde{W}(y, x) = 0$ in the latter. If $0 \leqq y \leqq \beta$ but y is not exceptional, define $\tilde{W}(y, x) = W(y, x)$. If $y > \beta$, define $\tilde{W}(y, x) = \tilde{W}(\beta, x)$. Finally, set $\tilde{W}(-y, x) = \tilde{W}(y, x)$. The function $\tilde{W}$ so defined is symmetric in y, nondecreasing in y for $y \geqq 0$, and has the property that

(5.18) $$W(y, z) \geqq \tilde{W}(y, x) - \epsilon'/4 \qquad \text{for } |x - z| \leqq \epsilon_1 \text{ and all } y,$$

and also that

(5.19) $$\int_{-\beta}^{\beta} \tilde{W}(y, x)q(y, x(1 - x) + h)\, dy \geqq \alpha - \epsilon'/2.$$

Now, let $N = N(\epsilon', x, k)$ be such that, for $n > N$ and with $\epsilon'' = \epsilon'/4(\alpha + 1)$, the conclusion (5.16) holds with the λ-interval including the interval $(-\beta, \beta)$, and such that $|d(k, n, x) - 1| < \epsilon'$ for $n > N$. Write $\bar{\mu}$ for the random variable defined by putting $T_{ki}^{(n)}/n$ for y_i in the definition of μ. Since $\bar{\mu}$ tends to zero in probability (according to P_x^*) as $n \to \infty$, we may also suppose N to be such that, for $n > N$, $P_x^*\{|\bar{\mu}| + \beta/\sqrt{n} < \epsilon_1\} \geqq 1 - \epsilon''$. Next, we recall the statement made immediately following the statement of Lemma 1, that for $n = 1$ the conclusion of Lemma 1 holds if the normal probability density is replaced by one truncated at $(-\beta, \beta)$. We also note that the integral (with respect to λ) of this truncated density multiplied by $\tilde{W}(\rho - \lambda, x)$ is easily seen (by an argument like that used to deduce (4.11) from (4.9)) to be minimized at $\rho = 0$. We note, as in previous sections, that if (5.12) is true under the restriction to nonrandomized ϕ (in the definition of r^*), then (5.12) is a fortiori true without this restriction. Thus, from (5.2), (5.16), (5.17), (5.18), and (5.19), we have for $n > N$ that, with P_x^*-probability at least

$$(5.20) \qquad 1 - 2\epsilon'' > 1 - \epsilon',$$

$T_k^{(n)}$ will be such that

$$\begin{aligned}
&r_{kn}^*(x, T_k^{(n)}) \\
&\quad \geqq \inf_\rho \int_{-\beta}^{\beta} W(\rho - \lambda, x + \mu + \lambda/\sqrt{n})(1 - \epsilon'')q(\lambda, x(1-x) + h)\, d\lambda \\
(5.21) \quad &\quad \geqq (1 - \epsilon'') \inf_\rho \int_{-\beta}^{\beta} [\bar{W}(\rho - \lambda, x) - \epsilon'/4] q(\lambda, x(1-x) + h)\, d\lambda \\
&\quad = (1 - \epsilon'') \int_{-\beta}^{\beta} [\bar{W}(-\lambda, x) - \epsilon'/4] q(\lambda, x(1-x) + h)\, d\lambda \\
&\quad \geqq (1 - \epsilon'')(\alpha - 3\epsilon'/4) > \alpha - \epsilon'.
\end{aligned}$$

This completes the proof of (5.12) and thus of Theorem 5.

We have not stated a corollary to Theorem 5 of the type given after Theorems 3 and 4. For $F \varepsilon \mathfrak{F} - \mathfrak{F}_c$, a weight function of the form (5.1) seems less meaningful because the loss contributed at a saltus x of F is measured by $W(y, z)$, where $z = F(x + 0)$. There are also certain technical difficulties in that the numerator of (5.6) need no longer be the same if $\mathfrak{F}_c$ is replaced by $\mathfrak{F}$. We shall not bother with the circumlocutions (e.g., additional restrictions on W) necessary to obtain a corollary from Theorem 5 in the same trivial manner as such corollaries were obtained from Theorems 3 and 4.

Theorem 5 implies certain much weaker results which, for special forms of W, may also be obtained from results obtained by Aggarwal [6]. He considers only the class C_n of procedures which with probability one set $g(x) = c_j^{(n)}$ for $Z_j^{(n)} \leqq x < Z_{j+1}^{(n)}$, where the $\{Z_j^{(n)}\}$ are the ordered values of the $\{X_j^{(n)}\}$. (Such procedures have constant risk for $F \varepsilon \mathfrak{F}_c$ and W_n^* of the form (5.1).) For the special functions $W(y, z) = |y|^r$ and $W(y, z) = |y|^r/z(1 - z)$ (r a positive integer), he obtains the best $c_j^{(n)}$ explicitly in a few cases and in the other cases characterizes them as the solutions of certain equations. In the former cases ϕ_n^* may be seen to be asymptotically best in C_n. This result is an immediate consequence of Theorem 5, where the result is proved for the class D_n of all procedures, of which the class C_n is a small subclass.

6. Other loss functions; multinomial estimation problems. The results obtained in the previous three sections may be extended to a more general class of loss functions to which the same methods of proof may be seen to apply. Thus, for example, in Sections 3 and 4 we could consider the maximum deviation over a set of x values for which $F(x)$ is in a specified *sub*set of the unit interval (this will involve techniques like those used in Section 5); the formulation of Theorem 5 already includes weight functions which may (e.g.) vanish for certain values of $F(x)$, and other modifications (e.g., to consider a finite set of points) are mentioned in the paragraph preceding Theorem 5. We may also consider (in Section 4) loss functions such as $W_1(r_1) + W_2(r_2)$ where r_1 and r_2 are the maximum deviations over two (not necessarily disjoint) sets of the type mentioned above and W_1 and W_2 are functions of the type considered in Section 4. Linear combinations of loss functions of this last type and the type considered in Section 5 may similarly be treated. In all of the above we may replace $\sup_x |g(x) - F(x)|$ by $\sup_x [|g(x) - F(x)|h(F(x))]$, where h is any nonnegative function (suitably regular), without any difficulty; this includes as a special case maximization over a subset as described above.

Thus, it appears that our results hold for a very general class of weight functions. It would of course be of interest to subsume all cases under one unified criterion and one method of proof. In the portion of Section 7 which is devoted to heuristic remarks, such a criterion (symmetry and convexity of a certain functional) is indicated; unfortunately, it does not include all cases treated above (e.g., the result of Section 3, which is apparently somewhat deeper), some of which will be seen in Section 7 to be slightly more difficult to handle than the symmetric convex functionals. A more general class Ω of monotone functionals for which (perhaps under slight regularity conditions) our results would seem likely to hold, and which includes the weight functions of Sections 3, 4, and 5 as well as those of the previous paragraph, is also indicated in Section 7. In the present context, this class consists of nonnegative functionals W of the function $|\delta|$ defined by $\delta(y) = g(F^{-1}(y)) - y$, $0 \leqq y \leqq 1$ (where we suppose for simplicity that the possible c.d.f.'s F under consideration are members of $\mathfrak{F}_c$ which are for each F strictly increasing for $\sup F^{-1}(0) \leqq y \leqq \inf F^{-1}(1)$) for which $W(|\delta_1(y)|) \leqq W(|\delta_2(y)|)$ whenever $|\delta_1(y)| \leqq |\delta_2(y)|$ for $0 \leqq y \leqq 1$. However, at this writing it is not evident how to give a rigorous *unified* proof (as distinguished from the heuristic one of Section 7) even for the class of weight functions which are convex symmetric functionals (of δ, in the present context), let alone to give one for the class Ω.

Another modification is to consider $\sup_x |g(x) - F(x)|h(x)$ above instead of $\sup_x |g(x) - F(x)|h(F(x))$. In this case ϕ_n^* will not have constant risk over $\mathfrak{F}_c$. However, this case is easily treated as follows: suppose for simplicity that h is continuous and bounded (the unbounded case is trivial and may be treated by a similar argument). Let J be an interval in which h is entirely within a prescribed $\epsilon > 0$ of $\sup_x h(x)$. We may for simplicity suppose J to be the unit interval. Then the risk function of ϕ_n^* will attain a value close to its maximum for $F = U$. The argument of Sections 3 and 4 may now be applied. In a similar manner we may consider in Section 5 loss functions for which the risk function of ϕ_n^* is not a constant; for example, (5.1) could be replaced by

$$W_n^*(F, g) = \int_{-\infty}^{\infty} W(\sqrt{n}[g(x) - F(x)], x)\, d\mu(x) \tag{6.1}$$

for a specified function W and measure μ satisfying certain regularity conditions.

An interesting question is whether or not our results can be extended to yield a *sequential* asymptotic minimax character, e.g., in the sense of Wald [7]. This is too large a topic to be discussed thoroughly in this paragraph, but a few indicative comments are in order. An essential idea present in the form of the ξ_{kn} of Sections 3, 4, and 5 is that, when k is large, a certain multinomial estimation problem is *almost as difficult* as the problem of estimating F. This suggests that, when the weight function considered here is such that the corresponding multinomial problem has (perhaps only asymptotically) a fixed sample-size minimax estimator (among all sequential estimators), then we may conclude that the fixed sample-size procedure ϕ_n^* is asymptotically minimax among all *sequential* procedures. An examination of [7] shows that such an asymptotic sequential minimax property for the multinomial problem will often be easy to prove using methods like Wald's.

Finally, the methods of this paper (without the limit considerations as $k \to \infty$) may be used to prove certain asymptotic minimax results for the estimation of the parameter π of the multinomial distribution (3.4) as $n \to \infty$, for any fixed k. To see this, we note that, under fairly general conditions of monotonicity and symmetry of the weight function (similar to those of Sections 3 to 5), the limiting risk function of $T_k^{(n)}/n$ as $n \to \infty$ will be continuous in a neighborhood of the point of B_k at which its maximum is achieved. Hence, for any $\epsilon > 0$, there will exist an interior point V_k^* of B_k in a neighborhood of which the limiting risk function of $T_k^{(n)}/n$ is continuous and at which point the limiting risk of $T_k^{(n)}/n$ is within ϵ of its maximum. One can then find a sequence $\{\xi_{kn}\}$ of a priori distributions on B_k (similar to the sequence used in Sections 3, 4, and 5) which assigns to any neighborhood of V_k^* a probability approaching one as $n \to \infty$ and which "shrinks down" on V_k^* at a slow enough rate (see the remarks of the paragraph preceding that containing (3.6)) to make the a posteriori probability distribution of the $\sqrt{n}(p_j - T_{kj}^{(n)}/n)$ normal with mean 0 so that $T_k^{(n)}/n$ is asymptotically Bayes with respect to $\{\xi_{kn}\}$, with integrated risk approaching the limiting risk of $T_k^{(n)}/n$ at V_k^*. The asymptotic minimax character of $T_k^{(n)}/n$ follows. We need not detail the wide variety of weight functions for which this optimum asymptotic property of the classical multinomial estimator $T_k^{(n)}/n$ follows from the methods and the results of the three previous sections as well as of the present section. It is perhaps worth while to remark that, although the results in Sections 3, 4, and 5 are stated in terms of deviations of sums $\psi_{n1} + \psi_{n2} + \cdots + \psi_{nj}$ of components ψ_{ni} of the estimator ψ_n from $p_1 + p_2 + \cdots + p_j$ $(1 \leqq j \leqq k+1)$, the given proofs apply with only trivial modifications to weight functions depending on differences $\psi_{nj} - p_j$. Thus, for example, for any set of numbers $c_j > 0$, the asymptotic minimax character of $T_k^{(n)}/n$ for estimating $\pi \varepsilon B_k$ for the risk function

$$r_n(\pi, \psi_n) = 1 - P_\pi\{|\psi_{nj} - p_j| \leqq c_j/\sqrt{n}, (1 \leqq j \leqq k+1)\} \tag{6.2}$$

follows from the asymptotic normality of the a posteriori distribution, noted above, and from the convexity and symmetry about ψ_n of the set of π (in R^{k+1}, not B_k) satisfying the inequalities in brackets in (6.2). (It is clear from this example that V_k^* need not be the V_k of Section 3.) The result for other risk functions follows similarly, using the methods and results of Sections 4, 5, and 6.

These asymptotic results for the multinomial estimation problem do not seem to have appeared previously in the literature. As indicated two paragraphs above, some of these multinomial results may also be extended to sequential problems.

7. Convex functionals and monotone functionals of stochastic processes; heuristic considerations. The first part of this section will be devoted to some simple remarks concerning convex symmetric functionals of random elements; these remarks will then be applied to give a short heuristic argument for many of the results obtained in previous sections.

Let $B = \{b\}$ be a linear space (or system) and ζ a random element with range in B and having a symmetric distribution, i.e., such that whenever A is a measurable subset of B so is $-A$ and $P\{\zeta \in A\} = P\{\zeta \in -A\}$. Let ω be a measurable real-valued convex functional on B which is symmetric ($\omega(b) = \omega(-b)$ for $b \in B$) and convex ($\omega(\lambda b_1 + (1-\lambda)b_2) \leqq \lambda\omega(b_1) + (1-\lambda)\omega(b_2)$ for $0 < \lambda < 1$ and $b_1, b_2 \in B$). We now note that, since $\min_b \omega(b) = \omega(0) > -\infty$ so that the expected value $E\omega(\zeta)$ is always defined, we may conclude that

$$E\omega(\zeta) = \min_{b \in B} E\omega(\zeta + b) \tag{7.1}$$

from the equation (implied by symmetry of P)

$$E\omega(\zeta + b) = E\omega(-\zeta + b) = \tfrac{1}{2}E\{\omega(\zeta + b) + \omega(-\zeta + b)\} \tag{7.2}$$

and the equation (implied by symmetry and convexity of ω)

$$\omega(\zeta + b) + \omega(-\zeta + b) = \omega(\zeta + b) + \omega(\zeta - b) \geqq 2\omega(\zeta). \tag{7.3}$$

We shall now apply (7.1) to the "tied-down" Wiener process (see [2]). B is now the space of continuous functions $b(t)$ on the unit interval $0 \leqq t \leqq 1$. The probability measure P assigns probability one to the subset B_0 of elements b of B satisfying $b(0) = b(1) = 0$. The measurable sets are generated by all sets of the form $\{b \mid b(t_0) < a_0\}$ for $0 \leqq t_0 \leqq 1$ and a_0 real. The joint distribution of $\zeta(t_1), \cdots, \zeta(t_n)$ for any $0 \leqq t_1 \leqq \cdots \leqq t_n \leqq 1$ is normal with $E\zeta(s) = 0$ and $E\{\zeta(s)\zeta(t)\} = \min(s, t) - st$ for $0 \leqq s, t \leqq 1$. Note that the distribution of ζ is symmetric.

Let W be any symmetric real-valued convex function on R^1. Then

$$W(\max_t |b(t)|)$$

is a convex functional of b and (7.1) implies that

$$EW(\max_t |\zeta(t)|) \leqq EW(\max_t |\zeta(t) + \rho(t)|) \tag{7.4}$$

for all continuous functions ρ. Generalizations of this result to the case where $\max_t$ is replaced by $\max_t h(t)$ in the manner of Section 6, or where ρ is allowed to be of a more general class than the continuous functions, are easily achieved by adjoining additional functions to B. One may also note that, for every $r > 0$,

$$P\{\max_t |\zeta(t)| > r\} \leqq P\{\max_t |\zeta(t) + \rho(t)| > r\} \tag{7.5}$$

for all continuous (or more general, as noted above) functions ρ. However, this cannot be proved in the same manner as (7.4), since the characteristic function of the subset of B for which $\max_t |b(t)| > r$ is not a convex functional on B. The validity of (7.5) follows, however, from (2.4) and Lemma 1. This strong result of a domination of an entire distribution function in the sense of (7.5) is deeper than the result (7.4); for (7.5) requires (in the proof of Lemma 1 in [8]) not merely the symmetry of the probability distribution, but also the convexity for every $u > 0$ of the set where the joint density of $\zeta(t_1), \cdots, \zeta(t_k)$ (for any k and $t_1, \cdots, t_k$) is $\geqq u$. (Note, for example, that it is not necessarily true for a symmetrically distributed real-valued random variable X that $P\{|X| > r\} \leqq P\{|X + \rho| > r\}$ for all real ρ.) Similarly, the result (7.4) for real functions $W(z)$ on the nonnegative reals which are nondecreasing in z for $z \geqq 0$ (but not necessarily convex) is a consequence of (7.5) but cannot be proved directly in the manner of (7.4) for convex W. Thus, to summarize, (7.4) for convex W follows from the symmetry of the probability measure, while in proving (7.4) for nondecreasing W (and, in particular, (7.5)) we use the additional assumption on the probability measure which is used to prove Lemma 1. We note that it has not been necessary to assume any integrability condition here.

It is interesting to note that, for the special case of a linear function $\rho(t) = c + dt$, the right side of (7.5) is given by formula (4.3) of [2] with $a = r - c - d$, $b = r - c$, $\alpha = r + c + d$, $\beta = r + c$ (unless $a \geqq 0$, $b > 0$, $\alpha \geqq 0$, $\beta > 0$, the probability in question is unity; our $\zeta(t)$ is Doob's $X(t)$). It does not seem completely apparent from the form of (4.3) of [2] that this expression, with the above substitutions, is a minimum for $c = d = 0$. The same is true of expectations with respect to the d.f. (4.3) of [2] of functions W of the type considered above.

Next, let the real-valued function $W(y, z)$ be symmetric and convex in y for each z ($-\infty < y < \infty$, $0 \leqq z \leqq 1$) and satisfy obvious measurability conditions. Let μ be any measure on the unit interval. Then

$$\omega(b) = \int_0^1 W(b(t), t)\, d\mu(t) \tag{7.6}$$

is a convex functional on B and hence (7.1) holds. In this case the result for the case where $W(y, z)$ is symmetric in y and nondecreasing in y for $y \geqq 0$ (but not necessarily convex for each z) is not much more difficult, although it cannot be handled by using (7.1): we need only apply Lemma 1 for $n = 1$ for each fixed z in this case in order to obtain the desired result.

More general convex functionals (such as combinations of the two varieties as treated in Section 6) or nonconvex functionals with certain monotonicity properties may be handled, similarly, by using (7.1) or consequences of Lemma 1 similar to (7.5), respectively. It is possible that the conclusion (7.1) holds for the class Ω of (not necessarily convex) functionals ω which are nonnegative, for which $\omega(\zeta) = \omega(|\zeta|)$, and for which $\omega(|\zeta_1|) \leqq \omega(|\zeta_2|)$ whenever $|\zeta_1(t)| \leqq |\zeta_2(t)|$ for all t. Similarly, results on processes other than the tied-down Wiener process, whose distributions are symmetrical or also satisfy the property which (as mentioned above) is used in [8] in proving the more general form of Lemma 1, may be obtained by using (7.1) or the generalization of Lemma 1 in [8], respectively;

We now turn to a heuristic argument for the results obtained in previous sections (except for certain results of Section 6, as noted below). This discussion may also be thought of as an outline of one intuitive explanation of why these results hold, the epsilontics and use of Bayes solutions in the previous sections supplying the needed rigor. However, the discussion which follows does not use Bayes solutions, and it would certainly be worth while to obtain an independent argument which would show that in the limit one need only consider "limiting" decision procedures of the type considered below, and thus to conclude that the argument which follows can be made rigorous by means of only brief additions.

In the previous sections we were concerned with estimating (for various weight functions) an unknown element F of $\mathfrak{F}_c$. Denote by $g_n(x; X^{(n)})$ such an estimator of $F(x)$ based on $X^{(n)}$ (for notational simplicity, we have considered a nonrandomized $\phi_n \varepsilon D_n$). Suppose we could show that for our considerations, at least asymptotically, it is only necessary to consider functions g_n of the form

$$g_n(x; X^{(n)}) = \psi_n(S_n(x)),$$

i.e., procedures in the class C_n mentioned in the last paragraph of Section 5 (this is one of two crucial gaps in our heuristic argument, for it is not obvious how to give a short proof, which in no way depends on the results of Sections 3 to 5, of this supposition). Procedures in C_n will have constant risk for $F \varepsilon \mathfrak{F}_c$ and for any of the weight functions of Sections 3, 4, and 5 (and those of Section 6 for which ϕ_n^* was not remarked to have constant risk). Thus, we may consider the distribution of the random function $\psi_n(S_n(t)) - t$ for $0 \leqq t \leqq 1$ where $F = U$. Write $\psi_n(z) = z + \rho_n(z)/\sqrt{n}$. Then

$$\sqrt{n}[\psi_n(S_n(t)) - t] = \sqrt{n}[S_n(t) - t] + \rho_n(S_n(t)).$$

If we could now suppose (and this is the other crucial nonrigorous development) that there is a sequence $\{\psi_n\}$ of minimax procedures (for $n = 1, 2, \cdots$) such that the corresponding sequence $\{\rho_n(z)\}$ has a continuous limit $\rho(z)$ uniformly in z as $n \to \infty$, and note that $\rho_n(S_n(t))$ would then be bounded (for n sufficiently large) and would tend to $\rho(t)$ with probability one as $n \to \infty$, then by [2] and [3] our consideration of $\sqrt{n}[\psi_n(S_n(t)) - t]$ would be reduced, asymptotically, to that of $\zeta(t) + \rho(t)$. The earlier comments of this section would then yield the desired asymptotic minimax properties of ϕ_n^*.

REFERENCES

[1] A. N. Kolmogorov, "Sulla determinazione empirica di una legge di distribuzione," *Inst. Ital. Atti. Giorn.*, Vol. 4 (1933), pp. 83–91.

[2] J. L. Doob, "Heuristic approach to the Kolmogorov-Smirnov theorems," *Ann. Math. Stat.*, Vol. 20 (1949), pp. 393–403.

[3] M. Donsker, "Justification and extension of Doob's heuristic approach to the Kolmogorov-Smirnov theorems," *Ann. Math. Stat.*, Vol. 23 (1952), pp. 277–281.

[4] PAUL LÉVY, *Théorie de L'Addition des Variables Aléatoires*, Gauthier-Villars, Paris (1937).

[5] J. V. USPENSKY, *Introduction to Mathematical Probability*, McGraw-Hill Book Co., New York (1937).

[6] O. P. AGGARWAL, "Some minimax invariant procedures for estimating a c.d.f.," *Ann. Math. Stat.*, Vol. 26 (1955), pp. 450–463.

[7] A. WALD, "Asymptotic minimax solutions of sequential point estimation problems," *Proceedings of the Second Berkeley Symposium on Mathematical Statistics and Probability*, University of California Press (1951), pp. 1–11.

[8] T. W. ANDERSON, "The integral of a symmetric unimodal function," *Proc. Amer. Math. Soc.*, Vol. 6 (1955), pp. 170–176.

[9] N. V. SMIRNOFF, "Approach of empiric distribution functions," *Uspyekhi Matem. Nauk.*, Vol. 10 (1944), pp. 179–206.

Reprinted from THE ANNALS OF MATHEMATICAL STATISTICS
Vol. 27, No. 4, December, 1956

CONSISTENCY OF THE MAXIMUM LIKELIHOOD ESTIMATOR IN THE PRESENCE OF INFINITELY MANY INCIDENTAL PARAMETERS

BY J. KIEFER[1] AND J. WOLFOWITZ[2]

Cornell University

Summary. It is shown that, under usual regularity conditions, the maximum likelihood estimator of a structural parameter is strongly consistent, when the (infinitely many) incidental parameters are independently distributed chance variables with a common unknown distribution function. The latter is also consistently estimated although it is not assumed to belong to a parametric class. Application is made to several problems, in particular to the problem of estimating a straight line with both variables subject to error, which thus after all has a maximum likelihood solution.

1. Introduction. Let $\{X_{ij}\}$, $i = 1, \cdots, n$, $j = 1, \cdots, k$, be chance variables such that the frequency function of $X_{i1}, \cdots, X_{ik}$ is $f(x \mid \theta, \alpha_i)$ when θ and α_i are given, and thus depends upon the unknown (to the statistician) parameters θ and α_i. The parameter θ, upon which all the distributions depend, is called "structural"; the parameters $\{\alpha_i\}$ are called "incidental". Throughout this paper we shall assume that the X_{ij} are independently distributed when $\theta, \alpha_1, \cdots, \alpha_n$, are given, and shall consider the problem of consistently estimating θ (as $n \to \infty$). The chance variables $\{X_{ij}\}$ and the parameters θ and $\{\alpha_i\}$ may be vectors. However, for simplicity of exposition we shall throughout this paper, except in Example 2, assume that they are scalars. Obvious changes will suffice to treat the vector case.

Very many interesting problems are subsumed under the above formulation. Among these is the following:

$$f(x \mid \theta, \alpha_i) = (2\pi\theta)^{-k/2} \exp\left\{\frac{-\sum_j (x_{ij} - \alpha_i)^2}{2\theta}\right\}. \tag{1.1}$$

Suppose now that the $\{\alpha_i\}$ are considered as unknown constants and we form in the usual manner the likelihood function

$$(2\pi\theta)^{-kn/2} \exp\left\{-\frac{1}{2\theta}\sum_{i,j} (X_{ij} - \alpha_i)^2\right\} \tag{1.2}$$

corresponding to (1.1). Then the maximum likelihood (m.l.) estimator of θ is

$$\frac{\sum_{i,j} (X_{ij} - \bar{X}_i)^2}{kn} \tag{1.3}$$

Received September 28, 1955.

[1] Research sponsored by the Office of Naval Research.

[2] The research of this author was supported in part by the United States Air Force under Contract AF 18(600)-685, monitored by the Office of Scientific Research.

with $\bar{X}_i = k^{-1}\sum_j X_{ij}$, and is obviously not consistent. This example is due to Neyman and Scott [1], who used it to prove that the m.l. estimator[3] need not be consistent when there are infinitely many incidental parameters (constants). The latter authors, to whom the names "structural" and "incidental" are due, seem to have been the first to formulate the general problem. Special forms of the problem, like Example 2 below, had been studied for a long time (e.g., Wald [2] and the literature cited there).

The general fact that, when the $\{\alpha_i\}$ are unknown constants, the m.l. estimator of θ need not be consistent, is certainly basically connected with the fact that, since there are only a constant number of observations which involve a particular α_i, it is in general impossible to estimate the $\{\alpha_i\}$ consistently. Now there are many meaningful and practical statistical problems where the $\{\alpha_i\}$ are not arbitrary constants but independently and identically distributed chance variables with distribution function (df) G_0 (unknown to the statistician). The question then arises whether the m.l. method, which does not always yield a consistent estimator when there are infinitely many incidental constants, and does yield consistent estimators in the classical parametric case where there are no incidental parameters, will give a consistent estimator in this case, where the $\{\alpha_i\}$ are independent chance variables with the common df G_0. This note is devoted to this question.

The answer is affirmative. Not only is the m.l. estimator of θ strongly consistent (i.e., converges to θ with probability one) under reasonable regularity conditions, but also the m.l. estimator of G_0 converges to G_0 at every point of continuity of the latter, with probability one (w.p.1). This is the more striking when one recalls that G_0 does not belong to a parametric class, i.e., a set of df's indexed by a finite number of parameters. (If G_0 were a member of such a given class, the problem would fall completely in the domain of classical maximum likelihood.) The interest of the present authors was originally in estimating θ. That G can also be estimated by the m.l. method is a felicitous by-product of our investigation. A heuristic explanation of the present result may be this: A sequence of chance variables is more "regular" than an arbitrary sequence of numbers. In the present procedure one does not attempt to determine the particular values of the chance variables $\{\alpha_i\}$, but only their distribution function; thus, we seek the m.l. estimator of the "parameter" $\gamma = (\theta, G)$ based on a sequence of independent random variables whose common distribution function is indexed by γ.

In sections 3, 4, and 5, the results are applied to three problems which seem to be of interest per se. Among these is the problem of fitting a straight line with both variables subject to normal error. This problem has a very long history and has been the subject of many investigations (see, for example [2], [7], [4], and the literature cited there); it seems interesting that it can, after all, be treated by the m.l. method. The verification of the regularity assumptions or the formulation of not too onerous conditions for them to be verified is sometimes not entirely ob-

[3] Throughout this paper, for the sake of brevity, we use the term "estimator" to mean "sequence of estimators for $n = 1, 2, \cdots$."

vious, and the verification of these assumptions (in the form used in Section 2) constitutes the main difficulty of the paper. As is explained in detail below, the fact that these assumptions imply the general consistency result of Section 2 follows from a modification of the proof of [5]. Professor Herbert Robbins has kindly called our attention to his abstract in *Ann. Math. Stat.*, vol. 21 (1950), p. 314, Abstract 35, which states that the m.l. estimator of G is consistent. Since nothing further has appeared on this subject, the intended restrictions under which the statement is true, and the intended method of proof, are unknown to the present authors. This seems to be the second instance in the literature where the m.l. estimator has been used to estimate an entire df which is not assumed to belong to a class depending only on a finite number of real parameters. The first instance of the employment of such an estimator is the classical estimation of a df by its empiric df (shown to be asymptotically optimal in [3]), which is its m.l. estimator (see the paragraph preceding the lemma in Section 2). The only other instance of the estimation of a df in the nonparametric case seems to be that of the estimation of identifiable df's in stochastic structures such as those of the present paper by means of the minimum distance method [4].[4] (The latter requires regularity conditions weaker than those of the present paper. Compare, for example, [4] with Example 2 below; see also Example 3a.)

In connection with these examples, and also in Section 6, we give some examples which illustrate the fact that the classical m.l. estimator may not be consistent, even in parametric examples which lack the pathological discontinuity sometimes present in hitherto published examples.

Section 6 also contains the statement of a simple device which can be used in the classical parametric case as well as in the case studied in this paper, to prove consistency of the m.l. estimator in some cases where the assumptions used in published proofs of consistency are not satisfied.

The proof in Section 2 is a modification of Wald's [5], and its fundamental ideas are to be found in [5]; for this reason some of its details will be omitted. Wald states in his paper that his method applies more generally when his Assumption 9 is fulfilled. However, this assumption is not fulfilled in our problem *ab initio* and some technical modifications have to be made. One obstacle to extending Wald's proof to our problem is in establishing an analogue of (16) in [5]; one "neighborhood of infinity" does not always seem to suffice. Also some changes in the assumptions are made necessary by the nature of our problem. The results of the present paper can be extended in the usual manner to abstract spaces, but we forego this. It should also be remarked that in [6] Wald studied the present problem of estimating a structural parameter.

The attitude towards the $\{\alpha_i\}$, i.e., whether they are to be regarded as unknown constants or identically and independently distributed chance variables or something else, seems to vary with the author and sometimes even within the

[4] A paper entitled "The minimum distance method," which gives the details and proofs of the results announced in [4], is scheduled for publication in a forthcoming issue of these Annals.

publications of the same author. For example, Wald [2], in his treatment of the problem of fitting a straight line mentioned above, considers the $\{\alpha_i\}$ as unknown constants; and Neyman and Scott, in their general formulation of the problem given in [1] and described at the beginning of the present section, also consider the $\{\alpha_i\}$ as unknown constants. On the other hand, Neyman in his treatment [7] of the straight line problem treats the α_i as independently and identically distributed chance variables. Also Neyman and Scott [8] criticize Wald's solution [2] on the ground that the conditions he postulates on the sequence of constants $\{\alpha_i\}$ are such that they are unlikely to be satisfied when the $\{\alpha_i\}$ are independently and identically distributed chance variables. Our own point of view and perhaps also that of the other writers cited, is that one need not insist on any one formulation to the exclusion of all others. There are certainly reasonable statistical problems where the $\{\alpha_i\}$ may be looked upon as independently and identically distributed chance variables, and consequently the problem of the present paper is statistically meaningful and interesting. This is also the attitude implicit in [4] and [9].

2. Proof of consistency. As we have stated earlier, the essential idea of the proof comes from [5]. A compactification device has to be employed because the space Γ defined below may not be compact.

We postulate that the following assumptions are fulfilled (see also the paragraph preceding the lemma at the end of this section):

Assumption 1: $f(x \mid \theta, \alpha)$ is a density with respect to a σ-finite measure μ on a Euclidean space of which x is the generic point. (This is also Wald's Assumption 1.)

Let Ω be the space of possible values of θ, and let A be the space of values which α_i can take. (Both Ω and A are measurable subsets of Euclidean spaces, f is jointly measurable in x and α for each θ, and we hereafter denote by $\theta_i^{(s)}$ $(1 \leq s \leq r)$ the components of a point θ_i in Ω and by $|\alpha|$ the Euclidean distance from the origin of a point $\alpha \in A$; τ will denote Lebesgue measure on A.) Let $\Gamma = \{G\}$ be a given space of (cumulative) distributions of α_i. Let θ_0, G_0 be, respectively, the "true" value of the parameter θ and the "true" distribution of α_i. It is assumed that $\theta_0 \in \Omega$ and $G_0 \in \Gamma$. Let $\gamma = (\theta, G)$ be the generic point in $\Omega \times \Gamma$. We define

$$f(x \mid \gamma) = \int_A f(x \mid \theta, z)\, dG(z) \tag{2.1}$$

and $\gamma_0 = (\theta_0, G_0)$. In the space $\Omega \times \Gamma$ we define the metric

$$\begin{aligned}\delta(\gamma_1, \gamma_2) &= \delta([\theta_1, G_1], [\theta_2, G_2]) \\ &= \sum_{s=1}^{r} |\operatorname{arc\,tan} \theta_1^{(s)} - \operatorname{arc\,tan} \theta_2^{(s)}| \\ &\qquad + \int_A |G_1(z) - G_2(z)|\, e^{-|z|}\, d\tau(z).\end{aligned} \tag{2.2}$$

Let $\bar{\Omega} \times \bar{\Gamma}$ be the completed space of $\Omega \times \Gamma$ (the space together with the limits of its Cauchy sequences in the sense of the metric (2.2)). Then $\bar{\Omega} \times \bar{\Gamma}$ is compact.

ASSUMPTION 2 (Continuity Assumption): It is possible to extend the definition of $f(x \mid \gamma)$ so that the range of γ will be $\bar{\Omega} \times \bar{\Gamma}$ and so that, for any $\{\gamma_i\}$ and γ^* in $\bar{\Omega} \times \bar{\Gamma}$, $\gamma_i \to \gamma^*$ implies

$$f(x \mid \gamma_i) \to f(x \mid \gamma^*), \tag{2.3}$$

except perhaps on a set of x whose probability is 0 according to the probability density $f(x \mid \gamma_0)$. (The exceptional x-set may depend on γ^*; $f(x \mid \gamma^*)$ need not be a probability density function.) (This assumption corresponds to Wald's continuity Assumptions 3 and 5.)

ASSUMPTION 3: For any γ in $\bar{\Omega} \times \bar{\Gamma}$ and any $\rho > 0$, $w(x \mid \gamma, \rho)$ is a measurable function of x, where

$$w(x \mid \gamma, \rho) = \sup f(x \mid \gamma'),$$

the supremum being taken over all γ' in $\bar{\Omega} \times \bar{\Gamma}$ for which $\delta(\gamma, \gamma') < \rho$. (This assumption is made for the reasons given by Wald. See his remarks following Assumption 8 in [5].)

ASSUMPTION 4 (Identifiability Assumption): If γ_1 in $\bar{\Omega} \times \bar{\Gamma}$ is different from γ_0, then, for at least one y,

$$\int_{-\infty}^{y} f(x \mid \gamma_1)\, d\mu \neq \int_{-\infty}^{y} f(x \mid \gamma_0)\, d\mu, \tag{2.4}$$

the integral being over those x all of whose components are $\leqq$ the corresponding components of y. (This is the same as Wald's Assumption 4.)

Let X be a chance variable with density $f(x \mid \gamma_0)$. *The operator E will always denote expectation under γ_0*; γ_0 will always, of course, be a member of $\Omega \times \Gamma$.

ASSUMPTION 5 (Integrability Assumption): For any γ in $\bar{\Omega} \times \bar{\Gamma}$ we have, as $\rho \downarrow 0$,

$$\lim E\left[\log \frac{w(X \mid \gamma, \rho)}{f(X \mid \gamma_0)}\right]^+ < \infty. \tag{2.5}$$

(This assumption is implied by assumptions corresponding to Wald's Assumptions 2 and 6.)

For any γ in $\bar{\Omega} \times \bar{\Gamma}$ other than γ_0, define $v = \log f(X, \gamma) - \log f(X, \gamma_0)$. We begin the proof of consistency by showing that

$$Ev < 0. \tag{2.6}$$

First, if γ is in $\Omega \times \Gamma$, $Ee^v \leqq 1$. Hence from (2.3) and Fatou's lemma it follows that, for any γ in $\bar{\Omega} \times \bar{\Gamma}$,

$$Ev \leqq Ee^v \leqq 1. \tag{2.7}$$

If v is $-\infty$ with probability one according to $f(x \mid \gamma_0)$, then (2.6) is obvious. Suppose therefore that $v > -\infty$ with positive probability according to

$f(x \mid \gamma_0)$. Then, by Jensen's inequality and (2.7),

$$Ev \leqq \log Ee^v \leqq 0, \tag{2.8}$$

and the first equality sign can hold only if v is a constant c with probability one according to $f(x \mid \gamma_0)$. If the first equality sign does not hold (2.6) follows at once. Consider, therefore, the constant c. If $c < 0$ then (2.6) holds. If $c > 0$ then (2.8) is violated. We cannot have $c = 0$ because of Assumption 4. This proves (2.6).

Now, as $\rho \downarrow 0$, for $\gamma' \neq \gamma_0$,

$$\lim E\left[\log \frac{w(X \mid \gamma, \rho)}{f(X \mid \gamma_0)}\right]^+ = E\left[\log \frac{f(X \mid \gamma)}{f(X \mid \gamma_0)}\right]^+ \tag{2.9}$$

by (2.3), (2.5), and Lebesgue's dominated convergence theorem. Also,

$$\lim E\left[\log \frac{w(X \mid \gamma, \rho)}{f(X \mid \gamma_0)}\right]^- = E\left[\log \frac{f(X \mid \gamma)}{f(X \mid \gamma_0)}\right]^-, \tag{2.10}$$

since the integrand of the left member decreases monotonically to the integrand of the right member. Hence, as $\rho \rightarrow 0$,

$$\lim E\left[\log \frac{w(X \mid \gamma, \rho)}{f(X \mid \gamma_0)}\right] = E \log \frac{f(X \mid \gamma)}{f(X \mid \gamma_0)} < 0 \tag{2.11}$$

by (2.6). Just as in [5] (see also [12]) it may then be shown that, for any positive ρ, there exists an $h(\rho), 0 < h(\rho) < 1$, such that the probability is one that, for all n sufficiently large,

$$\sup \left\{ \frac{\prod_{i=1}^{n} f(X_i \mid \gamma)}{\prod_{i=1}^{n} f(X_i \mid \gamma_0)} \right\} < h^n, \tag{2.12}$$

the supremum being taken over all γ in $\bar{\Omega} \times \bar{\Gamma}$ for which $\delta(\gamma, \gamma_0) > \rho$, and where $X_1, X_2, \cdots$ are independent chance variables with the common density $f(x \mid \gamma_0)$.

Let $L(x_1, \cdots, x_n \mid \gamma) = \prod_1^n f(x_i \mid \gamma)$. A *modified m.l. estimator* is defined to be a sequence of μ-measurable functions $\{\hat{\gamma}_n\}$ such that

$$L(x_1, \cdots, x_n \mid \hat{\gamma}_n(x_1, \cdots, x_n)) \geqq c \sup_\gamma L(x_1, \cdots, x_n \mid \gamma)$$

for almost all (μ) $x_1, \cdots, x_n$ for each n, where c is a positive number (the supremem is over $\Omega \times \Gamma$); for $c = 1$, this of course defines an m.l. estimator. (We shall not be concerned in this paper with conditions which ensure the existence of such measurable functions, although reasonable conditions are not difficult to formulate.) We also define a *neighborhood m.l. estimator* to be a sequence of μ-measurable functions $\{\gamma_n^*\}$ such that there exists a sequence of positive numbers ϵ_n with $\lim_{n\to\infty} \epsilon_n = 0$ for which $\sup_{\gamma \in \Pi_n} L(x_1, \cdots, x_n \mid \gamma) = \sup_\gamma L(x_1, \cdots, x_n \mid \gamma)$ for almost all $(\mu) x_1, \cdots, x_n$, where Π_n is the set of all γ in $\Omega \times \Gamma$ for which

$\delta(\gamma, \gamma_n^*(x_1, \cdots, x_n)) < \epsilon_n$. (Thus, neighborhood m.l. estimators exist in some cases where m.l. and modified m.l. estimators do not; this will be useful in making clear certain examples below where the lack of consistency is not merely due, as it might seem, to the fact that no strict m.l. or modified m.l. estimator exists.)

The above result (2.12) implies the strong convergence of m.l., modified m.l., and neighborhood m.l. estimators (in the respective cases where they exist). The component of the estimator which estimates G_0 converges to it at all its points of continuity w.p.1.

We remark that the above proof actually demonstrates consistency if, in the definition of m.l. estimator (or its variants), the supremum is taken over $\bar{\Omega} \times \bar{\Gamma}$ instead of over $\Omega \times \Gamma$ or, in fact, over any subset of $\bar{\Omega} \times \bar{\Gamma}$ containing γ_0. This last fact implies that if consistency is verified in an example where $\Omega = \Omega_1$, $\Gamma = \Gamma_1$, then it automatically holds in the example where $\Omega = \Omega_2$, $\Gamma = \Gamma_2$, whenever $\Omega_2 \subset \Omega_1$ and $\Gamma_2 \subset \Gamma_1$. In particular, this remark applies to the examples of Sections 3, 4, and 5.

We remark that Assumption 1 is not really essential in the above proof. Let P_γ denote the probability measure of X when γ is the true parameter value, and let $d(x, \gamma, \gamma_0) = r(x, \gamma, \gamma_0)/[1 - r(x, \gamma, \gamma_0)]$, where $r(x, \gamma, \gamma_0)$ denotes a Radon-Nikodym derivative of P_γ with respect to $P_\gamma + P_{\gamma_0}$ at the point x. If, for each $\gamma_0 \varepsilon \Omega \times \Gamma$, Assumptions 2 and 3 are satisfied when $f(x \mid \gamma)$ is replaced by $d(x, \gamma, \gamma_0)$, if (2.4) is replaced by the condition that $d(x, \gamma, \gamma_0) = 1$ does not hold on a set of probability one under γ_0 for any γ, and if $f(x \mid \gamma)/f(x \mid \gamma_0)$ is replaced by $d(x, \gamma, \gamma_0)$ (with a similar replacement for $w(x \mid \gamma, \gamma_0)$) in Assumption 5 and in the argument of the section, then (2.12) (with the replacement noted above) will still hold. An m.l. estimator $\hat{\gamma}$ is now defined to be one for which $\sup_\gamma \prod_1^n d(X_i, \gamma, \hat{\gamma}) = 1$ (with an analogous definition of modified and neighborhood m.l. estimator). We have not stated our assumptions and result (2.12) in this more general setting above because the stated form of the assumptions will suffice in most applications and will be easier to verify than assumptions stated in terms of $d(x, \gamma, \gamma_0)$ (which must be verified for each γ_0). As an example of the use of the more general result just cited, consider the problem of estimating the df F of a sequence of independent identically distributed discrete random variables, it being assumed that the true probability measure P_F (corresponding to the df F) satisfies

$$\sum_x P_F(x) \log P_F(x) > -\infty,$$

the sum being over all points x for which $P_F(x) > 0$. Then the assumptions are easily seen to be satisfied, and we may conclude that the sample df, which is the m.l. estimator, is a consistent estimator of F, a well-known result which does not usually seem to be considered as an example of the consistency of the m.l. estimator. (Of course, even if no restrictions of discreteness or logarithmic summability are placed on P_F, the sample df is still consistent and, as pointed out in the introduction, this is the m.l. estimator. However, Assumption 5 is not satisfied in this case.)

Before proceeding to our examples in subsequent sections, we prove a simple lemma which will be useful later in verifying Assumption 5.

LEMMA. *If $f(z_1, \cdots, z_k)$ is a bounded probability density function with respect to Lebesgue measure μ on Euclidean k-space R^k, and if*

$$\int_{|z_i|>1} (\log |z_i|) f \, d\mu < \infty \qquad (1 \leq i \leq k), \tag{2.13}$$

then

$$-\int_{R^k} f \log f \, d\mu < \infty. \tag{2.14}$$

PROOF: If we prove that (2.13) implies (2.14) when f is replaced by cf in these equations, where $c > 0$, then the lemma is clearly proved. Thus, since f was assumed bounded, we may hereafter assume $f \leq (2e)^{-1}$. (The new f need not have integral unity.) Let

$$g(z_1, \cdots, z_k) = f(z_1, \cdots, z_k) + \prod_{i=1}^{k} (z_i^2 + 1)^{-1}. \tag{2.15}$$

Clearly, (2.13) is true with f replaced by g. Moreover, since $g(z_1, \cdots, z_k) < e^{-1}$ outside of a sufficiently large sphere about the origin, and since $-f \log f < -g \cdot \log g$ if $0 < f < g < e^{-1}$, it suffices to prove (2.14) with f replaced by g, assuming g bounded and (2.13) with f replaced by g. By (2.13), we have

$$\int_{R^k} g \log \prod_{i=1}^{k} (1 + z_i^2)^{\frac{1}{2}} \, d\mu < \infty. \tag{2.16}$$

Thus, it suffices to prove the finiteness of

$$\begin{aligned} &-\int_{R^k} g \log g \, d\mu - \int_{R^k} g \log \prod_{i=1}^{k} (1 + z_i^2)^{\frac{1}{2}} \, d\mu \\ &\qquad = \int_{R^k} g \log \prod_{i=1}^{k} (1 + z_i^2)^{\frac{1}{2}} \left\{ \frac{-\log [g \prod (1 + z_i^2)^{\frac{1}{2}}]}{\log \prod (1 + z_i^2)^{\frac{1}{2}}} \right\} d\mu. \end{aligned} \tag{2.17}$$

The fact that $g(z_1, \cdots, z_k) \geq \prod (z_i^2 + 1)^{-1}$ (see (2.15)) implies easily that the bracketed expression in (2.17) is ≤ 1; by (2.16), this completes the proof of the lemma.

3. Example 1. Structural location parameter, incidental scale parameter. Let k be a positive integer, let μ be Lebesgue measure on Euclidean k-space, let g be a univariate probability density function with respect to Lebesgue measure, and let

$$f(x_i \mid \theta, \alpha_i) = \frac{1}{\alpha_i^k} \prod_{j=1}^{k} g\left(\frac{x_{ij} - \theta}{\alpha_i}\right), \tag{3.1}$$

where $x_i = (x_{i1}, \cdots, x_{ik})$. (Thus, observations are taken in groups of $k \geqq 1$, the value of the incidental parameter being the same within each group. The (unconditional) density of $X_i = (X_{i1}, \cdots, X_{ik})$ is given by Equation (2.1). Thus, the X_i are independent, but, for fixed i, the $X_{ij}(j = 1, \cdots, k)$ need not be independent.) Here Ω is the real line. Some further assumptions on g will be made below; we remark here that the important case

$$g(x) = (2\pi)^{-\frac{1}{2}} e^{-(x^2/2)} \tag{3.2}$$

will satisfy our assumptions. (See also (3.4) below.)

The cases $k = 1$ and $k > 1$ are essentially different. In Example 1a the consistency of the m.l. estimator will be proved for $k = 1$ assuming that A is the set of values $\alpha \geqq c$ where c is a known positive constant, and it is pointed out that the property of consistency of the m.l. estimator *does not hold* without this assumption. The proof of consistency in Example 1a is actually carried out for $k \geqq 1$ since this requires little additional space and will save space in Example 1b where we may refer back to 1a for proofs. In Example 1b we prove consistency of the m.l. estimator in the case $k > 1$ without assuming $\alpha \geqq c > 0$.

Example 1a. We assume that $k \geqq 1$ and that A is the set of all real values $\alpha \geqq c$ where c is a known *positive* constant. In the case $k = 1$, this assumption on A can be weakened slightly to an assumption on the behavior of $G(\alpha)$ as $\alpha \to 0$; however, some such assumption is necessary for consistency, since the last example of Section 6 shows that, even in cases where Γ is restricted to a simple parametric class of df's on a set of positive reals which is *not* bounded away from zero, it can happen that no m.l. or modified m.l. estimator exists and that there are neighborhood m.l. estimators which are not consistent.

We now state our assumptions on g and G_0. They seem reasonable and are in a form which makes brief proofs possible; they undoubtedly can be weakened. (These last remarks apply also to Examples 2 and 3. See also the first part of Section 6 for one method by which we can prove the results of our examples under weaker conditions.) We hereafter assume

(a) $\sup_x g(x) < \infty$;

(b) g is lower semicontinuous and for every $\epsilon > 0$ there is a continuous function $h_\epsilon \geqq g$ for which $\int [h_\epsilon(x) - g(x)]\, dx < \epsilon$;

(c) $\lim_{|x| \to \infty} g(x) = 0$;

(3.3) (d) $-\int_{-\infty}^{\infty} g(x)[\log |x|]^+ \, dx > -\infty$;

(e) $\int_{-\infty}^{\infty} |x|^{it} g(x)\, dx \neq 0$ for almost all real t;

(f) $g(x) > 0$ for almost all x in some open interval whose closure contains the point $x = 0$.

We note that, in addition to being satisfied in the case (3.2), Assumption (3.3) is also satisfied in such important cases as

$$\text{(3.4)}\quad \begin{aligned} &\text{(a) } g(x) = 1/\pi(1 + x^2);\\ &\text{(b) } g(x) = 1 \text{ if } |x| < \tfrac{1}{2} \text{ and } g(x) = 0 \text{ otherwise};\\ &\text{(c) } g(x) = e^{-x} \text{ if } x > 0 \text{ and } g(x) = 0 \text{ otherwise.}\end{aligned}$$

Of course, if g does not satisfy (3.3) but if there is a function g^* satisfying (3.3) and for which $g(x) = g^*(x)$ almost everywhere, then we may carry out our considerations replacing g by g^*.

We assume that Γ consists of all G such that

$$\text{(3.5)}\qquad \int_c^\infty (\log \alpha)\, dG(\alpha) < \infty,$$

where c is the constant used before in the definition of A. For example, G belongs to Γ if, for some positive constants b and ϵ,

$$\text{(3.6)}\qquad 1 - G(\alpha) < \frac{b}{\log \alpha (\log\log \alpha)^{1+\epsilon}}$$

for $\alpha > e^e$; integration by parts will verify that (3.6) implies (3.5). Condition (3.5) is weaker than the requirement that any positive (not necessarily integral) movement of G be finite.

We now verify the assumptions of Section 2. We complete the definition of f for (θ, α) in $\bar{\Omega} \times \bar{A}$ by setting $f(x \mid \theta, \alpha) = 0$ whenever $\theta = \pm\infty$ or $\alpha = \infty$. For $(\theta, G) \, \varepsilon \, \bar{\Omega} \times \bar{\Gamma}$, we then define $f(x \mid \theta, G)$ by (2.1). (We remark that $\bar{\Gamma}$ obviously contains all df's on $\bar{A}$.) Assumption 1 is obviously satisfied. Assumption 3 follows from the fact that (3.3) implies that $f(x \mid \theta, G)$ is for each x lower semicontinuous in (θ, G) (in the sense of the metric δ) on $\bar{\Omega} \times \bar{\Gamma}$, and the fact that $\bar{\Omega} \times \bar{\Gamma}$ is separable. Write $h_\epsilon(x_i \mid \theta, \alpha) = \alpha^{-k} \prod_{j=1}^k h_\epsilon\, [(x_{ij} - \theta)/\alpha]$. In order to verify Assumption 2, we note that, by the lower semicontinuity in (θ, G) of $f(x \mid \theta, G)$ and by the Helly-Bray theorem, we have (assuming, as we may, that the h_ϵ of (3.3) (b) satisfies $\lim_{|x|\to\infty} h_\epsilon(x) = 0$) that $(\theta_i, G_i) \to (\theta^*, G^*)$ as $i \to \infty$ implies

$$\text{(3.7)}\qquad \begin{aligned} f(x \mid \theta^*, G^*) &\leqq \liminf_{i\to\infty} \int f(x \mid \theta_i, \alpha)\, dG_i \leqq \limsup_{i\to\infty} \int f(x \mid \theta_i, \alpha)\, dG_i \\ &\leqq \lim_{i\to\infty} \int h_\epsilon(x \mid \theta_i, \alpha)\, dG_i = \int h_\epsilon(x \mid \theta^*, \alpha)\, dG^*. \end{aligned}$$

Since the last member of (3.7) is greater than or equal to the first for all x and since their difference has integral $< \epsilon^k$ (with respect to μ), Assumption 2 follows at once.

In verifying Assumption 4, it clearly suffices to prove that, if $f(x \mid \theta_0, G_0) =$

$f(x \mid \theta_1, G_1)$ for almost all x, where $(\theta_i, G_i) \varepsilon \Omega \times \Gamma$ for $i = 0, 1$, then $(\theta_0, G_0) = (\theta_1, G_1)$. If an interval $0 < x < \epsilon$ satisfies (3.3) (f), there is a value β such that

$$P\{X_{1j} \leqq t \text{ for } 1 \leqq j \leqq k \mid \theta_0, G_0\} \leqq \beta$$

is satisfied (whatever be G_0) if and only if $t \leqq \theta_0$, a similar assertion holding if the interval $-\epsilon < x < 0$ satisfies (3.3) (f). Hence, it suffices to prove the above assertion when $\theta_0 = \theta_1$, since it cannot hold when $\theta_0 \neq \theta_1$. Let H_i be the df of the random variable $\log \alpha_1$ when G_i is the df of the random variable α_1; i.e., $H_i(t) = G_i(e^t)$. Then, putting $g^*(z) = e^z[g(e^z) + g(-e^z)]$ (g^* is the density of $\log |U|$ when g is the density of U), it suffices to prove that, if H_0 and H_1 are not identical, then $p_1(z_1, \cdots, z_k)$ and $p_2(z_1, \cdots, z_k)$ are not identical for almost all $(z_1, \cdots, z_k)$, where

$$p_i(z_1, \cdots, z_k) = \int_{-\infty}^{\infty} \prod_{j=1}^{k} g^*(z_j - \beta)\, dH_i(\beta). \tag{3.8}$$

Let g^{**} be the density function of $\sum_1^k Z_j/k$ when the Z_j are independent random variables with common density g^*. The above assertion is then implied by the assertion that the function

$$q(r) = \int_{-\infty}^{\infty} g^{**}(r - \beta)\, dH(\beta) \tag{3.9}$$

uniquely determines the df H. But if A, B, C are the characteristic functions of q, g^{**}, H, respectively, then $B(t) \neq 0$ for almost all t by (3.3) (e) and hence $C(t)$ is determined for those t for which $B(t) \neq 0$ by $C(t) = A(t)/B(t)$ and elsewhere by continuity. Thus, Assumption 4 is verified.

It remains to verify Assumption 5. Since $f(x \mid \theta, G)$ is uniformly bounded in x, θ, G, Assumption 5 will clearly be satisfied if

$$E \log f(X_1 \mid \theta_0, G_0) > -\infty. \tag{3.10}$$

Since the left side of (3.10) does not depend on θ_0, we may assume $\theta_0 = 0$. By (3.3) (d) and (3.5), we have

$$\begin{aligned} E[\log |X_{11}|]^+ &= E\left[\log \frac{|X_{11}|}{\alpha_1} + \log \alpha_1\right]^+ \\ &\leqq E\left[\log \frac{|X_{11}|}{\alpha_1}\right]^+ + E[\log \alpha_1]^+ < \infty; \end{aligned} \tag{3.11}$$

equation (3.10) is a consequence of (3.11) and the lemma at the end of Section 2.

This completes our verification of the fact that the assumptions of Section 2 are implied by (3.3) and (3.5).

Example 1b. We now assume $k > 1$. A is the set of all positive α, while Γ is the set of all df's G on A satisfying

$$\int_0^{\infty} |\log \alpha|\, dG(\alpha) < \infty. \tag{3.12}$$

We assume that g satisfies (3.3) (some alterations could be made here but, for the sake of brevity, we forego making them) and also that

$$\text{(3.13)} \quad \begin{aligned} &\text{(a)} \quad \lim_{|x|\to\infty} xg(x) = 0; \\ &\text{(b)} \quad \sup_{x_1} [\min_{r<j} | x_{1r} - x_{1j} |]^k \prod_{j=1}^k g(x_{1j}) < \infty. \end{aligned}$$

Assumption (3.13) is easily verified, for example, in cases (3.2) and (3.4).

We now verify the assumptions of Section 2. We define $f(x \mid \theta, \alpha) = 0$ whenever $\theta = \pm\infty$ or $\alpha = 0$ or ∞; $f(x \mid \theta, G)$ is then defined by (2.1) for $(\theta, G) \; \varepsilon \; \bar{\Omega} \times \bar{\Gamma}$. Assumptions 1, 3, and 4 are verified exactly as in Example 1a. In verifying Assumption 2, we may follow the demonstration of Example 1a, noting only that the h_ϵ of (3.3) (b) may (because of (3.13) (a)) clearly be assumed to satisfy $\lim_{|x|\to\infty} xh_\epsilon(x) = 0$, so that for every x none of whose components is θ^*,

$$\text{(3.14)} \qquad \lim_{\substack{i\to\infty \\ \alpha\to 0}} h_\epsilon(x \mid \theta_i, \alpha) = 0;$$

thus, for almost all (μ) x, the Helly-Bray theorem may still be used at the last step of (3.7), no difficulty being caused by the possibility that $\lim \inf_{i\to\infty} G_i(0) < G^*(0)$.

It remains to verify Assumption 5. Now, $f(x \mid \theta, G)$ is no longer uniformly bounded as it was in Example 1a. However, by (3.13) (b), there is a constant B such that, for all $x_1 = (x_{11}, \cdots, x_{1k})$ none of whose components are equal, every $\theta \; \varepsilon \; \Omega$, and every $\alpha \; \varepsilon \; A$,

$$\text{(3.15)} \quad \begin{aligned} f(x_1 \mid \theta, \alpha) &= [\min_{r<s} | x_{1r} - x_{1s} |]^{-k} \left\{ [\min_{r<s} | y_{1r} - y_{1s} |]^k \prod_{j=1}^k g(y_j) \right\} \\ &\leqq B[\min_{r<s} | x_{1r} - x_{1s} |]^{-k}, \end{aligned}$$

where $y_{1r} = (x_{1r} - \theta)/\alpha$. Hence, for almost all x_1,

$$\text{(3.16)} \quad \begin{aligned} \sup_{\substack{\theta\varepsilon\Omega \\ \alpha\varepsilon A}} \log f(x_1 \mid \theta, \alpha) &\leqq \log B + k \max_{r<s} \log [1/ | x_{1r} - x_{1s} |] \\ &\leqq \log B + k \sum_{\substack{r,s \\ r<s}} [\log (1/| x_{1r} - x_{1s} |)]^+. \end{aligned}$$

Now, by (3.3) (a), there is a value B' such that $g(z) \leqq B'$ for all z. Hence, by (3.12), B_1 denoting a finite constant, we have

$$\text{(3.17)} \quad \begin{aligned} E[\log(1/ | X_{11} - X_{12} |)]^+ &\leqq E[\log 1/\alpha_1]^+ + E[\log (\alpha_1/ | X_{11} - X_{12} |)]^+ \\ &\leqq B_1 - 2 \int_{-\infty}^{\infty} g(z_2) \int_{z_2}^{z_2+1} B' \log (z_1 - z_2) \, dz_1 \, dz_2 \\ &= B_1 + 2B' < \infty. \end{aligned}$$

From (3.16) and (3.17), we obtain

$$E \sup_{\gamma \in \bar{\Omega} \times \bar{\Gamma}} \log f(X_1 \mid \gamma) < \infty. \tag{3.18}$$

Assumption 5 is a consequence of (3.18) and of (3.10), the latter of which is proved exactly as in Example 1a. This completes the verification of the assumptions of Section 2 in Example 1b.

The discrete analogue of Example 1 can be carried out similarly by letting x, θ, α take on only rational values; this is, however, of less practical importance. The multivariate extension of Example 1 (X_{ij} a vector) may also be carried out similarly.

4. Example 2. The straight line with both variables subject to error.

In this section we shall treat the case $k = 1$ of fitting a straight line with both variables subject to normal error, a famous problem with a long history.

We consider a system $\{(X_{i1}, X_{i2})\}$, $i = 1, 2, \cdots$, of independent chance 2-vectors (the two components X_{i1}, X_{i2} need not be independent for fixed i). We have $\theta = (\theta_1, \theta_2)$, Ω the entire plane, $\theta_0 = (\theta_{10}, \theta_{20})$, A the entire line. Γ is the totality of all non-normal (univariate) distributions G (a chance variable which is constant with probability one is to be considered normally distributed with variance zero) which satisfy

$$\int (\log |\alpha|)^+ \, dG(\alpha) < \infty.$$

It is known to the statistician that

$$X_{i1} = \alpha_i + u_i,$$

$$X_{i2} = \theta_{10} + \theta_{20}\alpha_i + v_i,$$

where (u_i, v_i) are jointly normally distributed chance variables with means zero, each pair (u_i, v_i) distributed independently of every other pair and of the independent chance variables $\{\alpha_i\}$, with a common covariance matrix which is unknown to the statistician.

It is known (see [10]) that the distribution of (X_{i1}, X_{i2}) then determines θ_0 uniquely, but in general not G_0, the "true" df of α_i, or the "true" covariance matrix

$$\begin{Bmatrix} d_{11}^0 & d_{12}^0 \\ d_{12}^0 & d_{22}^0 \end{Bmatrix}$$

of (u_i, v_i). However, a "canonical" complex is determined. (See [4].)

Complete the spaces Ω, A, and Γ to obtain $\bar{\Omega}$, $\bar{A}$ and $\bar{\Gamma}$. The space $\bar{\Gamma}$ contains all normal distributions on A, but this will cause us no trouble in estimating θ_0, as we shall soon see.

Let D be the space of all triples (d_{11}, d_{12}, d_{22}) such that

$$d_{11} \geqq \lambda_{11} > 0, \quad d_{22} \geqq \lambda_{22} > 0,$$
$$d_{11} d_{22} - d_{12}^2 \geqq \lambda_{12} > 0,$$

where $\lambda_{11}, \lambda_{12}, \lambda_{22}$, are given positive numbers. (This will be discussed further below.) We define a metric in D in the same way that one is defined on Ω. Let $\bar{D}$ be the completed space. We shall assume that the "true" triple $d_{11}^0, d_{12}^0, d_{22}^0$ is in D.

The place of $\bar{\Omega} \times \bar{\Gamma}$ in Section 2 and in Example 1 will now be taken by $\bar{\Omega} \times \bar{\Gamma} \times \bar{D}$. We therefore define

$$\gamma = (\theta_1, \theta_2, G, d_{11}, d_{12}, d_{22})$$

as the generic point in $\bar{\Omega} \times \bar{\Gamma} \times \bar{D}$.

Let $f(x_1, x_2 \mid \theta_1, \theta_2, \alpha, d_{11}, d_{12}, d_{22})$ be the joint density function of (X_{i1}, X_{i2}) when $\theta = (\theta_1, \theta_2)$, $\alpha_i = \alpha$, and the covariance matrix of (u_i, v_i) is

$$\begin{Bmatrix} d_{11} & d_{12} \\ d_{12} & d_{22} \end{Bmatrix}$$

(μ is Lebesgue measure in the plane). If, in the above, θ is in $\bar{\Omega} - \Omega$ or α is in $\bar{A} - A$ or (d_{11}, d_{12}, d_{22}) is in $\bar{D} - D$, we define f to be zero. Finally we define

$$f(x_1, x_2 \mid \gamma) = \int_{\bar{A}} f(x_1, x_2 \mid \theta_1, \theta_2, \alpha, d_{11}, d_{12}, d_{12})\, dG(\alpha).$$

It is known ([10] and [4]) that all γ in the same canonical class, and only such, define the same $f(x_1, x_2 \mid \gamma)$ (of course, to within a set of μ-measure zero). Two members of the same canonical class have the same $\theta = (\theta_1, \theta_2)$ but different G's and d_{ij}'s. We shall estimate only θ_0. For an estimator of the entire canonical complex by the minimum distance method under neccssary assumptions only, see [4].[5] In Section 5 below will be found an explanation of why the entire canonical complex cannot be estimated by the m.l. method.

From the definition of $f(x_1, x_2 \mid \gamma)$ it follows immediately that Assumptions 1, 2, and 3 of Section 2 are satisfied. Since we are estimating only θ_0, it is sufficient to verify Assumption 4 only for θ_0 and $\theta^* \neq \theta_0$, i.e., if we write the γ_0 and γ_1 of (2.4) as

$$\gamma_0 = (\theta_{10}, \theta_{20}, G_0, d_{11}^0, d_{12}^0, d_{22}^0),$$
$$\gamma_1 = (\theta_1^*, \theta_2^*, G_1, d_{11}, d_{12}, d_{22}),$$

only $\theta_0 = (\theta_{10}, \theta_{20})$ has to be different from the corresponding $\theta^* = (\theta_1^*, \theta_2^*)$. Now we know that G_0 is in Γ, hence is not normal and assigns probability one to A. If G_1 is also in Γ then Assumption 4 follows at once from the results of

[5] See footnote 4.

Reiersøl [10] or from [11]. If G_1 assigns probability less than one to A, $f(x \mid \gamma_1)$ assigns probability less than one to the Euclidean plane of (x_1, x_2). If G_1 is normal and assigns probability one to A, then (X_{i1}, X_{i2}) are jointly normal under γ_1, but not under γ_0. Thus Assumption 4 is always satisfied.

To verify Assumption 5 we proceed essentially as in Example 1, and use the lemma at the end of Section 2. Assumption 5 is satisfied if

$$E \log f(X_{i1}, X_{i2} \mid \gamma_0) > -\infty.$$

By the lemma this will follow if we prove

$$E\{\log |X_{ij}|\}^+ < \infty$$

for $j = 1, 2$. Now

$$\begin{aligned} E\{\log |X_{i1}|\}^+ &\leqq E\{\log [\,|X_{i1} - \alpha_i| + |\alpha_i|\,]\}^+ \\ &\leqq E\{\log [\,|X_{i1} - \alpha_i| + 1]\} + E\{\log |\alpha_i|\}^+ \\ &= E\{\log [\,|u_i| + 1]\} + E\{\log |\alpha_i|\}^+ \\ &< \infty. \end{aligned}$$

Similarly,

$$\begin{aligned} E\{\log |X_{i2}|\}^+ &\leqq E\{\log [\,|X_{i2} - \theta_{10} - \theta_{20}\alpha_i| + |\theta_{10} + \theta_{20}\alpha_i|\,]\}^+ \\ &\leqq E \log [\,|X_{i2} - \theta_{10} - \theta_{20}\alpha_i| + 1] + E\{\log |\theta_{10} + \theta_{20}\alpha_i|\}^+ \\ &\leqq E \log [\,|v_i| + 1] + \{\log |\theta_{10}|\}^+ + \mathrm{E} \log [1 + |\theta_{20}\alpha_i|\,] \\ &< \infty. \end{aligned}$$

Thus we have shown, under our assumptions on Γ and D, that Assumptions 1 through 5 of Section 2 are satisfied, so that the m.l. estimator of θ_0 converges strongly to θ_0 as $n \to \infty$.

The assumption on D (that d_{11}, d_{22}, and $d_{11} d_{22} - d_{12}^2$ are bounded away from zero) cannot be entirely dispensed with. For if D consists of all triples for which d_{11}, d_{22}, and $d_{11} d_{22} - d_{12}^2$ are positive, if S_n is the sample df of $x_{11}, \cdots, x_{n1}$, and if $\hat{\gamma}_\epsilon$ is the complex $(0, 0, S_n, \epsilon, 0, \sum_1^n x_{i2}^2)$, then it is easily verified that $\lim_{\epsilon \to 0} L((x_{11}, x_{12}), \cdots, (x_{n1}, x_{n2}) \mid \hat{\gamma}_\epsilon) = \infty$; thus, no m.l. or modified m.l. estimator exists, and there are neighborhood m.l. estimators which are not consistent (for θ).

The case $k > 1$ is much simpler to treat than the above case. It is easy to see that then the covariance matrix of (u_i, v_i) is uniquely determined, and from this it follows easily that the whole complex γ is uniquely determined. The problem can be treated in a manner similar to that of Examples 1b and 3b.

The problem of this section with the distribution of (u_i, v_i) other than normal may also be treated by the m.l. method, as in Examples 1 and 3. The last paragraph of Section 3 applies also to the present example.

5. Example 3. Structural scale parameter, incidental location parameter. We consider here the case of a structural scale parameter and an incidental location parameter; this reverses the roles of the two parameters of Example 1. Thus, we suppose μ to be Lebesgue measure on R^k and

$$f(x_i \mid \theta, \alpha) = \frac{1}{\theta^k} \prod_{j=1}^{k} g\left(\frac{x_{ij} - \alpha}{\theta}\right). \tag{5.1}$$

The cases $k = 1$ and $k > 1$ are essentially different, and we consider them separately.

Example 3a. *The case* $k = 1$. This example is another simple one where no m.l. estimator is consistent, and also shows, in a simpler setting, why in Example 2 the m.l. method was incapable of estimating the components of the canonical complex other than θ. Since Example 3a is intended to illustrate the *failure* of the m.l. method in certain situations, we shall for simplicity assume that g is given by (3.2); examples with other g (e.g., (3.4)) may be treated similarly. Ω may be taken to be any specified set of positive numbers containing more than one point; for the sake of brevity, we assume that Ω contains its greatest lower bound c (say) (and thus, that $c > 0$), but it is easy to carry through a similar demonstration (with modified or neighborhood m.l. estimators in place of m.l. estimators) when $c \not\in \Omega$. Γ is taken to be the class of all df's G on the real line for which $\int [\log |\alpha|]^+ \, dG(\alpha) < \infty$ and such that G has no normal component; i.e., no G in Γ can be represented as the convolution of two df's, one of which is normal with positive variance. (Γ may be further restricted, e.g., by the condition that for each G there is a bounded set outside of which G has no variation.)

All assumptions of Section 2 are easily verified except Assumption 4; there is no difficulty of identifiability in $\Omega \times \Gamma$, but there clearly *is* in $\bar{\Omega} \times \bar{\Gamma}$. Consider now the expression

$$\prod_{i=1}^{n} \int_{-\infty}^{\infty} \frac{1}{(2\pi)^{\frac{1}{2}} c} e^{-(1/2c^2)(x_i - s)^2} \, dM(s). \tag{5.2}$$

It is clear that the maximum of (5.2) with respect to M can be achieved only by an M which assigns probability one to the interval (min $(x_1, \cdots, x_n)$, max $(x_1, \cdots, x_n)$) and hence which has no normal component. This discussion of the expression (5.2) shows that, for every n, any m.l. estimator (the fact that the maximum is attained is easily verified) of (θ, G) subject to our assumption $\theta \geqq c$ *always* estimates θ to be c. Thus, no m.l. estimator of (θ, G) is consistent (unless $\theta = c$).

To summarize the result of this example, then, the m.l. method is incapable of estimating consistently the normal component of the df of the sequence $\{X_i\}$ of independent identically distributed random variables because, in every neighborhood of a point (θ, G) with $\theta > c$, there are points with $\theta = c$ (and for which the likelihood is larger).

Let N_σ denote the normal df with mean 0 and variance σ^2, and let $H_1 * H_2$ denote the convolution of the two df's H_1 and H_2.

It is interesting to note that, without any assumption on Γ (except the necessary identifiability assumption that G_0 has no normal component), the minimum distance method is capable of estimating (θ_0, G_0) consistently [4]. The difficulty noted above for the m.l. estimator is avoided by noting the *rate* at which the sample df S_n converges to the df $N_{\theta_0} * G_0$ of X_1 and estimating θ_0 *not* by the value t for which $N_t * H$ is closest to S_n for some normal-free H (this would encounter the same difficulty as the m.l. estimator, since, the smaller t is taken, the closer can $N_t * H$ be made to approximate S_n), but as the *largest* value for which there is an $N_t * H$ suitably close to S_n ("suitably" is connected with the rate mentioned above.)

One could modify the example as considered above so as not to require G_0 to have no normal component, and try then to escape the difficulty of non-identifiability by asking for an estimator of the canonical representation of (θ, G), this representation consisting of two df's, the normal and nonnormal components of $N_\theta * G$. The previous demonstration then shows that no m.l. estimator of the canonical representation estimates it consistently, and thus illustrates, in a simpler setting than that of Example 2 with $k = 1$, why the m.l. estimator could not be used in Example 2 to estimate the components of the canonical complex other than θ.

We remark that it is easy in many cases such as that of the present example to prove a result such as the one that, (t_n, H_n) denoting an m.l. estimator of (θ_0, G_0) after n observations, the df $N_{t_n} * H_n$ converges w.p.1 to $N_{\theta_0} * G_0$ as $n \to \infty$. Such a property is much weaker than that of the consistency of the m.l. estimator, and does not lie much deeper than the Glivenko-Cantelli theorem.

Example 3b. *The case* $k > 1$. We assume f to be given by (5.1) with $k > 1$. The function g is assumed to satisfy the conditions (a), (b), (c), and (d) of (3.3); conditions (a) and (b) of (3.13), and

$$\int_{-\infty}^{\infty} e^{itx} g(x)\, dx \neq 0 \quad \text{for almost all real } t. \tag{5.3}$$

(As in Example 1a, weaker conditions could be assumed here if we assumed also $\theta \geqq c > 0$; the above conditions are analogous to those of Example 1b.) Thus, for example, (3.2) and (3.4) satisfy these assumptions. Ω is the set of all values $\theta > 0$, while A is the real line and Γ is the set of all df's G on A for which

$$\int_{-\infty}^{\infty} [\log |\alpha|]^+ \, dG(\alpha) < \infty. \tag{5.4}$$

We now verify the assumptions of Section 2. We define $f(x \mid \theta, \alpha) = 0$ when $\theta = 0$ or ∞ or $\alpha = \pm\infty$. The definition of $f(x \mid \theta, G)$ for $(\theta, G) \in \bar{\Omega} \times \bar{\Gamma}$ is then given by (2.1). Assumptions 1, 2, and 3 are now verified as in Example 1b, interchanging the roles of θ and α in the latter (including the definition of $h_\epsilon(x \mid \theta, \alpha)$) and noting that (3.14)) still holds for almost all (μ) x, with this interchange. In order to verify Assumption 4, we note, for $(\theta, G) \varepsilon \Omega \times \Gamma$, that θ is determined by the density function of $X_{11} - X_{12}$ and that, for almost all real t, the characteristic function of G is then given by $B(t/k, \cdots, t/k)/[C(\theta t/k)]^k$ where $B(t_1, \cdots, t_k)$ is the characteristic function of $X_{11}, \cdots, X_{1k}$ and $C(t)$ is the characteristic function of g.

Finally, Assumption 5 is a consequence of equation (3.18), which is proved in the present case exactly as in Example 1b (using (3.15), (3.16), and (3.17), with α_1 replaced by θ in the latter), and of equation (3.10) (with f defined by (5.1)). Equation (3.10) in the present example is a consequence of the lemma at the end of Section 2 and of

$$\begin{aligned} E\{\log |X_{11}|\}^+ &\leqq E\{\log [\, |X_{11} - \alpha_1| + |\alpha_1|\,]\}^+ \\ &\leqq E \log [\, |X_{11} - \alpha_1| + 1] + E\{\log |\alpha_1|\}^+ < \infty. \end{aligned} \tag{5.5}$$

This completes the verification of the assumptions of Section 2 in Example 3b. The last paragraph of Section 3 applies also to the present example.

6. The Classical case. Miscellaneous remarks. It does not seem to have been noticed in the literature that a simple device exists for proving consistency of the m.l. estimator in certain cases where the regularity conditions of published proofs fail. This device may be used in the case studied in the present paper (to prove consistency in the examples under weaker conditions than those stated) as well as in the classical parametric case. We now illustrate this device in an example of the latter case.

When Γ consists only of distributions which give probability one to a single point, the problem of the present paper becomes the classical problem of estimating the parameter θ and the parameter σ (say) to which G_0 gives probability one. If θ may be any real value and σ any positive value, then the function $(1/\sigma)g((x - \theta)/\sigma)$ of Section 3 does not satisfy Wald's integrability condition or the corresponding condition of any other published proof; one verifies easily that (2.5) is not satisfied for any point in the (θ, σ) half-plane which lies on the line $\sigma = 0$. (The line $\sigma = 0$ has to be added to Ω in the process of forming $\bar{\Omega}$. As in earlier sections, we assume the true σ_0 to be >0.) Often, however, when the observations are considered as if they were taken in groups of two or more, the integrability condition will be satisfied. Such is the case, for example, with the density function

$$\frac{1}{\pi} \frac{\sigma}{\sigma^2 + (x_1 - \theta)^2} \cdot \frac{1}{\pi} \frac{\sigma}{\sigma^2 + (x_2 - \theta)^2}$$

and the normal density function

$$\frac{1}{(2\pi)^{\frac{1}{2}}\sigma} \exp\left\{-\frac{1}{2}\frac{(x_1 - \theta)^2}{\sigma^2}\right\} \cdot \frac{1}{(2\pi)^{\frac{1}{2}}\sigma} \exp\left\{-\frac{1}{2}\frac{(x_2 - \theta)^2}{\sigma^2}\right\}.$$

(Of course the estimator from the normal distribution is known to be consistent, but this does not alter the validity of the example.) In such cases it follows from Wald's proof [5] (using the compactification device used above) or from the result of Example 1b that the m.l. sequence of estimators considered only after an even number of observations in consistent, and from this it is an easy matter to show that the entire m.l. sequence of estimators is consistent.

We shall now discuss the integrability conditions of [5] and of the present paper. The integrability condition (2.5) involves the difference of two logarithms; the integrability condition as given by Wald in [5] requires the finiteness of the expected value of each logarithm. The form (2.5) is satisfied whenever the condition of [5] is, and has one other advantage which we shall now illustrate by an example. Let the observed chance variable X have density function $\theta e^{-\theta x}$ for $x > 0$ and zero elsewhere. The parameter θ is unknown and Ω is the positive half-line, so that $\bar{\Omega}$ contains the point $\theta = 0$. One verifies easily that the condition of [5], and hence (2.5), are satisfied. Suppose now that, instead of observing X, one observes $Y = e^{(e^X)}$, which therefore has the density function

$$\frac{\theta}{x} (\log x)^{-\theta-1}$$

for $x > e$, and zero elsewhere. One readily verifies that, when $\theta < 1$,

$$E \log \left\{ \frac{\theta}{Y} (\log Y)^{-\theta-1} \right\} = -\infty,$$

so that the condition of [5] is not satisfied when $0 < \theta_0 < 1$. Thus, whether the condition of [5] is satisfied depends in this instance on whether one observes X or Y; this is an unfortunate circumstance, since the estimation problems are in simple correspondence. On the other hand, condition (2.5) is invariant under one-to-one transformation of the observed chance variable because the numerator and denominator of the ratio in (2.5) are multiplied by the same Jacobian. (In particular, therefore, the chance variable Y satisfies (2.5).)

Without resorting to artificial or pathologic examples as is sometimes done in the literature, it is still easy to give instances where the m.l. method does not give consistent estimators in the classical parametric case. For example, consider the density function

$$\frac{1}{2(2\pi)^{\frac{1}{2}}} \exp\left\{-\tfrac{1}{2}(x-\theta)^2\right\} + \frac{1}{2(2\pi)^{\frac{1}{2}}\sigma} \exp\left\{-\frac{1}{2}\frac{(x-\theta)^2}{\sigma^2}\right\}$$

of the sequence of independent and identically distributed chance variables $X_1, X_2, \cdots$. Here θ and σ are the unknown parameters, θ may be any real number and σ any positive number. It is easy to see that the supremum of the likelihood function is almost always infinite, no m.l. or modified m.l. estimator exists, and there are neighborhood m.l. estimators (where θ_0 is estimated by X_1, say) which are obviously not consistent.

REFERENCES

[1] J. Neyman and E. L. Scott, "Consistent estimates based on partially consistent observations" *Econometrica*, Vol. 16 (1948), pp. 1–32.

[2] A. WALD, "The fitting of straight lines if both variables are subject to error," *Ann. Math. Stat.*, Vol. 11 (1940), pp. 284–300.

[3] A. DVORETZKY, J. KIEFER AND J. WOLFOWITZ, "Asymptotic minimax character of the sample distribution function and of the classical multinomial estimator," *Ann. Math. Stat.*, Vol. 27 (1956), pp. 642–669.

[4] J. WOLFOWITZ, "Estimation of the components of stochastic structures," *Proc. Nat. Acad. Sci., U.S.A.*, Vol. 40, No. 7 (1954), pp. 602–606.

[5] A. WALD, "Note on the consistency of the maximum likelihood estimate," *Ann. Math. Stat.*, Vol. 20 (1949), pp. 595–601.

[6] A. WALD, "Estimation of a parameter when the number of unknown parameters increases indefinitely with the number of observations," *Ann. Math. Stat.*, Vol. 19 (1948), pp. 220–227.

[7] J. NEYMAN, "Existence of consistent estimates of the directional parameter in a linear structural relation between two variables," *Ann. Math. Stat.* Vol. 22 (1951), pp. 497–512.

[8] J. NEYMAN AND E. L. SCOTT, "On certain methods of estimating the linear structural relation between two variables," *Ann. Math. Stat.* Vol. 22 (1951), pp. 352–361.

[9] J. WOLFOWITZ, "Estimation by the minimum distance method," *Ann. Inst. Stat. Math.* Tokyo, Vol. 5 (1953), pp. 9–23.

[10] O. REIERSØL, "Identifiability of a linear relation between variables which are subject to error," *Econometrica*, Vol. 18 (1950), pp. 375–389.

[11] J. WOLFOWITZ, "Consistent estimators of the parameters of a linear structural relation," *Skand. Aktuarietids.* (1952), pp. 132–151.

[12] J. WOLFOWITZ, "On Wald's proof of the consistency of the maximum likelihood estimate," *Ann. Math. Stat.*, Vol. 20 (1949), pp. 602–603.

SEQUENTIAL TESTS OF HYPOTHESES ABOUT THE MEAN OCCURRENCE TIME OF A CONTINUOUS PARAMETER POISSON PROCESS

J. Kiefer[1] and J. Wolfowitz[1*]
Cornell University

This paper presents tables which give the operating characteristic and average sample time functions of sequential probability ratio tests concerning the mean of a Poisson process and analogous discrete time results for the exponential distribution.

1. INTRODUCTION AND SUMMARY

In [1] the study of sequential tests of hypotheses about stochastic processes with continuous time parameter was initiated. Among other results, formulas were derived for the OC (operating characteristic) and AST (average sample time) functions of sequential probability ratio tests concerning the mean of a Poisson process. This process is a familiar one and arises in many practical situations. It is generally assumed to obtain when one studies meson incidence on a Geiger counter or such phenomena as the initiation of telephone calls and the expiration of light bulbs in life testing, or when one counts bacteria in milk on a slide ("time" is replaced by one coordinate of the slide). The purpose of the present paper is to give tables which will give the OC and AST functions of a wide variety of such tests.

The OC and AST functions computed from these tables will also prove useful as approximations to the OC and ASN (average sample number) functions for the now classical Wald (discrete time) sequential probability ratio test (see [2]) for testing hypotheses concerning the mean of a sequence of independent observations with common Poisson distribution. (See [1] for a discussion of the relationship between the classical tests of sequential analysis and the continuous time tests of [1].) The tables also yield exact results for discrete time problems involving the exponential distribution.

2. THE TESTS

Let $X(t)$ $(t \geq 0)$ be a Poisson process with stationary independent increments and mean occurrence time $1/\lambda$ (i.e., $EX(1) = \lambda$ is the expected number of occurrences per unit time[2]), so that the probability that $X(T) = k$ is $e^{-\lambda T} (\lambda T)^k/k!$ for $k = 0, 1, 2, \ldots$. It is desired to test the hypothesis $H_1: \lambda = \lambda_1$ against $H_2: \lambda = \lambda^* \lambda_1$ where $\lambda_1 > 0$, $\lambda^* > 1$. Let α_1, α_2 be

---

[1]Under contract with the Office of Naval Research.

[2]We mention the following typographical errors on p. 259 of [1]: the mean occurrence time was stated as λ instead of $1/\lambda$, and a minus sign was printed as a plus sign in equation (4.2) of [1].

*Manuscript received 5 March 1956

positive numbers with $\alpha_1 + \alpha_2 < 1$. The optimal test procedure (in the sense described in [1]) which satisfies the condition that the probability of a wrong decision be α_i (or even $\leqq \alpha_i$) when H_i is true is described by two real numbers, $J > 0$ and $K < 0$, as follows: Stop the first instant that

(2.1) $$X(t) - \frac{\lambda_1}{r^*} t$$

is $\leqq K$ or $\geqq J$, accepting H_1 or H_2, respectively. (We may take $X(t)$ to be continuous on the right, and thus such a first instant to exist, with probability one.) Here

(2.2) $$r^* = \frac{\log \lambda^*}{\lambda^* - 1},$$

K is given by

(2.3) $$K = \left(\log \frac{\alpha_2}{1-\alpha_1}\right) \Big/ \log \lambda^*,$$

and J is a number satisfying

(2.4) $$J \leqq \left(\log \frac{1-\alpha_2}{\alpha_1}\right) \Big/ \log \lambda^* \leqq J + 1 .$$

The value of J may be obtained exactly as the solution of either of the equations

(2.5) $$p(\lambda^* ; J, K) = 1 - \alpha_1$$

or

(2.6) $$p(1/\lambda^* ; J, K) = \alpha_2$$

(substituting herein for λ^*, K, and α_1 or α_2 as prescribed above). Here the function p (which is tabulated below) is given by putting (for $\lambda > 0$)

(2.7) $$\begin{cases} r(\lambda) = (\log \lambda)/(\lambda - 1), & \lambda \neq 1 \\ r(1) = 1, \end{cases}$$

(2.8) $$f(r(\lambda); J, K) = p(\lambda; J, K),$$

and, for $r > 0$,

(2.9) $$f(r; J, K) = e^{Kr} \cdot \frac{\sum_{i < J} [-(J-i)re^{-r}]^i / i\,!}{\sum_{i < J-K} [-(J-K-i)re^{-r}]^i / i\,!}$$

If the tables which follow are to be used to obtain J, a good check is to use one of the equations (2.5) or (2.6) to obtain J and then to make sure the other is satisfied. For certain parameter values one equation may be superior to the other for interpolating in the tables, as will be clear from inspection, and this will dictate which equation to use to obtain J.

It will often be the case that the user of the test does not care about the exact values of α_1 and α_2, but only about their order of magnitude. In this event one can quickly scan the tables of p and find (tabled) values of J and K which give values of α_1 and α_2 (read off the tables at once as indicated by (2.5) and (2.6)) which are reasonable for his use, avoiding interpolation entirely.

The OC function $L(\lambda)$ = probability of accepting H_2 when $EX(1) = \lambda$ is obtained for the above test from the same function p (and its tables) through the formula

$$1 - L(\lambda) = f(\lambda r^*/\lambda_1; J, K) = p(q(\lambda r^*/\lambda_1); J, K), \tag{2.10}$$

where q is the inverse function of r; i.e., $q(\lambda r^*/\lambda_1)$ (written as q for short in the equation below) is defined by

$$\frac{\lambda r^*}{\lambda_1} = \frac{\log q}{q-1}.$$

We may also write (2.10) as

$$p(q, J, K) = 1 - L\left(\frac{\lambda_1 \log q}{r^*(q-1)}\right). \tag{2.11}$$

This will usually be the most rapid and convenient form for sketching an OC curve from the tables.

The AST function, $A(\lambda)$ = expected duration of test when $EX(1) = \lambda$, is given by

$$\begin{aligned} A(\lambda) &= \frac{1}{\lambda} h(q(\lambda r^*/\lambda_1); J, K) \\ &= \frac{1}{\lambda} s(\lambda r^*/\lambda_1; J, K), \end{aligned} \tag{2.12}$$

where h, defined by (2.12), is tabulated below, and s is defined for positive r by

$$\begin{aligned} s(r; J, K) = {} & I(J+1) - I(J-K+1) f(r; J, K) \\ & + e^{r(J-1)} \sum_{i<J-K-1} e^{-ri} \sum_{j \le i} (-r)^j \{e^{-Kr} \\ & \cdot f(r; J, K)(J-K-i-1)^j - m(J-i-1, j)\}/j!, \end{aligned} \tag{2.13}$$

where $I(t)$ is the greatest integer $< t$ and $m(y, j) = y^j$ if $y > 0$ and $= 0$ if $y \le 0$, j being an integer ≥ 0. Corresponding to (2.11), the most convenient form for sketching the AST curve is

$$A\left(\frac{\lambda_1 \log q}{r^* (q-1)}\right) = \frac{r^* (q-1)}{\lambda_1 \log q} h(q; J, K). \tag{2.14}$$

3. THE TABLES

The functions p and h defined in Section 2 are tabulated for values of λ, J, and K, which, it is hoped, will be useful in many applications. The values of J and -K chosen for tabulation were 2, 2.5, 3, 3.5, 4, 5, 6, 8, and 10. Not all possible pairs of (J, K) values in this range are tabulated, but rather those pairs where the ratio of the larger to the smaller of the numbers J and -K is not too large (usually <2). The tabulated range will thus include (for values of λ^* included in the table) such pairs of values of α_1 and α_2 as .01 and .05 (see (2.3) and (2.4)), but perhaps not .002 and .05 because the last pair would be less likely to be specified in an application.

For each pair (J, K), the values of λ for which p and h were tabulated were chosen according to the following criteria: (1) only values of λ between .05 and 20 were included (many practical cases will be covered by this range); (2) for fixed (J, K), the α_i being permitted to range from .05 to .001 (subject to the remark of the previous paragraph), whatever values of λ^* correspond to such α_1, α_2 for a given J, K, the tabulated values of λ should permit locating three points on the OC and AST curves between the values λ_1, λ_2. These criteria have, of course, resulted in the tabulation of values of λ which vary greatly with J and K. For example, for J = 2, K = -2, the λ values range from .05 to 20, with .4, 1.0, and 2.0 being the consecutive values which include 1; whereas for J = 10, K = -10, all eleven of the tabulated λ values lie between .5 and 2.0. The successive values of λ, J, and K were chosen with the object of making the jumps between successive p and h values regular and not too large. But the application of standard interpolation methods to these tables will not be too difficult, even though successive intervals between tabled values of λ, J, and K will often not be equal. As was remarked in Section 2, many users will be able to select an appropriate pair (J, K) (giving reasonable values of α_1, α_2) from a quick scan of the tables, without using interpolation.

Thirteen figures were used in the computations[3] (an amount necessary in some cases to obtain four or five significant figures for p and h), and the tables are accurate to within one unit in the last digit printed. As checks on the computation, each value of p was checked against the bounds (easily obtained as a consequence of (2.3), (2.4), and (2.5))

$$\begin{aligned} &\frac{\lambda^J - 1}{\lambda^J - \lambda^K} \leq p(\lambda; J, K) \leq \frac{\lambda^{J+1} - 1}{\lambda^{J+1} - K}, \lambda \neq 1, \\ &\frac{J}{J-K} \leq p(1; J, K) \leq \frac{J+1}{J-K+1}. \end{aligned} \tag{3.1}$$

Similarly, using Wald's equation (see [2] and equation (2.1) of [1]), it is easy to derive the following inequalities on h, which were used to check every computation of that function:

[3] The computations were performed on the Cornell Computing Center's CPC by Mr. Richard Lesser and Mrs. Dorothy Hartman, to whom the authors express their thanks.

$$\frac{J - (J - K)\, p(\lambda; J, K)}{1 - 1/r(\lambda)} \begin{Bmatrix} \leq \\ > \\ = \end{Bmatrix} h(\lambda; J, K) \begin{Bmatrix} < \\ \geq \\ = \end{Bmatrix} \frac{(J + 1) - (J + 1 - K)\, p(\lambda; J, K)}{1 - 1/r(\lambda)} \quad \text{if } \lambda \begin{Bmatrix} < \\ > \end{Bmatrix} 1,$$

$$J^2 + (K^2 - J^2)\, p(1; J, K) \leqq h(1; J, K) \tag{3.2}$$

$$\leqq (J + 1)^2 + [K^2 - (J + 1)^2]\, p(1; J, K).$$

It is interesting to note that the left side of each of the inequalities of (3.1) and (3.2) is the classical approximation of Wald which "neglects excess over the boundaries" and which is of common use in applications of sequential analysis (to discrete parameter problems). These approximations can be rather poor (in terms of relative error) for certain values of the parameters in the present continuous parameter problem. As one would expect, the approximation of the left side of (3.1) is worst when λ and J are small, the percentage of excess over the boundary being large with high probability in such cases. For large λ, however, the relative error in the approximation for $1 - p$ (which is pertinent; see (2.5)) may be even greater. For example, if the test $J = 2 = -K$ is used when $\lambda^* = 10$, the actual α_1 is .0046, whereas the approximation gives .010. When H_2 is true in this last example, the approximation underestimates $A(\lambda)$ by 20 percent. We need not recount the economic losses which can result from under- or over-estimation of α_1, α_2 and $A(\lambda)$.

On the other hand, for $J = 10 = -K$, the relative error of the approximation for α_1 is greatest when $\lambda^* = 2$, $\alpha_1 = .0008$, and the approximation gives .0010, while the relative error in the approximation for $A(\lambda)$ never exceeds 4 percent. This indicates that the approximations will be close enough in most practical applications requiring values of J and $-K$ beyond those tabulated. If values of J and $-K$ smaller than 2 are required, (2.9) and (2.13) can be computed directly in fairly short time.

The tables can also be used to obtain exact results for discrete time sequential problems involving the exponential distribution; see Eq. (3).

J = 2 K = -2			J = 2.5 K = -2			J = 3 K = -2		
λ	p	h	λ	p	h	λ	p	h
20.000	.99911	.3737	14.000	.99940	.5087	10.000	.99953	.6867
14.000	.99795	.5072	10.000	.99848	.6851	7.0000	.99848	.9561
10.000	.99548	.6810	7.0000	.99595	.9507	5.0000	.99546	1.330
7.0000	.98967	.9387	5.0000	.98981	1.314	3.5000	.98581	1.933
5.0000	.97771	1.282	3.5000	.97336	1.881	2.5000	.96001	2.806
3.5000	.95070	1.796	2.5000	.93642	2.660	2.0000	.92363	3.601
2.5000	.89949	2.459	2.0000	.89077	3.330	1.7000	.88142	4.290
2.0000	.84392	2.995	1.0000	.58614	5.688	1.0000	.62499	6.687
1.0000	.53894	4.701	.40000	.14983	6.166	.40000	.15360	7.341
.40000	.14392	5.028	.28571	.07946	5.763	.28571	.08047	6.823
.28571	.07761	4.731	.20000	.03959	5.325	.20000	.03982	6.280
.20000	.03911	4.398	.14286	.02033	4.964	.14286	.02038	5.843
.14286	.02020	4.119	.10000	.00998	4.652	.10000	.01000	5.470
.10000	.00995	3.876	.071429	.00510	4.415	.071429	.00510	5.190
.071429	.00509	3.693	.050000	.00250	4.214	.050000	.00250	4.956
.050000	.00250	3.539						

J = 3.5 K = -2			J = 2 K = -2.5			J = 2.5 K = -2.5		
λ	p	h	λ	p	h	λ	p	h
7.0000	.99943	.9584	20.000	.99911	.4672	14.000	.99940	.6360
5.0000	.99797	1.339	14.000	.99794	.6343	10.000	.99847	.8567
3.5000	.99242	1.965	10.000	.99545	.8520	7.0000	.99590	1.190
2.5000	.97479	2.910	7.0000	.98953	1.176	5.0000	.98958	1.646
2.0000	.94641	3.814	5.0000	.97720	1.609	3.5000	.97229	2.364
1.7000	.91038	4.632	3.5000	.94874	2.264	2.5000	.93227	3.360
1.0000	.65715	7.686	2.5000	.89317	3.117	2.0000	.88138	4.224
.40000	.15597	8.528	1.0000	.48318	5.872	1.0000	.53119	7.108
.28571	.08101	7.888	.50000	.14694	5.896	.50000	.15581	7.239
.20000	.03992	7.236	.40000	.09038	5.572	.40000	.09434	6.810
.14286	.02040	6.722	.28571	.04140	5.051	.28571	.04243	6.133
.10000	.01000	6.286	.20000	.01748	4.569	.20000	.01770	5.519
.071429	.00510	5.963	.14286	.00763	4.211	.14286	.00768	5.068
.050000	.00250	5.691	.10000	.00315	3.922	.10000	.00316	4.704
			.071429	.00136	3.717	.071429	.00136	4.442
			.050000	.00056	3.552	.050000	.00056	4.227

J = 3 K = -2.5			J = 3.5 K = -2.5			J = 4 K = -2.5		
λ	p	h	λ	p	h	λ	p	h
10.000	.99952	.8586	7.0000	.99942	1.198	5.0000	.99907	1.679
7.0000	.99846	1.196	5.0000	.99793	1.674	3.5000	.99578	2.482
5.0000	.99536	1.665	3.5000	.99211	2.461	2.5000	.98299	3.743
3.5000	.98523	2.425	2.5000	.97307	3.658	2.0000	.95877	5.013
2.5000	.95733	3.535	2.0000	.94148	4.813	1.7000	.92427	6.215
2.0000	.91679	4.555	1.7000	.90043	5.861	1.5000	.88270	7.291
1.7000	.86872	5.439	1.0000	.60528	9.606	1.0000	.63416	10.85
1.0000	.57142	8.357	.50000	.16643	10.03	.50000	.16949	11.45
.50000	.16207	8.624	.40000	.09847	9.362	.40000	.09947	10.65
.40000	.09687	8.081	.28571	.04329	8.352	.28571	.04345	9.461
.28571	.04299	7.241	.20000	.01785	7.476	.20000	.01787	8.450
.20000	.01781	6.497	.14286	.00771	6.848	.14286	.00771	7.732
.14286	.00770	5.958	.10000	.00316	6.350	.10000	.00316	7.163
.10000	.00316	5.527	.071429	.00136	5.996	.071429	.00136	6.759
.071429	.00136	5.220	.050000	.00056	5.708	.050000	.00056	6.430
.050000	.00056	4.970						

J = 2 K = -3			J = 2.5 K = -3			J = 3 K = -3		
λ	p	h	λ	p	h	λ	p	h
20.000	.99911	.5607	14.000	.99940	.7633	10.000	.99952	1.030
14.000	.99793	.7614	10.000	.99847	1.028	7.0000	.99845	1.435
10.000	.99544	1.023	7.0000	.99588	1.429	5.0000	.99531	2.000
7.0000	.98948	1.413	5.0000	.98947	1.979	3.5000	.98492	2.918
5.0000	.97697	1.938	3.5000	.97172	2.849	2.5000	.95564	4.270
3.5000	.94770	2.735	2.5000	.92967	4.068	2.0000	.91202	5.522
2.5000	.88922	3.785	2.0000	.87485	5.134	1.7000	.85922	6.612
1.0000	.43789	7.042	1.0000	.48566	8.527	1.0000	.52631	10.03
.50000	.10280	6.480	.50000	.10936	7.935	.50000	.11400	9.429
.40000	.05691	5.961	.40000	.05950	7.265	.40000	.06116	8.598
.28571	.02210	5.243	.28571	.02266	6.353	.28571	.02297	7.488
.20000	.00782	4.655	.20000	.00792	5.616	.20000	.00796	6.604
.14286	.00288	4.250	.14286	.00290	5.111	.14286	.00291	6.006
.10000	.00100	3.939	.10000	.00100	4.722	.10000	.00100	5.548
.071429	.00036	3.724	.071429	.00036	4.450	.071429	.00036	5.229

J = 3.5 K = -3			J = 4 K = -3			J = 5 K = -3		
λ	p	h	λ	p	h	λ	p	h
7.0000	.99942	1.438	5.0000	.99907	2.015	5.0000	.99981	2.019
5.0000	.99791	2.010	3.5000	.99570	2.982	3.5000	.99876	3.003
3.5000	.99195	2.958	2.5000	.98230	4.506	2.5000	.99290	4.617
2.5000	.97199	4.410	2.0000	.95631	6.055	2.0000	.97821	6.367
2.0000	.93803	5.823	1.7000	.91843	7.528	1.7000	.95241	8.159
1.7000	.89296	7.112	1.5000	.87204	8.848	1.5000	.91617	9.875
1.0000	.56099	11.53	1.0000	.59094	13.02	1.3000	.84853	12.12
.50000	.11725	10.94	.50000	.11954	12.46	1.0000	.64000	16.02
.40000	.06221	9.939	.40000	.06287	11.28	.50000	.12228	15.54
.28571	.02314	8.624	.28571	.02322	9.758	.40000	.06355	13.98
.20000	.00798	7.593	.20000	.00799	8.578	.28571	.02329	12.03
.14286	.00291	6.901	.14286	.00291	7.789	.20000	.00800	10.55
.10000	.00100	6.372	.10000	.00100	7.189	.14286	.00292	9.571
.071429	.00036	6.005	.071429	.00036	6.769	.10000	.00100	8.829
						.071429	.00036	8.314

J = 6 K = -3			J = 2 K = -3.5			J = 2.5 K = -3.5		
λ	p	h	λ	p	h	λ	p	h
3.5000	.99966	3.010	20.000	.99911	.6542	14.000	.99940	.8905
2.5000	.99717	4.668	14.000	.99793	.8885	10.000	.99847	1.200
2.0000	.98912	6.548	10.000	.99544	1.194	7.0000	.99587	1.668
1.7000	.97215	8.586	7.0000	.98946	1.651	5.0000	.98943	2.312
1.5000	.94477	10.66	5.0000	.97687	2.266	3.5000	.97141	3.335
1.3000	.88679	13.55	3.5000	.94714	3.207	2.5000	.92803	4.782
1.0000	.67857	19.02	2.5000	.88673	4.461	2.0000	.87030	6.057
.50000	.12364	18.64	1.0000	.40035	8.212	1.0000	.44732	9.946
.40000	.06382	16.69	.58824	.11618	7.463	.58824	.12574	9.146
.28571	.02332	14.30	.50000	.07216	6.940	.50000	.07692	8.478
.20000	.00800	12.52	.40000	.03589	6.235	.40000	.03756	7.583
.14286	.00292	11.35	.28571	.01181	5.358	.28571	.01211	6.483
.10000	.00100	10.47	.20000	.00349	4.697	.20000	.00354	5.663
.071429	.00036	9.858	.14286	.00109	4.266	.14286	.00110	5.129
			.10000	.00031	3.944	.10000	.00032	4.728

J = 3 K = -3.5			J = 3.5 K = -3.5			J = 4 K = -3.5		
λ	p	h	λ	p	h	λ	p	h
10.000	.99952	1.202	7.0000	.99942	1.678	5.0000	.99906	2.351
7.0000	.99845	1.675	5.0000	.99790	2.346	3.5000	.99565	3.481
5.0000	.99529	2.335	3.5000	.99187	3.456	2.5000	.98186	5.272
3.5000	.98475	3.411	2.5000	.97132	5.165	2.0000	.95455	7.104
2.5000	.95458	5.009	2.0000	.93563	6.843	1.7000	.91398	8.858
2.0000	.90868	6.501	1.7000	.88732	8.381	1.5000	.86348	10.43
1.7000	.85209	7.806	1.0000	.52276	13.44	1.0000	.55319	15.19
1.0000	.48781	11.70	.58824	.13852	12.63	.58824	.14268	14.41
.58824	.13302	10.88	.50000	.08270	11.64	.50000	.08437	13.24
.50000	.08032	10.05	.40000	.03932	10.34	.40000	.03974	11.72
.40000	.03864	8.957	.28571	.01237	8.782	.28571	.01241	9.928
.28571	.01228	7.632	.20000	.00357	7.650	.20000	.00357	8.638
.20000	.00356	6.656	.14286	.00110	6.922	.14286	.00110	7.811
.14286	.00110	6.026	.10000	.00032	6.380	.10000	.00032	7.195
.10000	.00032	5.554						

J = 5 K = -3.5			J = 6 K = -3.5			J = 2.5 K = -4		
λ	p	h	λ	p	h	λ	p	h
5.0000	.99981	2.355	3.5000	.99964	3.512	14.000	.99940	1.018
3.5000	.99875	3.504	2.5000	.99710	5.450	10.000	.99847	1.372
2.5000	.99274	5.394	2.0000	.98867	7.655	7.0000	.99587	1.907
2.0000	.97732	7.455	1.7000	.97054	10.06	5.0000	.98941	2.645
1.7000	.94972	9.576	1.5000	.94075	12.52	3.5000	.97125	3.822
1.5000	.91028	11.62	1.3000	.87660	15.95	2.5000	.92700	5.499
1.0000	.60377	18.69	1.0000	.64407	22.18	2.0000	.86712	6.991
.58824	.14826	18.02	.58824	.15151	21.68	1.0000	.41460	11.36
.50000	.08638	16.46	.50000	.08739	19.70	.58824	.09564	9.786
.40000	.04018	14.49	.40000	.04036	17.26	.50000	.05419	8.899
.28571	.01245	12.22	.28571	.01246	14.52	.40000	.02373	7.803
.20000	.00358	10.62	.20000	.00358	12.60	.28571	.00647	6.559
.14286	.00110	9.597	.14286	.00110	11.38	.20000	.00158	5.687
.10000	.00032	8.838	.10000	.00032	10.48	.14286	.00041	5.137

J = 3 K = -4			J = 3.5 K = -4			J = 4 K = -4		
λ	p	h	λ	p	h	λ	p	h
10.000	.99952	1.374	7.0000	.99942	1.918	5.0000	.99905	2.688
7.0000	.99845	1.915	5.0000	.99789	2.682	3.5000	.99561	3.981
5.0000	.99529	2.670	3.5000	.99181	3.954	2.5000	.98155	6.039
3.5000	.98467	3.905	2.5000	.97088	5.922	2.0000	.95331	8.159
2.5000	.95392	5.750	2.0000	.93391	7.869	1.7000	.91058	10.20
2.0000	.90635	7.488	1.7000	.88302	9.666	1.5000	.85661	12.04
1.7000	.84671	9.017	1.0000	.48938	15.36	1.0000	.52000	17.36
1.0000	.45456	13.36	.58824	.10572	13.47	.58824	.10902	15.35
.58824	.10138	11.62	.50000	.05837	12.18	.50000	.05958	13.83
.50000	.05665	10.53	.40000	.02485	10.61	.40000	.02513	12.01
.40000	.02442	9.205	.28571	.00661	8.872	.28571	.00663	10.03
.28571	.00656	7.716	.20000	.00160	7.677	.20000	.00160	8.666
.20000	.00159	6.682	.14286	.00042	6.931	.14286	.00042	7.819
.14286	.00042	6.034						

J = 5 K = -4			J = 6 K = -4			J = 8 K = -4		
λ	p	h	λ	p	h	λ	p	h
5.0000	.99982	2.692	3.5000	.99964	4.014	2.5000	.99953	6.270
3.5000	.99876	4.006	2.5000	.99705	6.231	2.0000	.99709	8.955
2.5000	.99263	6.171	2.0000	.98836	8.764	1.7000	.98941	12.12
2.0000	.97669	8.545	1.7000	.96930	11.54	1.5000	.97246	15.70
1.7000	.94766	11.00	1.5000	.93750	14.39	1.3000	.92401	21.35
1.5000	.90552	13.38	1.3000	.86785	18.37	1.2000	.87332	25.13
1.0000	.57143	21.36	1.0000	.61290	25.35	1.0000	.67568	33.31
.58824	.11346	19.14	.58824	.11606	22.98	.58824	.11846	30.74
.50000	.06104	17.16	.50000	.06177	20.50	.50000	.06232	27.21
.40000	.02541	14.83	.40000	.02552	17.65	.40000	.02559	23.29
.28571	.00666	12.34	.28571	.00666	14.65	.28571	.00666	19.27
.20000	.00160	10.65	.20000	.00160	12.64	.20000	.00160	16.61
.14286	.00042	9.608	.14286	.00042	11.39	.14286	.00042	14.97

J = 3 K = -5			J = 3.5 K = -5			J = 4 K = -5		
λ	p	h	λ	p	h	λ	p	h
10.000	.99952	1.718	7.0000	.99941	2.397	5.0000	.99906	3.360
7.0000	.99845	2.394	5.0000	.99789	3.354	3.5000	.99561	4.981
5.000C	.99527	3.340	3.5000	.99176	4.949	2.5000	.98132	7.576
3.5000	.98458	4.894	2.5000	.97042	7.440	2.0000	.95183	10.28
2.5000	.95319	7.239	2.0000	.93181	9.937	1.7000	.90604	12.92
2.0000	.90347	9.482	1.7000	.87722	12.27	1.5000	.84665	15.31
1.7000	.83943	11.48	1.0000	.43397	19.20	1.0000	.46428	21.70
1.0000	.39999	16.70	.66667	.10683	16.30	.66667	.11149	18.56
.66667	.10104	14.06	.58824	.06178	14.74	.58824	.06381	16.75
.58824	.05912	12.74	.50000	.02912	12.90	.50000	.02974	14.62
.50000	.02823	11.18	.40000	.00994	10.92	.40000	.01005	12.35
.40000	.00976	9.488	.28571	.00189	8.954	.28571	.00190	10.11
.28571	.00187	7.791	.20000	.00032	7.696	.20000	.00032	8.688
.20000	.00032	6.700						

J = 5 K = -5			J = 6 K = -5			J = 8 K = -5		
λ	p	h	λ	p	h	λ	p	h
5.0000	.99980	3.365	3.5000	.99964	5.018	2.5000	.99952	7.838
3.5000	.99871	5.009	2.5000	.99701	7.796	2.0000	.99699	11.20
2.5000	.99252	7.728	2.0000	.98797	10.99	1.7000	.98882	15.19
2.0000	.97593	10.73	1.7000	.96764	14.52	1.5000	.97026	19.75
1.7000	.94489	13.88	1.5000	.93272	18.18	1.3000	.91536	26.99
1.5000	.89856	16.95	1.3000	.85382	23.31	1.2000	.85647	31.79
1.0000	.51613	26.69	1.0000	.55882	31.68	1.0000	.62500	41.55
.66667	.11833	23.16	.66667	.12283	27.83	.66667	.12777	37.32
.58824	.06655	20.80	.58824	.06815	24.88	.58824	.06964	33.12
.50000	.03050	18.07	.50000	.03087	21.54	.50000	.03116	28.48
.40000	.01016	15.21	.40000	.01021	18.08	.40000	.01024	23.81
.28571	.00190	12.44	.28571	.00190	14.76	.28571	.00190	19.40
.20000	.00032	10.68	.20000	.00032	12.66	.20000	.00032	16.64

J = 10 K = -5			J = 3 K = -6			J = 3.5 K = -6		
λ	p	h	λ	p	h	λ	p	h
2.0000	.99925	11.27	10.000	.99952	2.061	7.0000	.99941	2.877
1.7000	.99613	15.48	7.0000	.99845	2.873	5.0000	.99789	4.026
1.5000	.98682	20.58	5.0000	.99527	4.010	3.5000	.99176	5.945
1.3000	.95055	29.57	3.5000	.98457	5.883	2.5000	.97025	8.961
1.2000	.90318	36.26	2.5000	.95293	8.731	2.0000	.93082	12.02
1.1000	.81567	44.18	2.0000	.90210	11.49	1.7000	.87390	14.92
1.0000	.67391	51.68	1.7000	.83529	13.99	1.0000	.38984	23.03
.66667	.12995	46.94	1.0000	.35714	20.03	.76923	.14217	20.27
.58824	.07016	41.39	.76923	.13219	17.49	.66667	.07054	17.71
.50000	.03123	35.43	.66667	.06658	15.30	.58824	.03620	15.59
.40000	.01024	29.55	.58824	.03461	13.50	.50000	.01454	13.31
.28571	.00190	24.05	.50000	.01410	11.55	.40000	.00397	11.06
.20000	.00032	20.61	.40000	.00390	9.619	.28571	.00054	8.981
			.28571	.00054	7.816			

J = 4 K = -6			J = 5 K = -6			J = 6 K = -6		
λ	p	h	λ	p	h	λ	p	h
5.0000	.99905	4.033	5.0000	.99981	4.038	3.5000	.99964	6.022
3.5000	.99559	5.981	3.5000	.99874	6.012	2.5000	.99699	9.360
2.5000	.98117	9.114	2.5000	.99247	9.285	2.0000	.98778	13.21
2.0000	.95109	12.41	2.0000	.97555	12.93	1.7000	.96667	17.51
1.7000	.90339	15.67	1.7000	.94327	16.79	1.5000	.92955	22.01
1.5000	.84014	18.66	1.5000	.89397	20.58	1.3000	.84333	28.36
1.0000	.41935	26.03	1.0000	.47059	32.02	1.0000	.51351	37.99
.76923	.15073	23.10	.76923	.16446	28.86	.76923	.17472	34.75
.66667	.07376	20.13	.66667	.07848	25.04	.66667	.08161	30.00
.58824	.03743	17.68	.58824	.03908	21.89	.58824	.04005	26.13
.50000	.01486	15.07	.50000	.01524	18.58	.50000	.01543	22.11
.40000	.00402	12.50	.40000	.00407	15.39	.40000	.00408	18.27
.28571	.00054	10.14	.28571	.00054	12.47	.28571	.00054	14.79

J = 8 K = -6			J = 10 K = -6			J = 4 K = -8		
λ	p	h	λ	p	h	λ	p	h
2.5000	.99952	9.408	2.0000	.99924	13.53	5.0000	.99905	5.378
2.0000	.99695	13.45	1.7000	.99601	18.59	3.5000	.99559	7.981
1.7000	.98848	18.28	1.5000	.98616	24.77	2.5000	.98111	12.19
1.5000	.96881	23.82	1.3000	.94657	35.75	2.0000	.95053	16.70
1.3000	.90882	32.72	1.2000	.89691	44.30	1.7000	.90093	21.24
1.2000	.84292	38.67	1.0000	.63265	62.02	1.5000	.83302	25.48
1.0000	.58140	50.02	.76923	.19611	59.26	1.0000	.35135	34.68
.76923	.18829	46.84	.66667	.08658	50.19	.76923	.08667	27.16
.66667	.08505	40.05	.58824	.04126	43.18	.66667	.03250	22.16
.58824	.04095	34.64	.50000	.01561	36.23	.58824	.01292	18.70
.50000	.01558	29.17	.40000	.00410	29.81	.50000	.00371	15.45
.40000	.00409	24.04	.28571	.00054	24.09	.40000	.00064	12.60
.28571	.00054	19.44						

J = 5 K = -8			J = 6 K = -8			J = 8 K = -8		
λ	p	h	λ	p	h	λ	p	h
5.0000	.99981	5.384	3.5000	.99964	8.030	2.5000	.99952	12.55
3.5000	.99874	8.019	2.5000	.99698	12.49	2.0000	.99691	17.96
2.5000	.99245	12.40	2.0000	.98764	17.67	1.7000	.98815	24.46
2.0000	.97527	17.32	1.7000	.96576	23.53	1.5000	.96720	32.02
1.7000	.94176	22.63	1.5000	.92607	29.77	1.3000	.90007	44.40
1.5000	.88893	27.96	1.3000	.82947	38.74	1.2000	.82303	52.71
1.0000	.40000	42.70	1.0000	.44186	50.70	1.0000	.51020	66.69
.76923	.09522	33.80	.76923	.10168	40.54	.76923	.11034	54.29
.66667	.03468	27.44	.66667	.03613	32.75	.66667	.03774	43.44
.58824	.01350	23.07	.58824	.01384	27.45	.58824	.01417	36.23
.50000	.00381	19.02	.50000	.00386	22.60	.50000	.00389	29.74
.40000	.00065	15.49	.40000	.00065	18.39	.40000	.00066	24.18

J = 10 K = -8			J = 5 K = -10			J = 6 K = -10		
λ	p	h	λ	p	h	λ	p	h
2.0000	.99923	18.04	5.0000	.99981	6.731	3.5000	.99964	10.04
1.7000	.99591	24.83	3.5000	.99874	10.02	2.5000	.99698	15.62
1.5000	.98544	33.17	2.5000	.99244	15.51	2.0000	.98760	22.13
1.3000	.94119	48.26	2.0000	.97520	21.73	1.7000	.96544	29.56
1.2000	.87912	59.65	1.7000	.94124	28.51	1.5000	.92453	37.61
1.0000	.56364	82.69	1.5000	.88671	35.45	1.3000	.82148	49.39
.76923	.11538	68.29	1.0000	.34783	53.37	1.0000	.38775	63.37
.66667	.03845	54.19	.83333	.10696	43.07	.83333	.11654	51.64
.58824	.01428	45.03	.76923	.05563	37.26	.76923	.05959	44.57
.50000	.00390	36.89	.66667	.01538	28.71	.66667	.01603	34.19
.40000	.00066	29.97	.58824	.00467	23.55	.58824	.00479	27.99
			.50000	.00095	19.15	.50000	.00096	22.74

J = 8 K = -10			J = 10 K = -10		
λ	p	h	λ	p	h
2.5000	.99952	15.68	2.0000	.99923	22.55
2.0000	.99690	22.46	1.7000	.99589	31.07
1.7000	.98805	30.64	1.5000	.98509	41.59
1.5000	.96649	40.26	1.3000	.93809	60.94
1.3000	.89496	56.29	1.2000	.86938	75.75
1.2000	.80979	67.18	1.0000	.50836	103.4
1.0000	.45455	83.36	.83333	.14041	87.16
.83333	.13079	69.19	.76923	.06805	74.39
.76923	.06492	59.37	.66667	.01708	56.2
.66667	.01676	45.18	.58824	.00494	45.76
.58824	.00490	36.86	.50000	.00098	37.09
.50000	.00097	29.91			

REFERENCES

[1] A. Dvoretzky, J. Kiefer, and J. Wolfowitz, "Sequential Decision Problems for Processes with Continuous Time Parameter. Testing Hypotheses." Ann. Math. Stat. 24 (1953), pp. 254-264.

[2] A. Wald, Sequential Analysis, John Wiley and Sons, 1947.

[3] G. E. Albert, "Accurate sequential tests on the mean of an exponetial distribution," Ann. Math. Stat. 27(1956), pp. 460-470.

* * *

Reprinted from
Naval Res. Logist. Quart. **3** (3), 205–219 (1956)

Reprinted from The Annals of Mathematical Statistics
Vol. 28, No. 1, March, 1957

SOME PROPERTIES OF GENERALIZED SEQUENTIAL PROBABILITY RATIO TESTS[1]

By J. Kiefer and Lionel Weiss

Cornell University and University of Oregon

0. Introduction and Summary. Generalized sequential probability ratio tests (hereafter abbreviated GSPRT's) for testing between two simple hypotheses have been defined in [1]. The present paper, divided into four sections, discusses certain properties of GSPRT's. In Section 1 it is shown that under certain conditions the distributions of the sample size under the two hypotheses uniquely determine a GSPRT. In the second section, the admissibility of GSPRT's is discussed, admissibility being defined in terms of the probabilities of the two types of error and the distributions of the sample size required to come to a decision; in particular, notwithstanding the result of Section 1, many GSPRT's are inadmissible. In Section 3 it is shown that, under certain monotonicity assumptions on the probability ratios, the GSPRT's are a complete class with respect to the probabilities of the two types of error and the average distribution of the sample size over a finite set of other distributions. In Section 4, finer characterizations are given of GSPRT's which minimize the expected sample size under a third distribution satisfying certain monotonicity properties relative to the other two distributions; these characterizations give monotonicity properties of the decision bounds.

1. Uniqueness of certain GSPRT's. In this section we identify a GSPRT with the two sequences of limits characterizing it. Using the same notation as in [1], we assume that the Lebesgue densities f_1 and f_2 satisfy the conditions in Section 2 of [1], even for c equal to zero or infinity (i.e., the probability ratio for any number of observations takes on no single value with positive probability), and are continuous (this last restriction is easily weakened; see also Remark 1 at the end of this section for further generalization).

First we make the transformation $Y_i = F_1(X_i)$. Under H_1, Y_i has a rectangular distribution; under H_2, the density of Y_i will be g (say), where $\int_0^1 g(y)\,dy = 1$.

Next we make the transformation $Z_i = \phi[g(Y_i)]$, where $\phi(u)$ is strictly increasing in u for $u \geqq 0$, $\phi(u)$ takes on no values outside the interval [0, 1], and also

$$\int_{\{y \mid \phi[g(y)] \leqq t\}} dy = t \quad \text{for} \quad 0 \leqq t \leqq 1.$$

Under H_1, the distribution of Z_i is rectangular, while under H_2, the density of Z_i is ϕ^{-1}; we note that $\phi^{-1}(z)$ is strictly increasing in z for $0 \leqq z \leqq 1$ and is 0 elsewhere.

Received December 30, 1955; revised October 2, 1956.

[1] Research sponsored by the Office of Naval Research.

Since we can always transform from $X_1, X_2, \cdots$ to $Z_1, Z_2, \cdots$ and carry out any GSPRT in terms of $Z_1, Z_2, \cdots$, from now on we assume that $f_1(x) = 1$ on [0, 1], and $f_2(x)$ is strictly increasing in x for $0 \leqq x \leqq 1$, with $\int_0^1 f_2(x)\, dx = 1$. We also assume f_2 piecewise differentiable (for the sake of obtaining the g_i below). Any GSPRT is carried out by seeing whether

$$b_m < \prod_{i=1}^{m} f_2(X_i) < a_m, \text{ etc.},$$

or, defining Q_i as $\log f_2(X_i)$, B_m as $\log b_m$, A_m as $\log a_m$, and W_m as $Q_1 + Q_2 + \cdots + Q_m$, whether $B_m < W_m < A_m$, etc.

Denoting by g_i the density function of Q_1 under H_i, we find that $g_2(q) = e^q \cdot g_1(q)$ identically in q.

Suppose the $m - 1$ pairs $(a_1, b_1), \cdots, (a_{m-1}, b_{m-1})$ are fixed. The joint conditional density function of $(Q_1, Q_2, \cdots, Q_m)$ under H_i, given that sampling continues beyond the $(m - 1)$st observation, is

$$h_i(q_1 \cdots, q_m) = \frac{\prod_{j=1}^{m} g_i(q_j)}{K_i}$$

in the region $\{b_j < w_j < a_j; j = 1, \cdots, m - 1\}$; $h_i(q_1, \cdots, q_m) = 0$ elsewhere. Here $K_i = P$ {sampling continues beyond $(m - 1)$ observations under H_i}, and we assume K_1 and K_2 are positive. Thus

$$h_2(q_1, \cdots, q_m) = \frac{K_1}{K_2} \cdot e^{w_m} \cdot h_1(q_1, \cdots, q_m).$$

Then, denoting the conditional density of W_m under H_i given that sampling continues beyond $m - 1$ observations by k_i, we have

$$k_2(w) = \frac{K_1}{K_2} \cdot k_1(w) \cdot e^w.$$

Now we make the following *Assumption* A: f_2 is such that $g_i(q) > 0$ for almost all $q (i = 1, 2,)$; thus, if S is any nondegenerate interval, $\int_S g_i(q)\, dq > 0$ for $i = 1, 2$. But this implies that if T is any nondegenerate interval, $\int_T k_i(w)\, dw > 0$ for $i = 1, 2$.

For any given positive numbers C, D, we now show that there is at most one solution (γ, δ) to the two equations

$$\int_\gamma^\delta k_1(w)\, dw = C, \qquad \int_\gamma^\delta k_2(w)\, dw = D.$$

For, given any γ which can be the first element of a solution, let $\delta(\gamma)$ be the uniquely determined value of δ which satisfies the first equation. Then it is easily verified that

$$\int_{\gamma}^{\delta(\gamma)} k_2(w)\, dw = \int_{\gamma}^{\delta(\gamma)} \frac{K_1}{K_2} \cdot k_1(w) \cdot e^{w}\, dw$$

is strictly increasing in γ. This proves that there is at most one solution (γ, δ) to the two equations.

Let $D_i(n; T)$ denote the probability that a decision is reached after no more than n observations when using the test T and H_i is true. The considerations above yield

THEOREM 1. *If Assumption* A *holds, and if T is a nontruncated* GSPRT, *then there is no* GSPRT *T' different from T and with $D_i(n; T') = D_i(n; T)$ for all n and for $i = 1, 2$. If Assumption* A *holds, and if T is a* GSPRT *truncated after m observations, while T' is another* GSPRT *with $D_i(n; T') = D_i(n; T)$ for all n and for $i = 1, 2$, then T and T' differ only in the terminal decision boundary at stage m.*

REMARK 1. If Assumption A is violated, there are two different GSPRT's, T and T', with $D_i(n; T') = D_i(n; T)$ for all n and for $i = 1, 2$. For we can find a GSPRT T whose first pair of limits is (B_1, A_1), such that for a positive ϵ,

$$\int_{A_1}^{A_1+\epsilon} g_1(q)\, dq = 0.$$

But then T' can be taken as the GSPRT whose first pair of limits is $(B_1, A_1 + \epsilon)$, the other limits agreeing with those for T. The inessential difference between T and T' in such a case as this will be evident to the reader: every sample sequence in a set of probability one under both H_i suffers the same fate under T' as under T. Similarly, in the discrete case (or where Q_1 can take on some constant value with positive probability), there is the aspect of randomization in which two tests with identical $D_i(n; T)$ may differ (e.g., if T and T' both always require at least 3 observations, for some value of $Q_1 + Q_2 + Q_3$, T may stop after 3 observations if $Q_1 < Q_2$, T' if $Q_2 < Q_1$). With these modifications in mind, it is evident that Theorem 1 is of broader validity than its stated form.

REMARK 2. If Assumption A holds, one can prove similarly that there is at most one GSPRT having given values for the elements of the following sequences of probabilities:

$$\{P(\text{accepting } H_1 \text{ under } H_2 \text{ at stage } n)\},$$

$$\{P(\text{accepting } H_2 \text{ under } H_1 \text{ at stage } n)\}.$$

Incidentally, it is easy by the methods of [1] or [2] to show that the GSPRT's (and k-decision problem analogues) form a complete class with respect to the generalized risk function consisting of such sequences. A similar remark applies if these sequences and the $D_i(n; T)$ are considered together as the risk function, etc.

2. Questions of admissibility.[2] We have proved in the previous section that under certain conditions there is at most one GSPRT corresponding to any two specified distributions of n (one under each H_i). This does not imply that all GSPRT's are admissible (letting α_i = probability of an incorrect decision under H_i and p_{in}^* = probability that the experiment terminates after more than n observations under H_i, a GSPRT is said to be admissible in this section if there is no second procedure for which all of the numbers α_i and p_{in}^* ($i = 1, 2$; $n = 0, 1, 2, \cdots$) are no greater, and at least one is less, than for the given procedure); as we shall see in a simple example below, the general question of how to characterize the admissible procedures seems quite difficult. Before turning to this example, we make a few remarks on admissibility. Firstly, it is clear that admissibility does not entail any simple monotone character of the constants a_n and b_n : on one hand, putting p_{in} = probability of terminating after n observations under H_i, on considering the minization of

$$\sum_{i=1}^{2} \xi_i \left[\alpha_i W_i + \sum_{n=1}^{\infty} D_{in} p_{in} \right] \text{where } D_{in} = \sum_{1}^{i} C_{jn}, \tag{2.1}$$

where the C_{in} are an increasing (resp., decreasing) sequence of positive numbers, by comparison with the case $C_{in} = 1$, it becomes clear that we may have admissible procedures with $a_n\uparrow$, $b_n\downarrow$ (resp., $a_n\downarrow$, $b_n\uparrow$); similarly, there are admissible procedures with $a_n\uparrow$, $b_n\uparrow$ or $a_n\downarrow$, $b_n\downarrow$. (In the case $C_{in}\uparrow$ (resp., $C_{in}\downarrow$), it is also interesting to note that $a^{(m)} \leqq a_m \leqq b_m \leqq b^{(m)}$ (resp., $a_m \leqq a^{(m)} \leqq b^{(m)} \leqq b_m$), where $a^{(m)}$ and $b^{(m)}$ are the constant bounds for which (2.1) is minimized when C_{in} is replaced by $C_{i,m+1}$ for all n.) On the other hand, by considering $C_{in} = (h_i + k_i) + (-1)^n h_i$, one may obviously obtain admissible procedures for which $a_n = a_{n+2}$, $b_n = b_{n+2}$, $a_{2n} < a_{2n+1} < b_{2n+1} < b_{2n}$ for all n. Other admissible procedures for which the a_n and b_n have no simple monotone character may be constructed similarly.

Secondly, one can sometimes give simple *necessary* conditions for admissibility (a sufficient condition such as that of being the essentially unique procedure which minimizes (2.1) for some choice of the constants ξ_i, W_i, C_{in} will usually be hard to verify). Suppose, as before, that every interval of positive values (but no single value) of $F_2(x)/f_1(x)$ has positive probability under both H_i. Let N be the smallest integer for which $a_N = b_N$ (if no such integer exists, write $N = \infty$). Then a necessary condition for admissibility of a GSPRT is

$$\sup_{n<N+1} a_n \leqq \inf_{n<N+1} b_n . \tag{2.2}$$

To see this, note that for any Bayes solution minimizing (2.1) the constants ξ_i, W_i must satisfy $a_n \leqq (W_1\xi_1/W_2\xi_2) \leqq b_n$ for all $n < N + 1$; thus, any Bayes solution must satisfy (2.2). Since the essentially complete class of procedures which is the closure in the sense of (Wald's) regular convergence (see [3], [2]) of

[2] In Section 2 the roles of the symbols a and b are reversed from what they are in the other sections.

this class of Bayes procedures satisfying (2.2) also satisfies (2.2), and since (see Section 1) this class must include all admissible GSPRT's, the necessity of (2.2) is established. As will be seen in the examples below, the condition (2.2) is not in general sufficient for admissibility.

We now give an example which will illustrate how complicated an explicit delimitation of the *admissible* GSPRT's seems to be. Consider all GSPRT's requiring at least 1 and at most 2 observations for testing $H_1:\theta = 1$ against $H_2:\theta = 2$ where

$$f_\theta^*(x) = \begin{cases} \theta e^{-\theta x}, & x \geqq 0 \\ 0, & x < 0. \end{cases} \tag{2.3}$$

Let X_1, X_2 be independent with density f_θ^*; write $Y_i = e^{-X_i}$. The hypotheses can then be rewritten as $H_1: Y_i$ have density $f_1(y)$ against $H_2: Y_i$ have density $f_2(y)$, where

$$\begin{aligned} f_1(y) &= \begin{cases} 1, & 0 < y < 1 \\ 0 & \text{otherwise} \end{cases} \\ f_2(y) &= \begin{cases} 2y, & 0 < y < 1 \\ 0 & \text{otherwise.} \end{cases} \end{aligned} \tag{2.4}$$

Thus, $\frac{1}{2}f_2(y)/f_1(y) = y$, and we may write the general form of the GSPRT as:

$$\begin{cases} \text{If } Y_1 \leqq a \text{ (resp., } \geqq b\text{), stop and accept } H_1 \text{ (resp., } H_2\text{);} \\ \text{If } a < Y_1 < b \text{ and } Y_1Y_2 \leqq k \text{ (resp., } >k\text{), accept } H_1 \text{ (resp., } H_2\text{).} \end{cases} \tag{2.5}$$

Here we may assume a, b, k lie between 0 and 1 inclusive. If $a = b$, k is of no importance. If $a < b$, we may suppose $0 < k < 1$, since $k = 0$ or $k = 1$ is clearly inadmissible (replace b by $b' = a$ or a by $a' = b$, respectively, to obtain better procedures); also, since $Y_1Y_2 \leqq Y_1$ with probability one under both H_1 and H_2, we may suppose $k \leqq a$, since $k > a$ is clearly inadmissible (replace a by $a' = \min(k, b)$ for a better procedure). All procedures with $a = b$ are admissible. To summarize, then, in investigating which tests are admissible, we may eliminate certain trivial cases mentioned above and hereafter assume

$$0 < k \leqq a < b \leqq 1. \tag{2.6}$$

The characteristics of any such procedure are easily computed and may be summarized in the risk vector of any such procedure, which is given by the quadruple

$$\begin{aligned} r(a, b, k) &= \{P_1 \text{ (accept } H_2),\ P_2 \text{ (accept } H_1),\ P_1(n = 2),\ P_2(n = 2)\} \\ &= \left\{1 - a - k \log \frac{b}{a},\ a^2 + 2k^2 \log \frac{b}{a},\ b - a,\ b^2 - a^2\right\}. \end{aligned} \tag{2.7}$$

The question of inadmissibility or admissibility of such a procedure is then that of whether or not there exists a test $(\bar{a}, \bar{b}, \bar{k})$ for which all components of $r(\bar{a}, \bar{b}, \bar{k})$ are $\leqq$ the corresponding ones of $r(a, b, k)$, with strict inequality for at least one component. Since no two different tests have identical risk functions (see Section 1), inadmissibility of (a, b, k) is equivalent to (I) the existence of $(\bar{a}, \bar{b}, \bar{k})$ not identical to (a, b, k) and with (II) all components of $r(\bar{a}, \bar{b}, \bar{k}) \leqq$ the corresponding ones of $r(a, b, k)$. The latter condition (II) may be written

$$(2.8)\quad \begin{aligned} &\text{(a)} && \bar{b} - \bar{a} && \leqq b - a \\ &\text{(b)} && \bar{b}^2 - \bar{a}^2 && \leqq b^2 - a^2 \\ &\text{(c)} && \bar{a} + \bar{k}\log(\bar{b}/\bar{a}) && \geqq a + k\log(b/a) \\ &\text{(d)} && \bar{a}^2 + 2\bar{k}^2\log(\bar{b}/\bar{a}) && \leqq a^2 + 2k^2\log(b/a). \end{aligned}$$

The possibility that $\bar{b} = \bar{a}$ may be eliminated in all that follows: if $\bar{b} = \bar{a}$, squaring both sides of (c) and comparing with (d) yields $(a + k\log(b/a))^2 \leqq a^2 + 2k^2 \log(b/a)$; i.e., $k\log(b/a) \leqq 2(k - a)$, which is impossible. Thus, we may hereafter assume $\bar{a} < \bar{b}$. Also, $\bar{a} \geqq \bar{k} > 0$ for admissibility. Thus, in particular, $0 < \log(\bar{b}/\bar{a}) < \infty$ in all that follows.

Combining (2.8) (c) and (d), we obtain

$$(2.9)\quad \left\{\max\left[0, \frac{a - \bar{a} + k\log(b/a)}{\log(\bar{b}/\bar{a})}\right]\right\}^2 \leqq \bar{k}^2 \leqq \frac{a^2 - \bar{a}^2 + 2k^2\log(b/a)}{2\log(\bar{b}/\bar{a})}.$$

In particular, the right-hand term of (2.9) must be >0. Thus, for a given $\bar{a}, \bar{b}$ with $0 < \bar{a} < \bar{b} \leqq 1$, (2.9) can be satisfied for some $\bar{k}$ with $0 < \bar{k} \leqq \bar{a} < \bar{b}$ if and only if either

$$(2.10)\quad \frac{a - \bar{a} + k\log(b/a)}{\log(\bar{b}/\bar{a})} \leqq 0 < \frac{a^2 - \bar{a}^2 + 2k^2\log(b/a)}{2\log(\bar{b}/\bar{a})}$$

or else

$$(2.11)\quad \begin{cases} \text{(a)} & 0 < \dfrac{a - \bar{a} + k\log(b/a)}{\log(\bar{b}/\bar{a})} \leqq \bar{a} \quad \text{and} \\ \text{(b)} & \left(\dfrac{a - \bar{a} + k\log(b/a)}{\log(\bar{b}/\bar{a})}\right)^2 \leqq \dfrac{a^2 - \bar{a}^2 + 2k^2\log(b/a)}{2\log(\bar{b}/\bar{a})}. \end{cases}$$

Equation (2.10) implies

$$(a + k\log(b/a))^2 \leqq \bar{a}^2 < a^2 + 2k^2\log(b/a),$$

the extreme members of which give $2a < 2k - k\log(b/a)$, an impossibility. Since the right side of (2.11) (b) is positive, we also see that the first inequality of (2.11) (a) is implied by (2.11) (b): otherwise, we would again have the contradiction $2a \leqq 2k - k\log(b/a)$. Thus, (2.8) (c) and (2.8) (d) may be satisfied for some $\bar{a}, \bar{k}, \bar{b}$ with $0 < \bar{k} \leqq \bar{a} < \bar{b} \leqq 1$ if and only if (2.11) (b) and the second half of (2.11) (a) may be satisfied for some $\bar{a}, \bar{b}$ with $0 < \bar{a} < \bar{b} \leqq 1$. Write

$$c = \log(b/a), \qquad \bar{c} = \log(\bar{b}/\bar{a}), \lambda = k/a, \qquad \gamma = \bar{a}/a.$$

Equations (2.8) (a), (b) may be written

$$(2.12)\qquad \begin{aligned} &\text{(a)} \qquad \bar{c} \leqq \log\,[1 + (e^c - 1)/\gamma], \\ &\text{(b)} \qquad \bar{c} \leqq \tfrac{1}{2}\log\,[1 + (e^{2c} - 1)/\gamma^2]. \end{aligned}$$

Since c, $\bar{c} > 0$, equation (2.12) (a) implies or is implied by (2.12) (b) according to whether $\gamma \leqq 1$ or $\gamma \geqq 1$. We may write the restriction $\bar{b} \leqq 1$ as

$$(2.13)\qquad \bar{c} \leqq c - \log b\gamma.$$

Equation (2.12) (a) implies (2.13) if $\gamma \leqq 1$, since $b \leqq 1$. If $\gamma > 1$, (2.12) (b) implies or is implied by (2.13) according to whether or not $\gamma < [1 + e^{2c}(b^{-2} - 1)]^{1/2}$.

To summarize, then, (2.11), (2.12), and (2.13) imply that a given (k, a, b) (and hence, (c, λ, b)) is inadmissible if and only if there exist positive numbers $\bar{c}$, γ with either $\bar{c} \neq c$ or $\gamma \neq 1$ (note from (2.9) that $\bar{c} = c$, $\gamma = 1$ imply $\bar{k} = k$ and hence $(\bar{a}, \bar{b}, \bar{k}) = (a, b, k)$) and satisfying

$$(2.14)\qquad \begin{aligned} &\text{(a)} \qquad f_1(\gamma) \leqq \bar{c} \leqq f_2(\gamma), \\ &\text{(b)} \qquad \bar{c} \geqq f_3(\gamma), \\ &\text{(c)} \qquad \gamma < \sqrt{1 + 2\lambda^2 c}, \end{aligned}$$

where

$$f_1(\gamma) = (1 + \lambda c - \gamma)/\gamma,$$

$$(2.15)\qquad f_2(\gamma) = \begin{cases} \log\,[1 + (e^c - 1)/\gamma] & \text{if} \quad \gamma \leqq 1, \\ \tfrac{1}{2}\log\,[1 + (e^{2c} - 1)/\gamma^2] & \text{if} \quad 1 < \gamma < [1 + e^{2c}(b^{-2} - 1)], \\ c - \log b\gamma & \text{if} \quad \gamma \geqq [1 + e^{2c}(b^{-2} - 1)]^{1/2}, \end{cases}$$

$$f_3(\gamma) = 2(1 - \gamma + \lambda c)^2 \,/\, (1 - \gamma^2 + 2\lambda^2 c),$$

and where condition (2.14) (c) merely expresses the positivity of the right side of (2.11) (b).

Suppose $\lambda < 1$ (the case $\lambda = 1$ can be treated easily). It is evident that $f_1(1) = \lambda c < c = f_2(1) = f_3(1)$ and that all points $(\gamma, \bar{c})$ with $|\gamma - 1|$ and $|\bar{c} - c|$ of sufficiently small magnitude and for which $\bar{c} \leqq f_2(\gamma)$, satisfy (2.14) (a) (and, obviously, (2.14)(c)). Hence, primes denoting derivatives, a *necessary* condition for (a, b, k) to be *admissible* is that

$$(2.16)\qquad f_2'(1-) \geqq f_3'(1) \geqq f_2'(1+).$$

Evaluating (2.16) from (2.15), we obtain

$$(2.17)\qquad 1 - e^{-c} \leqq (2\lambda - 1)\,/\,\lambda^2 \leqq \begin{cases} 1 - e^{-2c} & \text{if} \quad b < 1, \\ 1 & \text{if} \quad b = 1. \end{cases}$$

On the other hand, the necessary condition $\sup a_n \leqq \inf b_n$ of (2.2) may be written $1/2 \leqq \lambda \leqq e^c / 2$, or (since $\lambda < 1$)

$$(2.18) \qquad 0 \leqq (2\lambda - 1) / \lambda^2 \leqq \begin{cases} 4e^{-c}(1 - e^{-c}) & \text{if} \quad e^c < 2, \\ 1 & \text{if} \quad e^c \geqq 2. \end{cases}$$

Clearly, (2.18) includes many procedures not included in (2.17) (hence, not in the complement of the set of procedures described by (2.14)). Thus, equation (2.2) is *not sufficient* for admissibility.

We shall not consider this example further: it is already evident from (2.14) that, even in simple cases, the delimitation of the admissible procedures can become complicated.

3. Controlling the distribution of the sample size under distributions other than those being tested. In this section we shall characterize (under certain assumptions) an essentially complete class of tests (the risk function is given below) for testing $H_0 : f = f_{\theta_0}$ against $H_1 : f = f_{\theta_{k+1}}$ sequentially, where the test is based on independent random variables $X_1, X_2, \cdots$ with common density f_θ with respect to some σ-finite measure μ. There are specified values K and $\theta_1, \cdots, \theta_K$, as well as non-negative numbers $a_0, a_1, \cdots, a_{K+1}$ whose sum is unity. The "risk function" of a procedure consists of the vector

$$(3.1) \qquad \left\{ P_{\theta_0}\{\text{accept } H_1\}, P_{\theta_{K+1}}\{\text{accept } H_0\}, \sum_{i=0}^{K+1} a_i P_{\theta_i}\{n \geqq j\}, j = 1, 2, \cdots \right\},$$

where n is the (chance) number of observations required. We consider only procedures for which $P_{\theta_i}\{n < \infty\} = 1$ for all i. One procedure is said to be at least as good as a second one if each component of (3.1) for the first is no greater than the corresponding component for the second, and the notion of essential completeness is relative to this definition of "as good as."

We assume in this section that the $f_{\theta_i}(0 \leqq i \leqq K + 1)$ are finite everywhere and have the same region of positivity and, writing $p_{im}(x^{(m)}) = \prod_{j=1}^m f_{\theta_i}(x_j)$ with $x^{(m)} = (x_1, \cdots, x_m)$, that the functions $p_{im}(x^{(m)}) / p_{0m}(x^{(m)})$ and $p_{(K+1)m}(x^{(m)}) / p_{im}(x^{(m)})$ are for $1 \leqq i \leqq k$ strictly increasing functions of $p_{(K+1)m}(x^{(m)}) / p_{0m}(x^{(m)})$ on the domain of positivity of p_{0m} (of course, this means that either of the first two ratios can increase as the argument changes from one value to another, only if the last ratio also increases); thus, the results of this section apply to the case $f_\theta(x) = c(\theta)e^{\theta x}h(x)$ with $\theta_0 < \theta_1 < \cdots < \theta_{K+1}$. Our result is

THEOREM 3. *An essentially complete class for the above problem consists of those procedures which at the outset randomize between accepting H_0, accepting H_1, and taking a first observation, and which thereafter are* GSPRT's *for testing H_0 against H_1 (with appropriate randomization rules on the boundaries if μ is not atomless).*

PROOF: A trivial modification of the argument of LeCam [2] (there are two decisions, $K + 2$ states of nature here) shows that an essentially complete class can be obtained by taking the closure in the sense of regular convergence (see [2], [3]) of all Bayes strategies for the problem of minimizing

$$(3.2) \qquad \xi_0 a P_{\theta_0}(\text{accept } H_1) + \xi_{K+1} b P_{\theta_{K+1}}(\text{accept } H_0) + \sum_{i=0}^{K+1} \xi_i E_{\theta_i} C_i(n)$$

for all $a > 0$, $b > 0$, $C_i(m)$ strictly increasing in m and approaching infinity with m if i is such that $a_i > 0$ and $C_i(m) = 0$ if $a_i = 0$ (actually, C need not depend on i here, but must for the considerations of Remark 1 below) and all a priori probability measures $(\xi_0, \xi_1, \cdots, \xi_{K+1})$. Each such Bayes strategy is characterized by an initial randomization of the type described in the statement of the theorem and by a sequence $\{D_{im}\}$ $(i = 0, 1, m = 1, 2, \cdots)$ of closed convex subsets of the $(K + 1)$-simplex S (whose elements will be described by $K + 2$ nonnegative barycentric coordinates whose sum is unity) with the property that, Q_i denoting the point of S whose ith barycentric coordinate is unity, $Q_0 \,\varepsilon\, \text{int}\, D_{0m}$ (int denoting interior in the usual topology of S), $Q_{K+1} \,\varepsilon\, \text{int}\, D_{1m}$, these interiors are (for each m) disjoint, and for $1 \leqq j \leqq k$, $Q_j \,\varepsilon\, D_{0m} \cap D_{1m}$ and $Q_j \,\varepsilon\, \text{int}\, (D_{0m} \cup D_{1m})$ (see Chapter 4 of [3] or the paragraph following Lemma 4.1 of the present paper for details of arguments yielding these conclusions). The Bayes strategy relative to an a priori probability measure $\xi = (\xi_0, \cdots, \xi_{K+1})$ is then (after some initial randomization as described above) to compute the point $\xi^{(m)} = \xi^{(m)}(x^{(m)})$ of S whose jth component is $\xi_j p_{jm}(x^{(m)}) / \sum_{i=0}^{K+1} \xi_i p_{im}(x^{(m)})$ $(j = 0, \cdots, K + 1)$ after m observations and to accept H_0, accept H_1, or take another observation according to whether $\xi^{(m)} \,\varepsilon\, \text{int}\, D_{0m}$, $\xi^{(m)} \,\varepsilon\, \text{int}\, D_{1m}$, or $\xi^{(m)} \,\varepsilon\, S - D_{0m} - D_{1m}$, with some sort of randomization if $\xi^{(m)}$ is in the boundary of one or more D_{im} (under our assumptions, if μ is atomless, randomization is actually unnecessary).

Since the class of procedures described in the theorem is compact and closed in the sense of regular convergence (see [2]), the theorem will be proved if we show each Bayes strategy has the structure enunciated in the theorem. But if this is not true, there are values of ξ, a, b, the functions C_i, and a number $n > 0$ and values $x^{(n)}$, $y^{(n)}$ with $\xi^{(n)}(x^{(n)}) \neq \xi^{(n)}(y^{(n)})$, such that $\xi_0 > 0$ and $\xi_{K+1} > 0$ (otherwise, $P\{n = 0\} = 1$ for any Bayes strategy) and such that

$$(3.3) \qquad \begin{aligned} &\text{(a)} \quad \xi^{(n)}(x^{(n)}) \,\varepsilon\, D_{1n}, \\ &\text{(b)} \quad \xi^{(n)}(y^{(n)}) \not\varepsilon\, \text{int}\, D_{1n}, \\ &\text{(c)} \quad p_{(K+1)n}(y^{(n)}) / p_{0n}(y^{(n)}) > p_{(K+1)n}(x^{(n)}) / p_{0n}(x^{(n)}), \end{aligned}$$

or else there is a similar situation for D_{0n} (which is handled similarly). Now, the convex subset of S spanned by $\xi_n(x^{(n)})$, $Q_1, \cdots, Q_{K+1}$ is a subset of D_{1n} which consists of those points $w = (w_0, w_1, \cdots, w_{K+1})$ of S for which

$$(3.4) \qquad w_i t_0 \geqq w_0 t_i \qquad \text{for all} \qquad i > 0 \qquad \text{for which} \qquad t_i > 0,$$

where $\xi^{(n)}(x^{(n)}) = (t_0, t_1, \cdots, t_{K+1})$. Hence, writing $\xi^{(n)}(y^{(n)}) = (z_0, z_1, \cdots, z_{K+1})$, (3.3) (b) would imply that $z_i t_0 \leqq z_0 t_i$ for some $i > 0$ for which t_i (hence, ξ_i) > 0.

Thus, for that i (since also $\xi_0 > 0$), we have

$$p_{in}(y^{(n)}) \,/\, p_{0n}(y^{(n)}) \leqq p_{in}(x^{(n)}) \,/\, p_{0n}(x^{(n)}). \tag{3.5}$$

Since (3.3) (c) implies (by the assumption of this section) the negation of (3.5), we obtain a contradiction and the theorem is proved.

REMARKS: 1. Essentially the same proof works to show that the essentially complete class of Theorem 3 is essentially complete for the more general problem where the components of the risk function are

$$P_{\theta_0}\{\text{accept } H_1\}, P_{\theta_{K+1}}\{\text{accept } H_0\}, \sum_{i=0}^{K+1} a_{ir} P_{\theta_i}\{n \geqq j\} (r = 1, \cdots, S; j = 1, 2, \cdots)$$

the a_{ir} being given non-negative constants. One can also treat the case where the finite linear combination in (3.1) is to be replaced by $\int P_\theta\{n \geqq j\}\, dA(\theta)$ where A is a probability measure on a suitable family $\{P_\theta\}$ of probability measures. These considerations have obvious applications to practical problems where the a_i (or A) represent the probability distribution of the process parameter.

2. Theorem 3 can also be proved using the method of [1]; in fact, a proof of Theorem 3 can be obtained essentially by going through the proof in [1] and replacing P_1 and P_2 there by $\sum a_i P_{\theta_i}$ and making other obvious similar alterations.

3. In cases like those of Lemma 4.2 below other than (3), the assumptions of the present section are not satisfied; however, such cases can also be treated here with only minor modifications of the above analysis.

4. Procedures minimizing ***En*** **at a third point, etc.** Let f_{-1}, f_0, f_1 be three densities with respect to a σ-finite measure μ. We assume no two of the f_i are identical almost everywhere (μ). Let X_1, X_2, $\cdots$ be independently and identically distributed random variables with common density f with respect to μ. It is desired to test between the hypotheses $H_{-1}:f = f_{-1}$ and $H_1:f = f_1$. Let $\alpha_i(\delta)$ denote the probability that the procedure δ terminates with an incorrect decision when H_i is true ($i = \pm 1$). Let α_i^* be specified numbers satisfying $0 < \alpha_i^* < 1$ ($i = \pm 1$). Let $A_0(\delta)$ denote the expected value of n (the number of observations which have been taken at termination) when $f = f_0$ and δ is used. Our purpose here is to characterize procedures δ which, among all procedures satisfying

$$\alpha_i(\delta) \leqq \alpha_i^*(i = \pm 1), \tag{4.1}$$

minimize $A_0(\delta)$. Under suitable assumptions (those of Sections 3 and 4 differ), the class of procedures delimited in Theorem 3 will evidently contain the procedures which do this, but we shall obtain here a much finer characterization of them. For the remainder of this section we shall term such procedures "optimum." To avoid trivialities, we hereafter assume $\alpha_1^* + \alpha_2^* < 1$.

We first note a fairly obvious property of optimum procedures. Let Γ be the set of points in three-space of the form $(\alpha_{-1}(\delta), \alpha_1(\delta), A_0(\delta))$ for all possible δ

(not merely those satisfying (4.1)). Since one can randomize between two procedures at the outset, Γ is clearly convex. (The existence of points (a, b, c) with $c < \infty$ for any $a, b > 0$ follows from consideration of fixed-sample-size procedures. A convex combination of points giving positive weight to a point with $c = \infty$ will itself have $c = \infty$.) For any procedure δ satisfying (4.1) and with strict inequality for either $i = 1$ or $i = -1$, we may (by randomizing between δ and a procedure requiring no observations) obtain a δ' satisfying (4.1) and for which $A_0(\delta') < A_0(\delta)$; we may therefore restrict our search for optimum procedures to those δ for which equality holds in (4.1). Among the class of all such procedures there exists one minimizing $A_0(\delta)$, a consequence of Theorem 3.1 of Wald [3]. Let $e(\alpha_{-1}^*, \alpha_1^*) = \min_\delta A_0(\delta)$, the minimum being taken subject to (4.1) (with equality). For all $\epsilon > 0$ with $\epsilon < \min(\alpha_{-1}^*, \alpha_1^*)$ we have (recalling $\alpha_1^* + \alpha_{-1}^* < 1$) that $e(\alpha_{-1}^* - \epsilon, \alpha_1^* - \epsilon) > e(\alpha_{-1}^*, \alpha_1^*)$; for otherwise, if equality held, a randomization of the type noted parenthetically above would produce a δ satisfying (4.1) and for which $A_0(\delta) < e(\alpha_{-1}^*, \alpha_1^*)$, a contradiction. Since $e(\alpha_{-1}^* - \epsilon, \alpha_1^* - \epsilon) > e(\alpha_{-1}^*, \alpha_1^*)$, and since for any value $e > 0$ the points $(0, 1, e)$ and $(1, 0, e)$ are clearly in Γ, it is clear that Γ can not be supported at $(\alpha_{-1}^*, \alpha_1^*, e(\alpha_{-1}^*, \alpha_1^*))$ by a plane any of whose direction cosines is zero. Since Γ obviously can be supported at this point by a plane with non-negative direction cosines, we have

LEMMA 4.1. *Any optimum procedure must, for some positive* ξ_1, ξ_{-1}, ξ_0, *minimize*

$$\xi_1\alpha_1(\delta) + \xi_{-1}\alpha_{-1}(\delta) + \xi_0 A_0(\delta) \tag{4.2}$$

among all procedures δ. (*Conversely, any procedure minimizing* (4.2) *for some positive* ξ_i's *is obviously optimum for some* α_i^*'s.)

Thus, necessary conditions on optimum procedures may be found by characterizing "Bayes solutions" which minimize (for a given ξ_1, ξ_{-1}, ξ_0, all positive, and whose sum we may take to be unity) the "integrated risk" (4.2). Results like Theorems 4.8, 4.9, and 4.10 of Wald [3] (see also [4]) are easily seen to be valid in the present case (with the two values of the loss function and cost of experimentation altered from their unit values in (4.2) if desired). To summarize what we need of these results, *all* procedures minimizing (4.2) for all possible $\xi = (\xi_1, \xi_2, \xi_3)$ with $\xi_i \geqq 0$, $\sum \xi_i = 1$, are characterized in the 2-simplex in barycentric coordinates by two closed convex regions C_{-1} and C_1 as follows: after m observations ($m = 0, 1, 2, \cdots$) compute the a posteriori probability measure $\xi^{(m)}$ for the given a priori measure $\xi = \xi^{(0)}$ and the observed values of $X_1, \cdots, X_m$. Accept H_{-1}, accept H_1, or take another observation according to whether $\xi^{(m)}$ lies in the interior of C_{-1}, C_1, or the complement of $C_{-1} \cup C_1$; on the boundaries between regions, a Bayes solution may randomize in any way (depending on $X_1, \cdots, X_m$ if desired) between (or among) appropriate actions. We now describe the C_i. Let V_i be the point where $\xi_i = 1 (i = 0, \pm 1)$. A point ξ of C_i will be called an *interior* or *boundary* point of C_i according to whether or not *every* Bayes solution with respect to ξ immediately accepts H_i

with probability one. Clearly (see p. 121 of [3]) V_i is in the boundary of C_i for $i = \pm 1$, and a line segment V_0P (of positive length) of the line $\xi_1 = \xi_2$ is the intersection $C_1 \cap C_2$; the curve V_0PV_i is the boundary of C_i ; on the segment V_0P, except at the point P where one may randomize among accepting H_1 or H_{-1} or taking another observation, a Bayes solution must stop with probability one and randomize between accepting H_1 or H_{-1} (analogous to Theorem 4.9 of [3]). Of course, a necessary condition for a Bayes solution minimizing (4.2) to be optimum (for some α_i^*) is that it stop with probability one whenever $\xi^{(m)} = V_1$ or V_{-1} ; *we hereafter consider only Bayes solutions of this nature.*

In order to obtain a more detailed characterization, we now introduce certain assumptions. Write $x^{(m)} = (x_1, \cdots, x_m)$ and $p_{im}(x^{(m)}) = f_i(x_1)f_i(x_2) \cdots f_i(x_m)$ for $m > 0$ and $=1$ for $m = 0$.

Assumption A. For each m and $x^{(m)}$, $y^{(m)}$, if

$$p_{1m}(x^{(m)})p_{-1m}(y^{(m)}) \geqq p_{1m}(y^{(m)})p_{-1m}(x^{(m)}), \tag{4.3}$$

then

$$p_{0m}(x^{(m)})p_{-1m}(y^{(m)}) \geqq p_{0m}(y^{(m)})p_{-1m}(x^{(m)}) \tag{4.4}$$

and

$$p_{1m}(x^{(m)})p_{0m}(y^{(m)}) \geqq p_{1m}(y^{(m)})p_{0m}(x^{(m)}); \tag{4.5}$$

and strict inequality in (4.3) with both sides positive implies strict inequality in (4.4) and (4.5).

Assumption B. For each x_1, if

$$f_1(x_1) \geqq f_{-1}(x_1), \tag{4.6}$$

then

$$f_0(x_1) \geqq f_{-1}(x_1); \tag{4.7}$$

and, if

$$f_{-1}(x_1) \geqq f_1(x_1), \tag{4.8}$$

then

$$f_0(x_1) \geqq f_1(x_1). \tag{4.9}$$

Assumption C:

$$\limsup_{m\to\infty}\ \sup_{x^{(m)}} \frac{\min[p_{1m}(x^{(m)}), p_{-1m}(x^{(m)})]}{p_{0m}(x^{(m)})} = 0 \tag{4.10}$$

where the supremum is taken over those $x^{(m)}$ for which the denominator is positive.

Of course, if there is a value z_1 for which $f_{-1}(z_1) = f_1(z_1) = f_0(z_1) > 0$ (which will usually not be so), then Assumption B follows from Assumption A. Assumptions A and B are related to the monotone likelihood ratio assumption which occurs elsewhere in certain fixed-sample-size problems in statistics (e.g., [5]), but are not quite in that form, which (for a parametric class: put $\theta_j = 0, \pm 1$ in our case) states (for $n = 1$) $f_{\theta_2}(x)f_{\theta_1}(y) \geqq f_{\theta_1}(x)f_{\theta_2}(y)$ if $\theta_2 \geqq \theta_1$, $x \geqq y$. In Assumption A the $x^{(m)}$ are not necessarily simply-ordered (although a simple ordering of certain equivalence classes for the purpose of the present discussion will often be obvious), unlike the monotone likelihood ratio case; and, at least for fixed n, it is easy to see by examples that neither of Assumption B and the monotone assumption implies the other.

Before proceeding to the consequences of Assumptions A, B, and C, we note that they will be satisfied in many important cases (see also Remark 12 below):

LEMMA 4.2. *If* $f_i(x) = f(x, \theta_i)$ $(i = 0, \pm 1)$ *with* $\theta_{-1} < \theta_0 < \theta_1$ *(the inequalities may be reversed), then Assumptions* A, B, *and* C *are satisfied if* $f(x, \theta)$ *is (for example) of any of the following forms:*

$$(1)\quad f(x, \theta) = \begin{cases} e^{-(x-\theta)}, & x > \theta \\ 0, & x \leqq \theta \end{cases}$$

$(-\infty < \theta < \infty, \mu =$ *Lebesgue measure*);

$$(2)\quad f(x, \theta) = \begin{cases} \theta^{-1}, & 0 < x < \theta \\ 0 & \text{otherwise} \end{cases}$$

$(0 < \theta < \infty, \mu =$ *Lebesgue measure*);

$$(3)\quad f(x, \theta) = r(\theta)e^{\theta x} \text{ (Koopman-Darmois)}$$

(μ *any* σ*-finite measure not giving all measure to one point,* $r^{-1}(\theta) = \int e^{\theta x}\, d\mu(x)$, θ *any value for which* $r^{-1}(\theta) < \infty$).

We remark that the case $f(x, \theta) = \phi(\theta)e^{g(\theta)t(x)}$ with g strictly monotone can be reduced to case (3).

PROOF: Cases (1) and (2) are easy to verify directly (note that the last part of Assumption A is vacuous here). In case (3), we have $\prod_{i=1}^n f(x_i, \theta) / f(y_i, \theta) = e^{\theta z}$ where $z = \sum_1^n (x_i - y_i)$; Assumption A follows at once. Next, we note (differentiating under the integral sign where necessary, which is easily justified for any $\theta \varepsilon L$, where L is the interior of the interval of values θ for which $r^{-1}(\theta) < \infty$) that, for $\theta \varepsilon L$, we have $d^2 \log r(\theta) / d\theta^2 = (E_\theta X)^2 - E_\theta X^2 < 0$, where $E_\theta g(X)$ denotes the expected value of $g(X)$ when X has density $f(x, \theta)$. Hence, $-\log r(\theta)$ is *strictly* convex over the interval of θ for which $r^{-1}(\theta) < \infty$.

Putting $f_i(x) = f(x, \theta_i)$ with $\theta_{-1} < \theta_0 < \theta_1$, equation (4.6) is equivalent to

$$\frac{1}{\theta_1 - \theta_{-1}} \log \frac{r(\theta_1)}{r(\theta_{-1})} \geqq -x_1, \tag{4.11}$$

while (4.7) is equivalent to the expression obtained from (4.11) by substituting θ_0 for θ_1. Hence, we will have shown that (4.6) implies (4.7) if we show that $q(\theta) = (\theta - \theta_{-1})^{-1} \log r(\theta) / r(\theta_{-1})$ is monotonically nonincreasing in θ for $\theta > \theta_{-1}$. Thus, it suffices to show, for $\theta > \theta_{-1}$, that $b(\theta) \leqq 0$, where $b(\theta) = (\theta - \theta_{-1})^2 dq(\theta) / d\theta = [-\log r(\theta) / r(\theta_{-1})] + (\theta - \theta_{-1}) d \log r(\theta) / d\theta$. Since

$b(\theta_{-1}) = 0$, it suffices to show, for $\theta > \theta_{-1}$, that $0 \geqq db(\theta) / d\theta$. Since $db(\theta) / d\theta = (\theta - \theta_{-1}) \, d^2 \log r(\theta) / d\theta^2$, the desired result follows from that of the previous paragraph. The proof that (4.8) implies (4.9) is similar (or may be obtained from the preceding argument by replacing θ and x by $-\theta$ and $-x$).

Finally, let ρ be defined by $\theta_0 = \rho\theta_1 + (1 - \rho)\theta_{-1}$. Clearly, $0 < \rho < 1$. Because of the strict convexity of $-\log r(\theta)$ we have, for some number h with $0 < h < 1$ and for all x,

$$\left[\frac{f(x, \theta_1)}{f(x, \theta_0)}\right]^\rho \left[\frac{f(x, \theta_{-1})}{f(x, \theta_0)}\right]^{1-\rho} = \left[\frac{r(\theta_1)}{r(\theta_0)}\right]^\rho \left[\frac{r(\theta_{-1})}{r(\theta_0)}\right]^{1-\rho} = h. \tag{4.12}$$

Hence, for each m and all $x^{(m)}$,

$$\min \left[\frac{p_{1m}(x^{(m)})}{p_{0m}(x^{(m)})}, \quad \frac{p_{-1m}(x^{(m)})}{p_{0m}(x^{(m)})}\right] \leqq h^{m\epsilon} \tag{4.13}$$

where $\epsilon = \min [1/2\rho, 1/2(1 - \rho)]$. Equation (4.10) follows at once from equation (4.13). This completes the proof of Lemma 4.2.

We now prove some consequences of Assumptions A, B, and C. In what follows $\xi^{(0)}$ is a fixed a priori probability measure all of whose components are positive. The reader should recall the italicized statement a few lines above Assumption A. Write $D^{(0)} = \xi_{-1}^{(0)}/\xi_1^{(0)}$, where $\xi^{(0)} = (\xi_{-1}^{(0)}, \xi_0^{(0)}, \xi_1^{(0)})$.

LEMMA 4.3. *Under Assumption* A *there exist positive constants* b_m *and* a_m ($m = 0, 1, 2, \cdots$) *with* $b_m \leqq a_m$ *and such that*

(4.14) $\xi^{(m)} \in C_1$ *if and only if* $p_{1m}(x^{(m)}) / p_{-1m}(x^{(m)}) \geqq a_m$,

(4.15) $\xi^{(m)} \in C_{-1}$ *if and only if* $p_{1m}(x^{(m)}) / p_{-1m}(x^{(m)}) \leqq b_m$,

with strict inequality holding if and only if $\xi^{(m)}$ *is an interior point of the appropriate* C_i.

(Of course, the values a_m, b_m depend on $\xi^{(0)}$.)

PROOF: The method of proof is similar to that of Theorem 3. Let $x^{(m)}$ and $y^{(m)}$ be such that (4.3) holds, and let $\xi^{(m)}(x^{(m)})$, $\xi^{(m)}(y^{(m)})$ be the a posteriori probability measures corresponding to observed values $x^{(m)}$, $y^{(m)}$. Equation (4.15) will follow if we can show that $\xi^{(m)}(x^{(m)}) \in C_{-1}$ implies $\xi^{(m)}(y^{(m)}) \in C_{-1}$. (The reader will be aided in what follows if he draws a picture.) Now, (4.3) says that the line $V_0\xi^{(m)}(y^{(m)})$ lies toward V_{-1} from (or on) the line $V_0\xi^{(m)}(x^{(m)})$. Moreover, (4.3) implies (4.4) and (4.5), which say that the line $V_1\xi^{(m)}(y^{(m)})$ lies toward V_{-1} from (or on) the line $V_1\xi^{(m)}(x^{(m)})$ and that the line $V_{-1}\xi^{(m)}(y^{(m)})$ lies toward V_0 from (or on) the line $V_{-1}\xi^{(m)}(x^{(m)})$. Hence, $\xi^{(m)}(y^{(m)})$ lies inside or on the triangle T whose vertices are V_{-1}, $\xi^{(m)}(x^{(m)})$, and the intercept of $V_1\xi^{(m)}(x^{(m)})$ with $V_{-1}V_0$. Since T is contained in the triangle $V_0\xi^{(m)}(x^{(m)})V_{-1}$ which (by convexity) is contained in C_{-1}, (4.15) is proved. Moreover, since the last part of Assumption A implies that $\xi^{(m)}(y^{(m)})$ could lie on the line $V_{-1}\xi^{(m)}(x^{(m)})$ only if $\xi^{(m)}(y^{(m)})$ is V_{-1} or $\xi^{(m)}(x^{(m)})$, it is clear that $\xi^{(m)}(y^{(m)})$ is a boundary point of C_{-1} if and only if either $\xi^{(m)}(y^{(m)}) = V_{-1}$ (see the italicized remark a few lines above Assumption A in this case) or else $\xi^{(m)}(x^{(m)})$ is a boundary point of C_{-1} and $\xi^{(m)}(y^{(m)}) = \xi^{(m)}(x^{(m)})$; the latter implies equality in (4.3). Thus (modulo the italicized remark), defining b_m^* to be (for fixed $\xi^{(0)}$) the supremum (over $x^{(m)}$) of those values $p_{1m}(x^{(m)}) / p_{-1m}(x^{(m)})$ for which $\xi^{(m)}(x^{(m)}) \varepsilon C_{-1}$, and taking $b_m = b_m^*$ if $\xi^{(m)}(x^{(m)}) \neq V_{-1}$ is on the boundary of C_{-1} for some $x^{(m)}$ and $b_m^* < b_m <$ infimum of those $p_{1m}(x^{(m)}) / p_{-1m}(x^{(m)})$ for which $\xi^{(m)}(x^{(m)}) \not\varepsilon C_{-1}$ otherwise, we see from the previous sentence and the fact that $p_{1m}(y^{(m)}) / p_{-1m}(y^{(m)}) < p_{1m}(x^{(m)}) / p_{-1m}(x^{(m)})$ if $\xi^{(m)}(x^{(m)}) \neq \xi^{(m)}(y^{(m)}) \varepsilon T$, that the last part of the lemma as it applies to (4.15) is proved. Equation (4.14) (and the corresponding last part) is proved similarly.

LEMMA 4.4. *Under Assumptions* A *and* B, *there exist constants* $b_m \leqq D^{(0)} \leqq a_m$ *of Lemma* 4.3 *satisfying* $b_m \leqq b_{m+1}$, $a_m \geqq a_{m+1}$, *for* $m = 0, 1, 2, \cdots$.

PROOF: We shall prove the assertion regarding the b_m, a similar proof applying for the a_m. Keeping $\xi^{(0)}$ fixed as before, in order to prove $b_m \leqq b_{m+1}$ it clearly suffices to prove that $\xi^{(m)} \varepsilon C_{-1}$ and $p_{1,m+1}(x^{(m+1)}) / p_{-1,m+1}(x^{(m+1)}) \leqq p_{1m}(x^{(m)}) / p_{-1,m}(x^{(m)})$ imply that $\xi^{(m+1)} \varepsilon C_{-1}$ (the case where either ratio is 0/0 or where both are 0 is easily disposed of); i.e., that $\xi^{(m)} \varepsilon C_{-1}$ and $f_1(x_{m+1}) \leqq f_{-1}(x_{m+1})$ imply $\xi^{(m+1)} \varepsilon C_{-1}$. The last inequality says that the line $V_0\xi^{(m+1)}$ lies toward V_{-1} from (or on) the line $V_0\xi^{(m)}$; by (4.8) and (4.9), it implies that the line $V_{-1}\xi^{(m+1)}$ lies toward V^0 from (or on) the line $V_{-1}\xi^{(m)}$. Thus, $\xi^{(m+1)}$ lies in the triangle $V_0\xi^{(m)}V_{-1}$ and hence, by convexity, $\xi^{(m+1)} \varepsilon C_{-1}$. The remaining part of the lemma follows at once from the fact that $\xi_1^{(m)} \leqq \xi_{-1}^{(m)}$ is equivalent to $p_{1m} / p_{-1m} \leqq D^{(0)}$.

If $f_1(x) / f_{-1}(x)$ cannot take on a suitably dense set of values, the a_m and b_m might (for fixed $\xi^{(0)}$) not be unique and might correspond to $\xi^{(m)}$ in the interior of the C_i or the complement of $C_{-1} \cup C_1$. If this is not the case, the previous paragraph and the fact that $V_0PV_{-1} \varepsilon C_{-1}$ show that we can strengthen the weak inequality of Lemma 4.4. One possible formulation of this result is the following:

LEMMA 4.5. *If* $b_m < D^{(0)}$ *(resp.,* $a_m > D^{(0)}$*) and if for every open interval* J *containing* b_m *(resp.,* a_m*) the ratio* $p_{1m}(x^{(m)}) / p_{-1,m}(x^{(m)})$ *takes on values in* $J - \{b_m\}$ *(resp.,* $J - \{a_m\}$*) with positive probability under* H_1 *and* H_{-1} *(so that* b_m, a_m *are unique), and if for every open interval* J' *containing* $D^{(0)}$ *as a left (resp., right) end-point* $f_1(x) / f_{-1}(x)$ *takes on values in* J' *with positive probability under* H_1 *and* H_{-1}, *then* $b_m < b_{m+1}$*(resp.,* $a_m > a_{m+1}$*). In particular, in case* (3) *of Lemma* 4.2, *if* μ *is equivalent to Lebesgue measure (or if Lebesgue measure is absolutely continuous with respect to* μ*) on the real line, the* a_m *and* b_m *are unique and this last result holds.*

In fact, it remains only to prove the last assertion of the lemma, which follows at once from the fact that $e^{(\theta_1-\theta_{-1})x}$ takes on values in any interval of positive numbers with positive probability under H_1 and H_{-1}, if Lebesgue measure is absolutely continuous with respect to μ.

LEMMA 4.6. *Under Assumption* C, *for any* $\xi^{(0)}$ *all of whose components are (or in fact, for which* $\xi_0^{(0)}$ *is) positive, there is an integer* $N = N(\xi^{(0)})$ *such that every Bayes solution with respect to* $\xi^{(0)}$ *requires fewer than* N *observations with probability one under* f_1, f_{-1}, *and* f_0.

PROOF: Fix $\xi^{(0)}$. Since P is a positive distance from V_0, there is clearly a positive number c such that every Bayes solution must stop with probability one whenever either $p_{1m}(x^{(m)}) / p_{0m}(x^{(m)}) < c$ or else $p_{-1m}(x^{(m)}) / p_{0m}(x^{(m)}) < c$. The desired result now follows at once from (4.10). (Note again the remark made in italics just before Assumption A).

We may now summarize our results:

THEOREM 4. *Under Assumptions* A, B, *and* C (*in particular, under* (1), (2), *or* (3) *of Lemma* 4.2), *any procedure which minimizes* $A_0(\delta)$ *subject to* (4.1) *is a* GSPRT *of* H_1 *against* H_{-1} *with* $b_m \leqq b_{m+1} \leqq D^{(0)} \leqq a_{m+1} \leqq a_m$ *for* $m = 0, 1, 2, \cdots, N$ *and some* $D^{(0)}$, *which stops with probability one under* $f_i (i = 0, \pm 1)$ *after* N *or fewer observations. Under additional conditions specified in Lemma* 4.5 *the values* $a_m, b_m (m \geqq 1)$ *will be unique and* $a_m > a_{m+1}$ *or* $b_m < b_{m+1}$ *unless* $a_m = D^{(0)}$ *or* $b_m = D^{(0)}$, *where* $D^{(0)}$ *corresponds to the a priori distribution with respect to which the optimum procedure is Bayes.*

REMARKS, GENERALIZATIONS, ETC.

1. Of course, a GSPRT of Theorem 4 involves a randomization rule for all $m \leqq N$, including a possibly randomized starting rule ($m = 0$) if a_0 or $b_0 = D^{(0)}$. If μ is nonatomic, there will clearly exist an optimum GSPRT involving no randomization, except possibly in the starting rule. The lack of uniqueness of the a_m and b_m in cases not covered by Lemma 4.5 is of course inessential, reflecting only that certain intervals of values of p_{1m}/p_{-1m} have probability zero under all f_i.

2. In all of the above, the X_i are random elements whose range is immaterial (not necessarily real) as long as the appropriate assumptions are satisfied. To conserve space we have not included statements about the obvious sets of measure zero where various conditions may be permitted to fail.

3. As an example of what can happen when our assumptions are not satisfied, we mention briefly the following example: Suppose $f_j(x) = 1 / \pi[1 + (x - j)^2]$, $j = 0, \pm 1$, $\mu =$ Lebesgue measure. In this Cauchy case it is easy to see that, for $\xi^{(0)}$ with all components positive, the set of possible $\xi^{(1)}$ values is a simple closed curve minus the point $\xi^{(0)}$, and lies entirely in the interior of the triangle $V_0 V_1 V_{-1}$. Assumption A is not satisfied, and there is no reason why the result of Lemma 4.3. should be valid. Also, since (e.g.) $f_j(\sqrt{3/2}) f_j(-\sqrt{3/2})$ does not depend on j, there is no reason why the result of Lemma 4.6 should hold here.

4. Remarks analogous to those of Section 2 can be made here: for concave (resp., convex) nondecreasing $c(n)$, the minimization of a linear combination such as (4.2) with $A_0(\delta)$ replaced by $E_0 c(n)$ under δ may be compared in an obvious fashion to the minimization when $c(n)$ is replaced by the linear homogeneous cost function $c_m^*(n)$ passing through $(1, c(m + 1) - c(m))$ (i.e., to the solution of the problem we have considered): the stopping region will now change with m, being contained in (resp., containing) that fixed region for the problem concerning $c_m^*(n)$.

5. The methods used herein (especially the geometric type argument of

Theorem 3 and Lemma 4.3) may also be usefully applied to obtain structure theorems in other sequential decision problems under suitable regularity conditions. For example, the stopping rule for any Bayes solution (risk = expected loss + En) for the k-decision problem of choosing which of $\theta_1 < \theta_2 < \cdots < \theta_k$ is the true parameter value when the X_i have the density $f(x, \theta_j)$ of (3) of Lemma 4.2 for some j is easily seen for large n to approach that which says to stop if and only if each of $k - 1$ certain SPRT's of θ_j against θ_{j+1} $(1 \leqq j < k)$ says to stop.

6. For practical use, our results may be put into more convenient form. For example, if f_j is for some $\Delta > 0$ the normal density with mean Δj and unit variance, our results say that there are constants $A_1 > A_2 > \cdots > A_{N-1} > 1$ such that the (essentially unique) procedure with type I and type II errors α which minimizes $A_0(\delta)$ stops the first time $|\sum_1^n X_i| \geqq A_n$ (making the appropriate decision) and never takes more than N observations (assuming a first observation is taken with probability one). Similar characterizations in the space of the range of the sufficient statistic may be made in other cases of Lemma 4.2.

7. For given α_i^*, it is interesting to consider $A_0(\delta^*)$ where δ^* is the SPRT with $\alpha_i(\delta^*) = \alpha_i^*$ (which minimizes $E_j n$ for $j = \pm 1$). Let $M = M(\alpha_{-1}^*, \alpha_1^*)$ be the smallest integer such that (4.1) may be satisfied by a fixed-sample-size procedure δ requiring M observations. It is easy to give examples where $A_0(\delta^*) < M$ (e.g., let f_0 be close to f_{-1} or f_1) and where $M < A_0(\delta^*)$ (in the example of Remark 6 above, as $\alpha \to 0$, $A_0(\delta^*)$ is of order $(\log \alpha)^2 > M(\alpha, \alpha)$). It would be interesting to obtain useful inequalities and limiting formulas for $e(\alpha_{-1}, \alpha_1)$, as well as $e(\alpha_{-1}, \alpha_1) / A_0(\delta^*)$ and $e(\alpha_{-1}, \alpha_1) / M(\alpha_{-1}, \alpha_1)$, analogous to those which can be obtained in sequential analysis [6]. Of course, if for each $\xi^{(0)}$ one has a knowledge of an upper bound on N, one can compute the procedures of Theorem 4 (for all α_i^*) by "working backwards" as in [3], [4]. Without investigating these topics further, we mention an interesting suggestion of Wolfowitz (who is also to be thanked for suggesting the problem of this section): There is in Case (3) of Lemma 4.2 with μ equivalent to Lebesgue measure a one-parameter family C of tests of the form "stop the first time there is a violation of the inequality $h_1 + Sn < \sum_1^n X_i < h_2 + Sn$ (h_1, h_2, S constants)" and which satisfy $\alpha_i = \alpha_i^* (i = \pm 1)$. One of these other than the unique SPRT δ^* of f_1 against f_{-1} which is a member of C may minimize $A_0(\delta)$ among members of C and may reduce $A_0(\delta)$ considerably from its values for δ^*. Investigation now being undertaken shows that this improvement may be appreciable in practical examples and can often be achieved without modifying $E_i n$ greatly for $i = \pm 1$. We also remark that truncated SPRT's will often be much better than untruncated SPRT'S (e.g., in the example of Remark 6 for α small) in making $A_0(\delta)$ small subject to (4.1); some data on this are available in (e.g.) [7]. These remarks apply also to 9 and 10 below.

8. Our results may be extended in an obvious fashion to consideration of minimizing $A_0(\delta)$ subject to (4.1) for continuous time processes [8].

9. One may obtain a result similar to that of our Theorem 4 for the problem of minimizing subject to (4.1) a (probability) average of $E_\theta n$ over a set of θ between θ_{-1} and θ_1 in the cases of Lemma 4.2 (see also Section 3). This corresponds to the practical situation where θ may be thought of as having a known probability distribution (e.g., certain industrial problems).

10. In any of the cases of Lemma 4.2, one can obtain results on the problem of minimizing $\sup_{\theta_{-1} \leqq \theta \leqq \theta_1} E_\theta n$ subject to (4.1). This can be done by obvious application of the Bayes technique, using Remark 9. In some cases it will be easy to guess at a value θ_0 $(\theta_{-1} \leqq \theta_0 \leqq \theta_1)$ such that a procedure minimizing $A_0(\delta)$ subject to (4.1) has its maximum $E_\theta n$ at $\theta = \theta_0$. This procedure will then clearly minimize $\sup_\theta E_\theta n$.

11. Results like those of Theorem 4 and Remarks 9 and 10 can also be obtained if a restriction of the type $E_i n \leqq c_i (i = \pm 1)$ is imposed in addition to (4.1).

12. Lemma 4.2 can easily be extended to include many other cases, e.g., many cases arising in simple fashion from those of Lemma 4.2. For example, Lemma 4.2 also holds for $f(x, \theta) = (t + 1)x^t\theta^{-t-1}$ if $0 < x < \theta$ (and $= 0$ otherwise), where $t > -1$.

REFERENCES

[1] L. Weiss, "Testing one simple hypothesis against another," *Ann. Math. Stat.*, Vol. 24 (1953), pp. 273–281.

[2] L. LeCam, "Note on a theorem of Lionel Weiss," *Ann. Math. Stat.*, Vol. 25 (1954), pp. 791–794.

[3] A. Wald, *Statistical Decision Functions*, John Wiley & Sons, 1950.

[4] A. Wald and J. Wolfowitz, "Bayes solutions of sequential decision problems," *Ann. Math. Stat.*, Vol. 21 (1950), pp. 82–89.

[5] H. Rubin, "A complete class of decision procedures for distributions with monotone likelihood ratio," (abstract) *Ann. Math. Stat.*, Vol. 22 (1951), p. 608.

[6] A. Wald, *Sequential Analysis*, John Wiley & Sons.

[7] R. E. Bechhofer, J. Kiefer, and M. Sobel, "A Sequential Multiple Decision Procedure for Certain Identification and Ranking Problems," to be published.

[8] A. Dvoretzky, J. Kiefer, and J. Wolfowitz, "Sequential decision problems for processes with continuous time parameter. Testing hypotheses," *Ann. Math. Stat.*, Vol. 24 (1953) pp. 254–264.

Reprinted from THE ANNALS OF MATHEMATICAL STATISTICS
Vol. 28, No. 3, September, 1957

INVARIANCE, MINIMAX SEQUENTIAL ESTIMATION, AND CONTINUOUS TIME PROCESSES

BY J. KIEFER[1]

Cornell University

1. Introduction and summary. The main purpose of this paper is to prove, by the method of invariance, that in certain sequential decision problems (discrete and continuous time) there exists a minimax procedure δ^* among the class of all sequential decision functions such that δ^* observes the process for a constant length of time. In the course of proving these results a general invariance theorem will be proved (Sec. 3) under conditions which are easy to verify in many important examples (Sec. 2). A brief history of the invariance theory will be recounted in the next paragraph. The theorem of Sec. 3 is to be viewed only as a generalization of one due to Peisakoff [1]; the more general setting here (see Sec. 2; the assumptions of [1] are discussed under Condition 2b) is convenient for many applications, and some of the conditions of Sec. 2 (and the proofs that they imply the assumptions) are new; but the method of proof used in Sec. 3 is only a slight modification of that of [1]. The form of this extension of [1] in Secs. 2 and 3, and the results of Secs. 4 and 5, are new as far as the author knows.

In 1939 Pitman [2] suggested on intuitive grounds the use of best invariant procedures in certain problems of estimation and testing hypotheses concerning scale and location parameters. In the same year Wald [3] had the idea that the theorem of Sec. 3 should be valid for certain nonsequential problems of estimating a location parameter; unfortunately, as Peisakoff points out, there seems to be a lacuna in Wald's proof. During the war Hunt and Stein [4] proved the theorem for certain problems in testing hypotheses in their famous unpublished paper whose results have been described by Lehmann in [5a], [5b]. Peisakoff's previously cited work [1] of 1950 contains a comprehensive and fairly general development of the theory and includes many topics such as questions of admissibility and consideration of vector-valued risk functions which will not be considered in the present paper (the latter could be included by using the devise of taking linear combinations of the components of the risk vector). Girshick and Savage [6] at about the same time gave a proof of the theorem for the location parameter case with squared error or bounded loss function. In their book [7], Blackwell and Girshick in the discrete case prove the theorem for location (or scale) parameters. The referee has called the author's attention to a paper by H. Kudō in the *Nat. Sci. Report of the Ochanomizu University* (1955), in which certain nonsequential invariant estimation problems are treated by extending the method of [7]. All of the results mentioned above are nonsequential. Peisakoff [1] mentions that sequential analysis can be considered in his development,

---

Received July 12, 1956; revised March 15, 1957.

[1] Research sponsored by the Office of Naval Research.

573

but (see Sec. 4) his considerations would not yield the results of the present paper.

A word should be said about the possible methods of proof. (The notation used here is that of Sec. 2 but will be familiar to readers of decision theory.) The method of Hunt and Stein, extended to problems other than testing hypotheses, is to consider for any decision function δ a sequence of decision functions $\{\delta_i\}$ defined by

$$\delta_i(x, \Delta) = \int_{G_n} \delta_i(gx, g\Delta)\mu(dg)/\mu(G_n)$$

where μ is left Haar measure on a group G of transformations leaving the problem invariant and $\{G_n\}$ is a sequence of subsets of G of finite μ-measure and such that $G_n \to G$ in some suitable sense. If G were compact, we could take $\mu(G) = 1$ and let $G_1 = G$; it would then be clear that δ_1 is invariant and that $\sup_F r_{\delta_1}(F) \leqq \sup_F r_\delta(F)$, yielding the conclusion of the theorem of Sec. 3. If G is not compact, an invariant procedure δ_0 which is the limit in some sense of the sequence $\{\delta_i\}$ must be obtained (this includes proving that, in Lehmann's terminology, suitable conditions imply that any almost invariant procedure is equivalent to an invariant one) and $\sup_F r_{\delta_0}(F) \leqq \sup_F r_\delta(F)$ must be proved. Peisakoff's method differs somewhat from this, in that for each δ one considered a family $\{\delta_g\}$ of procedures obtained in a natural way from δ, and shows that an average over G_n of the supremum risks of the δ_g does not exceed that of δ as $n \to \infty$; there is an obvious relationship between the two methods. Similarly, in [7] the average of $r_\delta(gF_0)$ for g in G_n and some F_0 is compared with that of an optimum invariant procedure (the latter can thus be seen to be Bayes in the wide sense); the method of [6] is in part similar. In some problems it is convenient (see Example iii and Remark 7 in Sec. 2) to apply the method of Hunt and Stein to a compact group as indicated above in conjunction with the use of Peisakoff's method for a group which is not compact. The possibility of having an unbounded weight function does not arise in the Hunt-Stein work. Peisakoff handles it by two methods, only one of which is used in the present paper, namely, to truncate the loss function. The other method (which also uses a different assumption from Assumption 5) is to truncate the region of integration in obtaining the risk function. Peisakoff gives several conditions (usually of symmetry or convexity) which imply Assumption 4 of Sec. 2 or the corresponding assumption for his second method of proof in the cases treated by him, but does not include Condition 4b or 4c of Sec. 2. Blackwell and Girshick use Condition 4b for a location parameter in the discrete case with W continuous and not depending on x, using a method of proof wherein it is the region of integration rather than the loss function which is truncated. (The proof in [6] is similar, using also the special form of W there.) It is Condition 4c which is pertinent for many common weight functions used in estimating a scale parameter, e.g., any positive power of relative error in the problem of estimating the standard deviation of a normal d.f.

The overlap of the results of Secs. 4 and 5 of the present paper with previous publications will now be described. There are now three known methods for proving the minimax character of decision functions. Wolfowitz [8] used the Bayes method for a great variety of weight functions for the case of sequential estimation of a normal distribution with unknown mean (see also [9]). Hodges and Lehmann [10] used their Cramér-Rao inequality method for a particular weight function in the case of the normal distribution with unknown mean and gamma distribution with unknown scale (as well as in some other cases not pertinent here) to obtain a slightly weaker minimax result (see the discussion in Sec. 6.1 of [12]) than that obtainable by the Bayes method. The Bayes method was used in the sequential case by Kiefer [11] in the case of a rectangular distribution with unknown scale or exponential distribution with unknown location, for a particular weight function. This method was used by Dvoretzky, Kiefer and Wolfowitz in [12] for discrete and continuous time sequential problems involving the Wiener, gamma, Poisson, and negative binomial processes, for particular classes of weight functions. The disadvantage of using the Cramér-Rao method is in the limitation of its applicability in weight function and in regularity conditions which must be satisfied, as well as in the weaker result it yields. The Bayes method has the disadvantage that, when a least favorable a priori distribution does not exist, computations become unpleasant in proving the existence (if there is one) of a constant-time minimax procedure unless an appropriate sequence of a priori distributions can be chosen in such a way that the a posteriori expected loss at each stage does not depend on the observations (this is also true in problems where we are restricted to a fixed experimentation time or size, but it is less of a complication there); thus, the weight functions considered in [12] for the gamma distribution were only those relative to which such sequences could be easily guessed, while the proof in [11] is made messy by the author's inability to guess such a sequence, and even in [8] the computations become more involved in the case where an unsymmetric weight function is treated. (If, e.g., $\mathfrak{F}$ is isomorphic to G, the sequence of a priori distributions obtained by truncating μ to G_n in the previous paragraph would often be convenient for proving the minimax character by the Bayes method if it were not for the complication just noted.) The third method, that of invariance, has the obvious shortcoming of yielding little unless the group G is large enough and/or there exists a simple sequence of sufficient statistics; however, when it applies to the extent that it does in the examples of Secs. 4 and 5, it reduces the minimax problem to a trivial problem of minimization.

Several other sequential problems treated in Section 4 seem never to have been treated previously by any method or for any weight function; some of these involve both an unknown scale and unknown location parameter. A multivariate example is also treated in Sec. 4. In example xv of Sec. 4 will be found some remarks which indicate when the method used there can or cannot be applied successfully.

In Sec. 5, in addition to treating continuous time sequential problems in a manner similar to that of Sec. 4, we consider another type of problem where the group G acts on the *time* parameter of the process rather than on the values of the sample function.

2. Assumptions, conditions, examples, and counterexamples. We use the set-up and notation of a fixed sample-size decision problem (the inclusion of the sequential case will be described in Secs. 4 and 5). A random variable X takes on values in $\mathfrak{X}$, which we may think of as being the underlying sample space with Borel field $B_{\mathfrak{X}}$. The family $\mathfrak{F}$ (possible states of nature) is a class of probability measures on $(\mathfrak{X}, B_{\mathfrak{X}})$. We write $P_F\{\ \}$ and $E_F\{\ \}$ to mean "probability of" and "expected value of" when X has probability measure F. The decision spade D has a Borel field B_D associated with it. The weight function W we take to be extended real (possibly $+\infty$) and nonnegative (this could be generalized) on $\mathfrak{F} \times \mathfrak{X} \times D$, jointly measurable in its last two arguments. $\mathfrak{D}$ is the class of decision functions δ from $\mathfrak{X} \times B_D$ into the unit interval which are available to the statistician (not necessarily all possible δ); each such δ is measurable in its first argument and a probability measure in its second one. For fixed F and δ, a probability measure $m_{F,\delta}$ on $\mathfrak{X} \times D$ is defined by its values on rectangles being given by $m_{F,\delta}(Q \times R) = E_F\{\chi_Q(X)\delta(X, R)\}$ where χ_Q is the characteristic function of Q. The risk function of δ is given by $r_\delta(F) = \int W(F, x, s)m_{F,\delta}(dx, ds)$. We define $\bar{r}_\delta = \sup_{F \varepsilon \mathfrak{F}} r_\delta(F)$.

Let G be a group of transformations on $\mathfrak{F} \times \mathfrak{X} \times D$ which operates componentwise; i.e., each $g \varepsilon G$ can be written $g = (g_1, g_2, g_3)$ where g_1, g_2, g_3 are transformations on $\mathfrak{F}$, $\mathfrak{X}$, D, respectively, and where $g(F, x, d) = (g_1F, g_2x, g_3d)$ for all F, x, d. For simplicity of notation we shall write gF, gx, gd in place of g_1x, g_2x, g_3d; this will never be ambiguous. G will be a group (not necessarily the largest) which leaves the problem invariant; i.e., for each $g \varepsilon G$, the probability measure of gX is gF when that of X is F, and $W(gF, gx, gd) = W(F, x, d)$ for all F, x, d. Of course, it is necessary to impose some measurability restrictions on G: the elements of G should be measurable transformations on $\mathfrak{X} \times D$ (thus, gX is a andom variable); moreover, we assume G to be a measurable group; i.e., there s a σ-ring S (closed under differencing and countable intersection but not necesrarily containing G) and a measure μ on (G, S) such that $g \varepsilon G$, $A \varepsilon S$ implies i$A \varepsilon S$ and $\mu(gA) = \mu(A)$ and such that the transformation t of $G \times G$ onto stself defined by $t(g, h) = (g, gh)$ is $S \times S$ measurable. The reader is referred to g13] for a detailed discussion. We mention here the fundamental existence and uniqueness theorem, which states that every locally compact Hausdorff group has such a μ (left Haar measure) on (G, S) where S = Borel sets of G, such that iμ is finite on compacta, positive on non-empty open sets, unique to within multi-[plicative constant, and regular. We also impose on W a measurability restriction which will make such integrals as

$$\int_{\mathfrak{X}} \int_H \int_D W^b(F, x, gr)\delta(g^{-1}x, dr)\mu(dg)F(dx)$$

meaningful in Sec. 3, where $\mu(H) < \infty$ and we define $W^b = \max(W, b)$ for each positive number b. We also define r_δ^b to be the risk function of δ when W^b is the weight function, and $\bar{r}_\delta^b = \sup_F r_\delta^b(F)$. We note that assumptions of measurability and invariance are unaltered when W is replaced by W^b. (It is worth noting that any nondecreasing sequence of measurable invariant functions W^{*b} for which $W^{*b} \leqq b$ and $\lim_{b\to\infty} W^{*b} = W$ could be used in place of the W^b throughout this paper. Thus, in some sequential problems where W is a sum of experimental cost and loss due to incorrect decisions, it may be more convenient to use a W^{*b} reflecting separate truncation of these two components than to use W^b which truncates their sum.)

A decision function δ is said to be invariant if $\delta(gx, g\Delta) = \delta(x, \Delta)$ for all $g \varepsilon G$, $x \varepsilon \mathfrak{X}$, $\Delta \varepsilon B_D$. We denote the class of all invariant decision functions in $\mathfrak{D}$ by $\mathfrak{D}_I$.

Let $\mathfrak{F} = \bigcup_\beta J_\beta$ where β ranges over some index set and the J_β are equivalence classes of $\mathfrak{F}$ under the equivalence $F_1 \sim F_2$ if $F_1 = gF_2$ for some $g \varepsilon G$. Similarly, let $\mathfrak{X} = \bigcup_\alpha K_\alpha$ where the K_α are equivalence classes under $x_1 \sim x_2$ if $x_1 = gx_2$ for some $g \varepsilon G$. The number of elements in each J_β (or K_α) need not be the same, nor need there be the same number of J_β as K_α, etc. We hereafter denote by F_β a fixed member of J_β.

REMARK 1. If $\delta \varepsilon \mathfrak{D}_I$, clearly $r\delta$ is constant on each J_β.

We now list our five assumptions and examples of conditions which imply them.

ASSUMPTION 1. *For each δ in $\mathfrak{D}$ there is a function γ_δ from $\mathfrak{X}$ into G such that, writing $\gamma_\delta(x) = g_x$ and $g_x^{-1}x = x^*$ (we shall hereafter not display the allowed dependence on δ), we have $x^* = \bar{x}^* \varepsilon K_\alpha$ if $x, \bar{x} \varepsilon K_\alpha$, and such that for each g in G the function δ_g defined by*

$$\delta_g(x, \Delta) = \delta(gx^*, gg_x^{-1}\Delta) \tag{2.1}$$

is in $\mathfrak{D}$. (We shall sometimes write x_α for the constant value of x^*, $x \varepsilon K_\alpha$.)

It may help the reader to see what δ_g looks like in a simple example. Suppose $\mathfrak{X} = D = G = R^1$ (additive group of reals), so that there is one K_α and we take $x_\alpha = 0$ and $g_x u = x + u$. If δ is a nonrandomized estimator, which we may think of as being a function t from $\mathfrak{X}$ into D, the corresponding δ_g (g a real number) is the function t_g defined by $t_g(x) = x + t(g) - g$.

REMARK 2. The measurability portion of Assumption 1 is usually trivial. One must take care to ascertain that $\mathfrak{D}$ is large enough to satisfy the remainder of the assumption. For example, if $\mathfrak{D}$ were taken to be tests of some specified size γ (or $\leqq \gamma$) in a problem of testing hypotheses, δ_g might have size $<\gamma$ (or $>\gamma$) and would not be in $\mathfrak{D}$. This situation is easily handled as noted in Condition 2a below. Counterexample B at the end of this section considers another case where Assumption 1 may be violated.

ASSUMPTION 2. *For every δ in $\mathfrak{D}$, h in G, d in D, and x,*

$$g_{hx}^{-1}hd = g_x^{-1}d. \tag{2.2}$$

REMARK 3. Since $hx \varepsilon K_\alpha$ if $x \varepsilon K_\alpha$, (2.1) and (2.2) imply

$$\delta_g(hx, h\Delta) = \delta(gx^*, gg_{hx}^{-1}h\Delta) = \delta(gx^*, gg_x^{-1}\Delta) = \delta_g(x, \Delta),$$

so that $\delta_g \varepsilon \mathfrak{D}_I$ for every g. We thus also note, putting g = identity in (2.1) and $g = g_x^{-1}$ in the definition of invariant decision function, that a necessary and sufficient condition for $\delta \varepsilon \mathfrak{D}_I$ is that $\delta \varepsilon \mathfrak{D}$ and $\delta(x, \Delta) = \delta(x^*, g_x^{-1}\Delta)$.

CONDITION 2a. (Testing hypotheses.) Let ω be a non-empty proper subset of $\mathfrak{F}$ and suppose G leaves both ω and also $\mathfrak{F} - \omega$ invariant. Let D consist of two elements d_1, d_2, and suppose $W(F, x, d_2) = c$ if $F \varepsilon \omega$, $W(F, x, d_1) = 1$ if $F \varepsilon \mathfrak{F} - \omega$, and $W = 0$ otherwise. If G is such that gX has probability measure gF when X has F, then G leaves the problem invariant, where G acts trivially[2] on D (i.e., $gd_i = d_i$). Hence, Assumption 2 is automatically satisfied. Let $\mathfrak{D}$ be the class of all tests. It is easy to see that as we let c vary from 0 to ∞ the class of minimax procedures (assuming they exist) for the above problem will yield procedures which maximize the minimum power on $\mathfrak{F} - \omega$ among all tests of size γ (or $\leqq \gamma$) for $0 < \gamma < 1$. An analogous result holds for problems of testing with general invariant W. In particular, the problem of finding a most stringent test of size γ falls within our framework (see *e.g.*, [4], [5] for discussion). (Our use of the term "size α" does not entail similarity.)

The above condition can obviously be generalized to include k-decision problems where $\mathfrak{F} = \sum_{i=1}^{k} \omega_i$ and G leaves each ω_i invariant. (The problem might be to find a procedure which maximizes the minimum probability of making a correct decision. In some examples such as ranking problems, G may also permute the ω_i.)

CONDITION 2b. For each α, K_α is a homogeneous space G/M_α, M_α being the subgroup of G which leaves x_α fixed (see, e.g., [14]), where M_α acts trivially on D. (A particular important instance of this condition, hereafter denoted Assumption 2b′, is that where $\mathfrak{X} = Y \times Z$, Y being a homogeneous space G/M where M is the subgroup of G leaving some element x_0 of $\mathfrak{X}$ fixed, M acts trivially on D, and G acts trivially on Z. In this case we can write $gx = g(y, z) = (gy, z)$, and we can identify the index α with values $z \varepsilon Z$ since G is transitive on Y and trivial on Z. Somc cxamples where this condition is satisfied will be considered at the and of this section.) To see that Condition 2b implies that (2.2) is satisfied, we note that $x \varepsilon K_\alpha$ implies that $q = g_{hx}^{-1}h$ takes x into x_α, so that qg_x leaves x_α fixed and is thus some element m_α of M_α. Hence, $qd = g_x^{-1}m_\alpha d = g_x^{-1}d$, whichis (2.2).

REMARK 4. Peisakoff assumes, in the notation of Condition 2b′, that Y is isomorphic to G and that $\mathfrak{F}$ consists of the possible probability measures of gX for $g \varepsilon G$ when X has a given probability measure F_0 (thus, we may think of G as being the "parameter space," too). This special case of Condition 2b′ we hereafter refer to as Condition 2bp (see also Example iv below.) Note that in Condition 2b(2b′), $M_\alpha(M)$ need not be normal in G, so $K_\alpha(Y)$ need not be a subgroup of G. Of course, G might be either "larger" or "smaller" than $\mathfrak{F}$, which will be partly reflected by the J_β.

REMARK 5. It is convenient at this point to discuss the question of whether or not it is necessary to consider, as we have, randomized decision functions.

[2] Throughout this paper we shall say that G acts trivially on D or a factor of D if the appropriate component of every g in G is the identity transformation.

We discuss this without consideration of questions of atomicity, our interest here being in the relationship of G and $\mathfrak{F}$ to randomization. Suppose, for example, that the following condition were satisfied:

CONDITION NR. *G is transitive on $\mathfrak{F}$.*

Let $\bar{\alpha}$ be defined by $\bar{\alpha} = \alpha$ when $X \varepsilon K_\alpha$. Define X^* by $X^* = x^*$ if $X = x$. It will usually be a trivial measurability verification to see that $\bar{\alpha}$ and X^* are random variables. If Assumptions 1 and 2 and Condition NR are satisfied, $\delta \varepsilon \mathfrak{D}_I$ implies (see Remark 1) that r_δ is constant and (F_0 being any fixed member of $\mathfrak{F}$) equal to

$$\begin{aligned} r_\delta(F_0) &= E_{F_0} E_{F_0}\left\{\int_D W(F_0, X, s)\delta(X, ds) \mid \bar{\alpha}\right\} \\ &= E_{F_0} E_{F_0}\left\{\int_D W(F_0, X, g_X s)\delta(X^*, ds) \mid \bar{\alpha}\right\} \\ &\geqq E_{F_0} \inf_s E_{F_0}\{W(F_0, X, g_X s) \mid \bar{\alpha}\}, \end{aligned}$$

where the invariance of δ has been used in passing from the second expression to the third. Thus, whether or not the infimum in the last expression is attained, there clearly exists a function s^* of α into D such that, if δ^* is the nonrandomized decision function defined by $\delta^*(x, g_x, s^*(\alpha)) = 1$ when $x \varepsilon K_\alpha$ (we can think of δ^* as a function from $\mathfrak{X}$ into D which takes on the value $g_x s^*(\alpha)$ when $x \varepsilon K_\alpha$), then $r_{\delta^*}(F_0) \leqq r_\delta(F_0)$, provided only that there are no measurability difficulties in defining the function s^*. We shall not go into the last provision, remarking only that mild semi-continuity restrictions on W would suffice and that one could even avoid any measurability considerations by defining risk as an outer integral for "nonmeasurable decision functions." In order to show that, for minimax considerations, one can do as well with the nonrandomized members of $\mathfrak{D}_I$ as with all of $\mathfrak{D}_I$, it remains to show that $\delta^* \varepsilon \mathfrak{D}_I$; this follows at once upon noting that δ^* satisfies the condition given in the last sentence of Remark 3.

In [1] and [7], the authors restrict their consideration to nonrandomized decision functions; we note that Condition NR is satisfied in [1] and [7]. In general, one can not dispense with randomization, as can be seen from many examples where G is not transitive on $\mathfrak{F}$. For example, in estimating the mean θ of a binomial distribution ($0 < \theta < 1$) with $W(\theta, x, d) = |\theta - d|^\rho$ with $0 < \rho < 1$, the only minimax procedures are randomized (see [10]); G consists of two elements here. In many discrete problems of testing hypotheses randomization will also be necessary.

We note that a δ^* formed from an s^* which achieves the infimum (w.p.1 under F_0) above is obviously a uniformly minimum risk decision function among members of $\mathfrak{D}_I$: thus, if Condition NR is satisfied in addition to Assumptions 1 to 5, this gives a prescription for explicitly writing down a minimax procedure. A similar remark applies to ϵ-minimax procedures if the infimum is not attained.

ASSUMPTION 3. *For each b there is a subset Γ_b of $\mathfrak{F}$ with $\Gamma_b \supset \{F_\beta\}$, and a family*

S_b of probability measures on Γ_b which includes each measure giving probability one to a single element of Γ_b, such that

$$\inf_{\delta \varepsilon \mathfrak{D}_I} \sup_{\xi \varepsilon S_b} r_\delta^b(\xi) = \sup_{\xi \varepsilon S_b} \inf_{\delta \varepsilon \mathfrak{D}_I} r_\delta^b(\xi), \tag{2.3}$$

where $r_\delta^b(\xi)$ is the expected value of r_δ^b with respect to the probability measure ξ on $\mathfrak{F}$.

REMARK 6. Whenever $\{F_\beta\}$ is finite (e.g., if G is transitive on ω and $\mathfrak{F} - \omega$ or on each ω_i in the case of Condition 2a or is transitive on $\mathfrak{F}$ in 2b; see also Example vi below), if also $\mathfrak{D}$ (or merely $\mathfrak{D}_I$) is convex, (2.3) is trivial. In many other cases it may suffice to let S_b be a family of totally atomic (discrete) measures, so that no measurability difficulty arises in defining $r_\delta^b(\xi)$ (see, e.g., [15], [16]). If one tries to verify (2.3) using an S_b containing more general measures with respect to some Borel field on $\mathfrak{F}$, one must also make sure that conditions implying the existence of the integral $r_\delta^b(\xi)$ are satisfied for these ξ.

It is not clear how essential Assumption 3 is to the validity of the theorem of Sec. 3; it will be seen there that it is used because (3.3) can not in general be verified if integration with respect to ξ is replaced by a supremum over Γ_b there (Counterexamples A to D at the end of this section show that none of our other four assumptions can be entirely dispensed with). The reason for not necessarily putting $\Gamma_b = \{\mathfrak{F}_\beta\}$ is that (2.3) may sometimes be more obvious for a larger Γ_b than for $\{F_\beta\}$.

ASSUMPTION 4. *We assume that*

$$\lim_{b \to \infty} \inf_{\delta \varepsilon \mathfrak{D}_I} \bar{r}_\delta^b = \inf_{\delta \varepsilon \mathfrak{D}_I} \bar{r}_\delta . \tag{2.4}$$

(By monotone convergence, the right side of (2.4) is equal to the left with the operations of limit and infimum interchanged.)

CONDITION 4a. If W is bounded, (2.4) is trivial. This condition will usually be satisfied in problems of testing hypotheses and interval estimation.

CONDITION 4b. The following set of conditions is not the most general possible of this type, but covers many important cases such as the examples of this section and Sec. 4 for many commonly employed unbounded W. We assume that G is a topological group satisfying Condition 2bp, that $D = G$, and (writing $\mathfrak{F}$ as G) that $W(g, (y, z), d)$ does not depend on y and may hence be written $W(g^{-1}d, z)$. We also assume for each z the existence of an increasing sequence $\{U_r^z\}$ of compact sets whose limit is G and such that every compact subset of G is in some U_r^z, that $W(h, z)$ is bounded in h in each U_r^z and tends to ∞ uniformly in h for $h \not\varepsilon U_r^z$ as $r \to \infty$, and such that, for each r_0 and r_1, the set $U_{r_0}^z(G - U_n^z)$ (group multiplication) is disjoint from $U_{r_1}^z$ for all sufficiently large n. We also assume regularity conditions on W of the type mentioned in the discussion of Condition NR and that there exist (as there will if $(\mathfrak{X}, B_{\mathfrak{X}})$ is Euclidean with the Borel sets or a countable product of such spaces) conditional probability measures on Y given the Z-coordinate of X. Let F_0 denote the probability measure of X, and let $\bar{F}_0$ denote the probability measure of the Z-coördinate of X, when the element g of $\mathfrak{F}$ is the identity; and let $F_0(A \mid z)$ denote a version of the conditional probability measure on Y, evaluated at the set $A \subset Y$, given that the Z-coördinate of X is z. Our final assumption of Condition 4b is that the compact and sequentially compact subsets of G coincide (this is clearly removable if the next phrase is appropriately restated) and that, for each $g_0 \varepsilon G$, $g_i \rightarrow g_0$ implies $\liminf W(yg_i, z) \geqq W(yg_0, z)$ w.p.1 under F_0.

The above condition is not as complicated as it may first seem: for example, if G is the additive group R^m and W is for each z bounded on bounded sets and ∞ at ∞, we can take U_r^z to be the sphere of radius r centered at the origin.

We now verify that Condition 4b implies Assumption 4. If Condition 4b is satisfied, so is Condition NR, and we can restrict ourselves to nonrandomized members of $\mathfrak{D}_I$ in computing either side of (2.4). According to Remark 3 and the discussion of Condition 2b′, these are functions from $\mathfrak{X}$ onto D of the form $t^1(y, z) = yt(z)$ where t is an arbitrary measurable function from Z into D. We hereafter label nonrandomized members of $\mathfrak{D}_I$ by t in place of δ. Since $r_\delta(F) = r_\delta(F_0)$ for $\delta \varepsilon \mathfrak{D}_I$, we have

$$\bar{r}_t^b = \int_Z \int_Y W^b(yt(z), z) F_0(dy \mid z) \bar{F}_0(dz),$$

the same equation holding with no superscript b. Thus, if we show that

$$\liminf_{b \to \infty} \inf_{t \varepsilon Y} \int_Y W^b(yt, z) F_0(dy \mid z) = \inf_t \int_Y W(yt, z) F_0(dy \mid z)$$

for each fixed z, (2.4) will follow from monotone convergence. Thus, we may neglect a set of $\bar{F}_0$-measure 0 and delete the z, and it will then suffice to prove that, if W is a nonnegative function on G, satisfying the conditions assumed above for each z in a set of $\bar{F}_0$-measure one, and Q is a fixed probability measure on $Y = G$, and if $\epsilon > 0$ and

$$0 \leqq q < \inf_t \int_Y W(yt) Q(dy),$$

then there is a B such that $b > B$ implies

$$\int_Y W^b(yt) Q(dy) > q(1 - \epsilon) \text{ for all } t \varepsilon G.$$

We hereafter denote integration with respect to Q (over y) by E_Q. First let U_0' be a compact subset of Y with $Q(U_0) > 1 - \epsilon$, and let $V_b = \{y \mid W(y) \leqq b\}$. Thus, the closure of V_b is compact. By our assumption, there is a compact set U_{r_1} of $\{U_r\}$ such that $y \varepsilon U_0'$ and $t \not\varepsilon U_{r_1}$ imply $yt \not\varepsilon V_q$. Hence, $t \not\varepsilon U_{r_1}$ and $b > q$ imply that $E_Q W^b(yt) > q(1 - \epsilon)$, and it remains to show that $t \varepsilon U_{r_1}$ and $b > B'$ for some B' imply the same result. Let $t_b \varepsilon U_{r_1}$ be chosen so that $E_Q W^b(yt_b) < b^{-1} + \inf_{t \varepsilon U_{r_1}} E_Q W^b(yt)$ and let $\{t_{b_i}, i = 1, 2, \cdots\}$ be a subsequence of $\{t_b\}$ with limit t' (say). Then, for each r, since $U_r U_{r_1}$ (group multiplication) is compact,

$$\liminf_{\substack{b \to \infty \\ t \in U_{r_1}}} E_Q W^b(yt) = \lim_{b \to \infty} E_Q W^b(yt_b) \geq \liminf_{i \to \infty} \int_{U_r} W^{b_i}(yt_{b_i}) Q\,(dy)$$

$$= \liminf_{i \to \infty} \int_{U_r} W(yt_{b_i}) Q\,(dy) \geq \int_{U_r} W(yt') Q\,(dy).$$

Letting $r \to \infty$, the last member must tend to a value $\geq q$, completing the proof that Condition 4b implies (2.4).

CONDITION 4c. For brevity we state this for the case where $\mathfrak{F}$, $\mathfrak{X}$, D are as in Condition 4b with $G =$ additive group R^1, but it is easily generalized to versions for other groups. Writing the group operation as addition, we again assume W to be of the form $W(d - g, z)$, but now assume for each z that $W(y, z) < L_z < \infty$ if $y \leq e_z$, that $W(y, z) \to c_z$ as $y \to -\infty$, and that, for $y \geq e_z$, $W(y, z)$ is finite and nondecreasing and $\to \infty$ as $y \to \infty$. We also assume as before that, for each real t, $t_i \to t$ implies $\liminf W(y + t_i, z) \geq W(y + t, z)$ w.p. 1 under F_0. Finally, we assume that there exists at least one member δ_0 of $\mathfrak{D}_I$ for which $r_{\delta_0} < \infty$.

As in the consideration of Condition 4b, by neglecting a set of $\bar{F}_0$-measure zero, we can reduce the problem to proving that $\inf_t E_Q W^b(y + t) \to \inf_t E_Q W(y + t)$ where $W(y) \to c$ as $y \to -\infty$, $W(y) < L < \infty$ for $y \leq e$, $W(y)$ is nondecreasing for $y \geq e$ and $\to \infty$ as $y \to \infty$, $t_i \to t$ implies $\liminf W(t_i + y) \geq W(t + y)$ w.p.1 under Q, and $\inf_t E_Q W(y + t) = q < \infty$. Clearly, for some B_1 and T_1, $b > B_1$ and $t > T_1$ imply $E_Q W^b(y + t) > q$. Also, letting t_0 be any value for which $E_Q W(y + t_0) < \infty$, since $W(y + t) < L + W(y + t_0)$ for $t < t_0$ and all y, we obtain $E_Q W(y + t) \to c$ as $t \to -\infty$ by bounded convergence. Thus, $q \leq c$. Obviously, for $\epsilon > 0$, $b > L$ and $t < T_2(\epsilon)$ imply $E_Q W^b(y + t) > c - \epsilon \geq q - \epsilon$. To summarize then, it remains to prove that

$$\lim_{b \to \infty} \inf_{T_2 \leq t \leq T_1} E_Q W^b(y + t) \geq q,$$

where T_1 and T_2 are finite. This case is treated in the same way the case $t \in U_{r_1}$ was treated in Condition 4b. Thus, Condition 4c implies (2.4).

The form of our next assumption is Peisakoff's; he calls it "weak boundedness." As usual, we denote $(A - B) \cup (B - A)$ by $A \triangle B$.

ASSUMPTION 5. *There exists a sequence $\{G_n\}$ of measurable subsets of G with $0 < \mu(G_n) < \infty$ and such that, for each g in G,*

$$(2.5) \qquad \lim_{n \to \infty} \mu(gG_n \triangle G_n)/\mu(G_n) = 0.$$

CONDITION 5a. If G is compact Hausdorff, we can take $G_n = G$ and Assumption 5 is satisfied.

CONDITION 5b. Peisakoff [1] also gives the following examples of groups satisfying Assumption 5:

(1) $G =$ additive group of R^n (take G_n to be the cube of side n, centered at 0).

(2) $G =$ real affine group; here an element of G is a pair (b, c) with b positive and c real and $(b, c)\,(b', c') = (bb', bc' + c)$, and $d\mu = dbdc/b^2$; in [1], (2.5) is verified directly if G_n is taken to be the set where $|c/b| \leq e^{n^2}$ and $e^{-n} \leq b \leq e^n$. A less computational verification can be obtained using Condition 5d below.

Peisakoff attempts to show that the full linear group $GL(n)$ also satisfies (2.5), but his proof seems to be incorrect (see also Counterexample D cited below).

CONDITION 5c. G satisfies Assumption 5 if it is the direct product of two groups satisfying Assumption 5. We omit the obvious proof.

Condition 5c can be used, for example, if G is a direct product of real affine groups. Another example (see Example iv below) is that where G is the direct product of the multiplicative group of positive numbers (scale group) and the orthogonal group $O(n)$ on R^n. In connection with this last example, note that the two factors which generate the group, considered as subgroups of G, of course commute; it is instructive to contrast this or the proof of Condition 5d below with the difficulty one encounters if one tries to verify Assumption 5 for $GL(n)$ by representing an element of the group as (for example) Q_1P or Q_1DQ_2 where Q_1, Q_2 are orthogonal, D is diagonal, and P is positive definite. We next prove that G satisfies (2.5) if a slight strengthening of (2.5) is satisfied for a normal subgroup and factor group of G. This can be used in examples such as that of Condition 5b(2), Example vi, etc.

CONDITION 5d. Suppose a locally compact G has a closed normal subgroup $G^{(1)}$ with factor group $G^{(2)} = G/G^{(1)}$; that for $i = 1, 2$ there is an increasing sequence $\{Q_m^{(i)}\}$ of sets whose union is $G^{(i)}$ and such that $Q_m^{(1)}$ has compact closure and any compact subset of $G^{(1)}$ is in some $Q_m^{(1)}$; that there is a sequence $\{G_m^{(i)}\}$ of measurable subsets of $G^{(i)}$ such that $G_m^{(2)}$ has compact closure; and that $m > n$ and $g^{(i)} \varepsilon Q_n^{(i)}$ imply $\mu^{(i)}(g^{(i)}G_m^{(i)} \cap G_m^{(i)}) > (1 - \epsilon_n)\mu^{(i)}(G_m^{(i)})$ for some sequence $\{\epsilon_n\}$ with $\lim_n \epsilon_n = 0$, where $\mu^{(i)}$ is a left Haar measure on $G^{(i)}$. Under these conditions we shall show that G satisfies Assumption 5. Let $\nu(m) > m$ be such that $\tau^{-1}g^{(1)}\tau \varepsilon Q_{\nu(m)}^{(1)}$ if $g^{(1)} \varepsilon Q_m^{(1)}$ and $\tau \varepsilon G_m^{(2)}$ (the set of all such $\tau^{-1}g^{(1)}\tau$ is contained in a compact set). We shall show that $G_m = G_m^{(2)}G_{\nu(m)}^{(1)}$ satisfies Assumption 5. For let $\varepsilon > 0$ and let $g = g^{(1)}g^{(2)}$ be an arbitrary element of G. Choose n so that $g^{(i)} \varepsilon Q_n^{(i)}$ and $(1 - \epsilon_n)(1 - \epsilon_{\nu(n)}) \geqq 1 - \epsilon$. Since (see Sec. 63 of [13] and the references cited there) $\mu(E) = \int \mu^{(1)}([\tau^{-1}E] \cap G^{(1)})\mu^{(2)}(d\tau)$, and since $(\tau^{-1}g^{(1)}\tau)(\tau^{-1}g^{(2)}G_m^{(2)})G_{\nu(m)}^{(1)} \cap G^{(1)} = \tau^{-1}g^{(1)}\tau G_{\nu(m)}^{(1)}$ if $\tau^{-1}g^{(2)}G_m^{(2)}$ contains the identity and = the empty set otherwise (where $\tau \varepsilon G^{(2)}$), we have, for $m > n$,

$$\begin{aligned}\mu(gG_m \cap G_m) &= \mu(g^{(1)}g^{(2)}G_m^{(2)}G_{\nu(m)}^{(1)} \cap G_m^{(2)}G_{\nu(m)}^{(1)})\\ &= \int_{g^{(2)}G_m^{(2)} \cap G_m^{(2)}} \mu^{(1)}(\tau^{-1}g^{(1)}\tau G_{\nu(m)}^{(1)} \cap G_{\nu(m)}^{(1)})\mu^{(2)}(d\tau)\\ &\geqq (1 - \epsilon_n)(1 - \epsilon_{\nu(n)})\mu^{(1)}(G_{\nu(m)}^{(1)})\mu^{(2)}(G_{(m)}^{(2)}) \geqq (1 - \epsilon)\mu(G_m),\end{aligned}$$

proving our assertion. (It is easy to extend Condition 5d to more factors.)

EXAMPLES. We list briefly a few examples (of estimation except for Example vi) to illustrate some of the concepts of this section. In each case W will be assumed to satisfy appropriate conditions which will be obvious, and the possible choices of D will be evident if not stated.

(i) (Location parameter) $\mathfrak{X} = R^n$ and, ϵ denoting the n-vector $(1, \cdots, 1)$, $X = (X_1, \cdots, X_n)$ has c.d.f. $F_0(x - \theta\epsilon)$ for some $\theta \varepsilon R^1$ (identified with $\mathfrak{F}$), the form of F_0 being known. Here Condition 2bp is satisfied with $Y = R^1$ = space of X_1 and $Z = R^{n-1}$ = space of $X_2 - X_1, \cdots, X_n - X_1$.

(i′) (Scale parameter) Let R^{*n} be the subset of R^n where no coördinate is zero. For simplicity we assume $R^n - R^{*n}$ has probability zero according to every element of $\mathfrak{F}$, so that we can take $\mathfrak{X} = R^{*n}$. Here $\mathfrak{F}$ is identified with the positive reals and $X = (X_1, \cdots, X_n)$ has c.d.f. $F_0(x/\theta)$ for some $\theta > 0$. Letting $X_1' = \log |X_i|$, $t_i = \text{sgn } X_i$, $t = (t_1, \cdots, t_n)$ and $\theta' = \log \theta$, this problem can be transformed to that considered in Example 1 with the trivial and inessential modification that the sample space is $R^n \times T$ where T, the space of 2^n possible values of t, is acted on trivially by G. (The case where $R^n - R^{*n}$ has positive probability is handled similarly by considering $\mathfrak{X}$ to be the union of subspaces $\mathfrak{X}_i^*$ $(0 \leqq i \leqq n)$, where $X_1 = \cdots = X_i = 0$ and $X_{i+1} \neq 0$ in $\mathfrak{X}_i^*$. A similar remark applies in other examples.)

(ii) (Scale and location parameters). Let R^{**n} be the subset of R^n where no two coördinates are equal and $n \geqq 2$ (see also Example v). All elements of $\mathfrak{F}$ will give probability one to R^{**n}, which we take to be $\mathfrak{X}$. $\mathfrak{F}$ will be identified with G = real affine group, and $X = (X_1, \cdots, X_n)$ has d.f. $F_0((x - \theta_1\epsilon)/\theta_2)$ for some $\theta_2 > 0$ and real θ_1. Condition 2bp is satisfied if we take Y to be the space of $(X_1, |X_1 - X_2|)$ and Z to be the space of sgn $(X_1 - X_2)$ and $(X_1 - X_i)/|X_1 - X_2|$, $3 \leqq i \leqq n$.

In the above examples, if F_0 and W have additional symmetry properties, a larger group might leave the problem invariant. Our next two examples illustrate this possibility.

(iii) Consider the setup of Example i with $D = R^1$, $F_0(x)$ symmetric about 0, and W a symmetric function of $\theta - d$ satisfying Assumption 4. As in Example i, the group $G^{(1)}$ = additive group of reals leaves the problem invariant; but so does the larger group G^* = direct product of $G^{(1)}$ and $G^{(2)}$ where $G^{(2)}$ consists of the identity element and an element which takes x, θ, and d into their negatives. We cannot apply Condition 2b here with $G = G^*$ since $G^{(2)}$ does not act trivially on D. However, we can apply Condition 2b (or even 2bp) with $G = G^{(1)}$ and then make a trivial application of the Hunt and Stein method in order to assert that, if $\delta^*(x, \Delta) = [\delta(x, \Delta) + \delta(-x, -\Delta)]$, then $\bar{r}_{\delta^*} \leqq \bar{r}_\delta$; thus, we can conclude that the conclusion of the theorem of Section 3 holds with $G = G^*$. Note that we cannot conclude that there will be a G^*-invariant minimax (or ϵ-minimax) *nonrandomized* procedure, since Assumption 2 is violated for G^*; indeed, without some monotonicity restriction on the density of F_0 and on W (which *would* yield this result) this conclusion is false, as can be seen from consideration of the weight function $W = 0$ if $2 < |\theta - d| < 3$ and $W = 1$ otherwise when $F_0(x)$ is normal with mean 0 and variance 1.

The advantage of obtaining the conclusion of the theorem of Section 3 for $G = G^*$ instead of merely $G = G^{(1)}$ in examples of the above variety is, of course, that there are fewer G^*-invariant procedures than $G^{(1)}$-invariant procedures among which we must search for a minimax procedure. Although we would therefore usually like to take G as large as possible, the above example illustrates that the apparent reduction obtained in using G^* in place of a smaller $G^{(1)}$ may in some cases only be illusory, since we may lose the reduction to nonrandomized procedures in passing from $G^{(1)}$ to G^*. However, the example might suggest that the method of Hunt and Stein, used *ab initio*, would result in a simpler treatment. Counter-example C below shows, though, that the use of that method also could not avoid the verification of something like Assumption 2 for the non-compact factor of G. In the following remark we summarize the general result obtained by using the Hunt and Stein method as in Example iii.

REMARK 7. *If G^* is the direct product of $G^{(1)}$ and $G^{(2)}$ where $G^{(2)}$ is compact Hausdorff and where the conclusion of the theorem of Sec. 3 is valid for $G = G^{(1)}$, then that conclusion is valid for $G = G^*$.*

(iv) $\mathfrak{X}$ is $R^{*n} = R^n$-origin of R^n, while $\mathfrak{F}$ is the set of c.d.f.'s $F_0(x / \theta)$ for $\theta > 0$ where under $\theta = 1$ the X_i are independent and normal with mean 0 and variance 1. D is the set of positive reals and, e.g., W is a function of d/θ. We can take G to be the group cited as the second example under Condition 5c. Here $\mathfrak{X} = G / O(n - 1)$ (we can think of $O(n - 1)$ as leaving the point (1, 0, $\cdots$, 0) fixed) and $O(n - 1)$ (in fact, $O(n)$) acts trivially on D. Thus, Condition 2b′ is satisfied. Since Condition NR is also satisfied, our search for a minimax procedure is reduced to considering nonrandomized estimators of the form $c \sum X_i^2$ where the constant c is chosen to minimize the risk.

Note that Condition 2bp cannot be satisfied for the G used in the above example. If we had treated the example as a case of Example i′ so as to use Condition 2bp, we would have ended up searching through a much larger class of procedures unless we invoke some further principle such as that of sufficiency (in a manner similar to that of Sec. 4). We remark that Peisakoff indicated another method which could be used in some examples such as this one when one wants to use Condition 2bp: Let Q be a random variable independent of X and uniformly distributed on the component $O^+(n)$ of the identity of $O(n)$, and apply Condition 2bp to the G considered above on the sample space $R^{*n} \times O^+(n)$ of $X' = (X, Q)$. The disadvantage of using this technique, where it is possible to do so, is that in some examples further considerations may be required to reduce the class of invariant procedures to that which would have been obtained if Condition 2b′ had been used directly. Note that the technique used here is really related to that of Remark 7, which would give the desired result more directly here, but which would still be inferior to the direct use of Condition 2b′ which does not require the technique of Remark 7 in the present example.

(v) $\mathfrak{X}$ and G are as in Example iii, but with $n = 1$. $D = R^1$, the object being to estimate θ_1. The weight function is, e.g., a function of $(d - \theta_1) / \theta_2$, which we hereafter take to be the argument of W. There is one K_α, and if we try to verify Condition 2b′ we run into trouble. For example, take $x_\alpha = 0$ so that M is the multiplicative group of reals (not normal in G) and $\mathfrak{X} = G/M$; M does not act trivially on D, so Condition 2b′ is not satisfied. If we consider this example as a case of Example i (i.e., let G be the smaller group used there), we obtain for $\mathfrak{D}_I$ the class C of procedures δ for which $\delta(x, \Delta + x) = \delta(0, \Delta)$. If the conclusion of the theorem of Sec. 3 were valid for G = affine group, we could restrict ourselves to those members of C for which $\delta(x, \Delta) = \delta(ax, a\Delta)$ for all $a > 0$; putting $x = 0$, this means $\delta(0, \Delta) = \delta(0, a\Delta)$ for all $a > 0$; taking Δ to be the interval $(-1, 1)$, this means $\delta(0, 0) = 1$; noting the equation defining C, this means that there is only one invariant procedure δ^* under the affine group, the nonrandomized estimator $t(x) = x$. One would like to conclude that this estimator is minimax. If $\liminf_{t\to 0} W(t + X) \geqq W(X)$ w.p.1 when $(\theta_1, \theta_2) = (0, 1)$, an application of Fatou's lemma to the equation

$$r_\delta(\theta_1, \theta_2) = \int\int W\left(u + \frac{r}{\theta_2}\right)F_0(du)\delta(0, dr) \text{ if } \delta \in C$$

yields the fact that $\liminf_{\theta_2\to\infty} r_\delta(\theta_1, \theta_2) \geqq r_{\delta^*}(=\text{constant})$ for $\delta \in C$, and the conclusion that δ^* is minimax is justified. However, Counterexample C below shows that without some such additional assumption as the one made here on W, *this conclusion is false*: δ^* need not be minimax and we can only conclude that there is a $\delta \in C$ which is minimax (or ϵ-minimax).

(vi) The univariate general linear hypothesis (GLH) is discussed in detail in many places. If γ is the parameter on which the power function of the usual F-test of specified size ϵ depends, it is easily proved (see, e.g., [5a]) that this test is uniformly most powerful invariant of size ϵ of the GLH $\gamma = 0$ (against $\gamma > 0$). There are several ways to apply the theorem of the next section to conclude, *e.g.*, that this test is most stringent of size ϵ (first proved by Hunt and Stein). One is to consider for fixed $\gamma_0 > 0$ the problem of testing $\gamma = 0$ against $\gamma = \gamma_0$, to note that G is transitive on ω and on $\mathfrak{F} - \omega$ in this case so that Assumption 3 is satisfied (as are the other assumptions), and thus to conclude that the above test is most stringent of size ϵ; since this is true for every $\gamma_0 > 0$, it follows that the test is most stringent for the original GLH. Another method (better than the above in other problems where such a property uniform in γ_0 may not hold) is to verify Assumption 3 directly for GLH; we can do this easily by applying the theory of [15] to the present case. Alternatively, (2.3) can be verified by considering, on the right side of (2.3), a ξ assigning probability one to the set consisting of one point in ω and one point at which the power function of the F-test differs most from the envelope power function.

Counterexamples. We now list briefly four counterexamples to the conclusion of the theorem of Sec. 3, only the third of which is new, in order to indicate that Assumptions 1, 2, 4, and 5 cannot be entirely dispensed with.

(A) In [6] and also in [7] are given examples which show that the conclusion of the theorem of Sec. 3 is false if (in terms of the present treatment) Assumption 4 is violated. We note here also that if, in the notation of p. 313 of [7], the weight function is altered to $f(s) = 1$ if s is an integer and $f(s) = \max(s, 0)$ otherwise, then there exist invariant procedures with finite risk $(=1)$, but the conclusion of the theorem is still false; thus, we see that if in Condition 4c the condition that $W(y, z) \to c_z$ as $y \to -\infty$ and $\to \infty$ monotonically as $y \to \infty$ were dropped while maintaining the condition that a $\delta_0 \,\varepsilon\, \mathfrak{D}_I$ with finite risk exists, Assumption 4 would not be implied.

(B) As Peisakoff has pointed out, the invariance theory applies to the general sequential case only if we restrict $\mathfrak{D}$ to consist of procedures which take at least a first observation with probability one. In Section 4 we shall discuss this in more detail (there are cases where this restriction of $\mathfrak{D}$ is not necessary); for the moment, we give an example to demonstrate that the conclusion of the theorem of the next section would not generally be true without such a restriction. Suppose we are limited to taking a single observation or else no observation on a random variable whose distribution depends only on a location parameter θ which we desire to estimate (see Example (i)), the loss from estimating θ incorrectly being bounded by 1 and the cost of experimentation being 2 or 0 depending on whether or not we take an observation. Any minimax procedure in $\mathfrak{D}$ must clearly take no observation with probability $\geqq \frac{1}{2}$ (a similar remark applying for ϵ-minimax procedures); however, the only invariant procedures take a first observation with probability one (see Sec. 4 for further discussion). The difficulty here is that Assumption 1 is violated, since g_x must depend on the observation and thus, for a δ which requires no observations, the δ_g of (2.1) would require no observations but would depend on the observation, and would thus not be a legitimate decision function.

(C) As an example which shows that Assumption 2 cannot be entirely dispensed with, consider the setup of Example v with $F_0(x) = 0$ if $x < 0$ and $=1$ if $x \geqq 0$, and let $W = 1$ if $d = \theta$ and $=0$ otherwise. This is essentially a game where one player says "don't you name the real number I name" and then names a real number, while the only affine-invariant procedure for the other player is, on hearing the number, to name the same number. The procedure δ^* of Example v is in fact uniformly worst and is clearly not minimax, while there exist many minimax procedures in the class C. This example can be made into one where all members of $\mathfrak{F}$ have densities with respect to a fixed σ-finite measure by restricting $\mathfrak{X}$, D, and G to the rationals (of course, this changes μ, and Condition 5b(2) is no longer applicable), and can be made more probabilistic by letting F_0 assign probability $\frac{1}{3}$ to each of the values -1, 0, 1; but the phenomenon persists. See Example v for an example of a condition which eliminates the phenomenon encountered in Counterexample C.

(D) Stein [17] has announced an example in testing hypotheses where all our assumptions except Assumption 5 are satisfied and where the conclusion of the theorem is false. This example shows that the real projective group and $GL(2)$ do not satisfy Assumption 5.

3. Proof of invariance theorem. We now use a modification of the method of proof used in [1] under Condition 2bp and Assumptions 4 and 5, in order to prove the following theorem (see also Remark 7 of Sec. 2):

THEOREM. *If G leaves the problem invariant and if Assumptions* 1 *to* 5 *are satisfied, then for any $\delta \,\varepsilon\, \mathfrak{D}$ and $\epsilon > 0$ there is a $\delta' \,\varepsilon\, \mathfrak{D}_I$ such that $\bar{r}_{\delta'} \leqq \epsilon + \bar{r}_\delta$. In* *particular, if δ^* is minimax among procedures in $\mathfrak{D}_I$, then it is minimax among procedures in $\mathfrak{D}$.*

PROOF. Our first step is to prove (3.5) below. Denote right invariant measure on G by μ^{-1}; i.e., $\mu^{-1}(E) = \mu(E^{-1})$. Fix b and $\delta \varepsilon \mathfrak{D}$ and let $\{G_n\}$ be a sequence satisfying Assumption 5, and define

$$H_{F,x}(g) = \int_D W^b(F, x, gr)\delta(g^{-1}x, dr). \tag{3.1}$$

Then, for $\gamma \varepsilon G$,

$$\begin{aligned}\lim_{n\to\infty} \left| \int_{G_n^{-1}} [H_{F,x}(g^{-1}) - H_{F,x}(\gamma g^{-1})]\mu^{-1}(dg)/\mu(G_n) \right| \\ = \lim_{n\to\infty} \left| \int_{G_n} [H_{F,x}(h) - H_{F,x}(\gamma h)]\mu(dh)/\mu(G_n) \right| \\ \leq \lim_{n\to\infty} 2b\mu(\gamma G_n \Delta G_n)/\mu(G_n) = 0,\end{aligned} \tag{3.2}$$

by Assumption 5. Using (3.2) with $\gamma = g_x$ and bounded convergence, we obtain, for any fixed $\xi \varepsilon S_b$,

$$\lim_{n\to\infty} \int_{\Gamma_b} \xi(dF) \int_{\mathfrak{X}} F(dx) \int_{G_n^{-1}} [H_{F,x}(g^{-1}) - H_{F,x}(g_x g^{-1})]\mu^{-1}(dg)/\mu(G_n) = 0. \tag{3.3}$$

It will simplify notation if we define the operation L by

$$L = \liminf_{n\to\infty} \int_{\Gamma_b} \xi(dF) \int_{\mathfrak{X}} F(dx) \int_{G_n^{-1}} \mu^{-1}(dg)[\mu(G_n)]^{-1} \int_D \cdot \tag{3.4}$$

Using (3.3), a change of variables, and (2.1), we obtain

$$\begin{aligned}LW^b(F, x, g^{-1}r)\delta(gx, dr) &= LW^b(F, x, g_x g^{-1}r)\delta([g_x g^{-1}]^{-1}x, dr) \\ &= LW^b(F, x, u)\delta(gx^*, d_u g g_x^{-1}u) \\ &= LW^b(F, x, u)\delta_g(x, du).\end{aligned} \tag{3.5}$$

Let $\delta \varepsilon \mathfrak{D}$. Using the fact (Assumption 3) that S_b includes every measure giving probability one to a single element of $\Gamma_b \supset \{F_\beta\}$ and that gX has probability measure gF when X has measure F, we have for any fixed $\delta \varepsilon \mathfrak{D}$,

$$\begin{aligned}\bar{r}_\delta &= \sup_{F \varepsilon \mathfrak{F}} \sup_b \int_{\mathfrak{X}} F(dx) \int_D W^b(F, x, r)\delta(x, dr) \\ &= \sup_b \sup_{\xi \varepsilon S_b} \sup_{g \varepsilon G} \int_{\Gamma_b} \xi(dF) \int_{\mathfrak{X}} F(dx) \int_D W^b(gF, gx, r)\delta(gx, dr) \\ &\geq \sup_b \sup_{\xi \varepsilon S_b} \liminf_{n\to\infty} \int_{G_n^{-1}} \mu^{-1}(dg)[\mu(G_n)]^{-1} \\ &\qquad \cdot \int_{\Gamma_b} \xi(dF) \int_{\mathfrak{X}} F(dx) \int_D W^b(gF, gx, r)\delta(gx, dr),\end{aligned} \tag{3.6}$$

where the inequality follows from the fact that an average is no greater than a supremum. Using Fubini's theorem ($\mu^{-1}(dg)$ on G_n^{-1} and $\xi(dF)F(dx)$ on $\Gamma_b \times \mathfrak{X}$ are both finite) and the invariance of W (i.e., $W^b(gF, gx, r) = W^b(F, x, g^{-1}r)$), we see that the last member of (3.6) is equal to the supremum with respect to b and ξ of the first member of (3.5). On the other hand, again using Fubini's theorem, the supremum with respect to b and ξ of the last member of (3.5) is equal to

$$(3.7) \qquad \sup_b \sup_{\xi \varepsilon S_b} \liminf_{n \to \infty} \int_{G_n^{-1}} \mu^{-1}(dg) r^b_{\delta_g}(\xi)/\mu(G_n) \geqq \sup_b \sup_{\xi \varepsilon S_b} \inf_{\delta \varepsilon \mathfrak{D}_I} r^b_\delta(\xi),$$

the inequality following from the fact that an average is no less than an infimum and that $\delta_g \varepsilon \mathfrak{D}_I$ for $g \varepsilon G$ (see Remark 3). Using first Assumption 3 and the fact that $\sup_{\xi \varepsilon S_b} r^b_\delta(\xi) = \bar{r}^b_\delta$ if $\delta \varepsilon \mathfrak{D}_I$, and then using Assumption 4, we see that the right side of (3.7) is equal to

$$(3.8) \qquad \sup_b \inf_{\delta \varepsilon \mathfrak{D}_I} \bar{r}^b_\delta = \inf_{\delta \varepsilon \mathfrak{D}_I} \bar{r}_\delta .$$

Thus, for each $\delta \varepsilon \mathfrak{D}$, the first member of (3.6) is no less than the last member of (3.8), proving the theorem.

The above theorem does not, of course, treat the question of whether or not a minimax procedure exists, i.e., whether $\inf_{\delta \varepsilon \mathfrak{D}} \bar{r}_\delta$ is attained. Conditions for this may be found, e.g., in [15] and [16]; the same conditions will usually apply for both $\mathfrak{D}$ and $\mathfrak{D}_I$, so that the conclusion of our theorem can be strengthened by the additional remark that a minimax procedure exists in $\mathfrak{D}_I$ if one exists in $\mathfrak{D}$. Various conditions for the attainment of $\inf_\delta \bar{r}_\delta$ are also given in [1] and [4] (see [5a]). Of course, for suitably simple W one can often write down an explicit formula for a minimax invariant procedure in the manner discussed under Condition NR of Sec. 2; for example, by now this formula is well known in the case studied in [6].

It is of interest to note an observation of Peisakoff to the effect that his proof (under Condition 2bp) will go through in many cases where the elements of $\mathfrak{F}$ are not *all* the distributions gF_0 for $g \varepsilon G$, but only a suitably large subset of these: e.g., in Example i of Sec. 2, the restriction $\theta \geqq 0$ might be imposed. This extension can also be carried out under our assumptions in certain cases where the restricted class of elements g for which $gF_\beta \varepsilon \mathfrak{F}$ is not compact.

4. The sequential case. Our setup in this section is that of Secs. 2 and 3 with certain interpretations. For simplicity our description is specialized to handle the examples stated at the end of this section, although a more general setup is obvious. The space $\mathfrak{X}$ is a product space $\mathfrak{X}_1 \times \mathfrak{X}_2 \times \cdots$ with denumerably many factors or a trivial modification of such a space as in Example ii or iv of Sec. 2, and we write a point of $\mathfrak{X}$ as $x = (x_1, x_2, \cdots)$ and the random variable X as $(X_1, X_2, \cdots)$. In the examples we treat, the $\mathfrak{X}_i$ will be copies of the same Euclidean space and the X_i will be independent and identically distributed according to each $F \varepsilon \mathfrak{F}$. The group G will act componentwise on $\mathfrak{X}$, so we may write $gx = (gx_1, gx_2, \cdots)$. The space D will be a product space $D_1 \times E$ where the "terminal decision space" D_1 has the role the space D had in fixed sample-size problems and $B_D = B_{D_1} \times B_E$ where (B_{D_1}, D_1) is the Borel sets on a subset of a Euclidean space and B_E contains at least the countable subsets of E. The "experimental decision space" E consists of all ordered k-tuples of (not necessarily distinct) positive integers for $k = 0, 1, 2, \cdots$, as well as infinite sequences of positive integers; we represent an element of E by $e_k = (a_1, a_2, \cdots, a_k)$, such a k-tuple representing an experiment carried out in k stages, the ith of which consisted of a_i "observations," namely, on $X_{s_{i-1}+1}, \cdots, X_{s_i}$, where we write $s(e)$ for the sum of the integers in e and $s_k = s((a_1, \cdots, a_k))$; e_0 represents the taking of no observations, and we write $e_\infty = (a_1, a_2, \cdots)$ for an e where experimentation never ceases. The group G acts trivially on E, so that we may write $gd = g(d_1, e) = (gd_1, e)$ in the sequel. The weight function W can depend on F, d_1, and e; for simplicity of exposition, in this section the weight function W will be a sum of two non-negative parts:

$$W(F, x, (d_1, e)) = W_1(F, d_1) + W_2(e), \tag{4.1}$$

although the more general form $W(F, d_1, e)$ can be treated in similar fashion. Thus, W_1 takes the place of the W of the fixed sample-size case and must satisfy the invariance condition $W_1(gF, gd_1) = W_1(F, d_1)$ for all F, d_1 and g. The cost of experimentation $W_2(e_k)$ we assume to be non-negative and finite if $k < \infty$ and infinite if $k = \infty$ (the cases where $W_2(e_k)$ is permitted to be infinite for $k < \infty$ in some treatments of decision theory to reflect upper limits on sampling will be covered by restricting $\mathfrak{D}$ as indicated in Remark 8 below), and we assume the existence of a finite number q and a real nondecreasing function h tending to infinity with its non-negative argument and such that, for all $k < \infty$ and e,

$$\begin{aligned} W_2((a_1, \cdots, a_k, 1)) &- W_2((a_1, \cdots, a_k)) < q, \\ W_2(e) &> h(s(e)), \\ W_2((a_1, \cdots, a_k, a_{k+1})) &\geqq W_2((a_1, \cdots, a_k)); \end{aligned} \tag{4.2}$$

in other words, the cost of taking *one* additional observation at any stage is bounded, for any finite number M only finitely many different e's cost less than M, and additional observations always have non-negative cost. One often imposes on W_2 practical restrictions such as $W_2((a_1 + a_2)) \leqq W((a_1, a_2))$, but this is inessential for our considerations. Typical specializations of W_2 often encountered in practice are $W_2((a_1, \cdots, a_k)) = \sum_1^k W_2((a_i))$ or $= W_2((\sum_1^k a_i))$ the latter case with $W_2((t)) = ct$ being especially important.

Denote by B_n the Borel field of members of $B_{\mathfrak{X}}$ which are cylinder sets with base in $\mathfrak{X}_1 \times \cdots \times \mathfrak{X}_n$; i.e., a B_n-measurable real function of x is one which depends on x only through $(x_1, \cdots, x_n)$, the only B_0-measurable functions being constants. We denote by $\mathfrak{D}^0$ the class of all sequential decision functions δ, i.e., functions δ on $\mathfrak{X} \times B_D$ which are probability measures on D for each x (see also the discussion of the paragraph containing (4.3) below for interpretation) where, in addition to the measurability requirements of Section 2, each $\delta \in \mathfrak{D}^0$ is assumed to satisfy the restriction that if $e = (a_1, \cdots, a_k)$ with $s(e) = r$ and if $Q_{e,a}$ is the set of all elements e_∞ or e_k of E of the form $e_\infty = (a_1, \cdots, a_k, a, \cdots)$ or $e_j = (a_1, \cdots, a_k, a, a_{k+2}, \cdots, a_j)$ for all $j \geqq k + 1$ and all $a_{k+2}, \cdots$, then $\delta(x, \Delta_1 \times e)$ (for each $\Delta_1 \varepsilon B_{D_1}$) and $\delta(x, D_1 \times Q_{e,a})$ are B_r-measurable in x; that is, the decision to stop taking observations or to take a particular number of observations at the next stage depends only on observations which have already been taken. Let $\mathfrak{D}^i$ denote the class of all δ in $\mathfrak{D}^0$ for which $\delta(x, D_1 \times e) = 0$ whenever $s(e) < i$; i.e., which for each x observe at least $x_1, \cdots x_i$ w.p.1. For $i \geqq 0$, let $\mathfrak{D}_I^i$ denote the invariant procedures in $\mathfrak{D}^i$; of course, δ is invariant if $\delta(gx, g\Delta_1 \times e) = \delta(x, \Delta_1 \times e)$ for all g, x, Δ_1, e. We have already seen in Counterexample B of Sec. 2 that the theorem of Sec. 3 will not generally be true if $\mathfrak{D} = \mathfrak{D}^0$ because not all of the δ_g of Assumption 1 will be decision functions. Of course, if G were compact we could use the method of [4] directly as outlined in Sec. 1, without any difficulty. For the examples treated at the end of this section it will suffice to take $\mathfrak{D} = \mathfrak{D}^1$ or $\mathfrak{D}^2$. (The sequential considerations of [1] consist of briefly pointing out an example of the sequential setup of $\mathfrak{D}$ and the necessity of not taking $\mathfrak{D} = \mathfrak{D}^0$.)

The question arises, how much do we lose by restricting $\mathfrak{D}$ to be $\mathfrak{D}^1$ or $\mathfrak{D}^2$ rather than $\mathfrak{D}^0$? The answer will usually be easy to verify. For example, suppose D_1, G, $\mathfrak{F}$, and the X_i are as in Example i (or i′) of Sec. 2 (Examples vii to x of the present section) and that W_1, which we may think of as a function of $\theta - d_1$, tends to its supremum w (say) when its argument tends to ∞ (or, similarly, $-\infty$). Then any procedure δ which requires 0 observations w.p.1 clearly has $\bar{r}_\delta = w$. Since any member of $\mathfrak{D}^0$ can be written as a probability mixture of a procedure in $\mathfrak{D}^1$ and a procedure which requires 0 observations w.p.1, it is evident that either every procedure requiring 0 observations w.p.1 is minimax, or else there is a $\delta \varepsilon \mathfrak{D}^1$ which is minimax. Which of these is the case will be easy to verify in most practical examples. In particular, if $w = \infty$, the second is always the case.

The function δ as given above is (with a different notation) the function p defined in Eq. (1.3) of [15]; $\delta(x, \Delta_1 \times e)$ is the probability, when δ is used and $X = x$, that the experiment will terminate with experimental decision e and terminal decision an element of the subset Δ_1 of D_1. The usual representation of a sequential decision function is obtained by letting $\bar{D}$ be the union of D_1 with the space L of positive integers and writing, for each element e of E and subset $\bar{\Delta}$ of $\bar{D}$,

$$\delta(\bar{\Delta} \mid x, e) = \frac{\delta(x, \bar{\Delta}_1 \times e) + \delta(x, D_1 \times Q'_e)}{\delta(x, D_1 \times Q_e)}, \tag{4.3}$$

where Q_e is the set of all elements of the form $e_\infty = (a_1, \cdots, a_k, \cdots)$ or $e_j = (a_1, \cdots, a_k, \cdots, a_j)$ of E for all $j \geqq k$, when $e = (a_1, \cdots, a_k)$ (thus, Q_e is the union of all $Q_{e,a}$ for $a > 0$, together with e), while Q'_e is the union over $a \varepsilon \bar{\Delta} \cap L$ of the sets $Q_{e,a}$, and we let $\bar{\Delta}_1 = \bar{\Delta} \cap D_1$. If the denominator of the right side of (4.3) is 0, define $\delta(\bar{\Delta} \mid x, e) = 1$ or 0 according to whether or not $1 \in \bar{\Delta} \cap L$; the definition in this case is only for definiteness and could be made in many other ways. The left side of (4.3) represents the conditional probability, when δ is used and given that $X = x$ and that the experiment has already proceeded (if $e = (a_1, \cdots, a_k)$) through k stages of experimentation as represented by e, that a terminal decision in $\bar{\Delta}_1$ is made or that the next stage of the experiment consists of a number of observations in $\bar{\Delta}_2 = \bar{\Delta} \cap L$. Clearly, $\delta(\bar{\Delta} \mid x, e)$ is B_r-measurable in x if $s(e) = r$, and the functions $\delta(\bar{\Delta} \mid x, e)$ on $B_{\bar{D}} \times \mathfrak{X} \times E$ satisfying obvious restrictions are in 1-to-1 correspondence with the functions $\delta(x, \Delta)$ on $\mathfrak{X} \times B_D$ as described originally ($B_{\bar{D}}$ consists of every union of a set in B_{D_1} and a set in B_E). Moreover, in terms of our later description, δ is invariant if $\delta(\bar{\Delta} \mid x, e) = \delta(g\bar{\Delta} \mid gx, e)$, where $g\bar{\Delta} = g(\bar{\Delta}_1 \cup \bar{\Delta}_2) = (g\Delta_1) \cup \Delta_2$. We shall use this representation of $\mathfrak{D}_I^i$ below.

The problems we are going to consider are ones in which the difficulty encountered in Counterexample B can be avoided as indicated above, and in which there is a very simple sufficient sequence $\{T_i\}$ of functions on $\mathfrak{X}$, T_i being B_i-measurable (the range space of T_i is immaterial). If one does not employ the principle of sufficiency in the manner of this section the theorem of Sec. 3 will only yield the dependence of the stopping rule on $x_\alpha(=(x_2 - x_1, \cdots, x_n - x_1)$ after n observations in Example vii, for example), nothing like the result we obtain. Specifically, we assume (see Example xv for further remarks)

Assumption 6. *For some positive integer m, Assumptions 1 and 2 are satisfied for $\mathfrak{D} = \mathfrak{D}^m$ with g_x a B_m-measurable function of x. There exists a sequence $\{T_i\}$ of functions with T_i a B_i-measurable sufficient statistic for $[(X_1, \cdots, X_i), \mathfrak{F}]$, such that there exist conditional probability* d.f.'s

$$F_r(y_1, \cdots, y_r \mid t_r) = P\{g_x^{-1}(X_1, \cdots, X_r) \leqq (y_1, \cdots, y_r) \mid T_r(X) = t_r\}$$

for $r \geqq m$ with the property

$$F_r(y_1, \cdots, y_r \mid t_r) \textit{ does not depend on } t_r. \tag{4.4}$$

It will aid understanding to consider an example at this point, Example vii of this section. The X_i are normal with unknown mean and known variance, and $\mathfrak{X}_i = G = D_1 = R^1$. We also identify $\mathfrak{F}$ with R^1 in obvious fashion. We can let $n = 1$ and $g_x u = u + x_1$ for u in $\mathfrak{X}_i$ or D_1, and identify the indices α with sequences $x_\alpha = g_x^{-1}x = (0, x_2 - x_1, x_3 - x_1, \cdots)$. Let $T_i = \sum_{j=1}^i X_j$. Since $g_x^{-1}X_1 = 0$ and $g_x^{-1}(X_2, \cdots, X_r) = (X_2 - X_1, \cdots, X_r - X_1)$, the distribution of $g_x^{-1}(X_1, \cdots, X_i)$ given that $T_i(X) = t_i$ is multivariate normal with means and covariances independent of t_i, so that Assumption 6 is satisfied. Similarly, in Example xi with G the affine group and $\mathfrak{X}_i = R^1$, we put $g_x^{-1}x_i = (x_i - x_1)/(x_2 - x_1)$, etc.

Assumption 6 is related to a property cited in [5a] as being proved in [4] in certain regular cases, to the effect that we lose nothing in the validity of the theorem of Sec. 3 for problems considered in [4] if we first use the principle of sufficiency and then apply the invariance principle to the space of a correctly chosen sufficient statistic. Assumption 6 also includes an additional strong property in that (4.4) is obviously not implied by this result of [4] (see also Example xv below). This assumption is easily verified in Examples vii to xiv.

Denote by $Q(s)$ the infimum of $\bar{r}_\delta - W_2(e_1^{(s)})$ over all δ with $\delta(x, D_1 \times e_1^{(s)}) = 1$, where $e_1^{(s)} = (s)$, and by $Q_I(s)$ the infimum when δ is also restricted to be invariant; thus, $Q(s)$ and $Q_I(s)$ are the values of $\inf_\delta \bar{r}_\delta$ over all δ or all invariant δ for the fixed sample-size problem with sample size s when the weight function is W_1. We assume

ASSUMPTION 7. *Either $\bar{r}_\delta = \infty$ for all $\delta \varepsilon \mathfrak{D}^m$ or else there is an integer m' with $Q(m') < \infty$.*

This assumption is easy to verify in practical cases for the examples considered in this section, where one will usually know $Q_I(j) < \infty$ for some j. The assumption can be shown, in fact, to be implied by our other assumptions under mild regularity conditions, although for the sake of brevity we forego such considerations here.

The main remaining difficulty in applying the theorem of Sec. 2 to the present problem is the verification of Assumption 4, which would usually be difficult to verify directly in sequential problems. Our form of the theorem which follows reduces this verification to the much simpler nonsequential one of Sec. 2.

THEOREM. *If G leaves the problem invariant and Assumptions* 3, NR, 5, 6, 7, (4.1), *and* (4.2), *as well as Assumption* 4 *for W_1 in each fixed sample-size problem with sample size $\geqq m$, are satisfied, and if $\mathfrak{D} = \mathfrak{D}^m$, then for each $\epsilon > 0$ there exists a fixed sample-size invariant procedure δ^* (the sample perhaps being taken according to some grouping) with $\bar{r}_{\delta^*} \leqq \epsilon + \inf_{\delta \varepsilon \mathfrak{D}} \bar{r}_\delta$. Thus, if $Q_I(s(e)) + W_2(e)$ is minimized over $s(e) \geqq m$ by $e = e'$ and if δ^* is a minimax invariant procedure for the fixed sample-size problem with sample size $s(e')$ ignoring W_2, then a minimax procedure for the sequential problem is to take $s(e')$ observations according to the grouping e' (which minimizes $W_2(e)$ over e satisfying $s(e) = s(e')$) and then to use δ^*.*

REMARK 8. Before proving the theorem we remark that the first paragraph of the proof below can easily be altered to handle the case where $\mathfrak{D}$ is further restricted in some way such as bounding k or the a_i or $s(e_k)$ in $e_k = (a_1, \cdots, a_k)$, etc. We have already noted the fact that it will usually be easy to verify whether a minimax procedure of $\mathfrak{D}^m$ or a more trivial procedure is minimax in $\mathfrak{D}^0$. We also note that one can think of G as acting on T_r for $r \geqq m$ in the examples treated by us, so that the conclusion of the theorem could be phrased in terms of invariant functions of T_r.

PROOF OF THEOREM. We may assume $\rho = \inf_{\delta \varepsilon \mathfrak{D}} \bar{r}_\delta < \infty$, the theorem being trivial otherwise. By Assumption 7 there is an m' and a procedure δ^0 with $\delta^0(x, D_1 \times e_1^{(m')}) = 1$ and $\bar{r}_{\delta^0} - W_2(e_1^{(m')}) = C < \infty$. Since the X_i are independent and identically distributed we can clearly assume $m' \geqq m$. Let ϵ be a positive number. The second line of (4.2) implies the existence of a number $N' > m$ such that any procedure $\delta \varepsilon \mathfrak{D}^m$ with $\bar{r}_\delta < \rho + \epsilon$ must require fewer than N' observations with probability $>1 - \epsilon$ for all $F \varepsilon \mathfrak{F}$. For any such δ define the procedure δ' as one which proceeds like δ except that whenever experimentation has reached a stage e (including e_0) where $s(e) < N' \leqq s(e) + t$ for some t with $\delta(x, e_1^{(t)} \mid e) > 0$, δ' assigns the probability $\delta(x, e_1^{(t)} \mid e)$ which δ assigned to the taking of t observations at the next stage (there may of course be several such t) to the taking of exactly m' additional observations one-by-one and, if these observations are taken, uses δ^0 on these last m' observations to reach a terminal decision. Since the X_i are independent and identically distributed, by the first and last lines of (4.2) we clearly have $\bar{r}_{\delta'} < \bar{r}_\delta + \epsilon(C + qm')$ and $\delta' \varepsilon \mathfrak{D}^m$. Since $\epsilon > 0$ is arbitrary, we conclude that our theorem will be proved if we prove it for the case where $\mathfrak{D}$ is restricted to the class $\mathfrak{D}^{m,N}$ of procedures in $\mathfrak{D}^m$ for which $\delta(x, D_1 \times E_N) = 1$, where N is a fixed integer and E_k is the set of e for which $s(e) < k$. We hereafter assume $\mathfrak{D} = \mathfrak{D}^{m,N}$.

In order to apply the theorem of Sec. 3 to the present case, it remains to verify Assumption 4 when $\mathfrak{D} = \mathfrak{D}^{m,N}$. Let $Q_I^b(s)$ be the value of $Q_I(s)$ when W_1 is replaced by W_1^b. By Assumption 4 in the fixed sample-size case, we have

$$\lim_{b\to\infty} [W_2(e) + Q_I^b(s(e))] = W_2(e) + Q_I(s(e))$$

for each fixed e with $s(e) \geqq m$. Since there are only finitely many e with $m \leqq s(e) < N$, we obtain, for $\mathfrak{D} = \mathfrak{D}^{m,N}$,

$$\begin{aligned} \lim_{b\to\infty} \inf_{\delta \varepsilon \mathfrak{D}_I} \bar{r}_\delta^b &= \lim_{b\to\infty} \min_{m \leqq s(e) < N} [W_2(e) + Q_I^b(s(e))] \\ &= \min_{m \leqq s(e) < N} [W_2(e) + Q_I(s(e))] = \inf_{\delta \varepsilon \mathfrak{D}_I} \bar{r}_\delta, \end{aligned}$$

which is Assumption 4 for the present problem.

Applying, then, the theorem of Sec. 3, we obtain for any $\delta \varepsilon \mathfrak{D}^{m,N}$ and $\epsilon > 0$ an invariant procedure δ' with $\bar{r}_{\delta'} \leqq \bar{r}_\delta + \epsilon$. Since δ' is invariant, we have

$$\begin{aligned} \delta'(x, \Delta \mid e) &= \delta'(g_x^{-1}x, g_x^{-1}\Delta \mid e) \\ &= \delta'(g_x^{-1}x, g_x^{-1}\Delta_1 \mid e) + \delta'(g_x^{-1}x, \Delta \cap L \mid e). \end{aligned}$$

Define the procedure δ'' by

$$\begin{aligned} \delta''(x, \Delta \mid e) &= E\{\delta'(g_X^{-1}X, g_X^{-1}\Delta_1 \mid e) \mid T_{s(e)} = T_{s(e)}(x)\} \\ &\quad + \int \delta'(y, \Delta \cap L \mid e) F_{s(e)}(dy_1, \cdots, dy_{s(e)} \mid T_{s(e)}(x)). \end{aligned} \tag{4.5}$$

Since B_{D_1} = Borel sets on a Euclidean set, this defines a decision function for some version of the conditional expected value (see, e.g., [18]). Clearly $\delta''(x, D_1 \times E_k) = 0$ for $k = m$ and $= 1$ for $k = N$, so $\delta'' \varepsilon \mathfrak{D}^{m,N}$. Since $\{T_i\}$ is sufficient, $r_{\delta'} = r_{\delta''}$. But for each e and $\Delta_2 \subset L$, Assumption 6 implies that $\delta''(x, \Delta_2 \mid e)$ is a constant. Hence δ'' can be considered to be a member of the class ϕ of probability mixtures of fixed sample-size procedures of sample-sizes $s(e)$ with $m \leqq s(e) < N$, where the sample may be taken according to some grouping e (independent of X). It is easy to see that, under our assumptions, the result of the previous paragraph remains true if $\mathfrak{D}^{m,N}$ is replaced by ϕ and that Assumptions 1, 2, 3, and 5 remain satisfied; thus, the theorem of Sec. 3 is valid for $\mathfrak{D} = \phi$, so that there is a $\delta^* \varepsilon \phi_I \subset \mathfrak{D}_I^{m,N}$ with $\bar{r}_{\delta^*} \leqq \bar{r}_{\delta''} + \epsilon \leqq \bar{r}_\delta + 2\epsilon$. This completes the proof of the theorem, since condition NR implies the constancy of r_{δ^*} and hence the existence of a fixed sample-size $\delta^{\#} \varepsilon \phi_I$ with $r_{\delta^{\#}} \leqq r_{\delta^*}$.

We note that δ'' in the preceding paragraph can be proved invariant in our examples, for an appropriate version of the first term on the right in (4.5), but the proof as given seems just as short. The lack of dependence of W on x in (4.1) is of course used in invoking sufficiency.

EXAMPLES. We shall use the following notation in our examples, where z and θ_1 are real and θ_2 and γ are positive:

$$f_1(z; \theta_1, \theta_2) = \frac{1}{\sqrt{2\pi}\theta_2} e^{-(z-\theta_1)^2/2\theta_2^2},$$

$$f_2(z; \theta_1, \theta_2) = \begin{cases} 1/2\theta_2 & \text{if } |z - \theta_1| < \theta_2 \\ 0 & \text{otherwise,} \end{cases}$$

$$f_{3\gamma}(z; \theta_1, \theta_2) = \begin{cases} (z-\theta_1)^{\gamma-1} e^{-(z-\theta_1)/\theta_2} / \theta_2^\gamma \Gamma(\gamma) & \text{if } z > \theta_1 \\ 0 & \text{otherwise.} \end{cases}$$

In all the examples except xiv, the X_i will be independent real random variables whose common Lebesgue density will be assumed to be in some class of the above densities, which class we identify with $\mathfrak{F}$.

(vii) $\mathfrak{F}$ consists of the densities f_1 for $-\infty < \theta_1 < \infty$ with θ_2 assumed known and θ_1 to be estimated and hence $G = D_1 = \mathfrak{F}$ = additive group of R^1, W_1 being a function of $\theta_1 - d_1$. Note that in most practical examples Assumption 4 can be verified by applying Condition 4a or 4b or 4c, and the question of whether to use a procedure requiring no observations or one in $\mathfrak{D}^1$ will be easy to settle. Of course, we can take $T_i = \sum_{i=1}^j X_j$, $m = 1$, and $g_x u = u + x_1$, as previously mentioned. Thus, the conclusion of the theorem will be satisfied for most W_1 and W_2 encountered in practice. Of course, $Q_I(s)$ is easily computed in this case to be given by

$$Q_I(s) = \inf_h \int_{-\infty}^{\infty} W_1(h + u) f_1(u; 0, \theta_2 s^{-\frac{1}{2}})\, du; \tag{4.6}$$

and, if h_s achieves the minimum, a nonrandomized sequential minimax estimator will be given by taking $s(e')$ observations according to the grouping e' described in the statement of the theorem and then estimating θ_1 by $s(e')^{-1}T_{s(e')} + h_{s(e')}$.

(vii+) We mention several extensions of Example vii: (1) The form of the minimax (or an analogous ϵ-minimax) estimator above depends on θ_2 in such a way that if it were only known that θ_2 belonged to some set B (not necessarily the set of all positive numbers) and if W_1 were a function of $(d_1 - \theta_1)\theta_2^{-1}$ instead of $(d - \theta_1)$, then the estimator of the previous problem vii for the case $\theta_2 = 1$ would be minimax (or ϵ-minimax) here. (2) A second extension is to note that, for the original setup of Example vii, if W_1 is symmetric we can also apply the group of reflections as in Example iii of Sec. 2. If in addition W_1 is nondecreasing in $|\theta_1 - d_1|$, we obtain the sample mean ($h_s = 0$ in vii) as minimax estimator, a result first obtained in [8], with a special case in [9]. Note that the question of whether a procedure in $\mathfrak{D}^1$ or one requiring no observations is minimax is trivial in this case.

(viii) Same as vii, except that the possible distributions are the $f_{3\gamma}$ with $\gamma = 1$ and θ_2 known and θ_1 unknown with $-\infty < \theta_1 < \infty$. In this case $T_i = \min (X_1, \cdots, X_i)$ and the considerations and conclusions are as in vii with f_1 replaced by $f_{31}(u; 0, \theta_2 s^{-1})$ in (4.6), the minimax estimator being $T_{s(e)} + h_{s(e)}$. A very special case of this was obtained tediously in [11].

(ix) $\mathfrak{F}$ consists of the densities f_2 with θ_1 known and θ_2 unknown, $0 < \theta_2 < \infty$. Here θ_2 is to be estimated, so $D_1 = \mathfrak{F} = G^{(1)}$ = multiplicative group of positive reals. The weight function is a function only of θ_2/d_1. We can either take $G = G^{(1)}$ or can think of θ_1 is being 0 and let G = direct product of $G^{(1)}$ and $G^{(2)}$ where $G^{(2)}$ contains the identity and an element which multiplies X_i by -1 and leaves $\mathfrak{F}$ and D fixed. We have

$$m = 1 \quad \text{and} \quad T_i = \max (|X_1 - \theta_1|, \cdots, |X_i - \theta_1|)$$

and $g_x^{-1}u = u/(x_1 - \theta_1)$ if $G = G^{(1)}$, with an obvious modification if we let $G = G^{(1)} \times G^{(2)}$. In either case, Assumption 6 is satisfied. Of course, this problem is really the same as that of estimating θ_2 when X_i has density $1/\theta_2$ for $0 < x_i < \theta_2$ and 0 otherwise (put $X_1' = |X_i - \theta_1|$ above), and in this form the problem may be reduced to that of viii by a logarithmic transformation as in Example i′. The form of the analogue of (4.6) and of the minimax procedure are obvious. The special case $W_2 = (\theta_2 - d_1)^2/\theta_2^2$ was considered in [11]; Condition 4c is satisfied there.

(x) $\mathfrak{F}$ consists of the densities $f_{3\gamma}$ where γ and θ_1 are known and θ_2 is unknown, $0 < \theta_2 < \infty$. This is a scale parameter problem with G = the $G^{(1)}$ of ix, and we need only remark that the theorem applies with $T_i = \sum_{j=1}^{i} X_j$, the analogues of (4.6) and the form of the minimax procedure being obvious. This problem was treated for a particular γ and weight function in [10] and for a special class of weight functions in [12].

(x′) If X_i has symmetric density about known θ_1, the density of $|X_i - \theta_1|$ being that of Example x, the same considerations apply, using also the $G^{(2)}$ of ix. Similarly, the problem of estimating θ_2 when f_1 is the density and θ_1 is known can obviously be reduced to that of Example x.

The next three examples are similar in that, in each, there is both an unknown location parameter θ_1 and also an unknown scale parameter θ_2 with $-\infty < \theta_1 < \infty$ and $0 < \theta_2 < \infty$. In each case $m = 2$, G is the real affine group (see Example ii), and $g_x^{-1}x_i = (x_i - x_1)/(x_2 - x_1)$. There are three main types of problems in each example: (1) estimation of both θ_1 and θ_2, so that $D_1 = G$, $d_1 = (d_{11}, d_{12})$, W_1 is a function of $(\theta_1 - d_{11})/\theta_2$ and d_{12}/θ_2, and

$$g_x^{-1} d_1 = ((d_{11} - x_1)/(x_2 - x_1), d_{12}/(x_2 - x_1));$$

(2) estimation of θ_1, where $D_1 = R_1$, W_1 is a function of $(\theta_1 - d_1)/\theta_2$, $g_x^{-1} d_1 = (d_1 - x_1)/(x_2 - x_1)$; (3) estimation of θ_2, where D_1 = positive reals, W_1 is a func-

tion of d_1/θ_2, $g_x^{-1} d_1 = d_1/(x_2 - x_1)$; of course, (2) and (3) can really be considered as special cases of (1) where W_1 only depends on one of its two arguments. For each type and example it is simple to write down an analogue to (4.6) and the corresponding form of a minimax procedure. In each case the conditions of the theorem are easily verified for many commonly used W, and the verification of whether one should use a procedure in $\mathfrak{D}^2$ or one requiring one or no observations is also easy. The use of the Bayes method in these examples would of course be much more complicated than that in [8], [11], and [12].

(xi) $\mathfrak{F}$ consists of all densities f_1. Putting $\bar{X}^{(i)} = i^{-1}\sum_{j=1}^{i} X_j$, we have $T_i = (\bar{X}^{(1)}, \sum_{j=1}^{i}(X_j - \bar{X}^{(i)})^2)$ for $i \geqq 2$. Note that the problem of estimating θ_2, even for the appropriate weight function, cannot be obtained by the method of [10] without some modification, because of the nature of the Cramér-Rao bound.

(xii) $\mathfrak{F}$ consists of all densities f_2. Putting $U_i = \min(X_1, \cdots, X_i)$ and $V_i = \max(X_1, \cdots, X_i)$, we can take $T_i = (U_i, V_i)$ or $((U_i + V_i)/2, (V_i - U_i))$, for $i \geqq 2$. (The second form of T_i here and in the next example are pertinent to remarks made below in Example xv.)

(xiii) $\mathfrak{F}$ consists of all densities f_{31} (i.e., γ is known to be 1), and in the notation of the previous two examples we can take $T_i = (U_i, \bar{X}^{(i)})$ or $(U_i, \bar{X}^{(i)} - U_i)$ for $i \geqq 2$.

(xiv) As an example of a multivariate nature, suppose $\mathfrak{X}_i = R^J$ for some positive integer J, the X_i again being independent and identically distributed. Here $X_i = (X_{i1}, \cdots, X_{iJ})$, and we assume X_i has a multivariate normal distribution with the identity convariance matrix and unknown mean $\theta = (\theta_1, \cdots, \theta_J) \varepsilon R^J$. The problem is to estimate θ, so that $D_1 = \mathfrak{F} = G =$ additive group of R^J and W_1 is a function of the difference between the vectors d_1 and θ_1. Taking $m = 1$ and $T_i = \bar{X}^{(i)}$ and $g_x^{-1}u = u - x_1$ for $u \varepsilon R^J$, the theorem is applicable for many common weight functions. (Examples viii to xiii have similar multivariate analogues.)

(xiv+) We can extend Example xiv in the manner of vii+. In particular, if W_1 is an increasing function of the usual Euclidean *distance* between d_1 and θ, it is easy to see that $\bar{X}^{(s(c'))}$ is a minimax sequential estimator. The orthogonal group also leaves the problem invariant in this case, but this fact need not be used in obtaining the above form of the minimax estimator, it sufficing to apply a result of [19]. It is interesting to note that it is shown in [20] that, when W_1 is the squared length of the distance and $J > 2$, this estimator is not admissible.

(xv) As an example which illustrates the fact that the method of this section yields little if no T_i satisfy Assumption 6, consider the problem of estimating θ_1 when the X_i have density f_2 and θ_2 is known. This problem is considered for certain W in [15] and [21], and the minimax procedures obtained there are not fixed sample-size. As in Example xii, (U_i, V_i) is a minimal sufficient statistic. Assumption 6 cannot be satisfied for any sufficient T_i. The application of our method in this example would yield the form of the estimator obtained in [15] and [21], but would only yield the fact that the minimax stopping rule depends on $U_i - V_i$ at the ith stage; the stationary form of the minimax stopping rule seems to depend strongly on the particular nature of f_2. It will be noted that the previous examples differ from this one in that in the former, but not in the latter, *there is a natural version of T_i for $i \geqq m$ whose range is G and such that the problem in terms of the T_i is left invariant by the natural operation of G on the range of T_i.* This is the essence of the examples where the method of this section yields the conclusion of the theorem, although we have seen that G may be modified somewhat from what this statement indicates (see Examples vii+, ix, x′, and xiv+) to the case where the *range of T_i is a subgroup or homogeneous space of G.* We may add that, in most sequential *testing* problems, the invariance principle yields little, for reasons similar to those present in Example xv.

Remark 9. We end this section with a remark about other versions of the statistical problem, such as that of minimaxing the W_1 component of the risk subject to a bound on the W_2 component or vice versa. This includes such problems as the problem of finding optimum sequential estimators of bounded relative error of the scale parameters in Examples ix to xiv (in [7] there is some discussion of this problem but our results are not obtained) and that of obtaining optimum sequential interval estimators of prescribed length and confidence coefficient for the location parameters in Examples vii and viii. The latter problem is considered in [8] and [9] in the case of Example vii, while [8] considers also the problem of minimaxing one component of risk subject to inequalities on *two* others, etc. The discussion of [8], [21], and [12] shows at once on application of our theorem that results of all these types hold for appropriate fixed sample-size procedures, or probability mixtures thereof, in Examples vii to xiv.

5. Sequential problems with continuous time. In this section we will use the method developed in Secs. 3 and 4 to obtain certain sequential minimax results for decision problems concerned with stochastic processes with continuous time parameter. Two types of problems will be considered: in Part I of this section we treat problems where the invariance is present in the same form as in Sec. 4, while in Part II the invariance has to do with the time parameter.

I. *Extension of Section 4 to continuous time.* The problems we consider here will be continuous time analogues of certain of the problems of Sec. 4 (in fact, those of Sec. 4 can be considered as special cases of those here, in the manner of [12]). Since the proofs are essentially identical to those of Sec. 4, we shall not give them. In fact, rather than to state a general theorem, we shall merely list three examples. In each of these the separable process $\{X(t), t \geqq 0\}$ is one of independent and stationary increments which can be taken to be continuous on the right, and $X(T)$ is sufficient for $\{X(t), 0 \leqq t \leqq T\}$. As in Sec. 4, W can be a function of $\theta^{-1}d$ (θ being the unknown parameter) and of the experimentation decision, but for convenience of exposition we discuss the case where it is a sum $W_1 + W_2$. The cost of experimentation W_2 may either be taken to be of the form $W_2(T)$ if the process is observed continuously up to time T, or else the cost may be allowed to depend on the number and spacing of the instants at which the process is observed; a description of this and other modifications (such as the problem of having to give an estimate continuously), as well as a more detailed discussion of the nature of sequential decision functions in the case of continuous time, and of the processes considered, will be found in [12]. In all of the examples, assumptions on W_2 can be treated as in Sec. 4. The analogue here of the restriction to $\mathfrak{D}^1$ in Sec. 4 is that we must restrict ourselves to the union over all $\epsilon > 0$ of the classes $\mathfrak{D}^\epsilon$ of procedures which observe the process for at least $0 \leqq t \leqq \epsilon$ w.p.1 for all $F \varepsilon \mathfrak{F}$. When we consider $\mathfrak{D}^\epsilon$, the g_x is a function of $X(\epsilon)$. As in Sec. 4, it will be easy in most practical cases to decide whether there will be a minimax procedure in $\mathfrak{D}^\epsilon$ for some $\epsilon > 0$ or a minimax procedure which does not observe the process at all.

In each of the three examples, our result is, under assumptions on W like those of Sec. 4, that *there exists an invariant minimax or ϵ-minimax procedure which observes the process for a constant length of time* w.p.1 (or a minimax procedure which does not observe the process at all). Formulas for computing the minimax procedure can be given as in Sec. 4 or [12], and Remark 9 of Sec. 4 applies also to these examples.

(xvi) The process is the one-dimensional Wiener process with known variance per unit time and with $EX(t) = \theta_1 t$, the object being to estimate θ_1. Thus, G, $\mathfrak{F}$, D, and the form of W_1 are the same as in Example vii. In particular, in the special case of a symmetric monotone W_2, we obtain the result of Sec. 5 of [12].

(xvi′) For the Wiener process with unknown scale or unknown location and scale, it has been shown in [12] that the scale parameter can be estimated with arbitrarily high accuracy in arbitrarily short time. Hence, the only new practical problems that arise when the scale parameter is unknown do so because W_2 reflects the number of instants at which the process is observed. In this case, as indicated in [12], we obtain problems analogous to Example xi with G the affine group, or to Example x′ (see also the next example below). In either of these problems there will be an invariant minimax procedure which observes the process at a certain set of instants specified in advance of the experiment.

(xvii) The process is the Gamma process; i.e., $X(0) = 0$ and $X(1)$ has density function $f_{3\gamma}$ of Sec. 4 with $\theta_1 = 0$ and γ known, the object being to estimate the scale parameter θ_2. Here $\mathfrak{F}$, D, G, and W_1 are the same as in Example x of Sec. 4.

(xviii) Consider the J-variate Wiener process $X(t) = (X_1(t), \cdots, X_J(t))$ where the $X_i(t)$ are independent with known scale factors and $EX_i(t) = \theta_i t$, the θ_i being unknown, $-\infty < \theta_i < \infty$. This is the continuous time analogue of Example xiv, and the considerations there and in xiv+ carry over to the present example.

II. *Invariance in time.* We now consider a process $\{X(t), t \geqq 0\}$ with unknown parameter $\theta > 0$ and with the property that, if $\{X(t), t \geqq 0\}$ has probability law labeled θ, then the process $\{X_c(t), t \geqq 0\}$, defined by $X_c(t) = X(ct)$ where $c > 0$, has probability law labeled $c\theta$. The most familiar process of this kind is the Poisson process. Another such process is the gamma process with θ_2 known and γ unknown.

Suppose the weight function (for estimating θ) is a function only of d_1/θ and

T/θ, where d_1 is the terminal decision and T is the length of time experimentation is carried on (modifications of the type mentioned earlier in this section and discussed in [12] are also possible). Then clearly the multiplicative group of positive reals leaves the problem invariant, where we define $g(\{X(t)\}, \theta, (d_1, T)) = (\{X(gt)\}, g\theta, (gd_1, g^{-1}T))$, the group operation being ordinary multiplication. The difference here from previous problems is that G acts on the process by shifting the time argument of a sample function by a scale factor rather than by operating on the values of the sample function, and that G acts nontrivially on the experimental decision. The reason for allowing this last action and the accompanying dependence of W on T/θ rather than on T lies in the form of the result which this setup yields when one applies the invariance theorem and examines the invariant procedures.

The details here are slightly more delicate and lengthy than those in Part I, so we shall be content with sketching the main idea. Consider the Poisson process with right continuous sample functions. $X(\tau)$ is sufficient for $\{X(t), 0 \leqq t \leqq \tau\}$. Suppose we have a nonrandomized stopping function which depends on the sufficient statistic, i.e., a nonnegative functional T of the process with the property that the event $t_1 < T \leqq t_2$ is measurable with respect to the Borel field generated by $\{X(t), t_1 < t \leqq t_2\}$. For such a T to be invariant we must have $T(x) = cT(x_c)$ for all $c > 0$ and all sample functions x, where x_c is the sample function of X_c when x is the sample function of X. It is easy to see that such a stopping function as $T(x) =$ constant is not invariant, while $T_r(x) =$ first time t that $x(t) = r$, where r is a fixed positive integer, is. In the present problem we must restrict $\mathfrak{D}$ to decision functions which observe the process until at least the first time $X(t) = 1$ (that time gives g_x^{-1}). Under fairly general conditions one can verify whether or not a minimax procedure should observe the process at all and that, if it does, a stopping rule of the type T_r is minimax. Of course, an invariant nonrandomized estimator will be of the form constant/T_r. A special case of this result thus shows that the procedure suggested in Sec. 3 of [22] and which was asserted there to be minimax among all procedures using a particular stopping rule T_r (analogous to a fixed sample-size problem) actually has an optimum property among all sequential procedures: e.g., among all procedures which give at least the prescribed accuracy of estimation, this one minimaxes $E_\theta T/\theta$.

REFERENCES

[1] M. P. Peisakoff, "Transformation Parameters," Thesis, Princeton University, 1950.

[2] E. J. G. Pitman, "The estimation of location and scale parameters of a continuous population of any given form," *Biometrika*, Vol. 30 (1939); "Tests of hypotheses concerning location and scale parameters," *Biometrika*, Vol. 31 (1939), pp. 200–215.

[3] A. Wald, "Contributions to the theory of statistical estimation and testing hypotheses," *Ann. Math. Stat.*, Vol. 10 (1939), pp. 299–326.

[4] G. A. Hunt and C. Stein, "Most stringent tests of statistical hypotheses," unpublished.

[5a] E. L. Lehmann, Notes on testing hypotheses, University of California, Berkeley, 1949.

[5b] E. L. Lehmann, "Some principles of the theory of testing hypotheses," *Ann. Math. Stat.*, Vol. 21 (1950), pp. 1–26.

[6] M. A. Girshick and L. J. Savage, "Bayes and minimax estimates for quadratic loss functions," *Proceedings of the Second Berkeley Symposium on Mathematical Statistics and Probability*, University of California Press, 1951, pp. 53–74.

[7] D. Blackwell and M. A. Girshick, *Theory of Games and Statistical Decision Functions*, John Wiley and Sons, New York, 1954.

[8] J. Wolfowitz, "Minimax estimates of the mean of a normal distribution with known variance," *Ann. Math. Stat.*, Vol. 21 (1950), pp. 218–230.

[9] C. Stein and A. Wald, "Sequential confidence intervals for the mean of a normal distribution with known variance," *Ann. Math. Stat.*, Vol. 18 (1947), pp. 427–433.

[10] J. L. Hodges, Jr., and E. Lehmann, "Some applications of the Cramér-Rao inequality," *Proceedings of the Second Berkeley Symposium on Mathematical Statistics and Probability*, University of California Press, 1951, pp. 13–22.

[11] J. Kiefer, "Sequential minimax estimation for the rectangular distribution with unknown range," *Ann. Math. Stat.*, Vol. 23 (1952), pp. 586–593.

[12] A. Dvoretzky, J. Kiefer, and J. Wolfowitz, "Sequential decision problems for processes with continuous time parameter. Problems of estimation," *Ann. Math. Stat.*, Vol. 24, 1953, pp. 403–415.

[13] P. R. Halmos, *Measure Theory*, D. Van Nostrand Co., Inc., New York, 1950.

[14] C. Chevally, *Theory of Lie Groups*, Princeton University Press, Princeton, 1946.

[15] A. Wald, *Statistical Decision Functions*, John Wiley and Sons, New York, 1950.

[16] L. LeCam, "An extension of Wald's theory of statistical decision functions," *Ann. Math. Stat.*, Vol. 26 (1955), pp. 69–81.

[17] C. Stein, "On tests of certain hypotheses invariant under the full linear group" (abstract), *Ann. Math. Stat.*, Vol. 26 (1955), p. 769.

[18] R. R. Bahadur, "Sufficiency and statistical decision functions," *Ann. Math. Stat.*, Vol. 25 (1954), pp. 423–462.

[19] T. W. Anderson, "The integral of a symmetric unimodal function over a convex set and some probability inequalities," *Proc. Amer. Math. Soc.*, Vol. 6 (1955), pp. 170–176.

[20] C. Stein, "Inadmissibility of the usual estimate for the mean of a multivariate normal distribution" (abstract), *Ann. Math. Stat.*, Vol. 26 (1955), p. 157.

[21] C. R. Blyth, "On minimax statistical decision problems and their admissibility," *Ann. Math. Stat.*, Vol. 22 (1951), pp. 22–42.

[22] M. A. Girshick, H. Rubin, and R. Sitgreaves, "Estimates of bounded relative error in particle counting," *Ann. Math. Stat.*, Vol. 26 (1955), pp. 276–285.

J. Soc. Indust. Appl. Math.
Vol. 5, No. 3, September, 1957

OPTIMUM SEQUENTIAL SEARCH AND APPROXIMATION METHODS UNDER MINIMUM REGULARITY ASSUMPTIONS*

J. KIEFER[1]

1. Introduction and Summary

A. Introduction. Computers and experimenters frequently encounter problems where a function with a given domain is unknown, but where

* Received by the editors February 19, 1957 and in revised form July 31, 1957.
[1] Research sponsored by the Office of Naval Research.

values of the function may be obtained by inserting appropriate values of the argument into a machine or, in the case of an experimenter, by conducting experiments whose make-ups are determined by such values. On the basis of the knowledge thus obtained of a number of values of the function at various values of the argument, it is desired to make some estimate of a specified property of the function such as a value of the argument at which the function has a maximum or a zero, or the value of its integral over the domain, etc. Usually it is desired to make this estimate as accurate as possible subject to a specification of the number of values of the argument at which the function is to be computed (or observed), or to minimize the number of observations required to achieve a specified accuracy; this will be made precise later in this section. Thus, a typical practical problem is to estimate the percentages of various metals which should be present in an alloy in order to maximize its hardness; another typical problem is to find a zero of a function whose values can be obtained only through extensive machine computation.

The treatment of such problems in this paper differs from the classical methods in two respects. Firstly, the methods usually employed by computers and experimenters are *non-sequential*, i.e., the values of the argument at which the function is to be "observed" are all decided in advance of any computation and experimentation. Secondly, the assumption is often made that the function belongs to a rather limited class such as polynomials of a given degree, etc. In problems where the classical technique is not restricted in these two ways, one often finds that little attention has been given to considerations of how efficient the technique might be compared to other techniques; for example, Newton's method is really a sequential one (i.e., the argument at which each observation is taken depends on the values of previous observations) which can be used for a wide class of functions, but it is often not an efficient method, as we shall see in Section 2A.

The first discussion of efficient sequential methods for a problem of the type described above appeared in [1], the results of which are summarized in Section 2A. There is by now an extensive statistical literature stemming from the papers [2] and [3] which describe sequential techniques for the problems of finding a zero or a maximum when there are "errors" in the observations, and asymptotic efficiency properties of these techniques under appropriate assumptions are considered in [4] and [5]. In the present paper we shall be concerned only with non-statistical problems: *we assume there are never any errors present in the observed values of the function.* However, there are clearly many experimental problems where errors in the observations are so small that the techniques described herein, or slight modifications of them, may be usefully employed.

In each of the problems described, the class of functions considered is essentially the largest for which a non-trivial solution exists; if a class which is slightly larger is considered, all procedures are equally bad against the worst functions they encounter. However, the optimum procedures obtained are also optimum if the class of functions considered is reduced to a class which is sufficiently "dense" in an appropriate sense in the original class; we shall point out several instances of this situation in subsequent sections. We shall not give all of the results obtainable by methods similar to those used herein; for example, the considerations of Sections 3 and 4 can easily be carried over to the problems of Section 2B, but for the sake of brevity we shall forego this.

B. Nomenclature and notation. It will be convenient to give the nomenclature and notation we shall use before listing the contents of this paper, and we now do so.[2] Throughout this paper X and Y will denote given spaces and $\mathfrak{F}$ will denote a given class of functions from X into Y. g will denote a functional on $\mathfrak{F}$ taking on values in a space D. The meaning of these is the following: the computer or experimenter is presented with a function f about which he knows nothing except that it is a member of $\mathfrak{F}$. The property of f for which he wants to search, or which he wants to estimate, has value $g(f)$. (Thus, e.g., $D = X$ if $g(f)$ is the value where f has its maximum, or $D = R$ (the reals) if $g(f)$ is an integral of f.) In order to estimate this value, he may gain information about f by selecting in a manner which we shall now make precise values $x_1, x_2, \cdots$ and observing or computing the values $y_1 = f(x_1)$, $y_2 = f(x_2)$, $\cdots$. Specifically, let S_N denote the class of all "N-observation (nonrandomized) procedures s," such an s consisting of a collection $s = \{x_1, \delta_2, \delta_3, \cdots, \delta_N, d\}$ where x_1 is an element of X, δ_K is a function from the product space $X^{k-2} \times Y^{k-1}$ into $X(2 \leqq k \leqq N)$, and d is a function from $X^{N-1} \times Y^N$ into a given space D^* of *subsets* of D. Such a procedure is used as follows: One observes or computes in order $f(x_1), f(x_2), \cdots, f(x_N)$, with $x_k = \delta_k[x_2, \cdots, x_{k-1}, f(x_1), \cdots, f(x_{k-1})]$ for $2 \leqq k \leqq N (x_2 = \delta_2[f(x_1)])$ and then selects the subset $d[x_2, \cdots, x_N, f(x_1), \cdots, f(x_N)]$ of D (a member of D^*) as the final estimate (terminal decision) of a set containing $g(f)$. In some problems D^* may be the same as D and the terminal decision is merely a possible value

[2] Our formulation differs from that usually employed in games and statistics in that in the latter (in our setting) what we have called D would not be mentioned and D^* would be arbitrary; however, since our D is arbitrary, the formulation employed herein is just as general, and this formulation may help to emphasize the nature of the computational problems usually encountered in practice as ones of estimation. The function L could be permitted to take on values in a more general space, a possibility of interest mainly in connection with complete class theorems; for example, L might have two real components, reflecting inaccuracy and experimental cost (of course, one could also minimax the larger of these components).

of $g(f)$, while if D is for example the real line R, D^* might be the collection of intervals in R and the terminal decision is an *interval* of possible values of $g(f)$. Sometimes we shall restrict consideration to some subclass $\bar{S}_N$ of S_N, as will be seen in examples below.

In some problems the number N of observations will not be specified in advance, but will depend on the course of experimentation or computation. We let S^* denote the class of all (non-randomized) procedures s of the following form: $s = \{x_1, \delta_2, \delta_3, \cdots, d_1, d_2, \cdots\}$, where x_1 and the δ_i are as above except that δ_i now takes on values in the space X^* consisting of the union of X with a single additional point σ, and d_i is a function from $X^{i-1} \times Y_i$ into D^*. Such an s is used by computing, as before, $f(x_1), f(x_2), \cdots$, with $x_k = \delta_k[x_2, \cdots, x_{k-1}, f(x_1), \cdots, f(x_{k-1})]$, except that at the first time m that $\delta_{m+1} = \sigma$, one stops (after m observations) and makes terminal decision $d_m[x_2, \cdots, x_m, f(x_1), \cdots, f(x_m)]$. Thus, the δ_i include the *stopping rule* as well as the rule about where to observe next. We shall occasionally restrict our consideration to some subclass $\bar{S}^*$ of S^* (in fact, the S_N can be considered as such subclasses, and will be so considered in the sequel).

The final element of the problem which must be considered is the specification of a real function L on $\mathfrak{F} \times \bar{S}^*$. This is the "loss function," and $L(f, s)$ is the loss incurred when f is the function actually confronting the computer and s is the procedure he is using to estimate $g(f)$. The function L reflects the relative costs of various degrees of inaccuracy in the terminal decision, and of taking observations; the latter cost may depend, for example, on the values of the x_i and $f(x_i)$ and on whether several observations are taken simultaneously or whether observations are made one at a time (if, for example, δ_k does not depend on x_{k-1} and $f(x_{k-1})$, this can be taken to mean that the observation at x_k is made at the same time as that at x_{k-1}). In particular, a sequential procedure may cost more per observation than a nonsequential one which takes the same number of observations, all simultaneously. The loss function may also reflect such considerations as the following: In searching for a hard alloy, it may be that the hardest composition is too costly; thus, one may really want to estimate not the hardest make-up, but rather the most economical one in some sense. L may also reflect the lengths of time a computing machine requires for various procedures.

For $\epsilon > 0$ a procedure s^* is said to be ϵ-*minimax* (or simply *minimax* in the case $\epsilon = 0$) if

$$\sup_{f\in\mathfrak{F}} L(f, s^*) \leqq \epsilon + \inf_{s\in\bar{S}^*} \sup_{f\in\mathfrak{F}} L(f, s). \tag{1.1}$$

In some problems every procedure is minimax because the right side of (1.1) is infinite. In such cases (see Section 4) we shall often be interested in a procedure which is optimum in a stronger sense than this trivial one. For example, we may consider an increasing sequence $\{F_n\}$ of subsets of $\mathfrak{F}$ whose union is $\mathfrak{F}$, such that $q_n < \infty$, where

$$q_n = \inf_{s \in \bar{S}^*} \sup_{f \in F_n} L(f, s),$$

and such that a procedure s^* satisfies

$$\text{(1.2)} \qquad \lim_{n\to\infty} \sup_{f \in F_n} L(f, s^*)/q_n = 1.$$

Such an s^* will be called *asymptotically minimax* (relative to the sequence $\{F_n\}$).

Our formulation of the problem includes the possibility of minimaxing some measure of inaccuracy subject to some restriction on cost or number of observations, or of minimaxing the latter subject to some restriction on the former (either of these possibilities is achieved by restricting $\bar{S}^*$), as well as of minimaxing some combination of these two contributions to the loss. (A simple device which can sometimes be used to obtain solutions to the former two problems from a solution to the last one can be found, for example, in [6].) Also, accuracy can be specified in many ways; for example, in some problems of finding a zero of a real function of a real variable, it may be required to find a value within some specified ϵ of the zero, while in others it may be required to find a value x for which $|f(x)| < \epsilon$, and in some problems L may depend on f and s in a more complicated way.

Several other notions arising in the sequel will now be defined. A procedure s_1 is said to be *at least as good* as the procedure s_2 if $L(f, s_1) \leq L(f, s_2)$ for all f in $\mathfrak{F}$; if also strict inequality holds for at least one f, s_1 is *better* than s_2. For a given $\bar{S}^*$, a procedure s in $\bar{S}^*$ is said to be *admissible* if no s' in $\bar{S}^*$ is better than s. If P is a probability measure on $\mathfrak{F}$ (we shall omit the obvious measure-theoretic considerations), a procedure s^* is said to be *minimal* relative to P if

$$\int L(f, s^*)P(df) = \min_{s \in \bar{S}^*} \int L(f, s)P(df).$$

Throughout this paper, except in the next to last paragraph of this section and in Sections 2C and 3C, we shall consider only nonrandomized procedures as defined previously. In the exceptions just noted we shall have occasion also to mention *randomized* procedures (for practical considerations regarding these, see the next to last paragraph of this section); loosely speaking (which will suffice for our purpose—the reader can see any standard reference on sequential games or statistical problems for a precise definition), such a procedure specifies in place of δ_k a chance experiment determined by $x_2, \cdots, x_{k-1}, f(x_1), \cdots, f(x_{k-1})$ and whose outcome is the value to be taken for x_k (this outcome may also be σ in the case of S^*). For such procedures, the form of L will usually reflect an expected value of loss with respect to the probabilities of obtaining various results with the given procedure for each f. Of course, the class of randomized procedures includes the nonrandomized ones.

C. Summary. We now summarize the content of succeeding sections. Section 2A reviews the known minimax results for the problems of estimating the zero of a monotonic function on the unit interval I and of estimating the argument at which the maximum of a unimodal function on I is attained. Section 2B discusses the extension to "higher order" search problems and carries out the proof that a certain procedure is minimax for the problem of search for an inflection point. In Section 2C certain features of the procedures treated in 2A and 2B, such as the question of their admissibility, are discussed. Section 2D discusses some difficulties encountered in considering problems analogous to those of 2A and 2B for dimension greater than one.

In Section 3A, problems analogous to those of Section 2 are considered on a lattice. It is pointed out in Section 3B that, except for computational difficulties, the lattice treatment may be a useful formulation in practice, in which the main difficulty encountered in 2D is no longer present.

Section 4 treats the asymptotic theory for certain problems of Sections 2 and 3 (e.g., search for a zero of a monotone function on the whole real line).

In Section 5 the problem of approximating the integral of a real function on I is considered.

D. Remarks on methods of proof. We conclude the present section with a brief comparison of the methods used herein with other methods used in various problems in the literature of games and statistics in proving that certain procedures are minimax. The method most commonly used is (in our setting) to find a P and an s^* such that s^* is minimal relative to P and

$$\sup_f L(f, s^*) = \int L(f, s^*)P(df). \tag{1.3}$$

s^* is then easily seen to be minimax (sometimes no such P exists and a sequence of P's must be considered, but the idea is the same as above). In particular, if $L(f, s^*)$ is constant and s^* is minimal relative to some P, then s^* is minimax. One advantage of this method is that it proves that s^* is minimax among the class of all randomized procedures, whereas the methods employed in the present paper, and which are entirely different, demonstrate only the weaker result that the procedure under discussion is minimax among all *nonrandomized* procedures. The disadvantage of the former method is that in settings like ours where $\mathfrak{F}$ is such a large space, it is often computationally very messy to verify minimality with respect to some P (or sequence of P's); for an example of this nature in statistics, the reader may see [7]. In some of the examples cited in this paper, such as those of Section 2, it seems that one can carry out such a proof by the older method; however, in Sections 2C and 3A we shall see examples where this is not possible, since the suggested procedure is *not* minimax in the class of *randomized* procedures, but only in the class of *nonrandomized* ones. Since in practice one will often restrict consideration to nonrandomized procedures (so that one's final statement will be of the type "this is an estimate of $g(f)$ of error at most ϵ" rather than "this is an estimate whose *expected* error (at the outset of computation) was at most ϵ (the *actual* error may be more)"), we shall give proofs using methods like those employed in [1], which are also of interest per se, rather than to use the older method described above. Two other methods employed in the past in statistics are those of [8] (applicable in certain special problems with no counterpart here) and the method of invariance (see, e.g. [9]); in a problem like that of finding the maximum in Section 2A, the latter method would enable us to conclude that there is a minimax procedure which (like the one described there) uses only the *order* relationships of the $f(x_i)$ and not their actual values, but this would be of little aid in finding the explicit form of an actual minimax procedure.

The computation of the minimal procedures with respect to a given P (as distinguished from guessing a P or sequence of P's which can be employed in a minimax proof) is an easy task. It is sometimes of interest, aside from its use in minimax proofs, in obtaining a characterization of the admissible procedures or of a *complete class* of procedures (a class such that, for any procedure not in the class, there is one in the class which is better); the reader may consult any standard reference on statistical decision functions for details. Such considerations will not concern us, although occasionally the question of admissibility of a proposed procedure will be touched upon briefly.

2. Search Problems on Compact Continua

A. Review of previous results: first and second order search in one formulation (dimension one). (i) *First order search.* Among the simplest possible search problems is the following one, which, despite the trivial nature of the solution, will be useful to illustrate various points mentioned in Section 1: Let X be the unit interval I of real numbers $0 \leqq x \leqq 1$ (similar problems can be treated here and elsewhere in Section 2 where X is the circumference of a circle, etc.), let Y be the reals R, and let $\mathfrak{F}$ be the class of all f which are monotonically nondecreasing on I with exactly one zero in I; this zero is the $g(f)$ of the problem. (With no additional difficulty we could include f's which do not take on the value 0 but jump through it at $g(f)$ (e.g., $f(x) = -1$ or $+1$ according to whether $x < \frac{1}{2}$ or $x \geqq \frac{1}{2}$) and f's which take on the value 0 over an interval of values of which it is permitted to estimate any one.) Let D^* be the collection of all subintervals of I and suppose for fixed N that we are restricted to the class $\bar{S}_N$ of procedures in S_N for which d contains $g(f)$ for all f; that is, the interval finally named by the procedure after N observations must contain the zero of f. The length of this interval d, which of course depends on f and s, is the loss L which we want to minimax.

It is easily seen that a minimax procedure s_N^* is given as follows: $x_1 = \frac{1}{2}$; $\delta_2 = \frac{1}{4}$ if $f(x_1) > 0$ (so that we know $g(f) < \frac{1}{2}$), and similarly $\delta_2 = \frac{3}{4}$ if $f(x_1) < 0$, while if $f(x_1) = 0$ we can put $d = x_1$ and not even bother to observe further; continuing in this fashion, x_k will be midway between the largest previously observed x with $f(x) < 0$ (or 0 if there is no such x), say α_{k-1}, and the smallest for which $f(x) > 0$ (or 1 if there is none), say β_{k-1}. The interval d has α_N and β_N as endpoints (or is a single point x_r if $f(x_r) = 0$ for some r). $L(f, s^*)$ is then 2^{-N} for all f except those with $g(f) = q/2^N$ for some integer q (for which $L = 0$).

It is interesting to note that the description of $x_1, \delta_2, \cdots, \delta_k$ for s_N^* is the same for all $N \geqq k$; that is, it is unnecessary when taking the first k observations to specify how many observations N are to be taken, as long as it is known that $N \geqq k$. (This is in contrast to the problem discussed in Section 2A (ii), where x_1 differs for any two N.) There is a practical usefulness to this fact: one might decide to search for the zero of f with no advance specification of N, the number of observations being decided on the basis of how steep f seems to be in a neighborhood of its zero (ideally these considerations should be included in a precise specification of L, but experimenters do not always proceed in this way in practice); in such a situation it will nevertheless be possible to use this procedure which, *whatever* the number N of observations may turn out to be, would have been optimum even if that N had been specified in advance.

The way of looking at the procedure s_N^* which is a helpful point of view in examples where the property described in the previous paragraph is absent (see Sections 2A (ii) and 2B) is to think of s_1^* as being defined and then defining s_N^* inductively as follows: Supposing $s_1^*, s_2^*, \cdots, s_{N-1}^*$ have been defined, let the x_1 of s_N^* be $\frac{1}{2}$ and then put $h(x) = 2x$ or $2x - 1$ according to whether $f(x_1) > 0$ or $f(x_1) \leqq 0$, and let $f^*(y) = f(h^{-1}(y))$ for y in I; the rest of s_N^* is then described by prescribing that s_{N-1}^* be used on the function f^* and variable y in I, the d for f^* being translated back into one for f in an obvious manner. Thus, *every observation reduces the problem to the initial stage of the same problem with N reduced by* 1.

The reason we have called this "first order search" is because a single observation gives information regarding the direction from it in which $g(f)$ lies. In "kth order search" a set of k observations will yield such information, while fewer than k observations will give no such information. (The examples of Sections 2A (ii) and 2B are the cases $k = 2$ and $k = 3$.)

Extensions, generalizations, comparison with classical methods. It is easy to see that if $\mathfrak{F}$ is reduced to a suitably dense subclass of the $\mathfrak{F}$ used above (e.g., all polynomials in the above $\mathfrak{F}$, or all polygonal functions therein), the given procedure remains minimax. (Of course, if $\mathfrak{F}$ were reduced too much, e.g., to polynomials of degree $N - 1$, this result will obviously cease to hold.) On the other hand, if $\mathfrak{F}$ is enlarged to include too many non-monotonic functions, *every* procedure becomes trivially minimax ($\sup_f L(s, f) = 1$ for all s). If $\mathfrak{F}$ is enlarged only by the addition of monotonic functions with more than one zero, the definition of $g(f)$ determines whether or not the above procedure is minimax. For example, if $g(f)$ is the supremum of the zeros of f, the above procedure s_N^* is minimax if whenever $f(x_i) = 0$ we act as though $f(x_i) < 0$. On the other hand, if $g(f)$ is the center of the interval of zeros of f and if when $N = 2$ we have $x_1 = \frac{1}{2}$ and $f(x_1) = 0$, the best choice of x_2 (that which minimizes the maximum of L) is $\frac{1}{4}$ or $\frac{3}{4}$ (either value, or even a randomized choice between them, makes this maximum $\frac{3}{8}$), while the choice $x_1 = \frac{1}{3}$, $x_2 = \frac{2}{3}$ gives a procedure whose maximum loss over all of $\mathfrak{F}$ is $\frac{1}{3}$; thus, the x_1 and δ_2 of s_2^* do not yield a minimax procedure in this case.

Other loss functions could also be considered, as in Section 2C or in the first paragraph of [1], etc. Of course, the procedure given above will not be minimax for all L, but it or a slight modification of it will be minimax for many simple and reasonable forms of L.

We have mentioned in the introduction that Newton's method, which is a sequential method which works for many f, is not very efficient if $\mathfrak{F}$ is very large. Aside from the fact that *this method requires a knowledge* (or approximation) *of f' as well as of f* (this knowledge is of no real help in reducing $\inf_s \sup_f L(s, f)$ in our problem, since it is information of a local character), Newton's method is poor against many f: for example, if $f(x) = x^b$ with $b > 0$, Newton's method gives loss $x_1(1 - b^{-1})^{N-1}$ after N steps, which is many times greater than 2^{-N} if $b > 2$ and N is large.

(ii) *Second order search.* We now turn to the problem solved in [1]. The setup is that of the previous example, except that $\mathfrak{F}$ is now the class of all real unimodal functions f on the unit interval I; that is, every f in $\mathfrak{F}$ is either strictly increasing for $x \leq g(f)$ (the "mode") and strictly decreasing for $x > g(f)$, or else is strictly increasing for $x < g(f)$ and strictly decreasing for $x \geqq g(f)$. (Thus, the interval d must contain the true value $g(f)$, whatever f may be.) The ϵ-minimax procedure s_N^* of $\bar{S}_N$ described in [1] (no minimax procedure exists) for $N \geqq 2$ (every procedure is minimax for $N = 1$) is the following: Let U_n be the nth Fibonacci number ($U_0 = 0$, $U_1 = 1$, $U_n = U_{n-1} + U_{n-2}$ for $n \geqq 2$). s_2^* is defined by $x_1 = \frac{1}{2}$, $x_2 = \frac{1}{2} + \epsilon$, and $d = [0, x_2]$ or $[x_1, 1]$ according to whether $f(x_1) \geqq f(x_2)$ or $f(x_1) < f(x_2)$. Suppose s_{N-1} has been defined ($N \geqq 3$). We then define s_N^* as follows: $x_1 = U_{N-1}/U_{N+1}$, $x_2 = 1 - x_1 = U_N/U_{N+1}$. (The reader will be aided in visualizing s_N^* if he draws a diagram of the unit interval and x_1 and x_2. The relative magnitudes of $f(x_1)$ and $f(x_2)$ will determine either that $g(f)$ lies in the interval $[0, x_2]$, or else that it lies in $[x_1, 1]$. Subsequent observations can therefore be taken in the appropriate one of these two intervals, this interval being expanded linearly by the function h onto the unit interval, in the description which follows. At the same time, the function f on the appropriate interval becomes f^* on I. Thus, the next observation can be thought of as being taken in I in order to help find the maximum of the function f^*.) According to whether $f(x_1) \geqq f(x_2)$ or $f(x_1) < f(x_2)$, let $h(x) = xU_{N+1}/U_N$ or $h(x) = (-U_{N-1} + xU_{N+1})/U_N$, let $y = h(x)$ and $f^*(y) = f(h^{-1}(y))$ for $y \in I$, and define $y_2 = h(x_1)$ or $y_1 = h(x_2)$ in the respective cases. Thus, $y_2 = U_{N-1}/U_N$ or $y_1 = U_{N-2}/U_N$ Use s_{N-1}^* on the variable y and function f^* (for $y \in I$), noting that either $f^*(y_2)$ or $f^*(y_1)$ has already been observed. Obviously $s_N^* \in \bar{S}_N$.

Remarks. Unlike the minimax procedure in the previous problem, the one just described depends on N in the choice of x_1 and x_2. A procedure in S^* which is not minimax for any N but will often be useful in applications is the strategy s^* defined as follows: Let $x_2 = 1 - x_1 = -\frac{1}{2} + 5^{\frac{1}{2}}/2 = .618 = \mu$ (say). If $f(x_1) \geqq f(x_2)$, define $v(x) = x/\mu$ and $y = v(x)$, and $f^*(y) = f(v^{-1}(y))$ for $y \in I$. Putting $y_2 = v(x_1) = \mu$ and $y_1 = 1 - \mu$, we then use s^* on the variable y and function f^*, where we already have observed $f^*(y_2)$. (A similar procedure applies if $f(x_1) < f(x_2)$.) Continuing in this manner, at every stage we have the same geometric configuration, unlike the case of s_N^*. The advantage of this is that if the number of observations is not specified in advance but is determined after several values have been observed (e.g., more observations might be taken if f appears to be sharply peaked near its maximum), the use of any s_N^* (or sequence of s_N^*'s) can lead to great inefficiency if one decides after N observations to take more. When N is large, if s^* is used for N observations, the length of the final interval is about 1.17 times that of s_N^* (with $\epsilon \to 0$).

The procedure s_N^* is not admissible, and may be improved slightly for some f by noting that, under our assumptions on f, whenever the two largest observations are equal, $g(f)$ must lie in the interval whose endpoints are the corresponding two arguments. (See Section 2C for further discussion.)

We shall not detail here the counterparts to the discussion of Section 2A(i) regarding modifications of the result when $\mathfrak{F}$ is changed slightly, comparison with commonly used techniques, or consideration of other loss functions (see also Section 2C and the first paragraph of [1], regarding this last.)

B. Third and higher order search (dimension one). (i) *General remarks.* We have already indicated, in Section 2A(i), what kth order search problems are. We left unstated there the *form* of the "information" a set of k observations yields, and in general that would of course depend on $\mathfrak{F}$ and g. For example, when $k = 2$, suppose the $\mathfrak{F}$ of Section 2A(ii) is restricted further to include only those members of the original class $\mathfrak{F}$ which are, except at $x = g(f)$, differentiable with $|f'(x)| < 1$, and for which

$$f[g(f)-] = f[g(f)] = 2 + f[g(f)+];$$

then clearly any two observations, say at $x = a$ and $x = b$ with $a < b$, yield the information either that $g(f) < a$ or that $g(f) \geqq b$ or that $a \leqq g(f) < b$. This is quite different from the situation for the original $\mathfrak{F}$, where, except when $f(a) = f(b)$, the information yielded is either that $g(f) \leqq b$ or that $g(f) \geqq a$. Similarly, the larger k, the greater the variety of forms in which a set of k observations can yield information in kth order search problems. Perhaps the most natural of these possibilities is that for which, for the subclass of k-times continuously differentiable functions in $\mathfrak{F}$, $g(f)$ is the argument where the kth derivative is zero, and for which $\mathfrak{F}$ is large enough that a set of k observations will in the "worst cases" yield only the information that $g(f)$ is not to the left or right (respectively) of the most lefthand or righthand observation; it can be seen that the two examples of Section 2A satisfy this description.

Of course, the right side of (1.1) will increase with k (fixing X, D, and the form of L, as in Sections 2A(i) and 2A(ii)). Thus, for large N, the order of magnitude of the length of the final interval d when $k = 1$, 2, or 3 (the latter case will be treated next) is $.5^N$, $.618^N$, or $.707^N$, respectively.

(ii) *The case* $k = 3$. As an example of a higher order search problem of the type just described, we now consider the case $k = 3$, where the problem is one of finding an inflection point. The setup is exactly like that of Section 2A except that $\mathfrak{F}$ now consists of every real function f on I for which f is strictly convex (upwards) for $x \leqq g(f)$ and $-f$ is strictly convex for $x \geqq g(f)$. (f may be discontinuous at 0, 1, or the inflection point g, but is not linear on any interval of positive length; this could be relaxed in a manner similar to that which appeared in Section 2A(i) where an interval of zeros was permitted.) For all s in $\bar{S}_N$ the final interval d must contain $g(f)$, whatever f in $\mathfrak{F}$ may be the true one.

We shall now describe, for $N \geqq 4$, a procedure s_N^* and a procedure s_N^{**}, both of which will then be proved minimax (the superiority of s_N^{**} over s_N^* will be discussed in the next paragraph, as well as in that following the proof of minimaxity; s_N^* is considered anyway for expository reasons, because its structure closely parallels that of the minimax procedures previously discussed for $k = 1, 2$); the case $N = 3$ will be mentioned later, while for $N \leqq 2$ every procedure is trivially minimax. (The proof for the case $N \geqq 4$ will be seen to use methods similar to those employed in [1].) For any $f \in \mathfrak{F}$ and $0 < u < v < w < 1$, write $T(u, v, w, f) = L$ (for "left") or $= R$ (for "right") according to whether

$$\frac{f(v) - f(u)}{v - u} - \frac{f(w) - f(v)}{w - v}$$

is >0 or $\leqq 0$. Clearly, $T = L$ (resp., $T = R$) implies $g(f) < w$ (resp., $g(f) > u$). Let $V_3 = \frac{3}{2}$, $V_4 = 2$, $V_n = 2V_{n-2}$ for $n \geqq 5$ (thus, $V_{2n+2} = 2^n$, $V_{2n+5} = 3 \cdot 2^n$).

We first define s_n^*. s_4^* proceeds by taking $x_1 = \frac{1}{4}$, $x_2 = \frac{1}{2}$, $x_3 = \frac{3}{4}$, and then (having observed $f(x_1), f(x_2), f(x_3)$), $x_4 = \frac{3}{8}$ or $\frac{5}{8}$ according to whether $T(x_1, x_2, x_3, f) = L$ or R. After $f(x_4)$ is observed, define the final interval d by

$$d = \begin{cases} [0, \frac{1}{2}] \text{ if } T(x_1, x_2, x_3, f) = T(x_1, x_4, x_2, f) = L, \\ [\frac{1}{4}, \frac{3}{4}] \text{ if } T(x_1, x_2, x_3, f) = L \text{ and } T(x_1, x_4, x_2, f) = R, \\ [\frac{1}{4}, \frac{3}{4}] \text{ if } T(x_1, x_2, x_3, f) = R \text{ and } T(x_2, x_4, x_3, f) = L, \\ [\frac{1}{2}, 1] \text{ if } T(x_1, x_2, x_3, f) = T(x_2, x_4, x_3, f) = R. \end{cases}$$

Supposing that $S_{N-1}^*(N \geqq 5)$ has been defined, we define s_N^* as follows: $x_3 = 1 - x_1 = V_{N-1}/V_N (= \frac{2}{3}$ if N is odd, $\frac{3}{4}$ if N is even), $x_2 = \frac{1}{2}$. Having observed $f(x_1), f(x_2), f(x_3)$, let $h(x) = xV_N/V_{N-1}$ or $h(x) = 1 + (x - 1)V_N/V_{N-1}$ according to whether $T(x_1, x_2, x_3, f) = L$ or R. For $y \in I$, let $f^*(y) = f(h^{-1}(y))$, and let $y_1 = h(x_1)$, $y_2 = h(x_2)$ if $T = L$ and let $y_2 = h(x_2)$, $y_1 = h(x_3)$ if $T = R$. (Thus, if $T = L$, $y_1 = \frac{1}{2}$ and $y_2 = \frac{3}{4} = V_{N-2}/V_{N-1}$ if N is odd, while $y_2 = 1 - y_1 = \frac{2}{3} = V_{N-2}/V_{N-1}$ if N is even; if $T = R$, y_1 and y_2 are unity minus the above values.) s_N^* now proceeds by using s_{N-1}^* on the variable $y = h(x)$ and function f^* (for $y \in I$), noting that two of the three initial values $f^*(\frac{1}{2})$, $f^*(V_{N-2}/V_{N-1})$, $f^*(1 - V_{N-2}/V_{N-1})$ have already been observed (the subscripts on the y's may not correspond to those of x_1, x_2, x_3 in the definition of s_{N-1}^*). Continuing this process (the reader will be aided by drawing the original $X = I$ and noting how it is reduced in size by successive observations, the reduced interval being expanded linearly by the function h in the above description to become a unit interval) and transforming the final d into terms of the original variable, we see that $s_N^* \in \bar{S}_N$ and that $L(f, s_N^*) = 1/V_N$.

Before defining S_N^{**}, we note that S_N^* has the structure mentioned in Section 2A(i) according to which at each stage the problem is reduced to that for one fewer observations, with two of the three initial observations of a minimax procedure for the reduced problem already present. Moreover, there are only two possible configurations of the three initial observations, corresponding to odd and even N. We also note that there are many procedures in S_4 with maximum loss the same as that of s_4^* (namely, $\frac{1}{2}$); for example, x_1 and x_3 could be taken to be b and $1 - b$ with $\frac{1}{4} < b < \frac{1}{2}$, while x_4 is taken to be c or $1 - c$ with $b < c < \frac{1}{2}$ (the choice of the final d is obvious). In particular, define s_4^{**} in this manner with $b = \frac{1}{3}$ and $\frac{1}{3} < c < \frac{1}{2}$ (the exact choice of c does not matter at the moment). s_5^{**} takes $x_1 = \frac{1}{3}$, $x_2 = \frac{1}{2}$, $x_3 = \frac{2}{3}$, and after 3 observations with $T = L$ (a procedure symmetrical about $\frac{1}{2}$ to the one about to be described being used if $T = R$) takes $x_4 = b'$ with $\frac{1}{6} < b' < \frac{1}{3}$ and $x_5 = c'$ where $b' < c' < \frac{1}{3}$ or $\frac{1}{3} < c' < \frac{1}{2}$ according to whether $T(b', \frac{1}{3}, \frac{1}{2}, f) = L$ or R (the choice of the final d is obvious). For $N > 5$, if N is odd, s_N^{**} coincides with s_N^* until the problem has been reduced to that for five observations of which two are already present, at which stage s_5^{**} is used; if N is even, s_{N+1}^* is used for $N - 2$ observations and then s_4^* is used on the reduced problem with two observations present and two remaining to be taken. This construction yields an s_N^{**} which is in $\bar{S}_N$ and which has maximum loss $1/V_N$, and *which is better than* s_N^* (the set of f's for which $L(f, s_N^{**}) < 1/V_N$ depends on b and c or on b' and c'; for example, when $N = 4$, this set is smaller and the reduction in L is larger, the larger b is taken, while for fixed b there is no best choice of c, each c being inferior to a choice of c which is closer to b). Moreover, s_N^{**} has the property that *its first* $N - 2$ *observations* (this would be $N - 1$ if we let $b = \frac{1}{4}$, $b' = \frac{1}{6}$ above, but s_N^{**} would then only be as good as, not better than, s_N^*) *are the same as the first* $N - 2$ *observations of every* $s_{N'}^{**}$ *with* $N' > N$ (which is in contrast to the structure mentioned above of the s_N^* of this section and Section 2A(ii)). We shall return to this last fact in the paragraph following the proof below.

We shall now prove that (1.1) *holds with* $\epsilon = 0$ *for* $N \geqq 4$ *for the* s_N^* *and* s_N^{**} *just defined; i.e., that the right side of* (1.1) *with* $\epsilon = 0$ *equals* $1/V_N$ *for* $N \geqq 4$.

Let $\mathfrak{F}^*$ consist of those f in $\mathfrak{F}$ which have two continuous derivatives and for which $|f| \leqq 1$, $|f'| \leqq 1$, and $|f''| \leqq 1$. Let S_N^* consist of all s requiring N observations and for which the final interval d contains $g(f)$ for all f in $\mathfrak{F}^*$. The result (1.1) with $\epsilon = 0$ clearly follows from

$$\inf_{s \in S_N^*} \sup_{f \in \mathfrak{F}^*} L(f, s) = 1/V_N \tag{2.1}$$

for $N \geqq 4$.

For $N = 4, 5$, the result (2.1) may be verified by explicit calculation, it being only necessary to notice that when f and its first two derivatives are near zero for all x, the information that $f \in \mathfrak{F}^*$ helps little over the information that $f \in \mathfrak{F}$, and one can find a class of such functions which is large enough to insure that the left side of (2.1) is arbitrarily close to $1/V_N (N = 4, 5)$ when $\mathfrak{F}^*$ is replaced by that class. For $n \geqq 6$, assume (2.1) valid for $4 \leqq N < n$ but false for $N = n$, so that

$$\sup_{f \in \mathfrak{F}^*} L(D(f, \bar{s})) < \frac{1}{V_n} - 2\epsilon \tag{2.2}$$

for some $\bar{s} \in S_n^*$ and some $\epsilon > 0$. We shall derive a contradiction from (2.2).

We may suppose that, under $\bar{s}$, the functions δ_2 and δ_3 are constant. For, if not, use instead of $\bar{s}$ the procedure which uses the x_1 of $\bar{s}$ and thereafter uses $\bar{s}$ on the function f_1 defined by $f_1(x) = [f(x) - f(x_1)]/2$. Clearly, $g(f_1) = g(f)$ and $f_1 \in \mathfrak{F}^*$ if $f \in \mathfrak{F}^*$. This new procedure satisfies (2.2) if $\bar{s}$ does, and its δ_2 is the constant $\bar{\delta}_2(0)$ (where $\bar{\delta}_k$ refers to $\bar{s}$). Thus, we may assume $\bar{\delta}_2$ constant. If now $\bar{\delta}_2$, but not $\bar{\delta}_3$, is constant, use the procedure which, after observing $f(x_1)$ and $f(x_2)$ according to $\bar{s}$, thereafter uses $\bar{s}$ on the function f_2 defined by $f_2(x) = a[f(x) + bx + c]$ where $b = -[f(x_2) - f(x_1)]/(x_2 - x_1)$, $c = [x_1 f(x_2) - x_2 f(x_1)]/(x_2 - x_1)$, and $a = a(x_1, x_2, f(x_1), f(x_2)) > 0$ is chosen (as it may be) so that $f_2 \in \mathfrak{F}^*$. The constant $x_2 = \delta_2$ obtained by using $\bar{\delta}_2$ on f_1 or on f_2 is the same, and the δ_3 of the new procedure is the constant $\bar{\delta}_3(x_2; 0, 0)$; moreover, this procedure satisfies (2.2) if $\bar{s}$ does, completing the assertion at the outset of this paragraph. We can also assume x_1, x_2, x_3 are all different, or we would have a procedure in S_{n-1}^*, violating (2.1) for $N = n - 1$. For the procedure $\bar{s}$ we hereafter assume δ_2 and δ_3 constant and put $0 \leqq x_1 = \alpha < x_2 = \alpha + \beta < x_3 = \alpha + \beta + \gamma = 1 - \delta \leqq 1$.

Assuming (2.2), we shall now show $\alpha + \beta < \frac{1}{2}$; a similar demonstration shows $\gamma + \delta < \frac{1}{2}$, contradicting $\alpha + \beta + \gamma + \delta = 1$. We do this by constructing, under the hypothesis $\alpha + \beta \geqq \frac{1}{2}$, an $s' \in S_{N-2}^*$ for which

$$\sup_{f \in \mathfrak{F}^*} L(f, s') < 2/V_n = V_{n-2}, \tag{2.3}$$

violating the induction hypothesis.

Let h be a fixed function with two continuous derivatives, with $h'(x) \leqq 0$

for $\alpha + \beta - \epsilon/5 \leqq x \leqq \alpha + \beta$, with $h(\alpha + \beta - \epsilon/5) = 1$, and with

$$0 = h(\alpha + \beta) = h'(\alpha + \beta) = h''(\alpha + \beta)$$
$$= h'(\alpha + \beta - \epsilon/5) = h''(\alpha + \beta - \epsilon/5).$$

Let $b(x) = -(x - \alpha - \beta + \epsilon/5)^2/4$ and let c be a positive constant, to be chosen presently. For $f \in \mathfrak{F}^*$, define

$$\bar{f}(x) = \begin{cases} cf\left(\dfrac{x}{\alpha + \beta}\right) & \text{if } 0 \leqq x \leqq \alpha + \beta - \epsilon/5, \\ ch(x)f\left(\dfrac{x}{\alpha + \beta}\right) + b(x) & \text{if } \alpha + \beta - \epsilon/5 \leqq x \leqq \alpha + \beta, \\ b(x) & \text{if } \alpha + \beta \leqq x \leqq 1. \end{cases} \tag{2.4}$$

Noting that b and its first derivative are 0 at $\alpha + \beta - \epsilon/5$ and that its second derivative is $-\frac{1}{2}$ for all x, it is easy to verify that we can choose c small enough (depending on ϵ and h, but not on f) that $\bar{f} \in \mathfrak{F}^*$ if $f \in \mathfrak{F}^*$ and that $\bar{f}''(x) < 0$ if $x \geqq \alpha + \beta - \epsilon/5$. Thus, assuming c to be so chosen,

$$g(\bar{f}) = \begin{cases} (\alpha + \beta)g(f) & \text{if } g(f) \leqq 1 - \epsilon/5(\alpha + \beta), \\ \alpha + \beta - \epsilon/5 & \text{otherwise.} \end{cases}$$

In any event,

$$\left| g(f) - \frac{g(\bar{f})}{(\alpha + \beta)} \right| \leqq \epsilon/5(\alpha + \beta) \leqq 2\epsilon/5. \tag{2.5}$$

We now describe s': Use $\bar{s}$ on $\bar{f}$ by calculating $\bar{f}(=b)$ at $x = \alpha + \beta$ and $x = \alpha + \beta + \gamma$ for two of the first three "observations." Thus, f is really "observed" at only $n - 2$ points, corresponding to the dictates of $\bar{s}$ for the last $n - 2$ observations on $\bar{f}$. $\bar{s}$ is carried out on $\bar{f}$ in an obvious manner by translating observations on $\bar{f}$ into corresponding observations, at appropriate values of the argument, on f (if an observation on $\bar{f}$ is at an argument $\geqq \alpha + \beta$, we do not even have to observe f to obtain the observation on $\bar{f}$). Let $d(f, s) = [d_1(f, s), d_2(f, s)]$ be the final interval when f is the true function and s is used. s' is then defined to take $n - 2$ observations on f corresponding to the last $n - 2$ on $\bar{f}$ according to $\bar{s}$, and to put $d(f, s') = [\max\ (0, (\alpha + \beta)^{-1} d_1(\bar{f}, \bar{s}) - 2\epsilon/5), \min(1, (\alpha + \beta)^{-1} d_2(\bar{f}, \bar{s}) + 2\epsilon/5)]$. From (2.5) we have $s' \in S^*_{n-2}$; finally, from (2.2),

$$\sup_{f \in \mathfrak{F}^*} L(f, s') \leqq \frac{4\epsilon}{5} + (\alpha + \beta)^{-1} \sup_{f \in \mathfrak{F}^*} L(f, \bar{s}) \tag{2.6}$$
$$\leqq \frac{4\epsilon}{5} + \frac{2}{V_n} - \epsilon,$$

demonstrating (2.3) and thus completing the proof.

Remarks. We shall not discuss in detail all the questions analogous to those treated in Section 2A(i) for the first order case. We note here that in the case $N = 3$ an ϵ-minimax procedure can be obtained by taking observations at $\frac{1}{2}$, $\epsilon + \frac{1}{2}$, and $-\epsilon + \frac{1}{2}$, but that no minimax procedure exists, unlike the situation for $N > 3$. It is interesting to note that in the case $N = 4$ the maximum loss of a minimax procedure ($\frac{1}{2}$) is reduced only by an arbitrary $\epsilon > 0$ from the maximum loss of an ϵ-minimax procedure when $N = 3$. Also, whereas in the case $k = 2$ it was necessary at the outset to know N if an ϵ-minimax procedure was to be obtained for small ϵ, when $k = 3$ it is not necessary to know N in advance; for one can proceed in the manner in which all s_N^{**} start out and, after M observations, terminate experimentation with 2 (resp., 3) additional observations if M is even (resp., odd) by taking these observations as the last few according to s_N^{**} with $N = M + 2$ (resp., $M + 3$); the resulting procedure is then, for that N, even minimax compared to procedures in $\bar{S}_N$ where the same N is specified in advance. (One could also stop after the M observations or after one additional one, but this strong optimality property would then be lost.) Thus, N may be determined by the course of the experiment and an efficient procedure may still result in such cases where a precise specification of L is not present to determine the correct stopping rule (see Section 2A(i) for related discussion). A similar property holds if one specifies in advance only whether N is to be odd or even and uses the appropriate s_N^* (or, without that specification, one can use s_N^* with a loss of the optimality property if N turns out to be of the wrong parity). Thus, the procedure obtained above from the s_N^{**} seems preferable to that obtained from the s_N^*, for this usage where N is not determined in advance.

Just as the s_N^{**}'s all start out like the s_N^*'s with N odd, there is a class of procedures s_N^{***} (say) which proceed similarly but start out like the s_N^*'s with N even and which satisfy (1.1) for $N > 5$ (and, of course, $N = 4$). For $N = 5$, if the first three observations are taken at $\frac{1}{4}$, $\frac{1}{2}$, $\frac{3}{4}$, the procedure cannot be minimax.

We have seen that s_N^* is not admissible, at least for sufficiently large N; in Section 2C we shall see that for $N \geqq 4$ the procedures s_N^*, s_N^{**}, s_N^{***} are all inadmissible for a reason other than that involving the nonexistence of a best c in the discussion just preceding the minimax proof.

C. Admissibility, randomization, other formulations. We have remarked that the procedures of Sections 2A(ii) and 2B are inadmissible (the inadmissible analogue in the case of Section 2A(i) is that obtained by treating the possibilities $f(x_r) = 0$ and $f(x_r) < 0$ in the same way), and have indicated how the procedure of Section 2A(ii) can be improved when $f(x_i) = f(x_j)$ for some i and j. The procedures of Section 2B can also be improved, but now the opportunity for improvement will arise more frequently than in the previous problem, and the method of improvement will be more complicated; for example, if $N = 4$ and s_4^* is used (a similar remark applies for s_4^{**}) and $T(x_1, x_2, x_3, f) = R$, and if the straight line through $(\frac{5}{8}, f(\frac{5}{8}))$ and $(\frac{3}{4}, f(\frac{3}{4}))$ intersects that through $(\frac{1}{4}, f(\frac{1}{4}))$ and $(\frac{1}{2}, f(\frac{1}{2}))$ at a point with abscissa z (say) where $z > \frac{1}{2}$, it is easy to see that we can put $d = [z, 1]$ for the final interval instead of having a final interval of length $\frac{1}{2}$. If N is large and such a possibility occurs at the fourth or some other early observation, the description of a way to proceed subsequently which makes efficient use of this kind of information is very messy to write out in detail (the analogue in the case $k = 2$ when $f(x_1) = f(x_2)$ is more easily handled, e.g., by considering for $N - 2$ observations the problem of finding the maximum of f on the interval $[x_1, x_2]$), and hence an inadmissible procedure like s_N^* or s_N^{**} may be preferable for both hand and machine computations.

It seems intuitively clear that the procedures described in Sections 2A(ii) and 2B, as well as their analogues for larger k, are inadmissible only because ϵ and c can be reduced or because improvements of the type just described are possible for certain "lucky" f's, and that the procedures obtained by making alterations of this latter type are inadmissible only because of the possibility of reducing ϵ or c (i.e., *no* procedure is admissible, but the cited procedures cannot be improved upon by reducing L by very much). However, it appears tedious to write out a complete proof of this. The principal methods available for proving that a procedure s_N is admissible are to prove s_N the *unique* minimal procedure with respect to some P or, in the case where a sequence of P's is considered, to go through a fine calculation of L as is done in [6]. (Similar proofs could show a procedure is almost admissible in the sense noted above.) An easier task is merely to prove a minimality result (without uniqueness, etc.); as stated in Section 1, if (1.3) also holds (or its analogue for a sequence of P's), s_N is proved minimax among *randomized* procedures, too. Such a proof (or its analogue for ϵ-minimax procedures) is apparently not too difficult to obtain, but the admissibility question seems harder to treat.

We now consider a different formulation of the problem from that (L and $\bar{S}^*$) treated in Sections 2A and 2B. Let c be a specified number satisfying $0 < c < 1$, and let the class $\bar{S}^*$ of allowable procedures consist only of procedures for which the final interval d contains $g(f)$ and is of length $\leqq c$; the loss L which is to be minimaxed is the number of observations required before stopping (N is not specified in advance, but a limit on the accuracy is). *It is now not true that a minimax procedure among nonrandomized procedures is minimax among all procedures.* For example, with the $\mathfrak{F}$ and g of Section 2A(i), let $c = \frac{1}{5}$. Any minimax nonrandomized procedure s^* (there is a large number of them, many with L not constant) clearly has $\sup_f L(f, s^*) = 3$ On the other hand, let s' be the randomized procedure which puts $x_1 = \frac{4}{10}$ or $\frac{6}{10}$ with probability $\frac{1}{2}$ each and then, if $x_1 = \frac{4}{10}$ (a symmetrical description applying if $x_1 = \frac{6}{10}$), takes exactly one more observation at $x_2 = \frac{2}{10}$ if $f(x_1) > 0$, while if $f(x_1) \leqq 0$ a second observation is taken at $x_2 = \frac{6}{10}$ and observation ceases if $f(x_2) > 0$ while a third observation is taken at $x_3 = \frac{8}{10}$ if $f(x_2) \leqq 0$. If we denote by L' the *expected* number of observations (with respect to the random choice of x_1), it is easy to see that $\sup_f L'(f, s') = \frac{5}{2}$. This proves that no nonrandomized procedure is minimax among all procedures for this problem. However, the amount of reduction in $\sup_f L'(f, s)$ from using a randomized procedure instead of the procedure which is minimax among nonrandomized ones, will always be less than one. Thus, in practice one will often prefer to use a nonrandomized procedure which achieves slightly better accuracy at the expense of a fraction of an observation, for example, the procedure s_N^* for the smallest N for which $L(f, s_N^*) \leqq c$. (This situation will also be encountered in Section 3A.) The nonrandomized procedure just described indicates the duality between the two formulations in the nonrandomized case. In the randomized case we would have to weaken the demand of the second formulation to guarantee only that the *expected* length be $\leqq c$ in order to obtain this duality in which the solutions of the one problem yield those of the other; this last property is absent with the present formulations. The role of randomization will be discussed further in the last paragraph of Section 3A.

D. Multidimensional search problems. If the space X has dimension >1, the search problems analogous to those we have already considered become much harder with respect to obtaining a satisfactory notion of optimality and to the computation of optimum procedures. We shall illustrate these difficulties here for the problem of finding the maximum of a function of two variables: X is the unit square I_2 (we could similarly consider the unit disc, the surface of a sphere, etc.), the coördinate axes of which we denote by $x^{(1)}$ and $x^{(2)}$; Y is the space of reals; $\mathfrak{F}$ is a subclass of the functions f from X into Y with unique maximum (at $g(f)$), this subclass being further specified below; D^* is arbitrary, while L will be further specified below.

For $0 < \epsilon < \frac{1}{2}$ and $0 < t < 1$, let $\mathfrak{F}_{\epsilon,t}$ be the class of all functions with the following properties: The values $f(0, 0)$, $f(0, 1)$, $f(\epsilon, t)$, $f(2\epsilon, 0)$, and $f(2\epsilon, 1)$ are all different and $f(\epsilon, t)$ is the largest; in the region $x^{(1)} \leqq 2\epsilon$ the function f is linear on each of the four triangles (with (ϵ, t) a vertex) determined by the five points at which f has just been discussed, and is thus determined by these five values of f; in the region $x^{(1)} \geqq 2\epsilon$, f is linear and $\partial f/\partial x^{(1)}$ is negative; $-f$ is convex (upward) on I_2. Let $\mathfrak{F}'$ be the union of the $\mathfrak{F}_{\epsilon,t}$ over $0 < \epsilon < \frac{1}{2}$ and $0 < t < 1$. $\mathfrak{F}'$ is then a subclass of the class of concave functions which are constant on no line segment of positive length parallel to a coördinate axis and which have a unique maximum. Yet, if $\mathfrak{F} = \mathfrak{F}'$, it is easy to see that, even for this simple class of functions each of which is linear on each of five regions whose union is I_2, no procedure in S_N (for any N) can yield a final region d which always contains the true f and always has diameter <1. Thus, if L reflects the diameter of d, the problem in this form has no satisfactory solution, even for quite restricted $\mathfrak{F}$. (Of course, similar situations arise for $\mathfrak{F}$'s consisting only of polynomials, etc.)

Several alternatives are available to us in trying to alter the form of the problem to one with a satisfactory solution. Firstly, $\mathfrak{F}$ can be changed. Even if we restrict $\mathfrak{F}$ to functions which are unimodal on each linear section, or even which are strictly concave, with a known upper bound on the absolute values of the second derivatives in any case, we obtain nothing of interest unless we also impose some sort of lower bound; the resulting problem is then computationally much too messy to proceed very far in obtaining the explicit form of ϵ-minimax procedures (see Section 3B for further discussion of some of these difficulties; the additional difficulties in the present problem are even worse), although under enough restrictions it is easy to exhibit procedures for which the final d is small (e.g., by overwhelming the problem with a fine enough net of observations) and to obtain a trivial lower bound on the efficiency of such a procedure relative to that of a minimax procedure.

A second possibility is to let L reflect the area of d rather than its diameter. This also yields a very messy problem; but even worse is the fact that this is a formulation of little practical value, since in almost any problem a square of side .01 will be preferable as a final decision to a rectangle $.0001 \times 1$ (in practice one would not know in advance which coördinate axis might end up parallel to the longer side).

A third possibility is to restrict $\mathfrak{F}$ less than in the consideration two paragraphs above and, as in Section 2C, to specify a positive value $c < 1$ and ask for the procedure in S^* which minimaxes the number of observations required to obtain a final d of diameter $\leq c$. The difficulty here is that any $\mathfrak{F}$ which does not yield a nontrivial result two paragraphs above will have the same effect here, since $\sup_f L(f, s)$ will be infinite for every s. It is of interest for such large $\mathfrak{F}$ to look for asymptotically minimax procedures (see Sections 1 and 4) relative to some reasonable $\{F_n\}$, but this is at least as hard computationally as the problem two paragraphs above (put $\mathfrak{F} = F_n$ there with n large).

A remaining possibility will be considered in Section 3B, where we shall treat a discrete analogue (in X) which eliminates many difficulties encountered above and thus singles out the basic computational problem for consideration; we shall see that, even when the problem is simplified in this way, the form of the solution still seems very complex.

3. The Lattice Case

A. Dimension one. We now consider discrete analogues of the problems of Sections 2A-C. For the sake of brevity we shall treat only the case $k = 2$ (see Section 2A(ii)) here, although similar results hold for the other order search problems.

Let M be a positive integer, let X be the set of M integers $1, 2, \cdots, M$, let Y be the reals, and let $\mathfrak{F}$ be the unimodal function on X, i.e., for each f in $\mathfrak{F}$ there is a unique maximum at $g(f)$ (a set of consecutive argument values at which the maximum is attained could be considered similarly, as mentioned in Section 2A (ii)) and f is strictly increasing or decreasing to the left or right, respectively, of $g(f)$. There are again many possible choices of L, but we consider here only the two corresponding to those of Section 2A and Section 2C.

(i) *The formulation of Section* 2C. The analogue of the problem of Section 2C is to let C be a positive integer and ask for the procedure in $\bar{S}^*$ (again $\bar{S}^*$ is the class of procedures for which the final interval d always contains $g(f)$) for which the maximum number of observations required to obtain a final interval of length $\leqq C - 1$ (i.e., there are at most C integers in d) is minimized. For the sake of definiteness, let $C = 1$, that is, $g(f)$ must be located exactly. Let $q(M)$ be the smallest integer such that $M < U_{q(M)}$, where U_n is the nth Fibonacci number (see Section 2A(ii)). It is quite easy to prove that $N' = q(M) - 2$ is the smallest integer such that there exists a procedure in $\bar{S}^*$ (for $C = 1$) requiring at most N' observations for all f in $\mathfrak{F}$: on the one hand, the first N' observations of the procedure $s^*_{N'+1}$ of Section 2A(ii) on the function f_1 on I defined by $f_1(x) = f(x/(M+1))$ yield a procedure s' in the $\bar{S}^*$ of the present problem requiring N' observations when $M + 1 = U_{q(M)}$, and hence a procedure requiring at most N' observations exists whenever $M < U_{q(M)}$; on the other hand, if for some M there were a procedure s^* in $\bar{S}^*$ requiring at most $N'' = q(M) - 3$ observations, then for the problem of Section 2A(ii) if we use on f the procedure in $S_{N''+1}$ obtained by using s^* on the function f_2 defined by $f_2(x) = f([M+1]x)$ for the first N'' observations, and then take the $(N''+1)$st observation at an argument ϵ to the right of the value in I corresponding to the largest of the first N'' observations, we obtain a final interval of length at most $\epsilon + (M+1)^{-1} \leqq \epsilon + (U_{q(M)-1} + 1)^{-1} = \epsilon + (U_{N''+2} + 1)^{-1}$, contradicting the result of Section 2A(ii) for $N = N'' + 1$.

As in the case of Section 2C, the procedure s' defined above is not always minimax among the randomized procedures if $M + 1 < U_{q(M)}$. For example, when $M = 5$, if $x_1 = 3$ and if $x_2 = 2$ or 4 with probability $\frac{1}{2}$ each, with x_3 the other argument next to whichever of x_1 and x_2 yields the larger observation (and x_4 next to x_3, if necessary), we obtain a procedure in $\bar{S}^*$ for which the maximum expected number of observations is 3.5, compared with 4 for the best nonrandomized procedure.

The case $C > 1$ can be treated similarly.

(ii) *The formulation of Section* 2A. We now turn to the formulation of Section 2A(ii): Find the procedure in S_N which minimizes the maximum length (or the number of integers) in d. We shall not detail the solution, but remark that if $N \geqq q(M) - 2$ the solution is obvious from the discussion three paragraphs above; otherwise, if in the previous formulation $r(C, M)$ denotes the smallest integer such that there is a nonrandomized procedure $s_{C,M}$ (say) requiring this number of observations for which d always contains at most C points, then in the present formulation if C' is the smallest integer for which $r(C', M) \leqq N$, the minimax procedure may be obtained from $s_{C',M}$ and has at most C' points in d. The duality remarked on in the last paragraph Section 2C is thus also present here.

Unlike the case of Section 2A(ii), where randomization is evidently of no help, the minimax nonrandomized procedure here is not necessarily minimax among the randomized procedures. For example, if $M = 5$ and $N = 3$, a minimax nonrandomized procedure has a maximum of 2 points in the final d, while the randomized procedure whose three observations are taken in the same manner as the first three observations of the procedure described three paragraphs above, has 1.5 as the maximum expected number of points in the final d.

(iii) *Remarks on randomization.* The explanation of the role of randomization is this: In the formulation of Section 2C and the first formulation of the present section, it is the discreteness of the number of observations which in some cases (of c or M and C) makes any nonrandomized minimax procedure fail to be minimax among the randomized procedures (which by randomizing can achieve *fractional* expected numbers of observations); on the other hand, in the formulation of Section 2A(ii) and the second formulation of the present section, it is the discreteness of X in the latter case and the lack of it in the former which accounts for the improvement sometimes achievable by randomization in the one case but not in the other.

B. Dimension >1. We now discuss briefly the difficulties present in computing minimax procedures for multidimensional analogues of the problems of Section 3A. We shall again consider the problem of search for a maximum, but other problems would reflect the same difficulties. Other difficulties of the type encountered in Section 2D, where for many reasonable $\mathfrak{F}$'s there was no satisfactory solution, are absent here because in the present formulation X is finite. We shall discuss the 2-dimensional case here, letting M_1 and M_2 be specified integers >1 and X the set of points $x = (x^{(1)}, x^{(2)})$ where $x^{(i)}$ is any integer satisfying $1 \leqq x^{(i)} \leqq M_i$; thus, X is an $M_1 \times M_2$ lattice. (One can give a similar treatment for a hexagonal lattice, etc.) Y will be the reals and the assumptions on $\mathfrak{F}$ will be discussed below. We shall consider only the problem of finding a procedure which minimizes the maximum number of observations needed to locate that x for which $f(x)$ is a maximum, but the same difficulties encountered here are present in any of the other formulations.

If $\mathfrak{F}$ consists of every function which for each fixed $x^{(1)}$ satisfies, as a function of $x^{(2)}$, the assumptions on the $\mathfrak{F}$ of Section 3A, then clearly $M_1[q(M_2) - 2]$ (where q is as defined in Section 3A) is the number of observations required by the nonrandomized minimax procedure. If $\mathfrak{F}$ is further restricted to functions for which $\max_{x^{(2)}} f(x^{(1)}, x^{(2)})$ is unimodal in $x^{(1)}$, then the number of observations required is reduced to $[q(M_1) - 2]$ $[q(M_2) - 2]$. The interesting problems are usually those where $\mathfrak{F}$ is restricted much more; however, it will be obvious that the remark of the preceding sentence yields, in the case of most reasonable $\mathfrak{F}$, a procedure whose efficiency (= maximum number of observations for a minimax procedure divided by the number for the given procedure) is $\geqq [q(\min[M_1, M_2]) - 2]^{-1}$, which may be of some practical value when at least one M_i is small. (Usually it will be obvious how to improve this efficiency somewhat.)

The results one obtains under various further reasonable restrictions on $\mathfrak{F}$ show wide differences, but they all have structures which seem so hopelessly unsystematic that it is impossible to list the minimax procedures for various M_1, M_2 with any of the simplicity of the one-dimensional case. The reason for this is easily explained: In the one-dimensional case any set of observations yielded an interval in which $g(f)$ must lie and where further observations were taken, while in the present case (under any of a large variety of reasonable assumptions) the corresponding region can have quite a messy form. A typical assumption might be that every f in $\mathfrak{F}$ is concave, or that $\mathfrak{F}$ also satisfies the assumptions of the previous paragraph when $x^{(1)}$ and $x^{(2)}$ are interchanged, or that for any rectilinear border-to-border path which separates X into two regions, the maximum of f on the path must be greater than that on at least one of the two regions. Perhaps it is more natural to suppose the problem arises from that of Section 2D by restricting the X there to a sublattice of I_2 and hence to phrase the assumption in terms of the original f on I_2; or it may seem reasonable that if the assumption in terms of the lattice holds for a given f, then it should also hold for that f restricted to any sublattice (the last two of the three assumptions just listed can easily be seen by example to be deficient in this respect); however, we shall not dwell on such matters here, since they do not seem connected with the essential difficulty.

As an example which will illustrate the computational difficulties, then, suppose that $M_1 = M_2 = 4$ and that every f in $\mathfrak{F}$ satisfies both of the last two of the three assumptions mentioned in the previous paragraph (if, e.g., only the first of these two is assumed, considerations are even messier) and takes on different values at all points of X, and that $x_1 = (2, 2)$, $x_2 = (3, 2)$, $x_3 = (2, 3)$, $x_4 = (3, 3)$. Three cases must be considered (the other possibilities may be obtained from these by reflection and rotation). If $f(x_1) > f(x_2) > f(x_4) > f(x_3)$, then $g(f)$ cannot be (1, 3) or (4, 2) by unimodality and cannot be (1, 4), (2, 4), (3, 4), (4, 4), or (4, 3) by the path assumption; the remaining 5 points take 4 observations to search in the worst case. If $f(x_4) > f(x_1) > f(x_2) > f(x_3)$, we can similarly eliminate all points except (3, 4), (4, 3), (4, 4) as possible locations for $g(f)$ (other than x_4), so three additional observations are required in the worst case. Finally, if $f(x_2) > f(x_4) > f(x_1) > f(x_3)$, only the points (1, 2), (1, 3), (1, 4), (2, 4), and (3, 4) are eliminated, and five additional observations are evidently required to search the remaining seven points in the worst case (put $x_5 = (2, 1)$ and $x_6 = (4, 3)$; if $f(x_5) > f(x_6)$, the point (4, 4) is eliminated and an observation at (3, 1) must eliminate one other point; similarly if $f(x_6) > f(x_5)$). Because of the possible reflected and rotated configurations, it seems clear that there will be a minimax procedure with the above x_1, x_2, x_3, x_4, and hence that a maximum of 9 observations is required by the minimax procedure. Thus, we see that, even in an example where the M_i are so small, the detailed enumeration of all possibilities (in order to arrive at a minimax procedure) is messy and allows no simple systematization; for larger M_i it becomes prohibitively exhausting except by machine (and even by machine, it is very lengthy).

It would be desirable under reasonable assumptions at least to give explicitly a class of procedures which have asymptotically (M_1 and M_2 large) reasonably high efficiency (i.e., close to 1); but even this seems quite difficult.

4. Asymptotic Results

We shall consider here an asymptotic form of the problem of Section 2A (i). Other formulations can be considered similarly, but this example will illustrate the methods. In subsection C of this section, the corresponding result for the problem of Section 2A(ii) is mentioned; other problems yield similar results. Our treatment of the problem of Section 2A(i) will be divided into two subsections. In both of them X will be the nonnegative reals (the case where X is the space of all reals can be treated by taking one observation at $x = 0$ to determine whether $g(f) \geqq 0$ or <0, and then using the appropriate procedure of the present case, which thus yields the same asymptotic results); Y, $\mathfrak{F}$, and g will be the same as in Section 2A(i) with this modification of X. The space D will be the same as X, and D^* will be the intervals of D. The set F_n of (1.2) will be the nonnegative reals which are $\leqq n$. The class $\bar{S}^*$ will be different in the two treatments; in the second it will be the class S^{**} of all procedures for which the final interval d always contains the zero $g(f)$ and is of length $\leqq 1$; in the first it will be a subclass S^{***} of this class, and will be described below. In both cases $L(f, s)$ will be the number of observations required to locate $g(f)$ with the required precision when s is used. In addition to the asymptotic considerations of (1.2), we shall treat the finer result dealing with the error term in the approach of (1.2). The discrete case (X = the integers) obviously yields analogous results, as in Section 3.

A. Asymptotic minimax procedures based on the procedure of Section 2A(i). Let S^{***} be the class consisting of every procedure s in S^{**} (described above) which is characterized by some nondecreasing sequence of positive numbers $b_1, b_2, \cdots$ (depending on s) tending to ∞ and such that s proceeds by taking observations at $x_i = b_i$ until the first i, say i_0, for which $f(x_{i_0}) \geqq 0$; thereafter the interval $[b_{i_0-1}, b_{i_0}]$ is searched for $g(f)$ in the manner used in Section 2A(i), by expanding the scale there by a factor $b_{i_0} - b_{i_0-1}$, the number of observations required in this search being the least nonnegative integer $\geqq \log_2(b_{i_0} - b_{i_0-1})$. Thus, the number of observations required by such a procedure when $g(f) = \theta$ is[3]

$$L(\theta, s) = i + \log_2(b_i - b_{i-1}) + O(1) \quad (b_{i-1} < \theta \leqq b_i), \tag{4.1}$$

where throughout this section $O(1)$ denotes a function bounded in θ as $\theta \to \infty$.

The next paragraph shows that any asymptotically minimax procedure relative to S^{***} is also asymptotically minimax relative to S^{**} in the sense of (1.2), but we shall see in Section 4B that procedures in S^{***} are not as good as the best in S^{**} in the sense of the first order error term. (It should be noted, however, that procedures in S^{***} will be simpler to use in either hand or machine computation.) For the moment, we shall characterize the asymptotically minimax procedures in S^{***} in the sense of (1.2), and then in the finer sense of the error term.

(i) *Optimality in the sense of* (1.2). The rough characterization of asymptotic minimax procedures in S^{***} is trivial; it follows from the result of Section 2A (i) that $q_n \geqq \log_2 n$ (even if $\bar{S}^* = S^{**}$), while it is easy to exhibit many sequences $\{b_j\}$ for which $L(\theta, s)/\log_2 \theta \to 1$ as $\theta \to \infty$, and such sequences thus give asymptotically minimax procedures. In fact, writing

[3] We hereafter write $L(\theta, s)$ for $L(f, s)$ when $g(f) = \theta$ and $s \in S^{***}$. When $s \notin S^{***}$, L may depend on f other than through $g(f)$; in such cases, $L(\theta, s)$ will hereafter be written for the supremum of what we previously called $L(f, s)$, taken over all f for which $g(f) = \theta$.

$b_k = k^{k^{\alpha_k}}$, if $\alpha_k \geqq 1$ and if for some nondecreasing sequence $h_1, h_2, \cdots$ tending to ∞ we can write, at least for large k, $\alpha_k = k/h_k \log_2 k$ (it will be obvious how to weaken this assumption), then for $b_{k-1} < \theta \leqq b_k$ we have (denoting by $o(1)$ a function tending to 0 as $k \to \infty$),

$$(4.2) \quad \frac{L(\theta, s)}{\log_2 \theta} \leqq \frac{k + \log_2 (b_k - b_{k-1})}{\log_2 b_{k-1}} < \frac{k}{\log_2 b_{k-1}} + \frac{\log_2 b_k}{\log_2 b_{k-1}}$$
$$\leqq o(1) + [1 + o(1)]k^{\alpha_k}/(k-1)^{\alpha_{k-1}};$$

the logarithm of $k^{\alpha_k}/(k-1)^{\alpha_{k-1}}$ is

$$(4.3) \quad \begin{aligned} \alpha_k \log_2 k - \alpha_{k-1} \log_2 (k-1) &= \frac{k}{h_k} - \frac{(k-1)}{h_{k-1}} \\ &= \frac{k}{h_k}\left[1 - \frac{h_k}{h_{k-1}}\right] + \frac{1}{h_{k-1}} < \frac{1}{h_{k-1}} \end{aligned}$$

which tends to 0 as $k \to \infty$, proving our assertion. Typical choices of α_k are $\alpha_k =$ constant $\geqq 1$ and $\alpha_k = r + (\log_2 c - \log_2 \log_2 k)/\log_2 k$ where $c > 0$ and $r > 1$; the latter corresponds to $b_k = 2^{ck^r}$.

(ii) *A Finer optimality characterization.* We have seen that in order for s to be an asymptotically minimax procedure in S^{**}, it is necessary and sufficient that

$$(4.4) \quad \begin{aligned} &L(\theta, s) = \log_2\theta + o(\log_2\theta) \text{ for an unbounded set of } \theta, \\ &L(\theta, s) \leqq \log_2\theta + o(\log_2\theta) \text{ for all } \theta, \end{aligned}$$

where $o(q(\theta))$ is a function whose ratio to $q(\theta)$ tends to 0 as $\theta \to \infty$. (It is easy to construct procedures which violate the equation of the first line of (4.4) on an unbounded set of θ; for example, modify an asymptotically minimax procedure of the previous paragraph by taking an observation also at $b_k - 1$ for each k for which an observation is taken at b_k.) We shall now refine (4.4) and prove that every s in S^{***} (*not* S^{**}; see Section 4B) satisfies

$$(4.5) \quad L(\theta, s) \geqq \log_2\theta + (8 \log_2\theta)^{\frac{1}{2}} + o[(\log_2\theta)^{\frac{1}{2}}]$$

on an unbounded set of θ. This lower bound on L is attainable: it is easily verified that, for the sequence $b_k = 2^{k^2/2}$, $L(\theta, s)$ satisfies (4.5) with equality for all θ; on the other hand, most procedures which satisfy (4.4), e.g., $b_k = k^k$ or $b_k = 2^{ak^2}$ with $0 < a \neq \frac{1}{2}$, do not share this finer optimality property with the sequence $2^{k^2/2}$. We note that, in verifying such results as (4.5), one need only consider (for a given s in S^{***}) θ's of the form $b_{k-1} + 1$, since $L(\theta, s)$ is constant for $b_{k-1} < \theta \leqq b_k$.

We proceed to prove (4.5). We may clearly assume $b_k > 1$ for all k, so $\log_2 b_k > 0$. If (4.5) is not satisfied, using the remark at the end of the preceding paragraph we see that there is a sequence $\{b_k\}$ and an ϵ, $0 < \epsilon < 1$, such that

$$k + \log_2(b_k - b_{k-1}) \leqq \log_2 b_{k-1} + (1 - \epsilon)\,(8 \log_2 b_{k-1})^{\frac{1}{2}} \tag{4.6}$$

for all sufficiently large k. The inequality of (4.6) is easily rewritten as

$$b_k \leqq b_{k-1}[1 + 2^{(1-\epsilon)(8\log_2 b_{k-1})^{\frac{1}{2}}-k}]. \tag{4.7}$$

Iterating (4.7), we obtain for all large k,

$$b_k \leqq C \prod_{k=2}^{\infty} [1 + 2^{(1-\epsilon)(8\log_2 b_{k-1})^{\frac{1}{2}}-k}] \tag{4.8}$$

where C is a positive constant. Since $b_k \to \infty$ as $k \to \infty$, the product on the right side of (4.8) must be infinite, and hence

$$\sum_{k=1}^{\infty} 2^{(1-\epsilon)(8\log_2 b_k)^{\frac{1}{2}}-k} = \infty. \tag{4.9}$$

This implies that the exponent in (4.9) is $> -(1 + \epsilon)k/2$ for infinitely many k (otherwise the sum converges) i.e., that $\log_2 b_k > k^2/32$ for infinitely many k. We now show that

$$\log_2 b_k > k^2/32 \quad \text{for all sufficiently large } k. \tag{4.10}$$

In fact, the inequality being true for infinitely many k, if it were not true for all large k there would exist arbitrarily large k for which $\log_2 b_{k-1} \leqq (k - 1)^2/32$ and $\log_2 b_k > k^2/32$; substituting these inequalities into (4.7) is easily seen to lead to a contradiction for k sufficiently large. Thus, (4.10) holds. Next, writing $\log_2 b_k = 2c_k k^2$, (4.7) can be rewritten

$$2^{2(c_k - c_{k-1})k^2} \leqq 2^{2c_{k-1}(1-2k)} + 2^{2c_{k-1}(1-2k)+4(1-\epsilon)(k-1)c_{k-1}^{1/2}-k}. \tag{4.11}$$

Since $A > 0$ implies $-Ax^2 + 2Bx \leqq B^2/A$ for all real B and x, putting $x = \sqrt{c_{k-1}}$ we see that the exponent of the last term of (4.11) is

$$\leqq 2(1 - \epsilon)^2(k - 1)^2/(2k - 1) - k \leqq -\epsilon k.$$

Also, since by (4.10) $c_k > 1/64$ for k sufficiently large, we have $2c_k(1 - 2k) < -k/17$ for k large. The conclusions of the two preceding sentences together with (4.11) show that, for some $\epsilon' > 0$ and all sufficiently large k, $c_k - c_{k-1} \leqq -\epsilon'/k$; this implies that $c_k \to -\infty$ as $k \to \infty$, contradicting (4.10) and thus completing the proof of (4.5).

(iii) *Extensions.* One can similarly investigate the analogue of (4.5) with more terms in the expansion. We forego the details, but will now indicate the method by considering the second order error term. Put $c_k = (1 + \delta_k)/4$ with δ_k decreasing to 0 as $k \rightarrow \infty$ (it is easy to verify that δ_k increasing to 0 is worse). If L is then expanded as on the right side of (4.5), the o term in the expansion when $\theta = b_{k-1} + 1$ becomes $k^2[\delta_k - \delta_{k-1}]/2 + k\delta_{k-1}^2$ $[1 + 0(1)]/4 + O(1)$. If δ_k is too large (e.g., $\delta_k = (\log \log k)^{-1}$), this is positive, while $\delta_k = C/\log_e k$ is the largest (smooth) order for which the error term may be negative, namely, $(1 + o(1))\ (- C/2 + C^2/4)k/(\log_e k)^2$, whose minimum (at $C = 1$) is $- (1 + o(1))k/4\ (\log_e k)^2 = - (1 + o(1))$ $(2 \log_2\theta)^{\frac{1}{2}}/(\log_e 2)^2(\log_2\log_2\theta)^2$.

B. Improvement through modification of the procedure of Section 2A(i). We have seen that an asymptotically minimax procedure in S^{**} in the sense of (1.2) can already be found in S^{***}. The analogous finer result concerning the first order error term is false; we shall now show that every procedure s in S^{**} satisfies (in place of (4.5))

$$L(\theta, s) \geqq \log_2\theta + \log_2\log_2\theta + o(\log_2\log_2\theta) \tag{4.12}$$

on an unbounded set of values θ, and shall exhibit a procedure s^* which satisfies (4.12) with equality for all θ and is thus optimum in this finer sense. The reason this improvement over (4.5) is possible is that the procedures in S^{***}, because of the way in which they employ the procedure of Section 2A(i) in successively halving the interval $[b_{i-1}, b_i]$, have the largest error term at $\theta = b_{i-1} + 1$, and the attainment of (4.5) is geared to this fact; thus, by splitting the interval other than into halves at each stage, we can hope to do better. (As remarked previously, procedures in S^{***} will be easier to use in practice.)

Using reasoning of the type employed in Section 2, it is easily proved that for our asymptotic considerations it suffices to consider only those procedures in S^{**} which are described exactly like those in S^{***} except for the modification in splitting just described. Thus, we hereafter limit our discussion to procedures any of which is described by a sequence $\{b_i\}$ and an increasing differentiable function q which is convex downward (q' nonincreasing) and satisfies $q(b_i) = i$; the splitting of the interval $[b_{i-1}, b_i]$ (if θ lies in this interval) is then accomplished by taking the next observation at $q^{-1}(i - \frac{1}{2})$, the following one at $q^{-1}(i - \frac{1}{4})$ or $q^{-1}(i - \frac{3}{4})$, etc., until an interval of length $\leqq 1$ in which θ lies is determined. Since this may be looked upon as successive halving of the interval of q-values (from $q = i - 1$ to $q = i$) until an interval $[u, v]$ (say) of q-values is obtained such that $q^{-1}(u) < \theta \leqq q^{-1}(v)$ and $q^{-1}(v) - q^{-1}(u) \leqq 1$, and since the number of halvings needed to accomplish this is $-\log_2 q'(\theta) + O(1)$ (the fact that the b_j's must increase fairly rapidly to achieve (1.2) yields the error term), we see that

$$L(\theta, s) = q(\theta) - \log_2 q'(\theta) + O(1) \tag{4.13}$$
$$= q(\theta) - \gamma \log_e q'(\theta) + O(1)$$

where $\gamma = \log_2 e$. We shall show in the next paragraph that if p is positive increasing and convex downward with $p' < 1$ (it will be clear from the use in (4.15) of the result to be proved about p that this is the only case we need consider), it is *impossible* that

$$p(\theta) - \log_e p'(\theta) < \log_e \theta + \log_e \log_e \theta \tag{4.14}$$

for all sufficiently large θ. If $q = \gamma p$, noting that $\gamma \log_e \log_e \theta = \log_2 \log_2 \theta + O(1)$, we see that the impossibility of (4.14) for all large θ implies the impossibility of

$$q(\theta) - \log_2 q'(\theta) < \log_2 \theta + \log_2 \log_2 \theta + O(1) \tag{4.15}$$

for all large θ; thus, (4.12) will be proved.

Suppose, then, that (4.14) is satisfied for all sufficiently large θ. It is then satisfied when the term $p(\theta)$ on the left side of (4.14) is absent, and that implies $p'(\theta) > 1/\theta \log_e \theta$ and hence $p(\theta) > \log_e \log_e \theta - C_1$ for all large θ, where C_i's hereafter denote positive constants. This last inequality and (4.14) imply $p'(\theta) > C_2/\theta$ for θ large, and hence $p(\theta) > C_2 \log_e \theta - C_3$, which with (4.14) implies $-\log_e p'(\theta) < (1 - \epsilon) \log_e \theta$ for all large θ and some $\epsilon > 0$; this implies $p(\theta) > \theta^\epsilon/\epsilon - C_4$ for large θ, and since $p' < 1$ this clearly contradicts (4.14) for θ sufficiently large.

An example of a function q for which (4.12) is satisfied with equality is $C \log_d^{(k)} \theta$ for $k \geqq 3$ where $C > 0$, $d > 1$, and where $\log_d^{(k)} \theta = \log_d \log_d^{(k-1)} \theta$ with $\log_d^{(1)} \theta = \log_d \theta$. We shall not consider the next error term in (4.12), although it is clear from (4.15) that we have actually proved (4.12) with the o term replaced by a constant.

C. Other problems. Asymptotic results for other problems can be obtained similarly. We mention here as an example the analogue of (4.5) for the problem of Section 2A(ii). This problem is now formulated like that of Section 4A, except that $\mathfrak{F}$ consists of the unimodal functions on the positive reals with maximum at $g(f)$ and S^{***} consists of procedures characterized by increasing sequences $\{b_k\}$ and which proceed by letting $x_1 = b_1$, $x_2 = b_2$, etc., until $f(x_k) \leqq f(x_{k-1})$ for the first time; thereafter, the interval $[b_{k-2}, b_k]$ is searched by an appropriate method of Section 2A (ii). Writing $\nu = 1/\mu$ (see Section 2A(ii)), an optimum procedure in S^{**} and, in the finer sense of the first order error term, in S^{***}, is given by taking $b_K = \nu^{K^2/4}$, which attains the lower bound (analogous to (4.5)) satisfied by all s in S^{***}, namely,

$$L(\theta, s) \geqq \log_{\nu}\theta + 4(\log_{\nu}\theta)^{\frac{1}{2}}(1 + o(1)) \tag{4.16}$$

on an unbounded set of θ.

The analogue of (4.12) can be determined similarly.

5. Estimating Integrals

In this section we consider, under two sets of assumptions, the problem of computing an estimate of the integral of a real function on the unit interval, based on a specified total number of observations. All (nonrandomized) sequential schemes are under consideration. It turns out that, under either set of assumptions, there is in S_n a procedure which minimaxes the width of an estimating interval which is certain to contain the actual value $g(f)$ of the integral (or, equivalently, the error of a numerical estimate), and which is *nonsequential*, i.e., which specifies in advance all values of the argument at which the function is to be observed. Of the two sets of assumptions considered, the first assumes nothing about the differential properties of the function, only that it is monotone; the second assumes nothing regarding monotonicity, only a Lipschitz condition. Without some assumption such as one of these, it is evident that no procedure can guarantee better than trivial accuracy against the *worst* function it encounters. It is also easy to see, in both cases, that no procedure can guarantee a *relative* error of <1; we take L to be the width of the interval d, which must contain the value $g(f)$ (or, if d is taken to be a number, L is absolute error).

A. The solution for F monotone. Firstly then, let $\mathfrak{F}$ be the class of all real functions on I which are nondecreasing and for which $f(0) = 0$, $f(1) = 1$, and let $g(f)$ be the Riemann integral of f over I. A sequential procedure based on n observations is defined as before but with $g(f) \in d$ as the final condition on members of S_n. It is desired to find, for $n \geqq 1$, a procedure s^* in S_n for which (1.1) holds. A very short demonstration will now serve to show that the procedure s_n^* defined by taking $x_i = i/(n + 1)$ for $i \leqq i \leqq n$ and

$$d = \left[\frac{1}{n+1}\sum_{i=0}^{n} f\left(\frac{i}{n+1}\right), \frac{1}{n+1}\sum_{i=1}^{n+1} f\left(\frac{i}{n+1}\right)\right]$$

accomplishes this. In fact, let $\bar{s}$ be any procedure in S_n. Then, when $f(x) = x$, the procedure $\bar{s}$ will (perhaps sequentially) prescribe a certain set of values of the x_i, say $x_i = a_i(1 \leqq i \leqq n)$. Now, any other function $f \in \mathfrak{F}$ with $f(a_i) = a_i(1 \leqq i \leqq n)$ will prescribe the same values of the x_i and will result in the same set of observations $f(a_i) = a_i (1 \leqq i \leqq n)$. The functions for which this is so have values of $g(f)$ ranging from $\sum_0^n b_i(b_{i+1} - b_i)$ to $\sum_0^n b_{i+1}(b_{i+1} - b_i)$, where $b_0 = 0$, $b_{n+1} = 1$, and $b_1, \cdots, b_n$ are the a_i rearranged in nondecreasing order. The difference between these extremes is $\sum_0^n (b_{i+1} - b_i)^2 \geqq 1/(n+1)$. Hence, the supremum (over all $f \in \mathfrak{F}$) of $L(f, \bar{s})$ is $\geqq 1/(n+1)$. On the other hand, the s_n^* defined above is clearly in S_n and achieves $1/(n+1)$ for its maximum loss, proving (1.1).

Remarks. The problem treated above may also be formulated without the restriction $f(0) = 0$, $f(1) = 1$. Removing these restrictions on the definition of $\mathfrak{F}$ but supposing $f(0) < f(1)$, $L(f, s)$ is unbounded in f for every s. If one either asks for a procedure which minimaxes $L(f, s)/[f(1) - f(0)]$, or else asks, among all procedures which have the cogredient property of assigning final interval $[as + b, at + b]$ to the function $af + b$ whenever $[s, t]$ is assigned to f $(a > 0)$, for a procedure which for every fixed pair of values $c < d$ minimaxes $L(f, s)$ among all $f \in \mathfrak{F}$ with $f(0) = c$, $f(1) = d$ (many other similar formulations, including some in the manner of Section 4, yield the same conclusion), then such a procedure s'_{n+2} in S_{n+2} $(n \geqq 1)$ is that which takes n observations like s_n^* and the other two at $x = 0$ and $x = 1$. We remark that s_n^* and s'_{n+2} are not admissible for $n > 1$. For example, alter s_2^* by taking $x_1 = \frac{1}{3}$ and then $x_2 = \frac{2}{3}$ unless $f(\frac{1}{3}) > \frac{2}{3}$, in which case $x_2 = \frac{1}{6}$. The resulting procedure will be better than s_2^*, since whenever $f(\frac{1}{3}) > \frac{2}{3}$ it will yield a final interval of width $\frac{2}{3} - f(\frac{1}{3})/2 < \frac{1}{3}$, which last is the width for s_2^*. (Improvements on s_n^* can be made similarly, but a detailed description of admissible minimax procedures for general n would be long.)

B. The solution for F satisfying a Lipschitz condition. We now consider a second assumption on $\mathfrak{F}$, namely, that, for a specified value $C > 0$, it consists of all real functions f on I for which $|f(x) - f(y)| \leqq C|x - y|$ for all x, y. We may assume $C = 1$. S_n is as above. This time an s_n^* satisfying (1.1) will be shown to be given for $n \geqq 1$ by letting $x_i = (2i-1)/2n$ for $1 \leqq i \leqq n$ and $d = [e - \delta, e + \delta]$ where

$$e = \frac{1}{n}\sum_{i=1}^{n} f\left(\frac{2i-1}{2n}\right),$$

$$\delta = \frac{1}{4n} - \frac{1}{4}\sum_{i=1}^{n-1}\left[f\left(\frac{2i+1}{2n}\right) - f\left(\frac{2i-1}{2n}\right)\right]^2.$$

In fact, consider as before any $\bar{s} \in S_n$, but this time let a_i be the values taken on by the (sequentially determined) x_i when $f(x) = 0$, with a corresponding meaning for the b_i. Any $f \in \mathfrak{F}$ with $f(a_i) = 0$ $(1 \leqq i \leqq n)$ will give the same observations. The maximum f satisfying this and our Lipschitz condition is $b_1 - x$ for $x \leqq b_1$, $x - b_i$ for $b_i \leqq x \leqq (b_i + b_{i+1})/2$

(where $1 \leqq i \leqq n - 1$), $b_{i+1} - x$ for $(b_i + b_{i+1})/2 \leqq x \leqq b_{i+1}$ (where $1 \leqq i \leqq n - 1$), and $x - b_n$ for $x \geqq b_n$; the minimum possible f is the negative of this. The difference between the corresponding maximum and minimum possible values of $g(f)$ for $f \in \mathfrak{F}$ with $f(a_i) = 0$ for all i is

$$b_1^2 + (1 - b_n)^2 + \frac{1}{2}\sum_{i=1}^{n-1} (b_{i+1} - b_i)^2,$$

which is $\geqq \frac{1}{2}n$. Hence, $\sup_f L(f, \bar{s}) \geqq \frac{1}{2}n$. But $\sup L(f, s_n^*) = \frac{1}{2}n$, and it is easy to verify that, for specified values of $f((2i - 1)/2n)$ with $f \in \mathfrak{F}$, the maximum and minimum possible values of $g(f)$ are $e \pm \delta$. Thus, $s_n^* \in S_n$ and (1.1) is proved. We note that this s_n^* is not admissible for $n > 2$. For example, s_3^* is inferior to the procedure which takes $x_1 = \frac{1}{6} = 1 - x_2$ but takes x_3 at $\frac{1}{12}$ (or $\frac{11}{12}$) instead of $\frac{1}{2}$ if $x_2 - x_1 - | f(x_2) - f(x_1) |$ is sufficiently small.

C. Comparison with other procedures; generalisations. It is interesting to note that many commonly used approximation methods (e.g., that based on Gauss's points) which are geared to very restricted $\mathfrak{F}$ do not show up well for the $\mathfrak{F}$ considered here (or for a suitably large subset, for which the procedures s_n^* described above will often still be minimax). Results similar to those obtained here hold for integration on a circle or in higher dimensions, etc. Other linear functionals, i.e., integration with respect to other measures than Lebesgue measure, can also be treated; a particular case of interest is the estimation of a sum from a few of its terms, given some condition of regularity on the terms. Other formulations (e.g., to obtain an estimate of g of specified accuracy for various $\mathfrak{F}$) and various asymptotic problems can be treated similarly. There are several other interesting problems involving integrals, among which we mention only that of finding an x such that $F(x) = \alpha$ (a specified constant), where $F(0) = 0$ and $F' = f$ (here L might depend on the error in the estimate or on the corresponding error in F); this bears the same relationship to the problem of Section 2A (i) that the problem of Section 2A(i) bears to that of 2A(ii). In both cases, with a minor regularity restriction, the second problem could be reduced to the first if one could observe f' rather than f (of course, this reduction would be more useful for the case of second order search than for first order search).

REFERENCES

[1] J. KIEFER, *Sequential minimax search for a maximum*, Proc. Amer. Math. Soc., 4 (1953), pp. 502–506.

[2] S. MONRO AND H. ROBBINS, *A stochastic approximation method*, Ann. Math. Stat., 22 (1951), pp. 400–407.

[3] J. KIEFER AND J. WOLFOWITZ, *Stochastic estimation of the maximum of a regression function*, Ann. Math. Stat., 23 (1952), pp. 462–466.

[4] K. L. Chung, *On a stochastic approximation method*, Ann. Math. Stat., 25 (1954), pp. 463–483.

[5] J. Sacks, *Limit distributions of stochastic approximation methods*, to be published.

[6] C. Blyth, *On minimax statistical decision problems and their admissibility*, Ann. Math. Stat., 22 (1951), pp. 22–42.

[7] A. Dvoretzky, J. Kiefer, and J. Wolfowitz, *Asymptotic minimax character of the sample distribution function*, Ann. Math. Stat., 27 (1946), pp. 642–669.

[8] J. Hodges and E. L. Lehmann, *Some applications of the Cramer-Rao inequality*, Second Berk. Symp. on Prob. and Stat., Univ. of Cal. Press, Berkeley (1951), pp. 13–22.

[9] J. Kiefer, Invariance, sequential estimation, and continuous time processes, Ann. Math. Stat., 28 (1957), pp. 573–601.

Cornell University

ON THE DEVIATIONS OF THE EMPIRIC DISTRIBUTION FUNCTION OF VECTOR CHANCE VARIABLES

BY

J. KIEFER(1) AND J. WOLFOWITZ(2)

1. Introduction. Let F be a distribution function (d.f.) on Euclicean m-space, and let $X_1, X_2, \cdots, X_n$, be independent chance variables with the common d.f. F. The empiric d.f. S_n is a chance d.f. defined for any $x = (x_1, \cdots, x_m)$ as follows: $nS_n(x)$ is the number of X_i's, $i=1, \cdots, n$, such that, for $j=1, \cdots, m$, the jth component $X_i^{(j)}$ of X_i is less than x_j. Define

$$D_n = \sup_x \left| S_n(x) - F(x) \right|,$$

$$D_n^+ = \sup_x (S_n(x) - F(x)),$$

$$D_n^- = \sup_x (F(x) - S_n(x)),$$

and

$$G_n(r) = P\{ D_n < r/n^{1/2} \},$$
$$G_n^+(r) = P\{ D_n^+ < r/n^{1/2} \},$$
$$G_n^-(r) = P\{ D_n^- < r/n^{1/2} \}.$$

In this paper we prove two theorems, of which the first is the following:

THEOREM 1. *For each m there exist positive constants c_0 and c such that, for all n, all F, and all positive r,*

$$1 - G_n(r) < c_0 e^{-cr^2}, \tag{1.1}$$

$$1 - G_n^+(r) < c_0 e^{-cr^2}, \tag{1.2}$$

$$1 - G_n^-(r) < c_0 e^{-cr^2}. \tag{1.3}$$

The nub of the theorem is, of course, that it sets a minimum rate at which $G_n(r)$, $G_n^+(r)$, and $G_n^-(r)$ go to one as $r \to \infty$, independent of n and F. It is rather curious that a bound independent of F can be given, since the limits of G_n, G_n^+, and G_n^- (as $n \to \infty$) depend on F for $m > 1$. The limits of G_n, G_n^+, and G_n^- for $m=1$ are of course known [1] and independent of F when F is continuous. The limits for $m > 1$ are at present writing unknown.

Received by the editors September 6, 1956.

(1) Research under contract with the Office of Naval Research.

(2) This research was supported by the United States Air Force under Contract No. AF(600)-685 monitored by the Office of Scientific Research.

Theorem 1 for $m=1$ was proved in [2] by a method which took as its point of departure an exact expression for $G_n^+(r)$ due to Smirnov [3]. No such formula is known for the case $m>1$. The method of the present paper is entirely different and does not use the result of [2](3). The extension of the result from $m=1$ to $m=2$ presents difficulties; the extension of the result for $m=2$ to larger values of m by our method of proof is obvious, and proceeds by induction on m. Theorem 1 is used in proving Theorem 2.

The constants c_0 and c in general depend upon m. We make no attempt in this paper to obtain the best possible constants or even to perform some tedious calculations which would improve them. At the end of the proof of Theorem 1 we calculate possible values of c and give some suggestions for improving the constants (it is shown in [2] that 2 is the best value of c for $m=1$; we also show at the end of the proof of Theorem 1 that $c<2$ for $m>1$). We also point out that the supremum operation can be performed over a larger class of sets without affecting the result.

Before stating Theorem 2 we introduce some additional notation. For fixed F and positive integral k, write A_k^m for the subset consisting of every point in Euclidean m-space for which, for $1\leqq j\leqq m$, the jth coordinate w_j satifies $F_j(w_j)\leqq h_j/k\leqq F_j(w_j+0)$ for any integers h_j, where F_j is the (marginal) d.f. of $X_i^{(j)}$. Write

$$D_{n,k} = \max_{x\in A_k^m} | S_n(x) - F(x) | .$$

$$D_{n,k}^+ = \max_{x\in A_k^m} [S_n(x) - F(x)].$$

$$D_{n,k}^- = \max_{x\in A_k^m} [F(x) - S_n(x)].$$

as well as

$$G_{n,k}(r) = P\{ D_{n,k} < r/n^{1/2}\},$$

$$G_{n,k}^+(r) = P\{ D_{n,k}^+ < r/n^{1/2}\},$$

$$G_{n,k}^-(r) = P\{ D_{n,k}^- < r/n^{1/2}\},$$

and

$$H_n(r, r') = P\{ D_n^+ < r/n^{1/2}, D_n^- < r'/n^{1/2}\},$$

$$H_{n,k}(r, r') = P\{ D_{n,k}^+ < r/n^{1/2}, D_{n,k}^- < r'/n^{1/2}\}.$$

We shall also denote by

$$G_{\infty,k}, \quad G_{\infty,k}^+, \quad G_{\infty,k}^-, \quad H_{\infty,k}$$

(3) In a first, unpublished, version of [2], a weaker result than that mentioned below as appearing in [2] for the case $m=1$, was proved by a method which has points in common with the present proof of Theorem 1; one idea used in this method is due to P. Erdös.

the respective limits as $n\to\infty$ of the d.f.'s $G_{n,k}$, $G^+_{n,k}$, $G^-_{n,k}$, and $H_{n,k}$; the existence of these limits is a consequence of the multivariate central limit theorem.

Our second result is

THEOREM 2. *For every m and F, there exists a d.f. G (resp., G^+, G^-, H) such that the sequence of d.f.'s G_n (resp., G_n^+, G_n^-, H_n) converges to G (resp., G^+, G^-, H) at every continuity point of the latter as $n\to\infty$ and such that the sequence of d.f.'s $G_{\infty,k}$ (resp., $G^+_{\infty,k}$, $G^-_{\infty,k}$, $H_{\infty,k}$) converges to this d.f. G (resp., G^+, G^-, H) at every continuity point of the latter as $k\to\infty$.*

It is obvious that G, G^+, G^-, and H cannot be degenerate unless F is. Of course, as noted above, these d.f.'s depend on F.

Theorem 2 generalizes the result of Donsker [4] for the case $m=1$; we remark that our proof starts ab initio and does not make use of Donsker's result or method. Donsker's result is needed to justify Doob's [5] computation of G, G^+, G^-, and H in the case $m=1$, and Theorem 2 could perhaps prove of similar use in the more difficult problem of computing these limiting d.f.'s when $m>1$ if this is to be done by consideration of a Gaussian process (depending on F) with m-dimensional time. Donsker's result was also used in [2] in the case $m=1$ for proving certain asymptotic optimum properties of S_n in estimating F.

(*Added in proof*: In another paper we shall prove that analogous optimality results hold for S_n when $m>1$, *even though D_n is no longer distribution free and the distribution theory of D_n is unknown*. These results follow from those of the present paper, arguments like those of [1], and the fact that the integral of a continuous bounded function with respect to G_n converges to that with respect to G uniformly in continuous F; the latter result will also appear in another paper.)

Some generalizations of Theorem 2 are mentioned at the end of §3.

2. **Proof of Theorem** 1. We shall give a detailed proof for $m=2$. As we have remarked earlier, the proof for general m is by induction on m and is obvious to carry out. We shall indicate below the point where induction would be used. The result for the case $m=1$ can be obtained by an argument similar to but simpler than that used below. Alternatively, it can be obtained from Lemma 2 of [2].

Throughout this section c_0 and c will be a generic notation for positive constants which do not depend on n, r, and F. Hence these symbols in different places will not, in general, stand for the same numbers. No confusion will be caused by this.

We have

$$(1-G_n(r)) \leq (1-G_n^+(r)) + (1-G_n^-(r)).$$

We will content ourselves with proving (1.2). The proof of (1.3) follows in the same fashion, and (1.1) then follows from (1.2) and (1.3).

We shall assume that F is continuous. If this is not so (1.1), (1.2), and (1.3) hold a fortiori; the proof of this is the same as in the one-dimensional case and therefore obvious. Since F is continuous we may transform $X_i^{(1)}$ and $X_i^{(2)}$ separately so that the marginal d.f.'s F_j are uniform on $[0, 1]$ without changing $G_n(r)$, $G_n^+(r)$, or $G_n^-(r)$; we hereafter assume that the F_j are uniform on $[0, 1]$.

In the discussion which follows we shall always assume, to simplify the discussion, that, for any given number x_1, there is at most one i such that $X_i^{(1)} = x_1$. The probability that this be not so is zero.

In the course of the proof we shall always assume that $r < n^{1/2}$. The theorem is trivially true when this is not so.

If the theorem is true for all $r > R > 0$, it is true for all $r \geqq 0$. One has only to enlarge c_0, if necessary, so that $c_0 e^{-cR^2} > 1$; the inequalities (1.1), (1.2), and (1.3) are then trivially true. It will therefore be sufficient to prove the theorem for all r sufficiently large, say $> R > 3$. Then $n > R^2$.

Since n and r will be fixed in the present discussion we may allow ourselves the luxury of a notation simpler than that of the next section and not display all dependences on n and r. (We remind the reader that c and c_0 will not depend on n and r.) Define the events

$$L = \left\{ S_n(x_1, x_2) - F(x_1, x_2) > \frac{r}{n^{1/2}} \text{ for some } x_1 \leqq \frac{1}{2} \right\},$$

$$L' = \left\{ S_n(x_1, x_2) - F(x_1, x_2) < -\frac{r}{n^{1/2}} \text{ for some } x_1 \leqq \frac{1}{2} \right\},$$

$$B = \left\{ S_n\left(\frac{1}{2}, x_2\right) - F\left(\frac{1}{2}, x_2\right) > \frac{r}{4n^{1/2}} \text{ for some } x_2 \right\},$$

$$\bar{L} = \left\{ S_n(x_1, x_2) - F(x_1, x_2) > \frac{r}{n^{1/2}} \text{ for some } x_1 > \frac{1}{2} \right\},$$

$$L^* = \left\{ S_n(x_1, x_2) - F(x_1, x_2) > \frac{r}{n^{1/2}} \text{ for some } x_1 > \frac{1}{2}, \ x_2 > \frac{1}{2} \right\}.$$

Define the chance variables (z_1, z_2) when the event L occurs (we shall not need them when L does not occur) as follows: First, z_1 is the infimum of those values x_1 ($\leqq 1/2$) for which $\sup_{x_2}[S_n(x_1, x_2) - F(x_1, x_2)] > r/n^{1/2}$. There is then an i such that $X_i^{(1)} = z_1$. We define $z_2 = X_i^{(2)}$. We now define the event $L(x_1, x_2)$ for any pair x_1, x_2, $0 \leqq x_1 \leqq 1/2$, $0 \leqq x_2 \leqq 1$, as follows: $L(x_1, x_2)$ is the subset of L where $z_1 = x_1$, $z_2 = x_2$.

Define $r(x_1, x_2)$ as

$$\frac{1}{n^{1/2}} \left[(\text{least integer} > nF(x_1, x_2) + n^{1/2} r) - nF(x_1, x_2) \right].$$

Then, for almost all values (x_1, x_2) of (z_1, z_2), the event $L(x_1, x_2)$ implies the event

$$\left\{S_n(x_1, x_2) - F(x_1, x_2) = \frac{r(x_1, x_2)}{n^{1/2}}\right\}. \tag{2.1}$$

Define, for any x_2, $0 \leqq x_2 \leqq 1$, the events

$$B(x_2) = \left\{S_n\left(\frac{1}{2}, x_2\right) - F\left(\frac{1}{2}, x_2\right) > \frac{r}{4n^{1/2}}\right\}$$

and

$$J = \bigcup_{0 \leq x_1 \leq 1/2} \bigcup_{0 \leq x_2 \leq 1} L(x_1, x_2) B(x_2).$$

Obviously

$$1 - G_n^+(r) \leqq P\{L\} + P\{\bar{L}\}.$$

Our immediate goal will now be to prove

$$P\{L\} < c_0 e^{-cr^2} \tag{2.2}$$

for all F, and for r sufficiently large, say $> R$.

We have

$$P\{L\} \cdot \text{ess. inf. } P\{J \mid z_1, z_2\} \leqq \int_{0 \leqq x_1 \leqq 1/2; 0 \leqq x_2 \leqq 1} P\{J \mid x_1, x_2\} d_{x_1, x_2} P\{z_1 < x_1, z_2 < x_2\}$$
$$= P\{J \cap L\} \leqq P\{J\} \leqq P\{B\}.$$

Hence

$$P\{L\} \leqq \frac{P\{B\}}{\text{ess. inf. } P\{J \mid z_1, z_2\}}. \tag{2.3}$$

Our plan to prove (2.2) is as follows: First, we shall prove that

$$\text{ess. inf. } P\{J \mid z_1, z_2\} \geqq 1/2. \tag{2.4}$$

Then we shall prove that

$$P\{B\} < c_0 e^{-cr^2} \tag{2.5}$$

for $r > R$ and $n > r^2$.

Suppose the event $L(x_1, x_2)$ has occurred. Since there is exactly one i such that $X_i^{(1)} = x_1$, and since

$$\sup_{x_2} \left[S_n(x_1', x_2) - F(x_1', x_2)\right] \leqq \frac{r}{n^{1/2}}$$

for $x_1' < x_1$, we have

$$nS_n(x_1, 1) < nF(x_1, 1) + rn^{1/2} + 1. \tag{2.6}$$

Hence the number N of $X_1, \cdots, X_n$ which have first coordinate greater than x_1 is at least

$$M = n(1 - x_1) - rn^{1/2} - 1. \tag{2.7}$$

First suppose that $M \leqq 0$. Then

$$\frac{r}{n^{1/2}} \geqq -\frac{1}{n} - (1 + x_1). \tag{2.8}$$

Obviously $0 \leqq r(x_1, x_2) - r \leqq 1/n^{1/2}$. The event $B(x_2)$ occurs when

$$S_n\left(\frac{1}{2}, x_2\right) > F\left(\frac{1}{2}, x_2\right) + \frac{r}{4n^{1/2}} \cdot \tag{2.9}$$

From (2.1) and (2.8) we obtain that

$$\begin{aligned}
S_n\left(\frac{1}{2}, x_2\right) &\geqq S_n(x_1, x_2) \geqq F(x_1, x_2) + \frac{r}{n^{1/2}} \\
&\geqq F(x_1, x_2) + (1 - x_1) - \frac{1}{n} \\
&= F\left(\frac{1}{2}, x_2\right) - \left[F\left(\frac{1}{2}, x_2\right) - F(x_1, x_2)\right] + (1 - x_1) - \frac{1}{n} \qquad (2.10)\\
&\geqq F\left(\frac{1}{2}, x_2\right) - \left(\frac{1}{2} - x_1\right) + (1 - x_1) - \frac{1}{n} \\
&= F\left(\frac{1}{2}, x_2\right) + \frac{1}{2} - \frac{1}{n} \cdot
\end{aligned}$$

Since $r/4n^{1/2} < 1/4$, it follows that, for $n > R^2 > 9$ (which is all we need consider), (2.9) holds.

Suppose now that $M > 0$. Let R_0 be the region in the x_1', x_2' plane defined by the inequalities

$$x_1 < x_1' \leqq 1/2, \quad 0 \leqq x_2' \leqq x_2.$$

In order for $B(x_2)$ to occur it is sufficient, by (2.9) and (2.1), that, of the N chance variables among $X_1, \cdots, X_n$ whose first coordinates are greater than x_1, at least

$$n(F(1/2, x_2) - F(x_1, x_2)) - 3rn^{1/2}/4 \tag{2.11}$$

take on values in R_0. We shall compute a lower bound for the probability of this under the assumption that $N = M$ (>0). It will be easy to see that, if

$N > M$ or M is not an integer, the probability is a fortiori greater than this lower bound, which is $>1/2$ for $r > R$. This will prove (2.4).

If we define

$$p = \frac{F(1/2, x_2) - F(x_1, x_2)}{1 - x_1} \tag{2.12}$$

and

$$t = \frac{n(F(1/2, x_2) - F(x_1, x_2)) - 3rn^{1/2}/4 - Mp}{(Mp(1 - p))^{1/2}}, \tag{2.13}$$

we obtain that

$$t = \frac{r(p - 3/4) + p/n^{1/2}}{\left(\dfrac{Mp(1 - p)}{n}\right)^{1/2}}.$$

Since $p < 1/2$ and $M/n \leqq 1$ it follows that $t < -r/4$ for $r > R$. The probability in question is the probability that, of M independent Bernoulli chance variables with common probability p of a "success," the number N^* of "successes" satisfy the inequality

$$\frac{N^* - EN^*}{(E(N^* - EN^*)^2)^{1/2}} > t. \tag{2.14}$$

This probability is greater than

$$P\left\{\frac{N^* - EN^*}{(E(N^* - EN^*)^2)^{1/2}} > -\frac{r}{4}\right\} \tag{2.15}$$

which, by Chebyshev's inequality, is greater than $1 - 16/r^2$, which, for $r > R$, is $>1/2$, as was to be proved. This proves (2.4).

From [6, p. 288, Equation (96)], it follows that, for $r > R$,

$$P\left\{\left|S_n\left(\frac{1}{2}, 1\right) - \frac{1}{2}\right| > \frac{r}{16n^{1/2}}\right\} < c_0 e^{-cr^2}. \tag{2.16}$$

Suppose now that the event $\{|S_n(1/2, 1) - 1/2| \leqq r/16n^{1/2}\}$ occurs. Then the number n_1 of chance variables $X_1, \cdots, X_n$ with first coördinate not greater than $1/2$ satisfies

$$\frac{n}{2} - \frac{rn^{1/2}}{16} = n_2 < n_1 < n_3 = \frac{n}{2} + \frac{rn^{1/2}}{16} < \frac{9n}{16}.$$

Let $T_{n_1}(x_2)$ denote $1/n_1$ multiplied by the number of chance variables $X_1, \cdots, X_n$ whose first coordinates are less than $1/2$ and whose second coördinates are less than x_2. The relation

$$(2.17)\qquad nS_n\left(\frac{1}{2}, x_2\right) - nF\left(\frac{1}{2}, x_2\right) > \frac{rn^{1/2}}{4}$$

implies the relation

$$(2.18)\qquad n_1T_{n_1}(x_2) - n_1\left[2F\left(\frac{1}{2}, x_2\right)\right] > \frac{rn^{1/2}}{4} - \frac{rn^{1/2}}{8} F\left(\frac{1}{2}, x_2\right),$$

whose right member is not less than

$$(2.19)\qquad \frac{3rn^{1/2}}{16} > \frac{r(n_3)^{1/2}}{4} > \frac{r(n_1)^{1/2}}{4}.$$

The theorem for the case $m=1$ implies that

$$(2.20)\qquad P\left\{\sup_{x_2}\left[n_1T_{n_1}(x_2) - n_1\left(2F\left(\frac{1}{2}, x_2\right)\right)\right] > \frac{r(n_1)^{1/2}}{4}\right\} < c_0e^{-cr^2}.$$

Equations (2.16) to (2.20) prove (2.5) and hence (2.2).

In the proof of Theorem 1 for general m the induction on m would occur at this point. We have just used the theorem for $m=1$ to prove (2.5) for $m=2$. We can then use this to prove the result corresponding to (2.5) for $m=3$, and so on (x_2 represents all variables other than x_1 in this proof, when $m>2$).

Returning to the case $m=2$, the proof of

$$(2.21)\qquad P\{L'\} < c_0e^{-cr^2}$$

is practically the same as that of (2.2), and will be omitted. We shall henceforth assume that (2.21) holds, and use this fact to prove that

$$(2.22)\qquad P\{\bar{L}\} < c_0e^{-cr^2}.$$

First, applying the result (2.2) to the chance variables $X_1^*, X_2^*, \cdots$ defined by $X_i^* = (X_i^{(2)}, X_i^{(1)})$, we obtain for the original sequence $X_1, X_2, \cdots$ that

$$(2.23)\qquad P\left\{S_n(x_1, x_2) - F(x_1, x_2) > \frac{r}{n^{1/2}} \text{ for some } x_2 \leq \frac{1}{2}\right\} < c_0e^{-cr^2}.$$

For any pair (x_1, x_2), $1/2<x_1\leq 1$, $1/2<x_2\leq 1$, we define the following regions in the x_1', x_2' plane:

$$U_1(x_1, x_2) = \{x_1', x_2' \mid x_1 < x_1' \leq 1, x_2 < x_2' \leq 1\},$$
$$U_2(x_1, x_2) = \{x_1', x_2' \mid x_1 < x_1' \leq 1, 0 \leq x_2' < x_2\},$$
$$U_3(x_1, x_2) = \{x_1', x_2' \mid 0 \leq x_1' < x_1, x_2 < x_2' \leq 1\},$$

and the following events for $i=1, 2, 3$:

$$Q_i = \left\{\text{for some } (x_1, x_2), \frac{1}{2} < x_1 \leqq 1, \frac{1}{2} < x_2 \leqq 1, \text{the number of } (X_1, \cdots, X_n)\right.$$
$$\left.\text{in } U_i(x_1, x_2) \text{ minus expected number } < -\frac{rn^{1/2}}{3}\right\}.$$

Obviously

$$L^* \subset Q_1 \cup Q_2 \cup Q_3.$$

We shall prove

$$P\{Q_v\} < c_0 e^{-cr^2}, \qquad v = 1, 2, 3. \tag{2.24}$$

The result (2.24) follows for $v=1$ from the application of (2.21) to the sequence of chance variables $(1-X_i^{(2)}, 1-X_i^{(1)})$, for $v=2$ by the application of (2.21) to the sequence of chance variables $(1-X_i^{(1)}, X_i^{(2)})$, and for $v=3$ by the application of (2.21) (in the form (2.23)) to the sequence of chance variables $(X_i^{(1)}, 1-X_i^{(2)})$. Thus, (2.24) is proved, and this and (2.23) imply (2.22) and hence (1.2).

The proof of (1.3) is completed in a similar manner. Obviously (1.2) and (1.3) imply (1.1). This completes the proof of Theorem 1.

We shall now obtain explicit possible values for the constant c (c_0 could be obtained similarly, but this is of less interest). First consider the case $m=2$. In the definition of the set B, let us replace $r/4$ by $r/(2+\epsilon)$ with $\epsilon>0$; the proof of Theorem 1 then still holds, but will yield a larger value of c. Making appropriate changes in the argument, an analogue of (2.15) holds, as before. In (2.16) and what follows, put λ for 1/16. The constant c on the right side of (2.16) then becomes $2\lambda^2$. The displayed inequality on n_1 becomes

$$n\left(\frac{1}{2} - \lambda\right) < \frac{n}{2} - r\lambda n^{1/2} < n_1 < \frac{n}{2} + r\lambda n^{1/2} < n\left(\lambda + \frac{1}{2}\right). \tag{2.25}$$

Equation (2.17) (with $1/(2+\epsilon)$ for 1/4) and Equation (2.25) imply an analogue of (2.18) with (2.19) replaced by

$$rn^{1/2}\left(\frac{1}{2+\epsilon} - \lambda\right) > rn_1^{1/2}\left(\frac{1}{2+\epsilon} - \lambda\right) \Big/ \left(\lambda + \frac{1}{2}\right)^{1/2}.$$

The fact that we can take $c=2$ for $m=1$ implies, in place of (2.20),

$$P\left\{\sup_{x_2}\left| n_1 T_{n_1}(x_2) - 2n_1 F\left(\frac{1}{2}, x_2\right)\right| > rn^{1/2}\left(\frac{1}{2+\epsilon} - \lambda\right) \Big/ \left(\lambda + \frac{1}{2}\right)^{1/2}\right\}$$
$$< c_0 \exp\left(-2r^2\left(\frac{1}{2+\epsilon} - \lambda\right)^2 \Big/ \left(\lambda + \frac{1}{2}\right)\right). \tag{2.26}$$

The minimum of the coefficients of r^2 in the exponents of (2.26) and the analogue of (2.16) is maximized when ϵ is small by taking $\lambda \doteq .266$; this gives $c > .142 + o(1)$ as $\epsilon \to 0$, in (2.5). The same value of c may be obtained similarly for L' in (2.21), and also in (2.23). If this value of c is multiplied by 1/9, we obtain a value applicable in (2.24), (2.22), and (1.2); a similar argument applies for (1.1). Thus, we also obtain $c > .0157 - \epsilon'$ in (1.3) for $m = 2$, where ϵ' is an arbitrary positive value. Thus, $c = .0157$ is a possible value in (1.3) for $m = 2$.

For general m, we may similarly obtain a possible value for c. Let $2d_m^2$ be the value for c obtained by this argument for dimension m, with $d_m > 0$. For dimension m we then obtain $2d_{m-1}^2(1/2 - \lambda)^2$ for the coefficient of r^2 in (2.26) and thus the solution λ of the equation $2\lambda^2 = 2d_{m-1}^2(1/2 - \lambda)^2/(\lambda + 1/2)$ is the value of λ which maximizes the minimum of the coefficients of r^2 in (2.16) and (2.26). Rather than carry out the obvious analogue of the case $m = 2$ in terms of this inexplicit λ, we shall obtain explicitly a slightly smaller value of c. This value is suggested by the fact that d_{m-1}, and hence the above λ, is small for $m > 2$. Taking then for λ the value $d_{m-1}/2^{1/2}$, the two coefficients of r^2 are almost equal, the smaller (that of (2.26)) being

$$2d_{m-1}^2(1/2 - d_{m-1}/2^{1/2})^2/(1/2 + d_{m-1}/2^{1/2}).$$

The factor 1/9 above must be replaced by $(2^m - 1)^{-2}$. Thus, we obtain for possible values of d_m and c (the ϵ' no longer being needed):

$$c = 2d_m^2 \quad \text{(dimension } m\text{)},$$

$$(2.27) \qquad d_m = \frac{d_{m-1}(1/2 - d_{m-1}/2^{1/2})}{(1/2 + d_{m-1}/2^{1/2})^{1/2}(2^m - 1)} \qquad (d_2 = .088).$$

The above possible value for c is probably not a very good one ($c = .0157$ for $m = 2$ and $c = .000107$ for $m = 3$). It could be improved by considering $S_n(x) - F(x)$ at a large finite number of lines (in the case $m = 2$, for example) instead of just on the line $x_1 = 1/2$; but this would be at the expense of more tedious computations. The value $c = 2$ obtained in [2] for the case $m = 1$ is the best possible in the sense that (1.3) is clearly false for any $c > 2$ and any c_0. We next show that $c < 2$ for $m > 1$; i.e., (1.3) is false for $c = 2$ and any c_0 when $m > 1$ (in fact, this is so even if c_0 is permitted to depend on F).

In fact, consider the case $m = 2$ and suppose F_1 is the d.f. which distributes all probability uniformly on the line $L: x_1 + x_2 = 1$. Then $S_n(x, y) = 0$ w.p. 1 if $x + y \leq 1$, and for $x + y > 1$ we have $nS_n(x, y) =$ number of observations on L between $(1 - y, y)$ and $(x, 1 - x)$. Let $S_n^*(u) = S_n(u, 1)$. Of course, $S_n^*(u)$ is a univariate sample d.f. corresponding to the uniform d.f. on the unit interval. Denoting by D_n^+ and D_n^- the supremum positive and negative deviations of $S_n^*(u)$ from the function u, $0 \leq u \leq 1$, we have, w.p.1.,

$$\sup_{x,y} | S_n(x, y) - F_1(x, y) | = \sup_{u,v} | (S_n^*(u) - u) - (S_n^*(v) - v) | = D_n^+ + D_n^-.$$

From [5] we obtain for the limiting d.f. of $n^{1/2}(D_n^+ + D_n^-)$ (note, e.g., Equation (4.6) of [7], which gives the limiting d.f. of $n^{1/2}(D_n^+ + D_n^-)/2$), as $r \to \infty$,

$$\lim_{n\to\infty} P\{n^{1/2}(D_n^+ + D_n^-) > r\} \sim 8r^2 e^{-2r^2},$$

which demonstrates the impossibility of taking $c=2$. We note, in fact, that (1.3) cannot hold for $c=2$ and any c_0 and for all absolutely continuous F (or, instead, for all *discrete* F); this is obtained easily from the above result by taking a fixed r so large that $4r^2 > c_0$, a k so large that the above limiting probability for the case of the discrete approximation of F_1 is $> 6r^2e^{-2r^2}$, and an absolutely continuous d.f. F_2 whose probability is concentrated on such small spheres about the discrete points that the probability of a deviation $> r$ cannot be smaller for F_2 than for F_1.

The supremum operation involved in the definition of D_n, D_n^+, and D_n^- is over all sets of the form $x_j \leqq a_j$, $j=1, \cdots, m$, for all $a=(a_1, \cdots, a_m)$ in m-space. It is obvious that Theorem 1 applies also to the case when the supremum is taken over any of several larger classes of sets such as, for example, that which consists of all rectangular parallelepipeds with sides parallel to the coördinate planes. This will be of interest in statistical applications where it is often required or desired that the results be invariant under certain transformations of the chance variables, e.g., $X \to -X$.

3. **Proof of Theorem** 2. Let I^m denote the closed unit m-cell $\{x \mid 0 \leqq x_1, \cdots, x_m \leqq 1\}$. We shall first prove Theorem 2 for the case when F is continuous, and then, at the conclusion of the proof, we shall remark on how the proof proceeds for discontinuous F. As in §2, since F is now assumed continuous, it suffices to consider the case where all F_j's are uniform on $[0, 1]$, and we hereafter assume we are in this case. Write $Q_{k,0} = I^m$, and for $j>0$ let $Q_{k,j}$ be the subset of I^m whose first j coördinates are integral multiples of $1/k$ (thus, $Q_{k,m} = A_k^m$). Write

$$D_{n,k,j}^+ = \sup_{x \in Q_{k,j}} [S_n(x) - F(x)],$$

$$D_{n,k,j}^- = \sup_{x \in Q_{k,j}} [F(x) - S_n(x)].$$

For fixed $d>0$, r, and r', we shall show in the succeeding paragraphs that

$$\begin{aligned} &\limsup_{k\to\infty} \limsup_{n\to\infty} [P\{D_{n,k,j}^+ < r/n^{1/2};\ D_{n,k,j}^- < r'/n^{1/2}\} \\ (3.1)\qquad &- P\{D_{n,k,j-1}^+ < (r+d)/n^{1/2};\ D_{n,k,j-1}^- < (r'+d)/n^{1/2}\}] \leqq 0 \end{aligned}$$

for $1 \leqq j \leqq m$. Adding these inequalities over j yields

$$\limsup_{k\to\infty}\limsup_{n\to\infty}\left[H_{n,k}(r, r') - H_n(r + md, r' + md)\right] \leqq 0. \tag{3.2}$$

We have remarked in §1 that $\lim_n H_{n,k} = H_{\infty,k}$ exists. Hence, writing r for $r+md$ and r' for $r'+md$, (3.2) becomes

$$\limsup_{k\to\infty} H_{\infty,k}(r - md, r' - md) \leqq \liminf_{n\to\infty} H_n(r, r'). \tag{3.3}$$

Write $H^*(r, r') = \limsup_k H_{\infty,k}(r, r')$. Since obviously $H_n(r, r') \leqq H_{n,k}(r, r')$, we obtain from (3.3),

$$H^*(r - md, r' - md) \leqq \liminf_{n\to\infty} H_n(r, r') \leqq \limsup_{n\to\infty} H_n(r, r') \leqq H^*(r, r'). \tag{3.4}$$

Since H^* is clearly monotone and bounded, it is continuous except possibly on a denumerable set of lines parallel to the coördinate axes. Letting d tend to zero in (3.4) at continuity points (r, r') of H^*, we see that $\lim_n H_n(r, r')$ exists for all points (r, r') in the plane, except possibly on a denumerable set of lines parallel to the coördinate axes. This limit determines a left-continuous function H (say) which has variation one by Theorem 1 and which is clearly a d.f. Hence the sequence H_n converges to a d.f. H at every continuity point of the latter. Finally, since clearly we can also write, for all continuity points (r, r') of H,

$$\begin{aligned} H(r, r') &\leqq \liminf_{k\to\infty} H_{\infty,k}(r, r') \\ &\leqq \limsup_{k\to\infty} H_{\infty,k}(r, r') \leqq H(r + md, r' + md), \end{aligned} \tag{3.5}$$

letting $d\to 0$ shows that $\lim_k H_{\infty,k}(r, r') = H(r, r')$ at all continuity points of the latter, and hence that $H_{\infty,k}$ converges to H at every continuity point of the latter as $k\to\infty$. Thus, the theorem for H will be proved if we can show (3.1), and the result for G, G^+, and G^- can be obtained easily from this result or else can be proved directly in the same manner as the result for H.

We now prove (3.1). Fix $d>0$, r, r', and j. For h an integer, write $V_{k,j,h}$ for the subset of I^m where $h/k < x_j \leqq (h+1)/k$ and write $x^{(j)} = (x_1, \cdots, x_{j-1}, h/k, x_{j+1}, \cdots, x_m)$ if $x \in V_{k,j,h}$. Clearly, if the event

$$\Lambda_{n,k} = \left\{ \overset{+}{D}_{n,k,j} < r/n^{1/2},\ \overset{-}{D}_{n,k,j} < r'/n^{1/2} \right\}$$

occurs and the event

$$\Lambda^*_{n,k} = \left\{ \overset{+}{D}_{n,k,j-1} < (r + d)/n^{1/2},\ \overset{-}{D}_{n,k,j-1} < (r' + d)/n^{1/2} \right\}$$

does not occur, it is necessary that for some h with $0 \leqq h \leqq k-1$ the event

$$\Gamma_{n,k,h} = \left\{ \sup_{x\in V_{k,j,h}} \left| [S_n(x) - F(x)] - [S_n(x^{(j)}) - F(x^{(j)})] \right| > d/n^{1/2} \right\}$$

occurs. Write N_{nkjh} = number of X_i $(1 \leq i \leq n)$ whose values are in $V_{k,j,h}$. Now,

$$P\{\Gamma_{n,k,h}\} \leq P\left\{\sup_{V_{k,j,h}} \left| \frac{S_n(x) - S_n(x^{(j)})}{N_{nkjh}/n} - k\left[F(x) - F(x^{(j)})\right]\right| > dn^{1/2}/2N_{nkjh}\right\}$$

$$(3.6) \qquad + P\left\{\left|\frac{kN_{nkjh}}{n} - 1\right| \sup_{V_{k,j,h}} [F(x) - F(x^{(j)})] > d/2n^{1/2}\right\}$$

$$= P\{Y_{nkjh}\} + P\{Z_{nkjh}\} \text{ (say).}$$

Now, given that $N_{nkjh} = N$, the event Y_{nkjh} is a subset of the event that the maximum deviation of the empiric d.f. of N independent, identically distributed m-variate random variables from the corresponding theoretical d.f. is more than $(d/2)(n/N)^{1/2}/N^{1/2}$. Hence, by Theorem 1,

$$(3.7) \qquad P\{Y_{nkjh} \mid N_{nkjh} = N\} < c_0 \exp\{-cd^2n/N\}$$

where c_0 and c are positive constants. Also, by Chebyshev's inequality,

$$(3.8) \qquad P\{N_{nkjh} \leq n/k + n/k^{1/2}\} > 1 - 1/n.$$

From (3.7) and (3.8) we obtain

$$(3.9) \qquad P\{Y_{nkjh}\} < n^{-1} + c_0 \exp\{-cd^2k^{1/2}/2\}.$$

Also, an application of Chebyshev's inequality based on the fourth moment yields

$$(3.10) \qquad \begin{aligned} P\{Z_{nkjh}\} &= P\left\{\left|\frac{N_{nkjh}}{n} - \frac{1}{k}\right| > d/2n^{1/2}\right\} \\ &< \frac{3n^2/k^2 + n/k}{n^4[d^4/16n^2]} = \frac{16}{d^4}\left[\frac{3}{k^2} + \frac{1}{nk}\right] \end{aligned}$$

From (3.6), (3.9), and (3.10) we obtain

$$(3.11) \quad P\{\Lambda_{n,k} \vdash \Lambda^*_{n,k}\} < \frac{k}{n} + kc_0 \exp\{-cd^2k^{1/2}/2\} + \frac{16}{d^4}\left[\frac{3}{k} + \frac{1}{n}\right],$$

which proves (3.1) and, hence, Theorem 2 in the continuous case.

We now remark on the method of proving Theorem 2 when F has discontinuities. The conclusion follows by the same method as that used to prove Theorem 2 for continuous F, upon noting the manner in which any discontinuous F can be obtained from a continuous one by "lumping together" (in the same manner as that used to obtain Theorem 1 for discontinuous F) certain points in the domain of the latter.

Generalizations of the theorem may be obtained by noting that, as in the case of Theorem 1, the conclusion of Theorem 2 holds if the supremum of observed from theoretical frequency is taken over a larger class of sets than those of the form $x_j \leqq a_j$, $j=1, \cdots, m$ (for all a in m-space).

REFERENCES

1. A. N. Kolmogorov, *Sulla determinazione empirica di una legge di distribuzione*, Inst. Ital. Atti. Giorn. vol. 4 (1933) pp. 83–91.

2. A. Dvoretzky, J. Kiefer, and J. Wolfowitz, *Asymptotic minimax character of the sample distribution function and of the classical multinomial estimator*, Ann. Math. Stat. vol. 27 (1956) pp. 642–669.

3. N. V. Smirnov, *Approximation of the distribution law of a random variable by empirical data*, Uspehi Matematičeskih Nauk vol. 10 (1944) pp. 179–206.

4. M. Donsker, *Justification and extension of Doob's heuristic approach to the Kolmogorov-Smirnov theorems*, Ann. Math. Stat. vol. 23 (1952) pp. 277–281.

5. J. L. Doob, *Heuristic approach to the Kolmogorov-Smirnov theorems*, Ann. Math. Stat. vol. 20 (1949) pp. 393–403.

6. P. Lévy, *Théorie de l'addition des variables aléatoires*, Paris, Gauthier-Villars, 1937.

7. M. Kac, J. Kiefer and J. Wolfowitz, *On tests of normality and other goodness of fit based on distance methods*, Ann. Math. Stat. vol. 26 (1955) pp. 189–211.

CORNELL UNIVERSITY,
ITHACA, N. Y.

Reprinted from
Trans. Amer. Math. Soc. **87**, 173–186 (1958)

Reprinted from THE ANNALS OF MATHEMATICAL STATISTICS
Vol. 30, No. 2, June, 1959

K-SAMPLE ANALOGUES OF THE KOLMOGOROV-SMIRNOV AND CRAMÉR-V. MISES TESTS

J. KIEFER[1]

Cornell University

0. Summary. The main purpose of this paper is to obtain the limiting distribution of certain statistics described in the title. It was suggested by the author in [1] that these statistics might be useful for testing the homogeneity hypothesis H_1 that k random samples of real random variables have the same continuous probability law, or the goodness-of-fit hypothesis H_2 that all of them have some specified continuous probability law. Most tests of H_1 discussed in the existing literature, or at least all such tests known to the author before [1] in the case $k > 2$, have only been shown to have desirable consistency or power properties against limited classes of alternatives (see e.g., [2], [3], [4] for lists of references on these tests), while those suggested here are shown to be consistent against all alternatives and to have good power properties. Some test statistics whose distributions can be computed from known results are also listed.

1. Introduction. Let X_{ji} be independent random variables ($1 \leqq i \leqq n_j$, $1 \leqq j \leqq k$), X_{ji} having unknown continuous distribution function (d.f.)F_j. We are going to consider tests of two hypotheses, the homogeneity hypothesis

$$H_1{:}F_1 = F_2 = \cdots = F_k \tag{1.1}$$

and the goodness-of-fit hypothesis

$$H_2{:}F_1 = F_2 = \cdots = F_k = G, \tag{1.2}$$

where G is some specified continuous d.f. In the case of H_1, the hypothesis allows the common unknown d.f. to be any continuous d.f. The class of alternatives to H_1 or H_2 can be considered to be all sets $(F_1, \cdots, F_k)$ which violate (1.1) or (1.2), respectively; in discussing power under alternatives, continuity of the F_i is irrelevant.

Let

$$S_{n_j}^{(j)}(x) = n_j^{-1} \quad (\text{number of } X_{ji} \leqq x, 1 \leqq i \leqq n_j)$$

be the sample d.f. of the n_j observations in the jth set. We shall omit the subscript n_j whenever this causes no confusion. For $k = 1$ the Kolmogorov test [5] and Cramér-v. Mises ω^2 test [6] of H_2, and for $k = 2$ the Smirnov test [7] and the 2-sample analogue of the ω^2 test of H_1 considered by Lehmann [8] and Rosenblatt [9], may be thought of as test criteria based on simple measurements of distance between $S^{(1)}$ and G or between $S^{(1)}$ and $S^{(2)}$, respectively. (In this

---

Received August 15, 1955; revised June 26, 1958.

[1] Research under contract with the Office of Naval Research.

paper, the word "distance" is not used in the technical sense; see [23], following (5.1).) In [1], several analogous measurements of distance (dispersion) among the $S^{(j)}$ were suggested for testing H_1 or H_2 when k is larger than 2. For example, for testing H_1, some of the most obvious analogues are

$$U = \textstyle\sum_{q,r} \sup_x C_{q,r} \mid S^{(q)}(x) - S^{(r)}(x)|,$$

$$V = \sup_{q,r,x} C_{q,r} \mid S^{(q)}(x) - S^{(r)}(x)|,$$

$$T = \sup_x \textstyle\sum_j C_j[S^{(j)}(x) - \bar{S}(x)]^2,$$

$$W = \int_{-\infty}^{\infty} \textstyle\sum_j C_j[S^{(j)}(x) - \bar{S}(x)]^2 \, d\bar{S}(x),$$

$$Z = \max_j \int_{-\infty}^{\infty} C_j[S^{(j)}(x) - \bar{S}(x)]^2 \, d\bar{S}(x),$$

where $C_{q,r}$ and C_j are positive constants (see, however, the next paragraph) and $\bar{S}(x) = \sum_j n_j S_{n_j}^{(j)}(x)/\sum_j n_j$ is the sample d.f. of the pooled k samples. Similarly, for testing H_2, one might use corresponding statistics U', V', T', W' or Z', obtained from the above by writing G for $S^{(r)}$ or $\bar{S}$. Each of this last collection of statistics has a distribution which does not depend on G in the case that H_2 is true, and each of the first collection has a distribution which does not depend on what the common d.f. is when H_1 is true. In all cases, large values of the statistic lead to rejection of the hypothesis. It is clear that an appropriate choice of the C_j and $C_{q,r}$ in the case $k = 1$ of H_2 or the case $k = 2$ of H_1, reduces each of these tests to one of those previously mentioned for those cases in [5], [6], [7], [8], [9] (in the case of [8] and [9], the integrating measure is altered slightly, as discussed in connection with (2.8) below).

Many tests may be constructed along similar lines by allowing the C_j and $C_{q,r}$ to be functions (of the $S^{(j)}$ for H_1 and of $G(x)$ for H_2) as in the treatments of Kac [11] and Anderson and Darling [12] when $k = 1$, by using other measures of distance or dispersion, etc. In Section 5 we shall mention a few statistics whose limiting distributions are easy to obtain from those of the usual Kolmogorov-Smirnov and ω^2 statistics, but which are intuitively less appealing than those we have mentioned, especially from a practical point of view. In fact, the limiting distribution of V' or Z' (suitably normalized) is that of the maximum of multiples of k independent random variables with limiting Kolmogorov or ω^2 distributions, and is thus trivial to obtain from these latter distributions. From a practical point of view, the problem of testing H_2 may thus seem to be satisfactorily answered by these statistics.

Thus, our main goal is to obtain the limiting distribution under H_1 of appropriate statistics for testing that hypothesis, and the corresponding results we shall obtain for tests of H_2 are less important by-products of the investigation. Specifically, in Section 3 we shall obtain the limiting distribution of T (and T') for $C_j = n_j$, as the $n_j \to \infty$, while in Section 4 we obtain the limiting distribution of W (and W') under the same conditions. The limiting distributions of U, V, and Z seem more difficult to obtain, and the methods of this paper do not apply at all to those statistics.

Many different proofs of the Kolmogorov-Smirnov results [5] and [7] now exist. Combinatorial proofs such as those of Feller [10] and of several papers by Russian authors (such as Smirnov, Gnedenko, Korolyuk) seem inapplicable to the problem of obtaining the limiting distribution of the generalizations T and T' of the Kolmogorov-Smirnov statistics. The geometric aspects of Doob's proof [13] clearly cannot be directly generalized. However, the approach used by Kac in several papers since 1949, e.g., in [11], to obtain various results such as that of Kolmogorov, can be generalized with some slight technical modifications to give results on the Wiener process in dimensions >1 which can be used with an analogue of Donsker's result [14] to obtain the limiting distribution of T; such results for closely related problems have in fact been studied by Rosenblatt [17]. The method of Anderson and Darling [12] could also be used, but perhaps guessing the solution to the appropriate diffusion equation is more difficult than the approach used here.

In Section 2, therefore, we reduce the problem of finding the limiting distribution of T or T' to a calculation regarding a multidimensional Wiener process, and outline the steps to be carried out in performing this calculation. The solution is then obtained in Section 3. A similar method will work for the limiting distributions of W and W', but these may be obtained more easily by convolving the usual ω^2 distribution with itself an appropriate number of times (Section 4). In Section 5 the statistics mentioned three paragraphs above and whose distributions may be obtained from existing tables, are discussed. The power of the tests considered in this paper is discussed briefly in Section 6, where several other remarks are made. Finally, Section 7 contains tables of some of the limiting distributions obtained in the paper.

2. Reduction of the problem. We hereafter write N for the vector $(n_1, \cdots, n_k)$ and consider (now exhibiting the dependence on N)

$$T_N = \sup_x \sum_j n_j [S_{n_j}^{(j)}(x) - \bar{S}_N(x)]^2,$$

$$T_N' = \sup_x \sum_j n_j [S_{n_j}^{(j)}(x) - G(x)]^2,$$

$$W_N = \int_{-\infty}^{\infty} \sum_j n_j [S_{n_j}^{(j)}(x) - \bar{S}_N(x)]^2 \, d\bar{S}_N(x),$$

$$W_N' = \int_{-\infty}^{\infty} \sum_j n_j [S_{n_j}^{(j)}(x) - G(x)]^2 \, dG(x).$$

(We shall also consider extensions of W_N; see equation (2.8).) Since the distribution of each of these statistics does not depend on G (resp., on the common d.f.) if H_2 (resp., H_1) is true, we shall as usual perform our calculations under the assumption that G and all F_i are the uniform d.f. on the unit interval.

Let $Y_1, Y_2, \cdots, Y_h$ be h independent separable Gaussian processes whose sample functions are functions of the same "time" parameter t, $0 \leq t \leq 1$, and such that $EY_i(t) = 0$ and $EY_i(t)Y_i(s) = \min(s, t) - st$ for each i. Thus, the Y_i are independent "tied-down Wiener processes" which may be represented as $Y_i(t) = (1 - t)^{-1}w_i(t/(1 - t))$, where the w_i are independent Wiener processes of the usual variety; i.e., w_i is a separable Gaussian process of independent increments with $Ew_i(\tau) = 0$ and $Ew_i(\tau)w_i(\sigma) = \min(\tau, \sigma)$ for $0 \leqq \tau, \sigma < \infty$. The use of such processes in [11], [12], [13] to obtain the Kolmogorov-Smirnov results is well known. Let

$$A_h(a) = P\{ \max_{0 \leqq t \leqq 1} \textstyle\sum_{i=1}^{h} [Y_i(t)]^2 \leqq a\}. \tag{2.1}$$

and

$$B_h(a) = P\left\{ \int_0^1 \textstyle\sum_{i=1}^{h} [Y_i(t)]^2 \, dt \leqq a \right\}. \tag{2.2}$$

When G is the uniform d.f., the k random functions

$$\nu_{n_j}^{(j)}(t) = \sqrt{n_j}(S_{n_j}^{(j)}(t) - t), \qquad 0 \leqq t \leqq 1$$

are independent of each other and as $n_j \to \infty$ their behavior approaches that of the processes $Y_1, \cdots, Y_h$ with $h = k$. More precisely, an obvious extension of the argument of Donsker [14] or Theorem 2 of Kiefer and Wolfowitz [15] to the present case shows at once that, at all continuity points of the limit (which, we shall see, means for all a),

$$\lim_{\text{all } n_j \to \infty} P\{T_N' \leqq a\} = A_k(a) \tag{2.3}$$

and

$$\lim_{\text{all } n_j \to \infty} P\{W_N' \leqq a\} = B_k(a). \tag{2.4}$$

Similarly, let H be a $k \times k$ orthogonal matrix such that the jth element of the first row of H is $(n_j/\sum n_j)^{\frac{1}{2}}$ for $1 \leqq j \leqq k$, and write ν_N for the k-vector whose jth component is the random function $\nu_{n_j}^{(j)}$. We have already discussed the asymptotic behavior of ν_N as the $n_j \to \infty$. The extension of the results of Donsker [14] or Kiefer and Wolfowitz [15] to the present case shows, on considering the sum of squares of the last $k - 1$ components of $H\nu_N$, which sum is equal to $\sum_j n_j[S_{n_j}^{(j)}(t) - \bar{S}_N(t)]^2$, that

$$\lim_{\text{all } n_j \to \infty} P\{T_N \leqq a\} = A_{k-1}(a). \tag{2.5}$$

We remark that, as in the case $h = 1$, if F_1 is not continuous, the statistics T_N and T_N' are equivalent to statistics obtained for the case of continuous F_1 by taking the supremum over a restricted range; thus, the d.f. of T_N or T_N' in such a case is not larger than what it is for continuous F_1.

Next, we consider W_N. Since we need to prove statements which differ slightly from those of Rosenblatt [9], and since the partial integrations in [9] require some alterations, we shall carry out the required demonstration in full here rather than to refer elsewhere.[2] We shall actually prove without extra difficulty a more general result than that needed here, but one which is useful in reducing the calculation of the limiting distribution of other integral criteria in the same way that we reduce that of W_n. Our result is (roughly) that an integral criterion formed by integrating with respect to a consistent estimator of the common F_i has the same limiting distribution if the consistent estimator is replaced by F_1. The following statement of it is thus easily generalized:

LEMMA. *Let $D \geqq 0$ be a continuous function of $k-1$ real variables which is bounded on bounded sets and such that*

$$(2.6) \qquad \int_{-\infty}^{\infty} \cdots \int_{-\infty}^{\infty} D(t_1, \cdots, t_{k-1}) \, | t_1 \cdots t_{k-1} | \, e^{-(t_1^2 + \cdots + t_{k-1}^2)/2} \, dt_1 \cdots dt_{k-1} < \infty.$$

Then, for each j, when all F_i are uniform on [0, 1],

$$(2.7) \qquad \int_0^1 D(\nu_{n_1}^{(1)}(t) - \nu_{n_k}^{(k)}(t), \cdots, \nu_{n_{k-1}}^{(k-1)}(t) - \nu_{n_k}^{(k)}(t)) \, d(S_{n_j}^{(j)}(t) - t)$$

converges to 0 in probability as all $n_i \to \infty$.

PROOF: It was proved by Dvoretzky, Kiefer, and Wolfowitz [16] that $P\{\sup_t \nu_{n_j}^{(j)}(t) > r\} < ce^{-2r^2}$ for all n_j and r, where c is a positive constant. Hence, (2.6) implies that if in (2.7) we replace the function D by max (D, L), where L is a constant, (2.7) is altered by a quantity which goes to 0 in probability as the constant $L \to \infty$, uniformly in the n_j. Hence, it suffices to prove (2.7) assuming D is bounded and uniformly continuous, which we now assume. The proof of Theorem 2 of Kiefer and Wolfowitz [15] shows that for any $\epsilon > 0$ there is a value m such that the probability that

$$\sup_{i/m \leqq t \leqq (i+1)/m} | \nu_{n_j}^{(j)}(t) - \nu_{n_j}^{(j)}(i/m) | < \epsilon$$

for all $i(0 \leqq i \leqq m-1)$ is at least $1 - \epsilon$ for all sufficiently large n_j. Thus, given any $\epsilon' > 0$, we can choose ϵ (and thus m) with regard to the modulus of continuity of D, so that for all n_j sufficiently large the probability will be $>1 - \epsilon'$ that the value of the integrand of (2.7) varies over a range of length $< \epsilon'$ as t varies from i/m to $(i+1)/m$, simultaneously for all i. On the other hand, when the n_j are sufficiently large, $S_{n_j}^{(j)}$ assigns measure arbitrarily close to $1/m$ to each of the intervals $i/m \leqq t \leqq (i+1)/m$, with probability arbitrarily close to 1. Since we have seen that D may be assumed bounded, the assertion of the lemma now follows easily.

We conclude at once from the lemma and the use of the orthogonal transformation H discussed in connection with T_N that if $a_1, \cdots, a_k$ are real numbers with $\sum a_i = 1$, then

$$(2.8) \qquad \lim_{\text{all } n_j \to \infty} P\left\{\int_{-\infty}^{\infty} \sum_j n_j \, [S_{n_j}^{(j)}(x) - \bar{S}_N(x)]^2 \, d[\sum_i a_i S_{n_i}^{(i)}(x)] \leqq a\right\} = B_{k-1}(a);$$

[2] Professor Rosenblatt has informed the author that he has constructed another correct proof of the result of [9], and has indicated that some corrections to [17] will appear shortly.

in particular,

$$(2.9) \qquad \lim_{\text{all } n_j \to \infty} P\{W_N \leqq a\} = B_{k-1}(a).$$

The extension (2.8) of (2.9) includes, for example, integration with respect to $k^{-1} \sum_j S_{n_j}^{(j)}$, which is what is done in the case $k = 2$ by Rosenblatt [9]. It is easy to extend (2.8) to allow the a_i to vary slightly with N, etc.

We note that we nowhere require the ratios n_i/n_j to approach positive finite limits. This requirement, which is made in [7], [9], [10], and [13] in the case $k = 2$ of H_1, is inessential, and our remarks show that the results there hold without this restriction.

3. The limiting distribution of T_N and T_N'. In [17] Rosenblatt studies the distribution of a class of suitably regular functionals of the h-dimensional process $Y = (Y_1, \cdots, Y_h)$ on $0 \leqq t \leqq 1$. We shall only state briefly the results we need from [17] and Kac's paper [11]. In fact, writing

$$\Lambda_c(t) = [(Y_1(t) + ct)^2 + \sum_2^h (Y_i(t))^2]^{\frac{1}{2}}$$

for $c \geqq 0$, if one considers only nonnegative functions v of Λ_c which satisfy the regularity conditions of [17], then the analysis there may be shortened somewhat, and we now summarize the results we need in that briefer form; the reader may consult [11] or [17] for details.

For any h-vector x and $t > 0$, with primes denoting transposes, write

$$(3.1) \qquad Q_0(x, t) = (2\pi t)^{-h/2} e^{-x'x/2t}$$

and, for $n > 0$, with E^h denoting Euclidean h-space and $d\xi = d\xi_1 d\xi_2 \cdots d\xi_h$,

$$(3.2) \qquad Q_{n+1}(x, t) = \int_0^t \int_{E^h} Q_0(x - \xi, t - \tau) v([\xi'\xi]^{\frac{1}{2}}) Q_n(\xi, \tau)\, d\xi\, d\tau.$$

It is easy to see that Q_n depends on x only through $x'x = r^2$ (say), so that we can write $Q_n(x, t) = \bar{Q}_n(r, t)$. Define the generating function (in $u \geqq 0$)

$$(3.3) \qquad Q(r, t, u) = \sum_{n=0}^{\infty} (-u)^n \bar{Q}_n(r, t)$$

and, for $r > 0$, its transform (in $s \geqq 0$)

$$(3.4) \qquad \psi(r) \equiv \psi_{s,u}(r) = \int_0^\infty Q(r, t, u) e^{-st}\, dt.$$

Write

$$(3.5) \qquad \phi(r) \equiv \phi_{s,u}(r) = r^{(h-1)/2} \psi_{s,u}(r).$$

One proves easily that ψ is the unique solution of the ordinary differential equation (for $r > 0$)

$$(3.6) \qquad \psi''(r) + \frac{h-1}{r} \psi'(r) - [2s + 2uv(r)]\psi(r) = 0$$

which satisfies

$$(3.7)\qquad \begin{aligned}&\text{(a) } \psi(r) \to 0 \text{ as } r \to \infty;\\ &\text{(b) } \psi'(r) \text{ is continuous for } r > 0;\\ &\text{(c) as } r \to 0,\ \psi'(r) \sim -\Gamma(h/2)\pi^{-h/2}r^{1-h}.\end{aligned}$$

It is sometimes convenient to rewrite (3.6) and (3.7) in other terms. For example, for $h > 1$ and suitably regular v, we can obtain ϕ as the unique solution (for $r > 0$) of

$$(3.6a)\qquad \phi''(r) - \left[2s + \frac{(h-1)(h-3)}{4r^2} + 2uv(r)\right]\phi(r) = 0$$

which satisfies

$$(3.7a)\qquad \begin{aligned}&\text{(a) } \phi(r) \to 0 \text{ as } r \to \infty;\\ &\text{(b) } \phi'(r) \text{ is continuous for } r > 0;\\ &\text{(c) as } r \to 0,\quad \phi(r) \sim \begin{cases}-\pi^{-1}r^{\frac{1}{2}}\log r & \text{if } h = 2,\\ \Gamma(h/2)\pi^{-h/2}r^{(3-h)/2}/(h-2) & \text{if } h > 2.\end{cases}\end{aligned}$$

(Equation (3.6) is merely the reduction to an ordinary differential equation of the partial differential equation of [17, equation (1.14)] when v depends only on $x'x$; (3.7) for the case $h \geqq 1$ is the analogue of [11], equation (3.14), for the case $h = 1$.)

Let $(w_1, \cdots, w_h)$ be the h-dimensional Wiener process described just above (2.1). Let

$$(3.8)\qquad \begin{aligned}\zeta(t) &= \int_0^t v((\sum_i [w_i(\tau)]^2)^{\frac{1}{2}})\, d\tau,\\ \sigma(q;t) &= P\{\zeta(t) < q\}.\end{aligned}$$

The function Q in the case of more general v is studied by Rosenblatt [17] because, as in the case $h = 1$ of Kac [11], it is desired to compute σ, and

$$(3.9)\qquad \int_0^\infty e^{-uq}\, d_q\, \sigma(q,t) = \int_{E^h} Q([x'x]^{\frac{1}{2}}, t, u)\, dx.$$

But it can also be seen, as it was in [11], equation (6.16), when $h = 1$, that if

$$(3.10)\qquad \begin{aligned}\eta_c &= \int_0^1 v(\Lambda_c(t))\, dt,\\ p_c(q) &= P\{\eta_c < q\},\end{aligned}$$

then

$$(3.11)\qquad \int_0^\infty e^{-uq}\, d_q\, p_c(q) = (2\pi)^{k/2}e^{c^2/2}Q(c, 1, u).$$

This is the use of Q which concerns us in obtaining distributions like those of (2.1) and (2.2).

In Kac's paper [11] it is only necessary to consider η_0, since p_0 is what we actually want to determine. However, $\psi_{s,u}(0)$ is infinite when $h > 1$, so that we are forced to consider η_c, determine $\psi_{s,u}(c)$ for c near 0, invert this to obtain $Q(c, 1, u)$, and the let $c \to 0$ to obtain $Q(0, 1, u)$. This continuity in c of $Q(c, 1, u)$ is proved by Rosenblatt [17] (it is also evident from the probabilistic meaning of η_c); the particular case of interest to us here involves another limit operation and will be discussed in the next paragraph.

In order to obtain the function A_h of (2.1), we consider, as did Kac [11] for the case $h = 1$, the function

$$v(r) = \begin{cases} 0 & \text{if} \quad r < a, \\ 1 & \text{if} \quad r \geqq a, \end{cases} \tag{3.12}$$

where $a > 0$. From (3.10) and (3.11) we then have

$$P\{\max_{0 \leq t \leq 1} \Lambda_c(t) < a\} = (2\pi)^{h/2} e^{c^2/2} \lim_{u \to \infty} Q(c, 1, u). \tag{3.13}$$

It is convenient to interchange the order of inverting with respect to s and letting $u \to \infty$; i.e., by bounded convergence we have

$$\psi_{s,\infty}(c) \equiv \lim_{u \to \infty} \psi_{s,u}(c) = \int_0^\infty \lim_{u \to \infty} Q(c, t, u) e^{-st}\, dt, \tag{3.14}$$

so that we can invert $(2\pi)^{h/2} e^{c^2/2} \psi_{s,\infty}(c)$ with respect to s and set $t = 1$ to obtain the left side of (3.13) and then, from the probabilistic meaning of Λ_c, let $c \to 0$ and obtain, for $a \geqq 0$,

$$A_h(a^2) = \lim_{c \to 0} P\{\max_{0 \leq t \leq 1} \Lambda_c(t) < a\}. \tag{3.15}$$

For the v of (3.12), the solution of (3.6) satisfying the conditions (3.7) is easily obtained in terms of modified Bessel functions of the first and third kind ([18], Vol. 2, [20]). The solution is of the form $\phi(r) = r^{(h-1)/2}\psi(r) = C_1 r^{\frac{1}{2}} K_{(h-2)/2}(r(2s)^{\frac{1}{2}}) + C_2 r^{\frac{1}{2}} I_{(h-2)/2}(r(2s)^{\frac{1}{2}})$ for $0 < r < a$, and of the same form with s replaced by $s + u$ and with C_1 and C_2 replaced by C_1' and C_2' (say) for $r \geqq a$, where the C_i and C_i' depend on s and u. From (3.7)(a) or (3.7a)(a) we obtain $C_2' = 0$, and from (3.7)(c) or (3.7a)(c) we obtain

$$C_1 = 2(2s)^{(h-2)/4}(2\pi)^{-h/2}.$$

The other two constants are obtained from the continuity of ϕ and ϕ' at $r = a$. In particular, we obtain, writing $a(2s)^{\frac{1}{2}} = \alpha$ and $a(2s + 2u)^{\frac{1}{2}} = \beta$,

$$\frac{C_2}{C_1} = \frac{K_{h/2}(\alpha)K_{(h-2)/2}(\beta) - (\beta/\alpha)K_{(h-2)/2}(\alpha)K_{h/2}(\beta)}{I_{h/2}(\alpha)K_{(h-2)/2}(\beta) + (\beta/\alpha)I_{(h-2)/2}(\alpha)K_{h/2}(\beta)}. \tag{3.16}$$

When we let u (i.e., β) go to ∞, this ratio approaches the limit

$$-K_{(h-2)/2}(\alpha)/I_{(h-2)/2}(\alpha).$$

Thus, we have, for $0 < r < a$ and $h \geqq 1$,

$$(2\pi)^{h/2}\psi_{s,\infty}(r)$$

$$(3.17)\qquad = \frac{2(2s)^{(h-2)/4}}{r^{(h-2)/2}}\left\{K_{(h-2)/2}(r(2s)^{\frac{1}{2}}) - \frac{K_{(h-2)/2}(a(2s)^{\frac{1}{2}})I_{(h-2)/2}(r(2s)^{\frac{1}{2}})}{I_{(h-2)/2}(a(2s)^{\frac{1}{2}})}\right\}.$$

(The corresponding formula and subsequent inversion in [17] is incorrect,[2] due to a mistake in evaluating C_1).

To invert (3.17), we consider the Fourier-Bessel expansion of [18], Vol. 2, p. 104, equation (58):

$$(3.18)\qquad \frac{\pi J_\nu(xz)}{zJ_\nu(z)}[J_\nu(z)Y_\nu(Xz) - Y_\nu(z)J_\nu(Xz)] = \sum_{n=1}^{\infty}\frac{J_\nu(\gamma_{\nu,n}\,x)J_\nu(\gamma_{\nu,n}\,X)}{(z^2 - \gamma_{\nu,n}^2)[J_{\nu+1}(\gamma_{\nu,n})]^2},$$

where $\gamma_{\nu,n}(n = 1, 2, \cdots)$ are the positive zeros of J_ν, ν and z are arbitrary, and $0 < x \leqq X \leqq 1$. (A similar formula of Watson ([20], p. 499) seems incorrect, as can be seen in the case $\nu = \frac{1}{2}$, $z \to 0$ there.) Divide both sides of (3.8) by $J_\nu(xz)$ and let $x \to 0$, noting that $J_\nu(\gamma_{\nu,n}x)/J_\nu(xz) \to (\gamma_{\nu,n}/z)^\nu$; it is easy to justify taking the limit inside the sum. Put $z = ia(2s)^{\frac{1}{2}}$ and $X = r/a$. We then obtain, from (3.17), (3.18), and the relation of I and K to J and Y, where $\nu = (h - 2)/2$,

$$(3.19)\qquad (2\pi)^{h/2}\psi_{s,\infty}(r) = 4\sum_{n=1}^{\infty}\left(\frac{\gamma_{\nu,n}}{ar}\right)^\nu \frac{J_\nu(r\gamma_{\nu,n}/a)}{[J_{\nu+1}(\gamma_{\nu,n})]^2(2a^2s + \gamma_{\nu,n}^2)}.$$

It is easy to see that this series can be inverted term-by-term with respect to s; inverting and setting $t = 1$, we have from (3.14),

$$(3.20)\qquad P\{\max_{0\leq t\leq 1}\Lambda_r(t) < a\} = 2e^{r^2/2}\sum_{n=1}^{\infty}\left(\frac{\gamma_{\nu,n}}{ar}\right)^\nu \frac{J_\nu(r\gamma_{\nu,n}/a)e^{-\gamma_{\nu,n}^2/2a^2}}{[J_{\nu+1}(\gamma_{\nu,n})]^2a^2}.$$

Finally, letting $r \to 0$, we have, from (3.15) and (3.20),

THEOREM. For $h \geqq 1$ (*see also* (3.27) *and* (3.31)),

$$(3.21)\qquad A_h(a^2) = \frac{4}{\Gamma\left(\frac{h}{2}\right)2^{h/2}a^h}\sum_{n=1}^{\infty}\frac{(\gamma_{(h-2)/2,n})^{h-2}\exp[-(\gamma_{(h-2)/2,n})^2/2a^2]}{[J_{h/2}(\gamma_{(h-2)/2,n})]^2}.$$

Thus, writing $\Phi_k(x) = A_k(x^2)$ for $x > 0$ and $\Phi_k(x) = 0$ otherwise, Φ_{k-1} and Φ_k are the limiting d.f.'s of $\sqrt{T_N}$ and $\sqrt{T_N'}$, respectively.

The series converges rapidly (see also the discussion of the two succeeding paragraphs for large a), but reduces to an expression in terms of elementary functions only when $h = 1$ or $h = 3$. When $h = 1$, we have $\gamma_{-\frac{1}{2},n} = (2n - 1)\pi/2$ and thus $[J_{\frac{1}{2}}(\gamma_{-\frac{1}{2},n})]^2 = 4/(2n - 1)\pi^2$. Thus, for $a > 0$,

$$(3.22)\qquad A_1(a^2) = \frac{(2\pi)^{\frac{1}{2}}}{a}\sum_{n=1}^{\infty}e^{-(2n-1)^2\pi^2/8a^2},$$

which is Smirnov's result, since T_N is the square of the usual Smirnov statistic when $k = 2$. Similarly, for $h = 3$ we obtain, for $a > 0$,

$$(3.23) \qquad A_3(a^2) = \frac{2^{\frac{1}{2}}\pi^{\frac{5}{2}}}{a^3} \sum_{n=1}^{\infty} n^2 e^{-n^2\pi^2/2a^2}.$$

In these cases we can obtain alternative expressions which are more useful for computations when a is large. These may be obtained directly by using an appropriate transformation on a theta function, or by noting that (3.17) reduces to

$$\pi^{\frac{1}{2}} \sinh [(a - r)(2s)^{\frac{1}{2}}]/s^{\frac{1}{2}} \cosh [a(2s)^{\frac{1}{2}}]$$

when $h = 1$ and to

$$(2\pi)^{\frac{1}{2}} \sinh [(a - r)(2s)^{\frac{1}{2}}]/r \sinh [a(2s)^{\frac{1}{2}}]$$

when $h = 3$, and these are tabled as theta function transforms in [19], Vol. 1, p. 258, equations (34) and (31), the first of which is wrong in sign. For $h = 1$ we obtain, letting $r \to 0$, the more familiar form of A_1 for $a > 0$,

$$(3.24) \qquad A_1(a^2) = 1 + 2\sum_{n=1}^{\infty} (-1)^n e^{-2n^2a^2}.$$

(For $h = 1$, but not for $h = 3$, we could have let $r \to 0$ before inverting, and used [19], Vol. 1, p. 257, equation (24).) For $h = 3$, the inverse Laplace transform is given in terms of a derivative of the theta function θ_4 ; letting $r \to 0$ yields

$$(3.25) \qquad A_3(a^2) = 1 + 4\sum_{n=1}^{\infty} [\tfrac{1}{2} - 2n^2a^2]e^{-2n^2a^2}.$$

The existence of the two forms for A_1 and A_3 suggests that a form more useful than (3.21) for large a might be found. There seems to be no simple analogue of the theta function transformation for the series of (3.21), but in this and the next two paragraphs we mention other computational approaches which may prove useful. There are other Fourier-Bessel expansions which can be employed in inverting (3.17). For example, one series for $J_\nu(xz)/J_\nu(z)$ ([18], Vol. 2, p. 104, equation (59)) gives (writing ν for $(h - 2)/2$)

$$(3.26) \qquad (2\pi)^{h/2}\psi_{s,\infty}(r) = 2\frac{(2s)^{\nu/2}}{r^\nu}\left\{K_\nu(r(2s)^{\frac{1}{2}}) - 2\sum_{n=1}^{\infty} \frac{J_\nu(\gamma_{\nu,n}\, r/a)\gamma_{\nu,n}\, K_\nu(a(2s)^{\frac{1}{2}})}{J_{\nu+1}(\gamma_{\nu,n})(2a^2s + \gamma_{\nu,n}^2)}\right\}.$$

Now, by [19], Vol. 1, p. 283, equation (40), $2(2s)^{\nu/2}K_\nu(r(2s)^{\frac{1}{2}})/r^\nu$ is the transform of $t^{-\nu-1}e^{-r^2/2t}$, which becomes 1 at $t = 1$, $r \to 0$. Since $(2a^2s + \gamma^2)^{-1}$ is the transform of $e^{-\gamma^2 t/2a^2}/2a^2$ and $(a/r)^\nu J_\nu(\gamma r/a) \to \gamma^\nu/2^\nu\Gamma(\nu + 1)$ as $r \to 0$, we obtain

$$(3.27) \quad A_h(a^2) = 1 - \frac{1}{2^\nu a^2\Gamma(\nu+1)} \sum_{n=1}^{\infty} \frac{(\gamma_{\nu,n})^{\nu+1}}{J_{\nu+1}(\gamma_{\nu,n})} \int_0^1 t^{-\nu-1}e^{-a^2/2t}e^{-(\gamma_{\nu,n})^2(1-t)/2a^2}\, dt.$$

For computational purposes, this formula has the disadvantage of involving a numerical quadrature, but it has the advantage that the series converges rapidly for a large.

Another way of trying to obtain a more useful formula for large a is to try to use the theta function transformation on a function close to that of (3.21). The following is such an approach *when h is odd.* We again write $\nu = (h-2)/2$. Now, for large n we have $\gamma_{\nu,n} \sim \pi(4n+2\nu-1)/4$ (see, e.g., [18], Vol. 2, pp. 60 and 85) and $[J_{\nu+1}(\gamma_{\nu,n})]^2 \sim 2/\pi\gamma_{\nu,n}$. Thus, an approximation to the summand of (3.21) is

$$(3.28) \qquad f(\nu, n, a) = \frac{\pi^{2\nu+2}}{2}\left[n + \frac{2\nu-1}{4}\right]^{2\nu+1} e^{-\pi^2[n+(2\nu-1)/4]^2/2a^2}.$$

How good an approximation this is of course depends on the exponential term; but the form of (3.28) is suggestive of theta functions. In fact, the transformation $\theta_3(t^{-1}v \mid -t^{-1}) = (-it)^{\frac{1}{2}}e^{i\pi v^2/t}\theta_3(v/t)$ ([18], Vol. 2, p. 370), on putting $t = -1/i\pi x$, becomes

$$(3.29) \qquad (\pi x)^{\frac{1}{2}}\sum_{n=-\infty}^{\infty} e^{-\pi^2[n+v]^2 x} = e^{-2\pi^2v^2x}\sum_{n=-\infty}^{\infty} e^{i\pi vn}e^{-n^2/x},$$

so that, for $2v$ a nonnegative integer,

$$(3.30) \qquad \sum_{n=1}^{\infty} e^{-\pi^2[n+v]^2x} = \frac{e^{-2\pi^2v^2x}}{2(\pi x)^{\frac{1}{2}}}\sum_{n=-\infty}^{\infty} e^{i\pi vn - n^2/x} - \frac{1}{2}\sum_{n=-2v}^{0} e^{-\pi^2[n+v]^2x}$$
$$= q_1(x) - q_2(x)(\text{say}).$$

Putting $v = (2\nu-1)/4$, differentiating $(2\nu+1)/2$ times with respect to x, and denoting the summand of (3.17) by $g(\nu, n, a)$, we thus obtain for odd $h \geq 3$,

$$(3.31) \qquad \frac{\Gamma\left(\frac{h}{2}\right)2^{h/2}a^h}{4}A_h(a^2) = \sum_{n=1}^{\infty}[g(\nu,n,a) - f(\nu,n,a)]$$
$$+ \frac{(-1)^{(2\nu+1)/2}\pi}{2}\left(\frac{d}{dx}\right)^{(2\nu+1)/2}[q_1(x) - q_2(x)]\,|_{x=\frac{1}{2}a^2}.$$

When f is close to g, this will be a convenient formula, since q_1 converges rapidly as $a \to \infty$ and q_2 will contain only $2\nu + 1$ terms.

Another approach to obtain different expressions from (3.17) to invert, and which allows us to let $r \to 0$ before inverting with respect to s, is to note that although the Laplace transform ψ of $Q(r, t, u)$ is infinite for $r = 0$, the transform of $t^m Q(r, t, u)$ is finite there for m an integer $> h/2$. But this is just $d^m\psi_{s,u}(r)/ds^m$. Thus, performing such a differentiation and letting $u \to \infty$ and $r \to 0$, we obtain an expression whose inverse transform with respect to s at $t = 1$ give $(2\pi)^{-h/2}A_h(a^2)$.

Tables of the functions A_h will be found in Section 7. Even when h is even, the computation is not very difficult. For example, when $h = 2$ the denominator of the summand of (3.21) is approximately $2/\pi\gamma_{0,n}$, as we have seen, and the series is easy to work with. For the next odd h above those we have considered in detail, $h = 5$, the $\gamma_{\nu,n}$ are solutions of $\tan x = x$ and the summand of (3.21) is $\pi\gamma_{\nu,n}^2(\gamma_{\nu,n}^2 + 1)e^{-(\gamma_{\nu,n})^2/2a^2}/2$.

4. The limiting distribution of W_N and W_N'. The differential equation of (3.6) and (3.7) can be solved, when $v(r) = r^2$, in terms of a confluent hypergeometric function (specifically, by (3.7)(a), in terms of the Whittaker function $W_{\kappa,\mu}$); but a more direct approach is to note, on reversing the order of integration and summation in (2.2), that the distribution B_h is merely the h-fold convolution of B_1 with itself. In the case $h = 1$, it is well known that $(2\pi)^{\frac{1}{2}}Q(0, 1, u) = [(2u)^{\frac{1}{2}}/\sinh(2u)^{\frac{1}{2}}]^{\frac{1}{2}}$. Raising this to the hth power, we obtain $(2\pi)^{h/2}Q(0, 1, u)$ for general h. We can now follow a procedure like that of Anderson and Darling ([12], p. 201): we obtain, on integrating by parts,

$$\int_0^1 e^{-ua}B_h(a)\,da = u^{-1}[(2u)^{\frac{1}{2}}/\sinh(2u)^{\frac{1}{2}}]^{h/2}. \tag{4.1}$$

Using the binomial expansion on $[1 - e^{-2(2u)^{\frac{1}{2}}}]^{-h/2}$, (4.1) becomes

$$2^{3h/4}\sum_{j=0}^{\infty}\frac{\Gamma(j + h/2)}{j!\Gamma(h/2)}u^{-1+h/4}e^{-(8u)^{\frac{1}{2}}(j+h/4)}. \tag{4.2}$$

This series can be inverted term-by-term in terms of tabled transforms, without computations like those of [12]: from [19], Vol. 1, p. 246, equation (9), we find that $u^{-1+h/4}e^{-(8u)^{\frac{1}{2}}(j+h/4)}$ is the Laplace transform of

$$2^{(2-h)/4}\pi^{-\frac{1}{2}}t^{-h/4}e^{-(j+h/4)^2/t}D_{(h-2)/2}(2(j + h/4)t^{-\frac{1}{2}}),$$

where D is the parabolic cylinder function. Thus, inverting (4.2) with respect to u, we obtain, for $a > 0$,

$$B_h(a) = \frac{2^{(h+1)/2}}{\pi^{\frac{1}{2}}a^{h/4}}\sum_{j=0}^{\infty}\frac{\Gamma(j + h/2)}{j!\Gamma(h/2)}e^{-(j+h/4)^2/a}D_{(h-2)/2}((2j + h/2)/a^{\frac{1}{2}}). \tag{4.3}$$

Thus, B_k and B_{k-1} are the limiting d.f.'s of W_n and W_n', respectively.

B_h can be written in a more convenient form if h is even. In that case if we write H_n for the nth Hermite polynomial, i.e., $H_n(x) = (-1)^n e^{x^2} d^n e^{-x^2}/dx^n$, we obtain from the relation between D_n and H_n ([18], Vol. 2, p. 117), for $a > 0$ and h even,

$$B_h(a) = \frac{2^{(h+1)/2}}{\pi^{\frac{1}{2}}a^{h/4}}\sum_{j=0}^{\infty}\frac{\Gamma(j + h/2)}{j!\Gamma(h/2)}e^{-2(j+h/4)^2/a}H_{(h-2)/a}((2j + h/2)/(2a)^{\frac{1}{2}}). \tag{4.4}$$

When h is odd, (4.3) can be written in terms of the Bessel functions $K_{\frac{1}{4}}$ and $K_{\frac{3}{4}}$, as follows: Since ([18], Vol. 2, p. 119) $D_{-\frac{1}{2}}(z) = (z/2\pi)^{\frac{1}{2}}K_{\frac{1}{4}}(z^2/4)$ and $D_{\frac{1}{2}}(z) = -e^{z^2/4}d[e^{-z^2/4}D_{-\frac{1}{2}}(z)]/dz = \pi^{-\frac{1}{2}}(z/2)^{\frac{3}{2}}[K_{\frac{1}{4}}(z^2/4) + K_{\frac{3}{4}}(z^2/4)]$, successive use of the recursion relation $D_{\nu+1}(z) = zD_\nu(z) - \nu D_{\nu-1}(z)$ and the fact that $K_\nu = K_{-\nu}$ yields $D_{m-\frac{1}{2}}$, for m a positive integer, in terms of $K_{\frac{1}{4}}$ and $K_{\frac{3}{4}}$.

In the case $h = 1$, substitution of the formula for $D_{-\frac{1}{2}}$ in terms of $K_{\frac{1}{2}}$ gives the formula of [12], equation (4.35).

Tables of B_h will be found in Section 7.

5. Criteria whose distributions may be obtained from previously known results. We limit our discussion to criteria for testing H_1 ; analogues for testing H_2 are obvious, and some criteria have been mentioned in Section 1. We shall also limit our discussion to criteria of the Kolmogorov-Smirnov type, ones of the integral (ω^2-) type being obtained similarly. Symbols newly defined in this section need not have their earlier meaning.

One of the simplest tests whose size may be computed from previously known results is that based on the maximum of the $k - 1$ random variables

$$Y_j = C_j'' \sup_x | S_j(x) - \sum_{i<j} n_i S_i(x) / \sum_{i<j} n_i |, \qquad (2 \leq j \leq k)'$$

which are obviously independent under H_1 (since, for example, the conditional distribution of $\sup_x | S_1(x) - S_2(x)|$ given the value of the function $n_1 S_1 + n_2 S_2$ does not depend on the latter). Y_j is distributed like a multiple of the Smirnov 2-sample criterion for sample sizes n_j and $\sum_{i<j} n_i$; thus, the tables of Massey [21] may be used in an obvious way to compute the d.f. of $\max_j Y_j$. Of course, asymptotically one may use the Kolmogorov-Smirnov distribution $A_1(a^2)$.

This test may be made more symmetrical by choosing at random the indexing j of the k sets. Another method of symmetrizing is to subdivide each of the k original sets of observations into $k!$ subsets, form $k!$ collections each of which contains one subset of each original set, index the subsets in each collection in a different one of the $k!$ possible ways, compute the maximum of the Y_j for each collection, and take the maximum of these over all collections.

A test based upon the Y_j of the previous paragraph is a special case of the class of tests based on the $k - 1$ quantities $Z_j = \sup_x | R_j(x) | \; (j = 2, \cdots, k)$ where the R_j are any $k - 1$ orthogonal linear combinations of the S_j which are orthogonal to $\bar{S}$; however, the Z_j will in general be independently distributed only in the limit, not for finite n_j as with the Y_j .

For $k = 3$, the asymptotic behavior of max (Y_2 , Y_3) was also noted by Fisz [22]. For $k > 3$ Fisz suggests dividing the k samples into approximately $k/3$ collections of 3 or 2 samples each, computing the above or the Smirnov statistic from each collection, and then computing the maximum of these. The resulting test is clearly inferior to those we have considered: it is not even consistent, since it tests effectively only differences *within* the various collections.

Another simple test whose size may be computed from previously known results is the following: Let the n_j observations in the jth sample be divided at random into $k - 1$ subsets, each subset containing approximately the same number of observations, and call the sample d.f.'s of the observations in the $k - 1$ subsets of the jth samples $S_{jr}(x) (1 \leq r \leq k, r \neq j)$; for any j_1 , j_2 with $j_1 \neq j_2$, the distribution of $Z_{j_1 j_2} = C'_{j_1 j_2} \sup_x |S_{j_1 j_2}(x) - S_{j_2 j_1}(x)|$ (where $C'_{j_1 j_2}$ is a suitable normalizing constant) may again be obtained from Massey's tables [21], and the size of a test of H_1 based on such a statistic as $\max_{j_1, j_2} Z_{j_1, j_2}$ is again easily computed, since the $Z_{j_1 j_2}$ are independent.

Tests based on statistics like $\sum Y_j$ are less convenient to use, since the computation of size entails the convolution of the Kolmogorov-Smirnov d.f. $\Phi_1(x) = A_1(x^2)$ with itself. For example, a single convolution of Φ_1 with itself using term-by-term integration of (3.24) yields the d.f. G_2 given for $z > 0$ by a slowly converging double sum of terms involving the normal d.f., and this is extremely poor for computational purposes. It is in fact easier to obtain G_2 by numerical integration of the convolution formula, and this has been done to obtain a table of G_2 in Section 7.

6. Power; miscellaneous remarks. We again limit the discussion to tests of H_1, similar remarks applying for H_2. We use the notation of Section 1.

It is easily seen that, for the test of size (approximately) $\alpha > 0$ based on T, U, V, or any of the procedures listed in the previous section (excluding that of Fisz [22] for $k > 3$), for any $\beta < 1$ there is a value $\delta(\alpha, \beta)$ such that any of these tests has power $> \beta$ against all alternatives for which

$$\sup_{q,r,x}\{|F_q(x) - F_r(x)| \min(n_q^{\frac{1}{2}}, n_r^{\frac{1}{2}})\} > \delta(\alpha, \beta).$$

However, tests based on criteria such as Z or W cannot be guaranteed to have the property just cited; this may be demonstrated exactly as it was for ω^2-type tests in another problem in the paper by Kac, Kiefer, and Wolfowitz [23]. Similar results may be proved relative to other measures of distance of alternatives from H_1, as in [23]. Thus, distance tests of the Kolmogorov-Smirnov type seem preferable in applications to those of the ω^2-type.

We note that the distribution of A_r obtained in Section 3 gives an asymptotic computation of power for certain alternatives when T is used.

We remark that the methods of this paper may be modified along the lines of the papers by Darling [24] and Kac, Kiefer, and Wolfowitz [23] in parametric cases, e.g., to test the hypothesis H_1 under the assumption that the F_j are all normal, or to test that the F_j are equal *and* normal.

In the case $k = 3$ of H_1, when all n_j are equal, David [25] has used a clever device to compute the distribution of $\max_{j,x}[S_j(x) - S_{j+1}(x)]$, where the subscripts are taken mod 3. The method does not seem to generalize.

The use of "distance" criteria in various nonparametric multi-decision problems, e.g., problems of ranking or of classification, is to be recommended, but the appropriate distribution theory is more complicated.

The author plans to return in another paper to consideration of some of the limiting distributions discussed here using a method somewhat similar to that of Doob [13].

7. Tables. The functions A_h of Section 3 and B_h of Section 4 ($1 \leqq h \leqq 5$), and the function G_2 defined in Section 5, have been tabled by the Cornell Computing Center's 650. I am indebted to Miss Susan Litt, Miss Virginia Walbran, Mrs. Jane Wiegand, Professor R. J. Walker, and Mr. R. C. Lesser, for carrying out this work.

(Continued at the foot of p. 438)

TABLE 1

Tables of $\Phi_i(x) = A_i(x^2)$ for $i = 1, 2, 3, 4, 5$

x	$\Phi_1(x)$	$\Phi_2(x)$	$\Phi_3(x)$	$\Phi_4(x)$	$\Phi_5(x)$
0.37	.000826				
0.38	.001285				
0.39	.001929				
0.40	.002808				
0.41	.003972				
0.42	.005476				
0.43	.007377				
0.44	.009730				
0.45	.012589				
0.46	.016005				
0.47	.020022				
0.48	.024682				
0.49	.030017				
0.50	.036055				
0.51	.042814				
0.52	.050306				
0.53	.058534	.000894			
0.54	.067497	.001256			
0.55	.077183	.001731			
0.56	.087577	.002342			
0.57	.098656	.003115			
0.58	.110394	.004079			
0.59	.122760	.005262			
0.60	.135717	.006696			
0.61	.149229	.008412			
0.62	.163255	.010441			
0.63	.177752	.012816			
0.64	.192677	.015566			
0.65	.207987	.018720	.000762		
0.66	.223637	.022307	.001035		
0.67	.239582	.026350	.001383		
0.68	.255780	.030874	.001824		
0.69	.272188	.035897	.002373		
0.70	.288765	.041437	.003050		
0.71	.305470	.047507	.003874		
0.72	.322265	.054116	.004866		
0.73	.339114	.061271	.006050		
0.74	.355981	.068976	.007447		
0.75	.372833	.077230	.009081		
0.76	.389640	.086029	.010977	.000820	
0.77	.406372	.095367	.013159	.001080	
0.78	.423002	.105233	.015649	.001406	
0.79	.439505	.115614	.018472	.001810	
0.80	.455858	.126496	.021649	.002306	
0.81	.472039	.137859	.025201	.002907	
0.82	.488028	.149685	.029149	.003631	
0.83	.503809	.161950	.033510	.004493	
0.84	.519365	.174632	.038300	.005511	

TABLE 1—*Continued*

x	$\Phi_1(x)$	$\Phi_2(x)$	$\Phi_3(x)$	$\Phi_4(x)$	$\Phi_5(x)$
0.85	.534681	.187705	.043534	.006704	
0.86	.549745	.201142	.049223	.008092	.000897
0.87	.564545	.214917	.055378	.009694	.001157
0.88	.579071	.229001	.062006	.011530	.001476
0.89	.593315	.243366	.069112	.013621	.001867
0.90	.607269	.257982	.076699	.015986	.002340
0.91	.620928	.272822	.084766	.018645	.002908
0.92	.634285	.287855	.093313	.021618	.003584
0.93	.647337	.303054	.102333	.024924	.004382
0.94	.660081	.318390	.111821	.028579	.005317
0.95	.672514	.333834	.121767	.032600	.006407
0.96	.684636	.349361	.132160	.037004	.007666
0.97	.696445	.364942	.142988	.041802	.009113
0.98	.707941	.380554	.154236	.047009	.010765
0.99	.719126	.396169	.165887	.052634	.012639
1.00	.730000	.411765	.177923	.058687	.014754
1.01	.740566	.427319	.190326	.065174	.017127
1.02	.750825	.442809	.203074	.072101	.019777
1.03	.760781	.458214	.216146	.079471	.022720
1.04	.770436	.473514	.229521	.087284	.025972
1.05	.779794	.488690	.243174	.095541	.029551
1.06	.788860	.503725	.257083	.104239	.033471
1.07	.797637	.518603	.271223	.113372	.037747
1.08	.806130	.533308	.285569	.122935	.042390
1.09	.814343	.547826	.300099	.132919	.047414
1.10	.822282	.562143	.314786	.143314	.052828
1.11	.829951	.576248	.329607	.154110	.058642
1.12	.837356	.590130	.344538	.165291	.064862
1.13	.844502	.603779	.359554	.176846	.071495
1.14	.851395	.617184	.374632	.188756	.078545
1.15	.858040	.630340	.389749	.201006	.086015
1.16	.864443	.643237	.404883	.213577	.093904
1.17	.870610	.655871	.420012	.226450	.102213
1.18	.876546	.668235	.435114	.239605	.110938
1.19	.882258	.680325	.450170	.253023	.120075
1.20	.887750	.692137	.465159	.266681	.129619
1.21	.893030	.703668	.480064	.280558	.139562
1.22	.898102	.714916	.494865	.294632	.149895
1.23	.902973	.725879	.509546	.308881	.160607
1.24	.907648	.736555	.524090	.323283	.171687
1.25	.912134	.746946	.538483	.337815	.183121
1.26	.916435	.757050	.552710	.352455	.194895
1.27	.920557	.766869	.566758	.367181	.206993
1.28	.924506	.776403	.580613	.381971	.219400
1.29	.928288	.785655	.594266	.396804	.232097
1.30	.931908	.794626	.607703	.411658	.245067
1.31	.935371	.803319	.620917	.426513	.258290
1.32	.938682	.811737	.633898	.441348	.271746
1.33	.941847	.819883	.646638	.456145	.285417
1.34	.944871	.827761	.659129	.470884	.299281

TABLE 1—*Continued*

x	$\Phi_1(x)$	$\Phi_2(x)$	$\Phi_3(x)$	$\Phi_4(x)$	$\Phi_5(x)$
1.35	.947758	.835374	.671366	.485547	.313318
1.36	.950514	.842727	.683343	.500117	.327506
1.37	.953143	.849824	.695055	.514577	.341825
1.38	.955651	.856670	.706498	.528911	.356254
1.39	.958041	.863269	.717669	.543104	.370771
1.40	.960318	.869627	.728564	.557141	.385356
1.41	.962487	.875748	.739183	.571009	.399989
1.42	.964551	.881638	.749523	.584696	.414648
1.43	.966515	.887302	.759585	.598190	.429314
1.44	.968383	.892745	.769367	.611479	.443968
1.45	.970158	.897973	.778871	.624554	.458590
1.46	.971846	.902992	.788096	.637405	.473163
1.47	.973448	.907808	.797046	.650025	.487667
1.48	.974969	.912425	.805720	.662404	.502087
1.49	.976413	.916849	.814122	.674537	.516406
1.50	.977782	.921086	.822255	.686418	.530607
1.51	.979080	.925142	.830121	.698041	.544676
1.52	.980310	.929023	.837724	.709401	.558598
1.53	.981475	.932733	.845067	.720496	.572360
1.54	.982579	.936278	.852154	.731321	.585948
1.55	.983623	.939664	.858990	.741874	.599352
1.56	.984610	.942897	.865579	.752155	.612560
1.57	.985544	.945980	.871926	.762160	.625561
1.58	.986427	.948921	.878036	.771890	.638346
1.59	.987261	.951723	.883913	.781345	.650906
1.60	.988048	.954393	.889563	.790525	.663233
1.61	.988791	.956934	.894991	.799432	.675320
1.62	.989492	.959352	.900203	.808066	.687161
1.63	.990154	.961651	.905203	.816430	.698749
1.64	.990777	.963837	.909998	.824526	.710081
1.65	.991364	.965913	.914593	.832356	.721151
1.66	.991917	.967885	.918994	.839925	.731957
1.67	.992438	.969756	.923206	.847235	.742495
1.68	.992928	.971530	.927235	.854290	.752763
1.69	.993389	.973213	.931087	.861094	.762760
1.70	.993823	.974807	.934766	.867651	.772485
1.71	.994230	.976317	.938280	.873967	.781936
1.72	.994612	.977746	.941633	.880045	.791116
1.73	.994972	.979099	.944830	.885891	.800024
1.74	.995309	.980378	.947878	.891509	.808660
1.75	.995625	.981586	.950781	.896905	.817028
1.76	.995922	.982728	.953546	.902084	.825130
1.77	.996200	.983807	.956176	.907052	.832966
1.78	.996460	.984824	.958676	.911813	.840542
1.79	.996704	.985784	.961053	.916375	.847859
1.80	.996932	.986689	.963311	.920741	.854921
1.81	.997146	.987542	.965455	.924919	.861732
1.82	.997346	.988345	.967488	.928913	.868296
1.83	.997533	.989102	.969417	.932729	.874618
1.84	.997707	.989813	.971245	.936373	.880703

TABLE 1—*Continued*

x	$\Phi_1(x)$	$\Phi_2(x)$	$\Phi_3(x)$	$\Phi_4(x)$	$\Phi_5(x)$
1.85	.997870	.990483	.972976	.939851	.886554
1.86	.998023	.991112	.974615	.943167	.892177
1.87	.998165	.991703	.976166	.946328	.897578
1.88	.998297	.992259	.977633	.949338	.902760
1.89	.998421	.992780	.979019	.952204	.907731
1.90	.998536	.993269	.980329	.954931	.912494
1.91	.998644	.993728	.981566	.957524	.917056
1.92	.998744	.994158	.982733	.959987	.921423
1.93	.998837	.994560	.983833	.962326	.925599
1.94	.998924	.994938	.984871	.964547	.929591
1.95	.999004	.995291	.985848	.966653	.933404
1.96	.999079	.995621	.986769	.968649	.937044
1.97		.995930	.987635	.970541	.940517
1.98		.996219	.988450	.972332	.943827
1.99		.996489	.989216	.974027	.946981
2.00		.996741	.989936	.975631	.949984
2.01		.996976	.990612	.977146	.952842
2.02		.997195	.991247	.978578	.955560
2.03		.997400	.991843	.979930	.958142
2.04		.997591	.992402	.981206	.960595
2.05		.997768	.992925	.982409	.962924
2.06		.997934	.993416	.983543	.965133
2.07		.998088	.993875	.984612	.967227
2.08		.998231	.994305	.985618	.969211
2.09		.998364	.994707	.986565	.971090
2.10		.998488	.995083	.987455	.972868
2.11		.998603	.995434	.988292	.974549
2.12		.998710	.995702	.989079	.976139
2.13		.998809	.996069	.989817	.977640
2.14		.998901	.996355	.990511	.979058
2.15		.998987	.996621	.991161	.980396
2.16		.999066	.996870	.991770	.981657
2.17		.999139	.997101	.992342	.982846
2.18			.997317	.992877	.983966
2.19			.997518	.993377	.985020
2.20			.997704	.993846	.986012
2.21			.997878	.994284	.986945
2.22			.998039	.994693	.987821
2.23			.998189	.995075	.988645
2.24			.998328	.995432	.989418
2.25			.998458	.995765	.990143
2.26			.998577	.996076	.990823
2.27			.998688	.996366	.991460
2.28			.998791	.996635	.992057
2.29			.998887	.996887	.992616
2.30			.998975	.997120	.993139
2.31			.999057	.997338	.993628
2.32			.999132	.997540	.994085
2.33				.997728	.994512
2.34				.997902	.994910
2.35				.998064	.995282

TABLE 1—*Continued*

x	$\Phi_1(x)$	$\Phi_2(x)$	$\Phi_3(x)$	$\Phi_4(x)$	$\Phi_5(x)$
2.36				.998215	.995629
2.37				.998354	.995952
2.38				.998483	.996253
2.39				.998603	.996534
2.40				.998714	.996795
2.41				.998817	.997038
2.42				.998911	.997263
2.43				.998999	.997473
2.44				.999080	.997668
2.45				.999155	.997849
2.46					.998016
2.47					.998172
2.48					.998316
2.49					.998449
2.50					.998573
2.51					.998687
2.52					.998793
2.53					.998891
2.54					.998981
2.55					.999065
2.56					.999142

TABLE 2

Table of the inverses $\Phi_i^{-1}(p)$

p	$\Phi_1^{-1}(p)$	$\Phi_2^{-1}(p)$	$\Phi_3^{-1}(p)$	$\Phi_4^{-1}(p)$	$\Phi_5^{-1}(p)$
.25	0.67645	0.89456	1.05493	1.18776	1.30375
.50	0.82757	1.05751	1.22349	1.35992	1.47855
.75	1.01918	1.25299	1.42047	1.55788	1.67728
.80	1.07275	1.30614	1.47337	1.61065	1.72997
.85	1.13795	1.37025	1.53692	1.67388	1.79299
.90	1.22385	1.45399	1.61960	1.75593	1.87462
.95	1.35810	1.58379	1.74726	1.88226	2.00005
.98	1.51743	1.73699	1.89743	2.03053	2.14698
.99	1.62762	1.84273	2.00092	2.13257	2.24798
.995	1.73082	1.94172	2.09773	2.22797	2.34235
.999	1.94948	2.15162	2.30296	2.43009	2.54217
.9999	2.22530	2.41695	2.56244	2.68565	2.79481

$\Phi_h(x) = A_h(x^2)$ is tabled in Table 1 for $1 \leqq h \leqq 5$ and for x in steps of .01 from $\Phi_h^{-1}(.001)$ to $\Phi_h^{-1}(.999)$. Tables of $\Phi_h^{-1}(p)$ for various often used values of p are given in Table 2. Thus, in using the statistic T (resp., T') to test H_1 (resp., H_2) when the n_j are large, with a test of size α, one should reject the hypothesis when $\sqrt{T} > \Phi_{k-1}^{-1}(1 - \alpha)$ (resp., $\sqrt{T'} > \Phi_k^{-1}(1 - \alpha)$).

(*Continued on p. 444*)

TABLE 3

Tables of $B_i(x)$ for $i = 1, 2, 3, 4, 5$

x	$B_1(x)$	$B_2(x)$	$B_3(x)$	$B_4(x)$	$B_5(x)$
0.01	.000006				
0.02	.002892				
0.03	.023832				
0.04	.066851				
0.05	.123719	.000324			
0.06	.186020	.001566			
0.07	.248436	.004768			
0.08	.308145	.010891			
0.09	.363856	.020564			
0.10	.415127	.034001			
0.11	.461959	.051075	.000914		
0.12	.504575	.071420	.001966		
0.13	.543293	.094544	.003735		
0.14	.578461	.119910	.006438		
0.15	.610424	.146986	.010272		
0.16	.639507	.175283	.015396		
0.17	.666005	.204366	.021924	.000708	
0.18	.690186	.233862	.029920	.001249	
0.19	.712291	.263459	.039405	.002067	
0.20	.732530	.292900	.050357	.003240	
0.21	.751092	.321978	.062721	.004848	
0.22	.768144	.350530	.076413	.006971	
0.23	.783833	.378432	.091332	.009682	
0.24	.798290	.405587	.107364	.013049	.000675
0.25	.811630	.431928	.124383	.017130	.001043
0.26	.823958	.457406	.142264	.021971	.001566
0.27	.835364	.481991	.160881	.027605	.002274
0.28	.845930	.505668	.180110	.034056	.003184
0.29	.855730	.528431	.199832	.041333	.004359
0.30	.864829	.550283	.219937	.049437	.005830
0.31	.873285	.571236	.240320	.058356	.007632
0.32	.881153	.591305	.260885	.068071	.009813
0.33	.888478	.610511	.281544	.078555	.012394
0.34	.895305	.628877	.302218	.089771	.015414
0.35	.901673	.646428	.322835	.101682	.018906
0.36	.907617	.663191	.343331	.114243	.022887
0.37	.913168	.679193	.363651	.127406	.027378
0.38	.918358	.694464	.383745	.141122	.032397
0.39	.923211	.709031	.403570	.155340	.037951
0.40	.927753	.722922	.423088	.170007	.044054
0.41	.932006	.736166	.442268	.185074	.050702
0.42	.935990	.748790	.461084	.200488	.057898
0.43	.939724	.760820	.479514	.216199	.065629
0.44	.943226	.772283	.497538	.232160	.073892
0.45	.946512	.783203	.515144	.248323	.082674
0.46	.949595	.793605	.532320	.264643	.091955
0.47	.952490	.803513	.549056	.281078	.101720
0.48	.955210	.812950	.565349	.297587	.111948
0.49	.957765	.821936	.581193	.314133	.122617

TABLE 3—*Continued*

x	$B_1(x)$	$B_2(x)$	$B_3(x)$	$B_4(x)$	$B_5(x)$
0.50	.960167	.830494	.596590	.330680	.133701
0.51	.962425	.838642	.611537	.347194	.145177
0.52	.964549	.846400	.626039	.363646	.157017
0.53	.966547	.853787	.640097	.380006	.169195
0.54	.968427	.860819	.653717	.396248	.181679
0.55	.970197	.867515	.666904	.412349	.194449
0.56	.971864	.873889	.679663	.428287	.207471
0.57	.973433	.879957	.692004	.444042	.220721
0.58	.974912	.885734	.703933	.459597	.234170
0.59	.976305	.891233	.715458	.474935	.247790
0.60	.977618	.896468	.726589	.490043	.261557
0.61	.978855	.901451	.737333	.504908	.275444
0.62	.980022	.906195	.747701	.519519	.289426
0.63	.981122	.910710	.757702	.533868	.303480
0.64	.982159	.915008	.767344	.547945	.317582
0.65	.983138	.919100	.776639	.561745	.331712
0.66	.984061	.922995	.785596	.575262	.345847
0.67	.984932	.926702	.794224	.588492	.359967
0.68	.985754	.930231	.802533	.601431	.374053
0.69	.986530	.933590	.810532	.614076	.388088
0.70	.987262	.936787	.818232	.626427	.402054
0.71	.987954	.939830	.825641	.638482	.415937
0.72	.988607	.942727	.832769	.650242	.429721
0.73	.989224	.945485	.839624	.661707	.443394
0.74	.989806	.948110	.846217	.672878	.456943
0.75	.990356	.950608	.852555	.683757	.470349
0.76	.990876	.952986	.858647	.694347	.483607
0.77	.991367	.955250	.864502	.704649	.496713
0.78	.991831	.957405	.870127	.714668	.509646
0.79	.992270	.959455	.875532	.724407	.522402
0.80	.992684	.961408	.880723	.733869	.534981
0.81	.993076	.963266	.885707	.743059	.547361
0.82	.993447	.965035	.890494	.751980	.559556
0.83	.993797	.966718	.895090	.760639	.571546
0.84	.994128	.968321	.899501	.769038	.583319
0.85	.994441	.969846	.903735	.777183	.594903
0.86	.994737	.971298	.907797	.785079	.606259
0.87	.995017	.972680	.911696	.792732	.617411
0.88	.995282	.973995	.915436	.800145	.628332
0.89	.995532	.975248	.919024	.807326	.639045
0.90	.995769	.976439	.922465	.814278	.649538
0.91	.995993	.977574	.925765	.821007	.659801
0.92	.996205	.978654	.928930	.827519	.669848
0.93	.996406	.979681	.931964	.833819	.679675
0.94	.996596	.980660	.934874	.839912	.689284
0.95	.996776	.981591	.937663	.845803	.698668
0.96	.996946	.982477	.940336	.851499	.707832
0.97	.997107	.983321	.942898	.857003	.716780

TABLE 3—*Continued*

x	$B_1(x)$	$B_2(x)$	$B_3(x)$	$B_4(x)$	$B_5(x)$
0.98	.997259	.984124	.945353	.862321	.725508
0.99	.997403	.984889	.947706	.867459	.734026
1.00	.997540	.985616	.949960	.872421	.742332
1.01	.997669	.986309	.952120	.877213	.750424
1.02	.997791	.986968	.954190	.881839	.758311
1.03	.997907	.987596	.956172	.886304	.765992
1.04	.998017	.988193	.958070	.890614	.773472
1.05	.998121	.988761	.959889	.894771	.780754
1.06	.998219	.989302	.961630	.898782	.787834
1.07	.998312	.989817	.963298	.902651	.794727
1.08	.998400	.990308	.964895	.906382	.801427
1.09	.998484	.990775	.966425	.909979	.807943
1.10	.998563	.991219	.967888	.913447	.814272
1.11	.998638	.991642	.969291	.916790	.820424
1.12	.998709	.992044	.970632	.920011	.826397
1.13	.998776	.992427	.971916	.923115	.832199
1.14	.998840	.992792	.973146	.926106	.837833
1.15	.998900	.993139	.974322	.928986	.843298
1.16	.998957	.993469	.975448	.931761	.848602
1.17	.999011	.993784	.976525	.934433	.853750
1.18	.999063	.994083	.977557	.937006	.858742
1.19		.994368	.978544	.939484	.863580
1.20		.994639	.979488	.941868	.868274
1.21		.994897	.980391	.944164	.872821
1.22		.995143	.981256	.946373	.877227
1.23		.995377	.082082	.948499	.881497
1.24		.995599	.982873	.950544	.885630
1.25		.995811	.983630	.952512	.889635
1.26		.996013	.984354	.954405	.893515
1.27		.996205	.985047	.956226	.897208
1.28		.996388	.985708	.957977	.900902
1.29		.996562	.986341	.959661	.904419
1.30		.996727	.986947	.961281	.907818
1.31		.996885	.987526	.962837	.911110
1.32		.997035	.988080	.964334	.914292
1.33		.997178	.988610	.965773	.917370
1.34		.997313	.989116	.967156	.920346
1.35		.997443	.989600	.968485	.923223
1.36		.997566	.990063	.969762	.926004
1.37		.997683	.990506	.970989	.928692
1.38		.997795	.990929	.972169	.931287
1.39		.997901	.991334	.973302	.933797
1.40		.998002	.991721	.974390	.936220
1.41		.998098	.992091	.975435	.938560
1.42		.998190	.992444	.976439	.940821
1.43		.998277	.992782	.977404	.943003
1.44		.998360	.993104	.978330	.945110
1.45		.998439	.993413	.979219	.947145
1.46		.998514	.993708	.980073	.949108
1.47		.998586	.993990	.980893	.951002

TABLE 3—*Continued*

x	$B_1(x)$	$B_2(x)$	$B_3(x)$	$B_4(x)$	$B_5(x)$
1.48		.998654	.994259	.981680	.952831
1.49		.998718	.994517	.982436	.954595
1.50		.998780	.994763	.983161	.956298
1.51		.998839	.994998	.983857	.957937
1.52		.998895	.995223	.984526	.959519
1.53		.998948	.995437	.985167	.961044
1.54		.998999	.995643	.985782	.962520
1.55		.999047	.995839	.986373	.963941
1.56		.999093	.996026	.986939	.965311
1.57			.996205	.987483	.966629
1.58			.996376	.988005	.967897
1.59			.996539	.988505	.969129
1.60			.996695	.988985	.970307
1.61			.996844	.989445	.971452
1.62			.966987	.989887	.972538
1.63			.997123	.990311	.973602
1.64			.997253	.990717	.974615
1.65			.997377	.991106	.975598
1.66			.997495	.991480	.976544
1.67			.997608	.991838	.977450
1.68			.997717	.992182	.978329
1.69			.997820	.992511	.979165
1.70			.997919	.992827	.979979
1.71			.998013	.993129	.980765
1.72			.998103	.993420	.981511
1.73			.998189	.993698	.982239
1.74			.998271	.993964	.982932
1.75			.998349	.994220	.983606
1.76			.998424	.994465	.984252
1.77			.998496	.994700	.984865
1.78			.998564	.994925	.985462
1.79			.998629	.995140	.986040
1.80			.998692	.995347	.986590
1.81			.998751	.995545	.987123
1.82			.998808	.995734	.987635
1.83			.998862	.995916	.988124
1.84			.998914	.996090	.988597
1.85			.998963	.996257	.989056
1.86			.999011	.996417	.989493
1.87			.999056	.996570	.989915
1.88				.996717	.990315
1.89				.996857	.990709
1.90				.996992	.991077
1.91				.997121	.991439
1.92				.997244	.991781
1.93				.997363	.992111
1.94				.997476	.992431
1.95				.997584	.992742
1.96				.997688	.993039
1.97				.997788	.993321

TABLE 3—*Continued*

x	$B_1(x)$	$B_2(x)$	$B_3(x)$	$B_4(x)$	$B_5(x)$
1.98				.997883	.993593
1.99				.997974	.993853
2.00				.998061	.994107
2.01				.998145	.994346
2.02				.998225	.994577
2.03				.998302	.994802
2.04				.998375	.995014
2.05				.998445	.995219
2.06				.998513	.995417
2.07				.998577	.995605
2.08				.998639	.995787
2.09				.998698	.995963
2.10				.998754	.996132
2.11				.998808	.996290
2.12				.998860	.996445
2.13				.998909	.996596
2.14				.998957	.996737
2.15				.999002	.996873
2.16				.999046	.997004
2.17					.997131
2.18					.997252
2.19					.997367
2.20					.997479
2.21					.997584
2.22					.997687
2.23					.997787
2.24					.997882
2.25					.997971
2.26					.998059
2.27					.998143
2.28					.998224
2.29					.998298
2.30					.998373
2.31					.998446
2.32					.998512
2.33					.998578
2.34					.998637
2.35					.998699
2.36					.998756
2.37					.998812
2.38					.998866
2.39					.998916
2.40					.998962
2.41					.999012
2.42					.999055

The corresponding tables of B_h, the limiting d.f. of W (with $h = k - 1$) and of W' (with $h = k$), and of B_h^{-1}, are Tables 3 and 4.

Tables 1 and 2 were computed from equation (3.21), while Tables 3 and 4 were computed using the form of (4.3) given in (4.4) and the paragraph following (4.4). A program developed at Cornell was used to obtain the Bessel functions by power series or asymptotic series in appropriate regions.

As a check, the tables for $h = 1$ were compared with that of Φ_1 of Smirnov [26] and that of B_1 of Anderson and Darling [12]. In the case of Φ_1, the last tabled figure often differed slightly; wherever a discrepency was noted in the last *two* places, the tables were checked by differencing, and Smirnov's appeared to be in error. The table of [12] checked with that of B_1 here.

As mentioned in Section 5, the easiest way to compute tables of the convolution G_2 of Φ_1 with itself appeared to be by numerical integration, and Table 5 was computed in this way. Thus, for example, to test H_1 with size α when $k = 3$, one can use the statistic $Y_2 + Y_3$ of Section 5 with $C_2'' = [n_1n_2/(n_1 + n_2)]^{\frac{1}{2}}$ and $C_3'' = [n_3(n_1 + n_2)/(n_1 + n_2 + n_3)]^{\frac{1}{2}}$, rejecting the hypothesis for large n_i when $Y_2 + Y_3 > G_2^{-1}(1 - \alpha)$.

Added in proof: The author has recently learned that the following independently obtained results, which overlap some of those of this paper, appeared somewhat after [1] and the submission of earlier versions of the present paper: the limiting d.f. of T_N has been considered by J. J. Gichman in *Teorya Veryotnostei i yeyau primenyenya*, vol. 2 (1957), pp. 380–384, using an approach like that of [12], and two papers by L. C. Chang and M. Fisz in *Science Record*, vol. 1 (1957), pp. 335–346, consider tests like those discussed in the second and fourth paragraphs of Section 5.

TABLE 4

Table of the inverses $B_i^{-1}(p)$

p	$B_1^{-1}(p)$	$B_2^{-1}(p)$	$B_3^{-1}(p)$	$B_4^{-1}(p)$	$B_5^{-1}(p)$
.25	0.07026	0.18545	0.31472	0.45103	0.59161
.50	0.11888	0.27757	0.44138	0.60668	0.77253
.75	0.20939	0.42098	0.62227	0.81775	1.00947
.80	0.24124	0.46640	0.67691	0.87980	1.07785
.85	0.28406	0.52481	0.74592	0.95734	1.16268
.90	0.34730	0.60704	0.84116	1.06311	1.27748
.95	0.46136	0.74752	1.00018	1.23730	1.46466
.98	0.61981	0.93320	1.20561	1.45913	1.70028
.99	0.74346	1.07366	1.35861	1.62263	1.87215
.995	0.86939	1.21412	1.51010	1.78345	2.03935
.999	1.16786	1.54027	1.85773	2.14949	2.40774
.9999	1.60443	2.00691	2.3495	2.66130	2.825

TABLE 5

Table of $G_2(x)$

x	$G_2(x)$	x	$G_2(x)$	x	$G_2(x)$	x	$G_2(x)$
.92	.0008	1.42	.2005	1.92	.7157	2.42	.9531
.93	.0011	1.43	.2100	1.93	.7238	2.43	.9549
.94	.0013	1.44	.2197	1.94	.7319	2.44	.9569
.95	.0016	1.45	.2295	1.95	.7396	2.45	.9586
.96	.0020	1.46	.2396	1.96	.7474	2.46	.9605
.97	.0024	1.47	.2497	1.97	.7549	2.47	.9621
.98	.0028	1.48	.2601	1.98	.7624	2.48	.9638
.99	.0034	1.49	.2705	1.99	.7695	2.49	.9653
1.00	.0040	1.50	.2811	2.00	.7767	2.50	.9669
1.01	.0048	1.51	.2917	2.01	.7835	2.51	.9682
1.02	.0056	1.52	.3025	2.02	.7904	2.52	.9697
1.03	.0065	1.53	.3133	2.03	.7969	2.53	.9709
1.04	.0076	1.54	.3242	2.04	.8035	2.54	.9723
1.05	.0087	1.55	.3352	2.05	.8097	2.55	.9734
1.06	.0100	1.56	.3463	2.06	.8160	2.56	.9747
1.07	.0115	1.57	.3573	2.07	.8219	2.57	.9757
1.08	.0131	1.58	.3685	2.08	.8278	2.58	.9769
1.09	.0149	1.59	.3796	2.09	.8335	2.59	.9779
1.10	.0168	1.60	.3909	2.10	.8391	2.60	.9790
1.11	.0189	1.61	.4020	2.11	.8445	2.61	.9798
1.12	.0212	1.62	.4133	2.12	.8499	2.62	.9808
1.13	.0238	1.63	.4244	2.13	.8549	2.63	.9816
1.14	.0265	1.64	.4356	2.14	.8600	2.64	.9826
1.15	.0294	1.65	.4467	2.15	.8648	2.65	.9833
1.16	.0326	1.66	.4579	2.16	.8697	2.66	.9841
1.17	.0359	1.67	.4689	2.17	.8742	2.67	.9848
1.18	.0395	1.68	.4801	2.18	.8788	2.68	.9856
1.19	.0434	1.69	.4910	2.19	.8830	2.69	.9862
1.20	.0475	1.70	.5020	2.20	.8873	2.70	.9869
1.21	.0528	1.71	.5127	2.21	.8914	2.71	.9874
1.22	.0564	1.72	.5236	2.22	.8954	2.72	.9881
1.23	.0612	1.73	.5342	2.23	.8992	2.73	.9886
1.24	.0663	1.74	.5449	2.24	.9030	2.74	.9892
1.25	.0717	1.75	.5554	2.25	.9066	2.75	.9896
1.26	.0773	1.76	.5658	2.26	.9102	2.76	.9902
1.27	.0832	1.77	.5761	2.27	.9135	2.77	.9906
1.28	.0893	1.78	.5864	2.28	.9169	2.78	.9912
1.29	.0957	1.79	.5964	2.29	.9200	2.79	.9915
1.30	.1023	1.80	.6064	2.30	.9232	2.80	.9920
1.31	.1092	1.81	.6162	2.31	.9261	2.81	.9923
1.32	.1164	1.82	.6260	2.32	.9291	2.82	.9928
1.33	.1237	1.83	.6355	2.33	.9318	2.83	.9930
1.34	.1314	1.84	.6451	2.34	.9346	2.84	.9934
1.35	.1392	1.85	.6543	2.35	.9371	2.85	.9937
1.36	.1474	1.86	.6636	2.36	.9397	2.86	.9941
1.37	.1557	1.87	.6726	2.37	.9420	2.87	.9943
1.38	.1642	1.88	.6816	2.38	.9445	2.88	.9946
1.39	.1730	1.89	.6902	2.39	.9466	2.89	.9948
1.40	.1820	1.90	.6989	2.40	.9489	2.90	.9952
1.41	.1911	1.91	.7073	2.41	.9509	2.91	.9953

REFERENCES

[1] J. Kiefer, "Distance tests with good power for the nonparametric k-sample problem" (Abstract), *Ann. Math. Stat.*, Vol. 26 (1955), p. 775.

[2] W. H. Kruskal, "A nonparametric test for the several sample problem," *Ann. Math. Stat.*, Vol. 23 (1952), pp. 525–540.

[3] I. R. Savage, "Bibliography on nonparametric statistics and related topics," *J. Amer. Stat. Assn.*, Vol. 48 (1953), pp. 844–906.

[4] P. DeMunter, "Consistance des tests non-paramétriques pour la compardison d'échantillons" *Acad. Roy. Belgique Bull. Cl. Sci.*, (1954), pp. 1106–1119.

[5] A. N. Kolmogorov, "Sulla determinazione empirica delle leggi di probabilita," *Giorn. Ist. Ital. Attuari*, Vol. 4 (1933), pp. 1–11.

[6] R. von Mises, *Wahrscheinlichkeitsrechnung*, Deuticke, Vienna, 1931.

[7] N. Smirnov, "On the estimation of the discrepancy between empirical curves of distribution for two independent samples," *Bul. Math. de l'Universite de Moscou*, Vol 2 (1939), fasc. 2.

[8] E. L. Lehmann, "Consistency and unbiasedness of certain non-parametric tests," *Ann. Math. Stat.*, Vol. 22 (1956), pp. 165–179.

[9] M. Rosenblatt, "Limit theorems associated with variants of the von Mises statistic," *Ann. Math. Stat.*, Vol. 23 (1952), pp. 617–623.

[10] W. Feller, "On the Kolmogorov-Smirnov limit theorems for empirical distributions," *Ann. Math. Stat.*, Vol. 19 (1948), pp. 177–189.

[11] M. Kac, "On some connections between probability theory and differential and integral equations," *Proceedings of the Second Berkely Symposium of Mathematical Statistics and Probability*, University of California Press, 1951, pp. 180–215.

[12] T. W. Anderson and D. A. Darling, "Asymptotic theory of certain 'goodness of fit' criteria based on stochastic processes," *Ann. Math. Stat.*, Vol. 23 (1952), pp. 193–212.

[13] J. L. Doob, "Heuristic approach to the Kolmogorov-Smirnov theorems," *Ann. Math. Stat.*, Vol. 20 (1949), pp. 393–403.

[14] M. L. Donsker, "Justification and extension of Doob's heuristic approach to the Kolmogorov-Smirnov theorems," *Ann. Math. Stat.*, Vol. 23 (1952), pp. 277–281.

[15] J. Kiefer and J. Wolfowitz, "On the deviations of the empiric distribution function of vector chance variables," *Trans. Amer. Math. Soc.*, Vol. 87 (1958), pp. 173–186.

[16] A. Dvoretzky, J. Kiefer, and J. Wolfowitz, "Asymptotic minimax character of the sample distribution function and of the classical multinomial estimator," *Ann. Math. Stat.*, Vol. 27 (1956), pp. 642–669.

[17] M. Rosenblatt, "On a certain class of Markov processes," *Trans. Amer. Math. Soc.*, Vol. 71 (1951), pp. 120–135.

[18] A. Erdélyi et al, *Higher Transcendental Functions* (3 vols.), McGraw-Hill, New York, 1953–1955.

[19] A. Erdélyi et al, *Tables of Integral Transforms* (2 vols.), McGraw-Hill, New York, 1954.

[20] G. N. Watson, *A Treatise on the Theory of Bessel Functions*, 2nd edn., Cambridge University Press, Cambridge, 1944.

[21] F. J. Massey, Jr., "Distribution table for the deviation between two sample cumulatives," *Ann. Math. Stat.*, Vol. 23 (1952), pp. 435–441; also Vol. 22 (1951), pp. 125–128.

[22] M. Fisz, "A limit theorem for empirical distribution functions," *Bull. de l'Acad. Pol. des Sci.*, Vol. 5 (1957), pp. 695–698.

[23] M. Kac, J. Kiefer, and J. Wolfowitz, "On tests of normality and other tests of good

ness of fit based on distance methods," *Ann. Math. Stat.*, Vol. 26 (1955), pp. 189–211.

[24] D. A. Darling, "The Cramér-Smirnov test in the parametric case," *Ann. Math. Stat.*, Vol. 26 (1955), pp. 1–20.

[25] H. David, "A three-sample Kolmogoroff-Smirnov test," *Ann. Math. Stat.*, Vol. 29 (1958), pp. 842–851.

[26] N. Smirnov, "Table for estimating the goodness of fit of empirical distributions," *Ann. Math. Stat.*, Vol. 19 (1948), pp. 279–281.

Reprinted from The Annals of Mathematical Statistics
Vol. 30, No. 2, June, 1959

ASYMPTOTIC MINIMAX CHARACTER OF THE SAMPLE DISTRIBUTION FUNCTION FOR VECTOR CHANCE VARIABLES

By J. Kiefer[1] and J. Wolfowitz[2]

Cornell University

Summary. The purpose of this paper is to prove Theorem 1 stated in Section 1 below and Theorem 2 of Section 6 and the results of Section 7. These theorems are the generalizations to vector chance variables of Theorems 4 and 5 and Section 6 of [1], and state that the sample distribution function (d.f.) is asymptotically minimax for the large class of weight functions of the type described below. The main difficulties are embodied in the proof of Theorem 1 (Sections 2 to 5), where the loss function is a function of the maximum difference between estimated and true d.f. The proof utilizes the results of [2] and is not a straightforward extension of the result of [1], because the sample d.f. is no longer "distribution free" (even in the limit), and hence it is necessary to prove the uniformity of approach, to its limit, of the d.f. of the normalized maximum deviation between sample and population d.f.'s (for a certain class of d.f.'s). The latter fact enables us essentially to infer the existence of a uniformly (with the sample number) approximately least favorable (to the statistician) d.f., by means of which the proof of the theorem is achieved. Theorem 2 (Section 6) considers loss functions of integral type, and more general loss functions are treated in Section 7.

1. Introduction and preliminaries.

The problem of finding a reasonable estimator of an unknown distribution function (d.f.) F in one or more dimensions is an old one. In the one-dimensional case the first extensive optimality results were obtained in [1]. It was shown there that, although a minimax procedure for sample size n may depend on the weight function as well as on n, the sample d.f. ϕ_n^* is asymptotically minimax as $n \to \infty$ for a very large class of weight functions which includes almost any weight function of practical interest. Also, an exact minimax procedure is extremely tedious to calculate in most practical cases, and is less convenient to use in practice than is ϕ_n^*. Moreover, one can obtain from [1] a bound on the relative difference between the maximum losses which can be encountered from using ϕ_n^* or the actual minimax procedure, and for many common weight functions this bound indicates that ϕ_n^* is very close to being minimax for fairly small values of n.

For dimension $m > 1$ the minimax problem presents difficulties which are not present when $m = 1$. (An outline of the main ideas and difficulties encountered in the proofs when $m = 1$ or when $m > 1$ will be given in Section 4; the

Received April 21, 1958.

[1] The research of this author was under contract with the Office of Naval Research.

[2] The research of this author was supported by the U. S. Air Force under Contract No. AF 18(600)-685, monitored by the Office of Scientific Research.

proof there is completed in Section 5; additional considerations for various weight functions are outlined in Sections 6 and 7.) These difficulties stem from the fact that neither ϕ_n^* nor any other known procedure which seems a reasonable candidate for optimality, has the distribution-free property possessed by ϕ_n^* when $m = 1$. This fact has led investigators of the problem when $m > 1$ to try (unsuccessfully) to find reasonable distribution-free procedures. Such investigations now seem to have been aimed in the wrong direction; for the main result of the present paper is that ϕ_n^* *is still asymptotically minimax for a large class of weight functions, even though it is no longer distribution free.*

The proof of the result just stated presents new difficulties far greater than those encountered when $m = 1$. In order to describe these difficulties briefly, let us suppose for the moment that the risk function is the expected value (under the true F) of $n^{1/2}$ times the maximum absolute deviation between estimated and true d.f. The computation of this risk function or its limit as $n \to \infty$ for the sequence of procedures ϕ_n^* (or any other reasonable sequence procedures) is known to present formidable difficulties, even for very simple continuous F (e.g., the uniform distribution on the unit square when $m = 2$). Our method of proof circumvents such a computation by showing that, when n is suitably large, the risk function of ϕ_n^* is changed arbitrarily little from what it would be if the maximum deviation were taken over a large but *finite* set of points instead of over all of m-space (this uses a result of [2]). Thus, the problem is reduced to a multinomial problem, similar to the reduction of [1] when $m = 1$, and we can circumvent the explicit computation of the risk there in a manner like that used in the multinomial case in [1], and which will be described in Section 3 below. But there remains another difficulty: in order to use a Bayes technique like that of [1] to prove the asymptotic minimax character of ϕ_n^*, we must show that there is a d.f. F_δ at which the risk function of ϕ_n^* is almost a maximum for all sufficiently large n; i.e., that the location of some approximate maximum does not "wander around" too much with n. Because of the distribution-free nature of the chance loss (for many common loss functions) under ϕ_n^* when $m = 1$, the existence of such an F_δ was automatic there (any continuous d.f. could be used); for $m > 1$, our proof requires the result of Lemma 1 of Section 2 below to obtain the existence of such an F_δ, at least when F is restricted to belong to a class of d.f.'s which in Section 5 is seen to be dense enough in an appropriate sense to yield the desired result. Once such an F_δ is known to exist, a sequence of approximately least favorable a priori distributions can be constructed for the approximating multinomial problem in the manner of [1]; this will be described in Section 4.

Aside from the difficulties described in the previous paragraph, the proofs of minimax results when $m > 1$ are very similar to those when $m = 1$. Therefore, rather than to repeat all of the details of [1], in each of Sections 4, 6, and 7 we will first describe the idea of the proof and then will indicate the modifications needed in the proof of the corresponding section of [1] to make it apply when $m > 1$.

We now give the notation used in this paper. m will denote any positive integer, fixed throughout the sequel. $\mathfrak{F}$ denotes the class of all d.f.'s on Euclidean m-space R^m, and $\mathfrak{F}^c$ denotes the subclass of continuous members of $\mathfrak{F}$. Let D be any subclass of the space of real functions on R^m. For simplicity we assume $\mathfrak{F} \subset D$, although it is really only necessary that D contains every possible function of the form S_n (defined below), for all n and $z^{(n)}$. Let B be the smallest Borel field on D such that every element of $\mathfrak{F}$ is an element of B and such that, for every positive integer k, real numbers $a_1, \cdots, a_k$, and m-vectors $t_1, \cdots, t_k$, the set $\{g \mid g \varepsilon D; g(t_1) < a_1, \cdots, g(t_k) < a_k\}$ is in B. (For example, we might have $D = \mathfrak{F}$ and B the Borel sets of the usual metric topology.) Let $\mathfrak{D}_n$ be the class of all real functions ϕ_n on $B \times R^{mn}$ such that $\phi_n(\cdot\,; z)$ is a probability measure (B) on D for each z in R^{mn} and such that $\phi_n(\Delta\,;\cdot)$ is a Borel-measurable function on R^{mn} for each Δ in B.

We now describe the statistical problem. Let $Z_1, \cdots, Z_n$ be independently and identically distributed m-vectors, each distributed according to some d.f. F about which it is known only that $F \varepsilon \mathfrak{F}$ (or $\mathfrak{F}^c$ or some other suitably dense subclass of $\mathfrak{F}$). The statistician wants to estimate F. Write $Z^{(n)} = (Z_1, \cdots, Z_n)$ and $z^{(n)} = (z_1, \cdots, z_n)$, where $z_i \varepsilon R^m$. Having observed $Z^{(n)} = z^{(n)}$, the statistician uses some decision function ϕ_n (a member of $\mathfrak{D}_n$) as follows: a function $g \varepsilon D$ is selected by means of a randomization according to the probability measure $\phi_n(\cdot\,; z^{(n)})$ on D; the function g so selected (which need not even be a member of $\mathfrak{F}$) is then the statistician's estimate of the unknown F. It is desirable to select a procedure ϕ_n which may be expected to yield a g which will lie close to the true F, whatever it may be; the precise meaning of "close" will be reflected by a weight function $W_n(F, g)$ which measures the loss when F is the true distribution function and g is the estimate of it. The probability of making a decision in Δ when ϕ_n is used and F is the true d.f. is

$$\mu_{F,\phi_n}(\Delta) = \int \phi_n(\Delta, z^{(n)}) F(dz^{(n)}), \tag{1.1}$$

which, as a function of Δ, will be a probability measure on D (see the next paragraph). Denoting expectation of a function on D with respect to this measure by E_{F,ϕ_n} (the symbol P_{F,ϕ_n} is used analogously, and the subscript ϕ_n will be omitted when it is not relevant), the risk function of the procedure ϕ_n is defined by

$$r_n(F, \phi_n) = E_{F,\phi_n} W_n(F, g); \tag{1.2}$$

i.e., it is the expected loss when F is true and ϕ_n is used. A sequence $\{\phi_n'\}$ of procedures is said to be *asymptotically minimax* relative to a sequence W_n of weight functions and a subclass $\mathfrak{F}'$ of $\mathfrak{F}$ if

$$\lim_{n\to\infty} \frac{\sup_{F\varepsilon\mathfrak{F}'} r_n(F, \phi_n')}{\inf_{\phi_n\varepsilon\mathfrak{D}_n} \sup_{F\varepsilon\mathfrak{F}'} r_n(F, \phi_n)} = 1. \tag{1.3}$$

(We note that this is a stronger property than that obtained by suppressing the supreme operation in the numerator and asking that the upper limit as

$n \to \infty$ be $\leqq 1$ for each F; this latter asymptotic property is much easier to verify than (1.3).) A nonrandomized decision function is one which for each $z^{(n)}$ assigns probability one to a single element (depending on $z^{(n)}$) of D. By ϕ_n^* we denote the nonrandomized procedure which chooses as decision the "sample d.f." S_n defined by

$$S_n(z) = n^{-1} \text{ (number of } Z_i \leqq z, 1 \leqq i \leqq n),$$

where as usual $Z \leqq z$ means that each component of Z is $\leqq$ the corresponding component of z. We shall not explicitly display the dependence of the chance function S_n on $Z^{(n)}$.

Obvious measurability considerations arise in connection with (1.1), (1.2), etc. These are handled exactly as in Section 1 of [1].

We can now state the main result of this paper, whose proof will occupy the next four sections (modifications and extensions are considered in Sections 6 and 7).

THEOREM 1. *Suppose* $W_n(F, g) = W(n^{1/2} \sup_z |g(z) - F(z)|)$, *where, for* $r \geqq 0$, $W(r)$ *is continuous, nonnegative, monotonically nondecreasing, not identically zero, and satisfies*

$$\int_0^\infty rW(r)e^{-c_m' r^2}\, dr < \infty \tag{1.4}$$

where c_m' *is given by* (1.8). *Then* $\{\phi_n^*\}$ *is asymptotically minimax relative to* $\{W_n\}$ *and* $\mathfrak{F}$.

Before listing the results of [2] which will be used in the present paper, we introduce some additional notation. When Z_1 has d.f. F, define

$$D_n = \sup_{x \in R^m} |S_n(x) - F(x)|$$

and

$$G_n(r\,; F) = P_F\{D_n < r/n^{1/2}\}.$$

For k a positive integer, write A_k^m for the subset of $(k+1)^m$ points in the m-dimensional unit cube $I^m = \{x \mid 0 \leqq x \leqq 1, x \in R^m\}$ for which each coordinate is an integral multiple of $1/k$. Write

$$D_{n,k} = \sup_{x \in A_k^m} |S_n(x) - F(x)|$$

and

$$G_{n,k}(r; F) = P_F\left\{D_{n,k} < \frac{r}{n^{1/2}}\right\}$$

We also write

$$G_{\infty,k}(r; F) = \lim_{n \to \infty} G_{n,k}(r; F);$$

the existence of this limit follows from the multivariate central limit theorem. Finally, let $\mathfrak{F}^*$ be the class of d.f.'s F which are in $\mathfrak{F}^c$ and for which each one-dimensional marginal d.f. of F is uniform on I^1. Clearly, if Z_1 has d.f. F in $\mathfrak{F}^c$, we can perform continuous transformations on the components of Z_1, so as to make the result have a d.f. F^* in $\mathfrak{F}^*$, without changing G_n. This fact will be used in the sequel.

The results of [2] which will be used in the present paper are the following (some of these results hold with little or no modification for F in $\mathfrak{F}$, but we need them here only for F in $\mathfrak{F}^*$):

A. (Theorem 2 of [2].) For F in $\mathfrak{F}^*$, there is a d.f. $G(\cdot\,;F)$ such that

$$\lim_{n\to\infty} G_n(r;F) = G(r;F) \tag{1.5}$$

at every continuity point of the latter. Moreover, for F in $\mathfrak{F}^*$,

$$\lim_{k\to\infty} G_{\infty,k}(r;F) = G(r;F) \tag{1.6}$$

and (obviously)

$$\lim_{k\to\infty} G_{n,k}(r;F) = G_n(r;F). \tag{1.7}$$

B. (Theorem 1 of [2].) There are positive constants c_m^* and c_m' (independent of n, F, and r) such that, for F in $\mathfrak{F}^*$, all n, and all $r \geqq 0$,

$$1 - G_n(r;F) < c_m^* e^{-c_m' r^2}. \tag{1.8}$$

Further remarks on possible values of c_m' are contained in [1] and [2].

C. For each F in $\mathfrak{F}$ there is an F_1 in $\mathfrak{F}^*$ such that, for all n and r,

$$G_n(r;F_1) \leqq G_n(r;F). \tag{1.9}$$

(This is fairly obvious; see [2] for further discussion.)

Of course, (1.8) and (1.9) also hold in the limit; i.e., with the subscript n deleted.

D. (A consequence of (3.11) of [2].) For all F in $\mathfrak{F}^*$, and for each $d > 0$,

$$G_{n,k}(r;F) - G_n(r+d;F) < \frac{c_1 k}{n} + c_2 k \exp\{-c_3 d^2 k^{1/2}\} + \frac{c_4}{d^4}\left(\frac{1}{k} + \frac{1}{n}\right), \tag{1.10}$$

where the c_i are positive constants depending only on m.

A further result of [2] will be given in Lemma 2 of Section 2, after some additional notation has been introduced.

In most of the arguments of this paper we will be dealing with F's which are in $\mathfrak{F}^c$. To simplify the discussion in such cases, we shall always assume that, for every real number t and integer j, at most one Z_i has its jth coordinate equal to t. The probability that this be not so is zero.

2. Uniformity of approach of G_n to G in the subclass $\mathfrak{F}_\epsilon$. The purpose of this section is to prove Lemma 1 (stated below), which will be used in Section 4 to prove the existence of an F_δ with the properties described in Section 1, when F is restricted to belong to a suitable subclass $\mathfrak{F}_\epsilon'$ of $\mathfrak{F}$. This and the multinomial result of Section 3 will then be used in Section 4 to demonstrate Theorem 1 with $\mathfrak{F}$ replaced by $\mathfrak{F}'_\epsilon$. The proof of Theorem 1 is then completed in Section 5 by showing that $\mathfrak{F}'_\epsilon$ is suitably dense in $\mathfrak{F}$ as $\epsilon \to 0$. Thus, although by far the greatest amount of new effort needed to prove Theorem 1 when $m > 1$ over what is needed when $m = 1$, is contained in the arguments of the present section, the reader who is interested mainly in the ideas of the statistical proof may read the statement of Lemma 1 and then go on to Section 3.

We first introduce some notation which will be used in this and subsequent sections. Let ϵ be a small positive number and let r be a positive number, both of which will be fixed in the present section. Other ϵ's with subscripts will be used in this paper to denote positive variables which will approach zero. The symbol $o(1 \mid \epsilon_i)$ is to denote a quantity which, as ϵ_i approaches zero, approaches zero uniformly in all other relevant quantities. Sometimes the latter will be explicitly indicated. Thus $o(1 \mid \epsilon_i \mid n, F)$ denotes a quantity which approaches zero, uniformly for all n (sometimes for all large n) and for all F (either in $\mathfrak{F}$ or in some indicated subclass), as $\epsilon_i \to 0$. The symbol $o(1 \mid \epsilon_i, n \mid F)$ denotes a quantity which approaches zero as $\epsilon_i \to 0$, $n \to \infty$, uniformly in F (either in $\mathfrak{F}$ or some indicated subclass). The symbol $o(1 \mid n \mid F)$ denotes a quantity which approaches zero as $n \to \infty$, uniformly in F (either in $\mathfrak{F}$ or some indicated subclass). The symbol $o(1 \mid d, N(d) \mid \cdot, \cdot)$ is to mean a quantity which approaches zero as $d \to 0$ while n stays larger than a suitable function $N(d)$ of d (which may change in various appearances of the symbol, although we shall sometimes use N, N', etc., to denote several such symbols which arise in the proof of the same lemma), and the approach of this quantity to zero is uniform in all other relevant quantities, which may be indicated where the dots are. The symbols $o(1 \mid \epsilon_i, N(\epsilon_i) \mid \cdot, \cdot)$ and $o(1 \mid k, N(k) \mid \cdot, \cdot)$ (with $k \to \infty$) will be used similarly. Finally the symbol θ will always denote a generic quantity <1 in absolute value; two θ's in different places need not be the same. The quantity d will always be >0.

Let $\mathfrak{F}_\epsilon$ be the subclass of those d.f.'s F in $\mathfrak{F}^*$ which have a Lebesgue density f_F in the subset of all points in I^m where at least one coordinate is $\geqq 1 - \epsilon$, and such that $\frac{1}{2} \leqq f_F \leqq 2$ almost everywhere in this region. The proofs of this section actually hold when $\mathfrak{F}_\epsilon$ is replaced by a somewhat larger class; but this is of little importance, the main use of Lemma 1 being to prove Theorem 1. (The relationship of $\mathfrak{F}'_\epsilon$ to $\mathfrak{F}_\epsilon$ will be stated in Section 4.)

LEMMA 1. *We have, for each fixed m,*

$$|G_n(r; F) - G(r; F)| = o(1 \mid n \mid F \,\varepsilon\, \mathfrak{F}_\epsilon). \tag{2.1}$$

The proof of Lemma 1 will require several supplementary lemmas. The proofs for all $m > 1$ are essentially the same, but the proof is most easily written out and followed in the case $m = 2$. *Hence, throughout the remainder of this section we shall carry out all proofs in the case* $m = 2$. The modifications in the statements and proofs which are necessary when $m > 2$ will usually be completely obvious;

and we shall explicitly mention, at appropriate points in the argument, those modifications which are not completely obvious.

Thus, we can write in coordinates $Z_n = (X_n, Y_n)$ and $z = (x, y)$, throughout the remainder of the section. (In most of the corresponding arguments for the case of m components, x will stand for the first $m - 1$ components of z, and y will stand for the last component of z.)

The idea of the proof of Lemma 1 is that (1.10) should somehow be used to prove Lemma 5, which, by a suitable uniformity result (Lemma 7) on the approach of the multinomial distribution to its limit, will yield (2.1). What is needed to obtain Lemma 5 from (1.10) is Lemma 4, the idea of which is that if $n^{1/2}|S_n(z) - F(z)|$ attains the value r somewhere, then it is very likely to attain the value $r + d$ somewhere, if d is small; it is the structure of $\mathfrak{F}_\epsilon$ which is used, in (2.18), to prove this.

For $0 < \epsilon_1 < 1$ we define the events

$$L_1(\epsilon_1) = \{\sup_{\substack{0 \leqq x \leqq \epsilon_1 \\ 0 \leqq y \leqq 1}} |S_n(z) - F(z)| \geqq r/n^{1/2}\}, \tag{2.2}$$

and

$$L_2(\epsilon_1) = \{\sup_{\substack{1-\epsilon_1 \leqq x \leqq 1 \\ 1-\epsilon_1 \leqq y \leqq 1}} |S_n(z) - F(z)| \geqq r/n^{1/2}\}. \tag{2.3}$$

(For the case of vectors with m components, the supremum in (2.2) is taken over the set where at least one of the $m - 1$ components of x is $\leqq \epsilon_1$; in (2.3), it is taken over the set where all m components are $\geqq 1 - \epsilon_1$.)

The next two lemmas lead up to Lemma 4.

LEMMA 2. *We have*

$$P_F\{L_1(\epsilon_1)\} = o(1 \mid \epsilon_1, n \mid F \,\varepsilon\, \mathfrak{F}^*), \tag{2.4}$$

and

$$P_F\{L_2(\epsilon_1)\} = o(1 \mid \epsilon_1, n \mid F \,\varepsilon\, \mathfrak{F}^*). \tag{2.5}$$

PROOF. An upper bound on the probability of $L_1(\epsilon_1)$ can be obtained from equations (3.6), (3.9), and (3.10) of [2], if, in the latter, we set $h = 0, j = 1, k = 1/\epsilon_1$ (the relevant argument of [2] is valid even if k is not an integer), $d = r$. We obtain

$$\begin{aligned} P_F\{L_1(\epsilon_1)\} &< \frac{1}{n} + c_o \exp\{-cr^2/2\epsilon_1^{1/2}\} \\ &\qquad + \frac{16}{r^4}\left(3\epsilon_1^2 + \frac{\epsilon_1}{n}\right) = o(1 \mid \epsilon_1, n \mid F \,\varepsilon\, \mathfrak{F}^*). \end{aligned} \tag{2.6}$$

We shall now use an argument like that by which (2.22) of [2] was proved, in order to prove (2.5). The event $L_2(\epsilon_1)$ implies the occurrence of at least one of the following events:

$$L_2^1 = \left\{ \sup_{\substack{1-\epsilon_1 \leq x \leq 1 \\ 1-\epsilon_1 \leq y \leq 1}} \mid (\text{number of } Z_1, \cdots, Z_n \text{ which satisfy } 0 \leq X_i \leq x, y \leq Y_i \leq 1) - \text{expected number} \mid \geq \frac{rn^{1/2}}{3} \right\},$$

$$L_2^2 = \left\{ \sup_{\substack{1-\epsilon_1 \leq x \leq 1 \\ 1-\epsilon_1 \leq y \leq 1}} \mid (\text{number of } Z_1, \cdots, Z_n \text{ which satisfy } x \leq X_i \leq 1, y \leq Y_i \leq 1) - \text{expected number} \mid \geq \frac{rn^{1/2}}{3} \right\}, \tag{2.7}$$

$$L_2^3 = \left\{ \sup_{\substack{1-\epsilon_1 \leq x \leq 1 \\ 1-\epsilon_1 \leq y \leq 1}} \mid (\text{number of } Z_1, \cdots, Z_n \text{ which satisfy } x \leq X_i \leq 1, 0 \leq Y_i \leq y) - \text{expected number} \mid \geq \frac{rn^{1/2}}{3} \right\}.$$

The random variables in the original sequence $\{Z_j\}$ all have the same distribution as $Z_1 = (X_1, Y_1)$. Apply the argument by which (2.4) was obtained for sequences all of whose members have the same distribution as each of the following, in order: $(1 - Y_1, X_1)$, $(1 - X_1, 1 - Y_1)$, and $(1 - X_1, Y_1)$. We obtain that

$$P_F\{L_2^i\} = o(1 \mid \epsilon_1, n \mid F \, \varepsilon \, \mathfrak{F}^*), \qquad i = 1, 2, 3. \tag{2.8}$$

Hence (2.5) is verified.

Define the events

$$L_3(\epsilon_1) = \left\{ \sup_{0 \leq x \leq 1} |S_n(x, 1) - x| < \frac{1}{\epsilon_1 n^{1/2}} \right\} \tag{2.9}$$

and

$$L_4(\epsilon_1) = \left\{ \sup_{\substack{\epsilon_1 \leq x \leq 1 \\ 0 \leq y \leq 1-\epsilon_1}} |S_n(z) - F(z)| \geq \frac{r}{n^{1/2}} \right\}. \tag{2.10}$$

From (1.8) we obtain that

$$P_F\{L_3(\epsilon_1)\} = 1 - o(1 \,|\epsilon_1|\, n, F \, \varepsilon \, \mathfrak{F}^*). \tag{2.11}$$

Write $L(\epsilon_1) = L_3(\epsilon_1) \cap L_4(\epsilon_1)$. Whenever $L(\epsilon_1)$ occurs we can define chance variables H and T as follows: $H = h$, $\epsilon_1 \leq h \leq 1$, and $T = t$, $0 < t \leq 1 - \epsilon_1$, if

$$|S_n(h, t) - F(h, t)| \geq r/n^{1/2} \tag{2.12}$$

and

$$|S_n(h', t') - F(h', t')| < r/n^{1/2} \tag{2.13}$$

for $\epsilon_1 \leq h' \leq 1$, $0 \leq t' < t$, as well as for $\epsilon_1 \leq h' < h$, $t' = t$. (In the m-component case, h' has all $m - 1$ components $\geq \epsilon_1$ and h can be specified by any rule which does not depend on y for $y > t$, and such that (2.12) holds.) Thus, if a horizontal line $y = t'$ is swept upward starting at $t' = 0$, the line $y = t$ is the first for which (2.12) can hold, and h is a well-defined value such that it does.

LEMMA 3. *We have, for some* $N(d)$ *and* $\epsilon_1 = d^{1/4}$,

$$(2.14)\quad P_F\left\{\sup_{T<y\leq 1} |S_n(H,y) - F(H,y)| \geqq \frac{r+d}{n^{1/2}} \,\middle|\, L(\epsilon_1)\right\} = 1 - o(1 \mid d, N(d) \mid F \,\varepsilon\, \mathfrak{F}_\epsilon) .$$

PROOF. We suppose that

$$(2.15)\quad S_n(H, T) - F(H, T) = r/n^{1/2}$$

and we will prove that, conditional on $L(\epsilon_1)$ occurring, the probability that

$$(2.16)\quad S_n(H, y) - F(H, y) \geqq (r + d)/n^{1/2}$$

for some y, $T < y \leqq 1$, is $1 - o(1 \mid d, N(d) \mid F \,\varepsilon\, \mathfrak{F}_\epsilon)$. This will be enough to prove (2.14), for (a) if the left member of (2.15) is greater than $r/n^{1/2}$ the result we want to prove is a fortiori true, and (b) if the left member of (2.15) is $\leqq -r/n^{1/2}$, it is proved, in the same way as below, that the probability (conditional on $L(\epsilon_1)$ occurring) that the left member of (2.16) be $\leqq -(r + d)/n^{1/2}$ for some y, $T < y \leqq 1$, is $1 - o(1 \mid d, N(d) \mid F \,\varepsilon\, \mathfrak{F}_\epsilon)$.

Define

$$n_1 = n(S_n(H, 1) - S_n(H, T)),$$

$$(2.17)\quad \bar{y} = \frac{F(H, y) - F(H, T)}{H - F(H, T)},$$

$$n_2(\bar{y}) = n(S_n(H, y) - S_n(H, T)).$$

From (2.9), (2.10), (2.15), and the definition of $\mathfrak{F}_\epsilon$, we have, in $L(\epsilon_1)$, if $\epsilon_1 < \epsilon$,

$$(2.18)\quad n_1 = n(H - F(H, T)) + \frac{\theta n^{1/2}}{\epsilon_1} - rn^{1/2} > n\epsilon_1^2/2 - n^{1/2}\left(\frac{1}{\epsilon_1} + r\right),$$

which goes to ∞ as $n \to \infty$ (uniformly in H and $T \leqq 1 - \epsilon_1$), and is thus arbitrarily large for $n >$ some $N'(\epsilon_1)$. Using (2.15) we find that (2.16) is equivalent to

$$(2.19)\quad \frac{n_2(\bar{y})}{n_1} - \frac{n\bar{y}(H - F(H, T))}{n_1} \geqq \frac{dn^{1/2}}{n_1}.$$

From (2.18) we obtain that the probability that (2.19) occur for some $\bar{y}$, $0 \leqq \bar{y} \leqq 1$, is $\geqq$ the probability that, for some $\bar{y}$,

$$(2.20)\quad \frac{n_2(\bar{y})}{n_1} - \bar{y} \geqq \frac{2d\,\epsilon_1^{-1}}{n_1^{1/2}} + \frac{4\epsilon_1^{-2}\bar{y}}{n_1^{1/2}},$$

provided that ϵ_1 is small enough and $n >$ a suitable $N_1(\epsilon_1)$.

Now set

$$(2.21)\quad \epsilon_1 = d^{1/4}$$

and suppose that d is small enough that (2.20) holds when ϵ_1 is given by (2.21).

Let $\{W(t), 0 \leqq t < \infty\}$ be the separable (Wiener) process with independent, normally distributed increments, $W(o) = 0$, $E(W(t)) = 0$, $\mathrm{Var}\,(W(t)) = t$. Given H, T, and n_1, the left member of (2.20) clearly is distributed as the difference between a sample d.f. and the uniform d.f. on the one-dimensional interval $0 \leqq \bar{y} \leqq 1$, when the sample d.f. is that of n_1 independent, uniformly distributed random variables. It follows from [3] and [4] and the fact that $n_1 \to \infty$ as $n \to \infty$ that, under (2.21), the conditional probability (given that $L(\epsilon_1)$ occurs) that (2.20) hold for some $\bar{y}$ approaches, uniformly in $\mathfrak{F}_\epsilon$, as $n \to \infty$,

$$(2.22) \qquad P\{W(t) \geqq 2d^{3/4} + (2d^{3/4} + 4d^{-1/2})t \text{ for some } t > 0\}.$$

The latter is, by [4], equation (4.2),

$$(2.23) \qquad \exp\{-2(2d^{3/4})(2d^{3/4} + 4d^{-1/2})\}$$

which approaches one as $d \to 0$. Hence, for d sufficiently small and $n >$ some $N(d)$, the conditional probability (given that $L(\epsilon_1)$ occurs) that (2.16) holds for some $y > T$ is arbitrarily close to 1, uniformly in $\mathfrak{F}_\epsilon$. This proves (2.14).

LEMMA 4. *We have, for some $N(d)$,*

$$(2.24) \qquad G_n(r + d; F) - G_n(r; F) = o(1 \mid d, N(d) \mid F \,\varepsilon\, \mathfrak{F}_\epsilon).$$

PROOF. Substituting (2.21) into (2.2), (2.4), (2.9), (2.10) and (2.11) (none of which previously depended on d in any way), and using Lemma 3, we have

$$(2.25) \qquad P_F\{r \leqq \sup_{\substack{0 \leqq x \leqq 1 \\ 0 \leqq y \leqq 1 - d^{1/4}}} \sqrt{n}|S_n(z) - F(z)| \leqq r + d\} = o(1 \mid d, N(d) \mid F \,\varepsilon\, \mathfrak{F}_\epsilon),$$

where of course the $N(d)$ may differ from that of Lemma 3. Now, the definition of $\mathfrak{F}_\epsilon$ is such that, by interchanging the roles of x and y, we obtain, in the same way that (2.25) was obtained,

$$(2.26) \qquad P_F\{r \leqq \sup_{\substack{0 \leqq x \leqq 1 - d^{1/4} \\ 0 \leqq y \leqq 1}} \sqrt{n}|S_n(z) - F(z)| \leqq r + d\} = o(1 \mid d, N(d) \mid F \,\varepsilon\, \mathfrak{F}_\epsilon).$$

(In the case of vectors with m components, there are $m - 2$ additional analogues of (2.26).) Finally, substituting (2.21) into (2.5) and combining the result with (2.25) and (2.26), we obtain (2.24).

LEMMA 5. *We have*

$$(2.27) \qquad 0 \leqq G_{n,k}(r; F) - G_n(r; F) = o(1 \mid k, N'(k) \mid F \,\varepsilon\, \mathfrak{F}_\epsilon).$$

PROOF. The left side of (2.27) is trivial. Adding (1.10) and (2.24), we have

$$(2.28) \qquad \begin{aligned} G_{n,k}(r; F) - G_n(r; F) &\leqq o(1 \mid d, N(d) \mid F \,\varepsilon\, \mathfrak{F}_\epsilon) \\ &\quad + c_2 k \exp\{-c_3 d^2 k^{1/2}\} + c_4/d^4 k + c_1 k/n + c_4 / d^4 n. \end{aligned}$$

Let $\epsilon' > 0$ be given arbitrarily. Let $d_1 > 0$ be such that the first term on the right side of (2.28) is $< \epsilon'/3$ if $d = d_1$ and $n > N(d_1)$. Let k_1 be such that the sum of the next two terms on the right side of (2.28) is $<\epsilon'/3$ when $d = d_1$ and $k > k_1$. For $k > k_1$, let $N'(k)$ be $>N(d_1)$ and be such that the sum of the last two terms of (2.28) is $<\epsilon'/3$ when $d = d_1$ and $n > N'(k)$. Then, putting $d = d_1$, we have that the right side of (2.28) is $<\epsilon'$ when $k > k_1$ and $n > N'(k)$. Thus, Lemma 5 is proved.

The discussion which immediately follows, as well as Lemma 6, leads up to the proof of Lemma 7.

There are k^2 cells into which I^2 is divided by the lines $x = i/k$, $y = j/k$, $i, j = 0, 1, \cdots, k$. (There are, of course, k^m cells in the case of vectors with m components.) Number the cells as follows: The cell bounded by $x = (i-1)/k$, $x = i/k$, $y = (j-1)/k$, $y = j/k$, is to be called the (i, j) cell. Write

$$\pi_{Fij} = P_F\{Z_1 \varepsilon \text{ cell } (i, j)\}.$$

Write $(i', j') \leqq (i, j)$ if $i' \leqq i, j' \leqq j$, and write $(i', j') < (i, j)$ if $(i', j') \leqq (i, j)$ and either $i' < j$ or $j' < j$. Let $\bar{H}$ be any collection of cells. For any fixed (i_0, j_0) not in $\bar{H}$, there clearly exist integers c_{ij} (depending only on $\bar{H}$ and (i_0, j_0)) such that we can write

$$F(i_0/k, j_0/k) = \sum_{\substack{(i,j) \leqq (i_0, j_0) \\ (i,j) \varepsilon H}} c_{ij} F(i/k, j/k) + \sum_{\substack{(i,j) \leqq (i_0, j_0) \\ (i,j) \notin H}} c_{ij}\, \pi_{Fij}, \tag{2.29}$$

identically in F (i.e., in the π_{Fij}).

Let $\epsilon_2 > 0$ be given. Call the cell (i, j) regular if $\pi_{Fij} \geqq \epsilon_2$ and $(i, j) \neq (k, k)$. Call the cell (i, j) singular if $\pi_{Fij} < \epsilon_2$ and $(i, j) \neq (k, k)$. Let $\bar{H}_F$ be the collection of regular cells, let (i_0, j_0) be singular under F, and let the c_{ij} be as in (2.29). Denote a summation over the region $(i, j) \leqq (i_0, j_0)$, $(i, j) \varepsilon \bar{H}_F$ by $\sum^{(F, i_0, j_0)}$. Then, clearly,

$$|F(i_0/k, j_0/k) - \textstyle\sum^{(F, i_0, j_0)} c_{ij} F(i/k, j/k)| < h(k)\epsilon_2, \tag{2.30}$$

where h is a suitable positive function of k alone, which can be chosen so that (2.30) is valid for every ϵ_2, every F, and every (i_0, j_0) singular for such an F; here the $\bar{H}_F$ depends on the F and ϵ_2 being considered, but the c_{ij} depend on these quantities only through $\bar{H}_F$.

Define $Q_{F,n}(\epsilon_2)$ to be the probability that

$$|S_n(i/k, j/k) - F(i/k, j/k)| < r/n^{1/2} \text{ for all } (i, j) \text{ in } \bar{H}_F,$$

$$|\textstyle\sum^{(F, i_0, j_0)} c_{ij}[S_n(i/k, j/k) - F(i/k, j/k)]| < r/n^{1/2} \tag{2.31}$$

$$\text{for all } (i, j) \neq (k, k) \text{ and not in } \bar{H}_F.$$

The proof of the next lemma is actually valid when $\mathfrak{F}^*$ is replaced by the class of all d.f.'s on I^2.

LEMMA 6. *We have*

$$|Q_{F,n}(\epsilon_2) - G_{n,k}(r; F)| = o(1 \mid \epsilon_2, N(\epsilon_2) \mid F \varepsilon \mathfrak{F}^*). \tag{2.32}$$

PROOF. Define, for (i_0, j_0) singular,

$$U = S_n(i_0/k, j_0/k) \tag{2.33}$$

and

$$V = \Sigma^{(F, i_0, j_0)} c_{ij} S_n(i/k, j/k). \tag{2.34}$$

Let B be the event defined by

$$B = \{|(U - V) - E(U - V)| < \epsilon_2^{1/4}/n^{1/2}\}. \tag{2.35}$$

Now, $U - V$ is just the last sum of (2.29) with π_{Fij} replaced by n^{-1} (number of $Z_1, \cdots, Z_n$ falling in cell (i, j)). Hence, $U - V$ has variance $<h'(k)\epsilon_2/n$, where h' is a suitable positive function of k. Thus, by Chebyshev's inequality,

$$P_F\{B\} > 1 - h'(k)\epsilon_2^{1/2}. \tag{2.36}$$

Of course, the definition of B depends on F, was well as on (i_0, j_0); but, again, h' can be chosen so that (2.36) holds for all F. Consider the events

$$A_1 = \{|U - EU| \geqq r/n^{1/2}, |V - EV| < r/n^{1/2}\} \tag{2.37}$$

and

$$A_2 < \{|U - EU| < r/n^{1/2}, |V - EV| \geqq r/n^{1/2}\}. \tag{2.38}$$

The definition of these events also depends on F and (i_0, j_0). Define $P_{Ft} = P_F\{A_t\}$, $t = 1, 2$. We are first going to show that, for all F for which (i_0, j_0) is singular,

$$P_{Ft} = o(1 \mid \epsilon_2, N(\epsilon_2) \mid F \,\varepsilon\, \mathfrak{F}^*), \qquad \text{for } t = 1, 2. \tag{2.39}$$

In proving this, let W stand for U in the case $t = 1$ and for V in the case $t = 2$. Then W is n^{-1} times the sum of n independent, identically distributed random variables, each bounded in absolute value by some constant L (independent of F). Let σ^2 be the variance and β_3 the absolute third moment about its expected value of each summand (i.e., of $W - EW$ when $n = 1$). Now, if $\sigma^2 < \epsilon_2^{1/8}$, Chebyshev's inequality yields $P_{Ft} < r^{-2}\epsilon_2^{1/8}$, so that (2.39) is verified in that case. On the other hand, if $\sigma^2 \geqq \epsilon_2^{1/8}$, by (2.36) we have

$$\begin{aligned} P_{Ft} &\leqq P_F\{B \cap A_t\} + P_F\{\bar{B}\} < P_F\{B \cap A_t\} + h'(k)\epsilon_2^{1/2} \\ &\leqq P_F\{r \leqq n^{1/2}|W - EW| \leqq r + \epsilon_2^{1/4}\} + h'(k)\epsilon_2^{1/2} \\ &\leqq P_F\{r/\sigma \leqq n^{1/2}|W - EW| / \sigma \leqq r/\sigma + \epsilon_2^{3/16}\} + h'(k)\epsilon_2^{1/2}. \end{aligned} \tag{2.40}$$

By the Berry-Esseen estimate (see, e.g., [5]) and the fact that $\beta_3/\sigma^3 \leqq L/\sigma$, we have from (2.40) for all F for which $\sigma^2 \geqq \epsilon_2^{1/8}$,

$$P_{Ft} \leqq h'(k)\epsilon_2^{1/2} + \epsilon_2^{3/16} + c_5 L n^{-1/2}\epsilon_2^{-1/16}, \tag{2.41}$$

where c_5 is a positive constant. Thus, (2.39) is proved.

Lemma 6 follows at once from (2.39).

The proof of the next lemma is also valid when $\mathfrak{F}^*$ is replaced by the class of all d.f.'s on I^2.

LEMMA 7. *For any fixed positive integer k, we have*

$$|G_{\infty,k}(r; F) - G_{n,k}(r; F)| = o(1|n|F \varepsilon \mathfrak{F}^*). \tag{2.42}$$

PROOF. Let $\epsilon_3 > 0$ be given arbitrarily. Choose ϵ_2 so small and $N(\epsilon_2)$ so large that, for this value of ϵ_2, the left side of (2.32) is $\leqq \epsilon_3/4$ for all F when $n > N(\epsilon_2)$. We shall show below that, writing $Q_F(\epsilon_2) = \lim_{n\to\infty} Q_{F,n}(\epsilon_2)$, we have

$$|Q_F(\epsilon_2) - Q_{F,n}(\epsilon_2)| < \epsilon_3/2 \tag{2.43}$$

for n sufficiently large, uniformly in F. Hence, we shall have, for n sufficiently large, uniformly in F, that the left side of (2.42) is no greater than

$$\begin{aligned} &|G_{\infty,k}(r; F) - Q_F(\epsilon_2)| + |Q_F(\epsilon_2) - Q_{F,n}(\epsilon_2)| \\ &\qquad + |Q_{F,n}(\epsilon_2) - G_{n,k}(r; F)| < \epsilon_3/4 + \epsilon_3/2 + \epsilon_3/4 = \epsilon_3, \end{aligned} \tag{2.44}$$

and (2.42) will be proved.

We shall now fix $\bar{H}$ and prove that (2.43) holds, uniformly in all F for which $\bar{H}_F = \bar{H}$, for n sufficiently large. Since k is fixed, the number of possible choices of $\bar{H}$ is finite, so that Lemma 7 will be proved.

Consider the joint distribution of the $n^{1/2}(S_n(i/k, j/k) - F(i/k, j/k))$ for all regular (i, j), which, as $n \to \infty$, approaches a multivariate normal distribution. Since $\pi_{Fij} \geqq \epsilon_2$ for any regular point it follows that the determinant of the covariance matrix of the $n^{1/2}(S_n(i/k, j/k) - F(i/k, j/k))$ (for regular (i, j)) is bounded away from 0 (and, of course, from ∞ as well) by a function of ϵ_2, uniformly in all F for which $\bar{H}_F = \bar{H}$. It follows from [6], page 121, that the maximum of the absolute value of the difference between the joint d.f. of these $n^{1/2}(S_n(i/k, j/k) - F(i/k, j/k))$ and their limiting multivariate normal d.f. is less than $n^{-1/2}M(\epsilon_2)$, where M is a real function of ϵ_2 only. The maximum of the density of this limiting normal d.f. is a real function only of ϵ_2, say $M'(\epsilon_2)$. Thus, the statements in the last two sentences are uniform in all F for which $\bar{H}_F = \bar{H}$.

It follows from (1.8) that the probability of a sufficiently large cube C in the space of the $n^{1/2}(S_n(i/k, j/k) - F(i/k, j/k))$ (for all regular (i, j)) which is centered at the origin, is greater than $1 - \epsilon_3/12$ uniformly in F and n. Hence this is also true of the limiting multivariate normal d.f. of the $n^{1/2}(S_n(i/k, j/k) - F(i/k, j/k))$.

Consider the region R in the space of these $n^{1/2}(S_n(i/k, j/k) - F(i/k, j/k))$, which is defined by (2.31) and whose probability is $Q_{F,n}(\epsilon_2)$. The region $R \cap C$ is a bounded polyhedron and can be approximated from within by a finite union R_1 of "rectangles" with sides parallel to the coordinate planes, such that the volume of the region $R_2 = [(R \cap C) - R_1]$ is $< \epsilon_4$, where $\epsilon_4 > 0$ is such that $\epsilon_4 M'(\epsilon_2) < \epsilon_3/12$. The set R_2 can be covered by a finite union R_3 of rectangles with sides parallel to the coordinate planes whose total volume is $< 2\epsilon_4$. Let m_3 be the number of rectangles in R_3, and m_1 be the number of rectangles in R_1.

The probability of R_3 according to the limiting normal d.f. is less than

$$2\epsilon_4 M'(\epsilon_2) < \epsilon_3/6. \tag{2.45}$$

The probability $P_F\{R_2\}$ of the region R_2 according to F is $<P_F\{R_3\}$, which, by the aforementioned result of Bergstrom [6], differs from the probability of R_3 according to the limiting normal d.f. by less than $4m_3M(\epsilon_2)n^{-1/2}$. Hence,

$$P_F\{R_2\} < \epsilon_3/6 + 4m_3M(\epsilon_2)n^{-1/2}. \tag{2.46}$$

Also, by Bergstrom's result just cited, the probability of R_1 according to the limiting normal d.f. differs from $P_F\{R_1\}$ by less than $4m_1M(\epsilon_2)n^{-1/2}$. Since the sum of this and the second term in the right member of (2.46) can be made less than $\epsilon_3/6$ by making n sufficiently large, it follows from the present paragraph and the previous two paragraphs that (2.43) holds for n sufficiently large, uniformly in all F for which $\bar{H}_F = \bar{H}$. This completes the proof of Lemma 7.

PROOF OF LEMMA 1. Let $\epsilon_5 > 0$ be chosen arbitrarily. Choose k' such that the right side of (2.27) is $<\epsilon_5/3$ for $k = k'$ and $n > N'(k')$. In particular, $0 \leqq G_{\infty,k'}(r; F) - G(r; F) \leqq \epsilon_5/3$. Choose N to be $>N'(k')$ and such that, for $k = k'$, the left member of (2.42) is $<\epsilon_5/3$ for $n > N'$. Then, for $n > N'$ and all F in $\mathfrak{F}_\epsilon$, we have

$$\begin{aligned} |G_n(r; F) - G(r; F)| &\leqq |G_n(r; F) - G_{n,k'}(r; F)| \\ &\quad + |G_{n,k'}(r; F) - G_{\infty,k'}(r; F)| + |G_{\infty,k'}(r; F) - G(r; F)| < \epsilon_5 . \end{aligned} \tag{2.47}$$

Since ϵ_5 was arbitrary, Lemma 1 is proved.

3. The multinomial result. We have mentioned in Section 1 that the main results of this paper are obtained by approximating the original problem by an appropriate multinomial problem. In the present section we summarize the needed multinomial results which were obtained in [1], and sketch the ideas of the proofs, unencumbered by the tedious details of [1]. Actually, we do not need the full strength of the results of [1], which are broader than those of Lemma 8 below in that, in the derivation of Section 3 of [1], the calculations were carried out in fine detail in order to obtain an error term which can be used to calculate an upper bound on the departure of ϕ_n^* from minimax character (in view of the lack of knowledge about the distribution of D_n, it seems more difficult to obtain a useful bound of this kind when $m > 1$). In fact, if one does not bother to obtain an error term, it is obvious how to shorten considerably the proof of the multinomial result in Section 3 of [1], and we shall see that this simple multinomial result without error term rests mainly on a result of v. Mises ([7], especially pages 84–86) which is almost forty years old.

We now introduce the needed notation. Let h be a positive integer and let B_h be the family of $(h + 1)$-vectors $\pi = \{p_i, 1 \leqq i \leqq h + 1\}$ with real components satisfying $p_i \geqq 0$, $\sum p_i = 1$. Let B'_h be a specified subset of $B_h \cdot B'_h$ can actually be fairly arbitrary in structure; to avoid trivial circumlocutions, we shall suppose in this section that B'_h is the closure of an h-dimensional open subset of B_h, although it will be obvious that Lemmas 8, 9, and 10 hold much more generally. Let $T^{(n)} = \{T_i^{(n)}, 1 \leqq i \leqq h + 1\}$, a vector of $h + 1$ chance variables, have a multinomial probability function arising from n observations with $h + 1$ possible outcomes, according to some π in B_h'; i.e., for integers $x_i \geqq 0$ with $\Sigma_1^{h+1} x_i = n$,

$$(3.1) \qquad P_\pi\{T_i^{(n)} = x_i, 1 \leqq i \leqq h + 1\} = \frac{n!}{x_1! \cdots x_{h+1}!} p_1^{x_1} \cdots p_{h+1}^{x_{h+1}}.$$

Let L be a positive integer, let γ_i be an $(h + 1)$-vector, $1 \leqq i \leqq L$, and let $\rho_i = \gamma_i'\pi$ (scalar product) be corresponding linear functions of π, $1 \leqq i \leqq L$. To avoid trivialities, we assume at least one ρ_i is not constant on B_h'. Let $\mathcal{E}_n$ be the class of all (possibly randomized) vector estimators of $\rho = \{\rho_i, 1 \leqq i \leqq L\}$, the weight function (which depends on n) being the simple one for which the risk function of a procedure ψ_n in $\mathcal{E}_n$ is

$$(3.2) \qquad 1 - P_{\pi,\psi_n}\{|d_i - \rho_i| \leqq r/n^{1/2}, 1 \leqq i \leqq L\},$$

where r is a positive value and we have written $d = \{d_i, 1 \leqq i \leqq L\}$ for the vector of decisions. Let ψ_n^* be the nonrandomized estimator whose ith component is $\gamma_i'T^{(n)}/n$ (the allowable decisions may be restricted to $\gamma'\pi$ for π in B_h' with only trivial modifications in what follows). Finally, a point π in B_h' is called an interior point if all its components p_i are positive, and if it has a neighborhood (in B_h) which is a subset of B_h'. The required multinomial result is:

LEMMA 8. *For any interior point π^* of B_h' there is a sequence $\{\xi_n\}$ of a priori distributions on B_h' converging in distribution to the distribution which gives probability one to π^* and such that $\{\psi_n^*\}$ is asymptotically Bayes relative to $\{\xi_n\}$ as $n \to \infty$, uniformly for $0 \leqq r \leqq R$ for any $R < \infty$; i.e., such that, uniformly in such r,*

$$(3.3) \qquad \lim_{n\to\infty} \frac{\int P_\pi\{|\rho_i - \gamma_i'T^{(n)}/n| > r/n^{1/2}, 1 \leqq i \leqq L\}\, \xi_n(d\pi)}{\inf_{\psi_n \in \mathcal{E}_n} \int P_{\pi,\psi_n}\{|d_i - \rho_i| > r/n^{1/2}, 1 \leqq i \leqq L\}\, \xi_n(d\pi)} = 1.$$

Of course, continuity considerations show that the positive (since not all ρ_i are constant) limit of the numerator of (3.3) is obtained by putting $\pi = \pi^*$ instead of integrating with respect to ξ_n, and then using the multivariate central limit theorem to compute the limiting probability.

The idea of the proof of Lemma 8 is very simple. Let Γ^* be the (nonsingular) covariance matrix of the limiting h-variate normal distribution of $n^{1/2}(n^{-1}T_i^{(n)} - p_i^*)$, $1 \leqq i \leqq h$, when $\pi = \pi^*$. Let ϵ be a small positive value and let ξ be the uniform *a priori* distribution in the (solid) sphere of radius ϵ about π^* in B_h'. (ϵ is small enough that this sphere consists entirely of interior points.) According to the result [7] of v. Mises, for any π'' in this sphere, with probability one when $T^{(n)}$ is distributed according to π'', the *a posteriori* density function of $n^{1/2}(p_i - T_i^{(n)}/n)$, $1 \leqq i \leqq h$ (calculated assuming ξ to be the *a priori* distribution) will tend to the h-variate normal density with means 0 and covarance matrix Γ'' (corresponding to π'') as $n \to \infty$. If the *a posteriori* density were really normal with the stated parameters, it would follow at once from a result [8] of Anderson that the *a posteriori* probability of the event

$$\{n^{1/2}|\rho_i - d_i| \leqq r, 1 \leqq i \leqq L\} \tag{3.4}$$

(this probability is unity minus the *a posteriori* risk) is a maximum for $d = \gamma' T^{(n)}/n$, since the region (3.4) is for each d a convex symmetric (about a point depending on d) subset in the space of the h variables $n^{1/2}(p_i - T_i^{(n)}/n)$ (considering the latter to be unrestricted in magnitude). Since the actual *a posteriori* density is almost normal (with high probability as $n \to \infty$), ψ_n^* will be asymptotically Bayes. Finally, let ξ_n be the ξ just described when $\epsilon = \epsilon_n$, where ϵ_n goes to zero slowly enough that the above result still holds for ψ_n^* as $n \to \infty$. (For example, $\epsilon_n = n^{-\alpha}$ with $0 < \alpha < \frac{1}{2}$. The crucial consideration is that the radius $n^{1/2}\epsilon_n$ of the set of possible values of $n^{1/2}(\pi - \pi^*)$ approach infinity with n, as will therefore the radius of the set of possible values of $n^{1/2}(\pi - T^{(n)}/n)$ w.p.1 under ξ_n. The asymptotic problem is thus approximately one of estimating the mean of a multivariate normal distribution with known constant covariance matrix, when the mean can take on any value in an appropriate Euclidean space).

The actual proof—the precise handling of the approximations mentioned above, the uniformity in r, etc.—may be handled as in [1] or by complementing with appropriate estimates the argument of [7], but the main idea is really the simple one of [7].

The reason for wanting Lemma 8 in its stated form with the sequence $\{\xi_n\}$ shrinking down on π^* has to do with the problem of multinomial minimax estimation for the risk function (3.2). Let π^0 be the value of π at which the positive limit b (as $n \to \infty$) of the continuous risk function of ψ_n^* is a maximum. Since the ρ_i are not all constant, for any $\delta > 0$ there will, by continuity, be an interior point π^* of B_h' at which the limit of the risk function of ψ_n^* is at most $(1 + \delta)b$. From Lemma 8 and the sentence following (3.3) we conclude:

LEMMA 9. $\{\psi_n^*\}$ *is asymptotically minimax relative to* B_h' *and the risk function* (3.2).

We next consider a generalization of this result to other weight functions which are nondecreasing functions of $\max_i |d_i - \rho_i|$. Of course, the risk function is defined in the usual way. (A Bayes result analogous to Lemma 8 can be proved in the course of the demonstration, but we shall not bother to state it.) Let C_0 and C be positive constants such that

$$P_\pi\{n^{1/2} \max_i \gamma_i'|T^{(n)}/n - \pi| \geqq r\} < C_0 e^{-Cr^2} \tag{3.5}$$

for all r, all n, and all π in B_h'. The existence of such positive constants (which depend on h and the structure of B_h') follows from well known results on the multinomial (or, in fact, the binomial) distribution; in Section 4 we shall actually refer to (1.8) for appropriate values of these constants.

LEMMA 10. *Let $W(r)$ be a nondecreasing real function of r for $r \geqq 0$, not identically zero, and satisfying*

$$\int_0^\infty W(r) r e^{-Cr^2}\, dr < \infty. \tag{3.6}$$

Then $\{\psi_n^*\}$ is asymptotically minimax relative to B_h' and the weight functions

$$W_n(\pi, d) = W(n^{1/2} \max_i |\rho_i - d_i|). \tag{3.7}$$

The proof of Lemma 10 can be carried out, starting from scratch, along lines like those of Lemma 8. An easier proof, which was given in [1], rests upon the idea of reducing the proof essentially to that for the simple weight function already considered in Lemma 8. Specifically, if the *a posteriori* distribution of the variables $n^{1/2}(p_i - T_i^{(n)}/n)$ were actually normal with means 0 and the appropriate covarance matrix, then $d_i = \gamma_i' T^{(n)}/n$, $1 \leqq i \leqq L$, would minimize the *a posteriori* risk; for, if this choice of the d_i did not minimize the *a posteriori* risk and if H_1 and H_2 were respectively, the d.f.'s of $n^{1/2} \max_i |\rho_i - d_i|$ for the above choice of d_i and for a better choice, we would have

$$\int_0^\infty W(r)\, d[H_1(r) - H_2(r)] > 0,$$

which is easily seen to imply that $H_1(r') < H_2(r')$ for some r', contradicting Anderson's result cited previously (i.e., when the error terms are included, this contradicts the result of Lemma 8). The details of the proof are contained in Section 4 of [1].

We note that Lemma 10 exemplifies a principle which is of more general use in statistics: If one can verify suitable (asymptotic) Bayes results for an appropriate class of simple weight functions, the results will automatically hold for a general class of monotone weight functions.

We remark that Anderson's result can be used to prove the result of Lemma 10 for a larger class of weight functions, namely, every function of $n^{1/2}(d_i - \rho_i)$, $1 \leqq i \leqq L$, which is symmetric about the origin and which for each real value c has a convex (or empty) set for the domain where the function is $\leqq c$.

4. Proof of Theorem 1 when $\mathfrak{F}$ is replaced by $\mathfrak{F}_\epsilon'$. Define $\mathfrak{F}_\epsilon'$ to consist of every d.f. in $\mathfrak{F}^c$ which gives probability one to I^m and which can be realized as the d.f. of Z_1' (say) when Z_1 has a d.f. in $\mathfrak{F}_\epsilon$ and Z_1' is obtained from Z_1 by continuous monotonic transformations on the individual coordinate functions. Thus, $\mathfrak{F}_\epsilon' \supset \mathfrak{F}_\epsilon$, but $\mathfrak{F}_\epsilon'$ includes d.f.'s which are not in $\mathfrak{F}^*$. Clearly, for any F' in $\mathfrak{F}_\epsilon'$ there is an F in $\mathfrak{F}_\epsilon$ such that $G_n(r; F) = G_n(r; F')$ for all n and r.

In this section we use the results of Sections 2 and 3 to prove the following

LEMMA 11. *For m a positive integer, suppose that*

$$W_n(F, g) = W(n^{1/2} \sup_z |F(z) - g(z)|), \tag{4.1}$$

where $W(r)$ for $r \geqq 0$ is continuous, nonnegative, nondecreasing in r, not identically *zero, and satisfies* (1.4). *Then, for each* ϵ *with* $0 < \epsilon < 1$, $\{\phi_n^*\}$ *is asymptotically minimax relative to* $\{W_n\}$ *and* $\mathfrak{F}_\epsilon'$.

PROOF. We divide the proof into three paragraphs; ϵ is fixed in what follows.

1. By (1.4), (1.8), the last sentence of the first paragraph of this Section, and Lemma 1, the function $r_n(F, \phi_n^*)$ approaches a bounded limit as $n \to \infty$, *uniformly* for F in $\mathfrak{F}_\epsilon'$. (This limit is positive, by the known results in the case $m = 1$.) Hence, for any $\delta > 0$, there is a d.f. F_δ in $\mathfrak{F}_\epsilon$ and an integer N_δ such that

$$\sup_{F \varepsilon \mathfrak{F}_\epsilon'} r_n(F, \phi_n^*) < (1 + \delta) r_n(F_\delta, \phi_n^*) \tag{4.2}$$

for $n > N_\delta$. Define

$$r_{nk}(F, \phi_n) = E_{F,\phi_n} W(n^{1/2} \sup_{z \varepsilon A_k^m} |F(z) - g(z)|), \tag{4.3}$$

so that

$$r_{nk}(F, \phi_n^*) = \int_0^\infty W(r) d_r G_{n,k}(r; F). \tag{4.4}$$

Since $r_{nk} \leqq r_n$, it follows from (4.2) and the arbitrariness of δ that Lemma 11 will be proved if we show that

$$\liminf_{k \to \infty} \liminf_{n \to \infty} \inf_{\phi_n \varepsilon \mathfrak{D}_n} \sup_{F \varepsilon \mathfrak{F}_\epsilon'} r_{nk}(F, \phi_n) \geqq \lim_{n \to \infty} r_n(F_\delta, \phi_n^*). \tag{4.5}$$

2. Define

$$r_{\infty k}^* = \int_0^\infty W(r) d_r G_{\infty,k}(r; F_\delta) \tag{4.6}$$

and

$$r^* = \int_0^\infty W(r) d_r G(r; F_\delta). \tag{4.7}$$

Let $\mathfrak{F}_{\epsilon,k}'$ be the subset of $\mathfrak{F}_\epsilon'$ consisting of every absolutely continuous d.f. in $\mathfrak{F}_\epsilon'$ which has a density function which is a constant on each of the k^m open m-cubes of side $1/k$ in I^m whose corners are points of A_k^m. From equations (1.4) through (1.8) and the fact that $F_\delta \,\varepsilon\, \mathfrak{F}^*$, we have

$$\lim_{n \to \infty} r_{nk}(F_\delta, \phi_n^*) = r_{\infty k}^* \tag{4.8}$$

and

$$\lim_{n \to \infty} r_n(F_\delta, \phi_n^*) = r^* = \lim_{k \to \infty} r_{\infty k}^*. \tag{4.9}$$

Let $F_{\delta k}$ be that member of $\mathfrak{F}_\epsilon'$ for which $F_{\delta k}(z) = F_\delta(z)$ whenever $z \,\varepsilon\, A_k^m$. Clearly, for each k and n,

$$r_{nk}(F_{\delta k}, \phi_n^*) = r_{nk}(F_\delta, \phi_n^*). \tag{4.10}$$

From equations (4.6) through (4.10) and the fact that $\mathfrak{F}_{\epsilon k}' \subset \mathfrak{F}_\epsilon'$, we see that

(4.5) will be proved if we show that, for each fixed $k > 1$,

$$\liminf_{n\to\infty} \inf_{\phi_n \in \mathfrak{D}_n} \sup_{F \in \mathfrak{F}'_{\epsilon k}} r_{nk}(F, \phi_n) \geqq \lim_{n\to\infty} r_{nk}(F_{\delta k}, \phi_n^*). \tag{4.11}$$

Since a sufficient statistic for $\mathfrak{F}'_{\epsilon k}$ based on $Z^{(n)}$ is the collection $T^{(n,k)}$ of k^m real random variables which are equal to the number of components of $Z^{(n)}$ taking on values in each of the k^m cubes just described, we may replace $\mathfrak{D}_n$ in (4.11) by the class $\mathfrak{D}_{n,k}$ of decision functions depending only on $T^{(n,k)}$. But the definition of r_{nk} then shows that the left side of (4.11) may be viewed as the limiting minimax risk associated with the problem of estimating certain linear combinations of multinomial probabilities. If we put $h + 1 = k^m$ in Section 3 and think of the p_i as being assigned to the k^m cubes and think of the $L = (k + 1)^m$ quantities ρ_i as being the values of the unknown d.f. at the $(k + 1)^m$ points in A_k^m, then the left side of (4.11) without the limit in n may be identified with the minimax risk for a multinomial problem with the setup of Lemma 10. (We shall discuss B'_h and the C of (3.5) in the next paragraph.)

3. Fix $k > 1$. For any F in $\mathfrak{F}'_{\epsilon,k}$, let π_F be the associated multinomial probability vector whose components are the p_i described in the previous paragraph. Let B'_h be the set of all such π_F in $\mathfrak{F}'_{\epsilon,k}$. From the definitions of $\mathfrak{F}_\epsilon$ and $\mathfrak{F}'_\epsilon$ it is clear that B'_h is a closed convex h-dimensional subset of the h-dimensional set B_h, and thus satisfies the requirements of Lemma 10. For the ρ_i defined in the previous paragraph, we can clearly take the C and C_0 of (3.5) to be the c'_m and c_m^* of (1.8). Hence, from Lemma 10, for each k, we have for the multinomial problem of Section 3 where B'_h and the ρ_i are as described above and the function W is that given in the statement of Lemma 11,

$$\liminf_{n\to\infty} \inf_{\psi_n \in \mathcal{E}_n} \sup_{\pi \in B'_h} r'(\pi, \psi_n) = \lim_{n\to\infty} \sup_{\pi \in B'_h} r'(\pi, \psi_n^*), \tag{4.12}$$

where we have written r' for the risk function in the multinomial problem. Since $r'(\pi_F, \psi_n^*) = r_{nk}(F, \phi_n^*)$ and since the left sides of (4.11) and (4.12) are equal because of the correspondence of $\mathcal{E}_n$ to $\mathfrak{D}_{n,k}$, of r' to r, and of B'_h to $\mathfrak{F}'_{\epsilon,k}$, we see that (4.11) follows from (4.12). Thus, Lemma 11 is proved.

5. Completion of the proof of Theorem 1; passage to the limit with ϵ. We now complete the proof of Theorem 1 by showing that $\mathfrak{F}_\epsilon$ is suitably dense in $\mathfrak{F}^*$ (and hence that $\mathfrak{F}'_\epsilon$ is suitably dense in $\mathfrak{F}^c$) as $\epsilon \to 0$. We require two lemmas to do this.

As in Section 2, the proof of the next two lemmas is very similar for all $m > 1$, but is most briefly written out when $m = 2$. For simplicity of presentation, we shall therefore again write out the details only in the case $m = 2$, and shall state explicitly all modifications for the case $m > 2$ which are not completely obvious.

Let $\mathfrak{F}'$ denote the class of all d.f.'s on I^2 (in the general case, on I^m). For F in $\mathfrak{F}'$ and $0 < \epsilon < 1$, define

$$\begin{aligned} \bar{F}(x, y) &= (1 - \epsilon)F(x, y / (1 - \epsilon)), \quad y \leqq 1 - \epsilon; \\ \bar{F}(x, y) &= (1 - \epsilon)F(x, 1) + x(y - 1 + \epsilon), \quad y > 1 - \epsilon. \end{aligned} \tag{5.1}$$

We shall not display the dependence on ϵ of the bar operation defined by (5.1). If $F \varepsilon \mathfrak{F}^*$ and we perform the bar operation of (5.1) on F to obtain $\bar{F}$ and then, interchanging the roles of x and y, perform the bar operation on $\bar{F}$ to obtain F^* (say), we clearly have $F^* \varepsilon \mathfrak{F}_\epsilon$. (In the case of chance vectors with m components, F^* is obtained after m such steps.) Let $\bar{Z}_1, \cdots, \bar{Z}_n$ be independent chance vectors with the common d.f. $\bar{F}$, let $\bar{S}_n$ be their sample (empiric) d.f., and define

$$\bar{D}_n = \sup_{z \varepsilon I} |\bar{S}_n(z) - \bar{F}(z)|.$$

Also, define $m = m(n, \epsilon)$ to be the greatest integer $\leqq n(1 - \epsilon)$.

We now prove the following lemma:

LEMMA 12. *We have*

$$\begin{aligned} P_{\bar{F}}\{\bar{D}_n < r/n^{1/2}\} \leqq P_F\{D_m < [r(1 + \epsilon) + 7\epsilon^{1/4}] / m^{1/2}\} \\ + o(1 \mid \epsilon, N(\epsilon) \mid r, F \varepsilon \mathfrak{F}^I). \end{aligned} \tag{5.2}$$

PROOF. Let C^* be the event

$$\{|\bar{S}_n(1, 1 - \epsilon) - (1 - \epsilon)| < \epsilon^{1/4} n^{-1/2}\}.$$

From Chebyshev's inequality we obtain

$$P_F\{C^*\} = 1 + o(1 \mid \epsilon \mid n, F \varepsilon \mathfrak{F}^I). \tag{5.3}$$

For small ϵ we have

$$\left|1 - \frac{n(1 - \epsilon)}{n(1 - \epsilon) + \theta n^{1/2}\epsilon^{1/4}}\right| < 4\epsilon^{1/4}/n^{1/2}. \tag{5.4}$$

Hence, when C^* occurs and ϵ is small,

$$\left|1 - \frac{n(1 - \epsilon)}{n\bar{S}_n(1, 1 - \epsilon)}\right| < 4\epsilon^{1/4}/n^{1/2}. \tag{5.5}$$

Since

$$\left|\frac{(1 - \epsilon)\bar{S}_n(z)}{\bar{S}_n(1, 1 - \epsilon)} - \bar{S}_n(z)\right| \leqq \left|\frac{(1 - \epsilon)}{\bar{S}_n(1, 1 - \epsilon)} - 1\right|, \tag{5.6}$$

we have

$$\bar{D}_n \geqq \sup_{\substack{0 \leqq x \leqq 1 \\ 0 \leqq y \leqq 1 - \epsilon}} \left|\frac{(1 - \epsilon)\bar{S}_n(z)}{\bar{S}_n(1, 1 - \epsilon)} - \bar{F}(z)\right| - \left|1 - \frac{(1 - \epsilon)}{\bar{S}_n(1, 1 - \epsilon)}\right|. \tag{5.7}$$

Also we have, for $y \leqq 1 - \epsilon$,

$$E_{\bar{F}}\left\{\frac{n\bar{S}_n(z)}{m'} \middle| n\bar{S}_n(1, 1 - \epsilon) = m'\right\} = \bar{F}(z) / (1 - \epsilon) = F(x, y / (1 - \epsilon)). \tag{5.8}$$

Hence the conditional d.f. of the first term on the right side of (5.7), given that $n\bar{S}_n(1, 1 - \epsilon) = m'$, is the same as the d.f. of $(1 - \epsilon)D_{m'}$. In what follows define $M' = n\bar{S}_n(1, 1 - \epsilon)$.

If m_1 and m_2 are two positive integers with $m_1 < m_2$, we can think of S_{m_2} as being obtained by adjoining $(m_2 - m_1)$ random vectors Z_i to the set of m_1 random vectors Z_i which gave rise to a corresponding realization of S_{m_1}. Hence, θ' denoting a value with $0 \leqq \theta' \leqq 1$, the corresponding values of $S_{m_1}(z)$ and $S_{m_2}(z)$ differ for all z by no more than

$$|S_{m_2}(z) - S_{m_1}(z)| = \left|\frac{m_1 S_{m_1}(z) + \theta'(m_2 - m_1)}{m_2} - S_{m_1}(z)\right| \leqq \frac{(m_2 - m_1)}{m_2}.$$

Thus, in C^*, where $|M' - m| < n^{1/2}\epsilon^{1/4} + 1$, we have that for each possible value of D_m there is a corresponding set of values of $D_{M'}$ of the same probability (these sets corresponding to different values of D_m arising from disjoint sets in the space of sequences $\{Z_i\}$) with $D_m \leqq D_{M'} + 2\epsilon^{1/4} m^{-1/2}$, provided $m >$ some $M(\epsilon)$.

From (5.3), (5.5), (5.7), (5.8), and the discussion of the previous paragraph, we have

$$P_{\bar{F}}\{\bar{D}_n < r/n^{1/2}\}$$

$$(5.9)\qquad \leqq \sup_{|m'-m| < n^{1/2}\epsilon^{1/4}+1} P_F\{(1-\epsilon)D_{m'} < (r + 4\epsilon^{1/4})/n^{1/2}\} + P_{\bar{F}}\{C^*\}$$

$$\leqq P_F\left\{D_m < \frac{r(1+\epsilon) + 7\epsilon^{1/4}}{m^{1/2}}\right\} + o(1 \mid \epsilon, N(\epsilon) \mid r, F \varepsilon \mathfrak{F}^I),$$

which proves Lemma 12.

We now prove

LEMMA 13. *For W satisfying the assumptions of Theorem* 1, *we have*

$$(5.10)\qquad \sup_{F\varepsilon\mathfrak{F}^*} r_n(F, \phi_n^*) = \sup_{F\varepsilon\mathfrak{F}_\epsilon} r_n(F, \phi_n^*) + o(1 \mid \epsilon, N'(\epsilon)).$$

PROOF. Define $m' = m'(n, \epsilon)$ to be the greatest integer $\leqq (1-\epsilon)^2 n$. Using Lemma 12 a second time (with a trivial modification since $m'(n, \epsilon)$ may differ by unity from $m[m(n, \epsilon), \epsilon]$) to go from $\bar{F}$ to F^*, we have at once, for any W satisfying (1.4) and the other assumptions of Theorem 1,

$$(5.11)\qquad r_n(F^*, \phi_n^*) \geqq r_{m'}(F, \phi_{m'}^*) + o(1 \mid \epsilon, N(\epsilon) \mid F \varepsilon \mathfrak{F}^I).$$

From (5.11) and the fact that $F^* \varepsilon \mathfrak{F}_\epsilon$ if $F \varepsilon \mathfrak{F}^*$, we have

$$(5.12)\qquad \sup_{F\varepsilon\mathfrak{F}_\epsilon} r_n(F, \phi_n^*) \geqq \sup_{F\varepsilon\mathfrak{F}^*} r_{m'}(F, \phi_{m'}^*) + o(1 \mid \epsilon, N(\epsilon)).$$

Now, as in the first part of the proof of Lemma 11, we have that $r_n(F, \phi_n^*)$ approaches a bounded limit as $n \to \infty$, uniformly for F in $\mathfrak{F}_\epsilon$. Hence,

$$(5.13)\qquad \sup_{F\varepsilon\mathfrak{F}_\epsilon} r_n(F, \phi_n^*) = \sup_{F\varepsilon\mathfrak{F}_\epsilon} r_{m'}(F, \phi_{m'}^*) + o_\epsilon(1 \mid n),$$

where $o_\epsilon(1 \mid n)$ denotes a term which, for each ϵ, goes to 0 as $n \to \infty$ (not necessarily uniformly in ϵ). From (5.12), (5.13), and the fact that $\mathfrak{F}_\epsilon \subset \mathfrak{F}^*$, we obtain

$$(5.14)\qquad \sup_{F\varepsilon\mathfrak{F}^*} r_{m'}(F, \phi_{m'}^*) = \sup_{F\varepsilon\mathfrak{F}_\epsilon} r_{m'}(F, \phi_{m'}^*) + o(1 \mid \epsilon, N''(\epsilon)).$$

Since the possible values of m' for $n > N''(\epsilon)$ include all integers $>N''(\epsilon)(1-\epsilon)^2 - 1 = N'(\epsilon)$ (say), Lemma 13 follows from (5.14).

LEMMA 14. *The statement of Theorem* 1 *holds with* $\mathfrak{F}$ *replaced by* $\mathfrak{F}^c$.

PROOF. We have previously alluded to the fact that, if Z_1 has a d.f. F in $\mathfrak{F}^c$, then by appropriate monotonic transformations on the individual coordinates of Z_1 we can obtain a random vector Z_1' (say) such that Z_1' has d.f. F' (say) in $\mathfrak{F}^*$ and $G_n(r; F') = G_n(r; F)$ for all r and n. Hence,

$$\sup_{F \epsilon \mathfrak{F}^c} r_n(F, \phi_n^*) = \sup_{F \epsilon \mathfrak{F}^*} r_n(F, \phi_n^*). \tag{5.15}$$

Moreover, in the same way we have

$$\sup_{F \epsilon \mathfrak{F}_\epsilon} r_n(F, \phi_n^*) = \sup_{F \epsilon \mathfrak{F}_\epsilon'} r_n(F, \phi_n^*). \tag{5.16}$$

Lemma 14 now follows at once from Lemma 11, (5.16), Lemma 13, (5.15), and the fact that $\mathfrak{F}_\epsilon' \subset \mathfrak{F}^c$.

PROOF OF THEOREM 1. Theorem 1 now follows immediately from Lemma 14 and (1.9).

We remark that the proof of Theorem 1 is clearly valid when $\mathfrak{F}$ is replaced by a suitably large subset.

It is not really necessary to prove Theorem 1 by using (1.9) and proving the result first for $\mathfrak{F}^c$ (in Lemma 14). For Lemma 13 clearly holds if in (5.10) we replace $\mathfrak{F}^*$ by $\mathfrak{F}^I$ and $\mathfrak{F}_\epsilon$ by the class of d.f.'s obtained by substituting $\mathfrak{F}^I$ for $\mathfrak{F}^*$ in the definition of $\mathfrak{F}_\epsilon$; one can carry through the arguments of Sections 2 and 4 with this altered definition of $\mathfrak{F}_\epsilon$ (appropriate results from [1] still hold), and obvious analogues of (5.15) and (5.16) then yield Theorem 1.

In Section 7 we shall discuss various modifications of Theorem 1 obtained by altering the way in which W depends on $F(z) - g(z)$.

6. Integral weight functions. Since for $m > 1$ the procedure ϕ_n^* does not have constant risk for F in $\mathfrak{F}^c$ and any common weight functions of the form given in equation (5.1) of [1], there is no longer any special reason for considering weight functions for which the dependence on F of the integrand is of the form considered there. Therefore, to make the proof of this section as simple as possible, we shall consider here the analogue of the special case of Section 5 of [1] wherein $W(y, z)$ does not depend on z, relegating the consideration of more complicated weight functions to Section 7. Our result is

THEOREM 2. *Let* $W(r)$ *be a monotonically nondecreasing nonnegative real function of* r *for* $r \geqq 0$ *which is not identically zero and which satisfies*

$$\int_0^\infty W(r) r e^{-2r^2} \, dr < \infty. \tag{6.1}$$

Then $\{\phi_n^*\}$ *is asymptotically minimax relative to* $\mathfrak{F}^c$ *and the weight functions*

$$W_n(F, g) = \int W(n^{1/2} |F(x) - g(x)|) \, dF(x). \tag{6.2}$$

PROOF. As in Section 5 of [1], the proof of this theorem is essentially easier than that of Theorem 1, since it is centered about the one-dimensional asymptotic result (6.12) (for each z). The analytic details are often like corresponding ones of Section 5 of [1], to which we shall consequently sometimes refer. The proof will be conducted in four numbered paragraphs.

1. From (6.1) and the uniformity of approach to its continuous limit of the d.f. of $n^{1/2}[S_n(z) - F(z)]$ for all z for which $\frac{1}{2} - |F(z) - \frac{1}{2}| > \delta > 0$ and all F in $\mathfrak{F}^c$ (the F-measure of this set of z approaches 1 as $\delta \to 0$, uniformly in F), we conclude at once from (6.1) that $r_n(F, \phi_n^*)$ has a bounded limit uniformly for F in $\mathfrak{F}^c$, and thus that (4.2) is satisfied with $\mathfrak{F}_\epsilon'$ replaced by $\mathfrak{F}^c$, for some F_δ in $\mathfrak{F}^c$ (of course, r_n is now to be computed using (6.2)). We can clearly suppose, and hereafter do, that F_δ is a d.f. on I^m. Let $\mathfrak{F}_{0k}'$ denote the class of d.f.'s defined in paragraph 2 of the proof of Lemma 11, with $\epsilon = 0$; thus, the B_h' of paragraph 3 of that proof now coincides with the B_h there.

2. As in Section 5 of [1], we shall let $\{\xi_{kn}\}$ be a sequence of *a priori* probability measures on B_h (we shall think of $\mathfrak{F}_{0k}'$ and B_h interchangeably), and we shall write $P_z^*\{A\}$ for the probability of an event expressed in terms of $T^{(n,k)} = T^{(n)}$ when the latter has probability function

$$P\{T_i^{(n)} = t_i^{(n)}, 1 \leqq i \leqq h+1\}$$
$$= \frac{1}{d(k, n, z)} \int_{B_h} f(z, \pi) P_\pi\{T_i^{(n)} = t_i^{(n)}, 1 \leqq i \leqq h+1\}\, d\xi_{kn}(\pi); \tag{6.3}$$

here P_π is defined in (3.1) and $f(z, \pi)$ is the Lebesgue density at z (in I^m) of the d.f. $F(\cdot, \pi)$ in $\mathfrak{F}_{0k}'$ corresponding to a given π in B_h; $d(k, n, z)$ is chosen to make (6.3) a probability function. We take $f(z, \pi)$ to be constant on the interior of each of the k^m cubes in I^m; this determines (6.3) for all z with all irrational components (hereafter called irrational z), to which such z we may limit all further discussion. For each such z and possible value $t^{(n)}$ of $T^{(n,k)}$, we define

$$r_{kn}(z, \phi, t^{(n)}) = \int_{B_h} E_\phi W(n^{1/2}|g(z) - F(z, \pi)|)\, d_\pi \xi_{kn}^*(\pi, z, t^{(n)}), \tag{6.4}$$

where, for Borel subsets B of B_h,

$$\xi_{kn}^*(B, z, t^{(n)}) = \frac{\int_B f(z, \pi) P_\pi\{t^{(n)}\}\, d\xi_{kn}(\pi)}{\int_{B_h} f(z, \pi) P_\pi\{t^{(n)}\}\, d\xi_{kn}(\pi)}; \tag{6.5}$$

we have used $P_\pi\{t^{(n)}\}$ to denote the function of (3.1).

For each n and k, if F is restricted to be in $\mathfrak{F}_{0k}'$, we may, as in Section 4, restrict our consideration to procedures ϕ in $\mathfrak{D}_{n,k}$. Denoting expectation with respect to P_z^* by E_z^*, we have as in (5.10) of [1],

$$\int r_n(F, \phi)\, d\xi_{kn} = \int_{I^m} E_z^* r_{kn}(z, \phi, T^{(n,k)})\, d(k, n, z)\, dz, \tag{6.6}$$

where dz denotes the differential element of Lebesgue measure on I^m. For fixed n, k, z, and $t^{(n)}$, let $r^*_{kn}(z, t^{(n)})$ denote the infimum of (6.4) over $\mathfrak{D}_{n,k}$. In order to prove Theorem 2, according to (6.6) and the discussion of paragraph 1 of this proof, it clearly suffices to show that, for some $\{\xi_{kn}\}$,

$$\text{(6.7)} \qquad \lim_{k\to\infty} \lim_{n\to\infty} \int_{I^m} E^*_z r_{kn}(z, \phi, T^{(n,k)})\, d(k, n, z)\, dz \geqq \lim_{n\to\infty} r_n(F_\delta, \phi^*_n).$$

3. Fix k. Let $\pi_{\delta k}$ be such that $F(z, \pi_{\delta k}) = F_\delta(z)$ for z in A^m_k. We may assume $\pi_{\delta k}$ is an interior point of B_h; for, if $\pi_{\delta k}$ were not an interior point, letting $F'_\delta = (1 - \delta')F_\delta + \delta' U$ where U is the uniform d.f. on I^m, we see easily that the right side of (6.7) can be decreased by at most a quantity which approaches 0 as $\delta' \to 0$ if F_δ is replaced by F'_δ there; we could thus replace $\pi_{\delta k}$ by the interior point π corresponding to F'_δ (for δ' small but positive) in what follows. Let ξ_{kn}, $n = 1, 2, \cdots$, be a sequence of *a priori* measures on B_h which "shrink down" on $\pi_{\delta k}$ as the ξ_n of Lemma 8 shrink down on π^*; e.g., ξ_{kn} is uniform on a sphere of radius $n^{-1/4}$ about $\pi_{\delta k}$. It follows at once that

$$\text{(6.8)} \qquad \lim_{n\to\infty} d(k, n, z) = f(z, \pi_{\delta k})$$

at all irrational z. Suppose we show that, for any irrational z and any $\epsilon > 0$, there is an $N = N(\epsilon, z, k)$ such that, for $n > N$, P^*_z assigns probability at least $1 - \epsilon$ to a set of $T^{(n,k)}$ values for which

$$\text{(6.9)} \qquad r^*_{kn}(z, T^{(n,k)}) + \epsilon > \int_{-\infty}^{\infty} W(y) q(y, \sigma(z, k))\, dy,$$

where $q(y, \sigma) = (2\pi\sigma^2)^{-1/2} \exp(-y^2/2\sigma^2)$ and where $\sigma(z, k)$ is continuous in z and

$$\text{(6.10)} \qquad \sigma(z, k) = F(z, \pi_{\delta k})[1 - F(z, \pi_{\delta k})] + o(1 \mid k \mid z) \leqq \tfrac{1}{4}.$$

Then, writing $V(z, k)$ for the expression on the right side of (6.9), we will clearly have (from (6.10), (6.1), and the continuity of q)

$$\lim_{n\to\infty} r_n(F_\delta, \phi^*_n) = \int_{I^m} \lim_{k\to\infty} V(z, k)\, dF_\delta(z)$$

$$\text{(6.11)} \qquad = \lim_{k\to\infty} \int_{I^m} V(z, k)\, dF_\delta(z)$$

$$= \lim_{k\to\infty} \int_{I^m} V(z, k) f(z, \pi_{\delta k})\, dz.$$

Thus, an application of Fatou's lemma to the left side of (6.7) shows that (6.11) and (6.8) will imply (6.7). Thus, it remains to prove (6.9) for the appropriate values of the arguments there.

4. The proof of (6.9) is similar to that of Lemma 8. For fixed z, the expression of (6.5) is like the *a posteriori* probability measure of π when ξ_{kn} is the *a priori*

measure, except for the factor $f(z, \pi)$. In fact, by the shrinking property of ξ_{kn} as $n \to \infty$ and the nature of $f(z, \pi)$, one obtains in the manner of [7] (see [1] for details) that, for any $\epsilon' > 0$ and for n suitably large, with probability $>1 - \epsilon'$ under P_z^*, the joint density according to ξ_{kn}^* of the quantities $\bar{\gamma}_i = n^{1/2}(p_i - t_i^{(n)}/n)$, $1 \leqq i \leqq h$ (where we have written $\pi = (p_1, \cdots, p_{h+1})$), in a spherical region of probability $>1 - \epsilon'$ under ξ_{kn}^*, is at least $(1 - \epsilon')$ times the appropriate normal density for which the $\bar{\gamma}_i$ have means 0, var $\bar{\gamma}_i = p_{\delta i}(1 - p_{\delta i})$, $\text{cov}(\bar{\gamma}_i, \bar{\gamma}_j) = -p_{\delta i}p_{\delta j}$ (the $p_{\delta i}$ being the components of $\pi_{\delta k}$). For $\epsilon'' > 0$, an elementary computation (the details being like those of [1], p. 661, except that now $m > 1$) then shows that, for a fixed arbitrary irrational z, the corresponding distribution of $n^{1/2}[F(z, \pi) - J_{n,k}(z)]$, where $J_{n,k}(z)$ is the obvious best linear estimator in $\mathfrak{D}_{n,k}$ of $F(z, \pi)$ for π in $\mathfrak{F}'_{0k}$ (not in general $S_n(z)$, unless $z \in A_k^m$), has, with probability $>1 - \epsilon''$ under P_z^*, an absolutely continuous component the magnitude of whose Lebesgue density is at least

$$(1 - \epsilon'')\, q(y, \sigma(z, k)) \tag{6.12}$$

on the interval $-1/\epsilon'' < y < 1/\epsilon''$, where $\sigma(z, k)$ is continuous in z and satisfies (6.10). Since ϵ'' is arbitrary, (6.9) follows easily from (6.12) and the trivial one-dimensional case of [8] (see [1] for details; the argument here is easier, since we have not yet included the additional dependence of W on other quantities as in [1] and Section 7 below). Thus, Theorem 2 is proved.

It is clear that Theorem 2 remains valid if $\mathfrak{F}^c$ is replaced by a suitably large subset. Further generalizations will be discussed in the next section.

7. Other loss functions. We list a few of the extensions of Theorems 1 and 2 which may be proved by the same methods with only minor modifications and no essential new difficulties in the proof. In fact, our treatment of the case $m > 1$ (compared with the argument of [1]) has been concentrated on the difficulty engendered by the nonconstancy of $r_n(F, \phi_n^*)$, and that nonconstancy (in the counterpart of modification F, below) is the only real new difficulty in any of the corresponding generalizations of Section 6 of [1] (the difficulty is more trivial there, where $m = 1$ and the nonconstancy is easier to deal with than in Theorems 1 and 2 above).

A. In Theorem 2, the form of W may be extended. For $m = 1$, the more general form $W(n^{1/2}|F(z) - g(z)|, F(z))$ was considered in Section 5 of [1]. The same form can be considered here, but perhaps the dependence on the second variable is no longer so natural; it may be replaced or supplemented, for example, by a dependence on the value of the marginal d.f.'s at the point z. The regularity condition which must be imposed on W in order for our method of proof to hold is, in any event, exactly the obvious analogue of that of Section 5 of [1]. For example, continuity and an appropriate integrability condition (the analogue of (5.5) of [1]) is more than enough.

B. In Theorem 2, W can be replaced by a measure (rather than a density) in the second argument of the W of A above (or its replacements, just above).

For example, when $m = 2$, one might be interested only in the estimation of the deciles of the marginal d.f.'s F_1 and F_2 (say) and, at each decile r of F_1, the deciles of the d.f. $F(r, y) / F_1(r)$ (and its counterpart with x and y interchanged).

C. An analogue of Theorem 2 (with any of the modifications noted above) for $\mathfrak{F}$ rather than $\mathfrak{F}^c$ is perhaps not too natural (see [1] for further comments), but can be given under suitable assumptions. An analogue of Theorem 1 or Theorem 2 for the class of purely discrete d.f.'s (e.g., on R^m, or on the integral lattice points of R^m) can also be given; for example, the former essentially follows from the fact that there is a discrete d.f. at which ϕ_n^* has almost the same risk as at F_δ when n is large (see (1.4) through (1.8)).

D. In Theorem 1, one can replace D_n by $\sup_z[|g(z) - F(z)|h(F(z))]$, where h is a suitably regular nonnegative function whose dependence on $F(z)$ may be replaced, e.g., by a dependence on the marginal d.f.'s, as in A above; a linear combination of such functions can also be employed. If h takes on only the values 0 and 1, this modification amounts to taking the supremum of the deviation over a suitable subset of R^m whose description depends on F.

E. In Theorem 1, one could consider the *measures* P, Q_n, and g^* corresponding to F, S_n, and g, and could let W depend on $\sup_A|P(A) - g^*(A)|$ where the supremum is taken over a suitable family of sets, e.g., rectangles with sides parallel to the coordinate axes. This presents no new difficulties.

F. The function h of D above, the second argument of W in A above, and the integrating measure of (6.2), can all be changed so as to depend only on z and not on $F(z)$ (or they can depend on both). This requires no new arguments, only obvious regularity conditions as on p. 664 of [1]. It is again the existence of an F_δ which is the crucial point.

G. The remarks on the *sequential* asymptotic minimax character of ϕ_n^* for suitable weight functions, which are contained on pp. 664–665 of [1], hold here without change.

H. Obvious combinations of the types of dependence of W_n on F and z which occur in Theorems 1 and 2 and in the previous remarks can be considered with no essential new difficulty. In fact, the asymptotic minimax character of ϕ_n^* seems to hold for a very general class of weight functions. The discussion of p. 664 of [1] indicates the possible breadth of that class, but we are even further than we were in the case $m = 1$ of [1] from being able to give a single simple, unified proof.

REFERENCES

[1] A. Dvoretzky, J. Kiefer, and J. Wolfowitz, "Asymptotic minimax character of the sample distribution function and of the classical multinomial estimator," *Ann. Math. Stat.*, Vol. 27 (1956), pp. 642–669.

[2] J. Kiefer, and J. Wolfowitz, "On the deviations of the empiric distribution function of vector chance variables," *Trans. Amer. Math. Soc.*, Vol. 87, Jan. 1958, pp. 173–186.

[3] M. D. Donsker, "Justification and extension of Doob's heuristic approach to the Kolomogorov-Smirnov theorems," *Ann. Math. Stat.*, Vol. 23 (1952), pp. 277–281.

[4] J. L. Doob, "Heuristic approach to the Kolmogorov-Smirnov theorems," *Ann. Math. Stat.*, Vol. 20 (1949), pp. 393–402.

[5] C. G. ESSEEN, "Fourier analysis of distribution functions," *Acta. Math.*, Vol. 77 (1945), pp. 1–125.
[6] H. BERGSTROM, "On the central limit theorem in the space R^k, $k > 1$," *Skand. Aktuarietids*, (1949), pp. 106–127.
[7] R. v. MISES, "Fundamentalsätze der Wahrscheinlichkeitsrechnung," *Math. Zeit.*, Vol. 4 (1919), pp. 1–97.
[8] T. W. ANDERSON, "The integral of a symmetric unimodal function," *Proc. Amer. Math. Soc.*, Vol. 6 (1955), pp. 170–176.

Reprinted from the *Proceedings of the Cambridge Philosophical Society*,
Volume 55, Part 4, pp. 328–332, 1959.

A FUNCTIONAL EQUATION TECHNIQUE FOR OBTAINING WIENER PROCESS PROBABILITIES ASSOCIATED WITH THEOREMS OF KOLMOGOROV–SMIRNOV TYPE

BY J. KIEFER†

Communicated by D. V. LINDLEY

Received 6 February 1959

1. *Introduction.* Since the first proofs by Kolmogorov (13) and Smirnov ((14), (15)) of their well-known results on the limit distribution of the deviations of the sample distribution function, many alternative proofs of these results have been given. For example, we may cite the various approaches of Feller (4), Doob (3), Kac (8), Gnedenko and Korolyuk (7), and Anderson and Darling (1). The approaches of (3), (8) and (1) rest on a probabilistic computation regarding the Wiener process, and are justified by the paper of Donsker (2) (see also (11)). Of all these approaches, only those of (8) and (1) can be extended to obtain the limit distributions of the 'k-sample' generalizations of the Kolmogorov–Smirnov statistics suggested in (9), and the author ((9), (10)) and Gihman (6) carried out such proofs.

The object of the present note is to give another method of derivation of the Kolmogorov–Smirnov results (and other related results of (3)) which is at the same time extremely simple computationally (compared, for example, with the method of obtaining the appropriate Green's function, integrating it, and passing to the limit) and which is also capable of extension to other problems such as that mentioned at the end of the previous paragraph. Of the other approaches mentioned above, that of Doob, although not extendable to these other problems, is the one most closely related to the present approach; both obtain the desired result by considering the variation of an appropriate function of the *boundary* parameters. Of course, the functional equation approach in probability and physics is old, but its justification and use in the present context seem to be new.

The approach of the present paper leads to a parabolic partial differential equation on an unbounded region, and an appropriate uniqueness theorem must be proved. Such regions do not seem to have been considered previously in the literature, and uniqueness is obtained only by imposing rather precise asymptotic boundary conditions, which in turn are obtained from the solutions to simpler problems. This technique obviously has much more general applicability than its use herein. The standard device of imposing a condition at infinity, which is familiar in such problems as that of the diffusion equation on a half plane, does not yield uniqueness in the present case, where much finer boundary conditions are needed.

In this paper we shall consider only the computation of certain Wiener process probabilities; the reader is referred to (3) and (10) for statistical interpretations.

† Research sponsored by the U.S. Office of Naval Research.

2. *The Kolmogorov–Smirnov theorems.* We first consider the simple 'one-sided' absorption result. Let $Y(t)$, $t \geqslant 0$, be a separable Wiener process with $EY(t) = 0$ and $EY^2(t) = t$, and for $x \geqslant 0$ and $y > 0$ write

$$P_1(x, y) = \Pr\{\inf_{0 \leqslant t < \infty} [Y(t) + (yt + x)] \leqslant 0\}. \tag{1}$$

Since $-Y(t) \geqslant yt + x$ (for all $t \geqslant 0$) is equivalent to $-yY(t/y^2) \geqslant t + xy$ and since $yY(t/y^2)$ is distributed like $Y(t)$, we have $P_1(x, y) = P_1(xy, 1)$. We write $P(z) = P_1(z, 1)$ for $z \geqslant 0$.

We see at once that P is continuous (and of course bounded) for $z \geqslant 0$, and that $P(0) = 1, P(\infty) = 0$. Next we note that if $p(s, w; t, z)$ is the Green's function corresponding to the one-dimensional diffusion equation with absorbing boundary $-t$ for $t \geqslant 0$, then we obviously have, for $t > 0$ and $z > 0$,

$$1 - P(z) = \int_0^\infty [1 - P(u)] p(0, z; t, u - t)\, du; \tag{2}$$

the only use we make of this equation and the well-known regularity properties† of p for such a boundary (we do not need the explicit form of p) is to conclude that P is appropriately differentiable in the sequel. Now, from the well-known behaviour of $\Pr\{\inf_{0 \leqslant t \leqslant \Delta} Y(t) \leqslant -z\}$, we can write, for $z > 0$ and as $\Delta \to 0$,

$$P(z) = \int_{-\infty}^{\infty} P(z + \Delta - \epsilon)\, (2\pi\Delta)^{-\frac{1}{2}} e^{-\epsilon^2/2\Delta}\, d\epsilon + o(z^2/2\Delta), \tag{3}$$

from which we easily obtain

$$\frac{d^2P}{dz^2} + 2\frac{dP}{dz} = 0. \tag{4}$$

(This last equation can also be derived with slightly more difficult justification from the analogous difference equation for a corresponding random walk setup, or can be obtained from (2) if we use only rudimentary additional knowledge about the properties of p.) Of course, the unique solution to (4) which satisfies $P(0) = 1$ and $P(\infty) = 0$ is e^{-2z}; thus, we have proved the 'one-sided' result of (15):

THEOREM 1. *$P_1(a, b) = e^{-2ab}$ for $a, b \geqslant 0$.*

(In statistical applications, one has $a = b$.)

We now turn to the Kolmogorov–Smirnov 'two-sided' result. With $Y(t)$ as above, we write, for $a > 0$, $b \geqslant 0$, and $|b| < c$,

$$P_1^*(a, b, c) = \Pr\left\{\sup_{t \geqslant 0} \frac{|Y(t) + b|}{at + c} \geqslant 1\right\}. \tag{5}$$

Just as before, we have $P_1^*(a, b, c) = P_1^*(1, ab, ac)$. We write $P(u, v) = P_1^*(1, u, v)$ for $0 \leqslant |u| \leqslant v$. Proceeding as before, we obtain, in analogy to (4), for $0 \leqslant |u| \leqslant v$,

$$\frac{\partial^2 P}{\partial u^2} + 2\frac{\partial P}{\partial v} = 0, \tag{6}$$

† The required properties of p, as well as the justification of the relationship between Brownian motion probabilities and the diffusion equation, can be found in (5) and (1); a recent paper by Doob considers this justification in the n-dimensional case.

with the boundary conditions

$$\left.\begin{aligned}&\text{(i)}\quad P(\pm v, v) = 1,\\ &\text{(ii)}\quad \lim_{v,a\to\infty}\ \sup_{|u|<v-a} P(u,v) = 0,\\ &\text{(iii)}\quad \lim_{v\to\infty} P(\pm(v-h),v) = e^{-2h}, \text{ uniformly for } 0 \leqslant h \leqslant H;\end{aligned}\right\} \tag{7}$$

the second and third of these conditions are trivial probabilistic consequences of Theorem 1. (The third condition is assumed to hold for each $H < \infty$.)

It is clear that one solution to (6) which satisfies (7) is

$$P(u,v) = 1 - \sum_{n=-\infty}^{\infty} (-1)^n e^{-2n^2v+2nu}. \tag{8}$$

Thus, as soon as we prove that (7) entails uniqueness of the solution to (6), we will, in particular, have derived the Kolmogorov–Smirnov result,

THEOREM 2. *For* $z > 0$, $P_1^*(z,z,0) = 2\sum_{n=1}^{\infty}(-1)^{n+1}e^{-2n^2z^2}$.

It remains to prove the uniqueness result mentioned above. The method of proof, which uses a standard device, is obviously extendable to a large class of partial differential equations on unbounded regions (we shall not bother to state this as a theorem), the essential fact being that conditions (ii) and (iii) of (7) imply that the difference $Q(u,v)$ (say) of two solutions to (6) satisfying (7) approaches 0 as $v \to \infty$, uniformly in $|u| \leqslant v$. For then if Q has a positive maximum M at (u_0, v_0) and $D > v_0$ is such that $Q(u,D) < \frac{1}{2}M$, the function $R(u,v) = Q(u,v) + M(v-v_0)/2D$ is $< M$ on the boundary of the triangle $|u| \leqslant v \leqslant D$, while $R(u_0,v_0) = M$. It follows that the positive maximum of R on this triangle is attained at an interior point of the triangle, at which point we obtain the contradiction

$$0 \geqslant \partial^2 R/\partial u^2 + 2\,\partial R/\partial v = M/D.$$

It would be convenient in some applications of the method outlined here, if the conditions corresponding to (7) (iii) could be weakened somewhat; it is not clear to what extent this can be done, although it is easy to see that some such fairly precise boundary condition at ∞ is needed for uniqueness.

3. *Generalizations*. In the same way that (8) was obtained, we can obtain, more generally, Doob's expression for

$$P(a,b,\alpha,\beta) = \Pr\{\sup_{t\geqslant 0}[Y(t)-(at+b)] \geqslant 0 \text{ or } \inf_{t\geqslant 0}[Y(t)+(\alpha t+\beta)] \leqslant 0\}. \tag{9}$$

In fact, as before, we have

$$P(a,b,\alpha,\beta) = P(1, ab, \alpha/a, a\beta).$$

Writing

$$P^*(u,v) = P(1, v-u, 2s-1, v+u)$$

for fixed $s > \frac{1}{2}$, we obtain, for $|u| \leqslant v$,

$$\frac{\partial^2 P^*}{\partial u^2} + 2(s-1)\frac{\partial P^*}{\partial u} + 2s\frac{\partial P^*}{\partial v} = 0. \tag{10}$$

The boundary conditions are those of (7), except that half of (iii) is altered to become $\lim P^*(-v+h, v) = e^{-2h(2s-1)}$. The same uniqueness proof holds, and we obtain easily

$$P^*(u,v) = 1 + \sum_{n=-\infty}^{\infty} \exp\{-(8n^2+4n)sv + 4nsu\} \cdot [e^{-(4n+2)v+2u} - e^{4nv}] \tag{11}$$

and thus we have

$$P(a,b,\alpha,\beta) = 1 + \sum_{n=-\infty}^{\infty} e^{-2n^2(a+\alpha)(b+\beta)-2nb(a+\alpha)}[e^{-2nb(b+\beta)-2ab} - e^{2na(b+\beta)}]. \tag{12}$$

which is equivalent to equation (4·3) of (3).

Finally, we turn to a multidimensional analogue of (8). For k a positive integer, let $Y_1(t), \ldots, Y_k(t)$ be independent processes with the same law as $Y(t)$ above. For $0 \leqslant u \leqslant v$, write

$$P_k(u,v) = \Pr\left\{\sup_{t\geqslant 0} \frac{(Y_1(t)+u)^2 + \sum_2^k Y_i(t)^2}{(v+t)^2} \geqslant 1\right\}. \tag{13}$$

Thus, $P_k(u,v)$ is the probability that a k-dimensional Brownian motion whose initial distance from the origin is u, will ever reach a sphere of initial radius v, centred at the origin, the radius increasing at uniform rate of one per unit of time. We obtain, for $0 \leqslant u \leqslant v$,

$$\frac{\partial^2 P_k}{\partial u^2} + \frac{k-1}{u}\frac{\partial P_k}{\partial u} + 2\frac{\partial P_k}{\partial v} = 0. \tag{14}$$

This is most easily derived from the corresponding equation wherein u is replaced by the k-vector of initial position, in terms of which there is spherical symmetry; the symmetric solutions to the latter equation correspond to those solutions of (14) for which

$$\left.\frac{\partial P_k}{\partial u}\right|_{u=0} = 0. \tag{15}$$

We also have the boundary conditions of (7) for $u, v \geqslant 0$; the third of these follows easily upon inscribing in and superscribing about the expanding sphere, appropriate expanding rectangular parallelopipeds whose sides are near each other near a point of the sphere. The uniqueness proof of the previous paragraph is applicable to the problem in terms of initial *position* mentioned above. We conclude then, that (14) has a unique solution on $0 \leqslant u \leqslant v$ satisfying (15) and the part of (7) which is concerned with $u, v \geqslant 0$.

Thus, for example, when $k = 3$ it is easy to write down the solution in terms of the functions $(1-cv/u)\, e^{cu-c^2v/2}$, which satisfy (14) and have a simple form when $u = v$. We obtain for $0 \leqslant u \leqslant v$,

$$P_3(u,v) = 1 - \sum_{n=-\infty}^{\infty} \left(1 - \frac{2nv}{u}\right) e^{2nu-2nv^2}. \tag{16}$$

Other applications of this method to certain more complicated problems will appear elsewhere.

REFERENCES

(1) ANDERSON, T. W. and DARLING, D. Asymptotic theory of certain 'goodness of fit' criteria based on stochastic processes. *Ann. Math. Stat.* 23 (1952), 193–212.

(2) DONSKER, M. Justification and extension of Doob's heuristic approach to the Kolmogorov–Smirnov theorems. *Ann. Math. Stat.* 23 (1953), 277–81.

(3) DOOB, J. L. Heuristic approach to the Kolmogorov–Smirnov theorems. *Ann. Math. Stat.* 20 (1949), 393–403.

(4) FELLER, W. On the Kolmogorov–Smirnov limit theorems for empirical distributions. *Ann. Math. Stat.* 19 (1948), 177–89.

(5) FORTET, R. Les Fonctions aléatoires du type de Markoff associées à certaines équations linéaires aux dérivées partielles du type parabolique. *J. Math. pures appl.* 22 (1943), 177–243.

(6) GIHMAN, J. J. A k-sample test of homogeneity. *Teorya Veryotnostei i yeyau primenyenya* (1957), pp. 380–4.

(7) GNEDENKO, B. V. and KOROLYUK, V. S. On the maximum discrepancy between the empirical distributions. *Dokl. Akad. Nuak. SSSR* (N.S.), 80 (1951), 525–9.

(8) KAC, M. On some connections between probability theory and differential and integral equations. *Proc. Second Berkeley Symp.* (Berkeley, 1951), 180–215.

(9) KIEFER, J. Distance tests with good power for the non-parametric k-sample problem. *Ann. Math. Stat.* 26 (1955), 775.

(10) KIEFER, J. K-sample analogues of the Kolmogorov–Smirnov and Cramer–v. Mises Tests. *Ann. Math. Stat.* 30 (1959).

(11) KIEFER, J. and WOLFOWITZ, J. On the deviations of the empiric distribution function of vector chance variables. *Trans. Amer. Math. Soc.* 87 (1958), 173–86.

(12) LÉVY, P. *Processus Stochastiques et Mouvement Brownien* (Paris, 1948).

(13) KOLMOGOROV, A. N. Sulla determionazione empirica della leggi di probilita. *Giorn. Ist. ital. Attuari*, 4 (1933), 1–11.

(14) SMIRNOV, N. V. On the estimation of the discrepancy between empirical curves of distribution of two independent samples. *Bull. Math. Univ. Moscou*, 2 (1939), fasc. 2.

(15) SMIRNOV, N. V. Approach of empirical distribution functions. *Uspyekhi Matem. Nauk.* 10 (1944), 179–206.

CORNELL AND OXFORD UNIVERSITIES

Reprinted from THE ANNALS OF MATHEMATICAL STATISTICS
Vol. 32, No. 2, June, 1961

DISTRIBUTION FREE TESTS OF INDEPENDENCE BASED ON THE SAMPLE DISTRIBUTION FUNCTION

BY J. R. BLUM,[1] J. KIEFER,[2] AND M. ROSENBLATT[3]

Sandia Corporation and Indiana University; Cornell University; and Brown University

0. Summary. Certain tests of independence based on the sample distribution function (d.f.) possess power properties superior to those of other tests of independence previously discussed in the literature. The characteristic functions of the limiting d.f.'s of a class of such test criteria are obtained, and the corresponding d.f. is tabled in the bivariate case, where the test is equivalent to one originally proposed by Hoeffding [4]. A discussion is included of the computational problems which arise in the inversion of characteristic functions of this type. Techniques for computing the statistics and for approximating the tail probabilities are considered.

1. Introduction. The idea of using various simple functionals of the sample d.f. of vector chance variables in order to test the independence of components, is a natural one. Only the difficult distribution theory prevents the use of such tests and the resulting achievement of improvement in power performance over all currently used tests. Specifically, let Ω be the class of continuous d.f.'s on m-dimensional Euclidean space R^m, and let ω be the subclass consisting of every member of Ω which is a product of its associated one-dimensional marginal d.f.'s. Let $X_1, \cdots, X_n$ be independent random m-vectors with common unknown d.f. F, a member of Ω, and suppose that it is desired to test the hypothesis $H_0: F \varepsilon \omega$ against the alternative $H_1: F \varepsilon \Omega - \omega$. Let S_n be the sample d.f. of $X_1, \cdots, X_n$; i.e., for x in R^m, $S_n(x)$ is n^{-1} times the number of X_i all of whose components are less than or equal to the corresponding components of x, i.e.,

$$S_n(r_1, r_2 \cdots, r_m) = \frac{1}{n} \sum_{j=1}^{n} \prod_{i=1}^{m} \phi_{r_i}(X_j^{(i)}),$$

where $X_j = (X_j^{(1)}, \cdots, X_j^{(m)})$ and

$$\phi_r(x) = \begin{cases} 1 & \text{if } x \leqq r, \\ 0 & \text{if } x > r. \end{cases}$$

Write S_{nj} for the marginal d.f. associated with the jth component of S_n (i.e., for the sample d.f. of the jth component of the X_i), and let

$$T_n(r) = S_n(r) - \prod_{j=1}^{m} S_{nj}(r_j). \tag{1.1}$$

Received April 24, 1960; revised September 15, 1960.

[1] Research sponsored by the Office of Ordnance Research, U. S. Army, under Contract No. DA-33-008-ORD-965.

[2] Research sponsored by the Office of Naval Research.

[3] Research sponsored by the Office of Naval Research.

Then many tests based on T_n will have good power properties (see Section 4) and will be similar on ω. For example, the critical region based on large values of

$$A_n = \sup |T_n(r)|,$$

a statistic constructed in the spirit of the Kolmogorov-Smirnov statistics, evidently has such properties. It follows from the results of [8] that the d.f. of $n^{\frac{1}{2}}A_n$ under H_0 differs from unity by less than $c_1 \exp(-c_2 z^2)$ for all n and all arguments $z > 0$, where the c_i are positive constants. It can be shown that the limiting d.f. of $n^{\frac{1}{2}}A_n$ exists (and hence has the same behavior with z); since the proof is somewhat long but uses mainly ideas like those of [8], it will not be given here. The calculation of this asymptotic distribution seems formidable; it is equivalent to the computation of the d.f. of the maximum of a particular Gaussian process with multidimensional *time* parameter. A corresponding calculation of exact (nonasymptotic) distributions for various values of n can, of course, be achieved numerically, but such calculations are extremely laborious even if done by machines for rather small n.

Another critical region, constructed in the spirit of the von Mises-Cramér tests, is that based on large values of

$$B_n = \int [T_n(r)]^2 \, dS_n(r). \tag{1.2}$$

Adapting the well known technique of Kac and Siegert [5] to the present setting (such a multidimensional computation was first carried out in [12]), we shall obtain the characteristic function of the asymptotic distribution of nB_n under H_0 when $m = 2$ (Section 2), in which case the test turns out to be equivalent to one constructed on other heuristic grounds by Hoeffding [4] (see Section 5 below for the form in which Hoeffding stated his test). Certain variants of nB_n in the case $m > 2$ will be considered in Section 3.

In Section 4 questions of distribution under H_1, power, and estimation, and certain modifications, will be taken up. A particularly simple and computationally convenient form of the tests is given in Section 5. In Section 6 an approximation is suggested to the tail of the limiting distribution, which is compared with the exact results; this idea clearly has useful applications in many other problems. Methods for computing distributions of weighted sums of chi-square variables, which are relevant for computing the asymptotic distribution of nB_n as well as many other important distributions in statistics, are discussed in Section 7. The asymptotic distribution of nB_n for the case $m = 2$ is tabulated in Section 8.

2. The case $m = 2$. The statistic B_n is clearly distribution-free for F in ω. As usual, we can therefore carry out our computations when F is the uniform distribution on the unit square I^2. Let $T(x, y)$ be a separable Gaussian process depending on the "time" parameter (x, y) for (x, y) in I^2, and with

$$\begin{aligned} ET(x, y) &= 0, \\ ET(x, y)T(u, v) &= [\min(x, u) - xu][\min(y, v) - yv]. \end{aligned} \tag{2.1}$$

A routine computation (most easily accomplished by writing

$$S_{n1}(x)S_{n2}(y) = xS_{n2}(y) + yS_{n1}(x) - xy + O_p(n^{-1})$$

shows that (2.1) gives the mean and the asymptotic covariance of the random function $n^{\frac{1}{2}}T_n$. It follows from the appropriate analogue in the present case of the corrected argument of [12] or of the argument of Section 2 of [7] (the proof being very similar here) that the asymptotic distribution of nB_n is the same as that of

$$B = \int_0^1 \int_0^1 T^2(x, y)\, dxdy.$$

Writing $s = (x, y)$, $t = (u, v)$, and $K(s, t)$ for the last member of (2.1), we consider the integral equation

$$\int_{I^2} K(s, t)\phi(t)\, dt = \lambda\phi(t). \tag{2.2}$$

It is easily seen that the eigenvalues and (complete set of) eigenfunctions of (2.2) are $1/\pi^4 j^2 k^2$ and $2 (\sin \pi jx) (\sin \pi ky)$; $j, k = 1, 2, \cdots$. Hence, exactly as in [5] and [12], we conclude that

$$Ee^{izB} = \prod_{j,k=1}^{\infty} (1 - 2iz/\pi^4 j^2 k^2)^{-\frac{1}{2}}. \tag{2.3}$$

An equivalent result was first stated by Hoeffding [4], who stated two other different methods for obtaining (2.3). The corresponding d.f. of B is tabled in Section 8.

It is obvious that, because of the factorizability of $K(s, t)$ we can similarly obtain the characteristic function of the limiting d.f. for the case where a weight function of the form $W(S_{n1}(r))W(S_{n2}(r))$ is inserted in the integrand in the expression for B_n; one has merely to use the corresponding one-dimensional results on weighted ω^2 statistics (see, e.g., [1], [5]) to obtain the eigenvalues.

3. The case $m > 2$. For the sake of brevity we shall discuss in detail only the case $m = 3$; the corresponding results for other cases require only obvious changes.

Suppose, then, that F is the uniform distribution on the unit cube. Another routine computation (most easily accomplished in a manner analogous to that suggested in Section 2) yields

$$\begin{aligned}\lim_{n\to\infty} nET_n(x, y, z)T_n(y, v, w) &= \min (x, u) \min (y, v) \min (z, w) \\ &- yzvw \min (x, u) - xzuw \min (y, v) - xyuv \min (z, w) + 2xyzuvw.\end{aligned} \tag{3.1}$$

This kernel does not permit the simple treatment which that of (2.1) did, and the eigenvalues are at present unknown. This suggests that we look for a function T_n' of S_n for which

$$\begin{aligned}&\lim_{n\to\infty} nET_n'(x, y, z)T_n'(u, v, w) \\ &= [\min (x, u) - xu] [\min (y, v) - yv] [\min (z, w) - zw].\end{aligned} \tag{3.2}$$

Denoting by S_{njk} the 2-dimensional marginal d.f. of S_n corresponding to the jth and kth coordinates (sample d.f. of the jth and kth components of the X_i), we easily verify that the function T_n' defined by

$$\begin{aligned}T_n'(x, y, z) = S_n(x, y, z) - S_{n1}(x)S_{n23}(y, z) - S_{n2}(y)S_{n13}(x, z) \\ - S_{n3}(z)S_{n12}(x, y) + 2S_{n1}(x)S_{n2}(y)S_{n3}(z)\end{aligned} \tag{3.3}$$

does in fact satisfy (3.2). It follows, in the manner of Section 2, that if

$$B_n' = \int [T_n'(r)]^2 \, dS_n(r),$$

then for F in ω we have

$$\lim_{n\to\infty} Ee^{iznB_n'} = \prod_{j_1,j_2,j_3=1}^{\infty} (1 - 2iz/\pi^6 j_1^2 j_2^2 j_3^2)^{-\frac{1}{2}}.$$

Thus, the asymptotic distribution of nB_n' can be tabulated in the manner of the tabulation of Section 8. However, a test for independence based only on the statistic B_n' is not to be recommended, since the power of any such test will be small for many alternatives which are far from ω; for example, it is clear that $ET_n'(r) = 0$ if F is of the form $F(x, y, z) = F_1(x)F_{23}(y, z)$. A solution to this difficulty can be found in the fact that, *if* the components of the X_i are *pair*wise independent, then $ET_n'(r) = 0$ for all r if and only if $F \varepsilon \omega$. Thus, the three 2-dimensional sample d.f.'s of the components of the X_i can be used to detect departure from pairwise independence, while B_n' detects other possible departures from independence. There are obviously many ways in which these two effects can be combined in constructing a test, and only one of them will be made explicit here. Let $T_{njk}(p, q) = S_{njk}(p, q) - S_{nj}(p)S_{nk}(q)$, and let

$$B_{njk} = \int [T_{njk}(r)]^2 \, dS_{njk}(r).$$

A computation of covariances readily shows that the functions $n^{\frac{1}{2}}T_{n12}$, $n^{\frac{1}{2}}T_{n13}$, $n^{\frac{1}{2}}T_{n23}$, and $n^{\frac{1}{2}}T_n'$ are asymptotically independent. Thus, arguing in the same manner as before, we conclude that the statistic C_n, defined by

$$C_n = n(B_{n12} + B_{n13} + B_{n23} + bB_n'), \tag{3.4}$$

where b is a positive constant, has the asymptotic distribution with characteristic function

$$\lim_{n\to\infty} Ee^{izC_n} = \prod_{j,k} (1 - 2iz/\pi^4 j^2 k^2)^{-3/2} \prod_{j_1,j_2,j_3} (1 - 2biz/\pi^6 j_1^2 j_2^2 j_3^2)^{-1/2}. \tag{3.5}$$

The corresponding asymptotic distribution can be tabulated in the manner of Section 8. The power properties of a critical region consisting of large values of C_n can be obtained as in Section 4.

4. Asymptotic distribution under H_1; power; estimation; modifications. We consider the case $m = 2$ throughout this section; the analogous results obviously all hold when $m > 2$.

If $F(x, y)$ is *not* of the form $G(x)H(y)$, where G and H are the two continuous marginal d.f.'s of X_1, the limiting d.f. of $n^{\frac{1}{2}}B_n$ can be obtained by noting that $n^{\frac{1}{2}}B_n$ is asymptotically

$$n^{\frac{1}{2}} \iint [EU_n(x, y)]^2 \, d[S_n(x, y) - F(x, y)]$$

$$+ n^{\frac{1}{2}} \iint [U_n(x, y) - EU_n(x, y)] \, dF(x, y) + o_p(1),$$

where

$$U_n(x, y) = S_n(x, y) - G(x)S_{n2}(y) - H(y)S_{n1}(x) + G(x)H(y).$$

Writing

$$\Delta(x, y) = F(x, y) - G(x)H(y),$$

$$\epsilon(x, y) = \iint [\phi_u(x) - G(u)][\phi_v(y) - H(v)]\Delta(u, v) \, dF(u, v),$$

we obtain that $n^{\frac{1}{2}}[B_n - \iint \Delta^2(x, y) \, dF(x, y)]$ *is asymptotically normal with mean* 0 *and the same variance as the random variable* $\Delta^2(X, Y) +. 2\epsilon(X, Y)$, *where* (X, Y) *is distributed according to* F. An equivalent form of this result was given by Hoeffding [4].

Of greater interest for most applications is the limiting d.f. of nB_n when we consider a sequence $F^{(n)}$ of alternatives on I^0 for which

$$n^{\frac{1}{2}}[F^{(n)}(x, y) - G^{(n)}(x)H^{(n)}(y)] \rightarrow q(x, y)$$

(finite and continuous) as $n \rightarrow \infty$. We obtain, using arguments similar to those of Section 2, that the limiting d.f. of $n\mathrm{B}_n$ is the same as the d.f. of

$$B' = \iint [T(x, y) + q(x, y)]^2 \, dxdy.$$

Recalling the eigenvalues and eigenfunctions of K obtained in Section 2, we can write

$$T(x, y) = \sum_{j,k=1}^{\infty} 2\pi^{-2}j^{-1}k^{-1} \, (\sin \pi xj) \, (\sin \pi ky)X_{jk},$$

where the X_{jk} are independent normal variates with means 0 and variances 1. Hence, writing

$$q_{jk} = \iint 2q(x, y)(\sin \pi jx)(\sin \pi ky) \, dxdy,$$

we obtain for the limiting characteristic function of nB_n,

$$Ee^{B'it} = \left\{ \prod_{j,k} \left(1 - \frac{2it}{\pi^4 j^2 k^2}\right)^{-\frac{1}{2}} \right\}$$
$$\cdot \exp\left\{ -\tfrac{1}{2} \sum_{j,k} q_{jk}^2 \pi^4 j^2 k^2 + \tfrac{1}{2} \sum_{j,k} q_{jk}^2 \pi^4 j^2 k^2 \left(1 - \frac{2it}{\pi^4 j^2 k^2}\right)^{-1} \right\}.$$

For simple $q(x, y)$'s (e.g., where all but a finite number of the q_{jk} are zero), one could easily compute tables of the power, in the manner of Section 8. Even for general $q(x, y)$, an argument like that of Section 6 would yield information.

Without obtaining such quantitative results, we can easily give a lower bound on the power. The power properties of tests based on the sample d.f. have been discussed in detail in [6] and [7], and it will suffice to state briefly the analogous results for the problems treated in the present paper. Such results will clearly apply for arbitrary m, and for the sake of clarity and brevity we shall only state them for the case $m = 2$, the extensions to $m > 2$ being obvious.

Let F be a d.f. on R^2 and let F_1 and F_2 be the corresponding marginal d.f.'s. Write

$$\delta_F = \sup_{x,y} |F(x, y) - F_1(x)F_2(y)|$$

and

$$\gamma_F = \left\{ \int_{R^2} [F(x, y) - F_1(x)F_2(y)]^2 \, dF_1(x) \, dF_2(y) \right\}^{\frac{1}{2}}.$$

(A similar treatment applies if the integrating measure is replaced by F in the definition of γ_F.) Then, for $0 < \alpha, \beta < 1$, there is a constant $C(\alpha, \beta)$ such that, for each $d > 0$, there is a critical region based on large values of A_n with $n < C(\alpha, \beta) d^{-2}$ and which has size $\leqq \alpha$ on ω and power $\geqq \beta$ for all alternatives F for which $\delta_F \geqq d$. Thus, the behavior of the required sample size as a function of d is of the same order as in common parametric (e.g., Gaussian) examples. The same conclusion for B_n holds if δ_F is replaced by γ_F in the above.

It is clear that this guaranteed behavior of the power function against *all* alternatives is far superior to that of the other nonparametric tests previously described in the literature (outside of [4]). Many of the latter have zero efficiency compared with tests based on A_n or B_n. Perhaps the best of these classical tests is the chi-square test with the observations divided into the k_n^2 classes determined by $k_n - 1$ equally spaced values of $S_{n1}(x)$ and $S_{n2}(y)$. The optimum choice of k_n has not been investigated, but it is reasonable to suppose that the power function for the optimum choice will behave no better, and possibly worse, than that of the best chi-square test of goodness of fit (see [10], [6]). If this is so, we would conclude that, if N observations are required by the test based on A_n (resp., B_n) to achieve a goal in terms of δ_F (resp., γ_F) like that described in the previous paragraph, then at least $\bar{C}(\alpha, \beta)N^{5/4}$ observations are required by the best chi-square test.

We remark that the relationship between δ_F and γ_F is easily seen to be $\delta_F \geqq \gamma_F \geqq C\delta_F^2$, where $C > 0$.

In many applications it is desirable not merely to test for dependence, but rather to estimate the type of dependence. There are many possible formulations of this problem. If it is desired to estimate the entire function $F - F_1F_2$, then, for almost any reasonable weight function, a modification of the arguments of [9] shows that $S_n - S_{n1}S_{n2}$ is asymptotically a minimax estimator (as $n \to \infty$). Similar results hold for the problem of estimating various functionals of F, F_1, F_2.

These results on power and estimation also apply under such obvious modifications as that of considering the probabilities and empiric frequencies in all rectangles instead of only in third quadrants, of inserting a weight function in the definition of A_n and B_n, etc. Also, as in [6], [7], [8], the results on size and minimum power are not materially affected if discontinuous distributions are admitted. We note also that, just as in [7], the results are unaffected if the integrating measure S_n is replaced by $S_{n1}S_{n2} \cdots S_{nm}$ in the definition of B_n (many other functions could be used, too); in fact, the limiting d.f. is exactly the same with this modification.

5. Computation of the statistics. The statistic B_n (or one of its variants, such as those mentioned at the end of Section 4) is rather unwieldy for practical computations in its form (1.2), even if the integral is rewritten as a sum to take account of the atomicity of the integrating measure. The form originally suggested by Hoeffding for his statistic (which differs slightly from B_n) for $n \geqq 5$ was

$$D_n = \frac{1}{4n(n-1)(n-2)(n-3)(n-4)} \tag{5.1}$$
$$\cdot \sum'' \prod_{j=1}^{2} [\phi_{X_{i_1}^{(j)}}(X_{i_2}^{(j)}) - \phi_{X_{i_1}^{(j)}}(X_{i_3}^{(j)})][\phi_{X_{i_1}^{(j)}}(X_{i_4}^{(j)}) - \phi_{X_{i_1}^{(j)}}(X_{i_5}^{(j)})],$$

where ϕ is defined as in Section 1 and $\sum''$ denotes the sum over all 5-tuples $(i_1, \cdots, i_5)$ of different integers, $1 \leqq i_q \leqq n$. Another form of D_n, for use in computations, was given by Hoeffding in Section 5 of his paper.

A more convenient form than (1.2) for computational purposes is obtained by noting that, when $m = 2$,

$$n^2 T_n(X_j^{(1)}, X_j^{(2)}) = N_1(j)N_4(j) - N_2(j)N_3(j),$$

where $N_1(j)$, $N_2(j)$, $N_3(j)$, $N_4(j)$ are the numbers of points lying, respectively, in the regions $\{(x, y) \mid x \leqq X_j, y \leqq Y_j\}$, $\{(x, y) \mid x > X_j, y \leqq Y_j\}$, $\{(x, y) \mid x \leqq X_j, y > Y_j\}$, $\{(x, y) \mid x > X_j, y > Y_j\}$. Thus, we have only to count the number of points lying in each of the four regions determined by the vertical and horizontal lines through $X_j = (X_j^{(1)}, X_j^{(2)})$, and compute

$$B_n = n^{-4} \sum_{j=1}^{n} [N_1(j)N_4(j) - N_2(j)N_3(j)]^2. \tag{5.2}$$

Similarly, when $m > 2$ a statistic such as that of (3.4) can easily be written in terms of the numbers of points in each of the 2^m orthants determined by the

m hyperplanes through $X_j = (X_j^{(1)}, \cdots, X_j^{(m)})$ and parallel to the coordinate hyperplanes. Thus, for $m = 3$ the statistic C_n can be written in terms of quantities $N_1(j), \cdots, N_8(j)$. We omit the details.

6. Approximation to the tail probabilities of the limiting distribution. We again limit ourselves to the case $m = 2$, although the discussion which follows even has obvious applications to problems outside this paper.

The Laplace transform of the asymptotic distribution of the test statistic nB_n under the null hypothesis is

$$(6.1) \qquad \prod_{j,k=1}^{\infty}\left[1 + \frac{2t}{\pi^4 j^2 k^2}\right]^{-\frac{1}{2}}.$$

The singularity of this expression in the complex t-plane which has largest real part is located at $t = -(\pi^4/2)$. In the neighborhood of $t = -(\pi^4/2)$ the expression (6.1) has the same behavior as

$$(6.2) \qquad \left(1 + \frac{2t}{\pi^4}\right)^{-\frac{1}{2}} \prod_{(j,k)\neq(1,1)}\left[1 - \frac{1}{j^2k^2}\right]^{-\frac{1}{2}}.$$

Making use of the relation $[(\sin z)/z] = \prod_{n=1}^{\infty}\{1 - [z^2/(\pi^2 n^2)]\}$ we see that

$$\prod_{(j,k)\neq(1,1)}\left[1 - \frac{1}{j^2k^2}\right]^{-\frac{1}{2}} = \sqrt{2}\prod_{n=2}^{\infty}\left(\frac{\pi/n}{\sin(\pi/n)}\right)^{\frac{1}{2}}.$$

We have been unable to invert (6.1) directly. However, some of the Tauberian theorems for Laplace transforms (see e.g., [2], p. 269) suggest that if we invert (6.2) we should approximate the tail of our distribution reasonably well. Thus we are led to approximate our distribution in the tail by

$$(6.3) \qquad 2^{\frac{1}{2}}\prod_{n=2}^{\infty}\left(\frac{\pi/n}{\sin(\pi/n)}\right)^{\frac{1}{2}} P\left\{\frac{X^2}{\pi^4} \geq t\right\},$$

where X is a normal random variable with mean zero and unit variance.

The tabulation which follows gives the exact value of $1 - F(y)$ and the corresponding tail approximation, where F is the limiting distribution of $\pi^4 nB_n/2$, tabled in Section 8:

y	1 − F(y)	Tail approximation
2	.145	.115
3	.0414	.0361
4	.0130	.0118
5	.00424	.00395
6	.00142	.00134
7	.00048	.00046

Thus, the agreement is quite good for even moderate values of the size. A similar approximate computation for the asymptotic distribution of the von Mises statistic tabled in [1] also gave good agreement.

7. Some remarks on computations. The computation of the d.f. of a weighted sum of independent chi-square variables, such as that whose characteristic function is given by (2.3) or by (3.5), arises too frequently to require further mention of examples. Unfortunately, the computational techniques now available in the literature for such problems are often extremely poor in applications. While the authors have no panacea to suggest, it does seem appropriate to make a few remarks whose content has proved helpful in considering the computations of the present and other papers (e.g., [6], [7]).

A. *Useful inequalities for estimating truncation error.* In inverting expressions like (2.3), it is usually convenient to work with a finite product, and it is therefore necessary to have a bound on the error introduced by truncating the infinite product. To this end, we consider the random variable

$$Z = \sum_{k=1}^{\infty} c_k Y_k, \tag{7.1}$$

where $c_i > 0$ and the Y_i are independent chi-square variables with one degree of freedom (it will be obvious that the case where Y_i has n_i degrees of freedom can be reduced to this case). We seek an upper bound on the quantity

$$p = P\{Z > \epsilon\}, \tag{7.2}$$

where $\epsilon > 0$. The usual Chebyshev inequality is not very good here, and any of several modifications yields great improvement. The details of one such modification will now be given. We have, for $0 < T < (2 \max_k c_k)^{-1}$,

$$\begin{aligned} p &= P\{e^{TZ} > e^{\epsilon T}\} < e^{-\epsilon T} E e^{TZ} \\ &= \exp\{-\epsilon T - \tfrac{1}{2} \sum_k \log(1 - 2Tc_k)\}. \end{aligned} \tag{7.3}$$

Thus, for given c_k and ϵ, the best bound of this type is achieved by minimizing the expression in braces with respect to T. It is easier to obtain an *explicit* bound by first invoking an inequality such as

$$-\log(1 - 2Tc_k) \leqq -(c_k/c^*) \log(1 - 2Tc^*), \tag{7.4}$$

where $c^* = \max_k c_k$. Substituting the expression on the right side of (7.4) into the last expression of (7.3) and then minimizing with respect to T, and writing $S_j = \sum_k c_k^j$ and $\epsilon = S_1(1 + \delta)$, we obtain

$$P\{Z > (1 + \delta)S_1\} < \exp\{-(S_1/2c^*)[\delta - \log(1 + \delta)]\}. \tag{7.5}$$

This can be improved by using a sharper inequality in place of (7.4). For example, the substitution

$$-\log(1 - 2Tc_k) \leqq Tc_k - (c_k^2/c^{*2})[\log(1 - 2Tc^*) + Tc^*] \tag{7.6}$$

yields, in place of (7.5), the better bound

$$P\{Z > (1 + \delta)S_1\} < \exp\left\{-\frac{S_2}{2c^*}\left[\delta - \frac{S_1}{c^* S_1} \log\left(1 + \frac{c^* S_1 \delta}{S_2}\right)\right]\right\}. \tag{7.7}$$

Further improvements can be made similarly. Of course, the usual Chebyshev inequality is

$$P\{Z > (1 + \delta)S_1\} \leqq 1/(1 + \delta). \tag{7.8}$$

As an example, suppose we want to truncate the product in (2.3) by considering only terms for which $jk \leqq 10$. To estimate the error involved in doing this, we seek an upper bound on p where the set $\{c_k\}$ consists of the λ_{jk} of Section 2 for which $jk > 10$. Routine computations yield $e^{-11\delta^2}$ and $e^{-43\delta^2}$ for the bounds of (7.5) and (7.7), respectively, when δ is small. In any event, we see that ϵ in (7.2) must be between S_1 and $2S_1$ for c_k of this sort, in order to make p fairly small. Since $S_1 = .0043$ and since $EB = .027$ (where B is as in (2.3)), we can only conclude that an approximate computation of the d.f. of B obtained by this truncation, at a value x of the argument, may actually yield the true value of the d.f. at a value as far away as $x + .2EB$, and this would probably be unsatisfactory. A larger truncation value is thus indicated. If the value 10 determining this truncation is increased to L, S_1 varies approximately inversely with L.

Since the ratio of S_1 to EB is the critical factor in determining the adequacy of a truncation in computations like that just mentioned, and since S_1 often decreases very slowly with increasing truncation value in such examples, a large number of terms in the product (2.3) will have to be used for even fair accuracy. An improvement would probably result from substituting for the ignored terms a multiple of a chi-square variable with appropriate low moments, but it seems difficult to *guarantee* an appreciable improvement in accuracy in this way. We shall return to these considerations in Section 7C.

B. Some methods of expansion and inversion. One of the most commonly used techniques for inverting characteristic functions of the form

$$\prod_{j=1}^{k} (1 - a_j it)^{-\frac{1}{2}m_j} \tag{7.9}$$

where the m_j are positive integers and the a_j are positive, is that of Pitman and Robbins [11]. Although this technique and variants of it which represent the solution in slightly different form are sometimes useful, these methods suffer from three defects in many problems: (1) the solution is given in the form of an infinite series which converges rather slowly; (2) the terms of the series are quantities such as incomplete gamma functions, which may not be convenient for some machine computations; (3) the methods do not distinguish simple cases for which a simple inversion in finite terms is possible. For a trivial example of (3), we note that, if $k = 2$, $m_1 = m_2 = 2$, $a_1 = 1$, and $a_2 = 2$, the distribution in question is immediately found by a routine convolution of two exponentials to be $2(e^{-x} - e^{-2x})$, whereas the method of [11] expresses the result as the sum of an infinite series of incomplete gamma functions.

This suggests that it will often be efficient to factor out of the expression (7.9) the corresponding expression wherein each $\frac{1}{2}m_j$ is replaced by its integral part n_j (say), to expand $t^{-1} \prod (1 - a_j it)^{-n_j}$ into partial fractions (the extra factor t^{-1} being introduced so as to give the Fourier transform of the d.f. rather than of the density), and then to invert term by term. Thus, for example, in inverting the expression discussed in the paragraph following (7.8), we can factor out and invert such an expression, leaving only the factors corresponding to λ_{11}, λ_{22}, and λ_{33}; the d.f. corresponding to these terms must then be found by other means and can then be convolved with the d.f. corresponding to the other terms. It should also be noted that the partial fraction technique will often be easy to apply in cases where (7.9) is replaced by an infinite product. For example, the expression $t^{-1}\prod_{j=1}^{\infty}(1+2t/\pi^2j^2)^{-1}$, which is the Laplace transform of the d.f. "B_2" which was computed by other means in Section 4 of [7], can easily be rewritten as $t^{-1}+\sum_{j=1}^{\infty}(-1)^j4/\pi^2j^2(1+2t/\pi^2j^2)$, which we can invert at once to give, for $z>0$,

$$B_2(z) = 1 + 2\sum_{j=1}^{\infty}(-1)^j e^{-\pi^2 j^2 z/2}.$$

(Incidentally, this proves the following interesting relationship: if W_1 and W_2 are independent and each is distributed according to the limiting $n\omega_n^2$ distribution, then $\pi(W_1+W_2)^{\frac{1}{2}}/2$ is distributed according to the limiting Kolmogorov-Smirnov distribution.)

We must still discuss the inversion of general expressions like (7.9) or, with the aid of a factorization like that just discussed, of expressions like (7.9) with all $m_j = 1$. There are many possible expansions akin to that of [11], and for the sake of brevity we shall illustrate only a few such possibilities in the simple case of (7.9) where $k = 2$, $a_1 = 1$, $a_2 = c^2$ with $0 < c < 1$, and $m_1 = m_2 = 1$. Writing t for $-it$ in (7.9) (i.e., working with the Laplace transform), this expression becomes $q(t) = (1+t)^{-\frac{1}{2}}(1+c^2t)^{-\frac{1}{2}}$. Factoring out $(1+ct)^{-1}$, $(1+c^2t)^{-1}$, or $[1+(1+c^2)t/2]^{-1}$, respectively, and then using the binomial expansion on the remaining factor, we obtain the three expressions for $q(t)$,

$$\text{(a)}\quad q(t) = \frac{1}{1+ct}\sum_{j=0}^{\infty}(-1)^j c_j \frac{(1-c)^{2j}t^j}{(1+ct)^{2j}},$$

$$\text{(7.10)}\qquad \text{(b)}\quad q(t) = \frac{c}{1+c^2t}\sum_{j=0}^{\infty} c_j \frac{(1-c^2)^j}{(1+c^2t)^j},$$

$$\text{(c)}\quad q(t) = \frac{1}{1+\left(\dfrac{1+c^2}{2}\right)t}\sum_{j=0}^{\infty} c_j \frac{\left(\dfrac{1-c^2}{4}\right)^j t^{2j}}{\left[1+\left(\dfrac{1+c^2}{2}\right)t\right]^{2j}},$$

where $c_j = (2j)!/2^{2j}(j!)^2$. The second of these corresponds to the method of [11]. Thus we see that various expansions are available which differ in speed of convergence and difficulty of inversion. If suitable partial fraction or other routines are available for inverting the individual terms, an expression like (a) might be useful for some values of c; in other cases, (b) might be satisfactory. Without giving detailed calculations of examples, we can see how ill-advised it is always to use, mechanically, the same routine in every case.

C. Other inversion techniques. Because of the large number of terms which must be kept in (2.3) in order to obtain reasonable accuracy (as discussed in Section 7A) when applying the techniques we have discussed, and because of the other shortcomings of these methods (see Section 7B), it is reasonable to investigate other inversion techniques. For example, in the problem of Section 2, if we first take the product with respect to k, we obtain $\prod_j \{\sinh [(2\pi)^{\frac{1}{2}}/\pi j]/[(2\pi)^{\frac{1}{2}}/\pi j]\}^{-\frac{1}{2}}$ for the Laplace transform, and one can try various manipulations with this expression. Another possibility, which seems more fruitful in this and many other problems, is that of direct numerical integration to invert the expression of (2.3).

In order to perform such an integration, one must first tabulate the function (2.3) for various values of the argument. A method which seems to be much more efficient than that of directly multiplying together an appropriately large number of terms of the product is to use the fact that, in a neighborhood of $v = 0$, we have

$$-\tfrac{1}{2} \sum_{i,j \geqq h} \log\left(1 + \frac{v}{i^2 j^2}\right) = \sum_{k=1}^{\infty} a_k v^k, \tag{7.11}$$

where $a_k = (-1)^k \left(\sum_{j \geqq h} j^{-2k}\right)^2/2k$ (these coefficients can be written in terms of Bernoulli numbers). On the basis of preliminary estimates of

$$g(v) = v^{-1} \prod_{i,j} (1 + v/i^2 j^2)^{-\frac{1}{2}}$$

on the proposed line of integration, the value of h can be chosen so as to make the series (7.11) convergent over that (finite) portion of the line where the integration will actually be performed. The series can then be evaluated for appropriate complex v, exponentiated, and the result multiplied by the remaining factor of $g(v)$, which can be expressed in terms of hyperbolic sines and of powers of linear functions of v. The numerical integration can then be performed. This was the method used to obtain the tables of Section 8.

A recent paper by Grenander, Pollak, and Slepian [3] discusses an interesting computational technique for obtaining an approximation to limiting distributions such as those discussed above by solving a set of linear equations whose solution approximates that of an integral equation for the limiting d.f. or c.f. The reader is referred to [3] for details and related discussion.

8. Tables. The inversion of (2.3) was carried out by the method outlined in the second paragraph of Section 7C, which was calculated to require much less machine time than any of the other available methods. The authors are grateful to Professor R. J. Walker for carying out the computations on the Cornell Computing Center's 220. Table I gives values (under H_0) of

$$F(x) = \lim_{n \to \infty} P\{\tfrac{1}{2}\pi^4 n B_n \leqq x\},$$

while Table II gives values of $F^{-1}(p)$.

It is not very difficult to program a computing machine to evaluate the statistic B_n or the modifications of it mentioned in Section 4. It may be worthwhile, especially for small n, to reduce the error introduced when using the

TABLE I

$$F(y) = \lim_{n\to\infty} P_{H_0}\{\tfrac{1}{2}\pi^4 nB_n \leqq y\}$$

y	$F(y)$	y	$F(y)$	y	$F(y)$
.30	.00000	2.10	.87275	3.90	.98546
.35	.00010	2.15	.88084	3.95	.98627
.40	.00086	2.20	.88835	4.00	.98702
.45	.00389	2.25	.89534	4.05	.98774
.50	.01158	2.30	.90185	4.10	.98841
.55	.02614	2.35	.90791	4.15	.98905
.60	.04867	2.40	.91357	4.20	.98965
.65	.07899	2.45	.91885	4.25	.99022
.70	.11594	2.50	.92377	4.30	.99075
.75	.15784	2.55	.92838	4.35	.99126
.80	.20293	2.60	.93268	4.40	.99174
.85	.24960	2.65	.93670	4.45	.99219
.90	.29652	2.70	.94047	4.50	.99261
.95	.34267	2.75	.94400	4.55	.99301
1.00	.38730	2.80	.94730	4.60	.99339
1.05	.42994	2.85	.95039	4.65	.99375
1.10	.47027	2.90	.95329	4.70	.99409
1.15	.50816	2.95	.95602	4.75	.99441
1.20	.54354	3.00	.95857	4.80	.99471
1.25	.57645	3.05	.96097	4.85	.99499
1.30	.60697	3.10	.96322	4.90	.99527
1.35	.63521	3.15	.96533	4.95	.99552
1.40	.66131	3.20	.96732	5.00	.99576
1.45	.68540	3.25	.96918		
1.50	.70763	3.30	.97094	5.50	.99755
1.55	.72813	3.35	.97259	6.00	.99858
1.60	.74704	3.40	.97414	6.50	.99918
1.65	.76449	3.45	.97561	7.00	.99952
1.70	.78060	3.50	.97698	7.50	.99972
1.75	.79547	3.55	.97828	8.00	.99983
1.80	.80922	3.60	.97949	8.50	.99990
1.85	.82193	3.65	.98064	9.00	.99994
1.90	.83369	3.70	.98172	9.50	.99997
1.95	.84459	3.75	.98274	10.00	.99998
2.00	.85469	3.80	.98370	10.50	.99999
2.05	.86406	3.85	.98461	11.00	1.00000

TABLE II

p	$F^{-1}(p)$	p	$F^{-1}(p)$
.9	2.286	.998	5.68
.95	2.844	.999	6.32
.98	3.622	.9995	6.96
.99	4.230	.9998	7.82
.995	4.851	.9999	8.47

limiting d.f. (in particular, in the limiting covariance function) by using $(n-1)B_n$ instead of nB_n.

REFERENCES

[1] T. W. ANDERSON AND D. A. DARLING, "Asymptotic theory of certain 'goodness of fit' criteria based on stochastic processes," *Ann. Math. Stat.*, Vol. 23 (1952), pp. 193–212.

[2] G. DOETSCH, *Theorie und Anwendung der Laplace Transformation*, Dover Pub., New York, 1943.

[3] U. GRENANDER, H. O. POLLAK, AND D. SLEPIAN, "The distribution of quadratic forms in normal variates," *J. S. I. A. M.*, Vol. 7 (1959), pp. 374–401.

[4] WASSILY HOEFFDING, "A nonparametric test of independence," *Ann Math. Stat.*, Vol. 19 (1948), pp. 546–557.

[5] M. KAC, "On some connections between probability theory and differential and integral equations," *Proceedings of the Second Berkely Symposium of Mathematical Statistics and Probability*, University of California Press, 1951, pp. 180–215.

[6] M. KAC, J. KIEFER, AND J. WOLFOWITZ, "On tests of normality and other tests of goodness of fit based on distance methods," *Ann. Math. Stat.*, Vol. 26 (1955), pp. 189–211.

[7] J. KIEFER, "*K*-sample analogues of the Kolmogorov-Smirnov and Cramér-v. Mises tests," *Ann. Math. Stat.*, Vol. 30 (1959), pp. 420–447.

[8] J. KIEFER AND J. WOLFOWITZ, "On the deviations of the empiric distribution function of vector chance variables," *Trans. Amer. Math. Soc.*, Vol. 87 (1958), pp. 173–186.

[9] J. KIEFER AND J. WOLFOWITZ, "Asymptotic minimax character of the sample distribution function for vector chance variables," *Ann Math. Stat.*, Vol. 30 (1959), pp. 463–489.

[10] H. B. MANN AND A. WALD, "On the choice of the number of class intervals in the application of the chi square test," *Ann. Math. Stat.*, Vol. 13 (1942), pp. 306–317.

[11] E. J. G. PITMAN AND HERBERT ROBBINS, "Application of the method of mixtures of quadratic forms in normal variates," *Ann. Math. Stat.*, Vol. 20 (1949), pp. 552–560.

[12] M. ROSENBLATT, "Limit theorems associated with variants of the von Mises statistic," *Ann. Math. Stat.*, Vol. 23 (1952), pp. 617–623.

ON LARGE DEVIATIONS OF THE EMPIRIC D.F. OF VECTOR CHANCE VARIABLES AND A LAW OF THE ITERATED LOGARITHM

J. KIEFER

1. Introduction and preliminaries. Let F be a distribution function (d.f.) on m-dimensional Euclidean space R^m, and let $X_1, \cdots, X_n$ be independent chance vectors with common d.f. F. The empiric d.f. S_n is a chance d.f. on R^m defined as follows: if $x = (x_1, \cdots, x_m)$, $nS_n(x)$ is the number of X_i's, $1 \leq i \leq n$, such that, for $j = 1, \cdots, m$, the jth component $X_i^{(j)}$ of X_i is less than or equal to x_i.

When $m = 1$, the distribution of $D_n = \sup_x |S_n(x) - F(x)|$ is the same for all continuous F, and Kolomogorov [5] first computed the limiting distribution of $n^{1/2}D_n$ as $n \to \infty$. Chung [1] gave a bound on the error term which was sharp enough to yield a law of the iterated logarithm for the empiric d.f. and, in fact, the more precise complete characterization of monotone functions of upper and lower class. (The more recent literature contains several asymptotic expansions of Kolmogorov's distribution.) It was proved by Dvoretzky, Kiefer and Wolfowitz [2] that there is a universal constant C such that, for all $n > 0$ and $r \geq 0$,

$$P\{n^{1/2}D_n \geq r\} \leq Ce^{-2r^2}; \tag{1.1}$$

since $\lim_n P\{n^{1/2}D_n \geq r\}$ is asymptotically $2e^{-2r^2}(1 + o(1))$ as $r \to \infty$, the estimate (1.1) cannot be improved upon in this general form.

Much less is known when $m > 1$. The limiting d.f. of $n^{1/2}D_n$ was proved to exist by Kiefer and Wolfowitz [4]; of course, its form depends on F (and is unknown except in a few trivial cases), unlike the case $m = 1$. It was also proved in [4] that there exist positive constants c_m and c'_m such that, for all F, $n > 0$, and $r \geq 0$,

$$P_F\{n^{1/2}D_n \geq r\} \leq c'_m e^{-c_m r^2}, \tag{1.2}$$

whereby $P_F\{A\}$ we denote the probability of the event A when X_1 has d.f. F. Possible choices of the constants c_m were shown to be

$$c_2 = .0157,\ c_3 = .000107,\ \cdots\ (\lim_m c_m = 0). \tag{1.3}$$

It was also shown in [4] that, for $m > 1$, one cannot have $c_m = 2$ in (1.2); specifically, if $m = 2$ and F^* distributes probability uniformly on the line segment $\{(x_1, x_2): x_1 \geq 0, x_2 \geq 0, x_1 + x_2 = 1\}$, then as $r \to \infty$ we have

$$\lim_n P_{F^*}\{n^{1/2}D_n \geq r\} = 8r^2e^{-2r^2}(1 + o(1)). \tag{1.4}$$

Received May 9, 1960. Research sponsored by the Office of Naval Research.

Thus, even for a single F we cannot hope to achieve (1.1) in the case $m > 1$.

The main object of the present paper is to prove the next best thing, namely,

Theorem 1-m. *For each m and $\varepsilon > 0$ there is a constant $c(\varepsilon, m)$ such that, for all F, $n > 0$, and $r \geqq 0$, we have*

$$P_F\{n^{1/2}D_n \geqq r\} \leqq c(\varepsilon, m)e^{-(2-\varepsilon)r^2} . \tag{1.5}$$

(The labeling of the Theorem as 1-m is to make clear the induction on m in the proof.)

The result (1.5) clearly represents a marked improvement over (1.3), and in view of (1.4) this result (1.5) is the best possible of this form. Whether or not the ε in the exponent can be improved, e.g., to a term like $cr^{-2} \log r$ when $m = 2$, as in (1.4), is not known, and the methods of the present paper do not seem capable of shedding any light on the subject.

The weaker result which is obtained by replacing the left side of (1.5) by its limit as $n \to \infty$ can be proved more quickly, and the reader will have no difficulty in recognizing how the proof of the present paper can be shortened if only that result on the corresponding limiting Gaussian process (see [4]) is desired.

Theorem 1 yields an easy proof of a law of the iterated logarithm:

Theorem 2. *For every m and every continuous F,*

$$P_F\{\limsup_{n\to\infty} n^{1/2}D_n / (2^{-1} \log \log n)^{1/2} = 1\} = 1 . \tag{1.6}$$

(The same result holds for $D_n^+ = \sup_x [S_n(x) - F(x)]$ or for $D_n^- = \sup_x [F(x) - S_n(x)]$.)

In fact, the conclusion that the upper limit is $\geqq 1$ follows at once on applying Chung's result to a one-dimensional marginal d.f. and empiric d.f. On the other hand, the proof that, if $\lambda > 2^{-1}$, the probability is zero that $n^{1/2}D_n \geqq (\lambda \log \log n)^{1/2}$ for infinitely many n, is proved in a classical way. For example, the proof on page 48 of [1] requires only trivial modifications to apply to the present case, using the estimate (1.5) for ε sufficiently small and the fact that for some positive constants b and c we have $P_F\{D_n \leqq bn^{-1/2}\} \geqq c > 0$ for all n; the latter is an immediate consequence of Theorem 1-m or of the results of [4]. It is unnecessary to give more details of the proof of Theorem 2.

Obviously, our estimate (1.5) is not precise enough to yield a finer result analogous to Chung's. In fact, it is clear that the value of λ which divides the functions $[2^{-1}\log \log n + \lambda \log \log \log n]^{1/2}$ into upper and lower classes depends on F. For example, when $m = 2$ and $F = F^*$, the d.f. of $n^{1/2}D_n$ is that of $n^{1/2}(D_n^+ + D_n^-)$ for $m = 1$ (see [4]) and, following Chung in obtaining an error term in the approach to the limiting distribution and in the characterization of the upper and lower classes, we obtain 3/2 for the critical value of λ. On the other hand, if F distributes probability uniformly on the main diagonal $x_1 = x_2$ of the unit square, D_n has the same distribution as in the case $m = 1$, and Chung's result then yields $\lambda = 1$ as the critical value.

The method used to prove Theorem 1-m is an improvement of the method used to prove Theorem 1 of [4], the line of argument being similar. Lemma 2 extends equation (2.4) of [4], where the case $k = 2$, $j = 1$ was considered. Lemma 3-m improves the estimate of equation (2.5) of [4]. Lemma 1 is needed in order to obtain, in Lemma 6-m, an improvement on the estimate of equation (2.24) of [4]. However, the present paper is self-contained, and we shall not make use of the results of [2] or [4] in the proof.

The idea of using an argument like that of Lemma 2 is well known in such a context as that of the study of the maximum of partial sums of independent summands, where it is of course much easier to apply than in the case of "tieddown" processes such as the random functions $S_n(x) - F(x)$ studied here. In fact, it is just as easy to obtain such results for processes with independent increments in the case of m-dimensional "time" as it is in the classical case of one-dimensional time. For example, if $X_{i_1, i_2, \cdots, i_m}$, $1 \leqq i_j \leqq n_j$, $1 \leqq j \leqq m$, are independent random variables which are symmetric about 0 (these assumptions are easily relaxed) and

$$S_{i_1, i_2, \cdots, i_m} = \sum_{\substack{t_j \leqq i_j \\ 1 \leqq j \leqq m}} X_{t_1, t_2, \cdots, t_m},$$

we obtain, on using the classical argument m times,

$$P\{\max_{\substack{1 \leqq i_j \leqq n_j \\ 1 \leqq j \leqq m}} S_{i_1, \cdots, i_m} \geqq r\} \leqq 2^m P\{S_{n_1, \cdots, n_m} \geqq r\}.$$

Similary, the standard semi-martingale inequality on $E\{I_A U_n\}$, where $U_n = \max_{i \leqq n} Y_i$ (with $\{Y_i\}$ a semi-martingale), $A = \{U_n \geqq r\}$, and I_A is the indicator of A, has an obvious analogue for semi-martingales with m-dimensional time. Such results also carry over to the case of m-dimensional continuous time; for example, if $Y(t_1, \cdots, t_m)$ is a separable Gaussian process (m-dimensional time) with mean 0, independent stationary increments for t in the positive orthant, and $Y(t) = 0$ if t is outside the positive orthant, we have

$$P\{\sup_{\substack{0 \leqq t_i \leqq T_i \\ 1 \leqq i \leqq m}} Y(t_1, \cdots, t_m) \geqq r\} \leqq 2^m P\{Y(T_1, \cdots, T_m) \geqq r\}.$$

The results of the present paper yield inequalities for other processes with m-dimensional time, which require somewhat greater effort.

In the proof of Theorem 1-m we require one rather elementary result on the tails of the binomial distribution. Let $X_1, X_2, \cdots$ be independent random variables with $P_p\{X_1 = 1\} = 1 - P_p\{X_1 = 0\} = p$, where $0 < p < 1$ and we now use the subscript p to designate the underlying probability law. Write $\bar{X}_n = \sum_{i \leq n} X_i/n$. It is well known that, for some $C > 0$,

$$P_p\{n^{1/2} |\bar{X}_n - p| \geqq r\} < Ce^{-2r^2} \tag{1.7}$$

for all $n > 1$, $r \geqq 0$, and p. In fact, (1.7) is much weaker than (1.1). The central limit theorem implies that, for p and r fixed, the limit of the left side of (1.7) as $n \to \infty$ is less than $C \exp\{-r^2/p(1-p)\}$, but it is well known (see the next paragraph or [6], p. 285) that the 2 in the exponent on the right side of (1.7) cannot be replaced by $c'/p(1-p)$ for all n, p, and r, where $c' > 0$. What we require is that the right side of (1.7) can be replaced by $C_p \exp\{-r^2 g(p)\}$, where $g(p) \to \infty$ as $p \to 0$.

We shall prove this inequality with $g(p) = \log(p^{-1}e^{-1})$ for $p < e^{-1}$. The factor e^{-1} in the expression for $g(p)$ can be improved slightly. However, the result cannot be improved by very much, since for $r = n^{1/2}(1-p)$ we obtain $\exp\{-r^2(\log p^{-1})(1-p)^{-2}\}$ for the left side of (1.7); thus, $g(p)$ cannot be taken to be $\varepsilon + \log p^{-1}$ with $\varepsilon > 0$. If $g(p)$ is allowed to depend upon r, it is easy to obtain a better bound. This is also true if one desires separate inequalities for the positive and negative deviations. But for $r \sim n^{1/2}(1-p)$ and $p \to 0$, it is again true that little improvement over our bound is possible, for the positive deviations.

LEMMA 1. *For $0 < p < e^{-1}$, all $n > 0$, and all $r \geqq 0$,*

$$P_p\{n^{1/2} |\bar{X}_n - p| \geqq r\} < C_p \exp\{-r^2[\log p^{-1} - 1]\}, \tag{1.8}$$

where C_p depends only on p.

Proof. We proceed along classical lines. In fact, for the negative deviations we need only note, putting $p' = 1 - p$, that

$$P_p\{n^{1/2}(\bar{X}_n - p) \leqq -r\} = P_{p'}\{n^{1/2}(\bar{X}_n - p') \geqq r\}, \tag{1.9}$$

and from [3], equations (VI. 10.12) and (VI. 3.5), we obtain easily that this last probability is bounded by

$$\begin{aligned} C'_p \exp\{-r^2/2(1-p')\} &= C'_p \exp\{-r^2/2p\} \\ &\leqq C'_p \exp\{-r^2[\log p^{-1} - 1]\}, \end{aligned} \tag{1.10}$$

where here and in the sequel C_p, C'_p, and C''_p depend only on p but may change meaning in different appearances. We need a slightly better bound than that just cited, for the positive deviations. Writing $b(K; n, p) = P_p\{n\bar{X}_n = k\}$, and letting the integer h be defined by $(n+1)p - 1 < h \leqq (n+1)p$, we have, for $k > h$,

$$(1.11)\quad \log \frac{b(k; n, p)}{b(h; n, p)} = \sum_{j=h+1}^{k} \log \frac{b(j; n, p)}{b(j-1; n, p)} = \sum_{j=h+1}^{k} \log \frac{p(n-j+1)}{(1-p)j}.$$

Since the summand is decreasing in j, we may bound the sum from above by integrating directly, with respect to j, from h to k. On combining terms, we obtain, for $n \geqq C'_p$ (which is all we need consider),

$$(1.12)\quad \log \frac{b(k; n, p)}{b(h; n, p)} < (k-h) \log \frac{p(n+1-h)}{(1-p)h}$$

$$+ (n+1-k) \log \frac{n+1-h}{n+1-k} - k \log \frac{k}{h} < C''_p + (k-h) - k \log \frac{k}{h}.$$

Consider the function

$$(1.13)\quad g(x) = x - (x + p_1) \log \left(1 + \frac{x}{p_1}\right) + \frac{\log (p_1^{-1} e^{-1+p_1})}{(1-p_1)^2} x^2$$

on the interval $[0, 1-p_1]$. We see easily that $g(0) = g(1-p_1) = 0$, that g is concave and decreasing for

$$0 \leqq x \leqq [(1-p_1)^2/2 \log (p_1^{-1} e^{-1+p_1})] - p_1,$$

and that g is convex beyond this last point. We conclude that $g(x) \leqq 0$ on $[0, 1-p_1]$. Putting $x = (k-h)/n$ and $p_1 = h/n$, we conclude that the expression (1.12) is no greater than

$$(1.14)\quad C''_p - \frac{[k-h]^2 \log (p_1^{-1} e^{-1})}{n}$$

for $h < k \leqq n$. Finally, since $|p_1 - p| < n^{-1}$, equation (1.14) yields

$$(1.15)\quad \log \frac{b(k; n, p)}{b(h; n, p)} < C''_p - \frac{[k-h]^2 \log (p^{-1} e^{-1})}{n}.$$

Equations (1.10) and (1.15), together with the well known estimate $\sum_{j \geqq 0} b(k+j; n, p) < b(k; n, p)(k+1)(1-p)/[k+1-(n+1)p]$ relating tail probabilities to individual terms ([5], equation (VI. 3.5)), immediately yield (1.8).

2. Proof of Theorem 1-*m*. As indicated in [4], for any discontinuous d.f. F there is a continuous $\bar{F}$ for which the left side of (1.5) is at least as large. Hence, we can and do assume that *F is continuous in all that follows.* We can then transform the coordinates one at a time, without changing the d.f. of D_n, so that *the marginal* (*one-dimensional*) d.f. *of* $X_1^{(j)}, 1 \leqq j \leqq m$, *is uniform on the interval* [0, 1]; we hereafter assume F to be of this form. We denote the unit m-cube, in which all X_i thus fall with probability one (w.p.1), by I^m.

In all that follows we need only consider values r for which $R(\varepsilon, m) < r \leqq n^{1/2}$, where $R(\varepsilon, m)$ is a fixed positive number. For $P_F(n^{1/2}D_n > n^{1/2}) = 0$, and if (1.5) is true for $r > R(\varepsilon, m)$ we can increase $c(\varepsilon, m)$ to make it hold for all $r \geqq 0$.

Let k be a real number $\geqq 2$ and let j be an integer, $1 \leqq j \leqq k-1$. Define

$$V_{jk} = \{(x_1, \cdots, x_m):\ (j-1)/k \leqq x_1 \leqq j/k\}\ ,$$
$$W_{jk} = \{(x_1, \cdots, x_m):\ x_1 = j/k\}\ .$$

We first show that if $S_n(x) - F(x)$ is $< rn^{-1/2}$ everywhere on $W_{j-1,k}$ and $\geqq rn^{-1/2}$ somewhere on the slab V_{jk}, then there is appreciable conditional probability that it is almost this large somewhere on the hyperplane W_{jk}. To this end, define the events

$$B_{jk}(r) = \left\{\sup_{V_{jk}}[S_n(x) - F(x)] \geqq rn^{-1/2}\right\}, \tag{2.1}$$
$$C_{jk}(r) = \left\{\sup_{W_{jk}}[S_n(x) - F(x)] \geqq rn^{-1/2}\right\}.$$

Denote the complement of an event C by $\bar{C}$.

LEMMA 2. *We have, for* $2 \leqq r \leqq n^{1/2}$, $k \geqq 2$, *and* $1 \leqq j \leqq k-1$,

$$P_F\left\{C_{jk}\left(r\left(1 - \frac{2}{k-j}\right)\right) \mid \bar{C}_{j-1,k}(r) B_{jk}(r)\right\} \geqq 1 - \frac{(k-j)^2}{r^2}\ . \tag{2.2}$$

Proof. If $B_{jk}(r)$ occurs, there is w.p.1 a smallest value x_1' of x_1 $(j-1 \leqq kx_1' \leqq j)$ for which $S_n(x) - F(x) \geqq rn^{-1/2}$, for some $x_2, \cdots, x_m$; for $x_1 = x_1'$, there is then a smallest value x_2' of x_2 for which this inequality is satisfied, and so on. Thus, we obtain a well-defined random vector $X' = (x_1', \cdots, x_m')$ for which $S_n(X') - F(X') \geqq rn^{-1/2}$ whenever $B_{jk}(r)$ occurs. Moreover, the event $\{x_1' = a_1, \cdots, x_m' = a_m\}$ depends only on $\{S_n(x), x'_1 \leqq a_1\}$. When $a = (a_1, \cdots, a_m)$, write $\bar{a} = (a_2, \cdots, a_m)$. We shall now prove that, for $j - 1 \leqq ka_1 \leqq j$,

$$P_F\left\{S_n(j/k, \bar{a}) - F(j/k, \bar{a}) \geqq r\left(1 - \frac{2}{k-j}\right)n^{-1/2} \mid X' = a;\ S_n(a) - F(a) \geqq rn^{-1/2};\ \bar{C}_{j-1,k}(r)\right\} \geqq 1 - \frac{(k-j)^2}{r}\ , \tag{2.3}$$

which clearly implies (2.2).

Suppose the event conditioning (2.3) occurs. Since $\bar{C}_{j-1,k}(r)$ occurs, a_1 is actually $> (j-1)/k$, and there is w.p.1 at most one X_i with first coordinate a_1. Hence, if $\bar{l}$ denotes a vector consisting of $m-1$ ones, we have

$$nS_n(a_1, \bar{l}) \leqq nF(a_1, \bar{l}) + rn^{1/2} + 1\,. \tag{2.4}$$

Hence, the number N of $X_1, \cdots, X_n$ which have first coordinate greater than a_1 is at least

$$M = n(1-a_1) - rn^{1/2} - 1\,. \tag{2.5}$$

If $M \leqq 0$, we thus have $rn^{-1/2} \geqq -n^{-1} + (k-j)/k$, from which we obtain, for $r \geqq 1$,

$$\begin{aligned} &S_n(j/k, \bar{a}) - F(j/k, \bar{a}) \\ &\geqq [S_n(a_1, \bar{a}) - F(a_1, \bar{a})] + [F(a_1, \bar{a}) - F(j/k, \bar{a})] \\ &\geqq rn^{-1/2} - k^{-1} \geqq \left(1 - \frac{1}{k-j}\right) rn^{-1/2} - n^{-1}(k-j)^{-1} \\ &\geqq \left(1 - \frac{2}{k-j}\right) rn^{-1/2}\,. \end{aligned} \tag{2.6}$$

On the other hand, if $M > 0$ we have $N > 0$. Let Q be the event that, of the N random variables X_i whose first coordinate is $> a_1$, at least

$$n[F(j/k, \bar{a}) - F(a_1, \bar{a})] - 2n^{1/2}(k-j)^{-1}r$$

take on values in the region

$$\{(x_1, \cdots, x_m) : a_1 < x_1 \leqq j/k;\ x_2 \leqq a_2, \cdots, x_m \leqq a_m\}\,.$$

Clearly, if we show that, for ν equal to any integer $\geqq M$,

$$\begin{aligned} P_F\{Q \mid X' = a;\ S_n(a) - F(a) \geqq rn^{-1/2};\ \bar{C}_{j-1,k}(r);\ N = \nu\} \\ \geqq 1 - \frac{(k-j)^2}{r^2}\,, \end{aligned} \tag{2.7}$$

this together with (2.6) yield (2.3).

Define

$$p = \frac{F(j/k, \bar{a}) - F(a_1, \bar{a})}{1 - a_1}\,.$$

If $p = 0$, (2.7) is trivial. We therefore assume $p > 0$, and define

$$t = \frac{n[F(j/k, \bar{a}) - F(a_1, \bar{a})] - 2n^{1/2}(k-j)^{-1}r - \nu p}{[\nu p(1-p)]^{1/2}}\,.$$

Since $\nu \geqq M$, (2.5) implies that

$$t \leqq \frac{r[p - 2(k-j)^{-1}] + pn^{-1/2}}{[\nu p(1-p)n^{-1}]^{1/2}}\,.$$

Since the event $X' = a$ depends only on $\{S_n(x), x_1 \leqq a_1\}$, the probability (2.7) is simply the probability that, of ν independent Bernoulli trials with probability p of a "success," the number Y of successes satisfies

$$\frac{Y - EY}{[E(Y-EY)^2]^{1/2}} \geqq t\,.$$

Since $p \leqq (k-j)^{-1}$ and $\nu/n \leqq 1$, it follows that $t \leqq -r(k-j)^{-1}$ for $r \geqq 2$. Hence, (2.7) follows at once from Chebyshev's inequality.

Next, let $B'_{jk}(r)$ and $C'_{jk}(r)$ be defined by replacing $S_n(x) - F(x)$ by $F(x) - S_n(x)$ on the right side of (2.1). In a manner almost identical to that used to prove Lemma 2, we obtain

LEMMA 2′. *We have, for* $r \geqq 2$,

$$P_F\left\{C'_{jk}\left(r\left(1 - \frac{2}{k-j}\right)\right) \middle| \bar{C}'_{j-1,k}(r); B'_{jk}(r)\right\} \geqq 1 - \frac{(k-j)^2}{r^2}\,. \tag{2.8}$$

In fact, this is even easier, since the case $N = 0$ now requires no calculation. We replace (2.5) by an upper bound on N, Q by the event that at most $n[F(j/k, \bar{a}) - F(a_1, \bar{a})] + 2n^{1/2}r(k-j)^{-1}$ observations fall in the region of probability p, obtain a positive lower bound on t, and thus a lower bound on $P\{Y - EY \leqq t\,[E(Y - EY^2]^{1/2}\}$.

Our next lemma requires an induction on m, which we exhibit in its number.

LEMMA 3-m. *For* $\varepsilon > 0$ *there is a number* $c_1(\varepsilon, m)$ *such that, for all* F, n, $r \geqq 0$, $k \geqq 2$, *and* $j \leqq k$,

$$P_F\{C_{jk}(r) \cup C'_{jk}(r)\} < c_1(\varepsilon, m)e^{-(2-\varepsilon)r^2}\,. \tag{2.9}$$

Proof. Of course, for $m = 1$ the result (2.9) is weaker than (1.7) (or (1.1)). For $m > 1$, consider $S_n - F$ on the set $W_{jk} \cup \{x : x_1 > j/k, x_2 = 1\}$. This clearly has the same distribution theory as an $(m-1)$-dimensional sample d.f. minus the corresponding continuous d.f. The desired result thus follows from Theorem 1-$(m-1)$.

Since

$$(2.10)\qquad P_F\{B_{jk}(\alpha)\} \leqq \frac{P_F\{C_{jk}(\beta)\}}{P_F\{C_{jk}(\beta)\,|\,C_{j-1,k}(\alpha);\,B_{jk}(\alpha)\}} + P_F\{C_{j-1\,k}(\alpha)\}\,.$$

we obtain at once from Lemmas 2, 2′, and 3-m,

LEMMA 4-m. *For $\varepsilon' > 0$ there is a number $c_2(\varepsilon', m)$ such that, for all F, n, $k \geqq 2$, $j \leqq k-1$, and $r \geqq 2(k-j)$,*

$$(2.11)\qquad P_F\Big\{\sup_{V_{jk}} |S_n(x) - F(x)| \geqq rn^{-1/2}\Big\}$$

$$\leqq c_2(\varepsilon', m)\exp\Big\{-r^2\Big(1-\frac{2}{k-j}\Big)^2(2-\varepsilon')\Big\}\,.$$

It now becomes evident that, by taking k large, we can prove Theorem 1-m by using the estimate (2.11) for those V_{jk} for which $k-j$ is large, provided we can also find an appropriate estimate for the region where $k-j$ is small, i.e., near $x_1 = 1$. Actually, we shall see that it suffices to find such an estimate for the region where x is close to $(1, 1, \cdots, 1)$, and the appropriate estimate is obtained in Lemma 6-m below. We first require a preliminary result which essentially improves the estimate (2.11) when j is small.

LEMMA 5-m. *There is a number $c_3(m)$ such that, for all F, $n \geqq 1$, $k \geqq 3$, and $r \geqq 2k[1-2/(k-1)]^{-m}$,*

$$(2.12)\qquad P_F\Big\{\sup_{V_{1k}} |S_n(x) - F(x)| \geqq rn^{-1/2}\Big\}$$

$$\leqq c_3(m)\exp\Big\{-r^2\Big[1-\frac{4m}{k-1}\Big]\log(ke^{-1})\Big\}\,.$$

Proof. Let $T_m = V_{1k}$, $T_{m-1} = W_{1k}$, $T_{m-q} = \{x\colon x_1 = k^{-1}, x_2 = 1, \cdots, x_q = 1\}$ for $q > 1$. As in the proof of Lemma 3-m, we see that $S_n - F$ on T_{m-q} has the same distribution theory as an $(m-q)$-dimensional sample d.f. minus the corresponding continuous d.f. with uniform marginal d.f.'s, on the subset of the $(m-q)$-cube where the first coordinate is $\leqq k^{-1}$. Hence, applying the argument (2.10) m times, where the last term on the right side of (2.10) is now zero, where in each successive application we use Lemma 2 or 2′ to obtain 3/4 as a lower bound on the denominator of the expression on the right side, and where in the qth application the right side numerator and left side of (2.10) are, respectively, probabilities of maximum deviations greater than or equal to $r[1-2/(k-1)]^q n^{-1/2}$ on T_{m-q} and greater than or equal to $r[1-2/(k-1)]^{q-1}n^{-1/2}$ on T_{m-q+1}, we obtain

$$(2.13)\quad P_F\left\{\sup_{V_{1k}} |S_N(x) - F(x)| \geqq rn^{-1/2}\right\} \leqq \left(\frac{4}{3}\right)^m \phi(k^{-1}, n, r[1 - 2/(k-1)]^m),$$

where $\phi(p, n, z)$ is the probability that the number Z of successes in n independent Bernoulli trials with probability p of success satisfies

$$|Z - np| \geqq zn^{1/2} .$$

Lemma 1 and (2.13) thus imply (2.12).

We are now ready to prove that large deviations of $|S_n(x) - F(x)|$ have suitably small probability if x lies in the region

$$U_\delta = \{x : 1 - \delta \leqq x_q \leqq 1,\ 1 \leqq q \leqq m\}$$

and δ is a small positive number.

LEMMA 6-m. *There is a number* $c_4(m)$ *such that, for all* F, n, $\delta \leqq 1/4$, *and* $r \geqq 2\delta^{-1}[1 - 3\delta]^{-m}$, *we have*

$$(2.14)\quad \begin{aligned} &P_F\left\{\sup_{U_\delta} |S_n(x) - F(x)| \geqq rn^{-1/2}\right\} \\ &\leqq c_4(m)\exp\{-r^2 4^{-m}(1 - 6m\delta)\log(\delta^{-1}e^{-1})\} . \end{aligned}$$

Proof. For a in U_δ, let $Q_a = \{x : x_q \leqq a_q,\ 1 \leqq q \leqq m\}$, and for any sequence $\sigma = (\sigma_1, \cdots, \sigma_m)$ of 1's and -1's, not all 1's, let

$$Q_a(\sigma) = \{x : x \in I^m;\ \sigma_q x_q < \sigma_q a_q,\ 1 \leqq q \leqq m\} .$$

Thus, Q_a and the $2^m - 1$ sets $Q_a(\sigma)$ are disjoint sets whose union is I^m minus the union of m hyperplanes $\{x : x_q = a_q\}$. Define the event

$$D_a(\sigma, s) = \left\{\left|[\text{number of } X_1, \cdots, X_n \text{ falling in } Q_a(\sigma)] - n\int_{Q_a(\sigma)} dF\right| \geqq n^{1/2}s\right\}.$$

If the event

$$B_\delta(r) = \left\{\sup_{U_\delta} |S_n(x) - F(x)| \geqq rn^{-1/2}\right\}$$

occurs, we define, in the manner of the definition of the random vector X' of the proof of Lemma 2, a random vector X'' such that $X'' \in U_\delta$ and $|S_n(X'') - F(X'')| \geqq rn^{-1/2}$. With probability one there is at most one X_i with any coordinate equal to the corresponding coordinate of X'', so that w.p. 1 the event $B_\delta(r) \cap \{X'' = a\}$ entails the event

$$(2.15)\quad \bigcup_\sigma D_a(\sigma, r/(2^m - 1)).$$

For any fixed σ, let $Y_i = (Y_i^{(1)}, \cdots, Y_i^{(m)})$ be defined by $Y_i^{(q)} = X_i^{(q)}$ (resp., $= 1 - X_i^{(q)}$) if $\sigma_q = 1$ (resp., $= -1$). Let G be the d.f. of Y_1 when F is the d.f. of X_1, and let S_n' be the sample d.f. of $Y_1, \cdots, Y_n$.

Let $b_q(\sigma, \delta) = 1 - \delta$ (rep., δ) if $\sigma_q = 1$ (resp., $= -1$), and let $V(\sigma, \delta) = \{x: 0 \leqq x_q \leqq b_q(\sigma, \delta),\ 1 \leqq q \leqq m\}$. The event $\bigcup_{a \in U_\delta} D_a(\sigma, s)$ is, w.p. 1, equivalent to the event

$$\left\{\sup_{V(\sigma,\delta)} |S'_n(x) - G(x)| \geqq s\right\}. \tag{2.16}$$

Since at least one σ_q is -1, at least one $b_q(\sigma, \delta)$ equals δ. Hence, if $k \geqq \delta^{-1}$, the set $V(\sigma, \delta)$ is a subset of a set obtained from V_{1k} by relabeling coordinates. Hence, by Lemma 5-m, the event (2.16) has probability no greater than

$$c_3(m)\exp\{-s^2[1 - 6m\delta]\log(\delta^{-1}e^{-1})\}\ .$$

The union (over σ) of $2^m - 1$ such events (with $s = r/(2^m - 1)$) is equivalent to the union over a in U_δ of the events (2.15), and thus contains the event $B_\delta(r)$, w.p. 1. This proves (2.14).

Proof of Theorem 1-m. (The proof which follows is valid for $m \geqq 1$, and does not require the use of (1.1), although the latter implies the desired result when $m = 1$.) Suppose $m \geqq 1$, and let $\varepsilon > 0$ be given. Choose δ to be the largest positive number which is $\leqq 1/6m$ and such that $4^{-m}(1 - 6m\delta)\log(\delta^{-1}e^{-1}) > 2$. Let k be the smallest positive number such that $k^{-1} < \delta/2$ and such that $2/(k - j) < \varepsilon/5$ for $j/k < 1 - \delta/2$, and let $\varepsilon' = \varepsilon/5$ in (2.11). The coefficients of $-r^2$ in the exponentials of (2.14) and (2.11) are thus $\geqq 2 - \varepsilon$ provided $j/k < 1 - \delta/2$. Thus, writing $V_q = \{x: x_q \leqq 1 - \delta\}$, we see that V_1 is contained in a union of fewer than k sets V_{jk} (namely, those for which $j/k < 1 - \delta/2$), for each of which the coefficient of $-r^2$ in (2.11) is $\geqq 2 - \varepsilon$. Thus, $r \geqq R(\varepsilon, m)$ implies

$$P_F\left\{\sup_{V_1} |S_n(x) - F(x)| \geqq rn^{-1/2}\right\} \leqq c_5(\varepsilon, m)e^{-r^2(2-\varepsilon)}\ , \tag{2.17}$$

where R and c_5 depend on ε and m but not on r, F, or n. Interchanging the roles of the first and qth coordinates, we obtain (2.17) with V_1 replaced by V_q. Since the union of U_δ and the V_q's ($1 \leqq q \leqq m$) is I^m, we obtain (1.5) for $r \geqq R'(\varepsilon, m)$ and thus (by possibly increasing $c(\varepsilon, m)$) for all $r \geqq 0$.

References

1. K. L. Chung, *An estimate concerning the Kolmogoroff limit distribution*, Trans. Amer. Math. Soc., **67** (1949), 36-50.
2. A. Dvoretzky, J. Kiefer, and J. Wolfowitz, *Asymptotic minimax character of the sample distribution function and of the classical multinomial estimator*, Ann. Math. Stat., **27** (1956), 642-669.
3. W. Feller, *An Introduction to Probability Theory and its Applications*, vol. 1, 2nd

edition, Wiley, New York (1957).
4. J. Kiefer and J. Wolfowitz, ***On the deviations of the empiric distribution function of vector chance variables***, Trans. Amer. Math. Soc., **87** (1958), 173-186.
5. A. N. Kolmogorov, ***Sulla determinrzione empirica di una legge di distribuzione***, Ist. Ital. Atti. Giorn., **4** (1933), 83-91.
6. P. Lévy, ***Théorie de L'Addition des Variales Aléatoires***, Gauthier-Villard, Paris (1937).

CORNELL UNIVERSITY

Reprinted from
Pacific J. Math. **11**, 649–660 (1961)

Reprinted from INFORMATION AND CONTROL, Volume 5, No. 1, March 1962

INFORMATION AND CONTROL 5, 44–54 (1962)

Channels with Arbitrarily Varying Channel Probability Functions

J. KIEFER* AND J. WOLFOWITZ†

Cornell University, Ithaca, N. Y.

I. INTRODUCTION

We begin by defining several terms whose significance will be apparent shortly. Let $D = \{1, 2, \cdots, d\}$ and $B = \{1, 2, \cdots, b\}$ be, respectively, the input and output alphabets. Let $S = \{w(\cdot \mid \cdot \mid i), i = 1, \cdots, c\}$ be c channel probability functions (c.p.f.'s). This means that, for $j = 1, \cdots, d$, $w(\cdot \mid j \mid i)$ is a nonnegative function with domain B such that $\sum_{k=1}^{b} w(k \mid j \mid i) = 1$.

Let n be an integer. Call any sequence of n elements, each a member of the set D (respectively, the set B) a transmitted (resp. received) n-sequence. Call any sequence of n elements, each one of $\{1, \cdots, c\}$, a channel n-sequence. Let

$$u_0 = (d_1, \cdots, d_n), \qquad v_0 = (b_1, \cdots, b_n), \tag{1.1}$$

and

$$\gamma_0 = (c_1, \cdots, c_n) \tag{1.2}$$

be, respectively, a transmitted n-sequence, a received n-sequence, and a channel n-sequence. Suppose u_0 is "sent (or transmitted) over the channel when the transmission is governed by γ_0." The chance received n-sequence

$$v(u_0) = (Y_1(u_0), \cdots, Y_n(u_0)) \tag{1.3}$$

is a sequence of independent chance variables such that

$$P\{v(u_0) = v_0 \mid \gamma_0\} = \prod_{s=1}^{n} w(b_s \mid d_s \mid c_s), \tag{1.4}$$

* The research of this author under contract with the Office of Naval Research.

† The research of this author was supported by the U. S. Air Force under contract No. AF 18 (600)-685, monitored by the Office of Scientific Research.

where the symbol on the left is the probability that $v(u_0) = v_0$ when the transmission is governed by γ_0. Thus the significance of $w(\cdot \mid \cdot \mid i)$, $i = 1, \cdots, c$, can be looked upon as follows: When the "letter" j is sent, and $w(\cdot \mid \cdot \mid i)$ governs its transmission, the probability that the letter k will be received is $w(k \mid j \mid i)$. The n letters of a received word are independently distributed. (For more detail about applications see, for example, Wolfowitz (1961).

A code (n, N, λ) for the present problem (when the c.p.f. varies arbitrarily) is a system

$$\{(u_1, A_1), \cdots, (u_N, A_N)\} \tag{1.5}$$

where $u_1, \cdots, u_N$ are transmitted n-sequences, $A_1, \cdots, A_N$ are disjoint sets of received n-sequences, and for *every* channel n-sequence γ_0 we have

$$P\{v(u_i) \in A_i \mid \gamma_0\} \geqq 1 - \lambda, i = 1, \cdots, N. \tag{1.6}$$

A number C is called the capacity (of the channel) if, for any $\epsilon > 0$ and λ, $0 < \lambda < 1$, there exists a code $(n, 2^{n(C-\epsilon)}, \lambda)$ for all sufficiently large n, and for all sufficiently large n there does not exist a code $(n, 2^{n(C+\epsilon)}, \lambda)$. (See also Wolfowitz (1961), Section 5.6.) A number $R \geqq 0$ is called a (possible) rate of transmission if, for any $\epsilon > 0$ and λ, $0 < \lambda < 1$, there exists, for all sufficiently large n, a code $(n, 2^{n(R-\epsilon)}, \lambda)$.

The code described in (1.5) can be more fully described as a code where neither the sender nor the receiver knows the channel sequence which governs the transmission of a word. Codes are described below which apply to the cases where the sender or the receiver or both know the channel sequence which governs the transmission of a word. For all four of these situations we give in the present paper necessary and sufficient conditions for the existence of a positive (possible) rate of transmission. In the case where both sender and receiver know the c.p.f. we actually determine the capacity.

The reader will find no difficulty in verifying that the existence of a positive rate for certain infinite collections of c.p.f.'s (of the nature of S of Section II) can be obtained from the methods and results of this paper.

The reader may also find it interesting to compare the results of Sections III and IV with those in the case of a compound channel (where the unknown c.p.f. is the same for each letter; see, for example, Wolfowitz (1961), Chapter 4).

The codes of the present paper, including the one described above, are all "nonrandomized." For the case where neither the sender nor the receiver knows the channel sequence which governs the transmission of a word, Blackwell, Breiman, and Thomasian (1960) studied the relation between the "nonrandomized" and "randomized" capacities. Their methods do not apply to the problems of the present paper, as they indicate in their discussion on page 566. This discussion gives an example where no positive rate of transmission exists and is included in the necessity condition of Theorem 1 below.

II. CASE WHERE NEITHER THE SENDER NOR THE RECEIVER KNOWS THE CHANNEL SEQUENCE

For each j in D consider the smallest convex body $T(j)$ which contains the c points of b-space

$$(w(1 \mid j \mid i), w(2 \mid j \mid i), \cdots, w(b \mid j \mid i)), i = 1, \cdots, c. \quad (2.1)$$

We shall now prove

THEOREM 1. *Necessary and sufficient for the existence of a rate of transmission greater than zero when neither sender nor receiver knows the channel sequence is that, among the convex bodies $T(1), \cdots, T(d)$, at least two be disjoint.*

PROOF OF NECESSITY. Assume that no two of $T(1), \cdots, T(d)$ are disjoint. Fix n and $\lambda < \frac{1}{2}$. We shall show that any code can contain only one member (i.e., $N = 1$). If δ is a distribution on channel n-sequences we define, for the sake of brevity,

$$P\{v(u_0) \in A \mid \delta\} = \sum_{\gamma_0} P\{v(u_0) \in A \mid \gamma_0\} \, \delta(\gamma_0).$$

The idea of the proof will be this: For any fixed n we will construct a distribution δ such that

$$P\{v(u_1) \in A_2 \mid \delta\} \geqq 1 - \lambda. \quad (2.2)$$

Of course, from (1.6),

$$P\{v(u_1) \in A_1 \mid \delta\} \geqq 1 - \lambda. \quad (2.3)$$

Since A_1 and A_2 are disjoint and $\lambda < \frac{1}{2}$, (2.2) and (2.3) yield a contradiction.

Suppose

$$u_1 = (x_1, x_2, \cdots, x_n)$$

$$u_2 = (y_1, y_2, \cdots, y_n).$$

Let g_i, $i = 1, \cdots, n$, be a point common to $T(x_i)$ and $T(y_i)$. Let $s^{(i)} = (s_1^{(i)}, s_2^{(i)}, \cdots, s_c^{(i)})$ and $t^{(i)} = (t_1^{(i)}, t_2^{(i)}, \cdots, t_c^{(i)})$ be the "barycentric" coordinates of g_i in the sets $T(x_i)$ and $T(y_i)$, respectively. (Of course, these sets have at most $\min(b, c)$ extreme points, and properly a point in one of them has as many barycentric coordinates as the number of extreme points. When, as above, we write the barycentric coordinates as being c in number, it is understood that a coordinate which corresponds to an inner (nonextreme) point is zero). Suppose that, when x_i (resp. y_i) is sent, the c.p.f. which governs its transmission were chosen at random, with probability $s_j^{(i)}$ (resp. $t_j^{(i)}$) that $w(\cdot \mid \cdot \mid j)$ would be chosen. It would follow then that, when either x_i or y_i is sent (under the above conditions), the probability that k ($k = 1, \cdots, b$) would be received is the same and equal to the kth Cartesian coordinate of g_i.

Now let δ (resp. δ^*) be the distribution on the channel n-sequences implied by the following: The elements of the channel n-sequence are independent chance variables, the distribution of the ith chance variable, $i = 1, \cdots, n$, being $s^{(i)}$ (resp. $t^{(i)}$). It follows that

$$P\{v(u_1) \in A_2 \mid \delta\} = P\{v(u_2) \in A_2 \mid \delta^*\}. \tag{2.4}$$

The right member of (2.4) is $\geqq 1 - \lambda$, by (1.6); this proves (2.2) and hence the necessity condition.

Proof of Sufficiency. We may suppose, without loss of generality, that $T(1)$ and $T(2)$ are disjoint. Then there is a plane in b-space which separates $T(1)$ and $T(2)$ and is disjoint from $T(1)$ and $T(2)$. Let $(l_1, \cdots, l_b, m)$ be its coordinates. Suppose an h-sequence (say z_1) consisting exclusively of ones or an h-sequence (say z_2) consisting exclusively of twos is sent over the channel, and let $N_i(z_j)$, $i = 1, \cdots, b$, $j = 1, 2$, be the number of elements i in the chance sequence $v(z_j)$. Let η, $0 < \eta < \frac{1}{16}$, be chosen arbitrarily. Now, reversing if necessary the indices 1 and 2, we may conclude from the law of large numbers that, when h is sufficiently large, the probability exceeds $1 - \eta$ that

$$\sum_{i=1}^{b} l_i N_i(z_1) < hm \text{ and } \sum_{i=1}^{b} l_i N_i(z_2) \geqq hm,$$

no matter what channel h-sequence governs the transmission of z_1 or z_2. From the above it follows that, if we construct the u's of the code (1.5) of consecutive blocks of h ones or h twos, each block can be "decoded correctly" with probability at least $1 - \eta$. The code whose existence we shall now demonstrate will have its sequences u so constructed. Since the result to be proved is one for large n there is no loss of generality in assuming that n is an integral multiple of h.

We now digress for a moment to describe a "t-error correcting" code. Take $d = b$. Then the code (1.5) is called t-error correcting if, for $i = 1, \cdots, N$, A_i consists of all n-sequences which differ from u_i in at most t places. (The condition (1.6) is no longer required. Since the A_i are disjoint it follows that any two u's of the code must differ in at least $(2t + 1)$ places.)

Now let $d = b = 2$. Suppose that S now contains a continuum of c.p.f.'s, each indexed by θ, where θ takes all values in the interval $[0, \eta]$. The c.p.f. $w(\cdot \mid \cdot \mid \theta)$ is defined as follows: $w(1 \mid 1 \mid \theta) = w(2 \mid 2 \mid \theta) = 1 - \theta$, $w(1 \mid 2 \mid \theta) = w(2 \mid 1 \mid \theta) = \theta$. It follows from the law of large numbers that, whatever be λ(fixed), $0 < \lambda < 1$, for n sufficiently large a $2\eta n$-error correcting code of length N is a code (n, N, λ) for the channel just described

$$(\text{i.e., } d = b = 2,\ S = \{w(\cdot \mid \cdot \mid \theta) \mid \theta \in [0, \eta]\}).$$

Since $\eta < \frac{1}{16}$ it follows from a result of Gilbert (1952, Theorem 1) that there is a positive r such that, for all n sufficiently large, there exists a $2\eta n$-error correcting code of length 2^{nr} for the channel described above.

To construct a code for the channel of our original problem when n is sufficiently large, we proceed as follows: A block of h ones (resp. twos) of the transmitted alphabet of the original problem corresponds to the symbol 1 (resp. 2) of the transmitted alphabet of the new problem for which an error correcting code will be used

$$(d = b = 2,\ S = \{w(\cdot \mid \cdot \mid \theta) \mid \theta \in [0, \eta]\}).$$

All blocks of h letters of the received alphabet of the original problem, which satisfy $l_i N_i < hm$ (resp. $l_i N_i \geqq hm$), are to correspond to the symbol 1 (resp. 2) of the received alphabet of the new problem; here N_i, $i = 1, \cdots, b$, is the number of elements i in the block of h letters. It follows that, whatever be λ, $0 < \lambda \leqq 1$, when n is sufficiently large there exists a code (n, N, λ) for our original problem with $\log N$ greater than the largest integer in nr/h. This completes the proof of sufficiency.

III. CASE WHERE THE CHANNEL SEQUENCE IS KNOWN TO THE RECEIVER BUT NOT TO THE SENDER

Let Γ_n be the totality of all channel n-sequences. For the case described in the title of this section a code (n, N, λ) is a system

$$(u_1, \{A_1(\gamma_0), \gamma_0 \in \Gamma_n\}), \cdots, (u_N, \{A_N(\gamma_0), \gamma_0 \in \Gamma_n\}) \tag{3.1}$$

where $u_1, \cdots, u_N$ are transmitted n-sequences, for each $\gamma_0 \in \Gamma_n$ the sets $A_1(\gamma_0), \cdots, A_N(\gamma_0)$ are disjoint sets of received n-sequences, and

$$P\{v(u_i) \in A_i(\gamma_0) \mid \gamma_0\} \geqq 1 - \lambda, \gamma_0 \in \Gamma_n, \quad i = 1, \cdots, N. \tag{3.2}$$

(For the application of such a code see, for example, Wolfowitz (1961, Chapters 3 and 4).) This section is devoted to a proof of the following:

THEOREM 2. *Necessary and sufficient for the existence of a positive rate of transmission when the receiver but not the sender knows the channel sequence is that, for some pair d_1, d_2 of elements of D,*

$$\sum_{i=1}^{b} | w(i \mid d_1 \mid j) - w(i \mid d_2 \mid j) | > 0, \quad j = 1, \cdots, c. \tag{3.3}$$

PROOF OF SUFFICIENCY. The proof will be similar to the proof of sufficiency in Theorem 1. Suppose (3.3) holds. For typographical simplicity assume $d_1 = 1$, $d_2 = 2$. Let η, $0 < \eta < \frac{1}{10}$, be chosen arbitrarily. From (3.3) and the law of large numbers it is not difficult to obtain the conclusion that there exists a positive integer h with the following property: Let $h^1 \geqq h$ be any integer. Suppose that a block of h^1 ones or a block of h^1 twos is sent over the channel with the transmission of every letter governed by the same c.p.f. $w(\cdot \mid \cdot \mid i)$, $i = 1, \cdots, c$. Then, no matter which is the c.p.f., known to the receiver, the latter can "correctly decode" which block (of ones or of twos) has been sent with probability at least $1 - \eta$. To put it more precisely: Let z_1 (resp. z_2) be the transmitted h^1-sequence which consists exclusively of ones (resp., of twos). Let $\gamma^{(i)}$, $i = 1, \cdots, c$, be the channel h^1-sequence which consists exclusively of elements i. There is a partition of the space of all received h^1-sequences into two disjoint sets $B_1^{(i)}$ and $B_2^{(i)}$ such that

$$P\{v(z_j) \in B_j^{(i)} \mid \gamma^{(i)}\} \geqq 1 - \eta, \; j = 1, 2; \; i = 1, \cdots, c.$$

We now proceed as in the last paragraph of the proof of sufficiency in Theorem 1. Instead of using blocks of h ones and h twos we use blocks of ch ones and ch twos. In any channel ch-sequence at least h elements must be the same, and the receiver knows which they are. He can therefore decode correctly the block of ch elements of the transmitted alphabet with probability at least $1 - \eta$. The remainder of the proof of sufficiency is as in Theorem 1.

PROOF OF NECESSITY. Assume that (3.3) does not hold. Then, for any pair a_1, a_2 of elements of D there exists an element of $1, \cdots, c$, say $c^*(a_1, a_2)$, such that

$$\sum_{i=1}^{b} | w(i \mid a_1 \mid c^*) - w(i \mid a_2 \mid c^*) | = 0.$$

Fix n and $\lambda < \frac{1}{2}$. We show that any code (3.1) can contain only one member (i.e., $N = 1$). Suppose

$$u_1 = (x_1, \cdots, x_n), \qquad u_2 = (y_1, \cdots, y_n).$$

Let γ^* be the channel n-sequence whose ith element, $i = 1, \cdots, n$ is $c^*(x_i, y_i)$. Then obviously

$$P\{v(u_1) \in A_2(\gamma^*) \mid \gamma^*\} = P\{v(u_2) \in A_2(\gamma^*) \mid \gamma^*\} \geqq 1 - \lambda.$$

Also, by (3.2),

$$P\{v(u_1) \in A_1(\gamma^*) \mid \gamma^*\} \geqq 1 - \lambda.$$

The last two statements are obviously in contradiction. Necessity is proved.

IV. CASE WHERE THE CHANNEL SEQUENCE IS KNOWN TO THE SENDER BUT NOT TO THE RECEIVER

We begin by describing a code (n, N, λ) for the case described in the title; our description will be, for the sake of brevity, a little informal but completely intelligible. The sets $A_1, \cdots, A_N$ are as in (1.5). When the sender wishes to send the ith word he no longer sends u_i, $i = 1, \cdots, N$. Instead he has a rule f_i which operates as follows: Let

$$\gamma_0 = (c_1, c_2, \cdots, c_n)$$

be the channel n-sequence which will govern the transmission of the word. This sequence γ_0 is known to the sender in the following way. When the sender is sending the jth letter, $j = 1, \cdots, n$, he knows

$(c_1, \cdots, c_j)$. The rule $f_i = (f_i^{(1)}, \cdots, f_i^{(n)})$ for sending the ith word tells the sender successively what each letter is to be. The jth letter, $j = 1, \cdots, n$, is given by the rule to be $f_i^{(j)}(c_1, \cdots, c_j)$ and is a function of the arguments exhibited. The place of $u_1, \cdots, u_N$ in the code (1.5) is now taken by $f_1, \cdots, f_N$. Of course, the analogue of (1.6) must hold.

For $k = 1, \cdots, c$, let $D(k)$ be the set of d points in b-space

$$\{D(i \mid k) = (w(1 \mid i \mid k), w(2 \mid i \mid k), \cdots, w(b \mid i \mid k)),\ i = 1, \cdots, d\}.$$

Now consider the totality of d^c sets $B^1(1), \cdots, B^1(d^c)$, each of which contains c points, one from each of $D(1), \cdots, D(c)$. Let $B(i)$ be the smallest convex body which contains the points of $B^1(i)$.

We now prove

THEOREM 3. *Necessary and sufficient for the existence of a positive rate of transmission when the sender but not the receiver knows the channel sequence is that at least two of* $B(1), \cdots, B(d^c)$ *be disjoint.*

PROOF OF SUFFICIENCY. Suppose $B(1)$ and $B(2)$, say, are disjoint, and $B(1)$ (resp. $B(2)$) contains $D(a_1(k) \mid k)$, $k = 1, \cdots, c$ (resp. $D(a_2(k) \mid k)$, $k = 1, \cdots, c$). The proof of sufficiency of Theorem 1 now applies with one difference. Instead of the sender sending long blocks of ones and long blocks of twos, he proceeds as follows: When he would wish to send a one (resp. a two) as part of a long block he sends the letter $a_1(k)$ (resp. $a_2(k)$) when he knows that $w(\cdot \mid \cdot \mid k)$ will be the c.p.f. according to which the received letter will be distributed. The complete proof is easy to supply after the model of the proof of Theorem 1.

PROOF OF NECESSITY. Suppose no two of $B(1), \cdots, B(d^c)$ are disjoint. Fix n and $\lambda < \frac{1}{2}$. Then the randomization argument of the proof of necessity of Theorem 1 can easily be applied to f_1 and f_2 to obtain the same contradiction as before.

V. CASE WHERE BOTH SENDER AND RECEIVER KNOW THE CHANNEL SEQUENCE

We now consider the case described in the title of this section. The receiver's knowing the channel n-sequence γ_0 (say) means that the sets A in the code corresponding to (1.5) are functions of γ_0, thus:

$$A_1(\gamma_0), \cdots, A_N(\gamma_0). \tag{5.1}$$

These sets are of course disjoint (for the same γ_0). There is such a system for every channel n-sequence. The words transmitted are defined by rules $f_1, \cdots, f_N$ as in Section IV.

Let $C(i)$, $i = 1, \cdots, c$, be the capacity of the c.p.f. $w(\cdot \mid \cdot \mid i)$ (i.e., of the discrete memoryless channel with (single) c.p.f. $w(\cdot \mid \cdot \mid i)$; see, e.g., Wolfowitz (1961, Chapter 3)).We will now show that the *capacity* of the channel of this section is the smallest of $C(1), \cdots, C(c)$, say C^*.

Clearly the capacity could not be greater than C^*. For, if $C^* = C(1)$, say, it would be enough to consider the channel sequence which consists entirely of ones to see that, for all sufficiently large n, there does not exist a code of length $2^{n(C^*+\epsilon)}$. It is therefore sufficient to prove that C^* is a possible rate of transmission for the present channel.

(As the channel has been defined above, the sender does not know the entire channel sequence in advance of sending a word (transmitted n-sequence); he knows only the c.p.f. for the letter he is sending and for the letters already sent. Whether the receiver knows the entire channel sequence in advance or not does not matter, since he does not "decode" the word (n-sequence) received until he has received the whole word. Suppose however, that the sender does know the entire channel sequence in advance of transmission. The argument of the preceding paragraph shows that the capacity of the channel (assuming that there is a capacity) could not be increased by this knowledge of the sender. On the other hand, this knowledge could not, obviously, decrease the capacity. It follows that, in determining the capacity of the channel of the present section we also determine the capacity of the channel modified so that the sender knows the entire channel sequence in advance of transmission of a word.)

This section is devoted to a proof of:

THEOREM 4. *The capacity of the channel of the present section, where the channel sequence is known to both sender and receiver, is C^*, the smallest of the capacities of the individual c.p.f.'s.*

As the earlier argument has shown, it is sufficient to show that C^* is a possible rate of transmission, which we now proceed to do. We shall write $\sqrt{n}$ as if it were always an integer, and leave to the reader the easy task of approximating it by an integer when that is necessary. Consider first another channel, say V, which is the same as the channel of the present section except that there are only c possible channel sequences (for each n) each one consisting of the same element repeated n times. Let $\epsilon > 0$ be fixed arbitrarily. Then (e.g., Wolfowitz (1961, Section 7.5)) there is a positive number α such that, when n is sufficiently large, there exists a code $((\sqrt{n}, 2^{\sqrt{n}(C^*-\epsilon)}, e^{-\alpha\sqrt{n}})$ for channel V. Let $z_1, \cdots, z_t$ be the elements u of this code $(t = 2^{\sqrt{n}(C^*-\epsilon)})$. We now construct a code $(n + s, t^{\sqrt{n}}, \sqrt{n}\, e^{-\alpha\sqrt{n}})$ for the channel of the present section, with $s = (\sqrt{n} - 1)c(c - 1) + (c - 1)$. Since $\sqrt{n}\, e^{-\alpha\sqrt{n}} \to 0$ as $n \to \infty$ it is easy to see that this proves the theorem.

Suppose the channel n-sequences of our problem were always of the following type: the first $\sqrt{n}$ elements are all the same, the second $\sqrt{n}$ elements are all the same, etc. Then we could construct the desired code for our problem as follows: Each element u is a succession of elements from $z_1, \cdots, z_t$, to a total of $\sqrt{n}$ elements in all. Each z sent can be correctly decoded with a probability at least $1 - e^{-\alpha\sqrt{n}}$, by the property of the code V. Hence the probability of error in the code just constructed is at most $\sqrt{n}\, e^{-\alpha\sqrt{n}}$. However, the channel n-sequences are not all of the above type.

We therefore proceed as follows: Suppose that in the case of the preceding paragraph we would have sent the sequence $z^{(1)}, z^{(2)}, \cdots, z^{(\sqrt{n})}$. Let $\gamma_0 = (c_1, c_2, \cdots, c_n)$ be the channel n-sequence which will govern the transmission of the word in our problem. The sender begins by sending the first element of $z^{(1)}$. Then, if $c_2 = c_1$ he sends the second element of $z^{(1)}$, and if $c_2 \neq c_1$ he sends the first element of $z^{(2)}$. The procedure at the third step may best be described by Table I. The procedure at the fourth and subsequent steps is now clear. As soon as a z has been entirely sent its place is taken by the next z whose transmission has not yet begun. The number of elements z which will be sent in this manner depends upon the sequence γ_0, because there will be "waste" at the end. However, it is clear that at least $(\sqrt{n} - c + 1)$ elements z will always be sent. Since the receiver knows the sequence γ_0 he can make each of the received symbols correspond to its own z.

Suppose $(c - 1)z$'s have not been sent. These could be in various positions (have various serial numbers), and both sender and receiver know their serial numbers after n symbols have been sent and received. Let the serial numbers be $\alpha_1 < \alpha_2 < \cdots < \alpha_{c-1}$. Beginning with the

TABLE I

$c_1 = c_2 = c_3$	Send third element of $z^{(1)}$
c_1, c_2, c_3 all different	Send first element of $z^{(3)}$
$c_1 \neq c_2 = c_3$	Send second element of $z^{(2)}$
$c_2 \neq c_1 = c_3$	Send second element of $z^{(1)}$
$c_1 = c_2 \neq c_3$	Send first element of $z^{(2)}$

$(n + 1)$st letter the sender sends $z^{(\alpha_1)}$ so that all its letters will be transmitted (not necessarily consecutively) under the same c.p.f.; at most $(\sqrt{n} - 1)\, c + 1$ letters will suffice for this. Then he sends $z^{(\alpha_2)}$ so that all its letters will be transmitted (not necessarily consecutively) under the same c.p.f., etc. At most $s = (\sqrt{n} - 1)\, c\, (c - 1) + (c - 1)$ letters will suffice to send all the missing z's. The sender sends exactly s letters; if fewer letters are needed the remainder can be any prearranged letters which are "ignored" by the receiver.

If fewer than $(c - 1)$ z's originally remained unsent the sender sends these as in the preceding paragraph, and then sends enough prearranged letters (which the receiver will "ignore") to make up a total of s letters. The receiver, who knows the channel sequence, knows the order in which all the z's have been sent, and knows which $\sqrt{n}$ received letters are to be used to decode any one z. Each z thus sent can be correctly decoded with a probability at least $1 - e^{-\alpha\sqrt{n}}$, by the property of the code V. Hence the probability of error in the code for our problem is at most $\sqrt{n}\; e^{-\alpha\sqrt{n}}$. This completes the proof of Theorem 4.

RECEIVED: October 9, 1961.

REFERENCES

BLACKWELL, D., BREIMAN, L., AND THOMASIAN, A. J., (1960), The capacities of certain channel classes under random coding. *Ann. Math. Stat.* **31**, No. 3, 558–567.

GILBERT, E. N., (1952), A comparison of signaling alphabets. *Bell System Tech. J.* **31**, 504–522.

WOLFOWITZ, J., (1961), "Coding Theorems of Information Theory." Springer, Berlin, Göttingen; Prentice-Hall, Englewood Cliffs, N. J.

Reprinted from THE ANNALS OF MATHEMATICAL STATISTICS
Vol. 34, No. 4, December, 1963

MINIMAX CHARACTER OF HOTELLING'S T^2 TEST IN THE SIMPLEST CASE

BY N. GIRI,[1] J. KIEFER,[2] AND C. STEIN[3]

Cornell University and Stanford University

Summary. In the first nontrivial case, dimension $p = 2$ and sample size $N = 3$, it is proved that Hotelling's T^2 test of level α maximizes, among all level α tests, the minimum power on each of the usual contours where the T^2 test has constant power. A corollary is that the T^2 test is most stringent of level α in this case.

1. Introduction. Let $X_1, \cdots, X_N$ be independent normal p-vectors with common mean vector ξ and common nonsingular covariance matrix Σ. Write $N\bar{X} = \sum_1^N X_i$ and $S = \sum_1^N (X_i - \bar{X})(X_i - \bar{X})'$. Let $\delta > 0$ (and finite) be specified. For testing the hypothesis $H_0: \xi = 0$ against $H_1: N\xi'\Sigma^{-1}\xi = \delta$ at significance level α, a commonly employed procedure is Hotelling's T^2 test, which rejects H_0 when $T^2 = N(N-1)\bar{X}'S^{-1}\bar{X} > C'$ or, equivalently, when $U = T^2/(T^2 + N - 1) > C$, where C (or C') is chosen so as to yield a test of level α. Throughout this paper $0 < \alpha < 1$, so that $0 < C < 1$.

In this paper we are interested in a minimax question regarding the T^2 test, namely, whether or not that test maximizes, among all level α tests, the minimum power under H_1. We succeed in proving that the answer is affirmative in the first nontrivial case, $p = 2$, $N = 3$ (for each possible choice of δ and α), although there are strong indications, mentioned at the end of this section, that the answer is also affirmative for general p and N. However, analytical difficulties make it seem most unlikely that our method of proof can be generalized to handle more than a few of these cases. What is worse is that this proof yields no real understanding of why the result holds, nor of what it is which distinguishes this problem from others where Stein has shown that the best invariant procedure under the real linear group (which, among procedures based on the sufficient statistic $(\bar{X}, S)$, the T^2-test is, here) is not minimax. (See Stein (1955), Lehmann (1959), pp. 231 and 338, and James and Stein (1960), p. 376.) We nevertheless publish the present result in the hope that it may interest others to attack the problem.

The results previously proved for the T^2 test include the best invariant character under the real linear group, as mentioned above. For testing H_0 against $H_1': \xi'\Sigma^{-1}\xi > 0$, Simaika (1941) proved this test to be uniformly most powerful

Received July 5, 1963.

[1] Research supported by ONR contract No. Nonr-401(03).

[2] Fellow of the John Simon Guggenheim Memorial Foundation; research supported in part by ONR contract No. Nonr-266(04) (NRO 47-005).

[3] Research supported by National Science Foundation Grant GP-40 at Stanford University.

of level α among tests whose power function depends only on $\xi'\Sigma^{-1}\xi$; this result also follows easily from the best invariant property. Stein (1956) showed that the test is admissible for testing H_0 against H_1', but the method used there yields nothing for the test of H_0 against H_1; the minimax proof of the present paper also yields no admissibility result for the problem of testing H_0 against H_1. Hsu (1945) showed that the T^2 test maximizes a certain integral over H_1 of the power, but, as he points out, this property is shared by many other tests, since the integral in question is infinite in value; thus, this result cannot be used to prove the desired minimax property. Of course, when $p = 1$ we have the usual properties of the symmetric Student's test, and when $N \leqq p$ it is easy to see that the infimum over H_1 of the power of every test equals the size of the test; hence, the case $p = 2$, $N = 3$ is the simplest one to be considered.

We now outline briefly our method of proof. We may restrict attention to the space of the minimal sufficient statistic $(\bar{X}, S)$. The examples of Stein mentioned above show that the Hunt-Stein theorem can not be applied for the real linear group G of $p \times p$ nonsingular matrices $(p \geqq 2)$ which leave the present problem invariant, operating as $(\bar{X}, S; \xi, \Sigma) \to (g\bar{X}, gSg'; g\xi, g\Sigma g')$. However, the theorem does apply for the smaller group G_T of nonsingular lower-triangular matrices (zero above the main diagonal), which is solvable. (See Kiefer (1957), Lehmann (1959), p. 345.) Thus, there is a test of level α which is almost invariant (hence, in the present problem, there is such a test which is invariant; see Lehmann (1957), p. 225) under G_T and which maximizes, among *all* level α tests, the minimum power over H_1. Whereas T^2 was a maximal invariant under G, with a single distribution under each of H_0 and H_1, the maximal invariant under G_T is a p-dimensional statistic $R = (R_1, \cdots, R_p)'$ with a single distribution under H_0 but with a distribution which depends continuously on a $(p-1)$-dimensional parameter $\Delta = (\delta_1, \cdots, \delta_p)'$, $\delta_i \geqq 0$, $\sum_1^p \delta_i = \delta$ (fixed), under H_1. Thus, when $N > p > 1$ there is no UMP invariant test under G_T as there was under G. We compute the Lebesgue densities f_Δ^* and f_0^* of R, under H_1 and H_0. Because of the compactness of the reduced parameter spaces $\{0\}$ and $\Gamma = \{(\delta_1, \cdots, \delta_p): \delta_i \geqq 0, \sum_1^p \delta_i = \delta\}$ and the continuity of f_Δ^* in Δ, it follows (see Wald (1950)) that every minimax test for the reduced problem in terms of R, is Bayes. In particular, Hotelling's test $U = \sum_1^p R_i > C$, which is G_T-invariant, maximizes the minimum power over H_1 if and only if there is a probability measure λ on Γ such that, for some constant K,

$$\int_\Gamma \frac{f_\Delta^*(r_1, \cdots, r_p)}{f_0^*(r_1, \cdots, r_p)} \lambda(d\Delta) \begin{Bmatrix} \geqq \\ < \end{Bmatrix} K \quad \text{according to whether} \quad \sum_1^p r_i \begin{Bmatrix} \geqq \\ < \end{Bmatrix} C, \tag{1.1}$$

except possibly for a set of measure zero. (Here C depends on the specified α, and λ and K may depend only on C and the specified value $\delta > 0$.) An examination of the integrand in (1.1) will allow us to replace (1.1) by the equivalent

$$\int_{\Gamma} \frac{f_{\Delta}^{*}(r_1, \cdots, r_p)}{f_0(r_1, \cdots, r_p)} \lambda(d\Delta) = K \quad \text{if} \sum_{1}^{p} r_i = C. \tag{1.2}$$

We are able to evaluate the unique value which K must take on in order that (1.2) can be satisfied, and are then faced with the question of whether or not there exists a probability measure λ satisfying the left half of (1.2). Writing $\lambda^*(A) = \lambda(\delta A)$, we show that λ^*, if it exists, depends on C and δ only through $C\delta$. The development thus far, which hold for general p and $N > p$, is carried out in Section 2. In Section 3 we then obtain a λ and carry out the proof that it satisfies the left half of (1.2) in the special case $p = 2, N = 3$.

The complexity of λ and of our proof that it satisfies the left half of (1.2) make it seem desirable to try other approaches for the general problem, but we have thus far succeeded with none of these. One attempt which must occur to most people who work on this problem is to consider instead the problem where, for fixed $\Sigma = H_{\Sigma}H_{\Sigma}'$ (say), the vector $\eta = H_{\Sigma}^{-1}\xi$ is uniformly distributed on the sphere $\eta'\eta = \delta$ under H_1; one can then use G_T on this modified problem, for which the presence of the minimax property for the T^2 test would imply its presence in the original problem; unfortunately, one obtains a test other than Hotelling's, and which is not minimax for the original problem.

As announced earlier in an abstract (Giri and Kiefer (1962)), it is easy to see that, for every α, N, and p, Hotelling's test has certain local and asymptotic minimax properties as $\delta \to 0$ and $\delta \to \infty$. This lends credence to our conjecture that the minimax result proved for T^2 in the present paper actually holds for all N and p.

The result for the test based on the multiple correlation coefficient R^2 when $p = 3$, $N = 4$, which is analogous to the result of the present paper, will be published elsewhere.

2. Reduction of the problem to (1.2). Since much of this development proceeds along standard lines, we shall omit some of the routine details. The reader may consult Lehmann (1959) for nomenclature and for a treatment of invariance and minimax theory in hypothesis testing.

We need only consider test functions which depend on the sufficient statistic $(\bar{X}, S)$, the Lebesgue density of which is

$$\begin{aligned} f_{\xi,\Sigma}(\bar{x}, s) &= c(\det \Sigma)^{-(N+p-1)/2} (\det s)^{(N-p-2)/2} \\ &\quad \times \exp\{-\tfrac{1}{2} \operatorname{tr} \Sigma^{-1}(s + N(\bar{x} - \xi)(\bar{x} - \xi)')\} \end{aligned} \tag{2.1}$$

where $c = N^{p/2}/2^{Np/2}\pi^{p(p+1)/4} \prod_{i=1}^{p} \Gamma((N-i)/2)$.

We can compute a maximal invariant of $(\bar{X}, S)$ under the action of the group G_T of nonsingular lower triangular matrices which leave the problem invariant (as described in Section 1) in the usual fashion: If a function ϕ is invariant, then $\phi(\bar{x}, s) = \phi(g\bar{x}, gsg')$ for all g, s, $\bar{x}$. We may consider the domain of S to be the positive definite symmetric matrices, which have probability one for all ξ, Σ; then there is an F in G_T such that $S = FF'$. Putting $g = LF^{-1}$ where L is any diagonal matrix with values ± 1 in any order on the diagonal, we see that ϕ is a function only of the vector $LF^{-1}\bar{X}$ and hence, because of the freedom in the choice of L, of $|F^{-1}\bar{X}|$, or, equivalently, of the vector whose ith element is the square Z_i of the ith component of $F^{-1}\bar{X}$. Write $b_{[i]}$ for the i-vector consisting of the first i components of a p-vector b, and $C_{[i]}$ for the upper left-hand $i \times i$ submatrix of a $p \times p$ matrix C. Because of the way in which inverses are formed in G_T, $(F_{[i]})^{-1} = (F^{-1})_{[i]}$, so that

$$Z_i = \bar{X}'_{[i]}(F'_{[i]})^{-1}(F_{[i]})^{-1}\bar{X}_{[i]} = \bar{X}'_{[i]}(S_{[i]})^{-1}\bar{X}_{[i]}.$$

The vector $Z = (Z_1, \cdots, Z_p)'$ is thus a maximal invariant if it is invariant, and it easily seen to be the latter. Z_i is essentially Hotelling's statistic computed from the first i coordinates. We shall find it more convenient to work with the equivalent statistic $R = (R_1, \cdots, R_p)'$ where

$$\sum_1^i R_j = NZ_i/(1 + NZ_i)$$

or

$$R_i = NZ_i/(1 + NZ_i) - NZ_{i-1}/(1 + NZ_{i-1}) \qquad (Z_{-1} = 0).$$

It is easily verified that $R_i \geqq 0$, $\sum_1^p R_i \leqq 1$, and of course $\sum_1^p R_i = U = T^2/(N - 1 + T^2)$.

A corresponding maximal invariant $\Delta = (\delta_1, \cdots, \delta_p)'$ in the parameter space of (μ, Σ) under G_T when H_1 is true is easily seen to be given by

$$\delta_i = N\xi'_{[i]}(\Sigma_{[i]})^{-1}\xi_{[i]} - N\xi'_{[i-1]}(\Sigma_{[i-1]})^{-1}\xi_{[i-1]} \qquad (\delta_1 = N\xi_1^2/\Sigma_{11}).$$

Here $\delta_i \geqq 0$ and $\sum_1^p \delta_i = \delta$. The corresponding maximal invariant under H_0 takes on the single value $0 = (0, \cdots, 0)$. The Lebesgue density function f_Δ^* of $(R_1, \cdots, R_p)$ depends on Δ under H_1, and is a fixed f_0^* under H_0.

We must now compute f_Δ^* and f_0^*. (Actually, we need only obtain f_Δ^*/f_0^* for use in (1.2), so we could proceed without keeping track of factors not depending on Δ in this derivation; however, it is not much extra work to keep track of these factors, so we shall do so.) We may put $\Sigma = I$ and $N^{\frac{1}{2}}\xi = (\delta_1^{\frac{1}{2}}, \delta_2^{\frac{1}{2}}, \cdots, \delta_p^{\frac{1}{2}})' = N^{\frac{1}{2}}\rho$ (say) in (2.1), since f_Δ^* depends only on ξ and Σ only through Δ. Let B be the unique lower triangular matrix with positive diagonal elements for which $BB' = S$, and let $V = B^{-1}\bar{X}$. One computes easily the Jacobians $\partial S/\partial B = 2^p \prod_1^p b_{ii}^{p+1-i}$ and $\partial \bar{X}/\partial V = \det B = \prod b_{ii}$, so that the joint density of V and B when $\Sigma = I$, $\xi = \rho$ is

$$h_{\rho,I}(v, b) = 2^p f_{\rho,I}(bv, bb') \prod b_{ii}^{p+2-i}.$$

Putting $W = (W_1, \cdots, W_p)'$ with $W_i = |V_i|$, and noting that the p-vector w with positive components can arise from any of the 2^p vectors $v = Mw$ where M is diagonal with diagonal entries ± 1, we can write $g = bM$ where g ranges over *all* matrices in G_T and obtain for the density of W

$$h^*_{\rho,I}(w) = 2^p \int f_{\rho,I}(gw, gg') \prod_i |g_{ii}|^{p+2-i} \prod_{i \geq j} dg_{ij}$$

$$(2.2) \qquad = 2^p c \int \exp\{-\tfrac{1}{2} \operatorname{tr}[gg' + N(gw - \rho)(gw - \rho)']\} \cdot \prod_i |g_{ii}|^{N-i} \prod_{i \geq j} dg_{ij},$$

the range of integration being from $-\infty$ to $+\infty$ in each variable. Let A be a lower triangular matrix for which $A(I + NWW')A' = I$. Then $A'A = (I + NWW')^{-1} = I - (1 + NW'W)^{-1}NWW'$, so that $NW'A'AW = NW'W/(1 + NW'W)$. Since $A_{[i]}(I_i + NW_{[i]}W'_{[i]})A'_{[i]} = I_i$, we obtain similarly

$$(2.3) \qquad \begin{aligned} NW'_{[i]}A'_{[i]}A_{[i]}W_{[i]} &= NW'_{[i]}W_{[i]}/(1 + NW'_{[i]}W_{[i]}) \\ &= NZ_i/(1 + NZ_i) = \sum_{j=1}^{i} R_j, \end{aligned}$$

so that $N^{\frac{1}{2}}AW$ is a vector whose ith element is $R_i^{\frac{1}{2}}$. Writing $gA^{-1} = q$, we have $\partial g/\partial q = \prod a_{ii}^{p-i+1}$. Also, tr $NgW\rho' = (N^{\frac{1}{2}}\rho)'q(N^{\frac{1}{2}}AW) = \sum_{i \geq j} (\delta_i R_j)^{\frac{1}{2}} q_{ij}$. From the equality of the second and fourth expressions of (2.3) we see that $W_i^2 = N^{-1}R_i/(1 - \sum_1^i R_j)(1 - \sum_1^{i-1} R_j)$, so that

$$\begin{aligned} \partial W/\partial R &= \prod_{i=1}^{p} [(\partial W_i^2/\partial R_i)/(\partial W_i^2/\partial W_i)] \\ &= N^{-p/2}2^{-p} \prod_{i=1}^{p} \left[R_i^{-\frac{1}{2}} \left(1 - \sum_1^{i-1} R_j\right)^{\frac{1}{2}} \left(1 - \sum_1^{i} R_j\right)^{-\frac{3}{2}} \right] \\ &= N^{-p/2}2^{-p} \left(1 - \sum_1^{p} R_j\right)^{-\frac{3}{2}} \prod_{i=1}^{p} R_i^{-\frac{1}{2}} \prod_{i=1}^{p-1} \left(1 - \sum_1^{i} R_j\right)^{-1}. \end{aligned}$$

Since $\prod_1^i a_{jj}^2 = \det(A'_{[i]}A_{[i]}) = 1/\det(I_i + NW_{[i]}W'_{[i]}) = 1/(1 + NW'_{[i]}W_{[i]}) = 1 - \sum_1^i R_j$ and $g_{ii} = a_{ii}q_{ii}$, (2.2) yields

$$\begin{aligned} f^*_\Delta(r) = h^*_{\rho,I}(w(r))\partial w/\partial r &= \left[cN^{-p/2} \left(1 - \sum_1^p r_j\right)^{(N-p-2)/2} e^{-\delta/2} \Big/ \prod_1^p r_i^{\frac{1}{2}} \right] \\ &\quad \times \int \exp\{-\tfrac{1}{2} \sum_{i \geq j} [q_{ij}^2 - 2(\delta_i r_j)^{\frac{1}{2}} q_{ij}]\} \prod_i |q_{ii}|^{N-i} \prod_{i \geq j} dq_{ij}, \end{aligned}$$

the integration again being from $-\infty$ to $+\infty$ in each variable. For $i > j$, integration with respect to q_{ij} yields a factor $(2\pi)^{\frac{1}{2}} \exp\{\delta_i r_j/2\}$. For $j = i$, we obtain a factor

$$\begin{aligned} (2\pi)^{\frac{1}{2}} e^{r_i\delta_i/2} E[\chi_1^2(r_i\delta_i)]^{(N-i)/2} \\ = 2^{(N-i+1)/2}\Gamma((N - i + 1)/2)\phi((N - i + 1)/2, \tfrac{1}{2}; r_i\delta_i/2), \end{aligned}$$

where $\chi_1^2(\beta)$ is a noncentral chi-square variable with one degree of freedom and noncentrality parameter $\beta\,(= E\chi_1^2(\beta) - 1)$, and where ϕ is the confluent hypergeometric function (sometimes denoted as ${}_1F_1$),

$$\phi(a, b; x) = \sum_{j=0}^{\infty} [\Gamma(a+j)\Gamma(b)/\Gamma(a)\Gamma(b+j)j!]x^j. \tag{2.4}$$

Thus, finally, for $r \,\varepsilon\, H = \{r: r_i > 0, 1 \leqq i \leqq p; \Sigma r_i < 1\}$, we have

$$\begin{aligned} f_\Delta^*(r) &= \left[\pi^{-p/2}\Gamma(N/2)\left(1 - \textstyle\sum_1^p r_j\right)^{(N-p-2)/2} \Big/ \Gamma((N-p)/2)\prod_1^p r_i^{\frac{1}{2}}\right] \\ &\quad \times \exp\left\{-\delta/2 + \sum_{j=1}^p r_j \sum_{i>j} \delta_i/2\right\} \prod_{i=1}^p \phi((N-i+1)/2, \tfrac{1}{2}; r_i\,\delta_i/2). \end{aligned} \tag{2.5}$$

Of course, $f_0^*(r)$ is just the expression preceding the exponential in (2.5), while $f_\Delta^*(r)/f_0^*(r)$ is the exponential and the product following it.

The continuity in Δ over its compact domain Γ is evident, so we can conclude that the minimax character of the critical region $U \geqq C$ is equivalent to the existence of a probability measure λ satisfying (1.1). Clearly (1.1) implies (1.2). On the other hand, if there are a λ and a K for which (1.2) is satisfied and if $r^* = (r_1^*, \cdots, r_p^*)'$ is such that $\sum r_i^* = C' > C$, writing $f = f_\Delta^*/f_0^*$ and $r^{**} = Cr^*/C'$, we see at once that $f(r^*) = f((C'/C)r^{**}) > f(r^{**}) = K$ because of the form of f_Δ^*/f_0^* and the fact that $C'/C > 1$ and $\sum r_i^{**} = C$. This and a similar argument for the case $C' < C$ show that (1.2) implies (1.1). (Of course, we do not assert that the left side of (1.2) still depends only on $\sum r_i$ if $\sum r_i \neq C$.)

The computation of the next section is somewhat simplified by the fact that, for fixed C and δ, we can at this point compute the unique value of K for which (1.2) can possibly be satisfied. Let $\hat{R} = (R_1, \cdots, R_{p-1})$, and write $f_\Delta^*(\hat{r} \mid u)$ for the version of the conditional Lebesgue density of $\hat{R}$ given that $\sum_1^p R_i = u$ which is continuous in $\hat{r}$ and u for $r_i > 0$, $\sum_1^{p-1} r_i < u < 1$, and is 0 elsewhere; write $f_\delta^{**}(u)$ for the denstiy of $U = \sum_1^p R_i$ which is continuous for $0 < u < 1$ and vanishes elsewhere (and which depends on Δ only through δ). Then (1.2) can be written as

$$\int f_\Delta^*(\hat{r} \mid C)\, d\lambda(\Delta) = \left[K\frac{f_0^{**}(C)}{f_\delta^{**}(C)}\right] f_0^*(\hat{r} \mid C) \quad \text{for} \quad r_i > 0, \sum_1^{p-1} r_i < C. \tag{2.6}$$

The integral of (2.6), being a probability mixture of probability densities, is itself a probability density in $\hat{r}$, as is $f_0^*(\hat{r} \mid C)$. Hence, the expression in square brackets equals one. It is well known that, for $0 < C < 1$,

$$f_\delta^{**}(C) = \frac{\Gamma(N/2)\, e^{-\delta/2}}{\Gamma(p/2)\Gamma((N-p)/2)}\, C^{(p-2)/2}(1-C)^{(N-p-2)/2}\phi(N/2, p/2; C\delta/2). \tag{2.7}$$

(See Anderson (1958) or use (2.5).) Hence, from (2.5), (1.2) becomes

$$\int_\Gamma \exp\left\{\sum_{j=1}^p r_j \sum_{i>j} \frac{\delta_i}{2}\right\} \prod_{i=1}^p \phi((N-i+1)/2, \tfrac{1}{2}; r_i\,\delta_i/2)\, d\lambda(\Delta) = \phi(N/2, p/2; C\delta/2) \tag{2.8}$$

for all r with $r_i > 0$, $\sum r_i = C$. Write Γ_1 for the unit $(p-1)$-simplex $\{(\beta_1, \cdots, \beta_p): \beta_i \geqq 0, \sum_i^p \beta_i = 1\}$. Writing $\gamma = C\delta$ and making the change of variables $\beta_i = \delta_i/\delta$, $t_i = \gamma r_i/C$, and writing λ^* for the measure on Γ_1 associated with λ on $\Gamma(\lambda^*(A) = \lambda(\delta A))$, (2.8) becomes

$$\int_{\Gamma_1} \exp\left\{\sum_{j=1}^{p} t_j \sum_{i>j} \frac{\beta_i}{2}\right\} \prod_{i=1}^{p} \phi((N-i+1)/2, 1/2; \beta_i t_i/2)\, d\lambda^*(\beta_1, \cdots, \beta_p) = \phi(N/2, p/2; \gamma/2) \tag{2.9}$$

for all $(t_1, \cdots, t_p)$ with $\sum t_i = \gamma$ and $t_i > 0$ (hence, by analyticity, for all $(t_1, \cdots, t_p)$ with $\sum t_i = \gamma$). Thus, λ^*, if it exists, depends on C and δ only through their product γ. (When $p = 1$, Γ_1 is a single point, but the dependence on γ is genuine in other cases.)

3. The case $p = 2$, $N = 3$. Representing the integration in (2.9) in terms of $\beta_2 (0 \leqq \beta_2 \leqq 1)$ and noting that $\phi(\frac{3}{2}, \frac{1}{2}; x/2) = (1+x)e^{x/2}$, we obtain from (2.9), on writing $t_1 = \gamma - t_2$, $\beta_1 = 1 - \beta_2$,

$$\int_0^1 [1 + (\gamma - t_2)(1-\beta_2)]\, \phi(1, \tfrac{1}{2}; \beta_2 t_2/2)\, d\lambda^*(\beta_2) = e^{(t_2-\gamma)/2} \phi(3/2, 1; \gamma/2). \tag{3.1}$$

One could presumably try to solve (3.1) for λ^* by using the theory of the Meijer transform (with kernel $\phi(1, \frac{1}{2}; x/2)$). We proceed instead by expanding both sides of (3.1) as power series in t_2. Writing $\mu_i = \int_0^1 \beta^i\, d\lambda^*(\beta)$, $0 \leqq i < \infty$ for the ith moment of λ^* and

$$B = e^{-\gamma/2} \phi(\tfrac{3}{2}, 1; \gamma/2), \tag{3.2}$$

we obtain the equations

$$\begin{aligned} &\text{(a) } 1 + \gamma - \gamma\mu_1 = B \\ &\text{(b) } -(2r-1)\mu_{r-1} + (2r+\gamma)\mu_r - \gamma\mu_{r+1} = B[\Gamma(r+\tfrac{1}{2})/r!\,\Gamma(\tfrac{1}{2})] \quad r \geqq 1 \end{aligned} \tag{3.3}$$

as equivalent to (3.1). (Of course, $\mu_0 = 1$ for λ^* to be a probability measure.) One could now try to show that the sequence $\{\mu_i\}$ defined by $\mu_0 = 1$ and (3.3) satisfies the classical necessary and sufficient condition for it to be the moment sequence of a probability measure on [0, 1] or, equivalently, that the Laplace transform $\sum_0^\infty \mu_j(-t)^j/j!$ is completely monotone on $[0, \infty)$, but we have been unable to proceed successfully in this way. Instead, we shall obtain, in the next paragraph, a function $m_\gamma(x)$ which we then prove, in the succeeding paragraphs below, to be the Lebesgue density $d\lambda^*(x)/dx$ of an absolutely continuous probability measure λ^* satisfying (3.3) (and, hence, (3.1)). That proof does not rely on the somewhat heuristic development of the next paragraph, but we nevertheless sketch that development to give the reader an idea of where the m_γ of (3.8) came from, rather than merely to pull it out of thin air.

The generating function $\psi(t) = \sum_{j=0}^\infty \mu_j t^j$ of the sequence $\{\mu_i\}$ satisfies a differential equation which is obtained in the usual fashion by multiplying (3.3)

(b) by t^{r-1} and summing from 1 to ∞:

$$\begin{aligned} 2t^2(1-t)\psi'(t) - (t^2 - \gamma t + \gamma)\psi(t) \\ = Bt[(1-t)^{-\frac{1}{2}} - 1] + \gamma[t(1-\mu_1) - 1] \\ = Bt(1-t)^{-\frac{1}{2}} - t - \gamma. \end{aligned} \tag{3.4}$$

(A corresponding use, instead, of the Laplace transform to obtain (3.8) below, is more involved.) This is solved by treatment of the corresponding homogeneous equation and by variation of parameter, to yield

$$\psi(t) = \frac{e^{-\gamma/2t}}{(1-t)^{\frac{1}{2}}} \int_0^t e^{\gamma/2T} \left[\frac{-1}{2T(1-T)^{\frac{1}{2}}} - \frac{\gamma}{2T^2(1-T)^{\frac{1}{2}}} + \frac{B}{2T(1-T)} \right] dT, \tag{3.5}$$

the integration being understood to start from the origin along the negative real axis of the complex plane. The constant of integration has been chosen to make ψ continuous at 0 with $\psi(0) = 1$, and (3.5) defines a single-valued function on the complex plane minus a cut along the real axis from 1 to ∞. In fact, the analyticity of ψ on this region can easily be demonstrated by considering the integral of ψ on a closed curve about 0 avoiding 0 and the cut, making the inversion $w = 1/t$, shrinking the path down to the cut $0 \leqq w \leqq 1$, and using (3.30) below. Now, if there did exist an absolutely continuous λ^* whose suitably regular derivative m_γ satisfied

$$\int_0^1 m_\lambda(x)/(1-tx)\, dx = \psi(t), \tag{3.6}$$

we could obtain m_γ by using the simple inversion formula

$$m_\gamma(x) = (2\pi i x)^{-1} \lim_{\epsilon \downarrow 0} [\psi(x^{-1} + i\epsilon) - \psi(x^{-1} - i\epsilon)]. \tag{3.7}$$

However, there is nothing in the theory of the Stieltjes transform which tells us that an m_γ satisfying (3.7) does satisfy (3.6) (and, hence, (3.1)), so we use (3.7) only as a formal device to obtain an m_γ which we shall then prove, in the remaining paragraphs, satisfies (3.1). From (3.5) and (3.7) we obtain, for $0 < x < 1$,

$$m_\gamma(x) = \frac{e^{-\gamma x/2}}{2\pi x^{\frac{1}{2}}(1-x)^{\frac{1}{2}}} \left\{ \int_0^\infty e^{-\gamma u/2} \left[\frac{B}{1+u} - \frac{u^{\frac{1}{2}}}{(1+u)^{3/2}} \right] + B \int_0^x \frac{e^{\gamma u/2}}{1-u} du \right\}. \tag{3.8}$$

In order to prove that $d\lambda^*(x) = m_\gamma(x)\, dx$ (with m_γ defined by (3.8)) satisfies (3.1) with λ^* a probability measure, we must show that

(3.9)

(a) $m_\gamma(x) \geqq 0$ for almost all x, $0 \leqq x \leqq 1$,

(b) $\int_0^1 m_\gamma(x)\, dx = 1$,

(c) $\mu_1 = \int_0^1 x m_\gamma(x)\, dx$ satisfies (3.3) (a)

(d) $\mu_r = \int_0^1 x^r m_\gamma(x)\, dx$ satisfies (3.3) (b) for $r \geqq 1$.

Condition (3.9) (a) follows at once from (3.8) and the fact that $B > 1$ and $u^{\frac{1}{2}}(1 + u) < (1 + u)^{3/2}$ for $u > 0$. To prove (3.9) (d), we note that m_γ as defined by (3.8) satisfies the differential equation

$$(3.10) \quad m'_\gamma(x) + m_\gamma(x)[\gamma/2 + (1 - 2x)/2x(1 - x)] = B/2\pi x^{\frac{1}{2}}(1 - x)^{3/2},$$

so that an integration by parts yields, for $r \geqq 1$,

$$(r + 1)\,\mu_r - r\mu_{r-1} = \int_0^1 [(r + 1)\,x^r - rx^{r-1}]\, m_\gamma(x)\, dx$$

$$(3.11) \quad = \int_0^1 (x^r - x^{r+1})\, m'_\gamma(x)\, dx = \mu_r(1 - \gamma/2) + \gamma\mu_{r+1}/2 - \mu_{r-1}/2$$

$$+ B\Gamma(r + \tfrac{1}{2})/2\pi^{\frac{1}{2}}\, r!$$

which is (3.3) (b). The proof of (3.9) (b) and (c) relies on certain identities involving hypergeometric functions. In the next paragraph we list some of the readily available properties of hypergeometric functions which will be used in the proof.

The material summarized in the present paragraph can be found, for example, in Erdélyi (1953), Chapter 6. The confluent hypergeometric function (2.4) has an integral representation when $c > a > 0$ given by

$$(3.12) \quad \phi(a, c; x) = \frac{\Gamma(c)}{\Gamma(a)\,\Gamma(c - a)} \int_0^1 e^{xt}\, t^{a-1}\, (1 - t)^{c-a-1}\, dt.$$

The associated solution ψ to the hypergeometric equation has the representation

$$(3.13) \quad \psi(a, c; x) = \frac{1}{\Gamma(a)} \int_0^\infty e^{-xt}\, t^{a-1}\, (1 + t)^{c-a-1}\, dt$$

if $a > 0$. We shall use the fact that the general definition of ψ, as used in what follows when $a = 0$, satisfies

$$(3.14) \quad \psi(0, c; x) = 1.$$

We shall also use the differential properties

$$(3.15) \quad \frac{d}{dx}\phi(a, c; x) = \left(\frac{a}{c} - 1\right)\phi(a, c + 1; x) + \phi(a, c; x)$$

$$(3.16) \quad \frac{d}{dx}\psi(a, c; x) = \psi(a, c; x) - \psi(a, c + 1; x),$$

$$(3.17) \quad \frac{d}{dx}\psi(a, c; x) = x^{-1} a\,[(a - c + 1)\,\psi\,(a + 1, c; x) - \psi\,(a, c; x)],$$

and the identities

$$(a - c + 1)\phi(a, c; x) - a\phi(a + 1, c; x) + (c - 1)\phi(a, c - 1; x) = 0, \tag{3.18}$$

$$c\phi(a, c; x) - c\phi(a - 1, c; x) - x\phi(a, c + 1; x) = 0, \tag{3.19}$$

$$\psi(a, c; x) - a\psi(a + 1, c; x) - \psi(a, c - 1; x) = 0, \tag{3.20}$$

$$(c - a)\psi(a, c; x) - x\psi(a, c + 1; x) + \psi(a - 1, c; x) = 0. \tag{3.21}$$

A useful integration formula (Erdélyi, op. cit., p. 285, equation 16) is, for $y > 0$,

$$\int_0^\infty e^{-x}(x + y)^{-1}\,\phi(\tfrac{1}{2}, 1; x)\,dx = \psi(\tfrac{1}{2}, 1; y) \tag{3.22}$$

The formula obtained by the (obviously permissible) differentiation under the integral sign of (3.22) with respect to y and use of (3.17) is

$$\int_0^\infty e^{-x}(x + y)^{-2}\,\phi(\tfrac{1}{2}, 1; x)\,dx = -[\Gamma(\tfrac{1}{2})/2y][\psi(\tfrac{3}{2}, 1; y)/2 - \psi(\tfrac{1}{2}, 1; y)]. \tag{3.23}$$

The function m_γ defined by (3.8) can be written, using (3.13), in terms of hypergeometric functions, for $0 < x < 1$, as

$$\begin{aligned}
m_\gamma(x) &= \frac{e^{-\gamma x/2}}{2\pi\,[x(1 - x)]^{\frac{1}{2}}}\left\{\int_0^\infty e^{-\gamma u/2}\,[e^{-\gamma/2}\varphi(\tfrac{3}{2}, 1; \gamma/2)(1 + u)^{-1}\right.\\
&\qquad \left. - u^{\frac{1}{2}}(1 + u)^{-3/2}]\,du + e^{-\gamma/2}\,\phi(\tfrac{3}{2}, 1; \gamma/2)\int_0^x e^{\gamma u/2}(1 - u)^{-1}\,du\right\}\\
&= \frac{e^{-\gamma x/2}}{2\pi\,[x(1 - x)]^{\frac{1}{2}}}\left\{e^{-\gamma/2}\,\phi(\tfrac{3}{2}, 1; \gamma/2)\psi(1, 1; \gamma/2) - \Gamma(\tfrac{3}{2})\psi(\tfrac{3}{2}, 1; \gamma/2)\right.\\
&\qquad \left. + \phi(\tfrac{3}{2}, 1; \gamma/2)\left[\int_{1-x}^\infty v^{-1}\,e^{-\gamma v)2}\,dv - \int_1^\infty v^{-1}\,e^{-\gamma v/2}\,dv\right]\right\}\\
&= \frac{1}{2\pi\,[x(1 - x)]^{\frac{1}{2}}}\,\{e^{-\gamma/2}\,\phi(\tfrac{3}{2}, 1; \gamma/2)\psi(1, 1; \gamma(1 - x)/2)\\
&\qquad - e^{-\gamma x/2}\Gamma(\tfrac{3}{2})\psi(\tfrac{3}{2}, 1; \gamma/2)\}.
\end{aligned} \tag{3.24}$$

We now prove (3.9) (b). From (3.13), (3.12), and (3.22), we have

$$\begin{aligned}
&\int_0^1 \frac{1}{2\pi\,[x(1 - x)]^{\frac{1}{2}}}\,\psi(1, 1; \gamma(1 - x)/2)\,dx\\
&\quad = \int_0^1 \frac{1}{2\pi\,[x(1 - x)]^{\frac{1}{2}}}\,\psi(1, 1; \gamma x/2)\,dx\\
&\quad = \int_0^1 \frac{dx}{2\pi\,[x(1 - x)]^{\frac{1}{2}}}\int_0^\infty (1 + t)^{-1}\,e^{-\gamma x t/2}\,dt\\
&\quad = \frac{1}{2}\int_0^\infty (1 + t)^{-1}\,\phi(\tfrac{1}{2}, 1; \gamma t/2)\;e^{-\gamma t/2}\,dt = \tfrac{1}{2}\,\Gamma(\tfrac{1}{2})\psi(\tfrac{1}{2}, 1; \gamma/2).
\end{aligned} \tag{3.25}$$

From this, (3.24), and (3.12), we have, putting $\gamma/2 = z$,

$$(3.26)\qquad H(z) \equiv \frac{4e^z}{\Gamma(\frac{1}{2})}\int_0^1 m_{2z}(x)\,dx = 2\phi(\tfrac{3}{2},1;z)\psi(\tfrac{1}{2},1;z) - \phi(\tfrac{1}{2},1;z)\psi(\tfrac{3}{2},1;z).$$

We shall show that

$$(3.27)\qquad H'(z) - H(z) = 0,$$

from which $H(z) = Ce^z$. (This identity is probably known, but we were unable to find it in the literature.) By direct evaluation in terms of elementary integrals when $\gamma = 0$ (or by using (3.26) and the expansion of ψ and ϕ near $z = 0$), we have $\int_0^1 m_0(x)\,dx = 1$; hence, (3.9) (b) follows from (3.27). To prove (3.27), we use (3.15) and (3.16), which yield (omitting everywhere the argument z)

$$(3.28)\qquad \begin{aligned} H' - H = {} & \phi(\tfrac{3}{2}, 2)\psi(\tfrac{1}{2}, 1) + 2\phi(\tfrac{3}{2}, 1)\psi(\tfrac{1}{2}, 1) - 2\phi(\tfrac{3}{2}, 1)\psi(\tfrac{1}{2}, 2) \\ & + \tfrac{1}{2}\phi(\tfrac{1}{2}, 2)\psi(\tfrac{3}{2}, 1) - \phi(\tfrac{1}{2}, 1)\psi(\tfrac{3}{2}, 1) + \phi(\tfrac{1}{2}, 1)\psi(\tfrac{3}{2}, 2). \end{aligned}$$

To this expression add the following four left hand side expressions, each of which equals zero (where a and c are the arguments as they appear in (3.12) and (3.13) and where, as in (3.28), we again omit display of the common argument z of ϕ and ψ):

$$(3.29)\qquad \begin{aligned} &\psi(\tfrac{3}{2}, 1) \text{ times (3.18) with } a = \tfrac{1}{2}, c = 2; \\ &\psi(\tfrac{3}{2}, 2) \text{ times (3.19) with } a = \tfrac{3}{2}, c = 1; \\ &2\phi(\tfrac{3}{2}, 1) \text{ times (3.20) with } a = \tfrac{1}{2}, c = 2; \\ &-\phi(\tfrac{3}{2}, 2) \text{ times (3.21) with } a = \tfrac{3}{2}, c = 1; \end{aligned}$$

one obtains $H' - H = 0$, as desired.

We now verify (3.9) (c). We first note, from (3.12) and from (3.19) with $a = \frac{3}{2}$, $c = 1$, that

$$(3.30)\qquad \int_0^1 \frac{(1 + \gamma y)\, e^{\gamma y/2}}{2\pi\,[y(1 - y)]^{\frac{1}{2}}}\,dy = \phi(\tfrac{1}{2}, 1;\gamma/2)/2 + \gamma\phi(\tfrac{3}{2}, 2;\gamma/2)/4 = \phi(\tfrac{3}{2},1;\gamma/2)/2$$

An alternative way of writing (3.9) (c) is, by (3.9) (b) (which we have just proved),

$$(3.31)\qquad 1 = [\phi(\tfrac{3}{2}, 1; \gamma/2)]^{-1}\, e^{\gamma/2} \int_0^1 [1 + \gamma(1 - x)]\, m_\gamma(x)\,dx.$$

The right side of (3.31) may be expressed, using (3.25) and (3.30), as

$$\frac{1}{\phi(\frac{3}{2}, 1; \gamma/2)} \int_0^1 \frac{(1 + \gamma y)}{2\pi[y(1 - y)]^{\frac{1}{2}}} \{\phi(\tfrac{3}{2}, 1; \gamma/2)\psi(1, 1; \gamma y/2) - e^{\gamma y/2}\Gamma(\tfrac{3}{2})\psi(\tfrac{3}{2}, 1; \gamma/2)\} \, dy$$

$$(3.32) \qquad = \int_0^1 \frac{\gamma y \psi(1, 1; \gamma y/2)}{2\pi[y(1 - y)]^{\frac{1}{2}}} \, dy + \int_0^1 \frac{\psi(1, 1; \gamma y/2)}{2\pi[y(1 - y)]^{\frac{1}{2}}} \, dy - \frac{\Gamma(\frac{3}{2})\psi(\frac{3}{2}, 1; \gamma/2)}{\phi(\frac{3}{2}, 1; \gamma/2)} \int_0^1 \frac{(1 + \gamma y)e^{\gamma y/2}}{2\pi[y(1 - y)]^{\frac{1}{2}}} \, dy$$

$$= \int_0^1 \frac{\gamma y \psi(1, 1; \gamma y/2)}{2\pi[y(1 - y)]^{\frac{1}{2}}} \, dy + \frac{\Gamma(\frac{1}{2})\psi(\frac{1}{2}, 1; \gamma/2)}{2} - \frac{\Gamma(\frac{3}{2})\psi(\frac{3}{2}, 1; \gamma/2)}{2}.$$

To evaluate the integral on the last line of (3.32), we use (3.21) with $a = 1$, $c = 0$ and (3.14), (3.13), (3.12), and (3.23), to write

$$\int_0^1 \frac{\gamma y \psi(1, 1; \gamma y/2)}{2\pi[y(1 - y)]^{\frac{1}{2}}} \, dy = \int_0^1 \frac{[\psi(0, 0; \gamma y/2) - \psi(1, 0; \gamma y/2)]}{\pi[y(1 - y)]^{\frac{1}{2}}} \, dy$$

$$(3.33) \qquad = 1 - \int_0^1 \frac{dy}{\pi[y(1 - y)]^{\frac{1}{2}}} \int_0^\infty (1 + t)^{-2} e^{-(\gamma y t/2)} \, dt$$

$$= 1 - \int_0^\infty (1 + t)^{-2} \phi(\tfrac{1}{2}, 1; \gamma t/2) e^{-\gamma t/2} \, dt$$

$$= 1 + \Gamma(\tfrac{1}{2})[\psi(\tfrac{3}{2}, 1; \gamma/2)/2 - \psi(\tfrac{1}{2}, 1; \gamma/2)]/2.$$

Thus, (3.32) and (3.33) imply (3.13) and, hence, (3.9) (c).

REFERENCES

ANDERSON, T. W. (1958). *Introduction to Multivariate Statistical Analysis*. Wiley, New York.

ERDÉLYI, A. (1953). (Editor) *Higher Transcendental Functions*, **1** McGraw-Hill, New York.

GIRI, N. and KIEFER, J. (1962). Minimax properties of Hotelling's and certain other multivariate tests, (abstract). *Ann. Math. Statist.* **33** 1490–1491.

HSU, P. L. (1945). On the power function of the E^2-test and the T^2-test. *Ann. Math. Statist.*, **16** 278–286.

JAMES, W. and STEIN, C. (1960). Estimation with quadratic loss *Proc. Fourth Berkely Symp. Math. Statist. Prob.* **1** 361–379.

KIEFER, J. (1957). Invariance, minimax sequential estimation, and continuous time processes. *Ann. Math. Statist.* **28** 573–601.

LEHMANN, E. L. (1959). *Testing Statistical Hypotheses*. Wiley, New York.

SIMAIKA, J. B. (1941). An optimum property of two statistical tests. *Biometrika* **32** 70–80.

STEIN, C. (1955). On tests of certain hypotheses invariant under the full linear group, (abstract). *Ann. Math. Statist.* **26** 769.

STEIN, C. (1956). The admissiblity of Hotelling's T^2-test. *Ann. Math. Statist.*, **27** 616–623.

WALD, A. (1950). *Statistical Decision Functions*. Wiley, New York.

Reprinted from THE ANNALS OF MATHEMATICAL STATISTICS
Vol. 34, No. 3, September, 1963

ASYMPTOTICALLY OPTIMUM SEQUENTIAL INFERENCE AND DESIGN

BY J. KIEFER[1] AND J. SACKS[2]

Cornell University and Northwestern University

0. Summary. In recent years the study of sequential procedures which are asymptotically optimum in an appropriate sense as the cost c per observation goes to zero has received considerable attention.

On the one hand, Schwarz (1962) has recently given an interesting theory of the asymptotic shape, as $c \to 0$, of the Bayes stopping region relative to an a priori distribution F, for testing sequentially between two composite hypotheses $\theta \leqq \theta_1$ and $\theta \geqq \theta_2$ concerning the real parameter θ of a distribution of exponential (Koopman-Darmois) type, with indifference region the open interval (θ_1, θ_2). (An example of Schwarz's considerations is described in connection with Figure 4.) One aim of the present paper is to generalize Schwarz's results to the case where (with or without indifference regions) the distributions have arbitrary form and there can be more than two decisions (Sections 2, 3, 4). In this general setting we obtain, under mild assumptions, a family $\{\delta_c\}$ of procedures whose integrated risk is asymptotically the same as the Bayes risk. (In fact, extending Schwarz's result, a family $\{\delta_c'\}$ can be constructed so as to possess this asymptotic Bayes property relative to all a priori distributions with the same support as F, or even with smaller indifference region support than F.) Procedures like our $\{\delta_c\}$ have already been suggested by Wald (1947) for use in tests of composite hypotheses (e.g., the sequential t-test), but his concern was differently inspired.

At the same time, we show how such multiple decision problems can be treated by using simultaneously a number of sequential tests for associated two-decision problems.

A second aim is to extend, strengthen, and somewhat simplify the asymptotic sequential design considerations originated by Chernoff (1959) and further developed by Albert (1961) and Bessler (1960) (Section 5). Our point of departure here is a device utilized by Wald (1951) in a simpler estimation setting, and which in the present setting amounts to taking a preliminary sample with predesignated choice of designs and such that, as $c \to 0$, the size of this preliminary sample tends to infinity, while its ratio to the total expected sample size tends to zero. The preliminary sample can then be used to "guess the true state of nature" and thus to choose the future design pattern once and for all rather than to have to reexamine the choice of design after subsequent observations. (In Wald's setting the only "design" problem was to pick the size of the second sample of his two-sample procedure.) The properties of the resulting procedure can then be inferred

Received August 8, 1962; revised March 5, 1963.
[1] Research sponsored by the Office of Naval Research.
[2] Research sponsored by the Office of Naval Research and by NSF Grant G24500.

from the considerations of Sections 2, 3, and 4, where there is no design problem but where most of the work in this paper is done; using Wald's idea, we thereby obtain procedures for the design problem fairly easily, once we have the (non-design) sequential inference structure to build upon. The family $\{\delta_c^*\}$ so obtained has the same asymptotic Bayes property as that described above for the family $\{\delta_c\}$ of the non-design problem. Furthermore, a family $\{\delta_c^{**}\}$ can be constructed so that, like $\{\delta_c'\}$ in the non-design problem, it is asymptotically Bayes for all a priori distributions with the same support. The value of the asymptotic Bayes risk of such a family is closely related to the lower bound which was obtained by Chernoff et al for the risk function of certain procedures, and which gives another form for the optimality statement.

The role of the sequential procedures considered by Donnelly (1957) and Anderson (1960) for hypothesis testing with an indifference region is indicated at the end of Section 1. Asymptotic solutions to the problem of Kiefer and Weiss (1957) are given.

An Appendix contains proofs of certain results on fluctuations of partial sums of independent random variables, which are used in the body of the paper.

1. Introduction and comments on related work. Chernoff (1959), whose work initiated the design investigations mentioned in the summary, has also given an introductory heuristic discussion which motivates these considerations. We therefore omit such a discussion, mentioning only the differences between the present and previous work.

As mentioned in the summary, most of the effort in the present paper is devoted to generalizing Schwarz's results. Since we no longer have his exponential type structure, it is not possible to obtain the concise proofs and elegant characterizations that he obtains. It should be noted that Chernoff's and Albert's papers can be regarded as considering, inter alia, this (nondesign) problem without indifference region, if one allows only one design in their treatments. The first of these considers finitely many states, while the latter obtains ϵ-optimum families of procedures for tests of hypotheses containing infinitely many states. (Albert mentions an indifference region in the early part of his paper, but does not consider it in the domain of his risk function in the later part, so that his procedures are really ϵ-optimum only for the problem without indifference region.) Bessler also considers finitely many states, but infinitely many experiments.

Since we cannot invoke the monotone likelihood ratio structure to obtain Schwarz's simple asymptotic reduction of a hypothesis like $\theta \leqq \theta_1$ to one like $\theta = \theta_1$, we use compactness and appropriate continuity in Section 2 to reduce the problem to one involving finitely many states. The final argument used to obtain optimality (Theorem 1) reduces the consideration to that of testing between two simple hypotheses, and a comparison with known properties of the Sequential Probability Ratio Test (SPRT). When we introduce an indifference region in Section 3, this comparison must be made with a Bayes sequential test between two simple hypotheses with a single indifference state.

In the absence of compactness of the space of states of nature, Remarks 1, 2, 3 and 5 of Section 2 state the compactification assumptions we require. Albert also uses a compactification device in this case, but his is associated with the type of maximum likelihood (ML) argument development used in ML consistency proofs by Wald (1949) and by Kiefer and Wolfowitz (1956). Our development, which is in terms of Bayes procedures (as is Schwarz's), seems somewhat simpler to us; Albert remarks on the possible complexity of procedures using ML estimates in this way, on page 798 of his paper. (Asymptotic relations between Bayes and ML procedures in simpler contexts are well known.)

The optimality results of Chernoff, Albert, and Bessler are all stated in the non-Bayesian language of our Corollary to Theorem 1, while Schwarz's results are obtained in the Bayesian terms of Theorem 1 itself. (Although Schwarz's discussion is mainly in terms of shapes of stopping regions, an earlier, mimeographed version of his paper described his result also in terms of the Bayes risk, itself; in our case the difference between the two descriptions is greater, and is essentially the difference between Lemma 4 and Theorem 1.) Schwarz mentions briefly (page 234) the relationship between these two forms of the optimality result in his case, and Theorem 2 (and its analogues in Theorems 4, 5, and 6) make the result precise in our case: *for a given set of possible states of nature, we obtain a family of procedures which is asymptotically Bayes for every a priori distribution whose support is the given set.* (When there are only finitely many states, Chernoff's procedure clearly achieves this.) One can easily find examples where the hypothesis of the Corollary to Theorem 1 is violated by an asymptotically Bayes family, for example, when the family is asymptotically Bayes relative to an a priori distribution whose support is a proper subset of the given set; the violation of hypothesis and conclusion are only to be expected in such a case, since one can not generally find a family which is simultaneously asymptotically Bayes against a priori distributions with different support. It would be interesting to obtain the conclusion to Theorem 2 for the procedure of Theorem 1, but we are unable to prove such a result.

In the presence of an indifference region I, the family of procedures we obtain has the additional property that *it is also asymptotically Bayes relative to every a priori distribution whose support* S' *satisfies* $S' = (S - I) \cup A$ *with* $A \subset I$, *where* S *is the given supporting set* ($S \supset I$).

Turning to our design considerations, the Summary and Section 5 describe the spirit of these. As for the results themselves, our considerations extend those of the previous papers in this area by considering both infinite sets of possible states and infinite sets of possible experiments, by including multiple decisions, indifference regions, and semi-indifference regions, by eliminating the ϵ in Albert's ϵ-optimality result (by use of a trivial device which could also be applied in his treatment), and by obtaining procedures with the strong uniform Bayes property mentioned in the two previous paragraphs. However, we regard it as more important than these extensions, that one sees that asymptotically optimum designs can be described and used in the simpler manner evolving from our extension of

Wald's method (wherein design prescriptions are decided at the outset and at one later stage, rather than after every stage), and that their properties can be verified largely by reference to the nondesign results. While no asymptotic theory of this type can contain any strict admissibility results, an examination of various nonasymptotic design problems leads us to a preference in applications and where c is not too small, for certain nonrandomized design choices over randomized ones which are demonstrably worse. Chernoff had indicated in his paper that such nonrandomized choices can be used, but the formal demonstrations of optimality in previous papers refer only to the random method (mentioned in Section 5) for choosing designs.

Most of our notation is similar to that of previous papers in the area, as is our model of independent, identically distributed observations in the nondesign work, and of identical possible choices of experiment at each stage in the design work. The reader is referred to Wald (1947) and Chernoff (1959) for asymptotic properties of the SPRT which we use. The fundamental role of the information numbers (whose use, as Albert mentions, was described by Wald (1947), although they subsequently came to be known as the Kullback-Leibler numbers) is well described in Chernoff's introductory comments.

Simultaneous tests. We continue this introduction by describing briefly the treatment of multiple decision problems through the use of simultaneous tests. This device has long been used by practical people, and was treated in detail by Lehmann (1957a, 1957b) in questions of nonasymptotic, nonsequential admissibility. In our asymptotic sequential context, considerations are easier, due to the fact that even a fairly large change in the stopping boundary does not alter asymptotic optimality. As an example of this last fact, we plot, in a three-state, three-decision problem, the asymptotically (as $c \to 0$) optimum boundaries given in Section 4 for the three-decision analogue of the SPRT; they are the lines

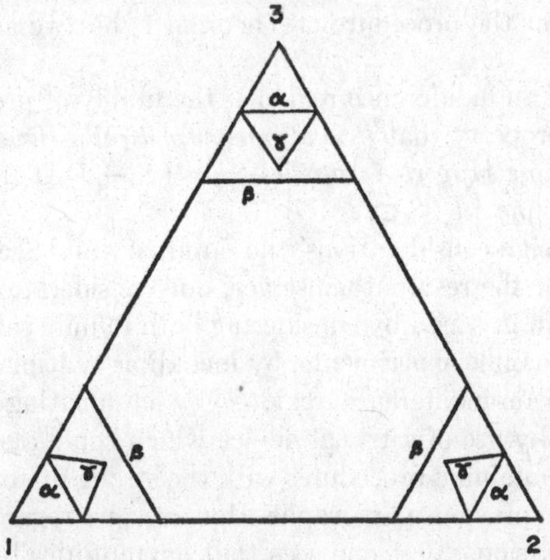

FIG. 1

α of Figure 1, namely (for $0 - 1$ loss) $\xi_{i,n} = 1 - c$ ($i = 1, 2, 3$), where $\xi_{i,n}$ is the a posteriori probability of hypothesis i after n observations. However, if these bounds are placed twice as far from the vertices, i.e., at $\xi_{i,n} = 1 - 2c$ (lines β), we change the expected sample size by a relative amount $\log 2/|\log c| = o(1)$, and still keep the probabilities of error to be of a smaller order of magnitude than the expected sample size. Thus, the second set of bounds is also asymptotically optimum. As Chernoff pointed out, the expected sample size is the overwhelming part of the asymptotic risk in these problems. As a consequence, the asymptotically optimum stopping bounds can vary greatly, and the almost impossible non-asymptotic problem of computing the best bounds disappears.

Now suppose that the statistician tried, in the classical tradition of solving new problems by using a conglomeration of old techniques, to carry on three SPRT's simultaneously, one between each of the three pairs of hypotheses and each with bounds $(1 - c)/c$ and $c/(1 - c)$ on the probability ratio, and that he stopped as soon as any two tests simultaneously dictated acceptance of the same hypothesis. Then his stopping bounds would be the broken lines γ, and these are again easily seen to be asymptotically optimum. It is not difficult to verify that the modification of this procedure which allows the tests to stop at different times (each one, as soon as possible) and makes a final decision as soon as two terminal decisions coincide or three are contradictory (in which case any decision can be made), is also asymptotically optimum, although this procedure cannot be represented so simply diagrammatically.

This technique of using simultaneous tests can also be used in more complicated cases. If one hypothesis consists of state 1 and a second consists of states 2 and 3, the procedure of Theorem 1 has the bounds $\alpha(\xi_{1,n} = 1 - c$ and $\xi_{2,n} + \xi_{3,n} = 1 - c)$ shown in Figure 2. If one instead tests 1 against 2 and 1 against 3 simultaneously with bounds $(1 - c)/c$ and $c/(1 - c)$ on the probability ratio, stopping as soon as both tests say to accept 1 or else either rejects 1, one obtains the bounds γ of Figure 2. The region where 1 is rejected is not even convex, but the procedure is asymptotically optimum; this again points up the extent to which small sample computational difficulties have become trivialized in the asymptotic theory. A modification like that indicated at the end of the previous paragraph is again possible.

As a final example, suppose we are testing between simple hypotheses 1 and 2, with 3 the simple indifference state. The procedure of Theorem 3 then stops as soon as either $\xi_{1,n}$ or $\xi_{2,n}$ is $<c$ (α, Figure 3). This rule can be described in terms of the tests of composite hypotheses without indifference region, of the previous paragraph: test $1 \cup 3$ against 2 and $2 \cup 3$ against 1, simultaneously. (If different critical constants were used, the lines α could be broken near the bottom line $\xi_{3n} = 0$.) Going all the way back to SPRT's, the boundaries γ of Figure 3, which are again asymptotically optimum, arise from using three simultaneous SPRT's and stopping as soon as either 1 or 2 is rejected by some test. The end of this section discusses these procedures further.

Further use of these ideas is contained in Sections 3 and 4. We summarize by

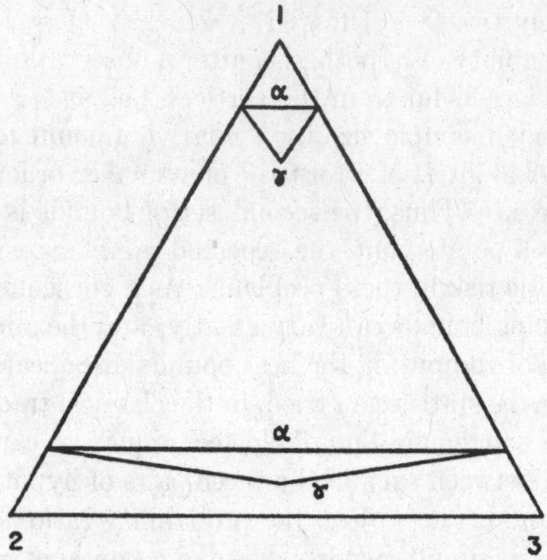

FIG. 2

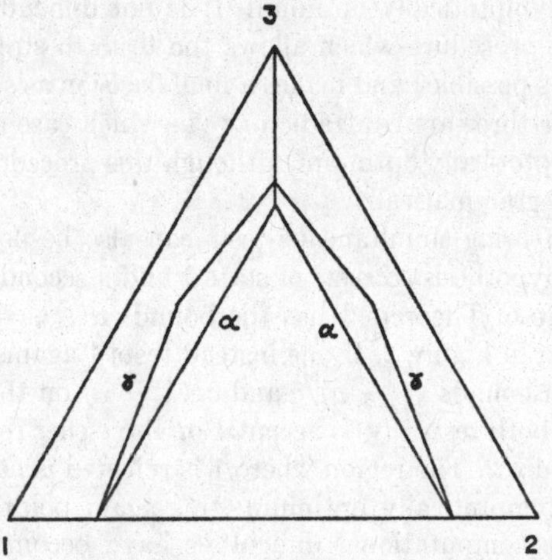

FIG. 3

remarking now that the whole development of asymptotically optimum multiple-decision procedures, with or without indifference and semi-indifference regions, can be carried out entirely in terms of such simultaneous tests. (If only SPRT's of *simple* hypotheses are used, this will, however, involve the use of appropriately fine coverings if the space of states of nature is infinite.)

The procedures of Donnelly, Anderson, and Schwarz. Let us consider the special application of the results of Section 3 to the problem where f_ω, h_α, and g_θ are normal with variance one and means -1, 0, and 1, respectively. In this case we can, with Schwarz, conveniently represent the asymptotic shape of procedures in the $(t, y) = (n/|\log c|, \sum_1^n X_i/|\log c|)$ plane. The solid lines β of Figure 4 indicate the stopping boundary corresponding to either α or γ of Figure 3 as $c \to 0$.

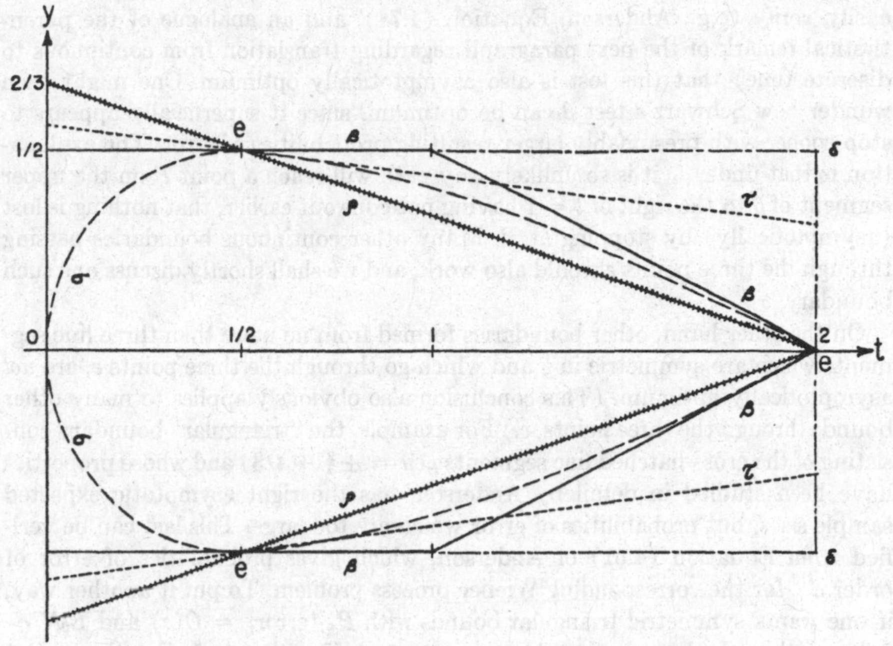

FIG. 4

This is Schwarz's famous pentagon. (The broken nature of β has nothing to do with that of γ, since α yields the same β; rather, the lower broken line β arises from the relation

$$\exp\left(\sum X_i - n/2\right)/\left[1 + \exp\left(\sum X_i - n/2\right) + \exp\left(-\sum X_i - n/2\right)\right] = c$$

which determines one half of α in Figure 3. It is the same shape which can be discerned more generally in the discussion of (3.16) of Section 3 in terms of coordinates $T_n/|\log c|$ and $S_n/|\log c|$ which correspond to $y - t$ and $-y - t$ in the present example.) The three points marked e are the intersection of the boundary with lines of "expected movement" under the three states ($\pm 45°$ lines under f_ω and g_θ, a horizontal line under h_α), and illustrate that the approximate expected sample size is $|\log c|/2$ under f_ω or g_θ and is $2\,|\log c|$ under h_α.

It is to be noted that, in the spirit of our discussion of simultaneous tests, this stopping rule can be obtained by drawing the well known parallel lines in terms of $(n, \sum X_i)$ for each of the three possible SPRT's with stopping constants c and $1/c$ for the probability ratio (it is impossible to reject h_α in both tests where it appears).

Other stopping boundaries of similar piecewise linear shape have been proposed by Donnelly (1957) and Anderson (1960). To obtain the same approximate sample sizes, such boundaries would also have to pass through the points e. One such boundary δ merely truncates, at $t = 2$, the SPRT of f_ω against g_θ. The idea of using such truncated sequential procedures goes back to Wald (1947). One can easily verify (e.g., Anderson, Equation (4.74), and an analogue of the parenthetical remark of the next paragraph regarding translation from continuous to discrete time) that this test is also asymptotically optimum. One might then wonder how Schwarz's test β can be optimum, since it superficially appears to stop sooner with presumably larger resulting probabilities of error. The explanation is that under f_ω it is so unlikely any path will reach a point P on the upper segment of β to the right of $t = 1$ having not gone out earlier, that nothing is lost (asymptotically) by stopping at P. Many other continuous boundaries passing through the three points e would also work, and we shall shortly discuss one such boundary, σ.

On the other hand, other boundaries formed from no more than three line segments which are symmetric in y and which go through the three points e, are *not* asymptotically optimum. (This conclusion also obviously applies to many other bounds through the three points, e.) For example, the "triangular" boundary consisting of the cross-hatched line segments $\rho(y = \pm\frac{2}{3} \mp t/3)$ and whose properties have been studied in detail by Anderson, has the right asymptotic expected sample sizes, but probabilities of error which are too large. This last can be verified from Equation (4.61) of Anderson, which gives probabilities of error of order $c^{8/9}$ for the corresponding Wiener process problem. To put it another way, if one wants symmetric triangular bounds with $P_{\pm 1}\{\text{error}\} = O(c)$ and $E_0 N \sim 2\,|\log c|$, then the bounds must be $y = \pm\, a \mp ta/2$ with $a \geqq 1$. For the best of these, $a = 1$, obtains $E_{\pm 1}N \sim \frac{2}{3}\,|\log c|$, efficiency $\frac{3}{4}$ compared with optimum bounds (like β or δ) for which $E_{\pm 1}N \sim \frac{1}{2}\,|\log c|$. (Translating the Wiener process results to the original discrete time normal problem, we see that by raising both lines slightly we obtain bounds ρ' for the Wiener process which still give order $c^{8/9+\epsilon}$ for probabilities of error, where $\epsilon < \frac{1}{9}$, and such that the probability that the corresponding discrete time process (obtained by examining the Wiener process at integral multiples of $t = 1/|\log c|$) with bounds ρ rejects f_ω when it is true, exceeds that for the Wiener process with bounds ρ' by $o(c)$; this last bound on the maximum difference between the two processes is standard.) A similar result holds for the procedure whose boundaries are the broken lines τ truncated at $t = 2$: we again obtain probabilities of error which are $c\,|\log c|/o(1)$, whereas the bounds β or δ yield a risk function of order $c\,|\log c|$.

We now discuss the particular boundary σ which is designated by a curved broken line in Figure 4 and whose equation is $y = \pm\, t \mp (2t)^{\frac{1}{2}}$ $(0 \leqq t \leqq 2)$. The boundary σ gives Schwarz's asymptotic shape of the Bayes stopping regions for the problem of testing $\theta \leqq -1$ against $\theta \geqq 1$ with indifference region $I = (-1, 1)$ when the support of the a priori distribution is the entire real line. For Schwarz's family of procedures (which is the same as the family $\{\delta_c^I\}$ of our Theorem 3) the asymptotic shape is given by σ when the a priori distribution is as described in the last sentence, and it is also easy to see that the same holds for the family $\{\delta_c'^I\}$ of our Theorem 4. Although $\{\delta_c^I\}$ and $\{\delta_c'^I\}$ have the same asymptotic shape, it is not clear that they enjoy the same asymptotic properties; e.g., in Theorem 4 we show (despite the lack of compactness in the parameter space, the methods of Section 3 will work because of the exponential structure of normal densities; the same will not generally be true in non-exponential situations) that $\{\delta_c'^{I}\}$, constructed for a given F, is asymptotically Bayes with respect to any G having the same support as F but we do not know if the same is true for $\{\delta_c^I\}$ (a discussion of this difference in the simpler context of Section 2 appears following the corollary to Theorem 1). The reason that the asymptotic shape does not seem to yield the asymptotic Bayes properties expressed in Theorem 4 is due to the necessity of knowing how P_θ {error} behaves asymptotically (for Theorem 4 we need P_θ {error} $= o(c\,|\log c|)$ uniformly for $\theta \not\in I$), and knowing the asymptotic shape of a family of procedures does not seem to yield precise enough knowledge about probabilities of error.

While knowledge of the boundary σ does not seem to yield the properties we obtain for $\{\delta_c'^{I}\}$, it is interesting to note that σ could be used to define a *new* family of procedures which will possess the desired properties. In fact, let $\bar{\delta}_c$ be the procedure which truncates at $2\,|\log c|$ observations, continues observing when $n < 2\,|\log c|$ and $n - (2\,|\log c|\,n)^{\frac{1}{2}} < S_n < (2\,|\log c|\,n)^{\frac{1}{2}} - n$, decides $\theta \geqq 1$ (resp. $\theta \leqq -1$) if S_n "goes out at the right" (resp. left) for $n < 2\,|\log c|$, and, if $2\,|\log c|$ observations are taken, $\bar{\delta}_c$ decides $\theta \geqq 1$ (resp. $\theta \leqq -1$) if $S_{2|\log c|} > 0$ (resp. < 0). By direct computation (which we omit), it can be shown that P_θ {wrong decision using $\bar{\delta}_c$} $= o(c\,|\log c|)$ uniformly for $|\theta| \geqq 1$, and that $E_\theta N(\bar{\delta}_c)$ has the "right" asymptotic value uniformly in θ, and this yields the result of Theorem 4. Thus, in this normal example, the family $\{\delta_c^I\}$ can be used to obtain an asymptotic shape σ from which we can go "back" and obtain a new family $\{\bar{\delta}_c\}$ (which is not the same as $\{\delta_c^I\}$) with asymptotic shape σ, which has the desired asymptotic Bayes properties, and which is simple to describe. For non-exponential problems we cannot generally expect to find simple asymptotic shapes which can be used to obtain such families $\{\bar{\delta}_c\}$.

Finally, we note that the $\{\delta_c'^{I}\}$ corresponding to the boundary σ (i.e., the $\{\delta_c'^{I}\}$ constructed for an F whose support is $(-\infty, \infty)$ is Bayes with respect to any G whose support contains the two points -1 and $+1$. That we can ignore that part of the support of the a priori distribution in $(-\infty, -1)$, or in $(1, \infty)$, is a consequence of the exponential structure of normal densities and is not, generally, possible in non-exponential situations. But, the reason we can "do away" with $(-1, 1)$ is generally true, and is stated in Theorem 4, where it is shown that we can replace the indifference region I by any subset I' of I and $\{\delta_c'^{I}\}$ remains asymptotically optimum. In particular, therefore, the $\{\delta_c'^{I}\}$ associated with σ is asymptotically optimum for our original three-state problem and is also asymptotically equivalent to the SPRT for testing between the simple hypotheses $\theta = -1$ and $\theta = +1$. Of course, the procedures which gave rise to the β and δ boundaries are also asymptotically optimum for testing $\theta = -1$ against $\theta = +1$ as well as for the original three-state problem; however, these procedures will not be asymptotically optimum in other problems for which the $\{\delta_c'^{I}\}$ corresponding to σ is asymptotically optimum.

Minimizing the maximum of EN subject to bounds on probabilities of error.

Kiefer and Weiss (1957) considered problems which, for the sake of brevity, we describe only in the special normal three-state setting just discussed. One problem is to minimize the maximum expected sample size, $\max(E_\omega N, E_\theta N, E_\alpha N)$, subject to specified upper bounds on probabilities of error when ω or θ is true. This is easily shown to be equivalent to minimizing

$$k_1 E_\omega N + k_2 E_\theta N + k_3 E_\alpha N + k_4 P_\omega\{\text{error}\} + k_5 P_\theta\{\text{error}\}$$

for some positive k_1, k_2, k_3, k_4, k_5. As the specified bounds on the probability of error go to zero, this problem can be seen to be equivalent to the asymptotic problem of Section 3. Thus, if we seek a family which asymptotically minimizes the maximum of EN subject to $P_\omega\{\text{error}\} \leqq q(c)$ and $P_\theta\{\text{error}\} \leqq q(c)$ where $q(c) = o(c\,|\log c|)$ and $q(c) = c/O(1)$ (for example, $q(c) = Kc$ where K is a positive constant), procedures corresponding to boundaries like the β and σ of Schwarz and the δ of Anderson (but not ρ or τ) are asymptotically optimum, as are many others. In fact, the fixed sample size procedure corresponding to the line $t = 2$ is asymptotically optimum for the present problem, although not for the previous one. (The normalization of q is for convenience.)

A problem auxiliary to the above, and of less intrinsic importance, is to minimize $E_\alpha N$ alone, subject to restrictions on $P_\theta\{\text{error}\}$ and $P_\omega\{\text{error}\}$. This problem really arises only because it is often true that its solution yields a solution to the more meaningful problem of the previous paragraph; nevertheless, we discuss it briefly here because of theoretical interest. This problem is equivalent to one of minimizing $k_3 E_\alpha N + k_4 P_\omega\{\text{error}\} + k_5 P_\theta\{\text{error}\}$, and is thus slightly different from that of the previous paragraph, so that the theorems of Section 3 do not lead immediately to an asymptotic solution. However from Lemma 5 of Section 3, we do obtain an asymptotic solution which is, in fact, the same one as for the first problem. Thus, in the normal example considered above, the procedures corresponding to β, σ, and δ, as well as many others, are asymptotically optimum in the sense of the present problem. It is interesting to note that, as a consequence of Lemma 5, any boundary which always guarantees stopping earlier than $t = 2$ cannot guarantee error probabilities of order $O(c)$.

L. Weiss has recently developed algorithms for the exact computation of procedures which solve this problem in certain cases (JASA, 1962). Actual computations by Weiss and D. Freeman indicate that the minimum $E_\alpha N$ must be very large before the asymptotic theory estimates it accurately.

Similar remarks apply to minimizing the maximum of EN over a larger indifference zone, etc. For example, in the normal example just discussed, the boundaries β, σ, and δ yield procedures which approximately minimize the maximum of EN over *all* normal distributions with unit variance, subject to the stated bounds on probabilities of error if the mean is $\leqq -1$ or $\geqq 1$. The corresponding k-decision problem can be treated in exactly the same manner.

2. Hypothesis testing without indifference region. In this section we prove the principal results for the two decision problem without indifference region; the latter modification is introduced in Section 3. The methods used in subsequent sections are very similar to those used here. Let Ω and Θ be two disjoint subsets of some Euclidean space (the restriction to Euclidean space is for convenience). Let $\{f_\omega ; \omega \varepsilon \Omega\}$, and $\{g_\theta ; \theta \varepsilon \Theta\}$ be two sets of densities, all absolutely continuous with respect to the same measure μ. The problem is to decide whether the true density is from Ω (Decision 1) or from Θ (Decision 2). Let $L_2(\omega)$ be the loss if $\omega \varepsilon \Omega$ is "true" and Decision 2 is made and let $L_1(\theta)$ be the loss if $\theta \varepsilon \Theta$ is "true" and Decision 1 is made. If the "correct" decision is made no loss is incurred. Let F be an a priori distribution on $\Omega \cup \Theta$ and denote that part of F which is concentrated on Ω by ξ and the part of F concentrated on Θ by η. We shall assume that $\Omega \cup \Theta$ is closed and that the support of F is $\Omega \cup \Theta$. This is for convenience in stating the results and the assumptions. To handle matters more generally all assumptions and results should be stated for the support of F (including, for example, the definitions in Assumption 2).

Independent and identically distributed observations may be taken sequentially with the cost of each observation being c with $0 < c < 1$. Let δ_c^* denote a Bayes solution with respect to F (the dependence on F of the various decision functions we define will be suppressed). Let δ_c be the decision function which, after n observations, stops and makes Decision 1 if

$$(2.1) \qquad \int f_\omega(x_1, \cdots, x_n)L_2(\omega)\xi(d\omega) > \frac{1}{c}\int g_\theta(x_1, \cdots, x_n)L_1(\theta)\eta(d\theta),$$

stops and makes Decision 2 if

$$(2.2) \qquad \int f_\omega(x_1, \cdots, x_n)L_2(\omega)\xi(d\omega) < c\int g_\theta(x_1, \cdots, x_n)L_1(\theta)\eta(d\theta),$$

and takes an $(n+1)$th observation if neither (2.1) nor (2.2) holds. (An asymptotically equivalent procedure, which is in the form of the k-decision procedure of Section 4, is to stop when the a posteriori risk is $< c$. This form also exhibits an invariance under changes in monetary scale which is not present in (2.1) and (2.2), although its absence in these last is of course irrelevant in asymptotic considerations.) The first result we wish to establish is that, under certain restrictions, $\lim_{c \to 0} r_c(F, \delta_c)/r_c(F, \delta_c^*) = 1$, i.e., that $\{\delta_c\}$ is "asymptotically Bayes" as the cost of observation goes to 0; r_c, of course, denotes the risk when c is the cost of observation.

ASSUMPTION 1. Put $\alpha_1 = \inf_\theta L_1(\theta)$, $\alpha_2 = \inf_\omega L_2(\omega)$, $\beta_1 = \sup_\theta L_1(\theta)$, $\beta_2 = \sup_\omega L_2(\omega)$. We assume that α_1 and α_2 are strictly positive and that β_1 and β_2 are finite.

ASSUMPTION 2. Put $\lambda_1(\omega, \theta) = E_\omega[\log f_\omega(X) - \log g_\theta(X)]$ and put $\lambda_2(\omega, \theta) = E_\theta[\log g_\theta(X) - \log f_\omega(X)]$. The indicated expectations are assumed to exist. Furthermore, we assume

(a) $\inf_\omega \inf_\theta \lambda_1(\omega, \theta) = \lambda_1 > 0$

(b) $\inf_\omega \inf_\theta \lambda_2(\omega, \theta) = \lambda_2 > 0$

(c) $\lambda_1(\omega, \theta)$ and $\lambda_1(\omega) = \inf_\theta \lambda_1(\omega, \theta)$ are both continuous in ω.

(d) $\lambda_2(\omega, \theta)$ and $\lambda_2(\theta) = \inf_\omega \lambda_2(\omega, \theta)$ are both continuous in θ.

REMARK. Assumption 2 guarantees that Ω and Θ are "separated."

ASSUMPTION 3. For each ω, θ

(a) $E_\omega[\log f_\omega(X) - \log g_\theta(X)]^2 < \infty$

$E_\theta[\log g_\theta(X) - \log f_\omega(X)]^2 < \infty$

(b) $\lim_{\rho\downarrow 0} E_\omega[\log \sup_{|\theta'-\theta| \leq \rho} g_{\theta'}(X) - \log g_\theta(X)]^2 = 0$

$\lim_{\rho\downarrow 0} E_\theta[\log \sup_{|\omega'-\omega| \leq \rho} f_{\omega'}(X) - \log f_\omega(X)]^2 = 0$

(c) $\lim_{\omega'\to\omega} E_\omega[\log f_{\omega'}(X) - \log f_\omega(X)]^2 = 0$

$\lim_{\theta'\to\theta} E_\theta[\log g_{\theta'}(X) - \log g_\theta(X)]^2 = 0.$

REMARK. (c) is slightly stronger than actually required. If (c) holds with the second moment replaced by the first moment and if the second moment is finite in some neighborhood of ω(or θ) then the argument below will proceed without change.

The first thing we do is to estimate the loss due to incorrect decision when δ_c is used. Let $P_\omega(c)$ be the probability of making Decision 2 when ω is the true state of nature and when δ_c is the decision function used. Let $Q_\theta(c)$ be the probability of making Decision 1 when θ is the true state of nature and δ_c is used.

LEMMA 1. $\int L_2(\omega)P_\omega(c)\xi(d\omega) + \int L_1(\theta)Q_\theta(c)\eta(d\theta) \leqq (\beta_1 + \beta_2)c.$

PROOF. Let A_j be the set of $(x_1, \cdots, x_j)$ where δ_c says stop after j observations and make Decision 2. Then

$$P_\omega(c) = \sum_{j=1}^{\infty} \int_{A_j} f_\omega(x_1, \cdots, x_j)\, d\mu^j.$$

Using (2.2) and Assumption 1 we have

$$\int_\Omega L_2(\omega)P_\omega(c)\xi(d\omega) \leqq c \sum_{j=1}^{\infty} \int_{A_j} \int_\Theta g_\theta(x_1, \cdots, x_j)L_1(\theta)\eta(d\theta)\, d\mu^j$$

$$= c \int (1 - Q_\theta(c))L_1(\theta)\eta(d\theta) \leqq c\beta_1.$$

The rest of the argument is obvious.

Lemma 2 which we now prove is the heart of the matter before us.

LEMMA 2. *Put* $\lambda_1(\omega) = \inf_\theta \lambda_1(\omega, \theta)$ *(see Assumption* 2*). Let* Θ *be a compact (therefore bounded) set in some Euclidean space. Let* $N(c)$ *be the number of observations required by* δ_c *to terminate. For each* ω_0,

$$E_{\omega_0}N(c) \leqq [1 + o(1)]\,[|\log c|/\lambda_1(\omega_0)] \tag{2.3}$$

as $c \to 0$ *(the* $o(1)$ *term may depend on* ω_0 *; this dependence is removed in Lemma* 3′*).*

PROOF. We will show (2.3) by proving it to be so for some random variable $N^*(c)$ (see (2.21) et seq) which is larger than $N(c)$. To obtain an $N^*(c)$ with which we can work we will suitably discretize Θ and alter (2.1). The first step is to obtain (2.8) below.

Fix ω_0 and put, for $\theta \in \Theta$,

$$M(\omega_0, \theta, \rho) = E_{\omega_0}[\log f_{\omega_0}(X) - \log \sup_{|\theta'-\theta|<\rho} g_{\theta'}(X)].$$

From Assumption 3, we obtain

$$(2.4)\qquad M(\omega_0, \theta, \rho) \text{ increases to } \lambda_1(\omega_0, \theta) \text{ as } \rho \text{ decreases to } 0.$$

Let $\{\epsilon_c ; c > 0\}$ be a set of positive numbers with $\epsilon_c \to 0$ as $c \to 0$. From (2.4) and Assumption 3, there is, for each c and each θ, a $\rho_c(\theta)$ such that

$$(2.5)\qquad \begin{aligned} &M(\omega_0, \theta, \rho_c(\theta)) \geqq \lambda_1(\omega_0, \theta) - \epsilon_c, \\ &E_{\omega_0}[\log f_{\omega_0}(X) - \log \sup_{|\theta'-\theta| \leqq \rho_c(\theta)} g_{\theta'}(X)]^2 < \infty. \end{aligned}$$

Let $U(\theta, \rho_c(\theta)) = \{\theta' \varepsilon \Theta | \, |\theta' - \theta| < \rho_c(\theta)\}$. Then $\{U(\theta, \rho_c(\theta)), \theta \varepsilon \Theta\}$ is a family of open sets which covers Θ and, since Θ is compact, we can extract a finite subcovering. Let $\{U(\theta_i, \rho_c(\theta_i)), i = 1, \cdots, T_c\}$ be such a finite subcovering of Θ and abbreviate $U(\theta_i, \rho_c(\theta_i))$ by U_i. The θ_i will also depend on c but we suppress this dependence. By (2.5), for each $i = 1, \cdots, T_c$,

$$(2.6)\qquad E_{\omega_0}[\log f_{\omega_0}(X) - \log \sup_{\theta \varepsilon U_i} g_\theta(X)] \geqq \lambda_1(\omega_0, \theta_i) - \epsilon_c.$$

From (c) of Assumption 3, we know that there is a γ_c such that, when $|\omega - \omega_0| < \gamma_c$,

$$(2.7)\qquad E_{\omega_0}[\log f_\omega(X) - \log f_{\omega_0}(X)] > -\epsilon_c.$$

Let $V_c = \{\omega \mid |\omega - \omega_0| < \gamma_c\}$. Then (2.6) and (2.7) yield

$$(2.8)\qquad E_{\omega_0}[\log f_\omega(X) - \log \sup_{U_i} g_\theta(X)] \geqq \lambda_1(\omega_0, \theta_i) - 2\epsilon_c$$

for all $\omega \varepsilon V_c$ and $i = 1, \cdots, T_c$. Since ω_0 is in the support of ξ (as is every point in Ω)

$$(2.9)\qquad \xi(V_c) > 0 \quad \text{for all} \quad c > 0.$$

We are now ready to work towards the definition of $N^*(c)$. First note that if $N_1(c)$ is the first n such that (2.1) holds then $N(c) \leqq N_1(c)$. Let $N_2(c)$ be the first n such that

$$(2.10)\qquad \int_{V_c} f_\omega(x_1, \cdots, x_n) L_2(\omega) \xi(d\omega) > \sup_\theta g_\theta(x_1, \cdots, x_n) \frac{\beta_1}{c}.$$

Clearly $N_2(c) \geqq N_1(c)$. Dividing both sides of (2.10) by $\xi(V_c)$ (which is positive by (2.9)) and taking logarithms we obtain an equivalent of (2.10), viz.,

$$(2.11)\qquad \begin{aligned} \log \int_{V_c} f_\omega(x_1, \cdots, x_n) L_2(\omega) \frac{\xi(d\omega)}{\xi(V_c)} &> |\log c| + \log \beta_1 \\ &+ |\log \xi(V_c)| + \log \sup_\theta g_\theta(x_1, \cdots, x_n). \end{aligned}$$

By the concavity of log and Jensen's Inequality

$$(2.12)\qquad \begin{aligned} \log \int_{V_c} f_\omega(x_1, \cdots, x_n) L_2(\omega) \frac{\xi(d\omega)}{\xi(V_c)} &\geqq \int_{V_c} \log f_\omega(x_1, \cdots, x_n) \frac{\xi(d\omega)}{\xi(V_c)} \\ + \int_{V_c} \log L_2(\omega) \frac{\xi(d\omega)}{\xi(V_c)} &= \sum_{j=1}^n \int_{V_c} \log f_\omega(x_j) \frac{\xi(d\omega)}{\xi(V_c)} \\ &+ \int_{V_c} \log L_2(\omega) \frac{\xi(d\omega)}{\xi(V_c)}. \end{aligned}$$

Recalling that $\{U_i\}$ covers θ,

$$\log \sup_\theta g_\theta(x_1, \cdots, x_n) = \log \sup_{i \leq T_c} \sup_{U_i} g_\theta(x_1, \cdots, x_n)$$

$$(2.13) \qquad = \sup_i \sup_{U_i} \sum_{j=1}^{n} \log g_\theta(x_j) \leq \sup_i \sum_{j=1}^{n} \sup_{U_i} \log g_\theta(x_j)$$

$$= \sup_i \sum_{j=1}^{n} \log \sup_{U_i} g_\theta(x_j).$$

Put

$$(2.14) \quad A_c = |\log \xi(V_c)| + \log \beta_1 + |\log c| - \int_{V_c} \log L_2(\omega) \frac{\xi(d\omega)}{\xi(V_c)}$$

$$(2.15) \quad Y_j = \int_{V_c} \log f_\omega(X_j) \frac{\xi(d\omega)}{\xi(V_c)} \qquad \text{for } j = 1, 2, \cdots$$

$$(2.16) \quad Z_j^i = \log \sup_{U_i} g_\theta(X_j) \qquad \text{for } i = 1, \cdots, T_c; j = 1, 2, \cdots.$$

Let $N_3(c)$ be the first n such that

$$(2.17) \qquad \sum_{j=1}^{n} Y_j - \sup_i \sum_{j=1}^{n} Z_j^i > A_c$$

or, equivalently, the first n such that

$$(2.18) \quad \sum_{j=1}^{n} [Y_j - \epsilon_c - Z_j^1] + \min_{1 \leq i \leq T_c} \sum_{j=1}^{n} [Z_j^1 + \epsilon_c - Z_j^i] > A_c.$$

(2.12)–(2.16) imply that $N_3(c) \geq N_2(c)$. From (2.8) we have

$$(2.19) \qquad E_{\omega_0}[Y_j - \epsilon_c - Z_j^i] \geq \lambda_1(\omega_0, \theta_i) - 3\epsilon_c \text{ for } i = 1, \cdots, T_c.$$

Suppose that $\{U_i\}$ are indexed so that the minimum (over i) of the left hand side of (2.19) occurs when $i = 1$. Then

$$(2.20) \qquad E_{\omega_0}[Y_j - \epsilon_c - Z_j^i] \geq \lambda_1(\omega_0) - 3\epsilon_c;$$
$$E_{\omega_0}[Z_j^1 + \epsilon_c - Z_j^i] \geq \epsilon_c \quad \text{for } i = 1, \cdots, T_c.$$

Let S_n denote the nth partial sum of the sequence $\{Y_j - \epsilon_c - Z_j^1\}$ and let B_n^i ($i = 1, \cdots, T_c$) denote the nth partial sum of the sequence $\{Z_j^1 + \epsilon_c - Z_j^i\}$. Since $\{X_j\}$ is a sequence of independent and identically distributed random variables the same is true for each of the sequences in the last sentence. Put $B_n = \min_{i \leq T_c} B_n^i$. Then (2.18) is equivalent to $S_n + B_n > A_c$. Let $N^*(c)$ be the first n such that, simultaneously,

$$(2.21) \qquad S_n > A_c \quad \text{and} \quad B_n \geq 0.$$

It is obvious that $N^*(c) \geq N_3(c)$. Our problem now is to show that the lemma holds for $N^*(c)$.

We hereafter assume, as we may, that $\lambda_1(\omega_0) - 3\epsilon_c > 0$. Let ν_1 be the first n such that $S_n > A_c$, ν_2 the second n such that $S_n > A_c$, etc. Let ϕ_t be the indicator function of the set where $B_{\nu_t} < 0, t = 1, 2, \cdots$. Then

$$N^*(c) = \nu_1 + \sum_{j=1}^{\infty} (\nu_{j+1} - \nu_j) \prod_{t=1}^{j} \phi_t .$$

Let $\nu_{j+1}^* - \nu_j$ be the first $m > 0$ such that $S_{m+\nu_j} - S_{\nu_j} > 0$. Since $S_{\nu_j} > A_c$, it follows that $S_{\nu^*_{j+1}} > A_c$ and, therefore, $\nu_{j+1}^* - \nu_j \geqq \nu_{j+1} - \nu_j$. Since $\nu_{j+1}^* - \nu_j$ depends on X's whose indices are greater than ν_j, it follows that $\nu_{j+1}^* - \nu_j$ is independent of $\phi_1, \cdots, \phi_j$. Consequently

$$E_{\omega_0} N^*(c) \leqq E_{\omega_0}\nu_1 + \sum_{j=1}^{\infty} E_{\omega_0}(\nu_{j+1}^* - \nu_j) E_{\omega_0} \prod_{t=1}^{j} \phi_t . \tag{2.22}$$

$\nu_{j+1}^* - \nu_j$ has the same distribution as the first n for which $S_n > 0$ so that, using (4.6) of Spitzer (1956), the Chebyshev inequality, (2.20), Assumption 3, and (2.5), we have

$$\begin{aligned} E_{\omega_0}(\nu_{j+1}^* - \nu_j) &= 1 + \sum_{k=1}^{\infty} P\{\max_{1 \leqq j \leqq k} S_j \leqq 0\} \\ &= \exp\left[\sum_{k=1}^{\infty} (1/k) P\{S_k \leqq 0\}\right] \\ &\leqq \exp\left\{[\mathrm{Var}_{\omega_0}(Y_1 - \epsilon_c - Z_1^1)/(\lambda_1(\omega_0) - 3\epsilon_c)^2] \sum_{k=1}^{\infty} (1/k^2)\right\} \\ &= D(c, \omega_0) \quad \text{(say)}. \end{aligned} \tag{2.23}$$

For the purpose of estimating $E_{\omega_0} \prod_{t=1}^{j} \phi_t$ we let $\sigma_i (i = 1, \cdots, T_c)$ be the *last* time $B_n^i < 0$ and then $\sigma = \max(\sigma_1, \cdots, \sigma_{T_c})$ is the last time $B_n < 0$. Observe that ν_j is at least as big as j so that

$$\begin{aligned} \sum_{j=1}^{\infty} E_{\omega_0} \prod_{t=1}^{j} \phi_t &= \sum_{j=1}^{\infty} P_{\omega_0}\{B_{\nu_1} < 0, \cdots, B_{\nu_j} < 0\} \leqq \sum_{j=1}^{\infty} P_{\omega_0}\{\sigma \geqq \nu_j\} \\ &\leqq \sum_{j=1}^{\infty} P_{\omega_0}\{\sigma \geqq j\} = E_{\omega_0}\sigma \leqq \sum_{i=1}^{T_c} E_{\omega_0}\sigma_i . \end{aligned} \tag{2.24}$$

We now apply Theorem D in the Appendix to $\{-B_k^i\}$ with $\mu_0 = \epsilon_c$ since, by (2.20), the summands in B_k^i have mean $\geqq \epsilon_c$, and, by (2.5), they have finite variance, so that

$$E_{\omega_0}\sigma_i \leqq -(2/\epsilon_c) E_{\omega_0}[\min_{k \geqq 0} B_k^i] + \sum_{k=1}^{\infty} P\{B_k^i - k\epsilon_c/2 < 0\}. \tag{2.25}$$

Since $B_k^i - k\epsilon_c/2$ is the kth partial sum of independent and identically distributed random variables whose mean is $\geqq \epsilon_c/2$ and whose variance is finite, we obtain from Erdös (1949) the fact that the last term in (2.25) is finite. The first term to the right of the inequality in (2.25) is finite because the summands in B_k^i have positive ($\geqq \epsilon_c$) mean and finite variance (this is well-known and can be seen in Theorem B of the Appendix). (2.24) and (2.25) yield

$$\sum_{j=1}^{\infty} E_{\omega_0} \prod_{t=1}^{j} \phi_t = h(c, \omega_0) \text{ (say) } < \infty. \tag{2.26}$$

The final estimate we need is the well-known fact from renewal theory that

$$E_{\omega_0} \nu_1 = \frac{[1 + o(1)]A_c}{E_{\omega_0}[Y_1 - \epsilon_c - Z_1^1]} \leqq \frac{[1 + o(1)]A_c}{\lambda_1(\omega_0) - 3\epsilon_c}. \tag{2.27}$$

Putting (2.23), (2.26), (2.27) into (2.22) we get

$$E_{\omega_0} N^*(c) \leqq [1 + o(1)]A_c/[\lambda_1(\omega_0) - 3\epsilon_c] + D(c, \omega_0)h(c, \omega_0).$$

The final step is to observe that ϵ_c may be chosen to go to 0 so slowly that the choices of V_c, $\{U_i\}$, etc. (which depend on c through ϵ_c) result in

$$|\log \xi(V_c)| = o(|\log c|), \qquad D(c, \omega_0)h(c, \omega_0) = o(|\log c|).$$

For, if this is the case,

$$A_c \leqq [1 + o(1)] \, |\log c| + \log \beta_1 + |\log \alpha_2| = [1 + o(1)] \, |\log c|$$

and then $E_{\omega_0} N^*(c) \leqq [1 + o(1)] \, |\log c|/\lambda_1(\omega_0)$, which completes the proof.

If we put $\lambda_2(\theta) = \inf_\omega \lambda_2(\omega, \theta)$ we can state

LEMMA 2′. *If Ω is compact, then for each θ_0*

$$E_{\theta_0} N(c) \leqq \frac{[1 + o(1)] \, |\log c|}{\lambda_2(\theta_0)} \tag{2.28}$$

as $c \to 0$ (the $o(1)$ term may depend on θ_0; see, however, Lemma 3′).

The proof of Lemma 2′ is, of course, the same as for Lemma 2.

REMARKS.

1. We may extend Lemmas 2 and 2′ to cases where Ω and Θ are not bounded subsets of a Euclidean space by using the following compactification assumption (in addition to assuming that our previously stated assumptions hold for compact subsets):

ASSUMPTION 4. There is for each ω a number r_ω such that

(a) $E_\omega[\log f_\omega(X) - \log \sup_{|\theta| \geqq r_\omega} g_\theta(X)]^2 < \infty$ and

(b) $E_\omega[\log f_\omega(X) - \log \sup_{|\theta| \geqq r_\omega} g_\theta(X)] \geqq \lambda_1(\omega)$.

The argument of Lemma 2 will go through quite easily by observing that we can take the $\{U_i\}$ there as a covering of $\Theta \cap \{\theta \mid |\theta| \leqq r_{\omega_0}\}$ and then we can take the remaining part of Θ as a neighborhood of infinity and add that to $\{U_i\}$ to give a covering of Θ. There is no problem in carrying out the remaining part of the argument. An analogous assumption will take care of Lemma 2′. These assumptions are satisfied in many common problems, for example, in ones of the univariate or multivariate exponential (*K-D*) or translation parameter type.

2. When, for any ω, we have

$$E_\omega[\log f_\omega(X) - \log \sup_{|\theta| \geqq r} g_\theta(X)] < \lambda_1(\omega) \tag{*}$$

for all r then the argument of Lemma 2 breaks down unless we have

ASSUMPTION 4′. If ω is such that (*) holds then there is a sequence $\{r_n\}$ (which may depend on ω) such that $r_n \to +\infty$ and

(a) $E_\omega[\log f_\omega(X) - \log \sup_{|\theta| \geqq r_n} g_\theta(X)]^2 < \infty$ for each n (boundedness in n is *not* assumed);

(b) The limit as $r \to +\infty$ of the left side of (*) is equal to $\lambda_1(\omega)$. (The limit exists because the l.h.s. of (*) is increasing in r).

To treat Lemma 2 in the presence of Assumption 4′ observe that, for fixed ω_0, we can take r_n sufficiently large *depending on* c so that $\Theta \cap \{\theta \mid |\theta| \geqq r_n\}$ can be added to $\{U_i\}$ to form a covering of Θ. Turning to (2.20) with the possibility now that $Z_j^1 = \log \sup_{|\theta| \geqq r_n} g_\theta(X_j)$ we see that all is well provided r_n is also large enough so that the l.h.s. of (*) with r replaced by r_n is larger than $\lambda_1(\omega_0) - \epsilon_c$.

For generalizing Lemma 2 to unbounded Ω and Θ we make the obvious remark that it is only necessary to have each $\omega \in \Omega$ satisfy either (a) or (b) of Assumption 4 or (a) or (b) of Assumption 4′.

3. The situations not covered by Remarks 1 and 2 include, of course, those where the second moment in (a) of each of the two assumptions fails to exist in the right way. In most regular situations the second moment will either exist for any choice of r or never exist. Here and in our other assumptions, finiteness of second moments can be weakened; recent work by R. H. Farrell replaces t^2 by convex $h(t)$ with $\lim_{t\to\infty} h(t)/|t| = \infty$ for some of these considerations. The possible pathology in the behavior of the first moment is that, for some ω,

$$\inf_\theta \lambda_1(\omega, \theta) = \lambda_1(\omega) > \lim_{r\to\infty} E_\omega[\log f_\omega(X) - \log \sup_{|\theta| \geqq r} g_\theta(X)]$$

indicating a lack of continuity at a point of infinity in Θ.

ASSUMPTION 5.

(a) $E_\omega[\log f_\omega(X) - \log g_\theta(X)]^2$ and $E_\omega[\log \sup_{|\theta'-\theta| \leqq \rho} g_{\theta'}(X) - \log g_\theta(X)]^2$ are continuous in ω; $E_\omega[\log f_{\omega'}(X) - \log f_\omega(X)]^2$ is jointly continuous in ω and ω'.

(b) Interchange ω and θ, ω' and θ', f_ω and g_θ in (a) and the assumption is continuity in θ and joint continuity in θ, θ'.

LEMMA 3. *If Ω and Θ are compact and Assumptions* 1, 2, 3 *and* 5 *are satisfied, then there is a constant M such that*

$$E_\omega N(c) \leqq M\,|\log c|, \qquad E_\theta N(c) \leqq M\,|\log c|$$

for all ω and θ.

PROOF. We will only consider $E_\omega N(c)$. Observe by the first part of the proof of Lemma 2 that, for each $\omega \in \Omega$, there is a family $U_1, \cdots, U_{T(\omega)}$ of open sets covering Θ such that for $i = 1, \cdots, T(\omega)$

$$E_\omega[\log f_\omega(X) - \log \sup_{U_i} g_\theta(X)] \geqq \lambda_1/2$$

$$E_\omega[\log f_\omega(X) - \log \sup_{U_i} g_\theta(X)]^2 < G_i(\omega) \text{ (say) } < \infty.$$

By making use of Assumptions 2, 3, and 5 we can obtain an open set $V(\omega)$ containing ω such that for any $\omega' \in V(\omega)$ and $\omega'' \in V(\omega)$

$$(2.29)\qquad E_{\omega'}[\log f_{\omega''}(X) - \log \sup_{U_i} g_\theta(X)] \geqq \lambda_1/4, \qquad i = 1, \cdots, T(\omega)$$

$$(2.30)\qquad E_{\omega'}[\log f_{\omega''}(X) - \log \sup_{U_i} g_\theta(X)]^2 \leqq G_1'(\omega) < \infty, \quad i = 1, \cdots, T(\omega)$$

$\{V(\omega), \omega \in \Omega\}$ covers Ω so we can extract a finite covering $V(\omega_1), \cdots, V(\omega_k)$. We will show that for each j there is a constant M_j such that

$$(2.31)\qquad \sup_{\omega \in V(\omega_j)} E_\omega N(c) \leqq M_j \,|\log c|$$

and then take $M = M_1 + \cdots + M_k$.

We will establish (2.31) for $j = 1$. Fix $\omega_0 \in V(\omega_1)$. We can now emulate the proof of Lemma 2 starting at (2.14), replacing V_c by $V(\omega_1)$ and ϵ_c by $\lambda_1/8$, and using (2.29) and (2.30) to conclude that

$$(2.32)\qquad E_{\omega_0} N(c) \leqq E_{\omega_0} \nu_1 + E_{\omega_0} \nu^* \sum_{i=1}^{T(\omega_1)} E_{\omega_0} \sigma_i$$

where ν_1 is the first time $S_n > A_c = |\log c| + \gamma$ (γ is some constant), S_n is the nth partial sum of independent and identically distributed random variables whose mean is $\geqq \lambda_1/8$ and whose variance is $\leqq \sum_{i=1}^{T(\omega_1)} G_i'(\omega_1) = G'$ (say) $< \infty$, ν^* has the same distribution as the first time $S_n > 0$, and σ_i is the last time $B_n^i < 0$ where B_n^i is the nth partial sum of independent and identically distributed random variables with mean larger than $\lambda_1/8$ and variance $\leqq 2G'$. From (2.23) it is immediate that

$$(2.33)\qquad E_{\omega_0} \nu^* \leqq \exp\left\{G'(8/\lambda_1)^2 \sum_{k=1}^{\infty} (1/k^2)\right\} = \rho_1 \quad \text{(say)}$$

for *all* $\omega_0 \in V(\omega_1)$. From (2.25) we obtain

$$E_{\omega_0} \sigma_i \leqq (-16/\lambda_1) E_{\omega_0}[\min_{n \geqq 0} B_n^i] + \sum_{k=1}^{\infty} P_{\omega_0}\{B_k^i - k\lambda_1/16 < 0\}.$$

The mean of the summands in B_k^i is always greater than $\lambda_1/8$ and the variance is bounded as long as $\omega_0 \in V(\omega_1)$ so we can use Theorems A and B in the Appendix to conclude that

$$(2.34)\qquad E_{\omega_0} \sigma_i \leqq \rho_2 \quad \text{(say)}, \qquad i = 1, \cdots, T(\omega_1), \text{ all } \omega_0 \in V(\omega_1).$$

Similarly, by Theorem C in the Appendix, we obtain

$$(2.35)\qquad E_{\omega_0} \nu_1 \leqq \rho_3 \,|\log c|, \qquad \text{all } \omega_0 \in V(\omega_1).$$

(2.33), (2.34), (2.35) used in (2.32) give (2.31) for $j = 1$ and we are finished.

REMARK 4. A refinement of the proof of Lemma 3 using the full strength of Theorem C in the Appendix can be used to show that the $o(1)$ term in Lemma 2 (and Lemma 2$'$) is uniform in $\omega_0(\theta_0)$. This refinement is necessary in the proof of Theorem 2 (to obtain (2.57)), and we state it as

LEMMA 3$'$. *Inequalities* (2.3) *and* (2.28) *hold with the* $o(1)$ *term independent of* ω_0 *and* θ_0.

REMARK 5. It will sometimes be possible to establish the conclusion of Lemmas 3 and 3$'$ even when Ω and Θ are not compact. If it can be shown that

$\sup_\Omega E_\omega N(c) = \sup_{\Omega_0} E_\omega N(c)$ where Ω_0 is a compact subset of Ω and similarly for Θ then Lemma 3 is clearly true. When Ω, Θ are disjoint intervals on the real line and f_ω, g_θ are exponential densities with respect to the same measure then such a result can be established. This is the context of the work of G. Schwarz (1962). Other examples are easy to give.

The next lemma is essentially due to Schwarz who proved it in the case of exponential densities. It will be seen that Lemma 4 is not used in the proof of Theorem 1 although a proof of Theorem 1 can be constructed using Lemma 4. We state it here because of its intrinsic interest and because of the intuitive understanding which it lends to the whole subject.

Let $S(c)$ be the set of points in the sample space where δ_c says to stop. Let $B(c)$ be the set of points where a Bayes solution with respect to F (and with c the cost of observation) says to stop.

LEMMA 4. *If Ω and Θ are compact and Assumptions* 1, 2, 3 *and* 5 *are satisfied, then there are positive constants σ and τ such that $S(\sigma c) \subset B(c) \subset S(\tau c\,|\log c|)$ for all $c \leq c_0(\tau)$ ($c_0(\tau)$ is some positive number).*

REMARK 6. What is needed here is the validity of the conclusions of Lemmas 2 and 2′ and the conclusion of Lemma 3.

PROOF. The first inclusion will follow from the fact that no Bayes solution would continue to take observations if the a posteriori loss is smaller than the cost of taking another observation. Indeed, if $x_1, \cdots, x_n$ have been observed, and we have

$$\int L_2(\omega) f_\omega(x_1, \cdots, x_n)\xi\,(d\omega) < \sigma c \int L_1(\theta) g_\theta(x_1, \cdots, x_n)\eta\,(d\theta)$$

then the a posteriori risk is smaller than

$$\frac{\sigma c \int L_1(\theta) g_\theta(x_1, \cdots, x_n)\eta\,(d\theta)}{\int g_\theta(x_1, \cdots, x_n)\eta\,(d\theta) + \int f_\omega(x_1, \cdots, x_n)\xi\,(d\theta)} \leq \sigma c \beta_1$$

with a similar relation if δ_c stops "at the other hypothesis." How to choose σ is apparent.

The second inclusion requires further argument. If $\delta_{\tau c|\log c|}$ says continue after observing $x_1, \cdots, x_n$ we have (abbreviating $f_\omega(x_1, \cdots, x_n)$ by f_ω^n and similarly with g_θ^n)

$$\tau c\,|\log c| \int L_1(\theta) g_\theta^n \eta\,(d\theta) < \int L_2(\omega) f_\omega^n\,\xi\,(d\omega) < \frac{1}{\tau c\,|\log c|} \int L_1(\theta) g_\theta^n \eta\,(d\theta).$$

Thus, the a posteriori loss when c is the cost, after observing $x_1, \cdots, x_n$, is greater than

$$\frac{\int L_2(\omega) f_\omega^n\,\xi\,(d\omega)}{\int f_\omega^n\,\xi\,(d\omega) + \int g_\theta^n \eta\,(d\theta)} \min\,[1, \tau c\,|\log c|],$$

and is also greater than

$$\frac{\int L_1(\theta) g_\theta^n \eta\,(d\theta)}{\int f_\omega^n \xi\,(d\omega) + \int g_\theta^n \eta\,(d\theta)} \min\,[1, \tau c\,|\log c|].$$

Consequently, the a posteriori risk is larger than

$$\tfrac{1}{2} \min\,[\alpha_1, \alpha_2] \min\,[1, \tau c\,|\log c|] \geqq \tau' c\,|\log c| \tag{2.36}$$

for c small enough (say $c \leqq c'$).

We will now show, if no τ yields the stated result, how to obtain a procedure which is an improvement of the Bayes procedure, and this gives a contradiction. Suppose B' is the set of sample points where the Bayes procedure stops and $\delta_{\tau c|\log c|}$ says continue. If $P_\omega(B')(P_\theta(B'))$ is the measure of B' induced by $f_\omega(g_\theta)$ then we might as well assume that

$$\pi(B') = \int P_\omega(B')\xi\,(d\omega) + \int P_\theta(B')\eta\,(d\theta) > 0.$$

If $x \,\varepsilon\, B'$ and n is the time the Bayes procedure says stop let us continue to take observations until the first ν such that

$$\frac{\int L_2(\omega) f_\omega(x_1, \cdots, x_{n+\nu})\xi\,(d\omega)}{\int L_1(\theta) g_\theta(x_1, \cdots, x_{n+\nu})\eta\,(d\theta)} > \frac{1}{c} \quad \text{or} \quad < c. \tag{2.37}$$

If we put $\xi^n(d\omega) = f_\omega^n \xi(d\omega)/(\int f_\omega^n \xi(d\omega) + \int g_\theta^n \eta(d\theta))$ and $\eta^n(d\theta) = g_\theta^n \eta(d\theta)/(\int f_\omega^n \xi(d\omega) + \int g_\theta^n \eta(d\theta))$ then to calculate the "conditional" properties (given $(x_1, \cdots, x_n)$) of (2.37) we need only turn to Lemmas 1 and 2 and substitute ξ^n for ξ. Thus the "conditional" loss (given $(x_1, \cdots, x_n)$) is smaller than $(\beta_1 + \beta_2)c$ and the "conditional" expected number of observations is, for each ω,

$$E_\omega \bar{N}(c) \leqq [1 + o(1)]\,|\log c|/\lambda_1(\omega) \tag{2.38}$$

and, for each θ,

$$E_\theta \bar{N}(c) \leqq [1 + o(1)]\,|\log c|/\lambda_2(\theta) \tag{2.39}$$

We are assuming here that ξ^n has Ω as its support and that η^n has Θ as its support. This is inessential and (2.38) and (2.39) will hold in any case as can be seen by applying Lemma 2 to the support of ξ^n and the support of η^n and concluding (2.38) with $\lambda_1(\omega)$ replaced by $\inf\{\lambda_1(\omega, \theta) \mid \theta \epsilon \text{ support of } \eta^n\}$ which is larger than $\lambda_1(\omega)$.

Let $\epsilon_1(\omega, x_1, \cdots, x_n, c)$ denote the $o(1)$ term in (2.38) and let $\epsilon_2(\theta, x_1, \cdots, x_n, c)$ denote the $o(1)$ term in (2.39)—the dependence on $x_1, \cdots, x_n$ is, of course, through ξ^n and η^n. From Lemma 3 we know that there is an $M(x_1, \cdots, x_n)$ such that $|\epsilon_1(\omega, x_1, \cdots, x_n, c)| + |\epsilon_2(\theta, x_1, \cdots, x_n, c)| \leqq M(x_1, \cdots, x_n)$

for all ω and all θ. Thus there is a number M_0 and a set $B^* \subset B'$ with $\pi(B^*) > 0$ such that for $x \varepsilon B^*$

$$|\epsilon_1(\omega, x_1, \cdots, x_n, c)| + |\epsilon_2(\theta, x_1, \cdots, x_n, c)| \leqq M_0$$

for all ω and all θ.

Now consider the procedure of (2.37) only for $x \varepsilon B^*$. We conclude from (2.38) and (2.39) that, for $\epsilon > 0$, there is a c_0 (which is the same for all $x \varepsilon B^*$) such that for all $c \leqq c_0$

$$\int E_\omega \bar{N}(c)\xi^n (d\omega) + \int E_\theta \bar{N}(c)\eta^n (d\theta) \leqq [1 + \epsilon] |\log c|/\min (\lambda_1, \lambda_2).$$

Thus if τ is chosen so that $\tau' > (1 + \epsilon)/\min (\lambda_1, \lambda_2)$ (see (2.36)) we can conclude that, for any $x \varepsilon B^*$, the a posteriori risk when the Bayes procedure says stop is larger than for the modified procedure. This contradiction establishes Lemma 4.

THEOREM 1. *If Ω and Θ are compact and Assumptions* 1, 2, 3, *and* 5 *are satisfied, then $\{\delta_c\}$ is asymptotically Bayes, i.e.,* $\lim_{c\to 0} r_c(F, \delta_c^*)/r_c(F, \delta_c) = 1$ *where δ_c^* is a Bayes solution when F is the a priori distribution and c the cost per observation.*

REMARK 7. Referring to Remarks 1–6 it is possible to describe more general conditions under which Theorem 1 is true.

PROOF. Let $\gamma_1 = \int_\Omega \xi(d\omega)/\lambda_1(\omega)$, $\gamma_2 = \int_\Theta \eta(d\theta)/\lambda_2(\theta)$. From Lemma 2 we know that, for each $\omega \varepsilon \Omega$,

$$\limsup_{c\to 0} E_\omega N(c)/|\log c| \leqq 1/\lambda_1(\omega)$$

and from Lemma 3 we have $E_\omega N(c) \leqq M |\log c|$ for all $\omega \varepsilon \Omega$. Consequently

$$\limsup_{c\to 0} \frac{1}{|\log c|} \int_\Omega E_\omega N(c)\xi (d\omega) \leqq \int_\Omega \limsup_{c\to 0} \frac{E_\omega N(c)}{|\log c|} \xi (d\omega) \leqq \gamma_1 .$$

Doing the same for Θ and letting $E_F N(c) = \int_\Omega E_\omega N(c)\xi(d\omega) + \int_\Theta E_\theta N(c)\eta(d\theta)$ we obtain

$$\limsup_{c\to 0} E_F N(c)/|\log c| \leqq \gamma_1 + \gamma_2 .$$

This together with Lemma 1 yields

$$\limsup_{c\to 0} r_c(F, \delta_c)/c\, |\log c| \leqq \gamma_1 + \gamma_2 . \tag{2.40}$$

If $\{\delta_c\}$ is *not* asymptotically Bayes we would have $\liminf_{c\to 0} r_c(F, \delta_c^*)/r_c(F, \delta_c) < 1 - 2\epsilon$ for some positive ϵ which implies, due to (2.40), that there is a sequence $\{c_i\}$ with $c_i \to 0$ such that

$$r_{c_i}(F, \delta_{c_i}^*) < (1 - 2\epsilon)c_i\, |\log c_i|\, (\gamma_1 + \gamma_2) \tag{2.41}$$

and, consequently,

$$E_F N^*(c_i) < (1 - 2\epsilon)\, |\log c_i|\, (\gamma_1 + \gamma_2) \tag{2.42}$$

where $N^*(c)$ is the number of observations required when δ_c^* is used. (2.42) implies that either $\int E_\omega N^*(c_i)\xi(d\omega) < (1-2\epsilon)\,|\log c_i|\,\gamma_1$ or $\int E_\theta N^*(c_i)\eta(d\theta) < (1-2\epsilon)\,|\log c_i|\,\gamma_2$. Suppose the former. Then

$$\int \left[\frac{E_\omega N^*(c_i)}{|\log c_i|} - \frac{1-\epsilon}{\lambda_1(\omega)}\right]\xi(d\omega) \leqq -\epsilon\gamma_1;$$

while from Lemma 3, Assumption 2, and the compactness of Ω,

$$\sup_{\omega,i}\left|\frac{E_\omega N^*(c_i)}{|\log c_i|} - \frac{1-\epsilon}{\lambda_1(\omega)}\right| = M_1 \quad \text{(say)},$$

where $M_1 < \infty$. Now, letting $\Omega_i = \{\omega \mid E_\omega N^*(c_i) - (1-\epsilon)\,|\log c_i|/\lambda_1(\omega) < 0\}$, we have

$$-M_1\xi(\Omega_i) \leqq \int_{\Omega_i}\left[\frac{E_\omega N^*(c_i)}{|\log c_i|} - \frac{(1-\epsilon)}{\lambda_1(\omega)}\right]\xi(d\omega) \leqq -\epsilon\gamma_1,$$

which implies that $\xi(\Omega_i) \geqq \epsilon\gamma_1/M_1 = \epsilon_1$ (say). Let $P_\omega^*(c)$ and $Q_\theta^*(c)$ be the probabilities of wrong decision when, respectively, ω and θ are true and δ_c^* is the decision function used. From (2.41) we know that

$$\int \frac{P_\omega^*(c_i)}{c_i\,|\log c_i|}\,\xi(d\omega) \leqq \frac{(\gamma_1+\gamma_2)}{\alpha_2}$$

(α_2 is defined in Assumption 1). Let $\epsilon_2 > 0$ and $\epsilon_1 > \epsilon_2$. It is possible to find a number K_1 such that $A_i = \{\omega \mid P_\omega^*(c_i) \leqq K_1 c_i |\log c_i|\}$ has the property that $\xi(A_i) \geqq \xi(\Omega) - \epsilon_2$; in fact, we can take $K_1 = (\gamma_1+\gamma_2)/(\alpha_2\epsilon_2)$. Let $B_i = \Omega_i \cap A_i$. Then $\xi(B_i) \geqq \epsilon_1 - \epsilon_2 > 0$ for each i. Since $\xi(\Omega) < \infty$ and the B_i have uniformly positive ξ-measure, it is easy to see that there is a subsequence $\{i_1, i_2, \cdots\}$ of the positive integers such that $B_{i_1} \cap B_{i_2} \cap \cdots \neq \phi$. Since (2.41) is valid for the subsequence $\{c_{i_j}\}$ of $\{c_i\}$ we might as well assume that $\{i_1, i_2, \cdots\}$ is the set of positive integers and, therefore, $\cap B_i \neq \phi$. Pick $\omega_0 \varepsilon \cap B_i$. By definition of Ω_i and A_i we have, for all i,

$$E_{\omega_0}N^*(c_i) \leqq (1-\epsilon)|\log c_i|/\lambda_1(\omega_0) \tag{2.43}$$

$$P_{\omega_0}^*(c_i) \leqq K_1 c_i\,|\log c_i|. \tag{2.44}$$

Let $\Theta_1 = \{\theta\,\varepsilon\,\Theta \mid (1-\epsilon)/\lambda_1(\omega_0) < (1-\epsilon/2)/\lambda_1(\omega_0,\theta)\}$. By the continuity of $\lambda_1(\omega,\theta)$ in θ we have Θ_1 non-empty and open and consequently $\eta(\Theta_1) > 0$. Just as the sets A_i were obtained above, we can find numbers K_2 and K_3 and sets G_i ($i = 1, 2, \cdots$) such that $G_i \subset \Theta_1$, $\eta(G_i) \geqq \epsilon_3 > 0$ for all i, and $E_\theta N^*(c_i) \leqq K_2\,|\log c_i|$, $Q_\theta^*(c_i) \leqq K_3\,|\log c_i|$ for $\theta\,\varepsilon\,G_i$. Again, as with $\{B_i\}$, we can find a subsequence of $\{G_i\}$ which has non-empty intersection and, for simplicity, we can assume $\cap G_i \neq \phi$. Choose $\theta_0\,\varepsilon\,G_i$ and we conclude that

$$E_{\theta_0}N^*(c_i) \leqq K_2\,|\log c_i| \tag{2.45}$$

$$Q_{\theta_0}^*(c_i) \leqq K_3\,|\log c_i| \tag{2.46}$$

and, furthermore, from the definition of Θ_1 and (2.43), we obtain

$$E_{\omega_0}N^*(c_i) < (1 - \epsilon/2)|\log c_i|/\lambda_1(\omega_0, \theta_0). \tag{2.47}$$

Consider the problem of testing ω_0 vs. θ_0 with cost of observation Hc_i ($H > 0$ and will be chosen below), 0-1 loss function, and a priori probabilities α and $1 - \alpha$ (to be chosen below) on ω_0 and θ_0 respectively. If δ_i' denotes the Bayes solution for this problem then, since δ_i' is a sequential probability ratio test (SPRT), it follows from the properties of SPRT's (which could be obtained from the previous results of this paper or from Wald (1947) or Chernoff (1959)) that

$$\begin{aligned}(2.48)\quad &\alpha r_{Hc_i}(\omega_0, \delta_i') + (1 - \alpha)r_{Hc_i}(\theta_0, \delta_i') \\ &\sim \alpha Hc_i|\log Hc_i|/\lambda_1(\omega_0, \theta_0) + (1 - \alpha)Hc_i|\log Hc_i|/\lambda_2(\omega_0, \theta_0).\end{aligned}$$

Considering $\delta_{c_i}^*$ as a test of these two simple hypotheses in the obvious way, we obtain from (2.44), (2.45), (2.46), and (2.47) that

$$\begin{aligned}(2.49)\quad &\alpha r_{Hc_i}(\omega_0, \delta_{c_i}^*) + (1 - \alpha)r_{Hc_i}(\theta_0, \delta_{c_i}^*) \leqq (\alpha K_1 + (1 - \alpha)K_3)c_i|\log c_i| \\ &+ Hc_i|\log c_i|[\alpha(1 - \epsilon/2)/\lambda_1(\omega_0, \theta_0) + (1 - \alpha)K_2].\end{aligned}$$

Now choose H such that $K_1 + (1 - \epsilon/2)H/\lambda_1(\omega_0, \theta_0) < (1 - \epsilon/4)H/\lambda_1(\omega_0, \theta_0)$, and then choose $\alpha < 1$ so that $(1 - \alpha)K_3 + (1 - \alpha)K_2H < (\alpha H\epsilon/8)/\lambda_1(\omega_0, \theta_0)$. This choice of H and α makes the right hand side of (2.49) smaller than

$$\frac{(1 - \epsilon/8)\alpha Hc_i|\log c_i|}{\lambda_1(\omega_0, \theta_0)} \sim \frac{(1 - \epsilon/8)\alpha Hc_i|\log Hc_i|}{\lambda_1(\omega_0, \theta_0)}$$

and this leads to a contradiction of (2.48).

COROLLARY. *If $\{\delta_c'\}$ is any family of procedures for which*

$$\sup_\omega r_c(\omega, \delta_c') + \sup_\theta r_c(\theta, \delta_c') \leqq Kc|\log c| \tag{2.50}$$

for some constant K, then, letting $N'(c)$ be the number of observations required by δ_c' to terminate,

$$E_\omega N'(c) \geqq [1 + o(1)]|\log c|/\lambda_1(\omega) \quad \textit{for all } \omega, \tag{2.51}$$

$$E_\theta N'(c) \geqq [1 + o(1)]|\log c|/\lambda_2(\theta) \quad \textit{for all } \theta, \tag{2.52}$$

which imply that

$$r_c(\omega, \delta_c') \geqq [1 + o(1)]|\log c|/\lambda_1(\omega) \tag{2.53}$$

and

$$r_c(\theta, \delta_c') \geqq [1 + o(1)]|\log c|/\lambda_2(\theta). \tag{2.54}$$

REMARK. 8. The uniformity in (2.50) is unnecessary; it is sufficient to have $r_c(\omega, \delta_c')$ and $r_c(\theta, \delta_c')$ of order $c|\log c|$ for each ω, θ.

PROOF. If (2.51) fails for ω_0 (say) then there is a positive number ϵ and a sequence $\{c_i\}$ with $c_i \to 0$ such that

$$E_{\omega_0} N'(c_i) < (1 - \epsilon)|\log c_i|/\lambda_1(\omega_0). \tag{2.55}$$

We can now argue as in Theorem 1 since (2.44), (2.45), and (2.46) follow immediately from (2.50) and (2.47) follows from (2.55).

It is important to note that the family of procedures $\{\delta_c\}$ depends on F and that the asymptotic Bayes property of $\{\delta_c\}$ is with respect to the same F. It is easily seen that $\{\delta_c\}$ is also asymptotically Bayes relative to any G whose support is the same as that of F and for which dG/dF is bounded. However it would be desirable to be able to state that $\{\delta_c\}$ is asymptotically Bayes with respect to *every* G having the same support (namely, $\Omega \cup \theta$) as F, without any further restrictions. To be able to state such a result it is sufficient to prove, in view of Lemma 3′ and Theorem 1, that

$$\sup_\omega P_\omega(c) + \sup_\theta Q_\theta(c) = o(c\,|\log c|). \tag{2.56}$$

Of course, if Ω and Θ are finite, then (2.56) follows trivially from Lemma 1, with $O(c)$ for $o(c\,|\log c|)$. Our methods do not seem sufficient to prove (2.56) in general, but we are able to achieve essentially the same end by obtaining the same result (see (2.59)) for a family $\{\delta_c'\}$ which still satisfies the conclusion of Lemma 3′ (see (2.60)) and which differs only slightly from $\{\delta_c\}$ (essentially by using slightly larger stopping bounds; Albert (p. 798, (c)) discusses a similar difficulty and device in his treatment). We require some slight further restrictions on $\{f_\omega\}$ and $\{g_\theta\}$ to obtain this result on the existence of a family $\{\delta_c'\}$ which satisfies this asymptotic Bayes property for all G having $\Omega \cup \Theta$ as its support. To this end let us assume

ASSUMPTION 6. For each $\omega \in \Omega$, $\theta \in \Theta$, $E_\omega[\sup_{|\theta'-\theta|<\rho} g_{\theta'}(X)/f_{\omega'}(X)]$ is finite for some $\rho > 0$ and all ω' in some neighborhood of ω. is continuous at $\rho = 0$, and is continuous in ω and ω'.

Since $E_\omega[g_\theta(X)/f_\omega(X)] = 1$ and $E_\omega[g_\theta(X)/f_\omega(X)]^h$ is convex and not constant in $h \in [0, 1]$, we know that for ϵ small enough there is an h_ϵ with $h_\epsilon \to 1$ as $\epsilon \to 0$ such that $E_\omega[g_\theta(X)/f_\omega(X)]^{h_\epsilon} \leqq 1 - 2\epsilon$. Then by Assumption 6 there is a ρ_ϵ and a neighborhood V_ϵ of ω such that

$$E_{\omega''}[\sup_{|\theta'-\theta|<\rho_\epsilon} g_{\theta'}(X)/f_{\omega'}(X)]^{h_\epsilon} \leqq 1 - \epsilon$$

for all $\omega' \in V_\epsilon$ and $\omega'' \in V_\epsilon$. Taking advantage of the compactness of Θ and Ω in the same way that we did in the proofs of Lemmas 2 and 3, we can find open sets $U_1, \cdots, U_{T_\epsilon}$ which cover Θ, and open sets $V_1, \cdots, V_{M_\epsilon}$ which cover Ω, and a number h_ϵ^* such that

$$\sup_{\omega' \in V_j, \omega'' \in V_j} E_{\omega''}[\sup_{U_i} g_\theta(X)/f_{\omega'}(X)]^{h_\epsilon^*} \leqq 1 - \epsilon$$

for $j = 1, \cdots, M_\epsilon$, $i = 1, \cdots, T_\epsilon$, and with $h_\epsilon^* \to 1$ as $\epsilon \to 0$.

If we define $\delta_{c,\epsilon}^* = \delta_{\bar c}$ where $\bar c = c^{1/h_\epsilon^*}$, then, from Lemma 3′, we have

$$E_\omega N(\delta^*_{c,\epsilon}) \leqq [1 + o(1)]|\log c|/\lambda_1(\omega)h^*_\epsilon \tag{2.57}$$

with the $o(1)$ term going to zero with c and uniformly in ω. A similar relation holds for $E_\theta N(\delta^*_{c,\epsilon})$. To calculate $P_\omega(\delta^*_{c,\epsilon})$ observe that, as in the proof of Lemma 2, we have, for all $\omega' \varepsilon V_j$,

$$P_{\omega'}(\delta^*_{c,\epsilon}) \leqq P_{\omega'}\left\{\sum_{k=1}^n \int_{V_j} \log f_\omega(X_k) \frac{\xi(d\omega)}{\xi(V_j)} - \sup_{i \leqq T_\epsilon} \sum_{k=1}^n \log \sup_{U_i} g_\theta(X_k) < \frac{\log c}{h^*_\epsilon} + |\log \xi(V_j)| \text{ for some } n\right\}.$$

Hence, putting

$$S_{in} = \sum_{k=1}^\infty \int_{V_j} [\log f_\omega(X_k) - \log \sup_{U_i} g_\theta(X_k)] \frac{\xi(d\omega)}{\xi(V_j)},$$

we have, for all $\omega' \varepsilon V_j$

$$\begin{aligned} P_{\omega'}(\delta^*_{c,\epsilon}) &\leqq \sum_{i=1}^{T_\epsilon} \sum_{n=1}^\infty P_{\omega'}\{S_{in} < \log c/h^*_\epsilon + |\log \xi(V_j)|\} \\ &= \sum_{i=1}^{T_\epsilon} \sum_{n=1}^\infty P_{\omega'}\{\exp(-h^*_\epsilon S_{in}) > (1/c) \exp[h^*_\epsilon \log \xi(V_j)]\}. \end{aligned} \tag{2.58}$$

Now,

$$E_{\omega'} \exp(-h^*_\epsilon S_{in}) = [E_{\omega'} \exp(-h^*_\epsilon S_{i1})]^n$$

and

$$\begin{aligned} E_{\omega'} \exp(-h^*_\epsilon S_{i1}) &\leqq E_{\omega'} \int_{V_j} \left[\frac{\sup_{U_i} g_\theta(X_1)}{f_\omega(X_1)}\right]^{h^*_\epsilon} \frac{\xi(d\omega)}{\xi(V_j)} \\ &\leqq \int_{V_j} (1-\epsilon) \frac{\xi(d\omega)}{\xi(V_j)} = (1-\epsilon); \end{aligned}$$

and this, used in (2.58) via the Chebyshev inequality, yields

$$P_{\omega'}(\delta^*_{c,\epsilon}) \leqq \sum_{i=1}^{T_\epsilon} \sum_{n=1}^\infty \{c/[\xi(V_j)]^{h^*_\epsilon}\}(1-\epsilon)^n = T_\epsilon c/\epsilon[\xi(V_j)]^{h^*_\epsilon}.$$

If we let ϵ depend on c (we will write ϵ_c) in such a way that $T_{\epsilon_c}/\epsilon_c[\xi(V_j)]^{h^*_{\epsilon_c}} = o(|\log c|)$ as $c \to 0$ and $\epsilon_c \to 0$ as $c \to 0$, then we obtain, for the procedures $\delta'_c = \delta^*_{c,\epsilon_c}$,

$$\sup_\omega P_\omega(\delta'_c) = o(c|\log c|); \tag{2.59}$$

and, since $h^*_{\epsilon_c} \to 1$ as $c \to 0$, we will also have from (2.57)

$$E_\omega N(\delta'_c) \leqq [1 + o(1)]|\log c|/\lambda_1(\omega) \tag{2.60}$$

with the $o(1)$ uniform in ω. A similar result for $Q_\theta(\delta'_c)$ and $E_\theta N(\delta'_c)$, when combined with the above, yield

THEOREM 2. *If Ω and Θ are compact and Assumptions* 1, 2, 3, 5, 6 *are satisfied, then, for a given a priori distribution F whose support is $\Omega \cup \Theta$, the family $\{\delta_c'\}$ just defined satisfies* (2.59), (2.60), *and their analogues in Θ, and is therefore asymptotically Bayes (as the cost $c \to 0$) with respect to every a priori distribution G having the same support as F.*

3. Indifference region, two decisions. With Ω, Θ as before let us introduce an indifference region I. For $\alpha \varepsilon I$ let h_α denote the density of an observation. Let F be an a priori distribution on $\Omega \cup \Theta \cup I$ and let ξ (resp. η, ψ) denote the restriction of F to Ω (resp. Θ, I). The loss function on Ω and Θ satisfies the assumptions of Section 2, and it is zero on I. (For the sake of notational simplicity, we will carry out the details below when the loss is zero-or-one, but the proof in general is the same except for obvious changes.) Let $\xi^{(n)} = \int_\Omega f_\omega(x_1, \cdots, x_n)\xi(d\omega)$, $\eta^{(n)} = \int_\Theta g_\theta(x_1, \cdots, x_n)\eta(d\theta)$, $\psi^{(n)} = \int_I h_\alpha(x_1, \cdots, x_n)\psi(d\alpha)$. Let $\lambda_{I1}(\alpha) = \inf_\omega E_\alpha[\log h_\alpha(X_1) - \log f_\omega(X_1)]$, $\lambda_{I2}(\alpha) = \inf_\theta E_\alpha[\log h_\alpha(X_1) - \log g_\theta(X_1)]$ and define $I_2 = \{\alpha \varepsilon I \mid \lambda_{I1}(\alpha) \geqq \lambda_{I2}(\alpha)\}$, $I_1 = \{\alpha \varepsilon I \mid \lambda_{I1}(\alpha) \leqq \lambda_{I2}(\alpha)\}$.

We may as well assume, and do, that F has $\Omega \cup \Theta \cup I$ as its support. Our regularity assumptions are the following:

ASSUMPTION IIIA. $\Omega \cup I_1$ and $\Theta \cup I_2$ are compact.

ASSUMPTION IIIB. The assumptions of Section 2 hold when (Ω, Θ) of Section 2 is replaced here by $(\Omega \cup I_1, \Theta)$ and also when (Ω, Θ) is replaced here by $(\Omega, \Theta \cup I_2)$. (In this replacement of Ω by $\Omega \cup I_1$, f_ω on Ω is replaced by f_ω on Ω together with h_α on I_1, etc.)

REMARK 9. As in Section 2, the compactness can often be weakened. The form of the assumptions as stated above is designed to include common cases where I is not closed but where its closure contains points of Ω and Θ. For example, in the case of normal random variables with variance one and mean μ, in testing $\Omega: a \leqq \mu \leqq -1$ against $\Theta: 1 \leqq \mu \leqq b$ with indifference region $-1 < \mu < 1$, we obtain $I_1 = \{-1 < \mu \leqq 0\}$ and $I_2 = \{0 \leqq \mu < 1\}$, neither of which is compact, but the above assumptions are satisfied.

Define δ_c^I as follows: take an $(n+1)$th observation if

$$(3.1) \quad \xi^{(n)}/[\xi^{(n)} + \eta^{(n)} + \psi^{(n)}] > c \quad \text{and} \quad \eta^{(n)}/[\xi^{(n)} + \eta^{(n)} + \psi^{(n)}] > c,$$

stop and choose Ω (resp., Θ) if

$$(3.2) \quad \eta^{(n)}/[\xi^{(n)} + \eta^{(n)} + \psi^{n}] \leqq c \text{ (resp., } \xi^{(n)}/[\xi^{(n)} + \eta^{(n)} + \psi^{(n)}] \leqq c)$$

and $\xi^{(n)} > \eta^{(n)}$ (resp., $\xi^{(n)} < \eta^{(n)}$), randomizing if $\xi^{(n)} = \eta^{(n)}$. The symmetric form of (3.2) is not really necessary; when both inequalities of (3.2) are satisfied, either decision could be made. (The form of (3.1) and (3.2) can be altered to reflect a posteriori *loss*, as in Section 2; again, this is a trivial alteration which is not even needed in view of the identical asymptotic behavior of the present and altered forms.)

Since we are only concerned with probabilities of error when the true state of nature is either in Ω or Θ, Lemma 1 is easily verified. Let $N_I(c)$ be the number of observation for δ_c^I to terminate. When $\omega \varepsilon \Omega$ is true, $E_\omega N_I(c) \leqq E_\omega N(c)$ where $N(c)$ is the first time $\eta^{(n)}/[\xi^{(n)} + \eta^{(n)}] \leqq c$ or, equivalently, $\xi^{(n)}/\eta^{(n)} \geqq 1/c - 1$. Now we can use Lemma 2 (which will work despite the fact that $\xi(\Omega) + \eta(\Theta) < 1$; it is only the positivity of $\xi(\Omega)$ and $\eta(\Theta)$ which is relevant) to conclude that

$$E_\omega N_I(c) \leqq [1 + o(1)] \, |\log c|/\lambda_1(\omega)$$

where λ_1 is as in Section 2. Similarly we get

$$E_\theta N_I(c) \leqq [1 + o(1)] \, |\log c|/\lambda_2(\theta).$$

For $\alpha \varepsilon I_2$, we have $E_\alpha N_I(c) \leqq E_\alpha N'(c)$ where $N'(c)$ is the first n such that $\xi^{(n)}/[\xi^{(n)} + \psi_2^{(n)}] \leqq c$, where $\psi_2^{(n)} = \int_{I_2} h_\alpha(x_1, \cdots, x_n)\psi(d\alpha)$. Again we use Lemma 2 to conclude that

$$E_\alpha N_I(c) \leqq [1 + o(1)] \, |\log c|/\lambda_{I2}(\alpha), \qquad \alpha \varepsilon I_2$$

and, similarly,

$$E_\alpha N_I(c) \leqq [1 + o(1)] \, |\log c|/\lambda_{I1}(\alpha), \qquad \alpha \varepsilon I_1.$$

Lemma 3 follows as before, noting again that we are always dealing with compact sets; for example, if the true state of nature is $\alpha \varepsilon I_1$, we are dealing with $\Omega \cup I_1$ and Θ. Lemma 4 also follows as before (but, again, it will not be used in proving the Theorem).

In the required modification of the proof of Theorem 1, we must now consider both the contingencies of Section 2 and also the possibility of $\int_I E_\alpha N^*(c_i)\psi(d\alpha)$ being too small, where $N^*(c_i)$ denotes the number of observations required by the Bayes solution to stop when c_i is the cost. As for the former, since δ_c^I is a test of Ω vs. Θ, we can conclude from Section 2 that

$$\text{(3.3)} \quad \int_\Omega E_\omega N_I^*(c)\xi(d\omega) + \int_\Theta E_\theta N_I^*(c)\eta(d\theta) \geqq [1 + o(1)] \, |\log c| \left[\int \frac{\xi(d\omega)}{\lambda_1(\omega)} + \int \frac{\eta(d\theta)}{\lambda_2(\theta)}\right].$$

As for the latter, to obtain the result

$$\text{(3.4)} \quad \int_I E_\alpha N_I^*(c)\psi(d\alpha) \geqq [1 + o(1)] \, |\log c| \int_I \frac{\psi(d\alpha)}{\max(\lambda_{I1}(\alpha), \lambda_{I2}(\alpha))}$$

we require Lemma 5 below. This lemma, which obtains the required results in the case where each of Ω, Θ, and I consists of one element, is used to obtain (3.4) in the same way that the analogous results concerning the SPRT were used to obtain (3.3) in Section 2. Putting these results together as in Section 2, we obtain

THEOREM 3. *Under the assumptions of the present section, with δ_c replaced by δ_c^I, the conclusion of Theorem 1 holds.*

We also obtain

COROLLARY. *Under the assumptions of the present section, if $\{\delta_c'\}$ is any family of* *procedures for which* $\sup_\omega r_c(\omega, \delta_c') + \sup_\theta r_c(\theta, \delta_c') + \sup_\alpha r_c(\alpha, \delta_c') \leqq Kc\,|\log c|$ *for some constant* K, *then* (2.51) *anc* (2.52) (*and thus* (2.53) *and* (2.54)) *hold, and also*

$$(3.5)\quad [1 + o(1)]\, c^{-1} \sup_\alpha r_c(\alpha, \delta_c') = \sup_\alpha E_\alpha N'(c) \geqq \frac{[1 + o(1)]\,|\log c|}{\max\,[\lambda_{f1}(\alpha), \lambda_{f2}(\alpha)]}.$$

In fact, all of the Corollary except for (3.5) follows from the Corollary to Theorem 1 (Section 2) and the fact that any test in the present context can be regarded as a test for the problem of Section 2. Lemma 5 yields (3.5) in the same way that properties of the SPRT yielded (2.51) and (2.52) in Section 2.

REMARK 10. The analogue of Remark 8 applies here.

We now turn to Lemma 5. We shall state and prove, for use in Section 4, a more general result than that needed in the present section, where we need consider only three states f, g, h. (We have dropped subscripts for convenience.) The specialization of Lemma 5 to the present case is then obtained by putting h for ω; f for θ_1, g for θ_2, "decision i" for the decision that θ_i is *not* the true state, and $m = k = 2$; Condition (3.7) is vacuous. We then obtain the desired conclusion that, for either δ_c^I or for the Bayes stopping rule,

$$(3.6)\qquad E_h N(c) \geqq [1 + o(1)]\, |\log c| / \max\,(\lambda_{f1}, \lambda_{f2}).$$

(The analogous results for f and g, which have already been stated in a more general context in the Corollary, followed from a comparison with the SPRT; they can also be obtained from Lemma 5, which generalizes to certain multiple decision problems the particular SPRT optimality results it states when $m = 1$, $k = 2$.)

LEMMA 5. *Consider any k-decision problem where the observations* $\{X_i\}$ *are taken sequentially and are independent and identically distributed. Let* $\omega, \theta_1, \cdots, \theta_m$ ($m \leqq k$) *be any* $m + 1$ (*not necessarily distinct*) *states of nature with corresponding densities* $f, g_1, \cdots, g_m$. *Suppose* $\lambda(\omega, \theta_i) = E_\omega[\log f(X) - \log g_i(X)] > 0$ *and* $E_\omega[\log f(X) - \log g_i(X)]^2 < \infty$ *for* $i = 1, \cdots, m$. *Put* $\lambda = \max_i \lambda(\omega, \theta_i)$. *Suppose* $\{\delta_j\,;\, j \geqq 1\}$ *is a sequence of decision functions satisfying*

$$(3.7)\qquad \sum_{t=m+1}^{k} P_\omega\{\delta_j \text{ makes decision } t\} = o(1) \qquad \text{as } j \to \infty$$

$$(3.8)\qquad P_{\theta_i}\{\delta_j \text{ makes decision } i\} \leqq A c_j\, |\log c_j|$$

for all i, *all* j, *where* A *is some positive constant and* $c_j \to 0$ *as* $j \to \infty$. ((3.7) *is empty if* $m = k$.) *Let* $N(j)$ *be the number of observations required by* δ_j *to terminate. Then* $E_\omega N(j) \geqq [1 + o(1)]\, |\log c_j| / \lambda$ as $j \to \infty$.

PROOF. Write N for $N(j)$ (there will be no confusion since j will be fixed for each calculation), and let $\epsilon > 0$. Put $S_{iN} = \sum_{r=1}^{N} [\log g_i(X_r) - \log f(X_r)]$, $D_i = \{S_{iN} \geqq -(1 - \epsilon)\, |\log c_j|\}$, $A_{in} = \{N = n, \delta_j \text{ selects decision } i\}$ ($i = 1, \cdots, m$). We are assuming here, merely for convenience, that each δ_j is non-randomized. Put $B_{in} = D_i \cap A_{in}$. The existence of $\lambda(\omega, \theta_1)$ implies that on the set where g_i is 0, f must be 0 *a. e.* (μ) so that, following Lemma 4 of Chernoff, we have

$$\begin{aligned} P_\omega\{\delta_j \text{ makes decision } i, D_i\} &= \sum_{n=1}^{\infty} \int_{B_{in}} \prod_{j=1}^{n} f(x_j)\, d\mu^{(n)} \\ &= \sum_{n=1}^{\infty} \int_{B_{in}} \prod_{j=1}^{n} [f(x_j)/g_i(x_j)] g_i(x_j)\, d\mu^{(n)} \\ &= \sum_{n=1}^{\infty} \int_{B_{in}} e^{-S_{iN}} \prod g_i(x_j)\, d\mu^{(n)} \\ &\leq c_j^{\epsilon-1} \sum_{n=1}^{\infty} P_{\theta_i}\{B_{in}\} \leq c_j^{\epsilon-1} P_{\theta_i}\{\delta_j \text{ makes decision } i\} \leq A c_j^{\epsilon} |\log c_j| \end{aligned}$$

where the last inequality follows from (3.8). Thus

$$(3.9) \qquad P_\omega\{\delta_j \text{ makes decision } i, D_i\} = o(1) \qquad \text{as } j \to \infty$$

We might as well assume that

$$(3.10) \qquad \sum_{t=1}^{k} P_\omega\{\delta_j \text{ makes decision } t\} = 1,$$

since, otherwise, we would have $E_\omega N = +\infty$.

Now

$$(3.11) \qquad \begin{aligned} P_\omega\{N < -(1-2\epsilon) \log c_j/\lambda\} \leq \sum_{i=1}^{m} P_\omega\{N < -(1-2\epsilon) - \log c_j/\lambda;& \\ S_{iN} < (1-\epsilon)\log c_j\} + P_\omega\{D_1 \cap \cdots \cap D_m\}.& \end{aligned}$$

From (3.10), (3.9), and (3.7),

$$(3.12) \qquad \begin{aligned} P_\omega\{D_1 \cap \cdots \cap D_m\} \leq \sum_{i=1}^{m} P_\omega\{D_i, \delta_j \text{ decides } i\}& \\ + \sum_{t=m+1}^{k} P_\omega\{\delta_j \text{ decides } t\} = o(1)& \end{aligned}$$

as $j \to \infty$. Also, for $r > 0$,

$$(3.13) \qquad \begin{aligned} &P_\omega\{N < -(1-\epsilon)\log c_j/\lambda,\ S_{iN} < (1-\epsilon)\log c_j\} \\ &\leq P_\omega\{\min_{1\leq n\leq r} S_{in} < (1-\epsilon)\log c_j\} \\ &\quad + P_\omega\{S_{iN}/N < -[(1-\epsilon)/(1-2\epsilon)]\,(\lambda(\omega,\theta_i));\ N > r\}. \end{aligned}$$

The last probability can be made arbitrarily small by taking r sufficiently large and using once more the result of Erdös (1949), while the first probability on the right side of (3.13) goes to zero for fixed r as $j \to \infty$. Hence,

$$(3.14) \quad P_\omega\{N < -[(1-2\epsilon)/\lambda]\log c_j;\ S_{iN} < (1-\epsilon)\log c_j\} = o(1)$$

as $j \to \infty$. Thus, using (3.14) and (3.12) in (3.11), we obtain

$$P_\omega\{N < -[(1-2\epsilon)/\lambda] \log c_j\} = o(1) \tag{3.15}$$

as $j \to \infty$, which proves Lemma 5.

The conclusion of Lemma 5 when $k = m = 2$ is related to a result of Hoeffding (1960) whose concern, in this situation, is to find lower bounds for $E_\omega N$, and one of his lower bounds (to be precise, (1.4) in Hoeffding's paper) has the asymptotic value given by Lemma 5 when (3.8) (of Lemma 5) holds. Hoeffding's conditions are stronger than ours so that his result does not include Lemma 5.

The reader may find it illuminating to consider the specialization of these results to the three-state problem discussed in Section 1 in connection with Figure 4. Writing $U_i = \log[f(X_i)/h(X_i)]$, $V_i = \log[g(X_i)/h(X_i)]$, $S_n = \sum_1^n U_i$, and $T_n = \sum_1^n V_i$ (with $S_0 = T_0 = 0$), the Bayes continuation region, by the analogue of Lemma 4, is easily seen to be

$$\max\{\,|S_n - T_n|, -S_n, -T_n\} < |\log c|\,(1 + o(1)), \tag{3.16}$$

for any f, h, g satisfying our conditions. If f, h, g are of exponential type with parameter values increasing in the order f, h, g, then S_n, T_n and n are linearly related, and the unbounded polygonal region (3.16) becomes Schwarz's (bounded) pentagon as exemplified by β of Figure 4 in the symmetric normal case.

Again, as in Section 2, we are able to establish the analogue of Theorem 2 for a family $\{\delta_c'^I\}$ obtained by modifying $\{\delta_c^I\}$, rather than for the original family $\{\delta_c^I\}$. In fact, there is now an additional difficulty: we cannot immediately refer to the argument preceding Theorem 2 because I is not necessarily separated from $\Omega \cup \Theta$ (see the example cited earlier in this section). We get around this by defining δ_{1c} to be the test of Section 2 with the same Ω but Θ replaced by $\Theta \cup I_2$, and define δ_{2c} to be the test of Section 2 with the same Θ but with Ω replaced by $\Omega \cup I_1$. Let $\bar{\delta}_c^I$ be the procedure which (1) continues observing if both δ_{1c} and δ_{2c} continue, (2) stops and selects Ω if δ_{2c} stops and selects $\Omega \cup I_1$ while δ_{1c} continues or stops and selects Ω, (3) stops and selects Θ if δ_{1c} stops and selects $\Theta \cup I_2$ while δ_{2c} continues or stops and selects Θ, and (4) stops and randomizes in all other situations. (In the case of finitely many possible states of nature this "simultaneous test" is essentially like δ_c^I or the procedure described in Figure 3; in the nonseparated case, it will differ from δ_c^I.) It is now possible to modify $\bar{\delta}_c^I$ (by modifying δ_{1c} and δ_{2c} in exactly the way δ_c was modified in the argument leading to Theorem 2) to obtain $\delta_c'^I$ with the desired properties.

We also observe that the family $\{\delta_c'^I\}$ whose construction was just described for a given F with support $\Omega \cup \Theta \cup I$ has an asymptotic risk function which, on any subset I' of I, is identical to that for the analogous asymptotically Bayes family relative to an F' with support $\Omega \cup \Theta \cup I'$. Thus, *the family* $\{\delta_c'^I\}$ *is also asymptotically optimum for the problem where* I *is replaced by a smaller* I' (although of course $\{\delta_c'^{I'}\}$ need not be optimum for $\Omega \cup \Theta \cup I$). This optimality of the given family for problems with reduced indifference region, which was discussed in the example illustrated by Figure 4, is to be contrasted with the loss of optimality which occurs if Ω or Θ is changed, except in certain cases such as those of exponential families as treated by Schwarz, and for which one family of procedures is optimum whether $\Omega = \{\omega : \omega \leq a\}$ or $\Omega = \{\omega : \omega = a\}$.

It follows from this remark on the effect of I that, in designing sequential procedures for large sample applications, we may as well take I to be so large as to include even the most remotely suspected indifferent possibilities, as long as our assumptions remain satisfied and computations remain tractable. "Robustness" can easily be built into indifference region performance from the outset, in the asymptotic theory. (Of course, the rate of approach to optimality may depend on the size of the chosen I.)

THEOREM 4. *The family* $\{\delta_c'^I\}$ *whose construction (with respect to any* F *with support* $\Omega \cup \Theta \cup I$*) was just described satisfies* P_ϕ {wrong decision} $= o(c\,|\log c|)$ *and*

$$E_\phi N'(c)/|\log c| = [1 + o(1)]r_c(\phi, \delta_c'^I)/c\,|\log c|$$

$$= \begin{cases} [1 + o(1)]/\lambda_1(\phi) & \text{if } \phi \varepsilon \Omega, \\ [1 + o(1)]/\lambda_2(\phi) & \text{if } \phi \varepsilon \Theta, \\ [1 + o(1)]/\max[\lambda_{I1}(\phi), \lambda_{I2}(\phi)] & \text{if } \phi \varepsilon I, \end{cases}$$

where the $o(1)$ *terms do not depend on* ϕ. *Hence,* $\{\delta_c'^I\}$ *is asymptotically Bayes relative to every a priori distribution* G *with support* $\Omega \cup \Theta \cup I'$ *where* $I' \subset I$.

4. k-decisions, with or without indifference and semi-indifference regions. The generalization of Section 2 and 3 to a k-decision problem is quite straightforward. First assume no indifference region, and let Ω_i, $i = 1, \cdots, k$ be k disjoint compact subsets of a Euclidean space and for $\omega \varepsilon \bigcup_{i=1}^k \Omega_i$ let f_ω denote the density of an observation. Let F be an a priori distribution on $\bigcup_{i=1}^k \Omega_i$ and let ξ_i be its restriction to Ω_i. For the sake of simplicity we shall describe the results when the loss is 0-1, the modifications needed for the more general case being obvious upon reference to Section 2. Let $\xi_{i,n,x}$ be the a posteriori probability of Ω_i when $x_1, \cdots, x_n$ is observed. Let δ_c be the procedure which stops at the first n for which $\xi_{i,n,x} > 1 - c$ for some i and selects that Ω_i for which $\xi_{i,n,x} > 1 - c$ (or randomizes in any way if there are several such i's, which could only occur if $c > 1/k$). The appropriate assumptions are obtained from those of Section 2 by replacing Ω and Θ by Ω_i and $\bigcup_{j \neq i} \Omega_j$, respectively, when the true state is an element of Ω_i (in place of Ω), for each $i = 1, 2, \cdots, k$. (For example we define $\lambda(\omega) = \inf_{\theta \notin \Omega_i} E_\omega \log [f_\omega(X_1)/f_\theta(X_1)]$ if $\omega \varepsilon \Omega_i$.) It can then be proved that $\{\delta_c\}$ is asymptotically Bayes, i.e., the conclusion of Theorem 1 holds for $\{\delta_c\}$. The key is to regard δ_c as a test of Ω_i vs $\bigcup_{j \neq i} \Omega_j$ when concerned with $E_\omega N(c)$ for $\omega \varepsilon \Omega_i$. In fact all arguments (except Lemma 1, which is trivial) can be reduced to those of Section 2. The Corollary and Theorem 2 follow similarly. The problem can also be treated through simultaneous tests, as exemplified in Figure 1.

When a single indifference region I is present (so that, for each i, if the true state is in Ω_i (resp., I), the loss is positive (resp., 0) if decision d_j is made, where $j \neq i$), the assumptions and methods of Section 3 carry over directly. There are now k sets $I_j = \{\alpha \varepsilon I \mid \lambda_{Ij}(\alpha) = \min_i \lambda_{Ii}(\alpha)\}$, etc.

For the more general situation when there may exist several "semi-indifference" regions, i.e., regions where there may be several possible correct decisions and also several possible incorrect decisions, we only need slightly more care to see that our methods carry over. Let Ω_0 denote the entire parameter space. For convenience we again take the loss function to be 0-1. For $1 \leqq i_1 < i_2 < \cdots < i_m \leqq k$, $1 \leqq m \leqq k$ define $\Omega_{i_1 \cdots i_m} = \{\omega \,\varepsilon\, \Omega_0 \mid L(\omega, i_1) = \cdots = L(\omega, i_m) = 0;\ L(\omega, j) = 1 \text{ if } j \not\varepsilon \{i_1, \cdots, i_m\}\}$. Thus, $\Omega_{i_1 \cdots i_m}$ is the set of ω's where the decisions $i_1, \cdots, i_m$ are correct and all others incorrect; it may or may not be empty. When $m = k$, $\Omega_{i_1 \cdots i_k} = \Omega_{1 \cdots k}$ is a "true" indifference region while, if $2 \leqq m < k$, $\Omega_{i_1 \cdots i_m}$ is a "semi-indifference" region. Let $B_i = \{\omega \mid L(\omega, i) = 0\}$, $i = 1, \cdots, k$. Let F be an a priori distribution on Ω_0. Let $F_{n,x}$ be the a posteriori distribution after n observations when $x = (x_1, \cdots, x_n)$ is observed. Let us define δ_c to be the procedure which stops as soon as $F_{n,x}(B_i) > 1 - c$ for some i and makes that decision i for which $F_{n,x}(B_i) > 1 - c$ or randomizes (in any way) among all decisions i for which $F_{n,x}(B_i) > 1 - c$. Let $N_i(c)$ be the first n such that $F_{n,x}(B_i) > 1 - c$ and $N_i(c) = \infty$ if no such n exists, i.e., $N_i(c)$ is the first n such that

$$\int_{B_i} f_\omega(x_1, \cdots, x_n) F\,(d\omega) \Big/ \int_{\Omega_0} f_\omega(x_1, \cdots, x_n) F\,(d\omega) > 1 - c. \tag{4.1}$$

Then, if $N(c)$ is the number of observations required by δ_c to terminate, we have

$$N(c) = \min\,(N_1\,(c), \cdots, N_k\,(c)). \tag{4.2}$$

For this procedure δ_c it is easy to check that the "loss" part of the risk must be $O(c)$, i.e., the analogue of Lemma 1 holds here. In fact, it follows, just as in Lemma 1, that

$$\int_{\Omega_0 - B_i} P_\omega\{\delta_c \text{ makes decision } i\} F\,(d\omega) \leqq c.$$

To proceed further we need to make assumptions and definitions that parallel those of Section 2.

ASSUMPTIONS.

1. $\lambda(\omega, \theta) = E_\omega\,[\log f_\omega\,(X) - \log f_\theta\,(X)]$ is continuous in both variables simultaneously. (f_ω (f_θ) denotes the density when the true state of nature is ω (θ).)
2. $\lambda(\omega, \theta) = 0$ if, and only if, $\omega = \theta$.
3. The obvious generalizations of Assumptions 3 and 5 of Section 2.
4. Ω_0 is compact, $\Omega_0 - B_i$ is compact for each i, and $F(\Omega_0 - B_i) > 0$ for each i.

REMARK 11. 1. and 2. are stronger than necessary; comparison with Assumptions 1 and 2 of Section 2 will provide weaker assumptions.

We define, for $\omega \,\varepsilon\, \Omega_{i_1 \cdots i_m}$,

$$\lambda(\omega) = \max_{1 \leqq j \leqq m} \min_{\theta \varepsilon \Omega_0 - B_{i_j}} \lambda(\omega, \theta) \tag{4.3}$$

Since $\{\Omega_{i_1 \cdots i_m}\}$ are disjoint sets there is no vagueness in this definition. Note that, when there are two possible decisions and a single indifference region as in Section 3, this definition is in agreement with the definition there of $\lambda(\alpha) = \max(\lambda_{I1}(\alpha), \lambda_{I2}(\alpha))$. The compactness of Ω_0 and $\Omega_0 - B_i$ and continuity of $\lambda(\omega, \theta)$ guarantees that $\min \lambda(\omega) > 0$.

From (4.2) we have, for $\omega \varepsilon \Omega_{i_1 \cdots i_m}$, that

$$E_\omega N(c) \leqq \min_{1 \leqq j \leqq m} E_\omega N_{i_j}(c). \tag{4.4}$$

From (4.1) it follows that $N_i(c)$ is no greater than the first n for which

$$\int_{B_i} f_\omega(x_1, \cdots x_n) F(d\omega) \Big/ \int_{\Omega_0 - B_i} f_\omega(x_1, \cdots x_n) F(d\omega) > \frac{1}{c}.$$

The methods of Lemma 2 enable us to conclude, therefore, that, for $\omega \varepsilon B_i$,

$$E_\omega N_i(c) \leqq [1 + o(1)] \, |\log c| / \min_{\theta \varepsilon \Omega_0 - B_i} \lambda(\omega, \theta). \tag{4.5}$$

Since $\omega \varepsilon \Omega_{i_1 \cdots i_m}$ implies that $\omega \varepsilon B_{i_1} \cap \cdots \cap B_{i_m}$, we can conclude from (4.4) and (4.5) that

$$E_\omega N(c) \leqq [1 + o(1)] \, |\log c| / \lambda(\omega) \tag{4.6}$$

as $c \to 0$, for each $\omega \varepsilon \Omega_{i_1 \cdots i_m}$. Thus the analogue of Lemma 2 is established. To obtain the analogues of Lemmas 3 and 3′ we can proceed as follows: Define

$$C_i = \{\omega \mid \lambda(\omega) = \min_{\theta \varepsilon \Omega_0 - B_i} \lambda(\omega, \theta)\}, \qquad i = 1, \cdots, k.$$

Some, but not all, C_i may be empty; the C_i's may not be disjoint; the union of the C_i's is Ω_0. Each C_i is closed (and therefore compact) because of the continuity of $\lambda(\omega, \theta)$. Since $\lambda(\omega) > 0$ for all ω we have $C_i \cap \Omega_0 - B_i$ empty for each i. Now argue, as in Lemma 3, by suitably covering C_i and $\Omega_0 - B_i$ as Ω and Θ were covered in proof of Lemma 3 to obtain $E_\omega N(c) \leqq M_i \, |\log c|$ for all $\omega \varepsilon C_i$. Taking $M = \max(M_1, \cdots, M_k)$ we obtain $E_\omega N(c) \leqq M \, |\log c|$ for all $\omega \varepsilon \Omega_0$.

To prove the result of Theorem 1, we note that it follows, as in Theorem 1, that if $\{\delta_c\}$ is not asymptotically Bayes then there is a sequence $\{c_j\}$ and a point $\omega_0 \varepsilon \Omega_{1 \cdots m}$ (or some $\Omega_{i_1 \cdots i_m}$ which for convenience of notation we take to be $\Omega_{1 \cdots m}$) such that, for the sequence of Bayes solutions $\{\delta^*_{c_j}\}$

$$E_{\omega_0} N^*(c_j) \leqq (1 - 2\epsilon) \, |\log c_j| / \lambda(\omega_0) \tag{4.7}$$

for some $\epsilon > 0$ and all j, and

$$\sum_{t=m+1}^{k} P_{\omega_0} \{\delta^*_{c_j} \text{ makes decision } t\} \leqq A c_j \, |\log c_j| \tag{4.8}$$

for all j. (If $\omega_0 \varepsilon \Omega_{1 \cdots k}$ then (4.8) is vacuous). Suppose for convenience that $\omega_0 \varepsilon C_1$ (ω_0 must be in $C_1 \cup \cdots \cup C_m$). Then, again as in the argument of Theorem 1, we can find $\theta_1 \varepsilon \Omega_0 - B_1$ such that $\lambda(\omega_0, \theta_1) \leqq [(1 - \epsilon)/(1 - 2\epsilon)] \lambda(\omega_0)$.

and $P_{\theta_1}\{\delta^*_{c_j}$ makes decision 1$\} < Ac_j\,|\log c_j|$ for all j. (Here and in (4.10) a subsequence of $\{c_j\}$ is rewritten, for convenience, as $\{c_j\}$, just as in the proof of Theorem 1.) We can continue and find, in addition to θ_1, (not necessarily distinct) values $\theta_2, \cdots, \theta_m$ with $\theta_i \,\varepsilon\, \Omega_0 - B_i$ such that

$$\lambda(\omega_0, \theta_i) \leqq [(1-\epsilon)/(1-2\epsilon)]\lambda(\omega_0) \tag{4.9}$$

$$P_{\theta_i}\{\delta^*_{c_j} \text{ makes decision } i\} \leqq Ac_j\,|\log c_j| \tag{4.10}$$

for all j and $1 \leqq i \leqq m$. We now apply Lemma 5 to $\delta^*_{c_j}$ with ω_0 playing the role of ω in the lemma (the conditions of the lemma are satisfied because of (4.7), (4.8), (4.10) and our assumptions) to obtain, with the help of (4.9),

$$E_{\omega_0}N^*(c_j) \geqq \frac{[1 + o(1)]|\log c_j|}{\max\limits_{1 \leqq i \leqq m} \lambda(\omega_0, \theta_i)} \geqq \frac{1 - 2\epsilon}{1 - \epsilon}[1 + o(1)]\frac{|\log c_j|}{\lambda(\omega_0)}.$$

This contradicts (4.7) and thereby establishes the result of Theorem 1 for $\{\delta_c\}$ as defined above (4.1).

The results of Theorems 2 and 4 can be obtained by a modification of $\{\delta_c\}$ like the one made in Section 3. In fact, for all pairs i, j ($i \leqq k$, $j \leqq k$) such that $C_i \cap B_j$ is not empty, let δ_{ijc} be the *test* of Section 2 of the closure $(\overline{C_i \cap B_j})$ of $C_i \cap B_j$ against $\Omega_0 - B_i$ (recall that C_i is defined after (4.6)). We now proceed by stopping the first time one of the δ_{ijc}'s terminates *and* accepts $\overline{C_i \cap B_j}$ and we make decision j (or randomize among those j's for which there are several pairs i, j).

THEOREM 5. *With* $\lambda(\omega)$ *as defined in* (4.3) *the results of Theorems* 1, 2, 3, *and* 4 *and their corollaries remain valid in k-decision problems with or without indifference and semi-indifference regions.*

The remarks in Section 3 concerning the dependence of the asymptotically optimum procedures on the support of the a priori distribution pertain here with the role of the indifference region played by $\Omega_{1\cdots k}$. The semi-indifference regions cannot be incorporated because (roughly) they affect the asymptotic value of $E_\omega N$ whereas the indifference region, as pointed out in Section 3, does not; and, in addition, the semi-indifference regions appear in the "loss" part of the risk whereas the indifference region does not.

We conclude the discussion in this section by considering the three-decision problem treated by Sobel and Wald (1949). The density functions are normal with (unknown) mean θ and variance 1. Numbers $\theta_1 < \theta_2 \leqq \theta_3 < \theta_4$ are specified and the problem is to decide whether $\theta \leqq \theta_1$, $\theta_2 \leqq \theta \leqq \theta_3$, or $\theta \geqq \theta_4$, with indifference between the first two decisions if the "true" $\theta \,\varepsilon\, (\theta_1, \theta_2)$, and with indifference between the second two decisions if $\theta \,\varepsilon\, (\theta_3, \theta_4)$. Thus, in the notation of this section, we have $\Omega_0 = (-\infty, +\infty)$, $\Omega_1 = (-\infty, \theta_1]$, $\Omega_2 = [\theta_2, \theta_3]$, $\Omega_3 = [\theta_4, \infty)$, $\Omega_{12} = (\theta_1, \theta_2)$, $\Omega_{23} = (\theta_3, \theta_4)$, and Ω_{13} and Ω_{123} are empty. According to (4.3) $\lambda(\theta) = (\theta - \theta_2)^2/2$ for $\theta \,\varepsilon\, \Omega_1$, $\lambda(\theta) = (\theta - \theta_3)^2/2$ for $\theta \,\varepsilon\, \Omega_3$, $\lambda(\theta) = \min\{(\theta - \theta_4)^2/2, (\theta - \theta_1)^2/2\}$ for $\theta \,\varepsilon\, \Omega_2$, etc. The procedure of Sobel and Wald is to test simultaneously (by sequential probability ratio tests) θ_1 against θ_2 and θ_3

against θ_4 and to decide that $\theta \leqq \theta_1$ if θ_1 and θ_3 are accepted, to decide $\theta_2 \leqq \theta \leqq \theta_3$ if θ_2 and θ_3 are accepted, and to decide $\theta \geqq \theta_4$ if θ_2 and θ_4 are accepted (it is impossible to accept θ_1 and θ_4 simultaneously under their restrictions). By comparing our asymptotic value of $E_\theta N(c)$ with that of their procedure (e.g. (6.8) or (6.9) of Sobel and Wald) it becomes clear that their procedure is asymptotically optimum if and only if $\theta_2 = \theta_3$ and the a priori distribution is concentrated on $\{\theta_1, \theta_2, \theta_4\}$. Moreover, when $\theta = (\theta_1 + \theta_2)/2$ their procedure gives $E_\theta N'(c)/|\log c| \to \infty$ where $N'(c)$ is the number of observations required by the Sobel-Wald procedure to terminate and give error probabilities of order $O(c)$. This last remark is analogous to the behavior of the SPRT for the problem of Figure 4.

5. The design problem. In this section we consider the problems of Sections 2, 3, and 4 when there are design questions developing from the additional feature that, at each stage of observation, the decision to take another observation is accompanied by a choice of an experiment (i.e., a design) to perform in order to make the additional observation. We shall always assume that the same design choices are available at each stage of observation. As mentioned in the summary and introduction, our aim here is to show that there are sequential procedures (each of which, of course, includes the choice of design at each stage) which can be described briefly, can be proved asymptotically optimum easily, and can be used in applications with a minimum of calculations and switching of designs. These procedures make use of an extension of the idea, first implemented by Wald (1951) in certain simpler estimation problems, of taking a preliminary sample which (when c is small) is large but is small relative to the total expected sample size, using this preliminary sample to estimate the "true" state of nature, and then deciding once and for all on the future course of experimentation. Before introducing detailed design considerations, let us make this idea more transparent. Here and throughout Section 5 we will give details only for the design problem associated with Section 2; the arguments in the context of Sections 3 and 4 will be quite similar.

Suppose first, to make the statement simple, that there are a finite number k of possible states of nature. For $1 \leqq i \leqq k$ suppose $\{\delta_c^i\}$ is a family of procedures (including choices of designs) such that, for all i and j,

$$\begin{aligned} E_j N(\delta_c^i) &= O(|\log c|), \\ E_i N(\delta_c^i) &= \mu_i |\log c| \, [1 + o(1)], \qquad (5.1) \\ P_j \{\text{wrong decision using } \delta_c^i\} &= o(c \log c). \end{aligned}$$

Let d_i be the "decision" that i is the true state (this is not necessarily the original decision space). Let $\{\delta_c^0\}$ be any family of procedures for which

$$\begin{aligned} E_j N(\delta_c^0) &= o(|\log c|), \\ P_i \{\delta_c^0 \text{ reaches decision } d_i\} &= 1 - o(1), \end{aligned} \qquad (5.2)$$

for $1 \leqq i, j \leqq k$. Consider the procedure δ_c^* defined as follows: First use δ_c^0. If "decision" d_i is made, then use δ_c^i (as though starting anew; we assume the same design choices are available at each stage). The final decision is reached using δ_c^i. Then, supposing for the moment that each observation costs the same amount c, (5.1) and (5.2) trivially yield, for the risk when i is true,

$$r_i(\delta_c^*) = cE_i N(\delta_c^*) [1 + o(1)] = \mu_i c|\log c| [1 + o(1)].$$

If μ_i can be shown to be the minimum possible value of $E_i N(\delta)$ as $c \to 0$ for any procedure which is Bayes relative to an F with support $\{1, 2, \cdots, k\}$, we will thus have, in $\{\delta_c^*\}$, an asymptotic Bayes procedure relative to any such F.

[Added in revision: L. R. Abramson, in his Columbia thesis, has, independently, employed this same technique in dealing with the case of two states of nature, two decisions, and k experiments, and his $\{\delta_c^0\}$ is a SPRT as is $\{\delta_c^i\}$, $i = 1, 2$. Under Abramson's restrictions, which imply that $\lambda^{e_j}(\theta) > 0$ for $\theta = \theta_1, \theta_2$ and $j = 1, \cdots k$ (see (5.9) and previous for the definition of λ^{e_j}), our construction of $\{\delta_c^i\}$ by any of the Methods I, II, III below shows that, in this special case, $\{\delta_c^i\}$ will be a SPRT for $i = 1, 2$ (this results from the fact that when there are two states of nature the maximin $\bar{e}$ (see (5.10) et seq.) is non-randomized), and one of the possibilities for our construction of $\{\delta_c^0\}$ (see the paragraph following (5.10)) is to take $\{\delta_c^0\}$ as a SPRT (the key here is that $\bar{e}^0$ can be taken, under Abramson's assumptions, as non-randomized).]

In the case where $\Omega \cup \Theta = \Phi$ consists of infinitely many states of nature we have to proceed with somewhat extra care. As in Section 2 let F be an a priori distribution on Φ and we suppose that the support of F is all of Φ. Suppose we can select designs and make observations $\{X_1, \cdots\}$ (not necessarily independent nor identically distributed) and, on the basis of these observations, suppose there is a sequence $\{t_n ; n \geqq 1\}$ of consistent estimators of the true parameter, i.e., for any $\epsilon > 0$

$$\lim_{n\to\infty} P_\phi \{ |t_n(X_1, \cdots, X_n) - \phi| < \epsilon\} = 1 \tag{5.3}$$

for each $\phi \varepsilon \Phi$. How to make the observations $X_1, \cdots$ in terms of the available designs is immaterial for the moment; we are merely supposing that it is possible to do so and to define $\{t_n\}$ which will satisfy (5.3). Let us further suppose that, for each $\gamma \varepsilon \Phi$ and each $\epsilon > 0$, there is a family $\{\delta_c^{\gamma,\epsilon}\}$ of decision procedures (which incorporates the choice of design at each stage) such that

$$\text{(a)} \quad \int_\Phi P_\phi\{\text{wrong decision using } \delta_c^{\gamma,\epsilon}\} F(d\phi) = O(c);$$

$$\text{(5.4)} \quad \text{(b)} \quad \sup_{\phi\varepsilon\Phi} E_\phi N(\delta_c^{\gamma,\epsilon}) = O(|\log c|)$$

$$\text{(c)} \quad E_{\phi'} N(\delta_c^{\gamma,\epsilon}) \leqq [1 + \epsilon + o(1)]\, |\log c|\, \mu(\phi'), \text{ for } |\phi' - \gamma| \leqq \epsilon'.$$

for some $\epsilon' > 0$ and where $\mu(\phi')$ is some positive number. Now define a family of procedures $\{\delta_c^{*\epsilon}\}$ as follows: Let $n(c)$ be a sequence of integers with $n(c) = o(|\log c|)$. Take $n(c)$ observations $x_1, \cdots, x_{n(c)}$ and compute $t_{n(c)}(x_1, \cdots, x_{n(c)})$. If $t_{n(c)} = \gamma$ use $\delta_c^{\gamma,\epsilon}$ and continue observation *independently of the $n(c)$ observations used to form $t_{n(c)}$* until $\delta_c^{\gamma,\epsilon}$ terminates and then make whatever decision $\delta_c^{\gamma,\epsilon}$ makes.

Thus, when c is the cost of observation,

$$\begin{aligned} r_c(\phi, \delta_c^{*\epsilon}) \leqq\ & E_\phi P_\phi\{\text{wrong decision using } \delta_c^{\gamma,\epsilon} \mid t_{n(c)} = \gamma\} + cn(c) \\ & + cO(|\log c|)P_\phi\{|t_{n(c)} - \phi| > \epsilon'\} \\ & + c\,|\log c|\,[1 + \epsilon + o(1)]\mu(\phi)P_\phi\{|t_{n(c)} - \phi| < \epsilon'\}. \end{aligned} \tag{5.5}$$

Now, integrating both sides with respect to F and using (b) of (5.4) we obtain

$$\begin{aligned} r_c(F, \delta_c^{*\epsilon}) &\leqq o(c\,|\log c|) + (1 + \epsilon)c\,|\log c| \int_\Phi \mu(\phi)F\,(d\phi) \\ &= [1 + \epsilon + o(1)]c\,|\log c| \int_\Phi \mu(\phi)F\,(d\phi) \end{aligned} \tag{5.6}$$

where the $o(1)$ term may depend on ϵ.

REMARK 12. If we assume that $\{t_n\}$ is uniformly consistent i.e., that the limit in (5.3) holds, for each $\epsilon > 0$, uniformly for $\phi \in \Phi$; if we assume that the $o(1)$ term in (c) of (5.4) is uniform in ϕ; and, finally, if we replace (a) of (5.4) by (a$'$)P_ϕ {wrong decision using $\delta_c^{\gamma,\epsilon}$} $= o(c\,|\log c|)$ uniformly in ϕ, we can obtain from (5.5)

$$r_c(\phi, \delta_c^{*\epsilon}) \leqq c\,|\log c|\,[1 + \epsilon + o(1)]\mu(\phi) \tag{5.7}$$

with the $o(1)$ term in (5.7) independent of ϕ but perhaps depending on ϵ. The difference between the uniform and non-uniform statements is related, as we shall see, to the difference between Theorem 1 and Theorem 2.

The $o(1)$ term in (5.6) depends on ϵ, but is $<\epsilon$ for $c \leqq C_\epsilon$ (say). Thus, for each ϵ, the bracketed expression in (5.6) is $<1 + 2\epsilon$ for $c \leqq C_\epsilon$, and if we define δ_c^* to be $\delta_c^{*\epsilon}$ for the smallest ϵ satisfying $C_\epsilon \geqq c$ (with the obvious modification if the infimum of such ϵ is not attained), we have

$$r_c(F, \delta_c^*) \leqq [1 + o(1)]c\,|\log c| \int_\Phi \mu(\phi)F\,(d\phi). \tag{5.8}$$

Our problem will be to show how to construct $\{t_n\}$ and $\{\delta_c^{\gamma,\epsilon}\}$ (resp. $\{\delta_c^0\}$ and $\{\delta_c'\}$) so that (5.3) and (5.4) (resp. (5.2) and (5.1)) are satisfied with $\mu(\phi)$ (resp. μ_i) in (5.4) (resp. (5.1)) the minimum possible value; it will then follow that $\{\delta_c^*\}$, as constructed in the above, is asymptotically Bayes with respect to F. This will yield the result of Theorem 1. To obtain the result of Theorem 2 we must construct $\{t_n\}$ and $\{\delta_c^{\gamma,\epsilon}\}$ so that the uniform versions (see Remark 12) of (5.3) and (5.4) hold. (When Φ is finite, Theorem 2 is the same as Theorem 1).

We are now ready for our design considerations. Let $\mathcal{E} = \{e\}$ be the set of one-observation experiments available at each stage. Each such experiment will be assumed to cost the same amount c, since the more general case can be treated without difficulty in the same way merely by dividing "information numbers" by experimental costs. For each k, given the outcomes of the first k experiments, the conditional distribution of the $(k + 1)$st observation depends only on the experiment chosen for this $(k + 1)$st stage. We alter the notation of Section 2 slightly and take f_θ^e to be the density of an observation when e is the experiment and $\theta \varepsilon \Phi$ is true, while $\lambda^e(\theta, \phi) = E_\theta \log [f_\theta^e(X_1)/f_\phi^e(X_1)]$ is the corresponding "information" concerning ϕ when θ is true; θ and ϕ are arbitrary elements in Φ. (In Section 2 we limited the notation so that $\theta \varepsilon \Theta$; this is no longer so.) For simplicity we can assume one measure μ applies for all e, although this is not really necessary. As in Chernoff (1959), Albert (1961), and Bessler (1960), we must also consider mixtures of experiments, i.e., probability measures $\bar{e}$ on $\mathcal{E}$. (We shall see that it always suffices, under the assumptions considered in the next paragraph, to limit consideration to measures $\bar{e}$ with finite support. It may nevertheless be convenient in applications to consider $\bar{e}$'s with infinite support. If $\mathcal{E}$ is nondenumerable, such $\bar{e}$'s are measures relative to a separable Borel field generated by open sets in a topology relative to which $\lambda^e(\theta, \phi)$ is continuous, etc.) Then any information number $\lambda^{\bar{e}}$ is defined to be the probabilistic mixture under $\bar{e}$ of λ^e's, etc.

The assumptions we require are similar to those of previous sections and will usually be easy to verify if f_θ^e is sufficiently regular in θ and e as the latter vary over domains which are assumed compact, although such compactness is not necessary. For brevity and simplicity, we assume the following:

ASSUMPTIONS.

(1) $\Omega, \Theta, \mathcal{E}$ are compact with Ω, Θ subsets of some Euclidean space.

(2) Assumption 1 of Section 2.

(3) Assumption 3 with a supremum over e inserted immediately to the left of each expectation sign.

(4) Assumption 2 is altered by requiring boundedness and continuity of $\lambda^e(\theta, \phi)$ in all three variables and

$$\sup_{\bar{e}} \inf_{\theta\varepsilon\Omega, \phi\varepsilon\Theta} \lambda^{\bar{e}}(\theta, \phi) > 0; \qquad \sup_{\bar{e}} \inf_{\theta\varepsilon\Theta, \phi\varepsilon\Omega} \lambda^{\bar{e}}(\theta, \phi) > 0. \tag{5.9}$$

(5) Assumption 5 with the continuity there uniform in e.

For the proof of the analogue of Theorem 2 we will also need

(6) Assumption 6 is altered with the insertion of a supremum over e to the left of the expectation sign and the continuity there is uniform in e.

We define

$$\lambda(\theta) = \sup_{\bar{e}} \inf_{\phi\varepsilon\Theta} \lambda^{\bar{e}}(\theta, \phi) \quad \text{if } \theta \varepsilon \Omega$$
$$= \sup_{\bar{e}} \inf_{\phi\varepsilon\Omega} \lambda^{\bar{e}}(\theta, \phi) \quad \text{if } \theta \varepsilon \Theta \tag{5.10}$$

Under the compactness and continuity Conditions (1) and (4) just mentioned, the game with kernel $\lambda^e(\theta, \phi)$ for each fixed $\theta \varepsilon \Omega$ is determined and has positive value $\lambda(\theta)$ which by (5.9) is bounded away from zero. The same holds for $\theta \varepsilon \Theta$. Moreover, for each $\epsilon > 0$ we can find a finite covering of Ω (or of Θ) such that, for each set S of the covering, there is an $\bar{e}_s$ with finite support and which is ϵ-maximin for all θ in S. Also, there is an $\bar{e}^0$ (say) with finite support and rational probabilities for which $\inf_{\theta\varepsilon\Omega,\phi\varepsilon\Theta} \lambda^{\bar{e}^0}(\theta, \phi) > 0$, and $\inf_{\theta\varepsilon\Theta,\phi\varepsilon\Omega} \lambda^{\bar{e}^0}(\theta, \phi) > 0$. These conclusions can be derived more generally from the work of Wald (1950) or

LeCam (1955) under assumptions analogous to Assumption 4, but our object here is brevity rather than greatest generality. $\lambda(\theta)$ turns out, as we shall see, to be the reciprocal of the minimum value of $\mu(\theta)$ for which (5.4) (or (5.1)) holds.

We now describe several simple ways of constructing asymptotically optimum designs. First, if Φ is finite, δ_c^0 can be constructed very easily: letting b be the denominator of the rational probabilities $\pi_1, \pi_2, \cdots, \pi_m$ (say) associated with $\bar{e}^0$, we need only take "blocks" of b experiments, $b\pi_j$ in each block being of the jth type making up $\bar{e}^0$; we can then view the vector of the outcomes from each of these blocks as a single "observation" and with such observations we take δ_c^0 to be the procedure $\delta_{q(c)}$ of Section 4 (without indifferences) with $q(c) = o(1)$ and $|\log q(c)| = o(|\log c|)$, e.g., $q(c) = |\log c|^{-1}$. Alternatively we could use the randomized experiment $\bar{e}^0$ at each stage.

When Φ is infinite there will often be a simple choice for $\{t_n\}$ using either the "block" method above or the randomized experiment $\bar{e}^0$ at each stage to obtain observations. In fact, under our assumptions, the sequence of maximum likelihood estimators, based on observations obtained as just described, is consistent (see Kiefer and Wolfowitz (1956)).

An alternative program, when Φ is infinite, is to construct $\{\delta_c^{*\epsilon}\}$ to satisfy (5.5) in a way analogous to the construction of $\{\delta_c^*\}$ from $\{\delta_c^0\}$ and $\{\delta_c^i\}$ following (5.2). We first construct, for each $\epsilon' > 0$, a family $\{\delta_c^{0,\epsilon'}\}$ (not to be confused with $\{\delta_c^{\gamma,\epsilon}\}$) as follows: Suppose that observations are taken either by the "block" method or by $\bar{e}^0$ at each stage. Let $U_1, \cdots, U_{k(\epsilon')}$ be a collection of open spheres of radius $\epsilon'/2$ which covers Φ and let $\gamma_1, \cdots, \gamma_{k(\epsilon')}$ be the centers of the spheres, so that we have $\gamma_i \varepsilon \Phi$ for $i = 1, \cdots, k(\epsilon')$. Consider the $k(\epsilon')$-decision problem with Φ the space of states of nature, decisions $d_1, \cdots, d_{k(\epsilon')}$, and loss function L satisfying $L(\phi, d_j) = 0$ if $\phi \varepsilon U_j$, $L(\phi, d_j) = 1$ if $\phi \not\varepsilon U_j$. We are now in the context of Section 4 (with indifferences and semi-indifferences) and we let $\delta_c^{0,\epsilon'}$ be the procedure there with the c in Section 4 replaced by $q(c)$ with $|\log q(c)| = o(|\log c|)$ and $q(c) = o(1)$, e.g., $q(c) = |\log c|^{-1}$. We now define $\{\delta_c^{*\epsilon}\}$ by taking ϵ' as in (5.4c), using $\delta_c^{0,\epsilon'}$ until it terminates, and if it leads to decision d_j we "estimate" the true parameter by γ_j and then use $\delta_c^{\gamma_j,\epsilon}$ until the latter terminates and we make the decision that $\delta_c^{\gamma_j,\epsilon}$ leads to. That (5.5) is satisfied for this definition of $\delta_c^{*\epsilon}$ follows because $E_\phi N(\delta_c^{0,\epsilon'}) = O(|\log q(c)|) = o(|\log c|)$ uniformly for $\phi \varepsilon \Phi$ and, if $\phi \varepsilon U_i$, $\sum' P_\phi\{\delta_c^{0,\epsilon'}$ makes decision $d_j\} = o(1)$ where $\sum'$ means summing over those j such that $U_i \cap U_j$ is empty.

We now turn to the construction of $\{\delta_c^{\gamma,\epsilon}\}$. We will give two methods, both of which can be used to obtain $\{\delta_c^i\}$. A third method which we present is appropriate when Φ is finite, i.e., for the construction of $\{\delta_c^i\}$. To describe the first two methods suppose that $\gamma \varepsilon \Omega$ and $\epsilon > 0$ are fixed. Let $\bar{e}_\gamma, \epsilon' > 0, \epsilon'' > 0$ be such that

$$\text{(5.11)} \quad \begin{aligned} &\text{(a)} \quad 0 < \epsilon'' < 2\epsilon \inf_{\theta\varepsilon\Omega} \lambda(\theta) \\ &\text{(b)} \quad \lambda(\theta) \leqq \inf_{\phi\varepsilon\Theta} \lambda^{\bar{e}_\gamma}(\theta, \phi) + \epsilon'' \quad \text{for } |\theta - \gamma| < \epsilon' \\ &\text{(c)} \quad \inf_{\theta\varepsilon\Omega, \phi\varepsilon\Theta} \lambda^{\bar{e}_\gamma}(\theta, \phi) > 0. \end{aligned}$$

Here $\bar{e}_\gamma$ may be chosen to have finite support and with rational probabilities because of the comments following (5.10). Condition (c) can be satisfied because we may always combine $\bar{e}_\gamma$ with a small multiple of $\bar{e}^0$. The fact that the remainder of (5.11) can be satisfied follows from our assumptions.

Method I. Define $\{\delta_c^{\gamma,\epsilon}\}$ by using the randomized experiment $\bar{e}_\gamma$ at each stage and the stopping and terminal decision rule of $\{\delta_c\}$ in Section 2 with the likelihood function there being replaced after n observations by $\prod_{j=1}^n f_\phi^{e_j}(x_j)$ where e_j is the actual (nonrandomized) experiment chosen at stage j (after randomization as prescribed by $\bar{e}_\gamma$). This is in the spirit of Chernoff's treatment, except that, once the "preliminary experiment" (associated with $\{t_n\}$) has been performed, we select the same $\bar{e}_\gamma$ to be used at every stage, rather than having to allow the randomized experiment to be changed continually with sequential calculations. Note that the role of the preliminary experiment is to obtain the "right" design. The relevant random walk considerations of Section 2 proceed as before with $\lambda(\theta, \phi)$ replaced by $\lambda^{\bar{e}_\gamma}(\theta, \phi)$ and without any additional design considerations, thus yielding (5.4) with (c) of (5.4) following from Lemma 2 and (a) of (5.11).

Method II. Instead of utilizing the randomized experiment $\bar{e}_\gamma$ we can use the "block" method described earlier for δ_c^0, replacing $\bar{e}^0$ in that description by $\bar{e}_\gamma$ and observing that the methods of Section 2 will yield

$$E_\theta\,\{\text{number of blocks}\} \sim [1 + \epsilon + o(1)]\,|\log c|/b\lambda(\theta)$$

since the appropriate information number when θ is true and $\phi \,\varepsilon\, \Phi$ is $b\lambda^{\bar{e}_\gamma}(\theta, \phi)$. This method thus yields a procedure with non-randomized choices of designs and which satisfies (5.4). (The possibility of using nonrandomized choices with the right asymptotic frequency was mentioned explicitly by Chernoff (1959) in the heuristic discussion of his "prototype" example.)

Method III. Here we assume that Φ is finite. The method will yield a direct construction of $\{\delta_c^i\}$ to satisfy (5.1). For each possible state i we specify a fixed sequence of experiments $e_{i1}, e_{i2}, \cdots$ such that the frequency of occurrences of each different e among $e_{i1}, \cdots, e_{in}$ tends as $n \to \infty$ to the probability assigned by a maximin $\bar{e}^i$ to e. (Since the set of possible states is finite, a maximin $\bar{e}^i$ with finite support is known to exist. Of course, if $\bar{e}^i$ has rational probabilities, we can consider blocks.) We assume here, for simplicity, that $\min_{\phi\neq\theta} \lambda^{\bar{e}^i}(\theta, \phi) > 0$ for each i, the modification needed otherwise being simple. Using a standard argument usually applied to identically distributed random variables, we again bound $E_\theta N$ by showing the smallness of

$$P_\theta\left\{\sum_{j=1}^{(1+\epsilon)|\log c|/\lambda(\theta)} \log\,[f_\theta^{e_{ij}}(X_j)/f_\phi^{e_{ij}}(X_j)] < |\log c|\right\}$$

by applying exponentiation and Chebyshev's inequality. It is important that the sequences $\{e_{ij}\}$ do not change with c.

To summarize, then, our methods reduce the construction of procedures and proofs in the design setting to those of Sections 2, 3, and 4 where there is no design problem, and we obtain (under Assumptions (1)–(5)) for a given a priori distribution F (whose support is Φ for convenience), a family $\{\delta_c^*\}$ such that (5.8) holds with $\mu(\phi) = 1/\lambda(\phi)$ where $\lambda(\phi)$ is defined in (5.10). By appropriate use of Remark 12, adding Assumption (6), and using the procedures $\{\delta_c'\}$ of Theorem 2, we can also obtain a family $\{\delta_c^{**}\}$ such that (5.8) holds with $\{\delta_c^*\}$ replaced by $\{\delta_c^{**}\}$ and F replaced by any G having the same support as F (without changing δ_c^{**}) with the $o(1)$ term in (5.8) uniform for all such G.

As the last step in our discussion we will show that, if $\{\delta_c^B\}$ is Bayes with respect to F, then

$$r_c(F, \delta_c^*)/r_c(F, \delta_c^B) \to 1 \tag{5.12}$$

i.e., $\{\delta_c^*\}$ is asymptotically Bayes as $c \to 0$.

To establish (5.12) we proceed as in Theorem 1 and conclude that, if (5.12) fails, there is a sequence $\{c_i\}$ and a $\theta \varepsilon \Omega$ (say) such that

$$\begin{aligned} E_\theta N(\delta_{c_i}^B) &\leqq (1 - 2\epsilon)\,|\log c_i|/\lambda(\theta) \\ P_\theta\{\text{wrong decision using } \delta_{c_i}^B\} &\leqq Kc_i\,|\log c_i|. \end{aligned} \tag{5.13}$$

By Assumption (4), we can find $\{\phi_1, \cdots, \phi_k\} \subset \Theta$ such that

$$0 < \frac{1 - 2\epsilon}{\lambda(\theta)} \leqq \frac{1 - \epsilon}{\sup_{\bar{e}} \inf_{1 \leqq j \leqq k} \lambda^{\bar{e}}(\theta, \phi_j)} < \infty \tag{5.14}$$

and $\{\phi_j\}$ can be selected so that, for a subsequence of $\{c_i\}$ (which we take to be $\{c_i\}$ for notational convenience),

$$P_{\phi_j}\{\text{wrong decision using } \delta_{c_i}^B\} \leqq Kc_i\,|\log c_i|$$

for all i and $1 \leqq j \leqq k$. Considering $\delta_{c_i}^B$ as a test of θ vs. $\phi_1, \cdots, \phi_k$ with available experiments $\mathcal{E}$ we see, by reference to Lemmas 4 and 5 of Chernoff (although Chernoff assumes $\mathcal{E}$ is finite his arguments remain valid under our assumptions on $\mathcal{E}$ and (5.14)), that

$$E_\theta N(\delta_{c_i}^B) \geqq \frac{[1 + o(1)]\,|\log c_i|}{\sup_{\bar{e}} \inf_{1 \leqq j \leqq k} \lambda^{\bar{e}}(\theta, \phi_j)}$$

which, together with (5.14), contradicts (5.13). Thus (5.12) is proved.

For the design problems in the context of Sections 3 and 4 the assumptions, definitions, and constructions are analogous to those described here for Section 2, and the verification of (5.12) proceeds by using the extension of Lemmas 4 and 5 of Chernoff to the multi-decision problem in the same way we used our Lemma 5 in Sections 3 and 4.

Interpreting the assumptions, definitions, and constructions in the appropriate manner to suit the context (Sections 2, 3, or 4) we can now state

THEOREM 6. *Under Assumptions* (1)–(5) *and with F the a priori distribution, the family $\{\delta_c^*\}$, constructed by any of the described methods, is asymptotically Bayes with respect to F for the problems of Sections* 2, 3, *or* 4 *when $\mathcal{E}$ is the set of available one-observation experiments. If Assumption* (6) *is added the family $\{\delta_c^{**}\}$ is asymptotically Bayes with respect to any G having the same support as F.*

The corollaries to Theorem 1 and Theorem 3 are also valid here; the corollary to Theorem 1 in the context of this section is, in fact, the optimality result of Chernoff, Albert, and Bessler. Less restrictive hypotheses, under which Theorem 6 remains valid can be obtained in analogy with Remark 7 in Section 2.

The dependence of the procedures on the support of the a priori distribution is subject to the same remarks as made in Section 3 and Section 4. Thus, for the design problem associated with Section 3, if the support F is $\Omega \cup \Theta \cup I$ then $\{\delta_c^{**}\}$ is asymptotically Bayes with respect to any G whose support is $\Omega \cup \Theta \cup I'$ for any subset I' of I.

The following is an example of these considerations (an example related to Bessler's Example 5): Suppose there are three possible states of nature and three corresponding decisions. The three possible states of nature specify (f, g, g), (g, f, g), and (g, g, f) as the possible vector of densities of three populations where f and g are specified. e_i is the experiment which takes an observation from population i. Bessler's solution to the design problem is to observe, at each stage, that population which, on the basis of previous observations, is the maximum likelihood estimator of the population with density f. A solution using Method II above is first to take $n(c) = o(|\log c|)$ (but with $n(c) \to \infty$) observations (e.g., $[|\log c|^{\frac{1}{2}}]$) from each population and, on the basis of these observations, find the maximum likelihood estimator of the population with density f. If this is population i we conduct the remaining experimentation from population i except that out of every $n(c)$ observations we reserve two observations one from each of the other two populations. (This modification is needed since λ^{e_i} can be zero.) Since the $o(1)$ term of the second line of (5.2) is then of order $e^{-n(c)}$ while the first line of (5.1) which it will multiply in the final expression for EN is of order $|\log c|\, n(c)$, it is easy to verify that this design is asymptotically optimum. (We have not satisfied (5.1), but have proceeded in a modified manner which is more expeditious.)

If f and g are normal with unit variance and means 1 and 0 respectively, Bessler's solution is to observe the population i for which $\sum_{j=1}^{n_i} x_{ij} - n_i/2$ is largest, where $\{x_{ij}, j = 1, \cdots, n_i\}$ are the previous observations from population i. However, for less simple f and g the calculations are more formidable, and for slightly more complicated problems, e.g., involving three different normal densities instead of two, Bessler shows how much more complicated his solution can become, with the necessity of prescribing randomization probabilities which vary from stage to stage.

APPENDIX

Let $\{\xi_i\}$ be a sequence of independent and identically distributed random variables with distribution G. Let $\mu(G) = E_G\xi_1$, $\sigma^2(G) = \operatorname{Var}_G \xi_1$, and for $\mu_0 > 0$, $0 < \sigma_0 < \infty$ put $\mathcal{G}(\mu_0, \sigma_0) = \{G \mid \mu(G) \leqq -\mu_0, \sigma^2(G) \leqq \sigma_0^2\}$, and for $\mu_0 > 0$ and $\rho_0 > 0$ let $\mathcal{G}^*(\mu_0, \rho_0) = \{G \mid \mu(G) \leqq -\mu_0, \mu(G)/\sigma(G) \leqq -\rho_0\}$. Let $S_k = \sum_1^k \xi_i$, $k = 1, 2, \cdots$; $S_0 = 0$.

THEOREM A.

$$\sup_{G \in \mathcal{G}(\mu_0, \sigma_0)} \sum_{k=1}^{\infty} P_G\{S_k > 0\} < \infty. \tag{A.1a}$$

(A.1b) $$\sup_{G \varepsilon \mathcal{G}^*(\mu_0, \rho_0)} \sum_{k=1}^{\infty} P_G\{S_k > 0\} < \infty.$$

PROOF. Since

$$P_G\{S_k > 0\} = P_G\{S_k - \mu(G)k > \mu(G)k\} \leqq P_G\{S_k - \mu(G)k > \mu_0 k\}$$

for all G we can dominate (A.1a) by

(A.2) $$\sup_{G \varepsilon \mathcal{G}'} \sum_{k=1}^{\infty} P_G\{S_k > k\}$$

where $\mathcal{G}' = \{G \mid \mu(G) = 0, \sigma^2(G) \leqq \sigma_0^2/\mu_0^2\}$. For fixed $G \varepsilon \mathcal{G}'$ the sum in (A.2) converges; see Erdös (1949). An inspection of Erdös' proof reveals that the only way the distribution of ξ_1 enters in the argument (outside of its mean being 0) is through its variance, and the use of Chebyshev's inequality in Erdös' argument shows that variance 1 (which is what he assumes) can be replaced by an upper bound on the variance. Thus (A.2) is finite and (A.1a) is thereby established.

(A.1b) follows from (A.1a) by observing that

$$\sup_{G \varepsilon \mathcal{G}^*(\mu_0, \rho_0)} \sum P_G\{S_k > 0\} = \sup_{G \varepsilon \mathcal{G}^*(\mu_0, \rho_0)} \sum P_G\{S_k/\sigma_0 > 0\} = \sup_{G \varepsilon \mathcal{G}(\rho_0, 1)} \sum P_G\{S_k > 0\}.$$

THEOREM B.

(A.3) $$\sup_{G \varepsilon \mathcal{G}(\mu_0, \sigma_0)} E_G[\max_{k \geqq 0} S_k] < \infty.$$

PROOF. It is well known that the expectation considered here is finite when $\mu(G) < 0$, $\sigma^2(G) < \infty$. To obtain the uniformity expressed in (A.3) we use Theorem 4.1 in Spitzer (1956) which states

$$E_G \max_{k \geqq 0} S_k = \sum_{k=1}^{\infty} (1/k) E_G S_k^+ .$$

Suppressing the dependence on G we can now write

$$E \max_{k \geqq 0} S_k = \sum_{k=1}^{\infty} \frac{1}{k} E S_k^+ = \sum_{k=1}^{\infty} \frac{1}{k} \int_{S_k > 0} (\xi_1 + \cdots + \xi_k)\, dP$$

$$= \sum_{k=1}^{\infty} \int_{S_k > 0} \xi_1\, dP \leqq \sum_{k=1}^{\infty} \int_0^{\infty} r\, d_r P\{\xi_1 < r, \xi_1 + \cdots + \xi_k > 0\}.$$

Since

$$P\{r' < \xi_1 < r, \xi_2 + \cdots + \xi_k > -\xi_1\}$$
$$\leqq P\{r' < \xi_1 < r, \xi_2 + \cdots + \xi_k > -r\} \leqq P\{r' < \xi_1 < r\},$$

we have (writing $\xi_k' = \xi_k - \mu/2$, $S_k' = S_k - k\mu/2$, $h = k - 1$)

$$E \max_{k \geqq 0} S_k \leqq \sum_{k=1}^{\infty} \int_0^{(-h\mu/2)} rP\left\{\xi_2 + \cdots + \xi_k > \frac{h\mu}{2}\right\} d_r P\{\xi_1 < r\}$$
$$+ \int_{(-h\mu/2)}^{\infty} r\, d_r P\{\xi_1 < r\}$$

$$\leqq \sum_{k=1}^{\infty} E\xi_1^+ P\{S'_{k-1} > 0\} + \int_0^\infty \sum_{h=0}^{[-2r/\mu+1]} r\, d_r P\{\xi_1 < r\}$$

$$\leqq E\xi_1^+ \sum_{k=1}^{\infty} P\{S'_{k-1} > 0\} + \int_0^\infty r(2 - 2r/\mu)\, d_r P\{\xi_1 < r\},$$

and this is finite by (A.1a) and the finiteness of $E\xi_1^+$ and $E\xi_1^2$.

THEOREM C. *Let N_t be the first k such that $S_k < -t$ $(t > 0)$. Then, for $G \varepsilon \mathcal{G}^*(\mu_0, \rho_0)$,*

$$E_G N_t \leqq -t/\mu(G) + \chi(t, G)$$

where

$$\sup_{G\varepsilon\mathcal{G}^*(\mu_0,\rho_0)} \chi(t, G) = o(t)$$

as $t \to +\infty$.

PROOF. Let $\{\epsilon_t\}$ be a sequence of positive numbers which will go to zero in a way to be chosen below. Then

$$EN_t = \sum_{k=0}^{\infty} P\{N_t > k\} = \sum_{k=0}^{\infty} P\{\min_{j\leqq k} S_j > -t\} \leqq \sum_{k=0}^{\infty} P\{S_k > -t\}$$

$$= \sum_{k\leqq(1+\epsilon_t)(-t/\mu)} P\{S_k > -t\} + \sum_{k>(1+\epsilon_t)(-t/\mu)} P\{S_k > -t\}$$

$$\leqq (1 + \epsilon_t)\frac{t}{-\mu} + \sum_{k>(1+\epsilon_t)(-t/\mu)} P\left\{S_k - \frac{k\mu}{1+\epsilon_t} > \frac{-k\mu}{1+\epsilon_t} - t\right\}$$

$$\leqq \frac{t}{-\mu} + \frac{\epsilon_t t}{-\mu} + \sum_{k=0}^{\infty} P\{S'_k(t) > 0\}$$

where $S'_k(t)$ is the kth partial sum of independent and identically distributed random variables with mean $\epsilon_t\mu(G)/(1 + \epsilon_t)$ and variance $\sigma^2(G)$ when G is the distribution of S_1. By (A.1b),

$$\sup_{G\varepsilon\mathcal{G}^*(\mu_0,\rho_0)} \sum_{k=0}^{\infty} P\{S'_k(t) > 0\} = B(t) \text{ (say)} < \infty.$$

Choose ϵ_t to go to zero so slowly that $B(t) = o(t)$. Since $\mu(G) \leqq -\mu_0$ for $G \varepsilon \mathcal{G}^*(\mu_0, \rho_0)$, we have

$$-\epsilon_t t/\mu + \sum_{k=0}^{\infty} P\{S'_k(t) > 0\} = o(t)$$

uniformly for $G \varepsilon \mathcal{G}^*(\mu_0, \rho_0)$, and thus Theorem C is proved.

THEOREM D. *Let ν be the last k such that $S_k > 0$. Then, for any $G \varepsilon \mathcal{G}(\mu_0, \sigma_0)$,*

$$E_G\nu \leqq (2/\mu_0)E_G[\max_{k\geqq0} S_k] + \sum_{k=1}^{\infty} P_G\{S_k + k\mu_0/2 > 0\}.$$

PROOF. Suppressing the dependence on G we have

$$Ev = \sum_{k=1}^{\infty} P\{\nu \leqq k\} = \sum_{k=1}^{\infty} P\{\max_{j \geqq k} S_j > 0\}$$

$$= \sum_{k=1}^{\infty} P\{\max_{j \geqq k} (S_j - S_k) + S_k > 0\} = \sum_{k=1}^{\infty} P\{M + S_k > 0\}$$

where M has the distribution of $\max_{j \geqq 0} S_j$ and is independent of $\{S_k\}$. Consequently, denoting by $[r]$ the greatest integer $\leqq r$,

$$\sum_{k=1}^{\infty} P\{M + S_k > 0\} = \sum_{k=1}^{\infty} \int P\{S_k > -m\}\, d_m P\{M \leqq m\}$$

$$= \sum_{k=1}^{[2m/\mu_0]} \int P\{S_k > -m\}\, d_m P\{M \leqq m\}$$

$$+ \sum_{k=[1+2m/\mu_0]} \int P\left\{S_k + \frac{\mu_0 k}{2} > \frac{\mu_0 k}{2} - m\right\} d_m P\{M \leqq m\}$$

$$\leqq \frac{2}{\mu_0} EM + \sum_{k=1}^{\infty} P\left\{S_k + \frac{\mu_0 k}{2} > 0\right\}$$

and Theorem D is proved.

REFERENCES

ALBERT, A. E. (1961). The sequential design of experiments for infinitely many states of nature. *Ann. Math. Statist.* **32** 774–799.

ANDERSON, T. W. (1960). A modification of the sequential probability ratio test to reduce the sample size. *Ann. Math. Statist.* **31** 165–197.

BESSLER, S. A. (1960). Theory and applications of the sequential design of experiments, k-actions, and infinitely many experiments. Technical Report no. 55, Department of Statistics, Stanford Univ.

CHERNOFF, H. (1959). Sequential design of experiments. *Ann. Math. Statist.* **30** 755–770.

DONNELLY, T. G. (1957). A family of truncated sequential tests. Doctoral dissertation, Univ. of North Carolina.

ERDÖS, P. (1949). On a theorem of Hsu and Robbins. *Ann. Math. Statist.* **20** 286–291.

HOEFFDING, W. (1960). Lower bounds for the expected sample size and the average risk of a sequential procedure. *Ann. Math. Statist.* **31** 352–368.

KIEFER, J. and WEISS, L. (1957). Some properties of generalized sequential probability ratio tests. *Ann. Math. Statist.* **28** 57–75.

KIEFER, J. and WOLFOWITZ, J. (1956). Consistency of the maximum likelihood estimator in the presence of infinitely many incidental parameters. *Ann. Math. Statist.* **27** 887–906.

LECAM, L. (1955). An extension of Wald's theory of statistical decision functions. *Ann. Math. Statist.* **26** 69–81.

LEHMANN, E. (1957a, 1957b). A theory of some multiple decision problems. I and II. *Ann. Math. Statist.* **28** 1–25 and 547–572.

SCHWARZ, G. (1962). Asymptotic shapes of Bayes sequential testing regions. *Ann. Math. Statist.* **33** 224–236.

SOBEL, M. and WALD, A. (1949). A sequential decision procedure for choosing one of three hypotheses concerning the unknown mean of a normal distribution. *Ann. Math. Statist.* **20** 502–522.

SPITZER, F. (1956). A combinatorial lemma and its application to probability theory. *Trans. Amer. Math. Soc.* **82** 332–339.

WALD, A. (1947). *Sequential Analysis*. Wiley, New York.

WALD, A. (1949). Note on the consistency of the maximum likelihood estimate. *Ann. Math. Statist.* **20** 595–601.

WALD, A. (1950). *Statistical Decision Functions*. Wiley, New York

WALD, A. (1951). Asymptotic minimax solutions of sequential point estimation problems. *Proc. Second Berkeley Symp. Math. Statist. Prob.* 1–11. Univ. of California Press.

Printed in the United States
By Bookmasters